Non-Equilibrium Thermodynamics and Statistical Mechanics

Non-Equilibrium Thermodynamics and Statistical Mechanics

Foundations and Applications

Phil Attard

OXFORD
UNIVERSITY PRESS

OXFORD
UNIVERSITY PRESS

Great Clarendon Street, Oxford, OX2 6DP,
United Kingdom

Oxford University Press is a department of the University of Oxford.
If furthers the University's objective of excellence in research, scholarship,
and education by publishing worldwide. Oxford is a registered trade mark of
Oxford University Press in the UK and in certain other countries

First Edition published in 2012

British Library Cataloguing in Publication Data
Data available

Library of Congress Cataloging in Publication Data
Data available

ISBN 978–0–19–966276–0

Preface

> Fate, Time, Occasion, Chance, and Change,
> To these all things are subject
>
> P. B. Shelley, 'Prometheus Unbound' (1820)

All things change in time

Motion and change are the nature of the world. The scale of motion ranges from the orbit of galaxies and planets, to the winds, ocean currents, and river flows, and even to the sub-microscopic, with the ceaseless movement of atoms and molecules. In the living world there is the locomotion of animals, the movement of limbs and muscles, and the flows of blood, sap, and other vital fluids. There is also human made motion in the form of transport by planes, trains, and automobiles, the mechanical movement of motorised tools, and controlled flows such as those of electricity, heat, and fluids.

There are as well other forms of change. Birth, growth, and death is a familiar trajectory for any individual living organism. But these also describe the evolution of a species as a whole, as well as that of the universe, planets, eco-systems, and even social structures. One can also include other physical phenomena that are the subject of more traditional scientific study such as the progress of chemical reactions, the development of heat and current fluxes, the formation of patterns by aerodynamic and hydrodynamic flows, the processes of self-assembly and physical organisation, and the dynamic deformation and response of materials. These of course occur in nature, industry, technology, and under controlled laboratory conditions.

The rôle of chance in our world should not be underestimated. When things occur, their exact trajectory, and their ultimate fate are not perfectly predictable. In general the complexity of systems, unknown initial conditions, and the influence of uncontrolled external forces all contribute a degree of randomness that leads to an uncertain future unconstrained by a strict fatalism. Chance means that changes in time occur not with pure determinism but rather with statistical probability.

Time changes everything

The point belaboured above is that time dependent phenomena are ubiquitous. The technical word for these is 'non-equilibrium'. It is a reflection of the chronological development of science that this, the most common class of systems, is described in the negative.

Initially, thermodynamics and statistical mechanics were developed for equilibrium systems, which do not change macroscopically with time. Thermodynamics is the science of macroscopic systems, and it provides universal laws and relationships that all equilibrium systems must obey. Statistical mechanics enables the probabilistic description of equilibrium systems at the molecular level. It gives the mathematical basis for the empirical laws of thermodynamics and it provides quantitative values for measured thermodynamic parameters.

It is one of the great ironies of science that the word 'dynamics' in 'thermodynamics', and the word 'mechanics' in 'statistical mechanics', both imply motion, when in fact both disciplines have been strictly formulated for static or equilibrium systems. Of course as an approximation they are often applied to time dependent systems, either instantaneously or else over time intervals small enough that any change is negligible, or they can be combined with an empirical theory such as hydrodynamics. But in terms of an exact treatment, thermodynamics and statistical mechanics are restricted to equilibrium systems. This raises the question: How does time change thermodynamics and statistical mechanics?

This book

This book seeks to answer that disarmingly simple question. A coherent formulation of non-equilibrium thermodynamics is given. The approach is based upon a particular form of entropy, and it has the advantage that almost all of the concepts of equilibrium thermodynamics carry over to the non-equilibrium field. It also enables a consistent derivation of most of the known non-equilibrium theorems and results, which exhibits their inter-relationships and places them in the context of a bigger picture.

The non-equilibrium probability distribution is also developed, and this provides a basis for the field of non-equilibrium statistical mechanics. Again, this enables a unified derivation of known and previously unknown theorems. Importantly, it also enables the development of computer simulation algorithms for non-equilibrium systems, which are used to test quantitatively the results and to illustrate them at the molecular level.

Because of the significance of time dependent phenomena, there are many books and scientific papers concerned with the formulation of non-equilibrium thermodynamics and non-equilibrium statistical mechanics, and with their application to specific systems. The selection of topics, underlying approach, and method of presentation vary enormously, although certain non-equilibrium theorems and results for which there is broad consensus commonly recur. Others results lie at the cutting edge of current research, and for these detailed justification and explanation are required.

As mentioned above, this book proceeds from the very fundamental principles that determine the optimum non-equilibrium thermodynamic state, and also from the equations of motion and probability distributions appropriate for non-equilibrium statistical mechanics. The strategy employed here is to set out the physical basis of the axioms, the close analogy between non-equilibrium and equilibrium principles, and, most importantly, the theorems and detailed results that follow as a consequence. In general an attempt is made to provide quantitative tests, experimental or computational, and detailed comparisons between different approaches, and alternative, independent derivations of the same result. It is hoped that such concrete evidence and the consistency of the approach will give some confidence in the fundamental principles that the book is based upon.

In the present book, most of the traditional topics in the non-equilibrium field are covered, and some new ones besides. What is perhaps unique here is that a single underlying approach suffices to derive and to describe all these results. The fields of non-equilibrium thermodynamics and non-equilibrium statistical mechanics are here regarded as a continuum that ranges from the macroscopic to the sub-microscopic, with Brownian motion and stochastic processes lying in the boundary region where they merge.

In a sense, this book is one long argument for non-equilibrium thermodynamics and statistical mechanics. There are several reasons why the reader may find the present approach useful and may have confidence in the results. First, is the simplicity of the concepts, examples, and equations. Stripping away all that is unnecessary removes the possibility of confusion masquerading as complexity, and displays the results in a clear and unambiguous light. Second, is the physical basis of the approach. Thermodynamics and statistical mechanics are derived from, and designed for the real world, and here is emphasised the physical basis and interpretation of all the terms that occur in each equation. This removes the likelihood of inadvertent non-physical behaviour due to artificial assumptions, it gives an intuitive feel to the equations and results, and it enables the common sense test to be readily applied. Third, is the coherence and self-consistency of the approach. Those theorems and results in non-equilibrium thermodynamics and statistical mechanics that are widely accepted are all derived here from a single approach based on entropy. This consilience gives some confidence in both the approach itself and the new results also generated by it. Fourth, the results of a number of computer simulations are given in the text, both to illustrate the procedures and to test quantitatively the results. In addition, certain experimental measurements are used, again quantitatively, to test predictions of the theory. Such tests should prove convincing, both of the individual results and of the formulation as a whole.

The fields of thermodynamics and statistical mechanics have grown over the years. This book is part of that evolution; it is intended to be timely rather than timeless. The principles for non-equilibrium thermodynamics and statistical mechanics set out herein consolidate the present state of knowledge and provide a basis for future growth and new applications.

Contents

Detailed Contents

Chapter 1

Prologue

This chapter gives an overview of the structure of the theory of non-equilibrium thermodynamics. The discussion of the Second Law of Thermodynamics in §1.2 emphasises the need to modify it to include time quantitatively for non-equilibrium systems. The conceptual nature of probability and randomness in the physical universe is canvassed in §1.3. Boltzmann's molecular interpretation of entropy provides the starting point for a general statistical treatment of entropy in §1.4 that focuses on its universal nature, its close connection with probability, and its appropriate formulation for non-equilibrium systems. The important concept of microscopic reversibility is introduced, which will be central to many of the non-equilibrium thermodynamic and statistical mechanical results in subsequent chapters. The chapter concludes with a discussion in §1.5 of the rôle of reservoirs in thermodynamics.

1.1 Entropy and the Second Law

The origin of equilibrium thermodynamics may be traced to the work of Clausius with his enunciation of the Second Law,[1]

Entropy increases during spontaneous changes
in the state of the total system. (1.1)

State here means the value of some unconserved variable, such as the spatial distribution of energy or molecules. The law may be formulated in various ways, some more precise than others, but all the same in their essence. At the time, entropy was purely a mathematical concept, the integral of the heat flow divided by temperature.[2] The Second Law itself was an empirical theorem based on many measurements and observations.

[1]Clausius, R. (1865), Ann. Phys. **125**, 353.

[2]Clausius, R. (1850), Ann. Phys. **79**, 368, 500. Clausius, R. (1851), Phil. Mag. **2**, 1, 102.

It was not until later that Boltzmann deduced the physical nature of entropy:[3] [4]

The entropy of a state is the logarithm of the number of molecular configurations in the state. (1.2)

As will be discussed in detail in coming chapters, the states are functions in phase space (the space of molecular positions and momenta), and so the number of molecular configurations is just the associated volume of phase space. It is an extraordinary achievement of intellect that at a time when the very existence of atoms and molecules was in dispute, Boltzmann had the insight to deduce the physical origins of the hitherto mathematical entropy.

On these two foundations rest the two towers that are thermodynamics and statistical mechanics. From them several important concepts may be immediately deduced. Boltzmann's insight showed the equivalence of entropy and probability, since, for equally likely molecular configurations, the probability of a state is just the number of molecular configurations in that state divided by the total number. This in turn is simply proportional to the exponential of the entropy of the state divided by Boltzmann's constant. The equivalence of entropy and probability introduced a stochastic element into thermodynamics, with the new interpretation of the Second Law being that spontaneous transitions are most likely to occur from less probable states to more probable states. (For a macroscopic system, the likelihood approaches certainty.) Superficially this is an intuitively obvious notion, and it appears that Boltzmann has provided both an explanation and a proof of the Second Law (but see below).

That the Second Law is really a probabilistic statement can be readily seen. For an equilibrium system (i.e. one that is macroscopically unchanging in time), fluctuations occur about the equilibrium or most probable state. These fluctuations are often immeasurably small experimentally, since their relative magnitude scales as the reciprocal of the square root of the size of the system, but they are nevertheless present due to the incessant motion of the molecules. This means that by definition transitions to states of lower entropy must occur, in apparent violation of the deterministic form of the Second Law. In fact, for these spontaneous fluctuations about equilibrium, for every transition that increases the entropy there must be an opposite transition that decreases the entropy. One can conclude from this that the Second Law of Thermodynamics is not designed for fluctuations, but rather for systems that have been prepared initially in an unlikely state that is far removed from the equilibrium state. In such circumstances one will, with probability approaching unity, observe transitions toward the state of greatest entropy.

---

[3]Boltzmann, L. (1872), Wien. Ber. **66**, 275. Boltzmann, L. (1877), Wien. Ber. **76**, 373.

[4]Strictly, this is in units of Boltzmann's constant, $k_\mathrm{B} = 1.38 \times 10^{-23}$ J/K. The logarithm makes entropy an extensive or additive quantity, like energy or size, whereas the number of configurations itself is multiplicative. Boltzmann's definition of entropy is valid for a total or isolated system, where the accessible molecular configurations all have the same energy and hence the same weight. It is often more useful to deal explicitly with the molecular configurations of a sub-system (e.g. when the total system consists of a sub-system and a reservoir), in which case 'number' must be replaced by 'weight', as will shortly become clear.

There is a more challenging issue in deciding to what extent Boltzmann's probabilistic interpretation of entropy proves the Second Law of Thermodynamics. Boltzmann's definition, Eq. (1.2), applies to the probability of a state. Clausius' law, Eq. (1.1), applies to the probability of transitions between states. These are not the same thing. It is directly relevant to what follows that equilibrium theory rests on Boltzmann's identification of entropy and probability for a state, whereas non-equilibrium theory rests on the nature of transition probabilities. Whether the latter are determined by Clausius' Second Law, and if so in what way, are now discussed.

It is intuitively appealing that transitions to a more probable state should themselves be more probable then transitions to a less probable state, but this needs to be made more precise. What can one deduce about the transition probability from the state probability, and how does this illuminate the Second Law?

Let $\wp(a,b|\tau)$ be the unconditional probability of observing the transition from the state b to the state a in a time interval $\tau > 0$, and let $\wp(b,a|\tau)$ be the probability of the reverse transition. In an equilibrium system there is no preferred direction of time and so one is just as likely to observe the forward transition as the reverse, $\wp(a,b|\tau) = \wp(b,a|\tau)$.[5] By the laws of probability, the unconditional probability is equal to the conditional probability times the probability of the conditioning statement, $\wp(a,b|\tau) = \wp(a|b,\tau)\wp(b)$. Here, $\wp(a|b,\tau)$ is the probability of the system being in the state a given that it was in the state b a time τ earlier, and $\wp(b)$ is the probability that the system is in the state b, which for an equilibrium system does not depend upon time. Now for a large enough time interval, the two states must be uncorrelated. In this case the unconditional probability is just the product of the singlet probabilities, $\wp(a,b|\tau)\to\wp(a)\wp(b)$, $\tau\to\infty$, and the conditional probability becomes independent of the initial state, $\wp(a|b,\tau)\to\wp(a)$, $\tau\to\infty$. If one now introduces a third state c, then in this limit

$$\frac{\wp(a|b,\tau)}{\wp(c|b,\tau)} \to \frac{\wp(a)}{\wp(c)}, \ \tau\to\infty. \tag{1.3}$$

If the state a is more probable than the state c (i.e. has more entropy, according to Boltzmann), then this says that a system in the state b is in the long term more likely to make the transition to a than to the state c. That is, transitions are more likely to be observed in the direction of increasing entropy. This is one sense in which Boltzmann's physical interpretation of entropy explains the Second Law of Thermodynamics.

The preceding argument cannot be the whole story, since it applies only in the long time limit. It is not clear to what extent the finite observations times upon which the Second Law of Thermodynamics is based qualify. A less rigorous but nevertheless illuminating argument for finite intervals goes as

[5]This assumes that the states are insensitive to the sign of the molecular velocities. Examples include the spatial distribution of energy or number. A slightly more complicated but in essence the same argument holds in the contrary case, such as the states representing the spatial distribution of momentum.

follows. Denote the value of the state by x, and let $S(x)$ be its entropy. Let $n_\pm(x)$ be the number of molecular transitions that increase or decrease the value of x. As mentioned above, for an equilibrium system, for every forward transition there is a reverse, $n_+(x) = n_-(x)$. The key assumption to make is that the number of such molecular transitions should be proportional to the number of molecular configurations, $n_\pm(x) \propto \exp S(x)/k_\mathrm{B}$. This says in essence that larger states (as measured by the number of molecular configurations) have a greater number of associated transitions. Considering a state to have a finite width Δ_x, then the excess number of forward transitions is $n_+(x + \Delta_x/2) - n_-(x - \Delta_x/2) \propto \Delta_x \mathrm{d}S(x)/\mathrm{d}x$. This says that if the entropy is an increasing function of x, then the forward transitions outweigh the backward transitions, and *vice versa*. This is in essence the Second Law of Thermodynamics.[6]

Although not rigorous, this particular argument serves to link Boltzmann's interpretation of entropy with Clausius' Second Law of Thermodynamics by relating the number of transitions between configurations to the number of configurations themselves. An essential ingredient of the argument is that the transitions are non-infinitesimal (i.e. it applies to steps of finite width), from which it follows that the driving force for the transitions is the gradient in the entropy.

Irrespective of the exact quantitative relationship between Boltzmann's Law (entropy represents the probability of a state) and Clausius' Law (transitions are in the direction of entropy increase), there are further important results that can be gleaned from the Second Law of Thermodynamics.

The Second Law says that an increase in entropy drives transitions. The corollary of this is that a system in which there are no spontaneous nett macroscopic transitions is in a state of maximum entropy. (The phrase 'nett macroscopic' recognises the existence of both molecular motion and macroscopic fluctuations.) This static state is called the equilibrium state, and by the probabilistic nature of entropy discovered by Boltzmann, it is the most likely state. Equilibrium thermodynamics always refers to the most likely state.

In practice, the major quantitative use of the Second Law of Thermodynamics is in determining the equilibrium state, by maximising the entropy, or equivalently by minimising the associated free energy. The Second Law tells the difference between possible constrained states in which the system is not-in-equilibrium. For example, of two possible crystal structures it gives the more stable one as the one with greater entropy.

1.2 Time Dependent Systems

The preceding discussion outlined the rôle of Clausius' Second Law and Boltzmann's identification of entropy for equilibrium systems. The issue now addressed is the appropriate generalisation of these two laws to time dependent

[6] See §1.4.4 for a more satisfactory presentation of this argument, and §7.6.2 for a rigorous statistical mechanical derivation.

systems. To do this, one has to identify the fundamental concepts that underly these two laws and to articulate their non-equilibrium analogue.

1.2.1 The Second Law is Timeless

The major qualitative outcome of the Second Law of Thermodynamics is that it gives the direction of the transitions between states. In common parlance, it gives the arrow of time: evolution forward in time is in the direction of increasing entropy. Note that this is a qualitative rather than a quantitative result. The Second Law is silent about the speed of motion, or the rate of transition.

This point deserves emphasis. As mentioned at the end of the preceding section, the quantitative outcome of the Second Law is limited to deciding the relative stability of constrained equilibrium states. A quantitative theory for non-equilibrium systems must be able to give the rate of transitions, or the speed of motion, or the value of material flows or fluxes, or the time at which events occur or their duration. The law says nothing quantitative about the transitions between them other than their direction. In short, the Second Law is timeless.

1.2.2 The Second Entropy

Extreme Dissipation

There are essentially two ways in which one might try to generalise the Second Law to include time quantitatively. One way—deprecated in this book—is to focus on the rate of change of entropy, which is also called the dissipation. Curiously enough, two axioms for the dissipation that are the exact opposite of each other have been proposed. One school of thought asserts that the rate of change of entropy is a maximum in the optimum non-equilibrium state, and the other school asserts that it is a minimum. Equivalently, it is claimed that spontaneous changes in a non-equilibrium system either increase or decrease the dissipation, respectively. Perhaps the best known advocate for such a principle is the Nobel laureate Prigogine, who at various times asserted one or the other, or even both.[7]

There are several reasons to doubt the validity of any approach to time dependent systems based upon finding the extreme value of the dissipation. The obvious one is that the proponents themselves cannot decide between the two contradictory propositions, which indicates that there is no compelling reason for either. Unlike Clausius' Second Law, there is no experimental or computational evidence to support any principle based upon the dissipation. Although one or other of the two proposed principles has been around for the better part of a century, it has not proved possible to erect upon either a coherent and complete theory for non-equilibrium thermodynamics, or to demonstrate their consistency with either known measurements or accepted theorems.

[7] Kondepudi, D. and Prigogine, I. (1998), *Modern Thermodynamics: From Heat Engines to Dissipative Structures*, (Wiley, Chichester).

States in Time

The approach taken in the present book to generalising Clausius' and Boltzmann's laws to non-equilibrium systems retains unchanged the key concepts from each law. From Boltzmann one takes the identification of entropy with the number (or weight) of molecular configurations of the state, and hence with the probability of the state. And from Clausius one takes the notion of constrained states, the increase in entropy during spontaneous transitions between such states, and the corollary that the entropy is a maximum in the optimum non-equilibrium state.

The new idea is to generalise the notion of state to a higher level that includes time. For example, in the equilibrium case the system is constrained to be in the state x, which is the value of a non-conserved variable such as the spatial distribution of energy or number. In the non-equilibrium case the relevant states are transitions in a certain time, for example $x \to x'$ in time τ, and the relevant entropy for a system constrained to have such a transition might be written $S^{(2)}(x', x|\tau)$. This is here called the second entropy; it could equally be named the transition entropy, or the two-time entropy. The unconditional transition probability is simply proportional to its exponential.

Conceptually, the second entropy is the same as the ordinary entropy; it is still the logarithm of the number (or weight) of molecular configurations in the state, except that now state refers to a transition, $x \xrightarrow{\tau} x'$, and the relevant molecular configurations are those that give x now and x' at a time τ in the future. What is new is that the non-equilibrium states now correspond to transitions between equilibrium states. This turns out to be the appropriate way to insert time into Boltzmann's and Clausius' laws.

The transition $x \to x'$ in time τ corresponds to a velocity, flux, or rate of change, namely $\overset{\circ}{x} \equiv (x' - x)/\tau$. It may also correspond to an event at a given time or over an interval, $x' = x(\tau|x)$. For simplicity, the word 'flux' will henceforth be used to denote generically all these time dependent objects. For technical reasons, it is useful for the time interval τ to be finite rather than infinitesimal (c.f. the discussion in §1.1 on p. 3). In the present view, the fluxes are the objects of constraint, and the second entropy for a given flux gives a quantitative answer to all the questions that could be asked of a non-equilibrium system.

The analogue of the Second Law for non-equilibrium thermodynamics may be formulated as

> The second entropy increases during
> spontaneous changes in the flux. (1.4)

This says that fluxes spontaneously evolve in time in the direction of increasing second entropy, and that the optimum or most likely flux, which may be called the stable non-equilibrium state, is the one with greatest second entropy. Note that even though flux can be written as conditional on the initial state, $\overset{\circ}{x}(x)$ or $x(\tau|x)$, strictly speaking this law applies to unconditional transitions, which corresponds to maximisation with respect to both x and x'.

As a concrete example, the flux under consideration could be taken as the heat flow in the presence of a fixed temperature gradient. In this case the steady state heat flow observed experimentally would correspond to the maximum of the second entropy with respect to all constrained heat flows in the presence of the given temperature gradient. If it were possible to turn on the temperature gradient, then the transient heat flux observed to increase from zero to its final steady state value would coincide with an increase in the second entropy. If the heat flow were to increase further past this optimum value, the second entropy would begin to decrease. If the temperature gradient is turned off, then the initial second entropy (i.e. that for zero temperature gradient but non-zero heat flux) would be lower than the second entropy for the final state of zero gradient and flux. (In this example, the heat flux is the $\overset{\circ}{x}$ and the temperature gradient of the sub-system is the x. The second entropy is a maximum with respect to x when the sub-system temperature gradient is equal to that applied by the reservoirs.)

In an analogous fashion, Boltzmann's molecular identification of entropy carries over to the second entropy,

The second entropy is the logarithm of the number of molecular configurations that give the flux. (1.5)

As will be discussed in much greater detail in the coming chapters, transitions or fluxes are functions in phase space, just like the equilibrium states themselves, and so the number of molecular configurations is again the associated volume of phase space.

Generalising the notion of states to pair transitions in a specified time is obviously the first in a hierarchy of such higher-order states. One can have the third entropy, for example, that gives the probability of three states, x_1, x_2, and x_3 at times t_1, t_2, and t_3, or, for systems homogeneous in time, at time intervals t_{21} and t_{32}. For higher-orders than this, one might refer to the path entropy. The pair level suffices for steady state non-equilibrium systems, and the higher-orders may be required for transient, harmonic, or otherwise time varying systems.

With this generalised view of transitions as states, the entire theory of non-equilibrium thermodynamics and non-equilibrium statistical mechanics becomes completely analogous to their equilibrium counterparts. In what follows of this chapter, the basic mathematical rules for entropy and second entropy are given. In later chapters these are applied to the physical problems that comprise non-equilibrium thermodynamics and non-equilibrium statistical mechanics.

1.3 Nature of Probability

To begin, a brief philosophical discussion of the concept of probability and the related notion of randomness is in order. Although thermodynamics and statistical mechanics are firmly rooted in the physical sciences with measurable outcomes and concrete consequences, it is not an empty exercise to discuss the conceptual basis that underlies these disciplines. The reason that this is important is that all scientists require an image or internal picture of their field that enables them to rationalise measurements, to reconcile results with intuition, to understand observed phenomena, and to forecast likely outcomes. In short, the conceptual basis of a field provides insight that can explain what is known, and that is generally a prerequisite for detailed calculation or measurement. Without physical insight one is *reduced* to mere mathematical proof: a logical derivation devoid of physical imagery may convict, but it does not convince.

Although it is important—indeed essential—to have mental images of natural phenomena, one also has to recognise that such are a representation of the phenomena, not the phenomena themselves. The fact that the image is a simplification of reality means that in all cases there must eventually arise a contradiction between the image and the fact. When this occurs one must be prepared to either augment the original picture, or else to replace it with a new way of thinking about the phenomenon. This can be more difficult than one would have thought, since so deeply rooted can these internal pictures become that one thinks of the phenomena only in their terms, and the normal reaction is to question the calculation or measurement rather than the concept itself.

In the case of probability, particularly in statistical mechanics, there are three philosophical positions that have been adopted: *frequency*, *credibility*, and *measure*. These can all be formulated so that their mathematical manipulation obeys the laws of probability, so mathematically there is no distinction between them. But as concepts that guide thinking about probability in the physical sciences, they are not equally useful.

1.3.1 Frequency

Probability interpreted as frequency holds that the probability of the outcome of an event is proportional to the number of times that it recurs in a long sequence of trials. This is the most widespread understanding of probability and is as likely to be found in departments of statistics as at horse races, casinos, and stock exchanges. In statistical mechanics, frequency as the conceptual basis for probability is implicit in the ensemble formulation of the subject.

The frequency interpretation of probability in the physical sciences is unsatisfactory for several reasons. Most tellingly, the number of times a particular outcome occurs differs every time the trial is repeated. Accordingly the frequency does not have a unique value for what is the same physical process, which is contrary to the essential requirement of a physical property. One concludes that the identification of frequency with probability confuses cause and effect: the reality is that frequency is a consequence of probability but is not identical to probability.

This confusion of rôles is also evidenced by the fact that frequency can only be calculated after the trial, whereas if probability is to have a physical reality it must exist independent of the trial. Once-only events (e.g. a horse race) deserve a probability, and yet one cannot physically carry out repeated trials of these. In these cases frequentists think of a fictitious trial of repetitious events and speak of probability as referring to the frequency of the outcome in this imaginary world. As mentioned above, in statistical mechanics the ensembles introduced by Boltzmann, and used by Maxwell and by Gibbs, share in common with the frequency interpretation of probability this fictitious replication of the original system.

1.3.2 Credibility

The second view of probability holds that it is the degree of reasonable belief that a particular outcome will occur. This subjectivist interpretation is widespread amongst Bayesian statisticians, and depends upon the connection between probability and inductive logic. A detailed presentation of this connection was given by Cox, who cast assertions in the language of set theory.[8] The difference between deductive and inductive reasoning is exemplified by the syllogisms:

Deductive	**Inductive**
If it rains, Bob carries an umbrella.	If it rains, Bob carries an umbrella.
It is raining.	Bob carries an umbrella.
Therefore, Bob carries an umbrella.	Therefore, it is likely raining.

The deductive reasoning says that rain makes an umbrella certain, $\wp(U|R) = 1$, whereas the inductive reasoning says that an umbrella increases the probability of rain, $\wp(R|U) > \wp(R)$. There are convincing arguments that inductive reasoning is more broadly applicable than deductive reasoning, that it is more applicable to real world problems, and that it more closely reflects the way people actually think and learn.

Inductive reasoning is closely related to the problem of inference, namely estimating the value of some parameter given limited information or data. In practice one communicates the probability distribution of the likely values of the parameter, which is interpreted as indicating one's belief or confidence in those values.

It is emphasised that the rules of inductive reasoning that follow from the set theoretical formulation of assertions are the same as the rules of probability. The issue then is not whether rational beliefs can be formulated as probability, but whether probability is solely a measure of reasonable belief.

In statistical mechanics, and in other branches of science, the subjectivist interpretation of probability underlies the so-called principle of maximum entropy. This is taken to have a particular meaning that is motivated by, but which is different from, the Second Law of Thermodynamics. It states that the probability distribution that should be used is the one that maximises the entropy constrained by the given information. A coherent formulation of statistical mechanics was given in terms of the principle of maximum entropy by

[8]Cox, R. T. (1961), *The Algebra of Probable Inference*, (Johns Hopkins Press, Baltimore, MD). Cox, R. T. (1978), *Of Inference and Inquiry: An Essay in Inductive Logic*, in *The Maximum Entropy Formalism*, Levine, R. D. and Tribus, M. (eds), (MIT Press, Cambridge, MA).

Jaynes.[9] For example, Jaynes asserts that the Maxwell-Boltzmann distribution results from maximising the entropy with respect to the probability, given the information that the system is constrained to have an average energy. Temperature, in this interpretation, is simply a Lagrange multiplier introduced to satisfy the constraint.

In the subjectivist view probability measures beliefs, which are modified by data or information, and entropy measures lack of information or ignorance. For this reason subjectivists often speak of the information entropy. The more information one has, the sharper and more constrained is the probability distribution and the lower is the entropy. The known information or data (e.g. the average number observed in a trial) acts as a constraint on the possible probability distributions. The principle of maximum entropy gives the recipe for finding the correct probability distribution given certain information. This distribution is the least biased distribution since any other distribution would either be inconsistent with the information or have a lower entropy, which implies the existence of additional constraints. Either these additional constraints are known, in which case they should be explicitly taken into account, or they are not known, in which case the lower entropy is a manifestation of an unjustified bias in one's beliefs. It is the laws of probability and the principle of maximum entropy that give the prescription for unbiased and rational beliefs.

In practice, the subjectivist viewpoint of probability and the related principle of maximum entropy have had relatively little impact on thermodynamics and statistical mechanics. There appear to be three main reasons for this. First is mathematical: as detailed in the following sections, there is a one-to-one relationship between the entropy of a state and the probability of a state. It therefore makes no mathematical sense to attempt to maximise the entropy with respect to the probability. Second, is that the approach divorces probability from the underlying physical causes. For example, viewing temperature as simply a Lagrange multiplier precludes the understanding gained from a lifetime of actual experience and a history of experimental measurement. Third, there is a deeply rooted philosophical or psychological resistance to the proposition that a measurement is just the belief of the experimentalist for the value of the measured quantity. Most scientists regard the world as having some objective reality, which it is their happy task to uncover, and the subjectivist viewpoint is alien to this philosophy. The dependence of the outcome of an experiment upon the knowledge of the observer, as in Jaynes' interpretation of statistical mechanics, or as in the general credibility interpretation of probability, is unsettling and disconcerting to those accustomed to dealing with physical phenomena. It is arguable that this interpretation over-emphasises the rôle of the observer to the exclusion of underlying physical causes. Many see it as peculiarly solipsistic to view entropy as a measure of the observer's uncertainty rather than of the physical disorder of the system.

---

[9] Jaynes, E. T. (1957), Phys. Rev. **106**, 620 and **108**, 171. Rosenkrantz, R. D. (1983), (Ed.), *E. T. Jaynes: Papers on Probability, Statistics, and Statistical Physics*, (D. Reidel, Dordrecht).

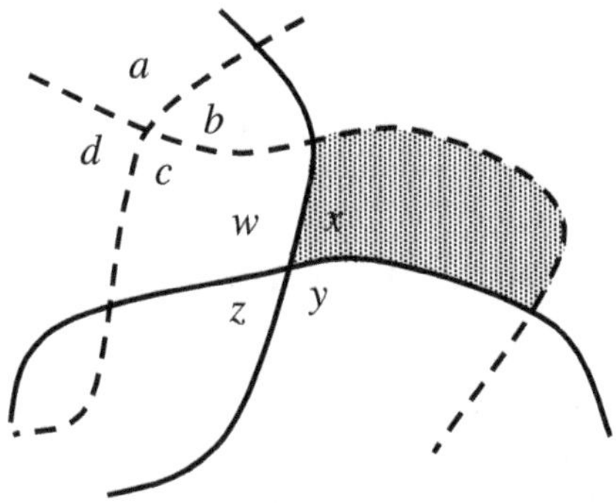

Figure 1.1: The space of conditions that lead to one collection of four mutually exclusive outcomes, $\{a, b, c, d\}$, (dashed boundaries), and to another collection of four mutually exclusive outcomes, $\{w, x, y, z\}$, (solid boundaries). The shaded region corresponds to the joint outcome cx. The weight of each outcome is proportional to the volume or measure of the conditions that cause it.

1.3.3 Measure

The third conceptual basis for probability, and the one adopted in this book, holds that it is a physical property of the system, namely it is the measure (or weight, or proportion) of states that give the designated outcome compared to the total weight of possible outcomes. Popper, in his objective philosophy of science, called it propensity.[10] The idea of measure differs from the frequency interpretation of probability in that it does not require a series of trials, real or imaginary, to obtain, but rather takes the probability to exist as a physical object that is a measurable or calculable from the underlying material properties of the system. In short, probability is taken as the cause, and frequency as an effect.

Figure 1.1 sketches this general idea for the physical basis of probability. The probability of an outcome is determined by its weight as measured in the space of conditions that influence the outcome relative to the weight of all possible outcomes. For example the collection of outcomes might be whether a coin lands head or tails, and the space of conditions might include the mass distribution of the coin, the initial orientation of the coin, the initial linear and angular momenta, the drag force of the wind, the elasticity and friction of the landing place, etc. In principle, by measurement or by calculation, one could map the multidimensional space of these conditions in terms of the particular outcome that they yield. There may be practical issues in quantifying the weight, including the possibility that different types of conditions have different dimensions, and that non-linear transformations of the conditions lead to different relative weights, but it may be argued that the prescription for quantitatively determining the weights in a particular problem is quite separate from the general notion

[10]Popper, K. R. (1959), Brit. J. Philos. Sci. **10**, 25.

of the existence of weight. Accepting that the relative weight of the conditions that lead to an outcome gives the probability of that outcome, it follows that the laws for combining and manipulating probabilities are the laws of set and measure theory. These are set out in §1.4 below.

1.3.4 Determination of Randomness

Any exploration of the physical basis of probability necessitates a discussion of the origin of randomness in the physical universe. This is particularly important in view of the historical development of statistical mechanics. Boltzmann's early identification of the entropy of a state with the number of molecular configurations in the state was made at a time when Newton's (equivalently Hamilton's) equations of motion were assumed to characterise the behaviour of atoms and molecules. These equations are both completely deterministic, which appears to contradict the randomness demanded by a probabilistic description, and time reversible, which appears to contradict the monotonic entropy increase embodied in Clausius' formulation of the Second Law of Thermodynamics.

To answer the first objection, namely how randomness arises from deterministic equations of motion, Boltzmann introduced the concept of an ensemble of systems, each member being an identical macroscopic copy of the original, but differing in their initial microscopic or molecular configuration. The ensemble of systems evolves in time, each member according to the deterministic equations of motion. The probability of a state at a given time is taken to be the proportion of the ensemble that is in the state. The randomness in this picture arises from the assigned distribution of the initial states of the members of the ensemble.

The ensemble picture is today more or less the standard view of probability in thermodynamics and statistical mechanics, but there are several weaknesses in it. First, it is no more than a mental image that has almost no relationship with actuality; the physical reality is not Avogadro's number of copies of identical systems, but rather a single system that is to be measured or characterised. As a mental artifice, ensemble theory is more akin to the frequency interpretation of probability than it is to the measure interpretation of probability as a physical object. Second, the ensemble picture does not give the prescription for the distribution of initial states, which itself determines the probability distribution, but this must be obtained by other physical considerations. Third, as will be discussed in detail in forthcoming chapters, it implies a conservation law for probability, which stems from a fixed number of ensemble members, that is not true in reality. As mentioned above, one must be careful to distinguish the physical reality, in this case probability, from the mental image of reality, in this case ensembles. That the number of ensemble members is conserved in this mental picture of probability does not make it true that probability is conserved in real physical systems. Fourth, ensembles are an artificial way of introducing indeterminacy in physical systems that is both unnecessary and overly complicated. The ensemble picture obscures an understanding of the true physical origin of randomness and precludes the full mathematical exploitation of randomness in statistical mechanics.

Although one can modify the ensemble picture to address each of these concerns, an arguably more satisfactory approach is to identify the actual origin of randomness in physical systems and to develop the notion and quantitative values of probability on that basis.

One possibility for randomness in the physical universe is the recognition that the world is ultimately quantum mechanical, which path was not available at the time of Boltzmann. However, this does not solve the problem because the vast majority of atoms and molecules behave classically due to their size, and for such systems one still has to account for indeterminacy. Quantum mechanics does not differ from classical mechanics for the motion of such atoms and molecules. Although there are systems for which quantum mechanics is essential, it is clear that randomness is more general than this and it occurs in both classical and quantum systems.

A second possibility is to consider that randomness arises from a lack of precision in the specifications of the initial conditions of the system, since it is known that the subsequent trajectory can be exquisitely sensitive to these. A related viewpoint attributes indeterminacy to the fact that the precise details of the intermolecular interactions are either unknown or are too complex to account for fully. Both these ideas tend toward the subjectivist view of probability. They take what are undoubtedly practical difficulties in the calculation of the behaviour of the system to be the physical principle that underlies randomness.

A third possibility, and the approach taken in this book, is based upon the so-called reservoir formalism, where the system of interest is actually a sub-system of a larger system with which it interacts. The molecular configuration of the sub-system is therefore a projection of the configuration of the total system, and the evolution of the sub-system is not determined by the sub-system configuration alone.[11] This indeterminacy in the sub-system evolution when the reservoir configuration is not treated explicitly is manifest in stochastic terms that are added to the classical picture of Hamilton's deterministic equations of motion, and this gives rise to a probabilistic treatment of the sub-system state. Quantitative treatments of the probability follow from the physical nature of the interaction between the sub-system and the reservoir, as will be given in later chapters.

This interpretation of randomness in the equations of motion—it arises from the projection of the total system onto the sub-system that is being treated explicitly—also accounts for the second objection that was made to Boltzmann's

[11] If $\mathbf{\Gamma}_{\rm s}$ denotes the molecular configuration of the sub-system, and $\mathbf{\Gamma}_{\rm r}$ that of the reservoir, then these determine the future configuration of the sub-system, $\mathbf{\Gamma}_{\rm s}' = \mathbf{\Gamma}(t|\mathbf{\Gamma}_{\rm s}, \mathbf{\Gamma}_{\rm r}')$. A different configuration of the reservoir gives a different evolution of the sub-system, $\mathbf{\Gamma}_{\rm s}'' = \mathbf{\Gamma}(t|\mathbf{\Gamma}_{\rm s}, \mathbf{\Gamma}_{\rm r}'')$. This means that if the reservoir coordinates are projected out of the problem, then the current sub-system configuration does not uniquely determine the evolution. This unpredictability is what is meant by saying that the equations of motion of the sub-system are stochastic: the probability distribution of the reservoir configuration determines the probability distribution of the sub-system evolution, and consequently the probability distribution of the sub-system itself. This projection operation has the same effect as what is often called a contracted description.

identification of entropy with the number of molecular configurations, namely that the equations of motion are time reversible but the Second Law of Thermodynamics gives a preferred direction to time. It turns out that the projection operation contributes both stochastic and dissipative terms to the equations of motion, and that these are time irreversible with the resultant tendency for entropy increase legislated by Clausius.

The criticism might be made that there is a degree of circularity in defining probability as weight or measure or even entropy. Whilst the self-referential nature of the definitions is undeniable, they nevertheless reflect the physical origin of probability, and they determine the mathematical rules for its manipulation. The physical interpretation of randomness as arising from the projection of external influences leads to the quantitative evaluation of probability. These laws of probability and its relation to entropy are now given in a general setting that abstracts from any specific physical system.

1.4 States, Entropy, and Probability

1.4.1 Macrostates and Microstates

The statistical basis of thermodynamics and statistical mechanics is here outlined. The ideas set out here that link entropy to probability are quite general and apply more broadly than the physical applications in the rest of the book. Conversely, the rôle of entropy in thermodynamics and statistical mechanics is largely independent of the precise physical or molecular details of the particular system under consideration.

The mathematical theory of entropy and probability is based on set theory (see Fig. 1.1 on p. 11). One needs the concepts of microstates, macrostates, and collectives. Basically, a collective contains all macrostates of a given type, and a macrostate is a set of microstates. At any instant the system may be characterised as being in a macrostate x_α of an x-collective. For example, this could be an energy collective, and x_α is a particular value of the energy of the system.[12] A *collective* is a complete set of disjoint states. This means that at any one time the system is in one, and only one, macrostate of a given collective. For the present, a discrete set of states is considered, $\alpha = 1, 2, 3 \ldots$. Only non-degenerate collectives are allowed, which means that there is a one-to-one relationship between the macrostate label and the value of the physical observable, $\alpha = \beta \Leftrightarrow x_\alpha = x_\beta$.

Macrostates belonging to different collectives are not disjoint, so that the system can simultaneously be in the macrostate x_α and the macrostate y_β, where, for example, x might represent the energy collective and y might represent the collective of number macrostates. This could also be denoted as the macrostate $\{x_\alpha, y_\beta\}$, or, if there is no ambiguity, simply $\alpha\beta$.

[12] For the present purposes, one should imagine that energy is not conserved. For example, this could be the energy in a specified half of the system. Or the 'system' could actually be a sub-system that can exchange energy with a reservoir.

There is a fundamental set of states called *microstates* that are complete, disjoint, and indivisible. For example, in classical statistical mechanics the microstates are taken to be the phase space of the atoms' positions and momenta. It is convenient to label the microstates by Roman letters and the macrostates by Greek letters. Each macrostate is a set of microstates. Because the macrostates in a collective comprise a disjoint and complete set, each microstate belongs to a unique macrostate of a given collective. A microstate belongs to a macrostate if the value of the physical observable when the system is in the microstate is equal to the value of the macrostate, $i \in \alpha \Leftrightarrow x_i = x_\alpha$. At any instant, the system is in one, and only one microstate, and in consequence it is in one, and only one macrostate of a given collective. It can of course be simultaneously in macrostates of different collectives.

The mathematical distinction between macrostates and microstates lies in the fact that microstates cannot be sub-divided into smaller states. But in practical physical applications, one never really has such an idealised mathematical indivisibility. For example, in taking classical phase space as the collective of microstates, in many cases one ignores electronic degrees of freedom, the rotations of non-spherical molecules, or the internal conformations of macromolecules. In almost every case one also ignores the configurations of the system or reservoir external to the sub-system of primary interest, so that in fact the microstates of the sub-system are actually macrostates of the total system.

For this reason the formal expression for entropy and probability that follows is invariant to the representation; it applies equally to macrostates as to microstates. Microstates may still be defined for any physical application, and they are to be interpreted as the finest level of description that will be used in the given problem. However, it is important to keep in mind that such microstates can themselves have internal states that are included as a whole rather than individually.

1.4.2 Weight and Probability

Each microstate has a non-negative weight, ω_i.[13] It follows that each macrostate has a weight that is the total weight of the microstates it contains,

$$\omega_\alpha = \sum_{i\in\alpha} \omega_i. \tag{1.6}$$

In the special case that the microstates have equal weight, it is often convenient to give this a value of unity, $\omega_i = 1$, in which case the weight of a macrostate is just the number of microstates that it contains, $n_\alpha = \sum_{i\in\alpha}$. The weight of the system is

$$W = \sum_\alpha \omega_\alpha = \sum_i \omega_i. \tag{1.7}$$

[13]The physical origin or value of this weight is not required here for the formal development of the theory. In practice, any internal degrees of freedom of the microstates can be expressed through the values of the weights.

Because the collectives are complete, the total weight of the system is the same whichever collective is used to obtain it.

The *probability* of a macrostate is the weight of the state divided by the total weight,

$$\wp_\alpha = \frac{\omega_\alpha}{W}. \tag{1.8}$$

This holds unchanged for microstates. The probability is obviously normalised to unity,

$$\sum_\alpha \wp_\alpha = 1. \tag{1.9}$$

The average of a function is defined as,

$$\langle f \rangle = \sum_\alpha \wp_\alpha f_\alpha. \tag{1.10}$$

This has been written for a function defined on a macrostate collective. An analogous formula holds for a function of the microstates.

The relationship between probability and weight is so direct that conceptually they must be regarded as the same thing. In particular, the laws of probability are a direct consequence of the linear additive nature of the weights and of the rules of set theory. Let α and β be macrostates in different collectives, and let $\alpha\beta$ denote the macrostate that the system is simultaneously in both, which can also be denoted by the intersection of the sets $\alpha \cap \beta$ (see Fig. 1.1 on p. 11). The weight of such a joint state is

$$\omega(\alpha\beta) = \sum_i \omega_i \delta(x_\alpha - x_i)\delta(y_\beta - y_i) = \sum_{i\in\alpha\cap\beta} \omega_i. \tag{1.11}$$

The Kronecker delta that appears here is defined such that $\delta(0) = 1$ and is zero otherwise. The sums are over microstates. If α and β belong to the same collective, $\omega(\alpha\beta) = \omega_\alpha \delta(\alpha - \beta)$. Because of the disjoint, complete nature of the collectives, one must have

$$\sum_\beta \omega(\alpha\beta) = \omega_\alpha, \tag{1.12}$$

which is a type of reduction condition, since the pair weight has been reduced to a singlet weight. It can also be regarded as a type of conservation law for weight. From this it follows

$$\sum_{\alpha,\beta} \omega(\alpha\beta) = W. \tag{1.13}$$

Accordingly the *unconditional* probability that the system is simultaneously in the macrostate α and the macrostate β is

$$\wp(\alpha\beta) = \frac{\omega(\alpha\beta)}{W}. \tag{1.14}$$

If the system is in the macrostate β, then the *conditional* probability that it is in the state α is

$$\wp(\alpha|\beta) = \frac{1}{\omega_\beta} \sum_{i \in \alpha\cap\beta} \omega_i = \frac{\omega(\alpha\beta)}{W} \frac{W}{\omega_\beta} = \frac{\wp(\alpha\beta)}{\wp(\beta)}. \tag{1.15}$$

The first equality follows because the conditioning statement that the system is in the macrostate β means that the total weight is ω_β, and the weight that is the numerator is that part of the total weight that is in the macrostate α. This is more usually written in the form of Bayes' theorem,

$$\wp(\alpha\beta) = \wp(\alpha|\beta)\wp(\beta). \tag{1.16}$$

In words this says that the probability that the system is simultaneously in the state α and the state β is equal to the probability that it is in the state α given that it is in the state β, times the probability that it is in the state β.

The remaining significant law of probability is that of disjunction,

$$\wp(\alpha \cup \beta) = \wp(\alpha) + \wp(\beta) - \wp(\alpha\beta), \tag{1.17}$$

where the union of the two sets appears on the left-hand side. This says that the probability that the system is in the state α, or that it is in the state β (which includes the possibility that it is in both), is equal to the probability that it is in the state α plus the probability that it is in the state β less the probability that it is in both. The last term is evidently a correction for double counting of the weight in the region of intersection.

1.4.3 Entropy

Entropy is conceptually equivalent to both weight and probability. The entropy of the total system is just the logarithm of the total weight,

$$S \equiv k_\mathrm{B} \ln W, \tag{1.18}$$

where Boltzmann's constant is $k_\mathrm{B} = 1.38 \times 10^{-23}$ J/K. The entropy of a macrostate is similarly defined as

$$S_\alpha \equiv k_\mathrm{B} \ln \omega_\alpha = k_\mathrm{B} \ln \sum_{i\in\alpha} \omega_i. \tag{1.19}$$

To avoid ambiguity in the case that one is dealing with more than one collective, the left-hand side would be best written $S(x_\alpha)$. As mentioned above, the formalism has been developed so that there is no real distinction between microstates and macrostates. This means that the entropy of a microstate is also the logarithm of its weight,

$$S_i \equiv k_\mathrm{B} \ln \omega_i. \tag{1.20}$$

In the event that one is dealing with equally weighted microstates, then without loss of generality one may give them unit weight, $\omega_i = 1$. In this case, the weight of a macrostate is just the number of microstates that it contains, $\omega_\alpha = n_\alpha \equiv \sum_{i\in\alpha}$, and the entropy of a macrostate is just the logarithm of the number of microstates that it contains, $S_\alpha = k_\mathrm{B} \ln n_\alpha$. This is Boltzmann's original definition of entropy.

From these definitions, the probability of a macrostate is just the exponential of its entropy,

$$\wp_\alpha = \frac{1}{W} e^{S_\alpha/k_\mathrm{B}}. \tag{1.21}$$

In view of its exceedingly simple definition, entropy is in concept the same as weight and probability. In most cases they could be used interchangeably. Because of its normalisation, probability contains less information than the other two.

Entropy does have one advantage over weight and probability, namely that it is a linear additive variable. The total weight of two independent macrostates is $W_\mathrm{total} = W_\alpha W_\beta$, whereas the total entropy, which is the logarithm of this, is $S_\mathrm{total} = S_\alpha + S_\beta$. Thermodynamics is largely predicated on the notion of linear additivity, which is why energy, number, volume, etc. play such a central rôle. By defining entropy in this logarithmic fashion, many thermodynamic formulae can be expressed more simply in terms of entropy than in terms of weight.

In physical terms, entropy is often interpreted as a measure of the disorder of a system. If one compares two macrostates, then the one with the greater number of microstates (or greater weight of microstates if they are not equally likely) is the one with greater entropy. But in the sense that ordered means predictable and disordered means unpredictable, the more microstates there are the less one is able to predict the specific microstate that the system is in. As another example, if one could place a few objects on a large grid, then the number of regular or ordered arrangements is very much less than the number of irregular arrangements, and since the logarithm of the number can be identified with the entropy, one sees that greater entropy corresponds to greater disorder. Similarly, in communications theory the entropy is often equated to the information content of a message: the less predictable the successive symbols in the message then the more informative it is said to be.[14]

One can rearrange the total entropy as

$$\begin{aligned} S &= k_\mathrm{B} \ln W \\ &= k_\mathrm{B} \sum_\alpha \wp_\alpha \ln \frac{\omega_\alpha W}{\omega_\alpha} \\ &= \sum_\alpha \wp_\alpha \left[S_\alpha - k_\mathrm{B} \ln \wp_\alpha \right]. \end{aligned} \tag{1.22}$$

The second equality follows because the sum over the macrostate probability is normalised to unity, and the third equality follows from the definitions, $S_\alpha \equiv$

[14]Information is here meant in a computational rather than a sociological sense; it is debatable whether a painting by Jackson Pollack is indeed more informative than one by, say, Piet Mondrian.

$k_{\mathrm{B}} \ln \omega_\alpha$ and $\wp_\alpha \equiv \omega_\alpha / W$. The interpretation of this is that the total entropy is the sum of two contributions: the $-k_{\mathrm{B}} \wp_\alpha \ln \wp_\alpha$ term represents the disorder associated with the breadth of the probability distribution (i.e. the uncertainty over which macrostate the system is in), and the $\wp_\alpha S_\alpha$ term represents the internal entropy of each macrostate.

This expression holds as a sum over any collective of macrostates, and it also holds as a sum over microstates. With this expression there is no preferred representation of the system; this is an invariant expression for the entropy of the system.

This invariance is exceedingly important. One often sees in both textbooks and the research literature the entropy written as[15]

$$\tilde{S} = -k_{\mathrm{B}} \sum_\alpha \wp_\alpha \ln \wp_\alpha. \tag{1.23}$$

Compared to the correct expression, Eq. (1.22), one sees that this is missing the internal entropy of the states. Equivalently, it sets this term to zero,

$$\sum_\alpha \wp_\alpha \tilde{S}_\alpha = 0. \tag{1.24}$$

It is often the case that only differences in entropy have physical meaning, and so it is always possible to add a constant to the macrostate entropy, $\tilde{S}_\alpha = S_\alpha + X$ (equivalently, a multiplicative factor for the macrostate weights), that makes the sum vanish, $X \equiv -\sum_\alpha \wp_\alpha S_\alpha$. However the side effect of this is that the total entropy $\tilde{S}$ would have a different value depending upon the collective used to evaluate it. Also, one would have to choose a different constant X for each collective, but this would mean that the macrostate weights were no longer the sum of weights of the microstates in each macrostate. Such a loss of the property of linear additivity is very severe, particularly in the case that the microstate weights have a quantitative value derived from some physical argument.

In the case of thermodynamics and statistical mechanics, which generally treat a sub-system and a reservoir, the full expression, Eq. (1.22), gives the total entropy of the sub-system and the reservoir, whether the sum is over microstates or macrostates. The variant expression, Eq. (1.23), gives the sub-system entropy alone if the sum is over microstates, and it gives a quantity that to leading-order is zero if the sum is over macrostates.

---

[15]This is essentially the H-function of Boltzmann. Gibbs called it the average of the index of probability, [Gibbs, J. W., (1902), *Elementary Principles in Statistical Mechanics Developed with Special Reference to the Rational Foundation of Thermodynamics*, (Yale Univ. Press, New Haven, CT). Reprinted by Dover, New York (1960)]. Shannon defined the so-called 'information entropy' by this expression in his theory of communications, [Shannon, C. E. and Weaver, W. (1949), *The Mathematical Theory of Communication*, (Univ. of Illinois Press, Urbana)]. Arguably, the most spirited proponent of this expression for the entropy has been Jaynes, who based his maximum entropy formulation of statistical mechanics and of probability theory upon it, (Jaynes, 1957; Rosenkrantz, 1983).

1.4.4 Transitions and the Second Entropy

Weight

In both the equilibrium and the non-equilibrium case, the microscopic state of the system changes in time. This is represented by transitions between microstates in a given time interval, $i \xrightarrow{\tau} j$. These microstate transitions may be deterministic, but, as for the microstate weights themselves, this is a mathematical abstraction that is generally not satisfied in the real world. Due to the 'internal' states that physical microstates often have (in some cases such internal states actually represent the state of the system external to the sub-system of interest), the microstate transitions generally have a stochastic element in addition to any underlying deterministic behaviour. Both extreme cases (fully deterministic and completely random), and everything in between, can be accommodated by a transition weight, $\omega(j, i|\tau)$, where the order of the arguments means $i \xrightarrow{\tau} j$.

Invoking a time interval implicitly assumes an equilibrium system which is to say a system that is homogeneous in time. This is consistent with the analysis of the preceding sections, where the probability of a state, $\wp_\alpha$ was explored, rather than the probability of a state at a given time, $\wp_\alpha(t)$. The present case that requires only a time interval is sufficient to account for the simplest non-equilibrium state, namely a steady state system. Shortly this will be generalised from a transition between a pair of microstates to a sequence of consecutive pair transitions, otherwise known as a path or a trajectory. In §8.3.1 the generic treatment will be given for time dependent weights, which is required for the most general non-equilibrium system.

Let α and β label macrostates in either the same or in different collectives. The weight of a transition between these two macrostates, $\alpha \xrightarrow{\tau} \beta$, is

$$\omega(\beta, \alpha|\tau) = \sum_{i\in\alpha} \sum_{j\in\beta} \omega(j, i|\tau). \tag{1.25}$$

Even if the microstate transitions are fully deterministic, the transitions between macrostates are always stochastic. This is not to say that they are completely random (in the sense of a uniform probability distribution), but rather that they are statistically rather than perfectly predictable. The stochastic character of macrostate transitions arises from the fact that specifying the macrostate does not specify the microstate that the system is in.

For each microstate i there is a conjugate microstate $i^\dagger$ such that

$$\omega(j, i|\tau) = \omega(i^\dagger, j^\dagger|\tau). \tag{1.26}$$

That is, the forward transition between the original states has the same weight as the reverse transition between the conjugate states. This is called *microscopic reversibility*. In a physical system, $i^\dagger$ represents the same microstate as i but with all the velocities reversed.[16] Microscopic reversibility must hold in

[16]If any magnetic fields or Coriolis forces are present, these should also be reversed, since both are derived from velocities.

a physical system for both deterministic and stochastic rules for the transitions between microstates. Note that in the sum over states, both i and $i^\dagger$ occur. For an equilibrium system, a microstate and its conjugate have the same weight, $\omega(i) = \omega(i^\dagger)$.

For each macrostate α in a particular collective there is also a conjugate macrostate, $\alpha^\dagger$. Again this has the physical interpretation of the macrostate with positions unchanged but velocities reversed. If the physical observable is an even function of the velocities, then $x_\alpha = x_{\alpha^\dagger}$, and, due to the non-degenerate nature of the macrostate labels, $\alpha = \alpha^\dagger$. If the physical observable is an odd function of the velocities, then $x_\alpha = -x_{\alpha^\dagger}$. If the collective is not an even function of the velocities (i.e. it has either odd or else mixed parity), then $\alpha \neq \alpha^\dagger$, and both occur once in any sum over macrostates. In view of this, the weight of the reverse transition between the conjugate macrostates satisfies

$$\begin{aligned}
\omega(\alpha^\dagger, \beta^\dagger|\tau) &= \sum_{i\in\alpha^\dagger}\sum_{j\in\beta^\dagger} \omega(i,j|\tau) \\
&= \sum_{i^\dagger\in\alpha}\sum_{j^\dagger\in\beta} \omega(i^\dagger,j^\dagger|\tau) \\
&= \sum_{i^\dagger\in\alpha}\sum_{j^\dagger\in\beta} \omega(j,i|\tau) \\
&= \omega(\beta,\alpha|\tau). \qquad (1.27)
\end{aligned}$$

The second equality uses the fact that $i \in \alpha \Leftrightarrow i^\dagger \in \alpha^\dagger$, and that $(\alpha^\dagger)^\dagger = \alpha$. The third equality uses microscopic reversibility. The final equality uses the fact that the $i^\dagger$ and $j^\dagger$ are dummy summation variables, and that each state and its conjugate occur once in the summation. One sees that microscopic reversibility implies macroscopic reversibility, which is to say that the forward and reverse transitions between macrostates have the same weight. This result holds for all macrostates: even, odd, and mixed. For macrostates of even parity, one has the stronger result $\omega(\alpha,\beta|\tau) = \omega(\beta,\alpha|\tau)$, but this is not required.

Macroscopic reversibility appears to say that there is no preferred direction of time, which would seem to contradict the Second Law of Thermodynamics. This question is addressed following the discussion of probability below.

For the case of simultaneous macrostates treated in §1.4.2, where $\omega(\alpha,\beta)$ was the weight attached to the system being in both the macrostate α and the macrostate β, a reduction condition was established as Eq. (1.12), $\sum_\beta \omega(\alpha,\beta) = \omega(\alpha)$. This followed from the disjoint, complete nature of the macrostates. There is obviously a close relationship between simultaneous macrostates and the present problem of transitions between macrostates, since the transition $i \stackrel{\tau}{\rightarrow} j$ can be regarded as the joint macrostate, i at t *and* j at $t+\tau$. Accordingly, since the set of target microstates is complete and disjoint, one must have a reduction rule for the transition,

$$\sum_j \omega(j,i|\tau) = \omega(i), \qquad (1.28)$$

where on the right-hand side is the weight of the microstate, which was denoted above as ω_i. This may be interpreted as a conservation law for weight: the original weight of microstate i is distributed unchanged by the transitions amongst all the target microstates j. In view of microscopic reversibility one must have

$$\sum_i \omega(j,i|\tau) = \sum_i \omega(i^\dagger, j^\dagger|\tau) = \omega(j^\dagger) = \omega(j). \tag{1.29}$$

Similar normalisations follow for the macrostate transitions,

$$\sum_\alpha \omega(\alpha,\beta|\tau) = \omega(\beta), \text{ and } \sum_\beta \omega(\alpha,\beta|\tau) = \omega(\alpha). \tag{1.30}$$

This result will be used to establish the very important reduction condition on the second entropy below.

In view of this normalisation, the total weight of transitions for the given time interval is

$$\sum_{i,j} \omega(j,i|\tau) = \sum_i \omega(i) = W. \tag{1.31}$$

This is the same as the total weight of the system, and is independent of the length of the time interval. There is no fundamental distinction between microstates and macrostates in these definitions, and the same total weight holds for the macrostate transitions,

$$\sum_{\alpha,\beta} \omega(\alpha,\beta|\tau) = \sum_\beta \omega(\beta) = W. \tag{1.32}$$

Probability

The unconditional transition probability is

$$\wp(\beta,\alpha|\tau) \equiv \frac{\omega(\beta,\alpha|\tau)}{W}. \tag{1.33}$$

The joint probability defined in §1.4.2, $\wp(\alpha\beta) \equiv \omega(\alpha\beta)/W$, is the $\tau \to 0$ of the present expression. The conditional transition probability is

$$\wp(\beta|\alpha,\tau) \equiv \frac{\wp(\beta,\alpha|\tau)}{\wp(\alpha)}. \tag{1.34}$$

This is the probability that the system will be in the macrostate β at a time τ in the future, given that it is currently in the macrostate α. These definitions apply as well to microstates.

The macroscopic reversibility for the weights yields

$$\wp(\alpha^\dagger, \beta^\dagger|\tau) = \wp(\beta,\alpha|\tau), \tag{1.35}$$

which says that one is just as likely to observe the forward transition between two macrostates as one is to observe the reverse transition between their conjugate states. For the conditional transition probability one obtains

$$\wp(\alpha^\dagger|\beta^\dagger,\tau) = \frac{\wp(\alpha^\dagger,\beta^\dagger|\tau)}{\wp(\beta^\dagger)} = \frac{\wp(\beta,\alpha|\tau)}{\wp(\beta)} = \wp(\beta|\alpha,\tau)\frac{\wp(\alpha)}{\wp(\beta)}. \tag{1.36}$$

This may be rearranged as

$$\frac{\wp(\alpha^\dagger|\beta^\dagger,\tau)}{\wp(\beta|\alpha,\tau)} = \frac{\wp(\alpha)}{\wp(\beta)} = e^{[S(\alpha)-S(\beta)]/k_{\rm B}}. \tag{1.37}$$

This says that the ratio of forward and reverse conditional transition probabilities is proportional to the exponential of the entropy difference of the two macrostates.

Boltzmann and Clausius Revisited

Before proceeding to the second entropy, it is worthwhile using this result for the conditional transition probability to revisit the discussion in §1.1 on p. 3 of the way in which Boltzmann's physical interpretation of entropy justifies the Second Law of Thermodynamics. In the case that α and β belong to the same collective of macrostates, and that γ is a macrostate approximately half way between the two of them, then the product of the ratio of the conditional transition probabilities from γ is

$$e^{[S(\alpha)-S(\beta)]/k_{\rm B}} = \frac{\wp(\alpha|\gamma,\tau)}{\wp(\gamma^\dagger|\alpha^\dagger,\tau)}\frac{\wp(\gamma|\beta,\tau)}{\wp(\beta^\dagger|\gamma^\dagger,\tau)} \approx \frac{\wp(\alpha|\gamma,\tau)^2}{\wp(\beta^\dagger|\gamma^\dagger,\tau)^2}. \tag{1.38}$$

The approximate equality follows because γ is the mid-macrostate, which implies that the transition $\gamma \to \alpha$ ought be approximately the same as the transition $\beta \to \gamma$, and the transition $\alpha^\dagger \to \gamma^\dagger$ ought be approximately the same as the transition $\gamma^\dagger \to \beta^\dagger$. For simplicity one can look at even parity macrostates, in which case this becomes

$$\frac{\wp(\alpha|\gamma,\tau)}{\wp(\beta|\gamma,\tau)} \approx e^{[S(\alpha)-S(\beta)]/2k_{\rm B}}. \tag{1.39}$$

This says that if the macrostates are ordered in terms of increasing entropy, then the likelihood of observing an entropy increasing transition from the current state over an entropy decreasing transition equals the exponential of half the difference in their entropy.[17] This is the connection between Clausius' law for spontaneous transitions and Boltzmann's physical identification of entropy.

Macroscopic reversibility for even parity states, $\wp(\alpha,\beta|\tau) = \wp(\beta,\alpha|\tau)$, says that the probability of the forward transition occurring is the same as that of the

[17] The factor of one half found here corrects its absence in the hand-waving argument in §1.1 on p. 3. One can invoke the mid-point theorem to assert that there must exist a γ between the two macrostates that makes the present result correct.

backward transition. Note that this is the unconditional probability. The result holds even if one state, say α, is far less likely than the other. Mathematically, the result says that if $\wp(\alpha) \ll \wp(\beta)$, then the conditional transition to the more likely state, $\alpha \xrightarrow{\tau} \beta$ is far more likely than the reverse, $\wp(\beta|\alpha,\tau) \gg \wp(\alpha|\beta,\tau)$, and the likelihood factors exactly balance. Physically, the result means that for every fluctuation *from* equilibrium, there is a reverse fluctuation back *to* equilibrium, where equilibrium means the most likely macrostate. This means that the unconditional transition from the unlikely state, $\wp(\beta,\alpha|\tau)$, which is in the direction toward equilibrium, must have been preceded by a fluctuation away from equilibrium, $\wp(\alpha,\beta|\tau)$.

Second Entropy

As for the macrostates themselves, for macrostate transitions one can define an entropy—the second entropy—that is the logarithm of the weight,

$$S^{(2)}(\alpha,\beta|\tau) \equiv k_{\mathrm{B}} \ln \omega(\alpha,\beta|\tau). \tag{1.40}$$

This could also be called the transition entropy, or the two-time entropy. The superscript (2) is actually redundant since the order of the entropy can be deduced from the number of its arguments. However, it does serve to differentiate the entropy of a transition from the ordinary entropy of multiple macrostates in different collectives. Also, by explicitly exhibiting the superscript as here it is emphasised that this is a distinct object from the more familiar ordinary entropy. One has an identical definition for the second entropy of microstate transitions.

By definition, the exponential of the second entropy gives essentially the unconditional transition probability,

$$\wp(\alpha,\beta|\tau) = \frac{1}{W} e^{S^{(2)}(\alpha,\beta|\tau)/k_{\mathrm{B}}}. \tag{1.41}$$

Similarly, the conditional transition probability is given by

$$\wp(\alpha|\beta,\tau) = e^{[S^{(2)}(\alpha,\beta|\tau)-S(\beta)]/k_{\mathrm{B}}}. \tag{1.42}$$

This suggests defining the conditional second entropy as

$$S^{(2)}(\alpha|\beta,\tau) \equiv S^{(2)}(\alpha,\beta|\tau) - S(\beta), \tag{1.43}$$

in terms of which the conditional transition probability is

$$\wp(\alpha|\beta,\tau) = e^{S^{(2)}(\alpha|\beta,\tau)/k_{\mathrm{B}}}. \tag{1.44}$$

The most likely state given the current state, $\overline{\alpha}(\tau|\beta)$, is the one that maximises the second entropy,

$$\left.\frac{\partial S^{(2)}(\alpha,\beta|\tau)}{\partial \alpha}\right|_{\alpha=\overline{\alpha}} = 0. \tag{1.45}$$

This definition, together with the normalisation condition on the transition weight, Eq. (1.30), may be used to establish the very important reduction condition,

$$\begin{aligned} S(\beta) &= k_{\mathrm{B}} \ln \omega(\beta) \\ &= k_{\mathrm{B}} \ln \sum_{\alpha} \omega^{(2)}(\alpha, \beta|\tau) \\ &\approx k_{\mathrm{B}} \ln \omega^{(2)}(\overline{\alpha}(\tau|\beta), \beta|\tau) \\ &= S^{(2)}(\overline{\alpha}(\tau|\beta), \beta|\tau). \end{aligned} \tag{1.46}$$

The third equality used the fact that the logarithm of a sum is approximately equal to the logarithm of the largest term in the sum, which is valid when the distribution of states α is sharply peaked about the most likely state, $\overline{\alpha}$. This reduction condition says that the maximum value of the second entropy in the future (or the past) is equal to the first entropy of the current state. This reduction condition will be generalised in §8.3.1 below to the case of time dependent weights, which is required for the most general non-equilibrium system.

1.4.5 The Continuum

The preceding formulation of macrostates, probability, and entropy has been in terms of discrete states, $\alpha = 1, 2, 3, \ldots$. In most cases, particularly in thermodynamics and classical statistical mechanics, the macrostates form a continuum. Energy is an example of such a continuous variable. The formalism is virtually unchanged in the continuum limit.

For example, if x is a real number that represents the value of a physical observable, then one defines the weight density $\omega(x)$ and the corresponding probability density, $\wp(x) = \omega(x)/W$. The latter represents the probability of finding the system within $\mathrm{d}x$ of x. The total weight is just

$$W = \int \mathrm{d}x\, \omega(x), \tag{1.47}$$

which is dimensionless. The average of a function of the continuous macrostate label is as usual

$$\langle f \rangle = \int \mathrm{d}x\, \wp(x) f(x). \tag{1.48}$$

There is one minor complication in defining the entropy for such a continuum of macrostates. Since it is preferable that arguments of logarithms be dimensionless, one introduces an arbitrary width $\Delta(x)$ that has the same dimensions as x. In terms of this the entropy of the macrostate x is defined to be

$$S(x) \equiv k_{\mathrm{B}} \ln[\omega(x)\Delta(x)]. \tag{1.49}$$

The probability density remains proportional to the exponential of the entropy, but the width also appears,

$$\wp(x) \equiv \frac{\omega(x)}{W} = \frac{e^{S(x)/k_{\mathrm{B}}}}{\Delta(x) W}. \tag{1.50}$$

It is emphasised that the width $\Delta(x)$ has no physical significance.[18] It can be taken to be constant, or it can be chosen to vary with x, or it can be neglected altogether. The explicit appearance of the width in the probability density cancels with its implicit contribution to the entropy, so that it in no way affects the probability density itself. Since the total weight is dimensionless, the explicit appearance of the width gives the probability density the correct dimensions of inverse x. For simplicity, it is probably best to take the width to be independent of x. In this case, differences in entropy or derivatives of the entropy do not depend upon the choice of its constant value. In practice, entropy is such a rapidly varying function of the physical observable that even if one has chosen the width to vary with x, this variation is negligible compared to that of the entropy itself.

As in the discrete case, the total entropy may be written as an average of the internal entropy and the probability,

$$\begin{aligned} S &= k_{\rm B} \ln W \\ &= \int {\rm d}x\, \wp(x) k_{\rm B} \ln W \\ &= \int {\rm d}x\, \wp(x) \left[S(x) - k_{\rm B} \ln \frac{e^{S(x)/k_{\rm B}}}{W} \right] \\ &= \int {\rm d}x\, \wp(x) \left[S(x) - k_{\rm B} \ln \{ \wp(x) \Delta(x) \} \right]. \end{aligned} \tag{1.51}$$

This is the continuum analogue of Eq. (1.22). Although the arbitrary width appears explicitly here, it cancels with the corresponding term in $S(x)$, and the total entropy is independent of the choice of $\Delta(x)$.

1.5 Reservoirs

In applications of thermodynamics and statistical mechanics one generally has in mind a particular system. The surroundings or rest of the universe are of interest only in so far as they affect the particular system, and it is almost always the case that such effects are indirect. If the details of such interactions were important, the relevant part of the surroundings would have to be included as part of the system of interest.

For example, if one is interested in the structure of a protein in solution, then most of the liquid can be ignored except possibly for those molecules immediately adjacent to the protein. The molecules of the liquid far removed from the protein need not be treated explicitly, but they appear indirectly, as a heat bath, for example, that sets the temperature of the protein and adjacent liquid. Similarly, if one is interested in the dynamics of protein folding, the distant

[18] In classical statistical mechanics, it is conventional to take the width or volume element of phase space to be proportional to Planck's constant to the power of the size of the subsystem. This is purely a convention that has no fundamental justification in terms of quantum mechanics or in terms of experimental measurement.

liquid molecules enter the equations of motion indirectly via a stochastic term derived from their temperature.

The use of reservoirs greatly simplifies thermodynamics and statistical mechanics since it avoids having to keep track of the full detail of the surroundings. This external influence is accounted for in a gross sense via a statistical probability that is derived from the thermodynamic state of the surroundings rather than from its exact molecular identity.

The reservoir formalism was originally developed by Gibbs (1902) to handle this situation. The formalism will be treated in explicit detail in later chapters, particularly for the non-equilibrium equations of motion. Here a brief outline of the general approach is described.

1.5.1 Equilibrium Systems

The total system consists of a sub-system and a reservoir, with the sub-system being of detailed interest and the reservoir being secondary. One generally identifies one or more conserved variables, the linear additive quantities such as the energy, number, volume, charge etc. that can exchange between the sub-system and the reservoir. Let x be the conserved quantity that is exchangeable between the reservoir and the sub-system,

$$x_{\text{total}} = x_{\text{s}} + x_{\text{r}}, \tag{1.52}$$

where the subscripts s and r stand for sub-system and reservoir, respectively. The total system is isolated from the rest of the universe, so that the total quantity does not change, $x_{\text{total}} = \text{const.}$ Consequently, changes in the sub-system are equal and opposite to those of the reservoir, $\Delta x_{\text{s}} = -\Delta x_{\text{r}}$.

The total entropy when the sub-system is in the macrostate x_{s} is the sum of that of the sub-system and that of the reservoir,

$$S_{\text{total}}(x_{\text{s}}|x_{\text{total}}) = S_{\text{s}}(x_{\text{s}}) + S_{\text{r}}(x_{\text{total}} - x_{\text{s}}). \tag{1.53}$$

Writing the entropy as the sum of that of the individual systems in the respective states is valid if the interaction between the two systems does not perturb significantly their individual natures. This is the case if the volume of the boundary region is very much less than the volume of either the sub-system or the reservoir. One further assumes that the reservoir is very much larger than the sub-system. Hence the reservoir formalism is predicated on the condition $|x_{\text{bdry}}| \ll |x_{\text{s}}| \ll |x_{\text{total}}|$. Since entropy is an extensive variable (this is the reason for defining it as a logarithm), it is of the same order as the macrostate itself. This enables a Taylor expansion of the reservoir entropy to be truncated after the linear term,

$$\begin{aligned} S_{\text{r}}(x_{\text{total}} - x_{\text{s}}) &= S_{\text{r}}(x_{\text{total}}) - x_{\text{s}} \frac{\mathrm{d}S_{\text{r}}(x_{\text{total}})}{\mathrm{d}x_{\text{total}}} + \mathcal{O}\left(\frac{x_{\text{s}}^2}{x_{\text{total}}}\right) \\ &= \text{const.} - x_{\text{s}}\lambda_{\text{r}}. \end{aligned} \tag{1.54}$$

The constant is independent of the sub-system and is of no interest or physical import. The specific field variable of the reservoir, which is the entropy derivative, depends upon the exchangeable variable; examples include the reservoir temperature, chemical potential, pressure, electrostatic potential etc. The first neglected term and higher are smaller than the retained term by factors of the size ratio of the sub-system to the reservoir, which vanishes in the present limit that the reservoir is infinitely larger than the sub-system. With this the sub-system dependent part of the total entropy is

$$S_{\text{total}}(x_{\text{s}}|\lambda_{\text{r}}) = S_{\text{s}}(x_{\text{s}}) - x_{\text{s}}\lambda_{\text{r}}. \tag{1.55}$$

Notice how now the reservoir only enters through its field variable.

This may be made more concrete by considering the canonical equilibrium system. In this case the sub-system can exchange energy, $x_{\text{s}} \Rightarrow E_{\text{s}}$ with a reservoir of (reciprocal) temperature, $\lambda_{\text{r}} \Rightarrow T_{\text{r}}^{-1} \equiv \partial S_{\text{r}}(E_{\text{r}}, N_{\text{r}}, V_{\text{r}})/\partial E_{\text{r}}$. The total entropy in this case is related to the Helmholtz free energy, as is discussed below.

The probability distribution for the macrostate of the sub-system is proportional to the exponential of the total entropy. That is

$$\wp(x_{\text{s}}|\lambda_{\text{r}}) = \frac{1}{W(\lambda_{\text{r}})} e^{S_{\text{s}}(x_{\text{s}})/k_{\text{B}}} e^{-x_{\text{s}}\lambda_{\text{r}}/k_{\text{B}}}, \tag{1.56}$$

where the total weight, also called the partition function, is

$$W(\lambda_{\text{r}}) = \int \mathrm{d}x_{\text{s}}\, e^{S_{\text{s}}(x_{\text{s}})/k_{\text{B}}} e^{-x_{\text{s}}\lambda_{\text{r}}/k_{\text{B}}}. \tag{1.57}$$

One could include the width of the macrostates to make this dimensionless, if one desired.

Again to make this concrete, for microstates of the sub-system, $\mathbf{\Gamma}_{\text{s}}$, which have no internal entropy, $S_{\text{s}}(\mathbf{\Gamma}_{\text{s}}) = 0$, for the canonical equilibrium system, $x_{\text{s}} \Rightarrow E_{\text{s}}(\mathbf{\Gamma}_{\text{s}})$ and $\lambda_{\text{r}} \Rightarrow T_{\text{r}}^{-1}$, this probability is $\wp(\mathbf{\Gamma}_{\text{s}}|T_{\text{r}}) = W(T_{\text{r}})^{-1} e^{-E_{\text{s}}(\mathbf{\Gamma}_{\text{s}})/k_{\text{B}}T_{\text{r}}}$. This may be recognised as the Maxwell-Boltzmann probability distribution.

The summed total entropy is $S_{\text{total}}(\lambda_{\text{r}}) = k_{\text{B}} \ln W(\lambda_{\text{r}})$. The word 'total' here means the sub-system entropy plus the reservoir entropy, and the word 'summed' means that it is summed over all sub-system macrostates. Here and below, reservoir entropy really means the sub-system dependent part of the reservoir entropy. In contrast there is the constrained sub-system entropy, $S_{\text{s}}(x_{\text{s}})$, the total constrained entropy, $S_{\text{total}}(x_{\text{s}}|\lambda_{\text{r}}) = S_{\text{s}}(x_{\text{s}}) - x_{\text{s}}\lambda_{\text{r}}$, and the maximal value of the total constrained entropy, $S_{\text{total}}(\overline{x}_{\text{s}}|\lambda_{\text{r}}) = S_{\text{s}}(\overline{x}_{\text{s}}) - \overline{x}_{\text{s}}\lambda_{\text{r}}$. The macrostate that gives the latter satisfies

$$\left.\frac{\partial S_{\text{total}}(x_{\text{s}}|\lambda_{\text{r}})}{\partial x_{\text{s}}}\right|_{x_{\text{s}}=\overline{x}_{\text{s}}} = 0, \text{ or } \left.\frac{\partial S_{\text{s}}(x_{\text{s}})}{\partial x_{\text{s}}}\right|_{x_{\text{s}}=\overline{x}_{\text{s}}} = \lambda_{\text{r}}. \tag{1.58}$$

Since the x derivative of the entropy is the field variable λ, this says that in the most likely macrostate there is equality between the field variables of the sub-system and the reservoir, $\overline{\lambda}_s \equiv \lambda_s(\overline{x}_s) = \lambda_r$. For example, if x were energy, then since the energy derivative of the entropy is equal to the reciprocal of the temperature, this says that in the most likely energy macrostate there is temperature equality between the sub-system and the reservoir.

Because the summed total entropy contains the maximal total entropy, one has the strict inequalities

$$S_{\text{total}}(\lambda_r) > S_{\text{total}}(\overline{x}_s|\lambda_r) \geq S_{\text{total}}(x_s|\lambda_r). \tag{1.59}$$

Thermodynamics always refers to the most likely state, and it is $S_{\text{total}}(\overline{x}_s|\lambda_r)$ that is generally meant.[19] Statistical mechanics generally refers to the sum over states, so in this case it is $S_{\text{total}}(\lambda_r)$. In practice the entropy is an exceedingly rapidly varying function of its argument, and so with practically negligible error one has $S_{\text{total}}(\lambda_r) \approx S_{\text{total}}(\overline{x}_s|\lambda_r)$. Finally, thermodynamics tends to deal with free energies rather than entropy. The relationship between the two is relatively trivial, since in general the free energy is minus the reservoir temperature times the total maximal constrained entropy,

$$F(\lambda_r) \equiv -T_r S_{\text{total}}(\overline{x}_s|\lambda_r) = T_r\lambda_r\overline{x}_s - T_r S_s(\overline{x}_s) \approx -k_B T_r \ln W(\lambda_r). \tag{1.60}$$

The final equality is the statistical mechanical formula for the free energy in terms of the partition function. For the case that the exchangeable variable is energy, one can make the identification $x_s \Leftrightarrow E_s$ and $\lambda_r \Leftrightarrow 1/T_r$, and this reads $F(T_r) = \overline{E}_s - T_r S_s(\overline{E}_s)$, which is the usual thermodynamic formula for the Helmholtz free energy.

The average value of the sub-system macrostate is

$$\begin{aligned} \langle x_s \rangle &\equiv \int \mathrm{d}x_s\, \wp(x_s|\lambda_r) x_s \\ &= \frac{1}{W(\lambda_r)} \int \mathrm{d}x_s\, e^{S_s(x_s)/k_B} e^{-x_s\lambda_r/k_B} x_s. \end{aligned} \tag{1.61}$$

It is an exercise to show that this can also be written as the derivative of the partition function,

$$\langle x_s \rangle = -k_B \frac{\partial \ln W(\lambda_r)}{\partial \lambda_r}. \tag{1.62}$$

In general the probability distribution is sharply peaked about its most likely value, which means that there is consistency between the thermodynamic formula and the statistical mechanical formula $\langle x_s \rangle \approx \overline{x}_s$.

1.5.2 Non-Equilibrium Steady State

A relatively small modification of the reservoir formalism suffices for one of the most common non-equilibrium systems, namely a steady thermodynamic flux.

[19] Most textbooks do not distinguish explicitly between the total entropy and the sub-system entropy.

Examples include heat flow due to a temperature gradient, shear flow due to a velocity gradient, and diffusion due to a chemical potential gradient. The specific case of steady heat flow will be treated in Ch. 4, and what follows is an abbreviated survey of the generic approach.

For the simplest case of one-dimensional flow in the z-direction, one considers the sub-system as sandwiched between two reservoirs with different values for the conjugate field variable, say $\lambda_{\mathrm{r},\pm}$. If the boundaries of the sub-system are at $z = \pm L_z/2$, then it turns out that the relevant variables are

$$\lambda_{\mathrm{r},0} \equiv \frac{\lambda_{\mathrm{r},+} + \lambda_{\mathrm{r},-}}{2}, \text{ and } \lambda_{\mathrm{r},1} \equiv \frac{\lambda_{\mathrm{r},+} - \lambda_{\mathrm{r},-}}{L_z}. \tag{1.63}$$

These have the interpretation of the mean value of the field variable, and the gradient of the field variable, respectively.

Let $\rho_x(z)$ be the volume density of the field variable that is exchangeable with the reservoirs (e.g. in the case of heat flow it is the energy density, and λ is the inverse temperature). Then the zeroth and first moments are

$$X_0 = A \int_{-L_z/2}^{L_z/2} \mathrm{d}z\, \rho_x(z), \text{ and } X_1 = A \int_{-L_z/2}^{L_z/2} \mathrm{d}z\, z\rho_x(z), \tag{1.64}$$

where A is the cross-sectional area. The entropy of the sub-system in the macrostate represented by these two moments is written $S_{\mathrm{s}}(X_0, X_1)$. It is straightforward to show that these moments are conjugate to the field variables defined above,

$$\frac{\partial S_{\mathrm{s}}(X_0, X_1)}{\partial X_0} = \lambda_{\mathrm{s},0}, \text{ and } \frac{\partial S_{\mathrm{s}}(X_0, X_1)}{\partial X_1} = \lambda_{\mathrm{s},1}. \tag{1.65}$$

That the first moment is conjugate to the gradient can be seen from the fact that their product has the dimensions of entropy.

In the optimum state, which in this case is the steady state, the field variables of the sub-system are equal to those applied by the reservoir,

$$\overline{\lambda}_{\mathrm{s},0} = \lambda_{\mathrm{r},0}, \text{ and } \overline{\lambda}_{\mathrm{s},1} = \lambda_{\mathrm{r},1}. \tag{1.66}$$

Hence for steady heat flow, the sub-system takes on a temperature gradient that is in essence equal to the temperature difference of the reservoirs divided by their separation.

The above results refer to structure, albeit the structure of a non-equilibrium system. However it is the flux that is the more obviously non-equilibrium quantity. It turns out that the flux in x, J_x, which is the amount of x crossing a plane per unit area per unit time, is related to the rate of change of the first moment of the sub-system,

$$J_x = \frac{1}{AL_z}\dot{X}_1^0. \tag{1.67}$$

This follows because the nett effect of a uniform flux inside the sub-system is to take an amount per unit time of $\Delta x = AJ_x$ from the boundary region at

$z = -L_z/2$, and to deposit it at the other boundary at $z = +L_z/2$, so that the internal rate of change of the moment is $-(-L_z/2)AJ_x + (L_z/2)AJ_x = L_zAJ_x$. Here the superscript 0 signifies the rate of change in the moment of the sub-system isolated from the reservoirs. In the next chapter the regression hypothesis will be derived, which states that the most likely flux for a given structure, X_1, is proportional to the derivative of the entropy,

$$\overline{\dot{X}_1^0} = C \frac{\partial S_s(X_0, X_1)}{\partial X_1} = C\lambda_{s,1}. \tag{1.68}$$

The constant C is a transport coefficient, such as the thermal conductivity, diffusion constant, etc.

This result holds for an isolated system. The reservoirs serve to fix the value of the moment, $\overline{X}_1$, via the implicit equation, $\lambda_{s,1}(\overline{X}_1) = \lambda_{r,1}$. The reservoirs also act as a source and sink for the variable in flux; they continually add and accept just the right amount of the material in flux at the boundaries so that the internal flux is cancelled and the first moment is maintained constant. That the flux in an isolated system is the same as in the presence of reservoirs will be true when the volume of the boundary region is negligible compared to the volume of the sub-system. One also needs the reservoirs to be so large that the amount of material exchanged has negligible effect on the value of their field variable, and also that the conductivity of the reservoirs be much greater than that of the sub-system, so that any gradient in the field variable in the reservoirs themselves can be neglected.

Chapter 2

Fluctuation Theory

This chapter develops fluctuation theory, primarily for equilibrium systems, but with the focus on transitions between thermodynamic states. It is these transitions that form the basis of non-equilibrium theory. Initially the analysis is restricted to macrostates of the same time parity, with behaviour of the time correlation function and the form of the second entropy being obtained by symmetry arguments. It is shown that the time correlation function decays exponentially in Markov systems. A small time expansion of the time correlation function that is valid in both Markov and non-Markov systems is carried out, and from it Onsager's regression hypothesis, Onsager's reciprocal relation, and the Green-Kubo expression for the transport coefficients are derived. The fluctuation analysis is then generalised to the case of systems of mixed time parity.

2.1 Gaussian Probability

In an equilibrium system fluctuations occur about the most likely state. In what follows, these will be taken to be Gaussian. Gaussian probability plays a central rôle in the physical universe for two reasons. First, the Gaussian probability distribution is stable, by which is meant that any variable that is the sum of other variables that each have a Gaussian probability distribution is itself Gaussian distributed. Second, the Gaussian probability distribution follows from the central limit theorem, which means that the probability distribution tends to the Gaussian distribution for any variable that is the sum of arbitrarily distributed variables as their number increases. Since the stochastic portion of the behaviour of all physical variables is due to multiple stochastic influences, at least classically, and since most variables of physical interest are linear additive variables, the Gaussian probability distribution is both necessary and sufficient to account for the statistical behaviour of reasonably likely states.

For a macrostate collective X, the fluctuations represent the departure from the equilibrium or most likely state, say $x = X - \overline{X}$. Consider M collectives of continuum macrostates, and let $\mathbf{x} = \{x_1, x_2, \ldots, x_M\}$ be the vector representing the state of the system as represented by fluctuations in these collectives. Obviously, $\overline{\mathbf{x}} = \mathbf{0}$.

The most important concepts emerge from the simplest case that each of the collectives is of pure parity, and all of them have the same parity. This means that either $\mathbf{x}^\dagger = \mathbf{x}$, or else $\mathbf{x}^\dagger = -\mathbf{x}$, which is to say that when the velocities of all the particles in the universe are reversed, either the value of the macrostate is unchanged or else it reverses sign. The more general case of mixed parity is treated in detail in §2.5.

The present system is taken to be isolated (no exchange with any reservoir). Eventually the influence of a reservoir will be treated in detail. For the present one can allude to one aspect of Onsager's regression hypothesis,[1] namely that the rate of regression back to equilibrium of a fluctuation in an isolated system is the same as though the initial departure from equilibrium had been induced by a reservoir.

For example, suppose the collective of interest is the energy difference between two halves of the isolated system. A non-zero value of this represents a fluctuation away from the equilibrium state of equal energy in the two halves. However, such an asymmetry in the energy distribution can also be induced by placing the system between two reservoirs of different temperatures and allowing energy exchange. The internal rate of relaxation of the energy asymmetry, which can be related to the heat flow across the mid-plane, is the same in both cases. This example illustrates why the transport properties of an isolated system are relevant to steady flows induced by externally applied gradients, which is contained in Onsager's regression hypothesis

In what follows, upper case Roman letters denote an $M \times M$ matrix. The juxtaposition of two vectors means a dyadic matrix, which has elements $\{\mathbf{xy}\}_{ij} = x_i y_j$; if there is no ambiguity with the scalar product, the symmetric dyadic may be written $\mathbf{x}^2 \equiv \mathbf{xx}$. Also, $\mathbf{z} = A\mathbf{x}$ is a vector formed by ordinary matrix multiplication, the scalar product of two vectors is $\mathbf{y} \cdot \mathbf{x} \equiv \mathbf{y}^{\mathrm{T}}\mathbf{x}$, where the superscript T means transpose, and the double scalar product can be written using the 'inside-out' convection, $A : \mathbf{xy} \equiv \mathbf{y}^{\mathrm{T}} A\mathbf{x} \equiv \mathbf{y} \cdot A\mathbf{x}$.

Define the time correlation matrix,

$$Q(\tau) \equiv k_{\mathrm{B}}^{-1} \langle \mathbf{x}(\tau)\mathbf{x}(0) \rangle \,. \tag{2.1}$$

The diagonal elements of $Q(\tau)$ are called autocorrelations, and the off-diagonal elements are called cross-correlations.

By the time homogeneity of an equilibrium system one has

$$\langle \mathbf{x}(\tau)\mathbf{x}(0) \rangle = \langle \mathbf{x}(0)\mathbf{x}(-\tau) \rangle \,, \text{ or } Q(\tau) = Q(-\tau)^{\mathrm{T}}. \tag{2.2}$$

By macroscopic reversibility one has

$$\langle \mathbf{x}(\tau)\mathbf{x}(0) \rangle = \left\langle \mathbf{x}(-\tau)^\dagger \mathbf{x}(0)^\dagger \right\rangle = \langle \mathbf{x}(-\tau)\mathbf{x}(0) \rangle \,, \text{ or } Q(\tau) = Q(-\tau). \tag{2.3}$$

[1] The regression hypothesis encompasses a number of concepts, including that the flux is linearly proportional to the thermodynamic force, that this force is the same whether the non-equilibrium state is due to a spontaneous fluctuation or is determined by an external reservoir or field, and that the flux is independent of time. It is derived in §2.3 and in §2.5.

These two mean that in the present pure parity case, the time correlation matrix is symmetric, $Q(\tau) = Q(\tau)^{\mathrm{T}}$.

One can perform an expansion of the entropy to second-order about the most likely state,

$$S(\mathbf{x}) = \frac{1}{2} S : \mathbf{xx}, \tag{2.4}$$

where

$$S \equiv \left. \frac{\partial^2 S(\mathbf{x})}{\partial \mathbf{x} \partial \mathbf{x}} \right|_{\mathbf{x}=\mathbf{0}} \tag{2.5}$$

may be called the fluctuation matrix. Evidently, the fluctuation matrix as a matrix of second derivatives is symmetric, $S^{\mathrm{T}} = S$. The main interest is in departures from the most likely state, and $S(\mathbf{0})$ has here been set to zero. The fluctuation matrix is negative definite, because the entropy of a fluctuation is by definition less than that of the most likely state.

In equilibrium thermodynamics, the second derivatives of the entropy that comprise a fluctuation matrix can be expressed as second derivatives of an appropriate free energy.[2] Examples of such quantities include the heat capacity, the thermal expansivity, and the isothermal compressibility. These material properties can be measured directly in the laboratory with macroscopic measurements, or they can be calculated with statistical mechanics from an intermolecular potential. A great deal of chemical engineering and materials science depends upon tabulated values of these coefficients.

The probability of the fluctuation is the exponential of this quadratic form for the entropy,

$$\wp(\mathbf{x}) = \frac{1}{W} \exp \frac{S : \mathbf{xx}}{2 k_{\mathrm{B}}}. \tag{2.6}$$

By design, this is a Gaussian probability. Normalisation gives the total entropy of the fluctuations as

$$W = (2\pi k_{\mathrm{B}})^{M/2} |-S|^{-1/2} = \left| -2\pi k_{\mathrm{B}} S^{-1} \right|^{1/2}, \tag{2.7}$$

where $|\ldots|$ denotes the determinant. Recall that the vector $\mathbf{x}$ has M components. The static correlation function, $Q_0 \equiv Q(0)$, is related to the fluctuation matrix by

$$Q_0 = -S^{-1}. \tag{2.8}$$

[2] Attard, P. (2002a), *Thermodynamics and Statistical Mechanics: Equilibrium by Entropy Maximisation*, (Academic Press, London).

This result follows from the definition,

$$\begin{aligned} Q_0 &= k_{\mathrm{B}}^{-1} \langle \mathbf{x}(0)\mathbf{x}(0) \rangle \\ &= \frac{1}{k_{\mathrm{B}} W} \int \mathrm{d}\mathbf{x}\, \mathbf{x}\mathbf{x} e^{S:\mathbf{x}\mathbf{x}/2k_{\mathrm{B}}} \\ &= \frac{1}{k_{\mathrm{B}} W} \int \mathrm{d}\mathbf{x}\, \mathbf{x} k_{\mathrm{B}} S^{-1} \frac{\partial}{\partial \mathbf{x}} e^{S:\mathbf{x}\mathbf{x}/2k_{\mathrm{B}}} \\ &= \frac{1}{W} \left\{ \left[\mathbf{x} S^{-1} e^{S:\mathbf{x}\mathbf{x}/2k_{\mathrm{B}}} \right]_{-\infty}^{\infty} - S^{-1} \int \mathrm{d}\mathbf{x}\, e^{S:\mathbf{x}\mathbf{x}/2k_{\mathrm{B}}} \right\} \\ &= -S^{-1}. \end{aligned} \tag{2.9}$$

In the fourth equality, an integration by parts has been used, with the integrated portion vanishing due to the negligible probability of extreme fluctuations.

The second entropy may be expanded similarly to quadratic order,

$$S^{(2)}(\mathbf{x}', \mathbf{x}|\tau) = \frac{1}{2} A(\tau) : \mathbf{x}\mathbf{x} + B(\tau) : \mathbf{x}'\mathbf{x} + \frac{1}{2} A'(\tau) : \mathbf{x}'\mathbf{x}'. \tag{2.10}$$

Here $\mathbf{x} \equiv \mathbf{x}(0)$ is the current state, and $\mathbf{x}' \equiv \mathbf{x}(\tau)$ is the future state. The matrices $A(\tau)$, $B(\tau)$, and $A'(\tau)$ are second derivative matrices,

$$A(\tau) \equiv \left. \frac{\partial^2 S^{(2)}(\mathbf{0}, \mathbf{x}|\tau)}{\partial \mathbf{x} \partial \mathbf{x}} \right|_{\mathbf{x}=\mathbf{0}}, \quad A'(\tau) \equiv \left. \frac{\partial^2 S^{(2)}(\mathbf{x}', \mathbf{0}|\tau)}{\partial \mathbf{x}' \partial \mathbf{x}'} \right|_{\mathbf{x}'=\mathbf{0}}, \tag{2.11}$$

and

$$B(\tau) \equiv \left. \frac{\partial^2 S^{(2)}(\mathbf{x}', \mathbf{x}|\tau)}{\partial \mathbf{x} \partial \mathbf{x}'} \right|_{\mathbf{x}'=\mathbf{x}=\mathbf{0}}. \tag{2.12}$$

Clearly the two 'pure' matrices are symmetric, $A(\tau) = A(\tau)^{\mathrm{T}}$ and $A'(\tau) = A'(\tau)^{\mathrm{T}}$ The second entropy must be negative with maximum $S^{(2)}(\mathbf{0}, \mathbf{0}|\tau) = 0$.

Time homogeneity means that $S^{(2)}(\mathbf{x}', \mathbf{x}|\tau) = S^{(2)}(\mathbf{x}, \mathbf{x}'|-\tau)$, which implies

$$A(-\tau) = A'(\tau), \text{ and } B(\tau) = B(-\tau)^{\mathrm{T}}. \tag{2.13}$$

Similarly, macroscopic reversibility gives the behaviour upon reversing the velocities, $S^{(2)}(\mathbf{x}', \mathbf{x}|\tau) = S^{(2)}(\mathbf{x}'^{\dagger}, \mathbf{x}^{\dagger}|-\tau) = S^{(2)}(\mathbf{x}', \mathbf{x}|-\tau)$. The final equality follows because each term in the quadratic form is unchanged when the sign of both vectors is changed, as in the pure odd case. Hence

$$A(\tau) = A'(\tau), \text{ and } B(\tau) = B(\tau)^{\mathrm{T}}. \tag{2.14}$$

The most likely future state given the present state is obtained by maximising the second entropy with respect to $\mathbf{x}'$. That is

$$\left. \frac{\partial S^{(2)}(\mathbf{x}', \mathbf{x}|\tau)}{\partial \mathbf{x}'} \right|_{\mathbf{x}'=\overline{\mathbf{x}'}} = 0. \tag{2.15}$$

This gives[3]

$$\overline{\mathbf{x}'} \equiv \overline{\mathbf{x}}(\tau, \mathbf{x}) = -A(\tau)^{-1}B(\tau)\mathbf{x}. \tag{2.16}$$

Similarly, the most likely present state that would lead to $\mathbf{x}'$ is $\overline{\mathbf{x}} \equiv \overline{\mathbf{x}}(-\tau, \mathbf{x}') = -A(\tau)^{-1}B(\tau)\mathbf{x}'$. The result that the most likely future position is linearly proportional to the current position is a formally exact consequence of the Gaussian form for the fluctuations. It justifies Onsager's original regression hypothesis, that the rate of return to equilibrium is linearly proportional to the gradient in the entropy. It represents a type of factorisation that splits the time dependence from the state dependence. However, until the time dependence of the fluctuation matrices is established, the full power of the hypothesis cannot be exploited, as will be discussed in greater detail in §2.3.

The second entropy may be rearranged in terms of the departure from the future most likely state,

$$\begin{aligned} S^{(2)}(\mathbf{x}', \mathbf{x}|\tau) &= \frac{1}{2}A(\tau) : \left[\mathbf{x}' + A(\tau)^{-1}B(\tau)\mathbf{x}\right]^2 \\ &\quad + \frac{1}{2}\left[A(\tau) - B(\tau)A(\tau)^{-1}B(\tau)\right] : \mathbf{x}\mathbf{x}, \end{aligned} \tag{2.17}$$

where the squared vector signifies the symmetric dyadic matrix.

The reduction condition was given as Eq. (1.46). This says that in the most likely state the second entropy must equal the first entropy,

$$S^{(2)}(\overline{\mathbf{x}}', \mathbf{x}|\tau) = S(\mathbf{x}). \tag{2.18}$$

A similar reduction holds for the most likely prior state, $S^{(2)}(\mathbf{x}', \overline{\mathbf{x}}|\tau) = S(\mathbf{x}')$. The physical origin of the reduction condition is that the probability distribution is sharply peaked about the most likely state such that fluctuations are relatively negligible. Using either of these and the preceding form for the second entropy establishes an exact relation between the two fluctuation matrices,

$$A(\tau) - B(\tau)A(\tau)^{-1}B(\tau) = S. \tag{2.19}$$

This result in essence means that there is only one independent fluctuation matrix.

The transition probability is the exponential of the second entropy,

$$\wp(\mathbf{x}', \mathbf{x}|\tau) = \frac{1}{W^{(2)}} e^{S^{(2)}(\mathbf{x}', \mathbf{x}|\tau)/k_{\mathrm{B}}}. \tag{2.20}$$

Obviously this is Gaussian, and from the rearranged form of the second entropy and the reduction condition one can see that the normalisation factor is

$$W^{(2)} = (2\pi k_{\mathrm{B}})^M \left|-A(\tau)\right|^{-1/2} \left|-S\right|^{-1/2}. \tag{2.21}$$

[3]There is a close relationship between the most likely state derived from the second (or higher-order) entropy and stochastic processes or Brownian motion, as is discussed in Chs 3, 10, and 11.

For Gaussian statistics, means equal modes, which allows the fluctuation matrices to be related to the time correlation matrix. One has

$$\begin{aligned} Q(\tau) &= k_{\mathrm{B}}^{-1} \langle \mathbf{x}'\mathbf{x} \rangle \\ &= k_{\mathrm{B}}^{-1} \langle \overline{\mathbf{x}}'\mathbf{x} \rangle \\ &= -A(\tau)^{-1}B(\tau)k_{\mathrm{B}}^{-1} \langle \mathbf{x}\mathbf{x} \rangle \\ &= A(\tau)^{-1}B(\tau)S^{-1}. \end{aligned} \tag{2.22}$$

Alternatively,

$$Q(\tau) = k_{\mathrm{B}}^{-1} \langle \mathbf{x}'\overline{\mathbf{x}} \rangle = S^{-1}B(\tau)A(\tau)^{-1}. \tag{2.23}$$

This is just the transpose of the preceding expression.

Using these, the generic regression hypothesis, Eq. (2.16), can be written in terms of the time correlation function,

$$\overline{\mathbf{x}}' = -A(\tau)^{-1}B(\tau)\mathbf{x} = -Q(\tau)S\mathbf{x}. \tag{2.24}$$

The term $S\mathbf{x} = \nabla S(\mathbf{x})$ is the thermodynamic driving force toward the equilibrium state. Post multiplying both sides of this by $\mathbf{x}^{\mathrm{T}}$ and taking the average yields an identity, since $\langle [\nabla S(\mathbf{x})]\mathbf{x}^{\mathrm{T}} \rangle = -k_{\mathrm{B}}\mathrm{I}$. The time dependence of the fluctuation matrices and of the time correlation function will be derived shortly, and the full power and utility of the regression hypothesis in non-equilibrium thermodynamics will be demonstrated.

The expressions for the time correlation function in terms of the fluctuation matrices can be inverted. Using the reduction condition, Eq. (2.19), and the macroscopic reversibility condition one has

$$A(\tau)^{-1}S = \mathrm{I} - A(\tau)^{-1}B(\tau)A(\tau)^{-1}B(\tau) = \mathrm{I} - Q(\tau)SQ(\tau)S. \tag{2.25}$$

Here and throughout I is the identity matrix. Hence

$$A(\tau) = [\mathrm{I} - SQ(\tau)SQ(\tau)]^{-1}\, S. \tag{2.26}$$

This is clearly symmetric, as it must be. Using this in the second form for the time correlation function and rearranging for the cross fluctuation matrix one obtains

$$B(\tau) = [\mathrm{I} - SQ(\tau)SQ(\tau)]^{-1}\, SQ(\tau)S = A(\tau)Q(\tau)S. \tag{2.27}$$

These show that the time correlation function completely determines the fluctuation coefficients, the second entropy, and the transition probability.

2.2 Exponential Decay in Markovian Systems

The preceding analysis provided a formally exact description of fluctuations in terms of the time correlation function. A key result was the generic regression hypothesis, Eq. (2.24), which showed that the most likely future state was linearly proportional to the current state. Now the aim is to derive the time dependence of the proportionality constant, which is the time correlation function.

Obtaining quantitatively the time correlation function from a given intermolecular potential is one of the objects of statistical mechanics. Thermodynamics, however, gives the generic behaviour of systems that is independent of such molecular details and that relies upon only a few macroscopically measured quantities. In this and the following section the time dependence of the fluctuation matrices and of the time correlation function are analysed with a view to expressing them in terms of a few macroscopic, hopefully universal, properties. In the first instance it will be shown that Markovian systems, which are a good approximation to some real world systems, have exponentially decaying time correlation functions. After that, it will be shown that in the small time limit, all systems, both Markovian and non-Markovian, have linearly decaying time correlation functions. The small time limit is all that is required to develop the non-equilibrium thermodynamics of steady state systems.

At very long times the initial and final macrostates must become uncorrelated. Hence the joint fluctuation in $\mathbf{x}'$ and $\mathbf{x}$ must be the same as the product of their independent fluctuations, $S^{(2)}(\mathbf{x}', \mathbf{x}|\tau) \rightarrow S(\mathbf{x}') + S(\mathbf{x})$, $\tau \rightarrow \infty$. This means that one must have

$$A(\tau) \rightarrow S, \; B(\tau) \rightarrow 0, \; \tau \rightarrow \infty. \tag{2.28}$$

These are consistent with the vanishing of the time correlation function in the long time limit, $Q(\tau) \rightarrow 0$, $\tau \rightarrow \infty$, as can be seen from Eq. (2.22).

At very short times, the most likely final state must be the same as the initial state, $\overline{\mathbf{x}}(\tau, \mathbf{x}) \rightarrow \mathbf{x}$, $\tau \rightarrow 0$. This follows because the system does not have time to leave its current macrostate. Using this in Eq. (2.16) shows that

$$A(\tau)^{-1}B(\tau) \rightarrow -\mathrm{I}, \; \tau \rightarrow 0. \tag{2.29}$$

Finally, for the present case of macrostates all with the same time parity, Eqs (2.13) and (2.14) show that the fluctuation matrices must be even functions of time,

$$A(\tau) = A(-\tau), \text{ and } B(\tau) = B(-\tau). \tag{2.30}$$

Accordingly the time correlation function must also be even, $Q(\tau) = Q(-\tau)$.

The three results just listed are formally exact, but one needs to establish additional behaviour to reduce the fluctuation matrices and the time correlation function to the few macroscopic parameters that are required for a non-equilibrium thermodynamic theory. To do this one has first to ascertain carefully the time regime over which such a universal theory is to be applied.

Figure 2.1 shows the time evolution of a fluctuation, $\overline{\mathbf{x}}(\tau, \mathbf{x})$. In this case the macrostate collective is the first energy moment, $\mathbf{x} \equiv E_1$, which tells the asymmetry of the energy distribution in the two halves of the system. Also the average value is used in place of the most likely value, but neither these

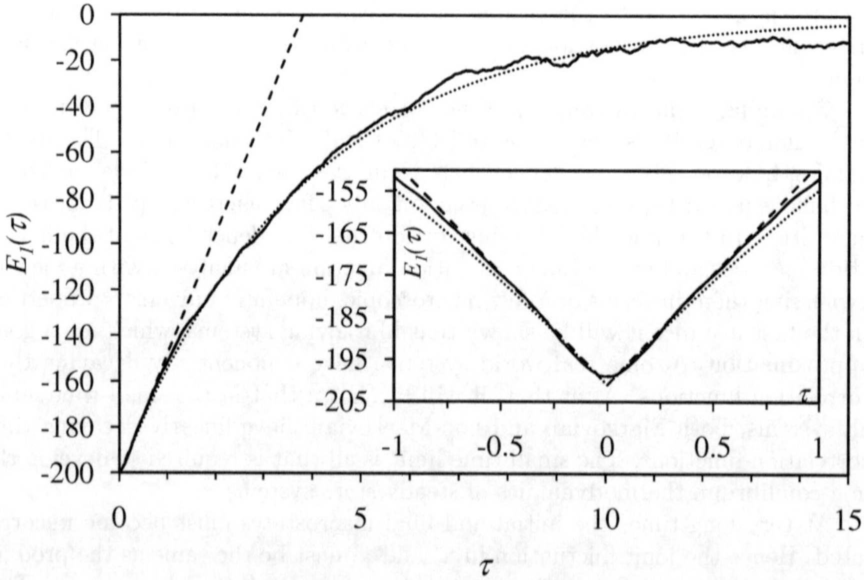

Figure 2.1: Regression of a fluctuation in the first energy moment. The isolated system at time $\tau = 0$ is in a dynamically disordered state, with non-zero energy moment Boltzmann-distributed according to a fixed temperature gradient. The evolution of the moment forwards and backwards in time (solid curve) is obtained by molecular dynamics simulations of a Lennard-Jones fluid using adiabatic Hamiltonian trajectories averaged over initial states. The dashed line is a fitted line, $E_1(\tau) = a + b|\tau|$, and the dotted curve is an exponential, $E_1(\tau) = a \exp[|\tau|b/a]$. The inset is a magnification for small times. More details may be found in P. Attard, J. Chem. Phys. **122**, 154101 (2005a).

nor the details of the simulation are important here. What is important is the generic behaviour shown in Fig. 2.1 that is typical of all fluctuations of macrostates of the same parity. That the regression in the fluctuation is even in time is true for macrostates and mixtures of macrostates that all have the same parity, either pure even or pure odd with respect to velocity reversal. In Fig. 2.1 one can identify three time regimes: the molecular or infinitesimal regime, the small time or linear or steady regime, and the long time regime. In this case the molecular regime is $|\tau| \lesssim 0.1$, and one can see that the fluctuation goes smoothly through the initial state, which means that it possesses a Taylor expansion, $\overline{\mathbf{x}}(\tau, \mathbf{x}) \sim \mathbf{x} + \mathbf{c}(\mathbf{x})\tau^2 + \mathbf{d}(\mathbf{x})\tau^4 + \ldots$. Only even powers of τ appear here. The small time regime is here $0.1 \lesssim |\tau| \lesssim 1$, and it is characterised by a linear regression in the fluctuation. Because the regression must be an even function of time, in the small time regime it must have the form, $\overline{\mathbf{x}}(\tau, \mathbf{x}) \sim \mathbf{a}(\mathbf{x}) + \mathbf{b}(\mathbf{x})|\tau|$ (with $\mathbf{a}(\mathbf{x}) \approx \mathbf{x}$ and terms of order τ^2, $|\tau|^3$ etc. neglected). Because the absolute value is not an analytic function, this is not a Taylor expansion, and this functional form cannot apply in the molecular regime, as the fits in the figure clearly show. One reason for nevertheless calling this the small time regime is that this must be a small-τ expansion, since this functional form cannot hold in the long time limit, where instead $\overline{\mathbf{x}}(\tau, \mathbf{x}) \rightarrow 0$, $|\tau| \rightarrow \infty$. Following the short time regime is what might be called the long time regime, here $|\tau| \gtrsim 1$. For this system the fluctuation decays exponentially, but such exponential behaviour does not occur in all systems. Indeed in some cases the approach to zero can be power-law, and in other cases it can even be non-monotonic.

The behaviour in the molecular regime depends upon the details of the intermolecular potential. It represents the time required for the system to come up to speed at the start of a regression before it settles into steady regression of the fluctuation. The small time or linear regime is the steady state regime in which the velocity or rate of change of the macrostate is constant. In what follows, the molecular regime will be referred to as infinitesimal τ, and the linear regime will be referred to as small τ. The rate of change or slope of the regression in this small time regime is related to the hydrodynamic transport coefficients, as will be explained shortly, and it is the regime to which non-equilibrium thermodynamics refers. The non-analytic form for the terms in the small-τ expansion arise from a resummation of an infinite number of terms in the Taylor expansion of the molecular regime. The upper limit on the small time regime is essentially that there cannot be a significant change in the state of the system, which means that $|\tau||\overline{\overset{\circ}{\mathbf{x}}}| \ll |\mathbf{x}|$. This follows because in essence the state provides the driving force for the velocity, and so the velocity cannot remain constant once the state has changed significantly.

A Markov system is one in which there is no memory. By this is meant that the future state is determined solely by the present state and not by the history of the system prior to the present state. The transition probability for three states factorises in a Markov system,

$$\begin{aligned} \wp(\mathbf{x}_3, \mathbf{x}_2, \mathbf{x}_1 | t_{32}; t_{21}) &= \wp(\mathbf{x}_3 | t_{32}, \mathbf{x}_2; t_{21}, \mathbf{x}_1) \wp(\mathbf{x}_2, \mathbf{x}_1 | t_{21}) \\ &\approx \wp(\mathbf{x}_3 | t_{32}, \mathbf{x}_2) \wp(\mathbf{x}_2, \mathbf{x}_1 | t_{21}). \end{aligned} \tag{2.31}$$

The first equality is formally exact, and expresses the probability of $\mathbf{x}_3$ conditional upon being in the preceding two states. The second equality represents the Markov approximation; it is true if the intermediate state $\mathbf{x}_2$ removes all influence or memory of the state $\mathbf{x}_1$ on the final state $\mathbf{x}_3$.

Not all systems display Markovian behaviour, but for those that do the generic regression hypothesis, Eq. (2.24), takes on a particularly simple form. The most likely transition $\mathbf{x}_1 \rightarrow \mathbf{x}_3$ in time t_{31} is

$$\overline{\mathbf{x}}_3 = -Q(t_{31}) S \mathbf{x}_1. \tag{2.32}$$

Mimicking the analysis of Fox and Uhlenbeck,[4] one can introduce an intermediate time, $t_1 < t_2 < t_3$, and write this as a pair of successive transitions,

$$\overline{\mathbf{x}}_3 = -Q(t_{32}) S \overline{\mathbf{x}}_2 = Q(t_{32}) S Q(t_{21}) S \mathbf{x}_1. \tag{2.33}$$

[4]Fox, R. F. and Uhlenbeck, G. E. (1970), Phys. Fluids, **13**, 1893, 2881.

The Markov approximation appears at the first equality, namely that the transition $\mathbf{x}_2 \rightarrow \mathbf{x}_3$ is not influenced directly by the prior state $\mathbf{x}_1$. Comparing these two expressions one must have

$$-Q(t_{31})S = Q(t_{32})SQ(t_{21})S, \tag{2.34}$$

which implies that

$$Q(\tau) = -e^{|\tau|\Lambda S/2}S^{-1}. \tag{2.35}$$

This is known as Doob's theorem, and it applies to any random process that is both Gaussian and Markov.[5]

One concludes that the time correlation functions for Markov systems are exponentially decaying. The decay rate is determined by a constant matrix Λ, which will turn out to be related to the transport coefficients of the system. Hence this result fulfils the thermodynamic goal of characterising the system by a few macroscopic parameters.

In the small time limit the exponential may be linearised to give,[6]

$$Q(\tau) = -S^{-1} - |\tau|\Lambda/2, \ |\tau||\Lambda S| \ll 1. \tag{2.36}$$

2.3 Small Time Expansion

The preceding analysis showed that in the Markov regime the time correlation function was linear in $|\tau|$ on scales much less than the exponential decay time. The absolute value arises because it is an even function of time. Since this is not an analytic function, one can conclude that this result cannot hold down to infinitesimal time scales, where instead one must have a Taylor expansion of the form

$$Q(\tau) = -S^{-1} + \frac{1}{2}\tau^2\ddot{Q}(0) + \ldots, \ \tau \rightarrow 0. \tag{2.37}$$

Hence one has to distinguish between on the one hand infinitesimal or molecular time scales, and on the other what will be called small time scales, which are defined as much less than the system decay time, but much greater than molecular time scales. The reason that the Markov regime cannot persist down to molecular time scales is that the system needs time to forget the preceding states.

In general (Markov and non-Markov), the influence of velocity decays relatively rapidly, and one can define a velocity relaxation time τ_{relax} such that $\overline{\mathbf{x}}(\tau|\mathbf{x},\dot{\mathbf{x}}) \approx \overline{\mathbf{x}}(\tau|\mathbf{x})$, $\tau \gtrsim \tau_{\text{relax}}$. For the small time expansion to be valid, one must have $|\tau| \ll |\mathbf{x}/\stackrel{\circ}{\overline{\mathbf{x}}}(\tau|\mathbf{x})|$. In a Markov system both of these can be simultaneously satisfied. By definition they cannot be simultaneously satisfied in a non-Markov system.

[5] Doob, J. L. (1942), Ann. Math. **43**, 351.

[6] The criterion for smallness might be better written $-|\tau|\Lambda S : \mathbf{xx} \ll \mathbf{x}\cdot\mathbf{x}$ for all $\mathbf{x}$.

In the general pure parity case the time correlation function must be an even function of τ, with a leading term $\mathcal{O}(\tau^2)$ in the molecular regime, and a leading term $\mathcal{O}(|\tau|)$ in the small time regime. This last result has just been proven for a Markov system using Doob's theorem for the exponential decay of correlations. Such exponential decay, either monotonic or damped sinusoid, is expected whenever the decay of the future state depends upon the magnitude of the current state, $\overset{\circ}{\overline{\mathbf{x}}}(\tau|\mathbf{x})$, and so one expects the result also to hold for non-Markov system in the small time regime. In §2.4.7, in which the transition $\mathbf{x}_1 \rightarrow \mathbf{x}_2 \rightarrow \mathbf{x}_3$ is briefly discussed, it is argued that non-Markov corrections to the Markov result are of higher-order in the time interval in the small time limit.

All of the following results apply to the small time regime, and it is assumed that the linearity in $|\tau|$ holds for all systems. In the non-Markov case, the time interval τ is not larger than the relaxation time, so that $\overline{\mathbf{x}}(\tau|\mathbf{x}) \neq \overline{\mathbf{x}}(\tau|\mathbf{x}, \dot{\mathbf{x}})$. Nevertheless the most likely conditional value $\overline{\mathbf{x}}(\tau|\mathbf{x})$ is well defined and corresponds to the most likely value depending on the starting position irrespective of the starting velocity (equivalently, summed over all starting velocities). Accordingly, expansions in powers of $|\tau|$ will now be obtained for the fluctuation matrices in the second entropy.

The second entropy in fluctuation approximation may be rearranged as

$$\begin{aligned} S^{(2)}(\mathbf{x}', \mathbf{x}|\tau) &= \frac{1}{2}A(\tau) : \mathbf{x}^2 + B(\tau) : \mathbf{x}'\mathbf{x} + \frac{1}{2}A(\tau) : \mathbf{x}'^2 \\ &= \frac{-1}{2}B(\tau) : [\mathbf{x}' - \mathbf{x}]^2 + \frac{1}{2}[A(\tau) + B(\tau)] : [\mathbf{x}'^2 + \mathbf{x}^2] . \end{aligned} \tag{2.38}$$

Recall the inside-out convention for the double scalar product, $A : \mathbf{xy} \equiv \mathbf{y} \cdot A\mathbf{x}$, and the square convention for the symmetric dyadic matrix, $\mathbf{x}^2 \equiv \mathbf{xx}$. This expression is restricted to fluctuations of variables all of the same parity, but no other restriction is placed on the system.

In the limit $\tau \rightarrow 0$, the future state must be the same as the present state, $\overline{\mathbf{x}}' = \mathbf{x}$. This means that the transition probability must become a delta-function, $\wp(\mathbf{x}', \mathbf{x}|\tau) \rightarrow \delta(\mathbf{x}' - \mathbf{x})\wp(x)$, $\tau \rightarrow 0$. In view of this, the first term in the second form for the second entropy must diverge as

$$B(\tau) = \frac{1}{|\tau|}\Lambda^{-1} + B + \mathcal{O}(\tau), \ |\tau| \rightarrow 0. \tag{2.39}$$

The notation for the coefficient of the leading $1/|\tau|$ term has been chosen to be consistent with that chosen for the Markov analysis, as will shortly become clear. The second term in the second form for the second entropy must remain finite, and so one must also have

$$A(\tau) = \frac{-1}{|\tau|}\Lambda^{-1} + A + \mathcal{O}(\tau), \ |\tau| \rightarrow 0. \tag{2.40}$$

To leading-order these give $A(\tau)^{-1}B(\tau) = -\mathrm{I}$, $|\tau| \rightarrow 0$, as demanded by Eq. (2.29). The expansions are even in time in this present pure parity case.

Using these two expansions in the reduction condition, Eq. (2.19), yields

$$\begin{aligned} S &= A(\tau) - B(\tau)A(\tau)^{-1}B(\tau) \\ &= \frac{-1}{|\tau|}\Lambda^{-1} + A - \left[\frac{1}{|\tau|}\Lambda^{-1} + B\right]\left[\frac{-1}{|\tau|}\Lambda^{-1} + A\right]^{-1}\left[\frac{1}{|\tau|}\Lambda^{-1} + B\right] \\ &= 2[A+B] + \mathcal{O}(\tau). \end{aligned} \tag{2.41}$$

Terms of linear order in the time interval have been neglected here. This determines the sum of the two constant contributions to the fluctuation matrices, which is all that will be required for the regression theorem. For a Markov system one can establish the stronger result that $B = 0$ and $A = S/2$, but this is not required here.

The most likely future state is

$$\begin{aligned} \overline{\mathbf{x}}(\tau, \mathbf{x}) &= -A(\tau)^{-1}B(\tau)\mathbf{x} \\ &= -\left[\frac{-1}{|\tau|}\Lambda^{-1} + A\right]^{-1}\left[\frac{1}{|\tau|}\Lambda^{-1} + B\right]\mathbf{x} \\ &= [\mathrm{I} + |\tau|\Lambda(A+B)]\,\mathbf{x} \\ &= \mathbf{x} + \frac{|\tau|}{2}\Lambda S\mathbf{x} + \mathcal{O}(\tau^2). \end{aligned} \tag{2.42}$$

This is the general form of the regression hypothesis that applies in both Markov and in non-Markov systems of pure parity. It shows that for small time intervals the fluctuation regresses linearly in the thermodynamic force and linearly in time. Post-multiplying by $k_{\mathrm{B}}^{-1}\mathbf{x}$ and taking the average yields the time correlation function, $Q(\tau) = -S^{-1} - |\tau|\Lambda/2$. This agrees with the Markov result above in the small time limit.

It should be made clear that one cannot extend these results for a non-Markov system for a small time interval to longer time scales by piecing together consecutive small time intervals, which is essentially what was done for the Markov system. Implicit in the second entropy $S^{(2)}(\mathbf{x}_2, \mathbf{x}_1|\tau)$ for the transition $\mathbf{x}_1 \stackrel{\tau}{\rightarrow} \mathbf{x}_2$, is that the initial macrostate $\mathbf{x}_1$ arose from a fluctuation and had no other special conditions. For the consecutive transitions $\mathbf{x}_1 \stackrel{t_{21}}{\rightarrow} \mathbf{x}_2 \stackrel{t_{32}}{\rightarrow} \mathbf{x}_3$, the intermediate state $\mathbf{x}_2$ does not arise from an unconditional fluctuation, but from the prior state $\mathbf{x}_1$. A non-Markov system retains a memory of this, and so one cannot break the third entropy down to the sum of two unconditional second entropies.

The second entropy in the small time limit is

$$\begin{aligned} S^{(2)}(\mathbf{x}', \mathbf{x}|\tau) &= \frac{-1}{2|\tau|}\Lambda^{-1} : [\mathbf{x}' - \overline{\mathbf{x}}(\tau, \mathbf{x})]^2 + \frac{1}{2}S : \mathbf{x}^2 \\ &= \frac{-1}{2|\tau|}\Lambda^{-1} : [\mathbf{x}' - \mathbf{x}]^2 + \frac{1}{2}S : \mathbf{x}'\mathbf{x} - \frac{|\tau|}{8}\mathbf{x} \cdot S\Lambda S\mathbf{x} \\ &= \frac{-1}{2|\tau|}\Lambda^{-1} : [\mathbf{x}' - \mathbf{x}]^2 + \frac{1}{4}S : \left[\mathbf{x}'^2 + \mathbf{x}^2\right] - \frac{|\tau|}{8}\mathbf{x} \cdot S\Lambda S\mathbf{x} \\ &\quad + \mathcal{O}(\tau^2). \end{aligned} \tag{2.43}$$

Here terms of order τ^2 have been neglected, including the term $-B : [\mathbf{x}' - \mathbf{x}]^2/2$. The first equality satisfies the reduction condition exactly, which is to say that the maximum value of the second entropy is equal to the first entropy,

$$\overline{S}^{(2)}(\tau, \mathbf{x}) \equiv S^{(2)}(\overline{\mathbf{x}}', \mathbf{x}|\tau) = \frac{1}{2} S : \mathbf{x}^2. \tag{2.44}$$

The final equality satisfies the reduction condition to order τ^2. One can in fact add a term of $\mathcal{O}(\tau^2)$ to the final equal that ensures that the reduction condition is satisfied exactly.

2.4 Results for Pure Parity Systems

2.4.1 Onsager Regression Hypothesis and Reciprocal Relations

Now several useful and well-known results are derived from the small time expansion. As has just been shown, the most likely future position to leading-order in τ is

$$\overline{\mathbf{x}}' \equiv \overline{\mathbf{x}}(\tau, \mathbf{x}) = -Q(\tau) S\mathbf{x} = \mathbf{x} + \frac{1}{2}|\tau| \Lambda S \mathbf{x}. \tag{2.45}$$

Hence the most likely coarse velocity, which is the average velocity over the interval, is

$$\overline{\overset{\circ}{\mathbf{x}}}(\mathbf{x}) \equiv \frac{\overline{\mathbf{x}}' - \mathbf{x}}{\tau} = \frac{\hat{\tau}}{2} \Lambda S \mathbf{x}. \tag{2.46}$$

Recall that $\hat{\tau} \equiv \text{sign}(\tau) = \tau/|\tau|$. The final two terms represent the thermodynamic force, $\mathbf{f}(\mathbf{x}) \equiv \partial S(\mathbf{x})/\partial \mathbf{x} = S\mathbf{x}$. Hence the result says that the most likely flux is linearly proportional to the thermodynamic driving force and is independent of the length of the time interval. It is this form that is known as the Onsager regression hypothesis (together with the identification of the internal force with the reservoir force); arguably, since it is here derived rather than assumed, it could be called a theorem. That the flux has opposite sign in the future, $\hat{\tau} > 0$, than in the past, $\hat{\tau} < 0$, reflects the fact that in the future the system most likely returns to equilibrium, and in the past the system most likely came from equilibrium.

The transport matrix is symmetric

$$\Lambda = \Lambda^{\mathrm{T}}, \tag{2.47}$$

because $A(\tau)$ and $B(\tau)$ are symmetric. This is the celebrated Onsager reciprocal relation.[7] This symmetry says that the cross-coupling transport coefficients are equal.

[7] Onsager, L. (1931), Phys. Rev. **37**, 405, and **38**, 2265.

As an example, take the case of isothermal diffusion in a multi-component mixture. In this case the regression theorem gives the mass fluxes in terms of the chemical potential gradients, which is known as Ficks law. The reciprocal relation says that the diffusion constant that gives the flux of species α in response to the gradient of chemical potential of species β, is equal to the diffusion constant that gives the flux of species β in response to the gradient of chemical potential of species α.

The coupling of electric and heat currents, which is known as a thermoelectric effect, provides another example. There is the Seebeck effect, which is the driving of an electric current by a temperature gradient, and the Peltier effect, which is heat flux driven by a voltage, otherwise known as an electrochemical gradient. The symmetry of the transport matrix Λ means that the proportionality constant for these two effects is the same.

The most widely used example of a type of reciprocal relation is for chemical reaction rates, where the rate coefficient for the forward reaction is known to be equal to that of the reverse reaction. This phenomenon originally motivated Onsager's Nobel prize-winning work.

Onsager's insight in deriving the reciprocal relations was that they were a consequence of macroscopic reversibility. This reversibility entered the present derivation in the symmetry of the cross fluctuation matrix, $B(\tau) = B(\tau)^{\mathrm{T}}$, and the equality of the 'pure' fluctuation matrices, $A(\tau) = A'(\tau)$. Equivalently, and closer to Onsager's original derivation, it follows from the symmetry of the time correlation function, $Q(\tau) = Q(\tau)^{\mathrm{T}}$. These specific symmetries only hold in the case that the macrostates all have the same parity; the more general case of mixed parity is treated in §2.5.

2.4.2 Green-Kubo Expression

The linear expansion for the time correlation function, Eq. (2.45), is actually equivalent to the Green-Kubo expressions for the transport coefficients.[8] This may be rearranged in more familiar form by writing

$$\begin{aligned}\Lambda &= \frac{-2}{|\tau|}\left[Q(\tau)+S^{-1}\right] \\ &= \frac{-2}{|\tau|}\left[k_{\mathrm{B}}^{-1}\left\langle \mathbf{x}(\tau)\mathbf{x}(0)\right\rangle - k_{\mathrm{B}}^{-1}\left\langle \mathbf{x}(0)\mathbf{x}(0)\right\rangle\right] \\ &= \frac{-2}{k_{\mathrm{B}}|\tau|}\int_0^\tau \mathrm{d}t' \left\langle \dot{\mathbf{x}}(t')\mathbf{x}(0)\right\rangle . \end{aligned} \tag{2.48}$$

The present expansion is valid for small but non-infinitesimal τ. In practice, Green-Kubo expressions such as this are evaluated in computer simulations for

[8] Green, M. S. (1954), J. Chem. Phys. **22**, 398. Kubo, R. (1966), Rep. Progr. Phys. **29**, 255. Kubo, R., Toda, M., and Hashitsume, N. (1978), *Statistical Physics II. Non-equilibrium Statistical Mechanics*, (Springer-Verlag, Berlin). Zwanzig, R. (2001), *Non-equilibrium Statistical Mechanics*, (Oxford University Press, Oxford). Actually, the relationship between the time correlation function and the transport coefficient was already given by Onsager in his papers on the reciprocal relations.

increasing values of τ, so that the left-hand side is actually $\tilde{\Lambda}(\tau)$. Once one goes beyond the infinitesimal regime, whose extent is not known *a priori*, there is a range of τ values over which $\tilde{\Lambda}(\tau)$ is constant, and this is taken as the value of the transport coefficient. Depending on the system, beyond the small time regime, $\tilde{\Lambda}(\tau)$ will decrease with increasing τ. Hence Λ is sometimes taken as the maximum value of $\tilde{\Lambda}(\tau)$, (see, for example, Fig. 10.2 on p. 338).

2.4.3 Physical Interpretation of the Second Entropy

An explicit expression for the second entropy can be obtained in terms of the transport matrix. Using the most likely future position, one has

$$\mathbf{x}' - \overline{\mathbf{x}}' = \tau \overset{\circ}{\mathbf{x}} - \frac{|\tau|}{2}\Lambda S\mathbf{x} + \mathcal{O}(\tau^2). \tag{2.49}$$

Hence to leading-order the second entropy is

$$\begin{aligned} S^{(2)}(\mathbf{x}',\mathbf{x}|\tau) &= S(\mathbf{x}) + \frac{1}{2}\left[\frac{-\Lambda^{-1}}{|\tau|} + \mathcal{O}(\tau^0)\right] : [\mathbf{x}' - \overline{\mathbf{x}}']^2 \\ &= S(\mathbf{x}) - \frac{|\tau|}{2}\Lambda^{-1} : \left[\overset{\circ}{\mathbf{x}} - \frac{\hat{\tau}}{2}\Lambda S\mathbf{x}\right]^2 + \mathcal{O}(\tau^2) \\ &= S(\mathbf{x}) - \frac{|\tau|}{2}\Lambda^{-1} : \overset{\circ}{\mathbf{x}}{}^2 + \frac{\tau}{2}\overset{\circ}{\mathbf{x}} \cdot S\mathbf{x} - \frac{|\tau|}{8}\mathbf{x} \cdot S\Lambda S\mathbf{x} + \mathcal{O}(\tau^2). \end{aligned} \tag{2.50}$$

The ordinary entropy is $S(\mathbf{x}) = S : \mathbf{x}^2/2$. The second entropy is here written explicitly for fluctuations in the flux. Unlike the position $\mathbf{x}'$, the flux $\overset{\circ}{\mathbf{x}}$ is an intensive function of time, which is to say that it should not depend upon the length of the time interval. The second entropy can be seen to be a linear or extensive function of the time interval.

The second entropy written in this form has a fairly transparent physical interpretation. The first term on the right-hand side is the ordinary entropy. It is all that remains once the flux has been optimised. In this sense the first entropy may be said to determine the structure of the non-equilibrium system. This is a direct consequence of the reduction condition. This term has great significance in the determination of non-equilibrium pattern formation, as will become evident in later chapters. The second term on the right-hand side of the final equality is quadratic in the flux, and, since it is always negative, it is unfavourable. This is the term that limits the flux. On general grounds a flux represents a dynamically ordered state, and so it must be entropically unfavourable, and, from symmetry, the decrease in entropy must depend upon the magnitude but not upon the direction of the flux, as the present quadratic form shows. The third term on the right-hand side of the final equality is positive when the flux is in the opposite direction to the fluctuation, $\overset{\circ}{\mathbf{x}} \cdot \mathbf{x} < 0$, since S is negative definite. That is, it is positive when the flux reduces the fluctuation. This term is half the rate of entropy production, since it is the flux times the thermodynamic force. This term is linear in the flux, as it must be from symmetry, and it is what drives the flux to be non-zero. The final term on the right-hand side of the final equality is negative, and depends explicitly only on the structure and not directly upon the flux. It is difficult to give a simple physical interpretation of this term, but mathematically it cancels with the second and third terms when the flux is optimum.

In summary, this expression for the second entropy shows that the optimum flux in a non-equilibrium system represents the balance of two competing effects: a positive effect related to the rate of entropy production that is linear in the flux and that dominates at small fluxes, and a negative effect representing the cost of the dynamic order of the flux that is quadratic in the flux and that dominates at large fluxes.

2.4.4 The Dissipation

This simple physical interpretation of the variational nature of the second entropy contrasts with the dissipation, which has been put forward by some as a basis for determining the behaviour of non-equilibrium systems, §1.2.2. The rate of entropy production for a general, not necessarily optimum, flux is

$$\dot{S}(\dot{\mathbf{x}}, \mathbf{x}) = \dot{\mathbf{x}} \cdot \frac{\partial S(\mathbf{x})}{\partial \mathbf{x}} = \dot{\mathbf{x}} \cdot S\mathbf{x}. \tag{2.51}$$

Using the above results for the most likely flux, the optimal rate of entropy production is therefore

$$\overline{\overset{\circ}{S}}(\mathbf{x}) = \frac{\hat{\tau}}{2}\mathbf{x} \cdot S\Lambda S\mathbf{x}. \tag{2.52}$$

The preceding analysis showed that the optimum flux is the one that maximises the second entropy. It neither maximises nor minimises the rate of entropy production. The maximum dissipation in Eq. (2.51) occurs when the flux is $\dot{x} = -\infty$, and the minimum dissipation occurs when the flux is $\dot{x} = +\infty$ (taking for simplicity a one component macrostate and $x > 0$). In contrast to the second entropy, the dissipation is only a bilinear form in the flux and the force, and so it cannot be made an extreme value. The most likely rate of entropy production, Eq. (2.52), is neither a maximum nor a minimum. This is why the rate of entropy production cannot be used as a variational principle for non-equilibrium systems (c.f. the remarks in §1.2.2).

2.4.5 Stability Theory

The first entropy of the fluctuations must be negative, and so the matrix S must be negative definite. This obviously ensures the existence of the Gaussian probability, and also the stability of the optimum state, $\overline{\mathbf{x}} = \mathbf{0}$.

The second entropy must also be negative, again since fluctuations measure the departure from the optimum state. Again this is necessary for the stability of the transitions. Hence the matrix $A(\tau)$ must be negative definite, as follows by setting $\mathbf{x} = 0$, for example. Similarly the matrix $A(\tau) + B(\tau)$ must be negative definite, as follows by setting $\mathbf{x} = \mathbf{x}'$. More generally, the double matrix $\begin{pmatrix} A & B \\ B & A \end{pmatrix}$ must be negative definite.

Because the transport matrix Λ is the negative of the leading-order term of $A(\tau)$, it must be positive definite. This imposes certain restrictions on the transport coefficients that it comprises. Such limits on the magnitudes of the cross-coupling coefficients can be useful in mixed component systems.

The rate of entropy production is determined by the matrix $S\Lambda S$. Writing $\mathbf{y} = S\mathbf{x}$, one has that $\mathbf{y} \cdot \Lambda\mathbf{y} \geq 0$ for all $\mathbf{y}$. Hence $S\Lambda S$ is positive definite. Accordingly, the result for the most likely coarse rate of entropy production, Eq. (2.52), implies that the entropy will most likely increase in the future, and that it most likely decreased in the past.

The first entropy for the fluctuation is negative, $S(\mathbf{x}) \leq 0$, with equality if and only if $\mathbf{x} = \overline{\mathbf{x}} = \mathbf{0}$. As just discussed, the most likely coarse rate of change of the first entropy, Eq. (2.52), is positive definite. These two conditions appear to satisfy the definition of a Lyapunov function, $\mathcal{L}(\mathbf{x})$ (in this case a negative Lyapunov function).[9] Such functions are associated with the stability of the optimum state of an equilibrium or of a non-equilibrium system. They can be thought of as a measure of the distance from the optimum state.

Some caution should however be exercised in either calling the first entropy a Lyapunov function, or else in using it for a trajectory. Generally a Lyapunov function is applied at each point on a trajectory (i.e. it is written $\mathcal{L}(\mathbf{x})$ rather than $\mathcal{L}(\mathbf{x}; \tau, \mathbf{x}_0)$ or $\mathcal{L}(\mathbf{x}; [\mathbf{x}])$). The conditions that a Lyapunov function must satisfy are often written $\mathcal{L}(\mathbf{x}) \leq 0$ and $\mathrm{d}\mathcal{L}(\mathbf{x})/\mathrm{d}t \geq 0$, and these are interpreted as implying that the trajectory monotonically approaches the optimum state, $\mathbf{x}(t) \to \overline{\mathbf{x}}$, (Keizer, 1987). The influence of the prior history of the trajectory is generally not included in the definition of the Lyapunov function.

For a Markov system, the most likely future value $\overline{\mathbf{x}}(t|\mathbf{x}_0)$ as a function of time t (even large times) approaches the optimum value exponentially, which is to say monotonically. Consequently, the first entropy also increases monotonically as a function of time (in a most likely sense). However for a non-Markov system neither of these two conditions necessarily holds, and one can have non-monotonic behaviour. The fact that $\overline{\overset{\circ}{S}} \geq 0$ is true for small time intervals and for no special conditions on the initial state $\mathbf{x}$. As discussed in the connection with the small-time, non-Markov expansion in §2.3 on p. 44, one cannot extend the second entropy analysis for non-Markov systems to long times by piecing together the results for small time intervals, because the intermediate states now have special conditions on them, namely that they arose from a specified prior state.

This point for non-Markov systems raises questions about the utility of Lyapunov functions that don't depend upon the prior history of the system. The two facts—the first entropy of a fluctuation is negative, and the most likely

[9]Keizer, J. (1987), *Statistical Thermodynamics of Non-equilibrium Processes*, (Springer-Verlag, New York), §7.2.

coarse rate of change of the first entropy immediately following a fluctuation is positive—do not tell the whole story about the behaviour of the first entropy on the whole trajectory of a fluctuation, because the rate of change of the first entropy can change sign on the most likely trajectory at long times for a non-Markov system. In this chapter, the stability of the most likely state and also the stability of the most likely transitions have been implicitly assumed throughout by postulating Gaussian fluctuations and negative definite quadratic forms for the first and second entropy, respectively.

2.4.6 Non-Reversibility of the Trajectory

A major consequence of the analysis is that the trajectory is not at all reversible. That is, if one goes backward from the final state then one does not return to the original initial state. Instead,

$$\begin{aligned}\overline{\mathbf{x}}(-\tau, \overline{\mathbf{x}}(\tau, \mathbf{x})) &= \left[\mathrm{I} + \frac{|\tau|}{2}\Lambda S\right]\left[\mathrm{I} + \frac{|\tau|}{2}\Lambda S\right]\mathbf{x} \\ &= \mathbf{x} + |\tau|\Lambda S\mathbf{x} + \mathcal{O}(\tau^2). \end{aligned} \tag{2.53}$$

Denote the three relevant states by $\mathbf{x}_1$, $\mathbf{x}_2 \equiv \overline{\mathbf{x}}(\tau, \mathbf{x}_1)$, and $\mathbf{x}_3 \equiv \overline{\mathbf{x}}(-\tau, \mathbf{x}_2)$. This says that in the future the system will move from $\mathbf{x}_1$ toward equilibrium, so that $S(\mathbf{x}_2) > S(\mathbf{x}_1)$. However, $\mathbf{x}_1$ is not as likely a prior state for $\mathbf{x}_2$ as is $\mathbf{x}_3$, since the latter is even closer to equilibrium, $S(\mathbf{x}_3) > S(\mathbf{x}_2) > S(\mathbf{x}_1)$.

2.4.7 Third Entropy

In the above only the singlet and pair fluctuations were analysed, which involved quadratic forms for the ordinary entropy and for the second entropy. For a Gaussian system, all orders of fluctuations can be expressed as sums of all possible pair-wise products. This means, for example, that the most general form for the third entropy is

$$\begin{aligned} S^{(3)}(\mathbf{x}_3, \mathbf{x}_2, \mathbf{x}_1|t_{32}, t_{21}) &= \frac{1}{2}C : \mathbf{x}_3^2 + \frac{1}{2}C' : \mathbf{x}_1^2 + \frac{1}{2}D : \mathbf{x}_2^2 \\ &\quad + E : \mathbf{x}_2\mathbf{x}_3 + E' : \mathbf{x}_2\mathbf{x}_1 + F : \mathbf{x}_1\mathbf{x}_3. \end{aligned} \tag{2.54}$$

The time dependence of the fluctuation matrices has not been shown. In the case that $t_{32} = t_{21}$, the primed matrices equal their unprimed counterparts. This could be rearranged to show explicitly the terms $[\mathbf{x}_3 - \mathbf{x}_2]^2$ and $[\mathbf{x}_2 - \mathbf{x}_1]^2$, and taking the limit $t_{31} \to 0$, one would deduce expansions with leading-order $|t|^{-1}$ for C, D, and E. The fluctuation matrix F, which contains the non-Markovian behaviour, is to leading-order $\mathcal{O}t^0$. For a Markovian system, $F = 0$, and the third entropy may be written as the sum of the second entropies for the individual transitions,

$$S^{(3)}(\mathbf{x}_3, \mathbf{x}_2, \mathbf{x}_1|t_{32}, t_{21}) = S^{(2)}(\mathbf{x}_3, \mathbf{x}_2|t_{32}) + S^{(2)}(\mathbf{x}_2, \mathbf{x}_1|t_{21}) - S(\mathbf{x}_2). \tag{2.55}$$

Subtracting the first entropy accounts for double counting of the intermediate state.

2.5 Fluctuations of Mixed Time Parity

2.5.1 Second Entropy and Time Correlation Functions

The preceding sections in this chapter treated transitions between macrostates all of which have the same time parity. That is, either $\mathbf{x}^\dagger = \mathbf{x}$, or else $\mathbf{x}^\dagger = -\mathbf{x}$, where the dagger denotes the conjugate state with all velocities reversed. In this section the mixed parity case, where neither of these hold, will be analysed.

It is convenient to deal with macrostates individually of pure parity. Without loss of generality, each of the x_i can be taken to be purely odd or purely even. If one is interested in a state of mixed parity, say y, then one can simply form its even and odd projections, $y_\pm \equiv [y \pm y^\dagger]/2$, and include these as two components of $\mathbf{x}$. Since $y = y_+ + y_-$, any property of y can be obtained by a linear combination of the equivalent property of $y_\pm$.

Define the diagonal parity matrix ε with elements $\varepsilon_{ij} = \pm\delta_{ij}$. That is, $\varepsilon_{ii} = 1$ if $x_i = x_i^\dagger$, and $\varepsilon_{ii} = -1$ if $x_i = -x_i^\dagger$. Hence $\mathbf{x}^\dagger = \varepsilon\mathbf{x}$. Also, the square of the parity matrix is the identity matrix, $\varepsilon\varepsilon = \mathrm{I}$.

The time correlation matrix was defined above as the equilibrium average of the dyadic matrix of the fluctuations, $Q(\tau) \equiv k_\mathrm{B}^{-1}\langle\mathbf{x}(\tau)\mathbf{x}(0)\rangle$. By the time homogeneity of an equilibrium system one has

$$\langle\mathbf{x}(\tau)\mathbf{x}(0)\rangle = \langle\mathbf{x}(0)\mathbf{x}(-\tau)\rangle\,, \text{ or } Q(\tau) = Q(-\tau)^\mathrm{T}. \tag{2.56}$$

By macroscopic reversibility one has

$$\langle\mathbf{x}(\tau)\mathbf{x}(0)\rangle = \left\langle\mathbf{x}(-\tau)^\dagger\mathbf{x}(0)^\dagger\right\rangle, \text{ or } Q(\tau) = \varepsilon Q(\tau)^\mathrm{T}\varepsilon. \tag{2.57}$$

The fluctuation expressions for the first entropy are unchanged from the pure parity case analysed in the preceding sections. One still has

$$S(\mathbf{x}) = \frac{1}{2}\mathbf{x}^\mathrm{T} S\mathbf{x}, \tag{2.58}$$

with the fluctuation matrix being

$$S \equiv \left.\frac{\partial^2 S(\mathbf{x})}{\partial\mathbf{x}\partial\mathbf{x}}\right|_{\mathbf{x}=\mathbf{0}} = -k_\mathrm{B}\langle\mathbf{x}\mathbf{x}\rangle^{-1} = -Q(0)^{-1}. \tag{2.59}$$

Since S is symmetric, so is $Q(0)$. Macroscopic reversibility in this case reads $Q(0) = \varepsilon Q(0)\varepsilon$, which implies that there is no instantaneous coupling between variables of different time parity,

$$\langle x_i(0)x_j(0)\rangle = 0, \text{ if } \varepsilon_{ii} \neq \varepsilon_{jj}. \tag{2.60}$$

It also follows that the parity matrix commutes with the fluctuation matrix, $\varepsilon S = S\varepsilon$, or $\varepsilon S\varepsilon = S$.

The most general quadratic form for the second entropy is

$$S^{(2)}(\mathbf{x}',\mathbf{x}|\tau) = \frac{1}{2}\mathbf{x}\cdot A(\tau)\mathbf{x} + \mathbf{x}\cdot B(\tau)\mathbf{x}' + \frac{1}{2}\mathbf{x}'\cdot C(\tau)\mathbf{x}'. \tag{2.61}$$

Here $\mathbf{x} \equiv \mathbf{x}(0)$ is the current state, and $\mathbf{x}' \equiv \mathbf{x}(\tau)$ is the future state. The matrices $A(\tau)$ and $C(\tau)$ are second derivative matrices of the same variable,

$$A(\tau) \equiv \left. \frac{\partial^2 S^{(2)}(\mathbf{0}, \mathbf{x}|\tau)}{\partial \mathbf{x} \partial \mathbf{x}} \right|_{\mathbf{x}=\mathbf{0}}, \text{ and } C(\tau) \equiv \left. \frac{\partial^2 S^{(2)}(\mathbf{x}', \mathbf{0}|\tau)}{\partial \mathbf{x}' \partial \mathbf{x}'} \right|_{\mathbf{x}'=\mathbf{0}}, \tag{2.62}$$

and as such are symmetric, $A(\tau) = A(\tau)^{\mathrm{T}}$ and $C(\tau) = C(\tau)^{\mathrm{T}}$. The matrix $B(\tau)$ is the cross second derivative,

$$B(\tau) \equiv \left. \frac{\partial^2 S^{(2)}(\mathbf{x}', \mathbf{x}|\tau)}{\partial \mathbf{x} \partial \mathbf{x}'} \right|_{\mathbf{x}'=\mathbf{x}=\mathbf{0}}, \tag{2.63}$$

and in general is neither symmetric nor asymmetric. The second entropy must be negative $S^{(2)}(\mathbf{x}', \mathbf{x}|\tau) \leq S^{(2)}(\mathbf{0}, \mathbf{0}|\tau) = 0$.

The symmetry properties of the second entropy are similar to those of the time correlation function. Time homogeneity means that $S^{(2)}(\mathbf{x}', \mathbf{x}|\tau) = S^{(2)}(\mathbf{x}, \mathbf{x}'|-\tau)$. Hence

$$A(-\tau) = C(\tau), \text{ and } B(\tau) = B(-\tau)^{\mathrm{T}}. \tag{2.64}$$

Similarly, macroscopic reversibility gives the behaviour upon reversing the velocities, $S^{(2)}(\mathbf{x}', \mathbf{x}|\tau) = S^{(2)}(\mathbf{x}'^{\dagger}, \mathbf{x}^{\dagger}|-\tau) = S^{(2)}(\varepsilon\mathbf{x}', \varepsilon\mathbf{x}|-\tau)$. Hence

$$\varepsilon A(\tau)\varepsilon = C(\tau), \text{ and } \varepsilon B(\tau)\varepsilon = B(\tau)^{\mathrm{T}}. \tag{2.65}$$

These mean that if all of the x_i are pure odd, or all of them are pure even, then $A(\tau) = C(\tau)$ and $B(\tau) = B(\tau)^{\mathrm{T}}$, as were invoked in the preceding sections of this chapter.

The most likely future state given the present state is obtained by maximising the second entropy with respect to $\mathbf{x}'$. One obtains

$$\overline{\mathbf{x}'} \equiv \overline{\mathbf{x}}(\tau, \mathbf{x}) = -C(\tau)^{-1} B(\tau)^{\mathrm{T}} \mathbf{x}. \tag{2.66}$$

Similarly, the most likely present state that would lead to $\mathbf{x}'$ is

$$\overline{\mathbf{x}} \equiv \overline{\mathbf{x}}(-\tau, \mathbf{x}') = -A(\tau)^{-1} B(\tau) \mathbf{x}'. \tag{2.67}$$

The second entropy may be rearranged in terms of the departure from the future most likely state,

$$\begin{aligned} S^{(2)}(\mathbf{x}', \mathbf{x}|\tau) &= \frac{1}{2} \left[\mathbf{x}' + C(\tau)^{-1} B(\tau)^{\mathrm{T}} \mathbf{x} \right]^{\mathrm{T}} C(\tau) \left[\mathbf{x}' + C(\tau)^{-1} B(\tau)^{\mathrm{T}} \mathbf{x} \right] \\ &\quad + \frac{1}{2} \mathbf{x}^{\mathrm{T}} \left[A(\tau) - B(\tau) C(\tau)^{-1} B(\tau)^{\mathrm{T}} \right] \mathbf{x}. \end{aligned} \tag{2.68}$$

It can also be written in terms of the departure from the prior most likely state,

$$\begin{aligned} S^{(2)}(\mathbf{x}', \mathbf{x}|\tau) &= \frac{1}{2} \left[\mathbf{x} + A(\tau)^{-1} B(\tau) \mathbf{x}' \right]^{\mathrm{T}} A(\tau) \left[\mathbf{x} + A(\tau)^{-1} B(\tau) \mathbf{x}' \right] \\ &\quad + \frac{1}{2} \mathbf{x}'^{\mathrm{T}} \left[C(\tau) - B(\tau)^{\mathrm{T}} A(\tau)^{-1} B(\tau) \right] \mathbf{x}'. \end{aligned} \tag{2.69}$$

The reduction condition (c.f. Eq. (1.46)) in which the second entropy reduces to the first entropy in the most likely state, obviously holds also for this mixed parity case. Explicitly,

$$S^{(2)}(\overline{\mathbf{x}}', \mathbf{x}|\tau) = S(\mathbf{x}), \text{ or } S^{(2)}(\mathbf{x}', \overline{\mathbf{x}}|\tau) = S(\mathbf{x}'). \tag{2.70}$$

Accordingly, the two preceding forms for the second entropy immediately allow one to conclude

$$A(\tau) - B(\tau)C(\tau)^{-1}B(\tau)^{\mathrm{T}} = S, \tag{2.71}$$

and

$$C(\tau) - B(\tau)^{\mathrm{T}}A(\tau)^{-1}B(\tau) = S. \tag{2.72}$$

These are actually equivalent, as can be seen by pre- and post-multiplying either one by the parity matrix, and invoking the macroscopic reversibility condition, Eq. (2.65).

For Gaussian statistics, means equal modes, which allows the fluctuation matrices to be related to the time correlation matrix. One has

$$\begin{aligned} Q(\tau) &= k_{\mathrm{B}}^{-1}\langle \mathbf{x}'\mathbf{x}\rangle \\ &= k_{\mathrm{B}}^{-1}\langle \overline{\mathbf{x}}'\mathbf{x}\rangle \\ &= -C(\tau)^{-1}B(\tau)^{\mathrm{T}}k_{\mathrm{B}}^{-1}\langle \mathbf{x}\mathbf{x}\rangle \\ &= C(\tau)^{-1}B(\tau)^{\mathrm{T}}S^{-1}. \end{aligned} \tag{2.73}$$

Alternatively,

$$\begin{aligned} Q(\tau) &= k_{\mathrm{B}}^{-1}\langle \mathbf{x}'\overline{\mathbf{x}}\rangle \\ &= -k_{\mathrm{B}}^{-1}\langle \mathbf{x}'\mathbf{x}'\rangle B(\tau)^{\mathrm{T}}A(\tau)^{-1} \\ &= S^{-1}B(\tau)^{\mathrm{T}}A(\tau)^{-1}. \end{aligned} \tag{2.74}$$

These two relations and the reduction condition allow the fluctuation matrices to be rewritten in terms of the correlation matrices. Using the first form of the reduction condition, Eq. (2.71), and the macroscopic reversibility condition, Eq. (2.65), one has

$$\begin{aligned} A(\tau)^{-1}S &= \mathrm{I} - A(\tau)^{-1}B(\tau)C(\tau)^{-1}B(\tau)^{\mathrm{T}} \\ &= \mathrm{I} - \left[\varepsilon C(\tau)^{-1}B(\tau)^{\mathrm{T}}\right]^2 \\ &= \mathrm{I} - \left[\varepsilon Q(\tau)S\right]^2, \end{aligned} \tag{2.75}$$

where the first form for the time correlation function has been used to obtain the final equality. With $\varepsilon Q(\tau) = Q(\tau)^{\mathrm{T}}\varepsilon$, this is

$$A(\tau) = \left[\mathrm{I} - SQ(\tau)^{\mathrm{T}}SQ(\tau)\right]^{-1} S. \tag{2.76}$$

This is clearly symmetric, as it must be. Using this in the second form for the time correlation function and rearranging for the cross fluctuation matrix one obtains

$$B(\tau) = \left[\mathrm{I} - SQ(\tau)^{\mathrm{T}} SQ(\tau)\right]^{-1} SQ(\tau)^{\mathrm{T}} S = A(\tau)Q(\tau)^{\mathrm{T}} S. \tag{2.77}$$

These show that the time correlation function completely determines the fluctuation coefficients, the second entropy, and the transition probability. In particular, with these the most likely future position, Eq. (2.66), may be written

$$\overline{\mathbf{x}}' = -Q(\tau)S\mathbf{x}. \tag{2.78}$$

This result is formally identical to the pure parity case, Eq. (2.24). The term $S\mathbf{x} = \partial S(\mathbf{x})/\partial \mathbf{x}$ is the thermodynamic driving force toward the equilibrium state, and as has been mentioned, post-multiplying both sides by $\mathbf{x}^{\mathrm{T}}$ and taking the average yields an identity, since $\langle [\nabla S(\mathbf{x})]\mathbf{x}^{\mathrm{T}} \rangle = -k_{\mathrm{B}}\mathrm{I}$.

2.5.2 Small Time Expansion for the General Case

The small time expansion for the fluctuations matrices and the time correlation function given in §2.3 was for a Markov or non-Markov, pure parity system. In this section the expansion is generalised to fluctuations of mixed parity. The small time expansion in this case is linear in $|\tau|$ and τ. (Recall the discussion at the start of §2.3 that distinguished between infinitesimal or molecular time scales and small or macroscopic time scales.)

In the limit $\tau \to 0$, the future state must be the same as the present state, $\overline{\mathbf{x}}' = \mathbf{x}$. This means that the transition probability must become a delta-function, $\wp(\mathbf{x}', \mathbf{x}|\tau) \to \delta(\mathbf{x}' - \mathbf{x})\wp(x)$, $\tau \to 0$. It follows that the fluctuation matrices must diverge as τ^{-1} in this limit. Hence one has expansions of the form,

$$A(\tau) = \frac{A_{-1}}{|\tau|} + \frac{A'_{-1}}{\tau} + A_0 + A'_0\hat{\tau} + \mathcal{O}(\tau), \tag{2.79}$$

and

$$B(\tau) = \frac{B_{-1}}{|\tau|} + \frac{B'_{-1}}{\tau} + B_0 + B'_0\hat{\tau} + \mathcal{O}(\tau). \tag{2.80}$$

Since $C(\tau) = A(-\tau)$, one also has

$$C(\tau) = \frac{A_{-1}}{|\tau|} - \frac{A'_{-1}}{\tau} + A_0 - A'_0\hat{\tau} + \mathcal{O}(\tau). \tag{2.81}$$

The matrices for A are symmetric. Since $B(-\tau) = B(\tau)^{\mathrm{T}}$, the unprimed coefficient matrices in its expansion are symmetric, and the primed ones are anti-symmetric.

In the pure parity case only powers of $|\tau|$ appeared, because the fluctuations were even in time. In the present mixed parity case, the fluctuations are not symmetric in time, and both $|\tau|$ and τ enter the expansion.

As mentioned previously, although this is a small time expansion, the time scales are longer than molecular time scales, and the non-analytic terms arise from resummation of the Taylor expansion that would apply on molecular time scales. By starting the expansion at τ^{-1}, the second entropy will turn out to be an extensive function of the time interval when written in terms of the flux, which is what one might expect.

In the $\tau \to 0$ limit one must have $A(\tau)^{-1}B(\tau) \to -\mathrm{I}$, Eq. (2.29). This means that the leading-order coefficients must satisfy

$$B_{-1} + \hat{\tau}B'_{-1} = -A_{-1} - \hat{\tau}A'_{-1}. \tag{2.82}$$

Since A'_{-1} is symmetric and B'_{-1} is antisymmetric, one must have $A'_{-1} = B'_{-1} = 0$. Replacing the leading coefficient by what will later become the symmetric part of the transport matrix, the expansion now is

$$A(\tau) = \frac{-\Lambda^{-1}}{|\tau|} + A_0 + A'_0\hat{\tau} + \mathcal{O}(\tau), \tag{2.83}$$

and

$$B(\tau) = \frac{\Lambda^{-1}}{|\tau|} + B_0 + B'_0\hat{\tau} + \mathcal{O}(\tau). \tag{2.84}$$

The relation between A_0 and B_0 will now be obtained by expanding the reduction condition to zeroth order in τ,

$$\begin{aligned}
S &= A(\tau) - B(\tau)C(\tau)^{-1}B(\tau)^{\mathrm{T}} \\
&= \left[-\Lambda^{-1}|\tau|^{-1} + A_0 + A'_0\hat{\tau}\right] - \left[\Lambda^{-1}|\tau|^{-1} + B_0 + B'_0\hat{\tau}\right] \\
&\quad \times \left[-\Lambda^{-1}|\tau|^{-1} + A_0 - A'_0\hat{\tau}\right]^{-1} \left[\Lambda^{-1}|\tau|^{-1} + B_0 - B'_0\hat{\tau}\right] \\
&= \left[-\Lambda^{-1}|\tau|^{-1} + A_0 + A'_0\hat{\tau}\right] - \left[\Lambda^{-1}|\tau|^{-1} + B_0 + B'_0\hat{\tau}\right] \\
&\quad \times \left[-\mathrm{I} + |\tau|\Lambda A_0 - \tau\Lambda A'_0\right]^{-1} \left[\mathrm{I} + |\tau|\Lambda B_0 - \tau\Lambda B'_0\right] \\
&= \left[-\Lambda^{-1}|\tau|^{-1} + A_0 + A'_0\hat{\tau}\right] - \left[\Lambda^{-1}|\tau|^{-1} + B_0 + B'_0\hat{\tau}\right] \\
&\quad \times \left[-\mathrm{I} + |\tau|\Lambda(A_0 + B_0) - \tau\Lambda(A'_0 + B'_0) + \mathcal{O}(\tau^2)\right] \\
&= \left[-\Lambda^{-1}|\tau|^{-1} + A_0 + A'_0\hat{\tau}\right] + \left[\Lambda^{-1}|\tau|^{-1} + A_0 + 2B_0 - A'_0\hat{\tau} + \mathcal{O}(\tau)\right] \\
&= 2(A_0 + B_0) + \mathcal{O}(\tau).
\end{aligned} \tag{2.85}$$

With this, the expansion for the time correlation function is

$$\begin{aligned}
Q(\tau) &= C(\tau)^{-1}B(\tau)^{\mathrm{T}}S^{-1} \\
&= \left[-\Lambda^{-1}|\tau|^{-1} + A_0 - A'_0\hat{\tau}\right]^{-1} \left[\Lambda^{-1}|\tau|^{-1} + B_0 - B'_0\hat{\tau}\right] S^{-1} \\
&= \left[\mathrm{I} - |\tau|\Lambda A_0 + \tau\Lambda A'_0\right]^{-1} \left[-\mathrm{I} - |\tau|\Lambda B_0 + \tau\Lambda B'_0\right] S^{-1} \\
&= \left[-\mathrm{I} - |\tau|\Lambda(A_0 + B_0) + \tau\Lambda(A'_0 + B'_0)\right] S^{-1} \\
&= -S^{-1} - \frac{|\tau|}{2}\Lambda + \tau\Theta + \mathcal{O}(\tau^2).
\end{aligned} \tag{2.86}$$

Since $Q(-\tau) = Q(\tau)^{\mathrm{T}}$, the coefficient of τ must be antisymmetric, and so one must have

$$\Theta \equiv \Lambda(A_0' + B_0')S^{-1} = -S^{-1}(A_0' - B_0')\Lambda. \tag{2.87}$$

The odd time fluctuation matrices can be written in terms of this antisymmetric transport matrix,

$$A_0' = \frac{1}{2}\left[\Lambda^{-1}\Theta S - S\Theta\Lambda^{-1}\right], \text{ and } B_0' = \frac{1}{2}\left[\Lambda^{-1}\Theta S + S\Theta\Lambda^{-1}\right]. \tag{2.88}$$

This means that one has essentially three coefficient matrices to determine the time correlation function $Q(\tau)$ to linear order in τ: S, Λ, and Θ. One requires one more relation between A_0 and B_0 to determine $A(\tau)$ and $B(\tau)$ to zeroth order in τ.

For the pure parity case, the transport matrix is symmetric. For the mixed parity case, the transport matrix would correspond to $L(\hat{\tau}) \equiv \Lambda - 2\hat{\tau}\Theta$, which evidently contains an antisymmetric part that is odd in time. By time homogeneity and macroscopic reversibility, $Q(-\tau) = \varepsilon Q(\tau)\varepsilon$, or in component form, $Q_{ij}(-\tau) = \varepsilon_{ii}\varepsilon_{jj}Q_{ij}(\tau)$. Recall that the parity matrix gives the time signature of the macrostate, $x_i^\dagger = \varepsilon_{ii}x_i$. Using this in the small time expansion one sees that the elements of the symmetric part of the transport matrix are zero for states of different parity,

$$\Lambda_{ij} = 0 \text{ if } \varepsilon_{ii}\varepsilon_{jj} = -1, \tag{2.89}$$

and that the elements of the antisymmetric part of the transport matrix are zero for states of the same parity,

$$\Theta_{ij} = 0 \text{ if } \varepsilon_{ii}\varepsilon_{jj} = 1. \tag{2.90}$$

Similarly $\{S^{-1}\}_{ij} = 0$ if $\varepsilon_{ii}\varepsilon_{jj} = -1$.

In view of these results it can be convenient to order the macrostates in $\mathbf{x}$ so that all those of even parity come before all those of odd parity. In this grouped representation the matrices sub-divide into four blocks: the matrices Λ, S^{-1} (and hence S) are symmetric and block diagonal, and the matrix Θ is antisymmetric and block adiagonal.

The optimum flux is

$$\overline{\overset{\circ}{\mathbf{x}}} = \frac{\hat{\tau}}{2}\left[\Lambda - 2\hat{\tau}\Theta\right] S\mathbf{x}. \tag{2.91}$$

This is the form of Onsager's regression hypothesis for a system of mixed parity.

The second entropy may be rewritten in terms of fluctuations in the flux,

$$\begin{aligned} S^{(2)}(\mathbf{x}',\mathbf{x}|\tau) &= \frac{1}{2}A(\tau):\mathbf{x}^2 + B(\tau):\mathbf{x}'\mathbf{x} + \frac{1}{2}C(\tau):\mathbf{x}'^2 \\ &= \frac{1}{2}C(\tau):[\mathbf{x}'-\mathbf{x}]^2 + [B(\tau)+C(\tau)]:\mathbf{x}'\mathbf{x} \\ &\quad + \frac{1}{2}[A(\tau)-C(\tau)]:\mathbf{x}^2 \\ &= \frac{1}{2}C(\tau):[\mathbf{x}'-\mathbf{x}]^2 + [\mathbf{x}'-\mathbf{x}]\cdot\left[B(\tau)^{\mathrm{T}}+C(\tau)\right]\mathbf{x} \\ &\quad + \frac{1}{2}\left[A(\tau)+2B(\tau)^{\mathrm{T}}+C(\tau)\right]:\mathbf{x}^2 \\ &= \frac{1}{2}S:\mathbf{x}^2 - \frac{|\tau|}{2}\Lambda^{-1}:\overset{\circ}{\mathbf{x}}{}^2 + \tau\,\overset{\circ}{\mathbf{x}}\cdot[A_0+B_0-\hat{\tau}(A_0'+B_0')]\,\mathbf{x} \\ &\quad - \frac{|\tau|}{2}\Lambda^{-1}:\overline{\overset{\circ}{\mathbf{x}}}(\mathbf{x};\hat{\tau})^2 + \mathcal{O}(\tau^2). \end{aligned} \tag{2.92}$$

The final term on the right-hand side ensures that the reduction condition is satisfied to linear order in τ. The third term contains what may be called the mixed parity thermodynamic force,

$$\mathbf{F}(\mathbf{x}) \equiv 2\,[A_0+B_0-\hat{\tau}(A_0'+B_0')]\,\mathbf{x} \equiv S\mathbf{x} - 2\hat{\tau}\Lambda^{-1}\Theta S\mathbf{x}. \tag{2.93}$$

The first term on the right-hand side is evidently the derivative of the entropy, and the second term is an additional term that only arises in the mixed parity case. Loosely speaking, the third term in the expression for the second entropy may be called half the rate of mixed parity entropy production. The terminology is not entirely satisfactory, because the additional mixed parity term does not contribute to the real rate of entropy production.

The physical interpretation of the second entropy written in terms of fluctuations in the flux is similar to that given for the pure parity case in §2.4.3. The first term on the right-hand side of the final equality is the ordinary entropy. It is the only term that survives when the flux is optimum, and it determines the structure of the non-equilibrium state for such an optimum flux. The second term is negative and quadratic in the flux. It represents the cost of dynamic order and it opposes any increase in the magnitude of the flux. The third term is half the rate of mixed parity entropy production. It drives the flux to be in the same direction as the thermodynamic force. From the definiteness of the matrices that is established below, this term favours the projection of the future flux in the opposite direction to $\mathbf{x}$ (i.e. to decrease $\mathbf{x}$), since this increases the second entropy, and it favours the alignment of the past flux and $\mathbf{x}$, both of which are expected on physical grounds. The fourth term is independent of the flux, at least explicitly, and is required to ensure the reduction condition for the optimum flux. The physical interpretation of the mixed parity term in the entropy production and of this flux-independent term is not clear.

The antisymmetric part of the transport matrix has an effect on the most likely future or past states. For the mixed parity case,

$$\overline{\mathbf{x}}(\tau,\mathbf{x}) - \overline{\mathbf{x}}(-\tau,\mathbf{x}) = -[Q(\tau)-Q(-\tau)]S\mathbf{x} = -2\tau\Theta S\mathbf{x} + \mathcal{O}(\tau^2). \tag{2.94}$$

This would be zero for the pure parity case. Similarly the fluctuation matrices now have a contribution that is odd in time.

The antisymmetric part of the transport matrix, Θ, is extremely important for the time development of a system at the molecular level. It will be seen that the adiabatic evolution of the system is contained in this term (c.f. the treatment of driven Brownian motion in §3.6.3, and the treatment of non-equilibrium statistical mechanics in Ch. 8).

The optimum rate of entropy production is unaffected by the antisymmetric contribution in the mixed parity case. One has

$$\begin{aligned} \overline{\overset{\circ}{S}}(\mathbf{x}) &= \frac{1}{\tau}\left[\overline{\mathbf{x}}(\tau,\mathbf{x}) - \mathbf{x}\right] \cdot S\mathbf{x} \\ &= \mathbf{x} \cdot \left[\frac{\hat{\tau}}{2} S\Lambda + S\Theta\right] S\mathbf{x} \\ &= \frac{\hat{\tau}}{2}\mathbf{x} \cdot S\Lambda S\mathbf{x} + \mathcal{O}(\tau). \end{aligned} \tag{2.95}$$

The final equality follows because the symmetric double scalar product of an antisymmetric matrix vanishes. This result for the dissipation is identical to the pure parity result, Eq. (2.52).

Since the second entropy must be less than or equal to zero, it follows that $A(\tau)$, $C(\tau)$, and $A(\tau) + C(\tau) \pm 2B(\tau)$ are all negative definite matrices. Since as $\tau \to 0$ the leading term dominates, it follows that Λ is a positive definite matrix. The leading contribution to the positive case of the third of these is $2A_0 + 2B_0 + 2\hat{\tau}B_0'$, and so one concludes that $A_0 + B_0 \pm B_0'$ must be a negative definite matrix. But only the symmetric part of a matrix contributes to its definiteness, so the term B_0' is irrelevant. Since $A_0 + B_0 = S/2$, one could have deduced directly that this is negative definite.

One could assume that the negative definiteness of the fluctuation matrix functions holds for the τ^0 term itself, arguing that there is a τ regime where this is large enough to dominate the τ^{-1} term, but where the neglected terms linear in τ are not yet significant. If this is the case then one has $A_0 \pm A_0'$, $A_0 \pm B_0$, and hence A_0 itself, are negative definite.

2.5.3 Magnetic Fields and Coriolis Forces

The operation of velocity reversal or conjugation, $\mathbf{x} \Rightarrow \mathbf{x}^\dagger$, is designed to reverse exactly the time evolution of the system. If the system is influenced by external forces or other interactions with external reservoirs, then the velocities of the atoms and molecules of these external systems also have to be reversed, otherwise the trajectory of the sub-system will not exactly traverse its former path.

Some forces, most notably magnetic forces, which arise from electronic motion or electric currents, and Coriolis forces, which arise from rotational motion, are velocity dependent. Hence implicit in the conjugation operation is that any magnetic field or rotational motion must also be reversed.

The time correlation function in an external magnetic field $\mathbf{B}$, for example, might be written $Q(\tau;\mathbf{B})$. In this case time homogeneity yields

$$\langle \mathbf{x}(\tau)\mathbf{x}(0)\rangle_{\mathbf{B}} = \langle \mathbf{x}(0)\mathbf{x}(-\tau)\rangle_{\mathbf{B}}, \text{ or } Q(\tau;\mathbf{B}) = Q(-\tau;\mathbf{B})^{\mathrm{T}}. \tag{2.96}$$

Macroscopic reversibility reads

$$\langle \mathbf{x}(\tau)\mathbf{x}(0)\rangle_{\mathbf{B}} = \left\langle \mathbf{x}(-\tau)^{\dagger}\mathbf{x}(0)^{\dagger}\right\rangle_{-\mathbf{B}}, \text{ or } Q(\tau;\mathbf{B}) = \varepsilon Q(-\tau;-\mathbf{B})\varepsilon. \tag{2.97}$$

For further discussion below, these two results can be combined in a form such that the sign of the time interval remains unchanged,

$$Q(\tau;\mathbf{B}) = \varepsilon Q(\tau;-\mathbf{B})^{\mathrm{T}}\varepsilon. \tag{2.98}$$

The matrices in the small time expansion of the correlation function now depend upon the magnetic field,

$$Q(\tau;\mathbf{B}) = -S(\mathbf{B})^{-1} - \frac{|\tau|}{2}\Lambda(\mathbf{B}) + \tau\Theta(\mathbf{B}) + \mathcal{O}(\tau^2). \tag{2.99}$$

The first two terms on the right-hand side are even in time, and the final term is odd. Hence time homogeneity, Eq. (2.96), implies that the entropy fluctuation matrix and the pure parity transport matrix are symmetric,

$$S(\mathbf{B}) = S(\mathbf{B})^{\mathrm{T}}, \text{ and } \Lambda(\mathbf{B}) = \Lambda(\mathbf{B})^{\mathrm{T}}, \tag{2.100}$$

and that the mixed parity transport matrix is antisymmetric,

$$\Theta(\mathbf{B}) = -\Theta(\mathbf{B})^{\mathrm{T}}. \tag{2.101}$$

Notice that the magnetic field need not, indeed must not, be reversed to ensure these symmetries. Macroscopic reversibility, Eq. (2.97), implies that the parity matrix commutes with the entropy fluctuation matrix and the pure parity transport matrix,

$$\varepsilon S(-\mathbf{B})\varepsilon = S(\mathbf{B}), \text{ and } \varepsilon\Lambda(-\mathbf{B})\varepsilon = \Lambda(\mathbf{B}), \tag{2.102}$$

and that it 'anti-commutes' with the mixed parity transport matrix,

$$\varepsilon\Theta(-\mathbf{B})\varepsilon = -\Theta(\mathbf{B}). \tag{2.103}$$

These hold if and only if the magnetic field is reversed. Similar results hold for Coriolis forces.

Useful, more explicit, forms of these results can be obtained in the small magnetic field limit. In this case a Taylor expansion about $\mathbf{B} = 0$ to quadratic order in the field can be made. One has

$$\Lambda_{ij}(\mathbf{B}) = \Lambda_{ij}^{(0)} + \mathcal{O}(\mathbf{B}^2), \text{ and } \Theta_{ij}(\mathbf{B}) = \mathbf{B}\cdot\Theta_{ij}^{(1)} + \mathcal{O}(\mathbf{B}^3),\ \varepsilon_{ii} = \varepsilon_{jj}, \tag{2.104}$$

and

$$\Lambda_{ij}(\mathbf{B}) = \mathbf{B}\cdot\mathbf{\Lambda}_{ij}^{(1)} + \mathcal{O}(\mathbf{B}^3), \text{ and } \Theta_{ij}(\mathbf{B}) = \Theta_{ij}^{(0)} + \mathcal{O}(\mathbf{B}^2),\ \varepsilon_{ii} \neq \varepsilon_{jj}. \tag{2.105}$$

Onsager's reciprocal relations (i.e. the symmetry of the transport matrix) were generalised to the case of magnetic fields and Coriolis forces by Casimir.[10] It should be noted that many books neglect the difference between the future and the past transport matrix (i.e. they do not avert to its dependence upon $\hat{\tau}$). Those forms for the reciprocal relations in the presence of magnetic fields when the time dependence is neglected can be reconciled with the results obtained here by invoking the symmetry of the time correlation that does not change the sign of the time interval, Eq. (2.98). The expressions given by de Groot and Mazur in their §§IV.3 and VII.4,[11] and the expression represented by Pottier's Eq. (2.8.5),[12] are consistent with Eq. (2.98).

[10]Casimir, H. B. G. (1945), Rev. Mod. Phys. **17**, 343.

[11]de Groot, S. R. and Mazur, P. (1984), *Non-equilibrium Thermodynamics*, (Dover, New York).

[12]Pottier, N. (2010), *Non-equilibrium Statistical Physics: Linear Irreversible Processes*, (Oxford University Press, Oxford).

Chapter 3

Brownian Motion

In the preceding chapter, both the fluctuations and the transitions in the macrostates of an equilibrium system were analysed in terms of Gaussian probabilities. This entailed a quadratic form for the second entropy, and from a small time expansion for the coefficients, several non-equilibrium theorems were established.

In this chapter, fluctuations are treated as a stochastic process. This serves to introduce some of the concepts and results for statistical mechanics that will be more formally derived from first principles in Chs 7 and 8. Readers primarily interested in non-equilibrium thermodynamics, or who wish to see an application of fluctuation theory in detail, may wish to skip directly to Chs 4–6 where conductive heat flow, fluctuation hydrodynamics, and non-equilibrium pattern formation in convective heat flow are treated.

For a stochastic process the system is viewed as following a random trajectory in time, $\mathbf{x}(t)$.[1] The prototypical stochastic process is Brownian motion, which originally referred to the erratic movement of small pollen grains in water recorded by Robert Brown (1828).[2] Einstein developed the first statistical analysis of the motion, and showed that the mean square displacement grew linearly with time.[3] A similar probabilistic analysis was given independently by Smoluchowski (1906).[4] Einstein's and Smoluchowski's approach was based upon the diffusion of the probability density, which was later formalised by Fokker and Planck.[5]

Langevin[6] formulated the problem by taking the acceleration of a particle to

[1] In many books the class of a stochastic process is denoted by a capital letter, $\mathbf{X}(t)$, and an actual realisation of the trajectory by a lower case letter, $\mathbf{x}(t)$. Here, however, this distinction will not be observed, and whether $\mathbf{x}(t)$ denotes the value of the macrostate at each time for a specific trajectory, or the set of all possible trajectories, must be gleaned from the context.

[2] Brown, R. (1828), Phil. Mag. **4**, 171. For a brief history of Brownian motion, see Haw, M. D. (2002), J. Phys.: Condens. Matter, **14**, 7769.

[3] Einstein, A. (1905), Ann. Phys. **17**, 549.

[4] von Smoluchowski, M. (1906), Ann. Phys. **21**, 756.

[5] Fokker, A. (1914), Ann. Phys. **43**, 810. Planck, M. (1916), Sitz. Ber. Preuss. Akad. Wiss. 324.

[6] Langevin, P. (1908), C. R. Acad. Sci. Paris, **146**, 530.

be due to deterministic and random forces. This generated a random walk from what is now called a stochastic differential equation. Langevin attributed the deterministic force to Stokes drag, a dissipative hydrodynamic force proportional to the velocity, and the random force to the fluctuating molecular motion of the solvent. Using the equipartition theorem, Langevin showed the magnitudes of these two forces were related. This has come to be called the fluctuation dissipation theorem and it has far greater generality than the original application to Brownian motion. Its derivation and application will play a central rôle in non-equilibrium statistical mechanics in subsequent chapters of this book.

The full treatment of Brownian motion and stochastic processes can only proceed from the basis of non-equilibrium statistical mechanics. Specifically, the fundamental form of the non-equilibrium probability distribution and the transition probability need to be established. However, in view of the historical primacy Brownian motion has played in the development of statistical mechanics, it seems worthwhile to give here an introductory treatment of the subject based on the second entropy theory of fluctuations that was elucidated in the preceding chapter.

What thermodynamic fluctuations and Brownian motion have in common is that the macrostate of interest evolves in time in a manner that has both deterministic and random elements. The analogy with thermodynamic fluctuations breaks down somewhat for the original application of Brownian motion to the movement of pollen grains in water, because in this case there is no restoring force acting on the particle, and no preferred position or direction of motion. This type of Brownian motion can also be called a random walk, or, a free Brownian particle. For this original case the analogy between a random walk and a thermodynamic fluctuation is not completely satisfactory, since thermodynamic fluctuations are localised about the equilibrium value of the macrostate. However one can apply an external field (e.g. an optical trap such as laser tweezers, or an electrostatic potential) that pins or directs the Brownian particle, and which more closely resembles a thermodynamic fluctuation. Free and trapped Brownian particles are treated in §§3.2 and 3.3, respectively.

Two important concepts emerge from this analysis: the fluctuation-dissipation theorem, which fixes the ratio of the magnitudes of the dissipative and stochastic forces, and the universal nature of the dissipative force, which is shown to be statistical in origin rather than hydrodynamic.[7]

Historically and conceptually, Brownian motion lies intermediate between thermodynamics and statistical mechanics. On the one hand it is a good quantitative model for thermodynamic fluctuations, and on the other hand, as in Einstein's original treatment, it can be usefully described by a probability distribution and a transition probability, which are the stuff of statistical mechanics. Position-velocity time correlation functions are given in the large drag regime in §3.5. In §3.6 an heuristic presentation of the probability distribution of a Brownian particle in a stationary and in a moving potential trap is given, also in the

[7]The dissipative force is consistent with Stoke's hydrodynamic drag, which is Langevin's original justification, but it is actually more fundamental. The functional form applies as well to non-hydrodynamic systems, as will be shown in Chs 7 and 8.

large drag regime. In §3.6.3 the probability distribution for a moving trap in the small drag regime is derived from the mixed parity fluctuation results of the preceding chapter. In §3.7 expressions for the evolution of the entropy and for the non-equilibrium probability distribution are obtained, and the Fokker-Planck equation and the limitations on Liouville's theorem are discussed.

3.1 Gaussian, Markov Processes

The state of the system at time t is denoted $\mathbf{x}(t)$ and is referred to generically as the position. This might be a three-dimensional vector representing the position of a single Brownian particle, or a multi-dimensional vector representing the positions of a number of Brownian particles, or it might be a set of fluctuating macrostates. For a stochastic process, the change in position along a trajectory has deterministic and stochastic contributions. The deterministic part is the most likely position, which, for Gaussian fluctuations for macrostates all of the same time parity in the small time limit was obtained above, Eq. (2.45). The stochastic part represents the Gaussian fluctuations about this most likely outcome. That is, the next step in the stochastic process can be written

$$\mathbf{x}(t+\tau) = \mathbf{x}(t) + \frac{|\tau|}{2}\Lambda S\mathbf{x}(t) + \tilde{\mathbf{R}}(t). \tag{3.1}$$

The most likely next position is this with $\tilde{\mathbf{R}} = \mathbf{0}$, $\overline{\mathbf{x}}(\tau|\mathbf{x}) \equiv \mathbf{x} + |\tau|\Lambda S\mathbf{x}/2$. The probability distribution for the random 'force' is just the exponential of the second entropy, Eq. (2.50),

$$\wp(\tilde{\mathbf{R}}) = \frac{|\Lambda|^{-1/2}}{(2\pi k_{\mathrm{B}}|\tau|)^{M/2}} \exp\left[\frac{-\Lambda^{-1} : \tilde{\mathbf{R}}\tilde{\mathbf{R}}}{2k_{\mathrm{B}}|\tau|}\right]. \tag{3.2}$$

Here M is the number of components in the position or macrostate vector $\mathbf{x}$, and k_{B} is Boltzmann's constant. For the Markov case, successive random forces are uncorrelated, $\left\langle \tilde{\mathbf{R}}(t+\tau)\tilde{\mathbf{R}}(t)\right\rangle = 0$. Their covariance is $\left\langle \tilde{\mathbf{R}}(t)\tilde{\mathbf{R}}(t)\right\rangle = k_{\mathrm{B}}|\tau|\Lambda$.

The stochastic process is written in the form of a difference equation rather than a differential equation. Recall that the theory is valid for time intervals $|\tau|$ that are larger than the molecular regime, but which are small enough that the change in state is relatively negligible. The random force $\tilde{\mathbf{R}}$ represents the sum total of the molecular influences on the evolution of the state during the non-infinitesimal time step τ. More precisely, it is the excess over the mean molecular influences, since the latter give the deterministic part of the evolution.

The simplest attempt to generate a trajectory based upon Eq. (3.1) is to use it recursively,

$$\mathbf{x}_{n+1} = \mathbf{x}_n + \frac{|\tau|}{2}\Lambda S\mathbf{x}_n + \tilde{\mathbf{R}}_n,\ n = 0, 1, 2, \ldots, \tag{3.3}$$

where $\mathbf{x}_n \equiv \mathbf{x}(t_n|\mathbf{x}_0)$, $t_n \equiv n\tau$, and the $\tilde{\mathbf{R}}_n$ are independent random variables, Gaussian distributed as above. This is only true for a Markov system, in which the value of the macrostate at each node depends only upon the value at the preceding node (c.f. the comment in §2.3 on p. 44). Gaussian, Markov processes are also called Ornstein-Uhlenbeck processes.

For a more general non-Markov process, the deterministic part of the process depends upon the values at two or more preceding nodes, the precise number required being dependent upon the rate of decay of the non-Markov influences. Such non-Markov stochastic process are treated in Ch. 10.

The Gaussian, Markov process described above is trivial for a deterministic force that is a linear function of the macrostate, $\partial S(\mathbf{x})/\partial \mathbf{x} = S\mathbf{x}$. This is the case for the thermodynamic fluctuations treated in the preceding chapter. The most likely trajectory can be immediately integrated to give exponential behaviour. More generally, the Brownian particle or stochastic process may occur in the presence of an arbitrary force, $\mathbf{F}(\mathbf{x}, t)$. In such cases either analytic or numeric solutions are required for the most likely trajectory and for averages of quantities over the trajectory.

In the physical theory of stochastic processes two thermodynamic variables appear: the thermodynamic force, $S\mathbf{x} = \partial S(\mathbf{x})/\partial \mathbf{x}$, and the transport matrix, Λ. The distribution of the random forces is completely determined by the transport matrix, and how the dissipative force is determined by the transport matrix and the thermodynamic force. This is the fluctuation dissipation theorem. The fluctuation part of the theorem refers to the random forces. The dissipative part refers to the deterministic forces, which terminology will be made clearer in the following example. The physical content of the fluctuation dissipation theorem is that there is a direct quantitative relationship between the stochastic forces and the deterministic forces. Whenever one has forces that arise from 'hidden' variables, such as the present resummation of molecular forces over the non-infinitesimal time step, then one always has a deterministic part and a stochastic part that are linearly proportional to each other and that are related by a fluctuation dissipation theorem. This point will prove important in developing equations of motion for the statistical mechanics of non-equilibrium systems. For the present, one can simply note that it is unphysical to have a deterministic dissipative contribution to the equations of motion whilst neglecting the stochastic contribution demanded by the fluctuation dissipation theorem.

3.2 Free Brownian Particle

As a concrete example, consider a free Brownian particle in a solvent. The stochastic motion of a Brownian particle is also called a random walk, and it is in many respects a good model for thermodynamic fluctuations in general. The macrostate of the system in the first case is taken to be the velocity $\mathbf{v}$. The Brownian particle has mass m, kinetic energy is $mv^2/2$, and entropy

$$S(\mathbf{v}) = \frac{-mv^2}{2T}, \tag{3.4}$$

where T is the temperature of the solvent, which acts as a heat reservoir. This is the total entropy, which is equal to the reservoir entropy because the Brownian particle has no internal entropy, since $\mathbf{v}$ represents a microstate of the sub-system. One can see that this is of Gaussian form, with the entropy matrix just $-m/T$ times the identity matrix. The variance of the velocity is therefore $\langle v^2 \rangle = 3k_{\rm B}T/m$. The thermodynamic force is $\partial S(\mathbf{v})/\partial \mathbf{v} = -m\mathbf{v}/T$, and hence the dissipative term in the difference equation is $(|\tau|\Lambda/2)\partial S(\mathbf{v})/\partial \mathbf{v} = -m|\tau|\lambda\mathbf{v}/2T$. Here the transport matrix is just $\Lambda = \lambda\mathrm{I}$, where λ is a scalar that will be related to the diffusion constant below. The stochastic process is then

$$\mathbf{v}(t+\tau) = \mathbf{v}(t) - \frac{|\tau|m\lambda}{2T}\mathbf{v}(t) + \tilde{\mathbf{R}}'(t), \tag{3.5}$$

or, with $\tilde{\mathbf{R}} \equiv m\tilde{\mathbf{R}}'/|\tau|$,

$$m\frac{\Delta\mathbf{v}(t)}{\Delta t} = -\gamma\mathbf{v}(t) + \tilde{\mathbf{R}}(t),\ \tau > 0. \tag{3.6}$$

The left-hand side can be interpreted as mass times acceleration, with $\Delta t \equiv \tau$. Hence the first term on the right-hand side is the drag or friction force, with the friction coefficient being

$$\gamma \equiv \frac{m^2\lambda}{2T}. \tag{3.7}$$

From Eq. (3.2), the variance of the random force is $\left\langle \tilde{\mathbf{R}}(t)\cdot\tilde{\mathbf{R}}(t) \right\rangle = 6k_{\rm B}T\gamma/|\tau|$. In the continuum limit, this becomes $\left\langle \tilde{\mathbf{R}}(t+s)\cdot\tilde{\mathbf{R}}(t) \right\rangle = 6k_{\rm B}T\gamma\delta(s)$. That the magnitude of the fluctuations is given by the friction coefficient is a manifestation of the fluctuation dissipation theorem.

Dropping the stochastic contribution, the most likely acceleration is

$$\dot{\overline{\mathbf{v}}}(t) = \frac{-\gamma}{m}\overline{\mathbf{v}}(t), \tag{3.8}$$

which evidently gives exponential decay of the velocity,

$$\overline{\mathbf{v}}(t) = e^{-\gamma|t|/m}\mathbf{v}(0). \tag{3.9}$$

For a Markov process, this exponential form of regression is exact. Accordingly one has

$$\begin{aligned} \langle \mathbf{v}(t)\cdot\mathbf{v}(0)\rangle &= \langle \overline{\mathbf{v}}(t)\cdot\mathbf{v}(0)\rangle \\ &= \frac{3k_{\rm B}T}{m}e^{-\gamma|t|/m}. \end{aligned} \tag{3.10}$$

One can relate the transport coefficient to the more usual diffusion constant by invoking Einstein's result that the variance of the position grows linearly with time,

$$\langle [\mathbf{r}(t)-\mathbf{r}(0)]\cdot[\mathbf{r}(t)-\mathbf{r}(0)]\rangle = 6|t|D, \tag{3.11}$$

where D is the diffusion constant. (This result will be derived in the following section.) Differentiating this with respect to time and invoking the exponential form of the regression theorem one obtains

$$\begin{aligned} D &= \frac{1}{3}\langle \mathbf{v}(t)\cdot[\mathbf{r}(t)-\mathbf{r}(0)]\rangle \\ &= \frac{1}{3}\int_0^t \mathrm{d}t' \, \langle \mathbf{v}(t)\cdot\mathbf{v}(t')\rangle \\ &= \frac{1}{3}\int_0^t \mathrm{d}t' \, \frac{3k_\mathrm{B}T}{m}e^{-\gamma|t-t'|/m} \\ &= \frac{k_\mathrm{B}T}{m}\frac{m}{\gamma}\left[1-e^{-\gamma t/m}\right] = \frac{k_\mathrm{B}T}{\gamma},\ t \gg m/\gamma. \end{aligned} \tag{3.12}$$

where the exponential term has been neglected for large times. This shows the relationship between the drag coefficient and the diffusion constant,

$$\gamma = k_\mathrm{B}T/D. \tag{3.13}$$

The diffusion constant is determined by the magnitude of the stochastic fluctuations in position, as will be seen next, and the friction coefficient is determined by the magnitude of the dissipative force. This explains the origin of the nomenclature fluctuation dissipation theorem.

3.3 Pinned Brownian Particle

Now the same Brownian particle can be treated from the point of view of position macrostates, $\mathbf{r}$. Suppose it experiences a potential energy in the form of a harmonic trap centred at the origin, $U(\mathbf{r}) = \kappa r^2/2$. Hence the entropy is

$$S(\mathbf{r}) = \frac{-\kappa}{2T}r^2, \tag{3.14}$$

the variance is $\langle r^2\rangle = 3k_\mathrm{B}T/\kappa$, and the thermodynamic force is $-\kappa\mathbf{r}/T$.

The stochastic process is then

$$\mathbf{r}(t+\tau) = \mathbf{r}(t) - \frac{|\tau|\kappa D}{k_\mathrm{B}T}\mathbf{r}(t) + \tilde{\mathbf{R}}_x(t), \tag{3.15}$$

where the transport matrix for this formulation is denoted $\Lambda \equiv 2D/k_\mathrm{B}$, where D will turn out to be the diffusion constant. This random force has covariance $\left\langle \tilde{\mathbf{R}}_x(t)\cdot\tilde{\mathbf{R}}_x(t)\right\rangle = k_\mathrm{B}|\tau|\Lambda = 2|\tau|D$. Taking the scalar product with $\mathbf{r}(t)$ yields

$$\langle[\mathbf{r}(t+\tau)-\mathbf{r}(t)]\cdot\mathbf{r}(t)\rangle = \frac{-|\tau|\kappa D}{k_\mathrm{B}T}\times\frac{3k_\mathrm{B}T}{\kappa} = -3D|\tau|, \tag{3.16}$$

since there is no correlation between the current position and the random force, $\left\langle \mathbf{r}(t)\tilde{\mathbf{R}}(t)\right\rangle = 0$. Note that this is negative, which says that both in the future and in the past the particle lies closer to the origin than it does now, which is what one expects because the origin is the equilibrium state. The right-hand side is independent of the spring constant, and so this expression also holds in the free particle limit, $\kappa \rightarrow 0$. Changing first τ to $-\tau$, and then t to $t+\tau$, shows that the left-hand side is also equal to $\langle [\mathbf{r}(t) - \mathbf{r}(t+\tau)] \cdot \mathbf{r}(t+\tau) \rangle$. Adding these together and taking the negative, one has

$$\langle [\mathbf{r}(t+\tau) - \mathbf{r}(t)] \cdot [\mathbf{r}(t+\tau) - \mathbf{r}(t)] \rangle = 6D|\tau|, \tag{3.17}$$

which is just Einstein's expression. This confirms the identification of D with the diffusion constant, and shows that it gives the magnitude of the fluctuations in position. The result is significant in that it predicts that the mean square displacement of a random walk, which is another name for Brownian motion, grows linearly in time, which contrasts with the quadratic growth that would be expected for deterministic motion. The present analysis for a pinned particle shows that the result is independent of the trap constant, but is restricted to the small time limit. For a free particle (i.e. a random walk in the absence of a trap), one might expect that the mean square displacement would grow linearly in time without limit. However, that is not such a good model for a thermodynamic fluctuation, which always has a restoring force that seeks to return it to the equilibrium state.

The analysis of velocity fluctuations invoked an exponential form for the time correlation function, which not only assumes Markovian behaviour, but it also invoked the long time limit, $|\tau| \gg m/\gamma$. However, the position analysis invoked the short time limit in the case of a non-zero trap constant, $|\tau| \ll k_{\mathrm{B}}T/\kappa D$. Hence the analyses are both valid in the regime

$$\frac{Dm}{k_{\mathrm{B}}T} \ll |\tau| \ll \frac{k_{\mathrm{B}}T}{\kappa D}. \tag{3.18}$$

This is satisfied in the small diffusion constant limit, or, equivalently, in the large drag limit.

One of the more interesting aspects of the above analysis is the emergence of the dissipative force in Eq. (3.6), $-\gamma\mathbf{v}$. This is the drag force that was originally invoked by Langevin, and rationalised on the basis of Stokes' law in hydrodynamics. In this the drag force on a sphere is linear in, and opposite to, the velocity, with the proportionality constant dependent upon the sphere radius, the fluid viscosity, and the hydrodynamic boundary conditions. Nowadays one would say that continuum hydrodynamics is not applicable on the length scales of Brownian particles, but nevertheless the dissipative force in this form is applicable, and the challenge is to justify it on the basis of non-equilibrium statistical mechanics. One compelling reason for seeking such a fundamental justification is that no such hydrodynamic argument exists for an analogous dissipative force for thermodynamic fluctuations, yet nevertheless these and other stochastic processes appear to behave identically to Brownian motion. A virtue of the present fluctuation approach is that the friction force law emerges from the regression theorem. That is, the form of the dissipative force is determined by the universal behaviour of thermodynamic fluctuations, rather than by a hydrodynamic result for a particular system.

3.4 Diffusion Equation

In a sense, the fundamental characteristic of a Brownian particle is that the average quadratic displacement, the variance, grows linearly with the time interval. One can construct a Gaussian transition probability from Eq. (3.17), namely

$$\wp(\mathbf{r}|\mathbf{r}',\tau) = \frac{e^{-[\mathbf{r}'-\mathbf{r}]\cdot[\mathbf{r}'-\mathbf{r}]/4D|\tau|}}{(4\pi D|\tau|)^{3/2}}, \quad \frac{Dm}{k_{\mathrm{B}}T} \ll |\tau| \ll \frac{k_{\mathrm{B}}T}{\kappa D}. \tag{3.19}$$

This is the probability that the particle will be at $\mathbf{r}$ at time $t+\tau$ given that it was at $\mathbf{r}'$ at time t. It is Gaussian because the total displacement over the time interval is the sum of many random minuscule displacements. The time interval cannot be too small, because the Einstein result only holds after long intervals such that the prior history of the particle has no influence. This is independent of the trap constant, so it applies as well to the free particle, $\kappa \to 0$.

Assume that the system is in a transient non-equilibrium state (due for example, to specified initial conditions) such that $\wp(\mathbf{r},t)$ is the probability of the Brownian particle being at $\mathbf{r}$ at time t. In this case the transition probability gives the evolution,

$$\wp(\mathbf{r},t+\tau) = \int \mathrm{d}\mathbf{r}'\ \wp(\mathbf{r}|\mathbf{r}',\tau)\wp(\mathbf{r}',t), \tag{3.20}$$

as follows from the laws of probability. Expanding $\wp(\mathbf{r}',t)$ to second-order about $\mathbf{r}$, this is

$$\begin{aligned}\wp(\mathbf{r},t+\tau) &= \int \mathrm{d}\mathbf{r}'\ \frac{e^{-[\mathbf{r}'-\mathbf{r}]\cdot[\mathbf{r}'-\mathbf{r}]/4D|\tau|}}{(4\pi D|\tau|)^{3/2}}\Big[\wp(\mathbf{r},t) + [\mathbf{r}'-\mathbf{r}]\cdot\nabla\wp(\mathbf{r},t) \\ &\qquad + \frac{1}{2}[\mathbf{r}'-\mathbf{r}]\,[\mathbf{r}'-\mathbf{r}] : \nabla\nabla\wp(\mathbf{r},t) + \ldots\Big] \\ &= \wp(\mathbf{r},t) + |\tau| D\nabla^2\wp(\mathbf{r},t). \end{aligned} \tag{3.21}$$

Assuming that the probability varies little over the time interval τ, this is

$$\frac{\partial\wp(\mathbf{r},t)}{\partial t} \approx \frac{\wp(\mathbf{r},t+\tau)-\wp(\mathbf{r},t)}{\tau} = \hat{\tau}D\nabla^2\wp(\mathbf{r},t), \tag{3.22}$$

where $\hat{\tau} \equiv \mathrm{sign}(\tau)$. Recall that τ cannot be made arbitrarily small because the Einstein result does not hold on infinitesimal time scales. The appearance of the sign of the time interval signifies the irreversibility inherent in thermodynamics and statistical mechanics.

This is known as the diffusion equation. It is a simplified version of the Fokker-Planck equation, which will be derived more rigorously in §3.7.3.

The solution of the diffusion equation is

$$\wp(\mathbf{r},t) = \frac{1}{(4\pi D|t|)^{3/2}} e^{-\mathbf{r}\cdot\mathbf{r}/4D|t|}. \tag{3.23}$$

This corresponds to the particle being at the origin initially, $\wp(\mathbf{r},0) = \delta(\mathbf{r})$. That this is a Gaussian centred on the origin reflects the fact that the random steps are equally likely positive as negative. The width of the Gaussian increases with time, which represents the diffusion over time of the particle from its starting point. The diffusion equation was originally given by Einstein and presented as a way to determine Avogradro's number. Such a measurement was successfully carried out by Perrin from the sedimentation of colloidal particles several years later.[8]

3.5 Time Correlation Functions

The simultaneous treatment of position and velocity fluctuations of the pinned Brownian particle is a mixed parity case that reveals certain self-consistency requirements. The general expression for the time correlation function was given in Ch. 2, Eq. (2.1). Focusing on one-dimensional motion for simplicity, the time correlation matrix for position-velocity macrostates is

$$Q(\tau) \equiv \begin{pmatrix} Q_{xx}(\tau) & Q_{xv}(\tau) \\ Q_{vx}(\tau) & Q_{vv}(\tau) \end{pmatrix} = k_{\mathrm{B}}^{-1} \begin{pmatrix} \langle x(\tau)x(0)\rangle & \langle x(\tau)v(0)\rangle \\ \langle v(\tau)x(0)\rangle & \langle v(\tau)v(0)\rangle \end{pmatrix}. \tag{3.24}$$

Here $x(t)$ is the position and $v(t) = \dot{x}(t)$ is the velocity, and $\langle x\rangle = \langle v\rangle = 0$. By macroscopic reversibility, the self-correlation functions $Q_{xx}(\tau)$ and $Q_{vv}(\tau)$ are even functions of the time interval, whereas the cross-correlation function is odd, $Q_{xv}(\tau) = -Q_{xv}(-\tau) = -Q_{vx}(\tau)$. Also, the velocity correlation functions are evidently the time-derivatives of the position correlation functions,

$$\dot{Q}_{xx}(\tau) = Q_{vx}(\tau), \text{ and } \dot{Q}_{xv}(\tau) = Q_{vv}(\tau). \tag{3.25}$$

Now $Q_{vv}(0) = \langle v^2\rangle / k_{\mathrm{B}} = T/m$ for this one-dimensional case. If, as above, one assumes Markovian behaviour, then

$$Q_{vv}(\tau) = \frac{T}{m} e^{-\gamma|\tau|/m}, \tag{3.26}$$

where, as above, γ is the friction coefficient. One does not linearise this, which is equivalent to the requirement that $|\tau| \gg m/\gamma$.

The off-diagonal element can be obtained by integration,

$$\begin{aligned} Q_{xv}(\tau) &= \int_0^\tau \mathrm{d}t\, Q_{vv}(t) \\ &= \hat{\tau}\frac{T}{\gamma}\left[1 - e^{-\gamma|\tau|/m}\right] \\ &= \hat{\tau}\frac{T}{\gamma}, \quad |\tau| \gg m/\gamma. \end{aligned} \tag{3.27}$$

[8]Perrin J. (1908), C. R. Acad. Sci. Paris, **146**, 967.

Notice that this is $\mathcal{O}(\tau^0)$, since $\hat{\tau} \equiv \text{sign}(\tau)$. Because this is proportional to the sign of the time interval, it formally vanishes when $|\tau| = 0$: $Q_{xv}(0) = [Q_{xv}(0^+) + Q_{xv}(0^-)]/2 = 0$. This is an exact requirement: there can be no instantaneous coupling between variables of opposite parity.

From the results given in the preceding section, $\langle r^2 \rangle = 3k_BT/\kappa$, and $\langle \mathbf{r}(t) \cdot [\mathbf{r}(t+\tau) - \mathbf{r}(t)] \rangle = -3D|\tau|$, for this one-dimensional case one can deduce that the position autocorrelation function is

$$Q_{xx}(\tau) = \frac{T}{\kappa} - \frac{D}{k_B}|\tau|, \tag{3.28}$$

where D is the diffusion constant. This is valid in the small time limit, $|\tau| \ll k_BT/2\kappa D$, which comes from insisting that the most likely change in the first entropy be relatively negligible, $[\overline{x}' - x]\partial S(x)/\partial x \ll |S(x)|$. The time derivative is $\dot{Q}_{xx}(\tau) = -D\hat{\tau}/k_B$. But since this must equal $Q_{vx}(\tau) = -Q_{xv}(\tau)$, comparison with the preceding result shows that $\gamma = k_BT/D$. This is the fluctuation dissipation theorem deduced in §3.2.

The most likely rate of regression of the Brownian particle for a given displacement is

$$\begin{aligned} \overline{v}(x) &= -Q_{vx}(\tau)S_{xx}x - Q_{vv}(\tau)S_{vv}\overline{v}(x) \\ &= \frac{\hat{\tau}T}{\gamma} \times \frac{-\kappa}{T}x + \frac{-Te^{-\gamma|\tau|/m}}{m} \times \frac{-m}{T}\overline{v}(x) \\ &= \frac{-\hat{\tau}\kappa}{\gamma}x, \quad |\tau| \gg m/\gamma. \end{aligned} \tag{3.29}$$

This says that for a given displacement, the fluctuation will regress in the future at a constant rate such that the drag force, $-\gamma v$, is equal and opposite to the spring force, $-\kappa x$. This is in accord with one's physical intuition.

All of the preceding results apply if

$$\frac{Dm}{k_BT} \ll |\tau| \ll \frac{k_BT}{2\kappa D}. \tag{3.30}$$

This is the small diffusion, or, equivalently, the large drag limit. That is, the velocity correlations decay much faster than the position correlations.

In the opposite limit of small drag, large diffusion, one linearises the exponential to obtain,

$$Q_{vv}(\tau) = \frac{T}{m} - \frac{\gamma T}{m^2}|\tau| + \mathcal{O}(\tau^2). \tag{3.31}$$

In this limit, $|\tau| \ll m/\gamma$. Integrating this, the off-diagonal element is

$$Q_{xv}(\tau) = \int_0^\tau \mathrm{d}t \left[\frac{T}{m} - \frac{\gamma T}{m^2}|t|\right] = \frac{T}{m}\tau - \frac{\gamma T}{2m^2}\hat{\tau}\tau^2 + \mathcal{O}(\tau^3). \tag{3.32}$$

Note that $Q_{xv}(0) = 0$. Setting $Q_{vx}(\tau) = -Q_{xv}(\tau)$ and integrating once more,

$$Q_{xx}(\tau) - Q_{xx}(0) = \int_0^\tau \mathrm{d}t \, Q_{vx}(t) = \frac{-T}{2m}\tau^2 + \frac{\gamma T}{6m^2}|\tau|^3. \tag{3.33}$$

Using $Q_{xx}(0) = T/\kappa$ this is

$$Q_{xx}(\tau) = \frac{T}{\kappa} - \frac{T}{2m}\tau^2 + \frac{\gamma T}{6m^2}|\tau|^3 + \mathcal{O}(\tau^4). \tag{3.34}$$

This shows that the Einstein result—that the mean square displacement grows linearly in time—does not hold in the small drag limit. This could also be called the ballistic regime.

3.6 Non-Equilibrium Probability Distribution

3.6.1 Stationary Trap

Although the full treatment must await the derivation of non-equilibrium statistical mechanics in Ch. 8, it is nevertheless possible to give a heuristic derivation of the probability distribution of a Brownian particle based on the above results.

With the potential trap located at $\mathbf{b}(t)$, the energy of the Brownian particle is the sum of its kinetic energy and the potential trap energy,

$$\mathcal{H}(\mathbf{x}, \mathbf{v}; t) = \frac{m}{2}\mathbf{v} \cdot \mathbf{v} + \frac{\kappa}{2}[\mathbf{x} - \mathbf{b}(t)] \cdot [\mathbf{x} - \mathbf{b}(t)]. \tag{3.35}$$

This assumes a spherical trap. One expects the Brownian particle to follow the minimum of the potential trap as it moves, and so this may be described as a driven Brownian particle.

If the trap is stationary, $\mathbf{b}(t) = \mathbf{b}$, then the usual equilibrium results apply. The entropy of the position-velocity macrostate is

$$S(\mathbf{x}, \mathbf{v}) = \frac{-m}{2T}\mathbf{v} \cdot \mathbf{v} - \frac{\kappa}{2T}[\mathbf{x} - \mathbf{b}] \cdot [\mathbf{x} - \mathbf{b}], \tag{3.36}$$

where T is the temperature. The probability distribution is simply the exponential of this,

$$\wp_{\rm MB}(\mathbf{x}, \mathbf{v}|\mathbf{b}) = \frac{(m\kappa)^{3/2}}{(2\pi k_{\rm B}T)^3} e^{-m\mathbf{v}\cdot\mathbf{v}/2k_{\rm B}T} e^{-\kappa[\mathbf{x}-\mathbf{b}]\cdot[\mathbf{x}-\mathbf{b}]/2k_{\rm B}T}, \tag{3.37}$$

where $k_{\rm B}$ is Boltzmann's constant. For this stationary trap, the most likely velocity is zero, $\overline{\mathbf{v}} = \mathbf{0}$. This equilibrium probability distribution is called the Maxwell-Boltzmann distribution.

Figure 3.1 shows four equally likely configurations of the Brownian particle in a fixed trap. Notice that the probability is unchanged by reflection of the particle position in the trap minimum or by reversal of the particle velocity. The fact that one has these two independent symmetries arises because the probability distribution is for position and velocity evaluated at the same time, $\wp(\mathbf{x}(t), \mathbf{v}(t)|\mathbf{b})$, and therefore there is no instantaneous coupling between them. If instead one asked for the probability of the position at one time and the velocity at another time, $\wp(\mathbf{x}(t), \mathbf{v}(t+\tau)|\mathbf{b})$, then the two upper configurations of Fig. 3.1 are equally likely, and the two lower configurations of Fig. 3.1 are

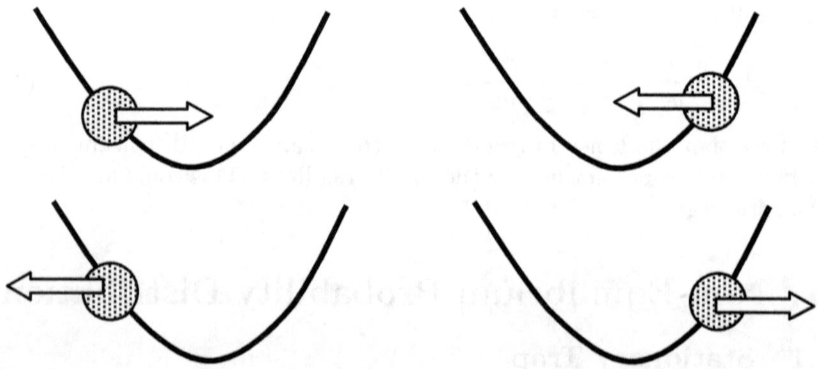

Figure 3.1: Four equally likely configurations of a Brownian particle in a stationary trap.

equally likely. Moreover, if the velocity is in the future, $\tau > 0$, then the upper configurations are more likely than the lower configurations, and *vice versa* for the past. This is Eq. (3.29) in pictures.

3.6.2 Uniformly Moving Trap

Now suppose that the potential trap is moving with uniform velocity, $\mathbf{b}(t) = \dot{\mathbf{b}}t$.[9] This gives rise to a steady state, in which the particle is fixed relative to the trap. That is, the most likely velocity of the particle must equal the trap velocity, $\overline{\mathbf{v}}(t) = \dot{\mathbf{b}}$. But by Eq. (3.29), in the steady velocity state, the drag force is equal and opposite to the trap force, and so the most likely displacement from the trap minimum must be $\overline{\mathbf{x}}(t) - \mathbf{b}(t) = -(\gamma/\kappa)\overline{\mathbf{v}}(t) = -\gamma\dot{\mathbf{b}}/\kappa$. Recall that the friction coefficient is related to the diffusion constant by $\gamma = k_\mathrm{B}T/D$. Both the most likely velocity and the most likely displacement are constant in time, as befits the steady state.

The most likely configurations are shown in Fig. 3.2 for two equal and opposite trap velocities. These configurations for a moving trap are obviously the same as the most likely future velocities of a stationary trap, the two upper configurations of Fig. 3.1.

The non-equilibrium probability distribution may be written $\wp(\mathbf{x}, \mathbf{v}|t, \dot{\mathbf{b}})$. Note that the position and the velocity are at the same time t. Although it is a steady state problem, the probability distribution is nevertheless dependent upon this time, and it is properly described as a non-equilibrium distribution. Based on the fluctuation theory, this probability distribution will be Gaussian. Since the most likely values of the position $\overline{\mathbf{x}}(t)$ and the velocity $\overline{\mathbf{v}}(t)$ have been given above, it remains to determine their variance.

---

[9]The results will also apply as an approximation when the variations in velocity are relatively negligible. In the more general case of arbitrary motion of the trap, memory effects must be accounted for, and one has to use the non-Markov analysis of Ch. 10.

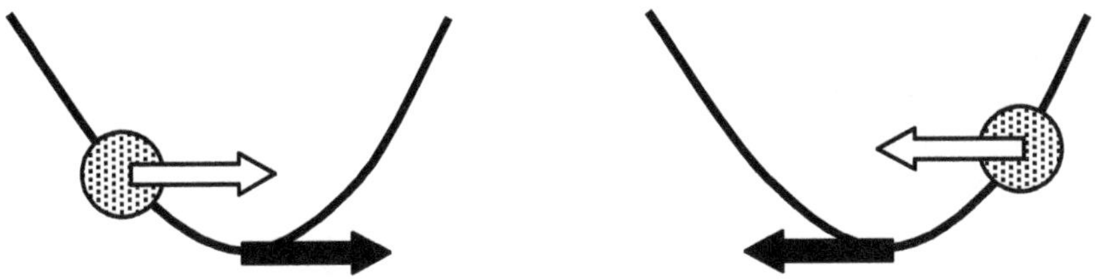

Figure 3.2: The respective most likely configurations of a Brownian particle in a trap moving to the right and to the left.

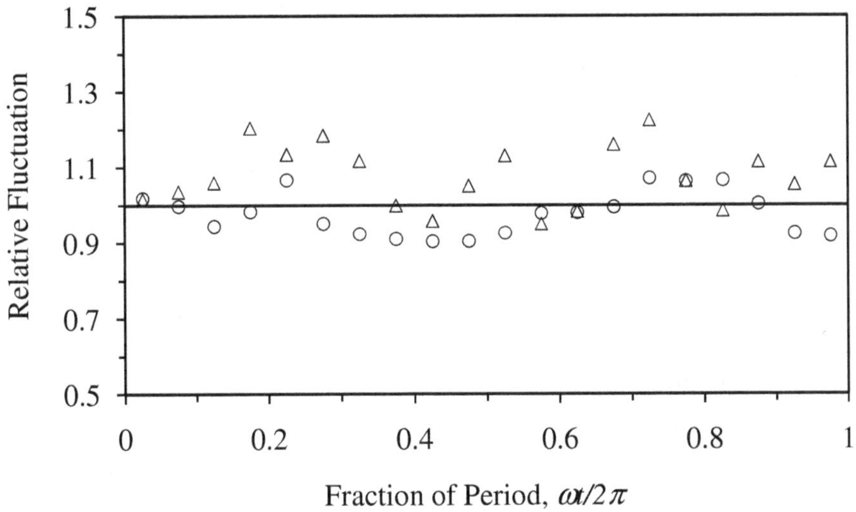

Figure 3.3: Relative fluctuations in a non-equilibrium system. The circles are for the position, $\left\langle [x(t) - \langle x(t) \rangle]^2 \right\rangle / (k_{\mathrm{B}}T/\kappa)$, and the triangles are for the velocity, $\left\langle [\dot{x}(t) - \langle \dot{x}(t) \rangle]^2 \right\rangle / (k_{\mathrm{B}}T/m)$, of the driven Brownian particle for $\omega = 0.7$. The line is a guide to the eye. The data were obtained by stochastic molecular dynamics simulations, [Attard, P. (2009a), J. Chem. Phys. **130**, 194113].

Within the frame of reference of the moving trap, fluctuations in position and velocity are identical to their equilibrium counterparts for a stationary trap. In other words, the motion of the trap is a macroscopic coordinate that does not directly contribute to the entropy, which is determined by the number of accessible microscopic configurations of the solvent. Hence the variance in this non-equilibrium case must be the same as in the equilibrium case.

This picture is confirmed by the non-equilibrium simulation data in Fig. 3.3, which shows the relative fluctuations about the average position and velocity for a Brownian particle in an oscillating parabolic trap in a soft-sphere solvent. The data were obtained by non-equilibrium stochastic molecular dynamics simulations (see Ch. 11). To the extent that the relative fluctuations equal unity, they confirm that the fluctuations in the non-equilibrium system are the same as those in the equilibrium system.

On the basis of these arguments and simulation data, one concludes that the non-equilibrium probability distribution is given by

$$\wp(\mathbf{x}, \mathbf{v}|t, \dot{\mathbf{b}}) = \frac{(m\kappa)^{3/2}}{(2\pi k_{\mathrm{B}}T)^3} e^{-m[\mathbf{v}-\overline{\mathbf{v}}(t)]^2/2k_{\mathrm{B}}T} e^{-\kappa[\mathbf{x}-\overline{\mathbf{x}}(t)]^2/2k_{\mathrm{B}}T}. \tag{3.38}$$

Recall $\overline{\mathbf{v}}(t) = \dot{\mathbf{b}}$ and $\overline{\mathbf{x}}(t) = \dot{\mathbf{b}}t - \gamma\dot{\mathbf{b}}/\kappa$. Here the notation $[\ldots]^2$ means the scalar product $[\ldots]\cdot[\ldots]$. Factors that are constant with respect to $\mathbf{x}$ and $\mathbf{v}$, but which may depend upon time, have been incorporated into the partition function.

This is the first non-equilibrium probability distribution given in the present book. In order to see the conceptual difference between this and the equilibrium Maxwell-Boltzmann distribution, use will be made of the conjugate microstate, $\{\mathbf{x}, \mathbf{v}\}^\dagger = \{\mathbf{x}, -\mathbf{v}\}$. Further, in what is commonly called time reversal, the effect of reversing the trap velocity, $\dot{\mathbf{b}} \Rightarrow -\dot{\mathbf{b}}$, will also be explored.

The equilibrium Maxwell-Boltzmann distribution for the stationary trap clearly obeys the symmetry

$$\wp_{\mathrm{MB}}(\{\mathbf{x}, \mathbf{v}\}^\dagger|\mathbf{b}(t)) = \wp_{\mathrm{MB}}(\{\mathbf{x}, \mathbf{v}\}|\mathbf{b}(t)). \tag{3.39}$$

Note that $\dot{\mathbf{b}}$ does not enter. This says that the Maxwell-Boltzmann distribution is insensitive to the direction of time.

In contrast, the non-equilibrium probability distribution changes upon velocity reversal,

$$\wp(\{\mathbf{x}, \mathbf{v}\}^\dagger|t, \dot{\mathbf{b}}) \neq \wp(\{\mathbf{x}, \mathbf{v}\}|t, \dot{\mathbf{b}}). \tag{3.40}$$

In fact, $\wp(\{\mathbf{x}, \mathbf{v}\}^\dagger|t, \dot{\mathbf{b}})/\wp(\{\mathbf{x}, \mathbf{v}\}|t, \dot{\mathbf{b}}) = e^{-2m\mathbf{v}\cdot\overline{\mathbf{v}}(t)/k_{\mathrm{B}}T}$. This dependence upon the arrow of time is the crucial difference between the Maxwell-Boltzmann distribution and a non-equilibrium probability distribution.

In general, the probability distribution is the exponential of the reservoir entropy, $\wp(\{\mathbf{x}, \mathbf{v}\}|t, \dot{\mathbf{b}}) = Z(t)^{-1} \exp S_{\mathrm{r}}(\{\mathbf{x}, \mathbf{v}\}|t, \dot{\mathbf{b}})/k_{\mathrm{B}}$. In the non-equilibrium case the reservoir entropy can be written as the instantaneous equilibrium expression or static part, plus a non-equilibrium excess or dynamic part,

$$S_{\mathrm{r}}(\{\mathbf{x}, \mathbf{v}\}|t, \dot{\mathbf{b}}) = S_{\mathrm{r,st}}(\{\mathbf{x}, \mathbf{v}\}|\mathbf{b}(t)) + S_{\mathrm{r,dyn}}(\{\mathbf{x}, \mathbf{v}\}|t, \dot{\mathbf{b}}). \tag{3.41}$$

The static part of this is just the usual Maxwell-Boltzmann factor,

$$S_{\mathrm{r,st}}(\{\mathbf{x}, \mathbf{v}\}|\mathbf{b}(t)) = \frac{-\kappa}{2T}[\mathbf{x} - \mathbf{b}(t)]^2 - \frac{m}{2T}\mathbf{v}^2. \tag{3.42}$$

Inserting this into the preceding equation for the reservoir entropy and equating it to the fluctuation expression given above, one sees that the non-equilibrium excess or dynamic part of the reservoir entropy must be equal to

$$S_{\mathrm{r,dyn}}(\mathbf{x}, \mathbf{v}|t, \dot{\mathbf{b}}) \equiv \frac{m}{T}\dot{\mathbf{b}} \cdot \mathbf{v} - \frac{\gamma}{T}\dot{\mathbf{b}} \cdot [\mathbf{x} - \mathbf{b}(t)]. \tag{3.43}$$

Here factors independent of $\mathbf{v}$ and $\mathbf{x}$ have been incorporated into the normalisation factor Z. This dynamic part of the entropy evidently depends upon the sign of the velocity. The full significance of this factorisation of the non-equilibrium probability distribution cannot be appreciated until the full presentation of non-equilibrium statistical mechanics in Ch. 8. Nevertheless a hint at the origin and interpretation of this term may be given as follows.

For a stationary trap, the position autocorrelation function was given as Eq. (3.28), $Q_{xx}(\tau) = T/\kappa - D|\tau|/k_{\mathrm{B}}$. This was valid in the short time limit, $|\tau| \ll k_{\mathrm{B}}T/2\kappa D$. Assuming Markovian behaviour, this may be extended to long times by exponentiation,

$$\begin{aligned} Q_{xx}(\tau) &= \frac{T}{\kappa} e^{-\kappa D|\tau|/k_{\mathrm{B}}T} \\ &= \frac{T}{\kappa} e^{-\kappa|\tau|/\gamma}, \end{aligned} \tag{3.44}$$

since $\gamma = k_{\mathrm{B}}T/D$. Differentiating this gives the velocity-position correlation function,

$$Q_{vx}(\tau) = \frac{-D\hat{\tau}}{k_{\mathrm{B}}} e^{-\kappa|\tau|/\gamma}. \tag{3.45}$$

Both these apply to a stationary trap in the long time limit.

These correlation matrices apply also to the fluctuations of a moving trap. Let $\mathbf{y}(t)$ denote the fluctuation, or departure from the most likely position at time t,

$$\mathbf{y}(t) \equiv \mathbf{x}(t) - \overline{\mathbf{x}}(t) = \mathbf{x}(t) - \mathbf{b}(t) + \gamma\dot{\mathbf{b}}/\kappa. \tag{3.46}$$

The departure velocity is similarly $\dot{\mathbf{y}}(t) \equiv \mathbf{v}(t) - \dot{\mathbf{b}}$. The most likely departure at time $t+\tau$ given the current departure is

$$\begin{aligned} \overline{\mathbf{y}}(t+\tau|\mathbf{y}(t),\dot{\mathbf{y}}(t)) &= -Q_{xx}(\tau)S_{xx}\mathbf{y}(t) - Q_{xv}(\tau)S_{vv}\dot{\mathbf{y}}(t) \\ &= \left[\mathbf{y}(t) + \frac{m\hat{\tau}}{\gamma}\dot{\mathbf{y}}(t)\right] e^{-\kappa|\tau|/\gamma}. \end{aligned} \tag{3.47}$$

The rate that the external potential does work on the sub-system is

$$\frac{\partial U(\mathbf{x},t)}{\partial t} = -\kappa\dot{\mathbf{b}}\cdot[\mathbf{x} - \mathbf{b}(t)] = -\kappa\dot{\mathbf{b}}\cdot\mathbf{y} + \text{const.} \tag{3.48}$$

Here and below, only terms that depend upon $\mathbf{x}$ and $\mathbf{v}$ are required for the probability distribution. The adiabatic rate of change of the static part of the reservoir entropy, is just $\dot{S}^0_{\mathrm{r,st}}(\{\mathbf{x},\mathbf{v}\}|\mathbf{b}(t)) = -T^{-1}\partial U(\mathbf{x},t)/\partial t$. It will be shown in Ch. 8 that the dynamic part of the reservoir entropy subtracts the total adiabatic contribution that occurred in reaching the current configuration from the current equilibrium value of the entropy. This can be estimated from the integral over the most likely trajectory leading to the present point from some time $t-s$ in the past. Discarding constants this is

$$\begin{aligned} S_{\rm r,dyn}(\mathbf{x},\mathbf{v}|t,\dot{\mathbf{b}}) &= -\int_{-s}^{0} \mathrm{d}\tau\ \dot{S}^0_{\rm r,st}(\overline{\mathbf{y}}(t+\tau),\dot{\overline{\mathbf{y}}}(t+\tau)|\mathbf{b}(t+\tau)) \\ &= \frac{-\kappa}{T}\dot{\mathbf{b}}\cdot\int_{-s}^{0} \mathrm{d}\tau\ \overline{\mathbf{y}}(t+\tau|\mathbf{y}(t),\dot{\mathbf{y}}(t)) \\ &= \frac{-\kappa}{T}\dot{\mathbf{b}}\cdot\int_{-s}^{0} \mathrm{d}\tau\ \left[\mathbf{y}(t)+\frac{m\hat{\tau}}{\gamma}\dot{\mathbf{y}}(t)\right]e^{-\kappa|\tau|/\gamma} \\ &= \frac{-\kappa}{T}\dot{\mathbf{b}}\cdot\left[\mathbf{y}(t)-\frac{m}{\gamma}\dot{\mathbf{y}}(t)\right]\frac{\gamma}{\kappa}\left[1-e^{-\kappa|s|/\gamma}\right] \\ &= \frac{-\gamma}{T}\dot{\mathbf{b}}\cdot[\mathbf{x}-\mathbf{b}(t)]+\frac{m}{T}\dot{\mathbf{b}}\cdot\mathbf{v}. \end{aligned} \tag{3.49}$$

In the final equality, it has been assumed that a sufficiently long trajectory has been used so that the exponential term can be neglected, which means that the result is independent of the lower limit of the integral. Also neglected have been terms quadratic in the trap velocity, which terms are independent of the current particle position and velocity. The final result, which agrees with Eq. (3.43) above, will be derived from first principles in the following section. On the basis of the present analysis one may conclude that the physical interpretation of the dynamic part of the reservoir entropy is that it corrects for double counting in the static part of the reservoir entropy by cancelling that part of the current value that arose from adiabatic rather than from heat processes.

3.6.3 Mixed Parity Formulation of the Moving Trap

The preceding subsection presented the non-equilibrium probability distribution for a Brownian particle in a uniformly moving trap in a somewhat heuristic fashion. This section gives a more rigorous derivation of the distribution. The notation and concepts foreshadow those to be used in the general non-equilibrium theory of Ch. 8. The present analysis also illustrates the connection between the non-equilibrium statistical mechanics and fluctuation theory of Ch. 2. In particular, the position and the velocity of the Brownian particle have opposite parity, and therefore the results for fluctuations of mixed parity, §2.5.1, are required for the full analysis.

There is no coupling between different Cartesian components, and so for simplicity the focus is on one-dimensional motion. The Brownian particle has position x and velocity $\dot{x}$. This represents the phase space of the sub-system, $X \equiv \{x,\dot{x}\}$ (the momentum $p=m\dot{x}$ is trivially related to the velocity), with the solvent being considered as the reservoir and being accounted for statistically. The time dependent external potential is of the form of a harmonic trap, $U(x,t)=\kappa[x-b(t)]^2/2$, in uniform motion, $b(t)=\dot{b}t$. The static part of the reservoir entropy is the instantaneous canonical form,

$$S_{\rm r,st}(X,t)=\frac{-1}{T}\mathcal{H}(X,t)=\frac{-m}{2T}\dot{x}^2-\frac{\kappa}{2T}[x-b(t)]^2. \tag{3.50}$$

As discussed in connection with Eq. (3.29), most likely the particle moves with the same velocity as the trap and is displaced from the minimum such that the drag force, $-\gamma\dot{\overline{x}}$, is equal and opposite to the restoring force, $-\kappa[\overline{x} - b(t)]$,

$$\dot{\overline{x}}(t) = \dot{b}, \text{ and } \overline{x}(t) = b(t) - \dot{b}\gamma/\kappa. \tag{3.51}$$

The friction coefficient is related to the drag force by $\gamma \equiv 1/\beta D$. One of the outcomes of the following analysis is a first principles derivation of this physically plausible result. For the initial analysis, it is not necessary to specify explicitly the most likely configuration, $\overline{X}(t) \equiv \{\overline{x}(t), \dot{\overline{x}}(t)\}$.

For the fluctuation $Y = \{y, \dot{y}\} = X(t) - \overline{X}(t)$, the reservoir entropy of the non-equilibrium system is[10]

$$S_{\mathrm{r}}(X,t) = \frac{\gamma \dot{b}^2}{T} t + \frac{1}{2} S : Y^2, \tag{3.52}$$

with the entropy matrix being the same as that of the equilibrium system,

$$S = \begin{pmatrix} -\kappa/T & 0 \\ 0 & -m/T \end{pmatrix}. \tag{3.53}$$

The justification for this result is that the entropy associated with the fluctuations in the non-equilibrium system (moving trap) are the same as those in the equilibrium system (stationary trap). The initial term that is independent of the fluctuation and that grows linearly with time represents the heat most likely dissipated to the reservoir by the drag force against the most likely motion of the particle.[11] This term has only trivial effects and could be incorporated into the partition function.

The mathematical derivation of this expression for the reservoir entropy is given in what follows of this subsection.

The following analysis is valid in the small drag regime, $|\tau| \ll m/\gamma$, in contrast to the analysis of the non-equilibrium probability distribution given in the preceding section, which was valid in the large drag regime, $m/\gamma \ll |\tau| \ll \gamma/2\kappa$ (equivalently, the small diffusion regime, $\beta m D \ll |\tau| \ll 1/2\beta\kappa D$, since $D = 1/\beta\gamma$). In the present case the velocity-velocity time correlation function decreases linearly with time,

$$\begin{aligned} Q_{\dot{x}\dot{x}}(\tau) &\equiv k_{\mathrm{B}}^{-1} \langle \dot{y}(t+\tau)\dot{y}(t)\rangle_X \\ &= \frac{T}{m} - \frac{\gamma T}{m^2}|\tau| + \mathcal{O}(\tau^2). \end{aligned} \tag{3.54}$$

[10]When it is necessary to be more precise, the reservoir entropy will be written $S_{\mathrm{r,Y}}(Y,t) \equiv (\gamma\dot{b}^2 t/T) - (\kappa/2T)y^2 - (m/2T)\dot{y}^2$, and as $S_{\mathrm{r,X}}(X,t) \equiv (\gamma\dot{b}^2 t/T) - (\kappa/2T)[x - \overline{x}(t)]^2 - (m/2T)[\dot{x} - \dot{\overline{x}}(t)]^2$. These are of course numerically equal, $S_{\mathrm{r,X}}(X,t) = S_{\mathrm{r,Y}}(Y,t)$. The partial time derivatives of these are not equal, $\partial S_{\mathrm{r,Y}}(Y,t)/\partial t \neq \partial S_{\mathrm{r,X}}(X,t)/\partial t$, because the derivative at constant configuration is not the same as the derivative at constant fluctuation.

[11] Strictly speaking, one should also add to this term $(\gamma^2\dot{b}^2/2\kappa T) + (m\dot{b}^2/2T)$. This is constant with respect to time and configuration, and so is immaterial.

Again this result is justified by the fact that the time correlation of the non-equilibrium fluctuations is identical to that of the equilibrium fluctuations. Integrating this, the position-velocity time correlation function is

$$\begin{aligned} Q_{x\dot{x}}(\tau) &\equiv k_{\rm B}^{-1} \langle y(t+\tau)\dot{y}(t)\rangle_X \\ &= \frac{T}{m}\tau - \frac{\gamma T}{2m^2}\hat{\tau}\tau^2 + \mathcal{O}(\tau^3), \end{aligned} \tag{3.55}$$

with $Q_{\dot{x}x}(\tau) = Q_{x\dot{x}}(-\tau)$. There is no instantaneous coupling between fluctuations of opposite parity, $Q_{x\dot{x}}(0) = k_{\rm B}^{-1}\langle y(t)\dot{y}(t)\rangle_X = 0$. Integrating once again, the position-position time correlation function is

$$\begin{aligned} Q_{xx}(\tau) &\equiv k_{\rm B}^{-1} \langle y(t+\tau)y(t)\rangle_X \\ &= \frac{T}{\kappa} - \frac{T}{2m}\tau^2 + \frac{\gamma T}{6m^3}|\tau|^3 + \mathcal{O}(\tau^4), \end{aligned} \tag{3.56}$$

with of course $Q_{xx}(0) = -S_{xx}^{-1}$. Normally one would neglect all terms $\mathcal{O}(\tau^2)$ and higher, but below the inverse of a matrix will be required, and it is necessary to retain the term $-T\tau^2/2m$ in $Q_{xx}(\tau)$ for this.

In what follows, the transport coefficient (here γ; more generally the Λ and the S) is taken to be independent of the configuration X, or, equivalently, the departure Y. In general it is a material property of the system, and it may depend upon time, both directly and indirectly by a dependence on the most likely state at the current time $\overline{X}(t)$. It is important to note that by definition the most likely state does not depend upon the current state but only upon time, $\partial\overline{X}(t)/\partial X(t) = 0$. This means that any potential non-linear effects are incorporated into the transport and fluctuation matrices via the current most likely state. The second entropy describes the fluctuations about the most likely state, and these in general are small enough not to effect the transport matrices. The independence of the transport matrix from the current state is significant in the presentation of the Fokker-Planck equation and the expression for the probability flux given below.

The most likely future position for the mixed parity case was given in terms of the time correlation function as Eq. (2.78),

$$\overline{Y}(t+\tau|Y,t) = -Q(\tau)SY, \tag{3.57}$$

or

$$\begin{aligned} \overline{y}(t+\tau|Y,t) &= \left[1 - \frac{\kappa}{2m}\tau^2 + \frac{\gamma\kappa}{6m^2}|\tau|^3\right] y + \left[\tau - \frac{\gamma}{2m}\hat{\tau}|\tau|^2\right]\dot{y} \\ &= y + \tau\dot{y} + \mathcal{O}(\tau^2), \\ \overline{\dot{y}}(t+\tau|Y,t) &= \left[-\frac{\kappa}{m}\tau + \frac{\gamma\kappa}{2m^2}\hat{\tau}|\tau|^2\right] y + \left[1 - \frac{\gamma}{m}|\tau|\right]\dot{y} \\ &= \dot{y} - \tau\frac{\kappa}{m}y - |\tau|\frac{\gamma}{m}\dot{y} + \mathcal{O}(\tau^2). \end{aligned} \tag{3.58}$$

These make sense. The second term on the right-hand side of each of these is just the adiabatic force, since $\dot{y}^0 = \dot{y}$ and $\ddot{y}^0 = \ddot{x}^0 - \ddot{\overline{x}}^0 = -\kappa y/m$. The third term on the right-hand side of the velocity equation can be interpreted as the friction force. As discussed below, one can add the zero-mean fluctuating force, $\tilde{R}$, to these to obtain the stochastic, dissipative equations.

The fluctuation matrices for the mixed parity case have the form given in Eqs (2.83) and (2.84),

$$A(\tau) = \frac{-1}{|\tau|}\Lambda^{-1} + A_0 + A_0'\hat{\tau} + \mathcal{O}(\tau), \tag{3.59}$$

and

$$B(\tau) = \frac{1}{|\tau|}\Lambda^{-1} + B_0 + B_0'\hat{\tau} + \mathcal{O}(\tau). \tag{3.60}$$

These coefficients were given in terms of the coefficients in the expansion for the time correlation function, Eq. (2.86), $Q(\tau) = -S^{-1} - |\tau|\Lambda/2 + \tau\Theta + \mathcal{O}(\tau^2)$. Comparing this to the explicit expression for the time correlation function given here, one has

$$\Lambda = \begin{pmatrix} T|\tau|/m & 0 \\ 0 & 2\gamma T/m^2 \end{pmatrix}, \quad \Theta = \begin{pmatrix} 0 & T/m \\ -T/m & 0 \end{pmatrix}. \tag{3.61}$$

Strictly speaking $\Lambda_{xx} = 0$, and the exhibited term belongs to a higher-order coefficient matrix. This term has been retained here because the matrix is singular without it. One has from Eq. (2.85),

$$A_0 + B_0 = \frac{S}{2} = \begin{pmatrix} -\kappa/2T & 0 \\ 0 & -m/2T \end{pmatrix}, \tag{3.62}$$

and from Eq. (2.88),

$$A_0' + B_0' = \Lambda^{-1}\Theta S = \begin{pmatrix} 0 & -m/T|\tau| \\ m\kappa/2\gamma T & 0 \end{pmatrix}. \tag{3.63}$$

Using these, the second entropy in the form of Eq. (2.92) is

$$\begin{aligned} & S^{(2)}(Y', t+\tau; Y, t) \\ &= S_{\mathrm{r}}(Y,t) - \frac{|\tau|}{2}\Lambda^{-1} : \overset{\circ}{Y}{}^2 + \tau \overset{\circ}{Y} \cdot [A_0 + B_0 \\ &\quad - \hat{\tau}(A_0' + B_0')]\, Y - \frac{|\tau|}{2}\Lambda^{-1} : \overline{\overset{\circ}{Y}}(Y;\hat{\tau})^2 + \mathcal{O}(\tau^2) \\ &= S_{\mathrm{r}}(Y,t) - \frac{m}{2T\tau^2}[y'-y]^2 - \frac{m^2}{4\gamma T|\tau|}[\dot{y}'-\dot{y}]^2 \\ &\quad - \frac{\kappa}{2T}[y'-y]y - \frac{m}{2T}[\dot{y}'-\dot{y}]\dot{y} + \frac{m}{T\tau}[y'-y]\dot{y} - \frac{\hat{\tau}m\kappa}{2\gamma T}[\dot{y}'-\dot{y}]y \\ &\quad - \frac{m}{2T\tau^2}[\overline{y'}-y]^2 - \frac{m^2}{4\gamma T|\tau|}[\overline{\dot{y}'}-\dot{y}]^2 + \mathcal{O}(\tau^2). \end{aligned} \tag{3.64}$$

Recall that the coarse velocity is $\overset{\circ}{Y} \equiv [Y'-Y]/\tau$. Also, $Y \equiv Y(t)$, $Y' \equiv Y(t+\tau)$, and $\overline{Y'} \equiv \overline{Y}(t+\tau|Y,t)$.

Maximising the second entropy with respect to y' and $\dot{y}'$ one obtains

$$\begin{aligned}\overline{y'} &= y + \tau\dot{y} + \mathcal{O}(\tau^2),\\ \overline{\dot{y}'} &= \dot{y} - \tau\frac{\kappa}{m}y - |\tau|\frac{\gamma}{m}\dot{y} + \mathcal{O}(\tau^2).\end{aligned} \tag{3.65}$$

These agree with the result obtained above directly from the time correlation function, $\overline{Y}(t+\tau|Y,t) = -Q(\tau)SY$. The most likely reservoir force is

$$\overline{R} = -|\tau|\frac{\gamma}{m}\dot{y} = \frac{|\tau|}{2}\Lambda_{\dot{x}\dot{x}}\frac{\partial S_{\rm r}(X,t)}{\partial \dot{y}}. \tag{3.66}$$

This has only a velocity component, because the most likely reservoir force for the position is $\mathcal{O}(\tau^2)$. The adiabatic force (equivalently acceleration) for the position and for the velocity are

$$\dot{y}^0 = \dot{y},\ \ddot{y}^0 = \frac{-\kappa}{\tau m}y. \tag{3.67}$$

This adiabatic force for the velocity came from the coupling of the position and velocity via $\{A_0' + B_0'\}_{x\dot{x}}$ (equivalently $\Theta_{x\dot{x}}$). Similarly, the adiabatic term in the position equation originated from $\{A_0' + B_0'\}_{\dot{x}x}$ (equivalently $\Theta_{\dot{x}x}$).

The stochastic part of the force can be added to these deterministic equations. From the result for the second entropy its variance is

$$\left\langle \tilde{R}\tilde{R}\right\rangle_{Y'} = |\tau|k_{\rm B}\Lambda = \begin{pmatrix} \dfrac{k_{\rm B}T\tau^2}{m} & 0 \\ 0 & \dfrac{2\gamma k_{\rm B}T|\tau|}{m^2}\end{pmatrix}. \tag{3.68}$$

Since $\left\langle \tilde{R}_y\tilde{R}_y\right\rangle_X \sim \mathcal{O}(\tau^2)$, the stochastic force on the position evolution can be neglected. Hence the stochastic dissipative equations become

$$\begin{aligned} y(t+\tau) &= y(t) + \tau\dot{y}(t) + \mathcal{O}(\tau^2),\\ \dot{y}(t+\tau) &= \dot{y}(t) - \tau\frac{\kappa}{m}y(t) - |\tau|\frac{\gamma}{m}\dot{y}(t) + \tilde{R} + \mathcal{O}(\tau^2).\end{aligned} \tag{3.69}$$

For brevity, the subscript denoting the velocity component is not shown on $\overline{R}$, $\tilde{R}$, or R.

Now the most likely configuration, $\overline{X}(t)$, is obtained. Recall that the fluctuation is $Y \equiv X - \overline{X}(t)$. The stochastic, dissipative equations of motion, transformed from the fluctuation Y to the actual configuration X, are

$$\begin{aligned} x(t+\tau) &= x(t) + \tau\frac{{\rm d}\overline{x}(t)}{{\rm d}t} + \tau\dot{x}(t) - \tau\overline{\dot{x}}(t)\\ &= x(t) + \tau\dot{x}(t),\\ \dot{x}(t+\tau) &= \dot{x}(t) + \tau\frac{{\rm d}\overline{\dot{x}}(t)}{{\rm d}t} - \tau\frac{\kappa}{m}\left[x(t) - \overline{x}(t)\right] - |\tau|\frac{\gamma}{m}\left[\dot{x}(t) - \overline{\dot{x}}(t)\right] + \tilde{R}\\ &= \dot{x}(t) + \tau\ddot{x}^0(t) + \overline{R}_{\rm X}(t) + \tilde{R}.\end{aligned} \tag{3.70}$$

This gives the trajectory of the particle if the most likely trajectory is known. The most likely reservoir force acting on the actual velocity is

$$\begin{aligned}\overline{R}_{\rm X}(t) &\equiv \overline{R}(t) + \tau\frac{{\rm d}\bar{\dot{x}}(t)}{{\rm d}t} - \tau\overline{\ddot{x}^0}(t)\\ &= -|\tau|\frac{\gamma}{m}[\dot{x} - \dot{b}] + 0 - \tau\frac{\gamma}{m}\dot{b}\\ &= -|\tau|\frac{\gamma}{m}\dot{x} - (\tau - |\tau|)\frac{\gamma}{m}\dot{b}. \qquad (3.71)\end{aligned}$$

The subscript X on the left-hand side indicates the most likely reservoir force acting on the actual coordinates, in this case the actual velocity; it does not indicate the position coordinate of the force vector. The final equality going forward in time, $\tau > 0$, is of the form of a traditional friction or drag force.

There are two expressions for the reservoir entropy. There is the fluctuation expression postulated above on the basis that fluctuations in a non-equilibrium system are the same as in an equilibrium system,

$$\begin{aligned}S_{\rm r}(X,t) &= \frac{\gamma\dot{b}^2 t}{T} - \frac{\kappa}{2T}y^2 - \frac{m}{2T}\dot{y}^2\\ &= \frac{\gamma\dot{b}^2 t}{T} - \frac{\kappa}{2T}[x - \overline{x}(t)]^2 - \frac{m}{2T}[\dot{x} - \bar{\dot{x}}(t)]^2, \qquad (3.72)\end{aligned}$$

with the initial, time dependent, constant representing the dissipation, as will shortly be justified, (see also footnote 11 on p. 77). And there is a formal decomposition of the reservoir entropy,

$$S_{\rm r}(X,t) = S_{\rm r,st}(X,t) + S_{\rm r,dyn}(X,t). \qquad (3.73)$$

The static part of the reservoir entropy is

$$S_{\rm r,st}(X,t) \equiv \frac{-\mathcal{H}(X,t)}{T} = \frac{-\kappa}{2T}[x - b(t)]^2 - \frac{m}{2T}\dot{x}(t)^2, \qquad (3.74)$$

and the non-equilibrium part of the reservoir entropy is

$$\begin{aligned}S_{\rm r,dyn}(X,t) &\equiv -\int_0^t {\rm d}t'\, \dot{S}^0_{\rm r,st}(\overline{X}(t'|X,t),t')\\ &= \frac{1}{T}\int_0^t {\rm d}t'\, \left.\frac{\partial\mathcal{H}(\overline{X}(t'|X,t),t'')}{\partial t''}\right|_{t''=t'}\\ &= \frac{-\kappa\dot{b}}{T}\int_0^t {\rm d}t'\, [\overline{x}(t'|X,t) - b(t')]. \qquad (3.75)\end{aligned}$$

The full justification for the form of this formal decomposition of the reservoir entropy will be given in Ch. 8. Here it may simply be noted that in a canonical system, the reservoir entropy changes by exchanging energy with the sub-system, $\Delta S_{\rm r} = -\Delta\mathcal{H}_{\rm s}/T$, and this is the only way that the sub-system energy can change. However, in a non-equilibrium system, the sub-system energy can also change by adiabatic evolution, $\dot{\mathcal{H}}^0_{\rm s} = \partial\mathcal{H}_{\rm s}/\partial t$, and so the change in reservoir entropy is the change in sub-system energy with this subtracted, $\Delta S_{\rm r} = -\left[\Delta \mathcal{H}_{\rm s} - \tau \dot{\mathcal{H}}_{\rm s}^0\right]/T$. Integrating this over the most likely backward trajectory from the present configuration gives the above expression.

The two expressions for the reservoir entropy must be equal, and equating them will yield the most likely configuration $\overline{X}(t)$. In order to do this, $S_{\rm r,dyn}(X,t)$ must first be evaluated. The deterministic form of the stochastic dissipative equations for the fluctuation can be rearranged as

$$\frac{\overline{Y}(t+\tau|Y,t) - Y(t)}{\tau} = MY(t),\ M \equiv \begin{pmatrix} 0 & 1 \\ -\kappa/m & -\hat{\tau}\gamma/m \end{pmatrix}. \tag{3.76}$$

Assuming Markovian behaviour, and taking the limit $\tau \to 0$, this has exponential solutions,

$$\overline{Y}(t'|Y,t) = e^{(t'-t)M} Y(t). \tag{3.77}$$

Inserting this into the non-equilibrium part of the reservoir entropy and integrating, one obtains

$$\begin{aligned} S_{\rm r,dyn}(X,t) &= \frac{-\kappa\dot{b}}{T}\int_0^t {\rm d}t' \left[\left\{e^{(t'-t)M}Y(t) + \overline{X}(t')\right\}_x - b(t')\right] \\ &= \frac{-\kappa\dot{b}}{T}\left\{M^{-1}Y(t)\right\}_x + c(t) \\ &= \frac{-\kappa\dot{b}}{T}\left[\frac{\gamma}{\kappa}[x-\overline{x}(t)] - \frac{m}{\kappa}[\dot{x} - \overline{\dot{x}}(t)]\right] + c(t), \end{aligned} \tag{3.78}$$

since $\hat{\tau} = -1$. The time dependent, configuration independent constant, $c(t) \equiv [-\kappa\dot{b}/T][-\dot{b}\gamma t/\kappa]$, will be discussed below.

With this result, the two expressions for the reservoir entropy, Eqs. (3.72) and (3.73), can be equated term by term. It is clear that the coefficients of the quadratic terms are equal. Equating the term linear in x yields

$$\frac{\kappa}{T} x\overline{x}(t) = \frac{\kappa}{T} x b(t) - \frac{\dot{b}\gamma}{T} x, \text{ or } \overline{x}(t) = b(t) - \dot{b}\gamma/\kappa. \tag{3.79}$$

Similarly the term linear in $\dot{x}$ yields

$$\frac{m}{T}\dot{x}\overline{\dot{x}}(t) = 0 + \frac{\dot{b}m}{T}\dot{x}, \text{ or } \overline{\dot{x}}(t) = \dot{b}. \tag{3.80}$$

These two result are physically reasonable and could have been written down directly. Most likely the particle moves with the same velocity as the trap, and it is most likely offset from the trap minimum by an amount such that the restoring force, $-\kappa[\overline{x}(t) - b(t)]$, is equal and opposite to the drag force, $-\gamma\overline{\dot{x}}(t)$.

Using this result for the most likely position, the time dependent part of the constant may now be obtained,

$$c(t) = \frac{-\kappa\dot{b}}{T}\int_0^t {\rm d}t' \left[\overline{x}(t') - b(t')\right] = \frac{\gamma\dot{b}^2 t}{T}. \tag{3.81}$$

This confirms in essence the most likely dissipation that was added to the fluctuation form for the second entropy above. In the most likely state, $\overline{Y} = 0$, the fluctuation expression, Eq. (3.72), is $S_r(\overline{X}(t), t) = \overline{S}_r(t)$. In that same state, $X = \overline{X}(t)$, the formal expression, Eq. (3.73), reduces to $S_r(\overline{X}(t), t) = S_{r,st}(\overline{X}(t), t) + S_{r,dyn}(\overline{X}(t), t) = (-m/2T)\dot{b}^2 - (\kappa/2T)(\gamma\dot{b}/\kappa)^2 + c(t)$. With these results, the two expressions for the reservoir entropy, Eqs (3.72) and (3.73), are equal (apart from an immaterial constant, $(\gamma^2\dot{b}^2/2\kappa T) + (m\dot{b}^2/2T)$; see footnote 11 on p. 77).

Using this explicit expression for the most likely position and velocity, the stochastic, dissipative equations of motion written in terms of the actual configuration are

$$\begin{aligned} x(t+\tau) &= x(t) + \tau\dot{x}(t) + \mathcal{O}(\tau^2), \qquad (3.82) \\ \dot{x}(t+\tau) &= \dot{x}(t) - \tau\frac{\kappa}{m}\left[x(t) - b(t) + \gamma\dot{b}/\kappa\right] - |\tau|\frac{\gamma}{m}\left[\dot{x}(t) - \dot{b}\right] + \tilde{R} \\ &= \dot{x}(t) - \tau\frac{\kappa}{m}\left[x(t) - b(t)\right] - |\tau|\frac{\gamma}{m}\dot{x}(t) + \tilde{R} - \frac{\gamma\dot{b}}{m}[\tau - |\tau|]. \end{aligned}$$

For $\tau > 0$, this has the form of the original Langevin equation, including a dissipative friction or drag force that is proportional to the actual velocity. For $\tau < 0$ there is an extra term that is missing in the original form of the equation. This term is essential for correct retrograde analysis.

3.7 Entropy, Probability, and their Evolution

This section derives expressions for the rates of change of entropy and probability. Most of the equations are structured so that the first equality is the generic result, and the subsequent equalities represent the specific application to the Brownian particle in the moving trap using the explicit results derived above. The Fokker-Planck equation and Liouville's theorem are discussed in the light of these general and specific analyses.

3.7.1 Time Evolution of the Entropy and Probability

The adiabatic rate of change of the reservoir entropy, expressed in terms of the fluctuation, is[12]

$$\begin{aligned} \dot{S}^0_{r,Y}(Y,t) &\equiv \frac{\partial S_{r,Y}(Y,t)}{\partial t} + \dot{Y}^0 \cdot \nabla S_{r,Y}(Y,t) \\ &= \frac{\gamma\dot{b}^2}{T} + \dot{y}\frac{-\kappa}{T}y + \frac{-\kappa}{m}y\frac{-m}{T}\dot{y} \\ &= \frac{\gamma\dot{b}^2}{T}. \qquad (3.83) \end{aligned}$$

[12]Recall, $S_{r,Y}(Y,t) \equiv (\gamma\dot{b}^2 t/T) - (\kappa/2T)y^2 - (m/2T)\dot{y}^2$, and $S_{r,X}(X,t) \equiv (\gamma\dot{b}^2 t/T) - (\kappa/2T)[x - \overline{x}(t)]^2 - (m/2T)[\dot{x} - \overline{\dot{x}}(t)]^2$, and that these are numerically equal.

Since $\dot{y}^0 = \dot{y}$ and $\ddot{y}^0 = \ddot{x}^0 - \overline{\ddot{x}^0}(t) = (-\kappa/m)[x - b(t) - \{\overline{x}(t) - b(t)\}] = -\kappa y/m$, one has $\dot{Y}^0 \cdot \nabla S_{\mathrm{r,Y}}(Y,t) = 0$. Hence only the partial time derivative contributes to $\dot{S}^0_{\mathrm{r,Y}}(Y,t)$. From the point of view of the actual configuration,

$$\begin{aligned}
\dot{S}^0_{\mathrm{r,X}}(X,t) &\equiv \frac{\partial S_{\mathrm{r,X}}(X,t)}{\partial t} + \dot{X}^0 \cdot \nabla S_{\mathrm{r,X}}(X,t) \\
&= \frac{\gamma \dot{b}^2}{T} + \frac{\kappa}{T}\dot{b}[x - \overline{x}(t)] + \dot{x}\frac{-\kappa}{T}[x - \overline{x}(t)] + \frac{-\kappa}{m}[x - b(t)]\frac{-m}{T}[\dot{x} - \dot{b}] \\
&= \frac{\gamma \dot{b}^2}{T} - \frac{\kappa}{T}\dot{y}y + \frac{\kappa}{T}[y - \gamma\dot{b}/\kappa]\dot{y} \\
&= \frac{\gamma \dot{b}^2}{T} - \frac{\gamma \dot{b}}{T}\dot{y}.
\end{aligned} \tag{3.84}$$

In passing, one notes that this invokes the result $\dot{X}^0 \cdot \nabla S_{\mathrm{r,st}}(X,t) = 0$. One sees that $\dot{S}^0_{\mathrm{r,X}}(X,t) \neq \dot{S}^0_{\mathrm{r,Y}}(Y,t)$. The reason for the difference is that the adiabatic development of the most likely configuration is not equal to the full development, $\overline{\dot{X}^0} \neq \dot{\overline{X}}$. Hence although the initial entropies are equal, $S_{\mathrm{r,X}}(X,t) = S_{\mathrm{r,Y}}(Y,t)$ when $Y = X - \overline{X}(t)$, the final ones aren't, $S_{\mathrm{r,X}}(X + \Delta_t \dot{X}^0, t + \Delta_t) \neq S_{\mathrm{r,Y}}(Y + \Delta_t \dot{Y}^0, t + \Delta_t)$, because $Y + \Delta_t \dot{Y}^0 \neq X + \Delta_t \dot{X}^0 - \overline{X}(t) - \Delta_t \dot{\overline{X}}(t)$.

The deterministic rate of change is the change on the most likely trajectory. For the reservoir entropy expressed in fluctuations it is (with $\overline{Y}' \equiv \overline{Y}(t+\tau|Y,t)$)

$$\begin{aligned}
\frac{\mathrm{d}^{\mathrm{det}} S_{\mathrm{r,Y}}(Y,t)}{\mathrm{d}t} &= \frac{1}{\tau}\left[S_{\mathrm{r,Y}}(\overline{Y}', t+\tau) - S_{\mathrm{r,Y}}(Y,t)\right] \\
&= \dot{S}^0_{\mathrm{r,Y}}(Y,t) + \frac{\overline{R}}{\tau}\frac{\partial S_{\mathrm{r,Y}}(Y,t)}{\partial \dot{y}} \\
&= \frac{\gamma \dot{b}^2}{T} + \hat{\tau}\frac{\gamma}{T}\dot{y}^2.
\end{aligned} \tag{3.85}$$

Using the fact that the most likely reservoir force acting on the actual velocity is $\overline{R}_{\mathrm{X}} = -|\tau|\gamma\dot{x}/m - (\gamma\dot{b}/m)(\tau - |\tau|) = \overline{R} - \gamma\dot{b}\tau/m$, the deterministic derivative of the reservoir entropy expressed in the actual configuration is

$$\begin{aligned}
\frac{\mathrm{d}^{\mathrm{det}} S_{\mathrm{r,X}}(X,t)}{\mathrm{d}t} &= \dot{S}^0_{\mathrm{r,X}}(X,t) + \frac{\overline{R}_{\mathrm{X}}}{\tau}\frac{\partial S_{\mathrm{r,X}}(X,t)}{\partial \dot{x}} \\
&= \frac{\gamma \dot{b}^2}{T} - \frac{\gamma \dot{b}}{T}\dot{y} + \frac{1}{\tau}\left[\frac{-|\tau|\gamma\dot{y}}{m} - \frac{\gamma\dot{b}}{m}\tau\right]\frac{-m\dot{y}}{T} \\
&= \frac{\gamma \dot{b}^2}{T} + \hat{\tau}\frac{\gamma}{T}\dot{y}^2.
\end{aligned} \tag{3.86}$$

The two deterministic derivatives agree, $\mathrm{d}^{\mathrm{det}} S_{\mathrm{r,X}}(X,t)/\mathrm{d}t = \mathrm{d}^{\mathrm{det}} S_{\mathrm{r,Y}}(Y,t)/\mathrm{d}t$, because the most likely trajectory is actually the deterministic trajectory. The fact that the forward, $\tau > 0$, and backward, $\tau < 0$, derivatives differ is a manifestation of the irreversibility of the equations of motion.

Using the formal decomposition of the reservoir entropy, the deterministic, backward, $\tau < 0$, derivative is

$$\begin{aligned}
\frac{\mathrm{d}^{\mathrm{det}} S_{\mathrm{r,X}}(X,t)}{\mathrm{d}t} &= \frac{\mathrm{d}^{\mathrm{det}} S_{\mathrm{r,st}}(X,t)}{\mathrm{d}t} + \frac{\mathrm{d}^{\mathrm{det}} S_{\mathrm{r,dyn}}(X,t)}{\mathrm{d}t} \\
&= \dot{S}^0_{\mathrm{r,st}}(X,t) + \frac{\overline{R}_{\mathrm{X}}}{\tau} \frac{\partial S_{\mathrm{r,st}}(X,t)}{\partial \dot{x}} - \dot{S}^0_{\mathrm{r,st}}(X,t) \\
&= \frac{1}{\tau}\left[\frac{-|\tau|\gamma\dot{y}}{m} - \frac{\gamma\dot{b}}{m}\tau\right]\frac{-m\dot{x}}{T} \\
&= \frac{-\gamma}{T}\left[\dot{x} - 2\dot{b}\right]\dot{x}. \qquad (3.87)
\end{aligned}$$

The second equality follows from the fundamental theorem of calculus, since the integral is over the same path as the derivative. This agrees with the two preceding results for the case $\tau < 0$. Notice that in order to obtain the agreement, it was necessary to retain the constant, time dependent term, $\gamma\dot{b}^2 t/T$, in the reservoir entropy. The significance of this result lies in the generality of the second equality,

$$\frac{\mathrm{d}^{\mathrm{det}} S_{\mathrm{r,X}}(X,t)}{\mathrm{d}t} = \frac{\overline{R}_{\mathrm{X}}}{\tau} \frac{\partial S_{\mathrm{r,st}}(X,t)}{\partial \dot{x}}, \quad \tau < 0. \qquad (3.88)$$

This is a general result that is demanded from the laws of material conservation, which are themselves the basis of Gibb's formulation of equilibrium thermodynamics. It provides the justification for, and the origin of, the formal expression for the entropy of a non-equilibrium system, Eq. (3.73).

The total rate of change of the reservoir entropy requires an expansion to quadratic order and an average over the stochastic force[13]

$$\begin{aligned}
\frac{\mathrm{d}S_{\mathrm{r,X}}(X,t)}{\mathrm{d}t} = \frac{\mathrm{d}S_{\mathrm{r,Y}}(Y,t)}{\mathrm{d}t} &= \frac{\mathrm{d}^{\mathrm{det}} S_{\mathrm{r,X}}(X,t)}{\mathrm{d}t} + \frac{1}{2\tau}\left\langle \tilde{R}\tilde{R}\right\rangle_{Y'} : \nabla\nabla S_{\mathrm{r,X}}(X,t) \\
&= \frac{\gamma\dot{b}^2}{T} + \hat{\tau}\frac{\gamma}{T}\dot{y}^2 - \hat{\tau}\frac{\gamma k_{\mathrm{B}}}{m}. \qquad (3.89)
\end{aligned}$$

By the equipartition theorem, $\langle \dot{y}^2 \rangle_X = k_{\mathrm{B}}T/m$, and so on average only the most likely dissipation, $\left\langle \dot{S}_{\mathrm{r,X}}(X,t)\right\rangle_X = \partial S_{\mathrm{r,Y}}(Y,t)/\partial t = \gamma\dot{b}^2/T$, contributes. The most likely dissipation arises on the most likely actual trajectory, $-\gamma\dot{\overline{x}}(t)^2/m = -\gamma\dot{b}^2/m$). The result says that the entropy created by the dissipation (due to the fluctuations, $-\gamma\dot{y}^2/m$), is on average cancelled by the fluctuations (of the fluctuations, $\gamma k_{\mathrm{B}}T/m^2$).[14] This exact cancellation of the stochastic and

[13]The discussion of the stochastic calculus in connection with the Fokker-Planck equation in footnote 18 on p. 89 below and in §11.3.3 is relevant to this entropy derivative and to the following probability derivative.

[14]The random force from the reservoir sometimes kicks the sub-system into a state of higher energy, and sometimes into a state of lower energy. Because of the non-linear potential, and because of the expansion to second-order, the energy increasing kicks exceed the energy decreasing ones, which corresponds to the reservoir losing energy via fluctuations going forward in time. Conversely, the drag force dissipates energy to the reservoir, increasing its entropy, again forward in time

the dissipative terms only occurs because the fluctuation dissipation theorem ensures that their magnitudes are equal.

The non-equilibrium probability density is essentially the exponential of the reservoir entropy,

$$\wp_Y(Y,t) = \frac{e^{S_{\mathrm{r,Y}}(Y,t)/k_\mathrm{B}}}{Z(t)} = \frac{e^{\beta\gamma \dot{b}^2 t} e^{-\beta\kappa y^2/2} e^{-\beta m\dot{y}^2/2}}{Z(t)}, \tag{3.90}$$

with $\beta = 1/k_\mathrm{B}T$, and the normalising partition function being $Z(t) = 2\pi e^{\beta\gamma\dot{b}^2 t} / \beta\sqrt{m\kappa}$. One can also write $\wp_X(X,t) = \wp_Y(Y,t)$, but this is not needed for what follows because the deterministic derivative gives the same result for both forms, as does the full derivative.

The deterministic derivative is

$$\begin{aligned}\frac{\mathrm{d}^{\mathrm{det}}\wp_Y(Y,t)}{\mathrm{d}t} &= \left\{ \frac{-\dot{Z}(t)}{Z(t)} + \frac{\mathrm{d}^{\mathrm{det}}S_{\mathrm{r,Y}}(Y,t)}{k_\mathrm{B}\mathrm{d}t} \right\} \wp_Y(Y,t) \\ &= \hat{\tau}\beta\gamma\dot{y}^2 \wp_Y(Y,t),\end{aligned} \tag{3.91}$$

and the full derivative, after expanding to quadratic order and averaging over the stochastic force, is

$$\begin{aligned}\frac{\mathrm{d}\wp_Y(Y,t)}{\mathrm{d}t} &= \frac{\mathrm{d}^{\mathrm{det}}\wp_Y(Y,t)}{\mathrm{d}t} + \frac{1}{2\tau}\frac{2\gamma|\tau|}{\beta m^2}\left[\left(\frac{\partial S_{\mathrm{r,Y}}(Y,t)}{k_\mathrm{B}\partial\dot{y}} \right)^2 \right. \\ &\quad \left. + \frac{\partial^2 S_{\mathrm{r,Y}}(Y,t)}{k_\mathrm{B}\partial\dot{y}^2} \right] \wp_Y(Y,t) \\ &= \hat{\tau}\beta\gamma\dot{y}^2\wp_Y(Y,t) + \frac{\gamma\hat{\tau}}{m}\left[\beta m\dot{y}^2 - 1\right]\wp_Y(Y,t).\end{aligned} \tag{3.92}$$

One can draw a very important conclusion from this result. The equipartition theorem shows that the stochastic term exactly vanishes when integrated over configuration space, $\int \mathrm{d}Y\, \wp_Y(Y,t)\left[\beta m\dot{y}^2 - 1\right] = \left\langle \beta m\dot{y}^2 - 1 \right\rangle_X = 0$. This means that to an approximation this term also vanishes point-wise,

$$\left[\beta m\dot{y}^2 - 1\right]\wp_Y(Y,t) \approx 0. \tag{3.93}$$

The reason that this is a reasonable approximation at every point in the configuration space is that on those points that are likely to occur, the equipartition theorem is satisfied, $\beta m\dot{y}^2 \approx 1$. Conversely, those points where the equipartition theorem is badly dissatisfied, $\beta m\dot{y}^2 \gg 1$, are unlikely to occur, $\wp_Y(Y,t) \approx 0$. (These approximations hold with increased accuracy for multiple particles.) This means that to a reasonable approximation the full derivative of the non-equilibrium probability density can be taken to be equal to the deterministic derivative

$$\frac{\mathrm{d}\wp_Y(Y,t)}{\mathrm{d}t} \approx \frac{\mathrm{d}^{\mathrm{det}}\wp_Y(Y,t)}{\mathrm{d}t}. \tag{3.94}$$

This result is more general that the present problem of Brownian motion; it is a result of the generalised equipartition theorem derived later in §3.7.4. However, it only holds for the non-equilibrium probability density; it does not hold for a general phase function, $\mathrm{d}f(X,t)/\mathrm{d}t \neq \mathrm{d}^{\mathrm{det}}f(X,t)/\mathrm{d}t$. The full significance and generality of the result can be better appreciated from the discussion of Liouville's theorem, later in §3.7.5.

3.7.2 Compressibility of the Equations of Motion

From the normalisation of the non-equilibrium probability density, one can obtain a relationship between the compressibility of the deterministic part of the equations of motion and the total time derivative of the probability density. With $Y' \equiv \overline{Y}(t+\tau|Y,t)$ the deterministic evolution of the fluctuation in a time step, one has[15]

$$\begin{aligned} 1 &= \int \mathrm{d}Y' \, \wp(Y', t+\tau) \\ &= \int \mathrm{d}Y \left[1 + \nabla \cdot [\tau \dot{Y}^0 + \overline{R}]\right] \left[\wp(Y,t) + \tau \frac{\mathrm{d}^{\mathrm{det}}\wp(Y,t)}{\mathrm{d}t}\right] \\ &= \int \mathrm{d}Y \, \wp(Y,t) + \int \mathrm{d}Y \left[(\nabla \cdot \overline{R})\wp(Y,t) + \tau \frac{\mathrm{d}^{\mathrm{det}}\wp(Y,t)}{\mathrm{d}t}\right] + \mathcal{O}(\tau^2). \end{aligned} \tag{3.95}$$

Since the compressibility of the adiabatic force vanishes, one has $\nabla \cdot [\tau \dot{Y}^0 + \overline{R}] = \nabla \cdot \overline{R}$. By normalisation, the first integral of the final equality is unity, and so the second integral must vanish. Hence in general, one must have that the average compressibility of the most likely trajectory must be equal and opposite to the total rate of change of the probability density,

$$\int \mathrm{d}Y \, \frac{\mathrm{d}\wp(Y,t)}{\mathrm{d}t} = \int \mathrm{d}Y \, \frac{\mathrm{d}^{\mathrm{det}}\wp(Y,t)}{\mathrm{d}t} = \frac{-1}{\tau} \left\langle \nabla \cdot \overline{R} \right\rangle_X . \tag{3.96}$$

Because the stochastic terms cancel in the total derivative of the probability density when integrated over phase space, as discussed in connection with Eq. (3.92), this result also holds for that case. This also means that the stochastic forces do not contribute to the compressibility.

For the particular case of driven Brownian motion, $\overline{R} = -\gamma|\tau|\dot{y}/m$, the compressibility is $\tau^{-1}\nabla \cdot \overline{R} = -\hat{\tau}\gamma/m$, a constant. Using Eq. (3.91) for the deterministic derivative of the probability distribution, and also the equipartition theorem, the left-hand side of this result is $\hat{\tau}\beta\gamma < \dot{y}^2 >_X = \hat{\tau}\gamma/m$, which confirms the validity of the result.

[15] For a general Cartesian vector space with volume element $\mathrm{d}r$, one has $\mathrm{d}r' = \prod_i \mathrm{d}r_i' = \prod_i \left[\mathrm{d}r_i + \tau \mathrm{d}r_i(\partial \overline{\dot{r}}_i/\partial r_i)\right] = \mathrm{d}r \left[1 + \tau \sum_i (\partial \overline{\dot{r}}_i/\partial r_i) + \mathcal{O}(\tau^2)\right] = \mathrm{d}r \left[1 + \tau \nabla \cdot \overline{\dot{r}}\right]$. The adiabatic motion does not contribute to this because Hamilton's equations are incompressible. Note that the stochastic contributions to the evolution of each vertex of the volume element, including the variance, cancel, and so they don't contribute to this. In this case the most likely value and the average value coincide.

The probability of a volume element, $P(Y,t) \equiv [\mathrm{d}Y\, \wp(Y,t)]$, has time derivative $\mathrm{d}P(Y,t)/\mathrm{d}t \equiv \tau^{-1}(\nabla \cdot \overline{R})[\mathrm{d}Y\, \wp(Y,t)] + \mathrm{d}Y\, \mathrm{d}\wp(Y,t)/\mathrm{d}t$. Its integral is therefore

$$\int \mathrm{d}Y \left[\frac{1}{\mathrm{d}Y}\frac{\mathrm{d}P(Y,t)}{\mathrm{d}t}\right] = \int \mathrm{d}Y \left[\frac{\mathrm{d}\wp(Y,t)}{\mathrm{d}t} + \frac{\wp(Y,t)}{\tau}\nabla \cdot \overline{R}\right] = 0. \tag{3.97}$$

Unlike the probability density, the total time derivative of the probability of a volume element is zero when integrated over all of phase space.

3.7.3 The Fokker-Planck Equation

The Fokker-Planck equation gives the expression for the partial time derivative of a probability density.[16] In the first place it will be derived in a fairly general setting, and then the explicit results for driven Brownian motion will be inserted.

As above, let $X \equiv \{x, \dot{x}\}$ be the configuration in phase space, let $Y = X(t) - \overline{X}(t)$ denote the fluctuation or departure from the most likely state, let $Y' = Y(t+\tau)$, and let $\overline{Y'} = \overline{Y}(t+\tau|Y,t)$ denote the conditional most likely fluctuation. The latter is given by the deterministic part of the stochastic, dissipative equations, $\overline{Y}(t+\tau|Y,t) = -Q(\tau)SY \equiv Y + \tau\dot{Y}^0 + \overline{R}$, with the force (more precisely, the acceleration) being

$$\dot{Y}^0 = \Theta SY, \text{ and } \overline{R} = -|\tau|\Lambda SY/2. \tag{3.98}$$

The first of these is the adiabatic contribution to the force (any possible non-adiabatic, asymmetric contributions from the reservoir have not been explicitly included), and the second term arises solely from the reservoir. The matrix ΘS couples only variables of opposite time parity, whereas the entropy matrix S couples only variables of the same time parity. This means that $\nabla \cdot \dot{Y}^0 = \Theta_{\alpha\gamma}S_{\gamma\alpha} = 0$. Hence the divergence of the total force is

$$\nabla \cdot [\tau\dot{Y}^0 + \overline{R}] = \frac{-|\tau|}{2}\mathrm{TR}(\Lambda S), \tag{3.99}$$

where TR denotes the trace.[17]

The transition probability is essentially the exponential of the second entropy. By the reduction condition, the maximum value of the second entropy occurs at $\overline{Y'}$ and equals the first entropy, $S^{(2)}(\overline{Y'}, t+\tau; Y, t) = S_\mathrm{r}(Y,t)$, as can be confirmed for the expression given explicitly as Eq. (3.64). Accordingly the conditional second entropy is defined as

$$\begin{aligned} S^{(2)}(Y', t+\tau|Y,t) &\equiv S^{(2)}(Y', t+\tau; Y, t) - S_\mathrm{r}(Y,t) \\ &= \frac{-1}{2|\tau|}\Lambda^{-1} : \left[Y' - \overline{Y'}\right]^2. \end{aligned} \tag{3.100}$$

[16] For a comprehensive account of the Fokker-Planck equation and Liouville's theorem, see Risken, H. (1984), *The Fokker-Planck Equation*, (Springer-Verlag, Berlin).

[17] Similarly, the antisymmetric nature of Θ means that it does not contribute to the rate of change of the reservoir entropy. In the general case this is $\mathrm{d}S_\mathrm{r}(Y,t)/\mathrm{d}t = \partial S_\mathrm{r}(Y,t)/\partial t + [\tau\dot{Y}^0 + \overline{R}] \cdot \nabla S_\mathrm{r}(Y,t) + \hat{\tau}k_\mathrm{B}\Lambda : \nabla\nabla S_\mathrm{r}(Y,t)/2 = \gamma\dot{b}^2/T + \hat{\tau}Y \cdot S\Lambda SY/2 + \hat{\tau}k_\mathrm{B}\Lambda : S/2$. The deterministic part of this is positive definite going forward in time. The final two terms cancel on average.

In terms of this, the exact conditional transition probability is a Gaussian,

$$\wp_{\rm ne}(Y',t+\tau|Y,t) = \frac{1}{Z^{(2)}} e^{S^{(2)}(Y',t+\tau|Y,t)/k_{\rm B}}. \tag{3.101}$$

Here the normalising partition function is $Z^{(2)} = (2\pi k_{\rm B}|\tau|)^{M/2}\sqrt{|\Lambda|}$ where M is the number of coordinates, which is two for the present case of Brownian motion.

Now the time evolution of an arbitrary probability density is given under the action of this, the actual transition probability. The partial time derivative of the probability density is

$$\begin{aligned}
&\frac{\partial \wp(Y',t)}{\partial t}\\
&= \frac{\wp(Y',t+\tau)-\wp(Y',t)}{\tau}\\
&= \frac{1}{\tau}\left\{-\wp(Y',t) + \int {\rm d}Y\, \wp(Y',t+\tau|Y,t)\wp(Y,t)\right\}\\
&= \frac{1}{\tau}\left\{-\wp(Y,t) + \int {\rm d}Y\, \frac{1}{Z^{(2)}} e^{-\Lambda^{-1}:\left[Y'-\overline{Y'}\right]^2/2k_{\rm B}|\tau|}\wp(Y,t)\right\}\\
&= \frac{1}{\tau}\left\{-\wp(Y,t) + \frac{1}{Z^{(2)}}\int {\rm d}\overline{Y'}\left[1-\nabla\cdot[\tau\dot{Y}^0+\overline{R}]\right] e^{-\Lambda^{-1}:\left[Y'-\overline{Y'}\right]^2/2k_{\rm B}|\tau|}\right.\\
&\qquad \left.\times\, \wp(Y'+Y-Y',t)\right\}\\
&= \frac{1}{\tau}\left\{-\wp(Y,t) + \frac{1}{Z^{(2)}}\int {\rm d}\overline{Y'}\left[1-\nabla\cdot[\tau\dot{Y}^0+\overline{R}]\right] e^{-\Lambda^{-1}:\left[Y'-\overline{Y'}\right]^2/2k_{\rm B}|\tau|}\right.\\
&\qquad \Big[\wp(Y',t) + (Y-\overline{Y'}+\overline{Y'}-Y')\cdot\nabla\wp(Y',t)\\
&\qquad \left.+\frac{1}{2}(Y-\overline{Y'}+\overline{Y'}-Y')^2:\nabla\nabla\wp(Y',t)\Big]\right\}\\
&= -\frac{1}{\tau}\nabla\cdot\Big[[\tau\dot{Y}^0+\overline{R}]\wp(Y',t)\Big] + \frac{\hat{\tau}k_{\rm B}}{2}\Lambda:\nabla\nabla\wp(Y',t) + \mathcal{O}(\tau).
\end{aligned} \tag{3.102}$$

In obtaining the final equality, use has been made of the facts that the weighted integral of $\overline{Y'}-Y'$ is zero, that $Y-\overline{Y'} = -[\tau\dot{Y}^0(Y)+\overline{R}(Y)] = -[\tau\dot{Y}^0(Y')+\overline{R}(Y')]+\mathcal{O}(\tau^2)$, and that the weighted integral of $(Y-\overline{Y'}+\overline{Y'}-Y')^2$ equals the weighted integral of $(\overline{Y'}-Y')^2$ plus $\mathcal{O}(\tau^2)$.

This is the Fokker-Planck equation. It has been assumed here that the transport and entropy matrices do not depend upon Y, (see the discussion in §3.6.3 on p. 78).[18]

[18]Derivatives and integrals of functions of stochastic variables generally require a stochastic calculus, either that due to Stratonovich or that due to Itô. The usual derivation of the Fokker-Planck equation proceeds from one or other of these: see Risken (1984), or Gardiner, C. W. (1983), *Handbook of Stochastic Methods*, (Springer, New York). In the case, and only in the case, that the variance of the random force, here denoted Λ, depends on phase

It is worthwhile giving this result in terms of the actual variable $X = Y + \overline{X}(t)$. Since $\overline{Y}(t+\tau|Y,t) = Y + \tau\dot{Y}^0 + \overline{R}$, one has $\overline{X}(t+\tau|X,t) = X + \tau\dot{X}^0 + \overline{R} + \tau \mathrm{d}\overline{X}(t)/\mathrm{d}t - \tau\dot{\overline{X}}{}^0 \equiv X + \tau\dot{X}^0 + \overline{R}_{\mathrm{X}}$. The Fokker-Planck equation for the probability distribution of the actual variable X is as above with $\wp(Y,t) \Rightarrow \wp(X,t)$ and $\overline{R} \Rightarrow \overline{R}_X$,

$$\begin{aligned}
\frac{\partial\wp(X,t)}{\partial t}& \\
&= \frac{-1}{\tau}\nabla\cdot\left[[\tau\dot{X}^0 + \overline{R}_{\mathrm{X}}(X,t)]\wp(X,t)\right] - \frac{\hat{\tau}k_{\mathrm{B}}}{2}\Lambda^{-1}:\nabla\nabla\wp(X,t) \\
&= -\left[\frac{\mathrm{d}\overline{X}(t)}{\mathrm{d}t} - \dot{\overline{X}}{}^0(t)\right]\cdot\nabla\wp(X,t) - \dot{X}^0\cdot\nabla\wp(X,t) \\
&\quad - \frac{1}{\tau}\nabla\cdot[\overline{R}(X,t)\wp(X,t)] - \frac{\hat{\tau}k_{\mathrm{B}}}{2}\Lambda:\nabla\nabla\wp(X,t).
\end{aligned} \tag{3.103}$$

Recall that $\wp(X,t)$ and $\wp(Y,t)$ are different functions of their arguments, which fact is signified by the different symbols used for their arguments. In the second equality, the first term on the right-hand side may be interpreted as a drift term.

These two expressions for the Fokker-Planck equation give the partial time derivative of an arbitrary (not necessarily the exact) probability distribution. The result invoked the exact conditional transition probability, which is equivalent to the exact stochastic, dissipative equations of motion.

For the particular case of driven Brownian motion that is a concern of this chapter, the Fokker-Planck equation is explicitly

$$\begin{aligned}
\frac{\partial\wp(X,t)}{\partial t} &= \frac{-\mathrm{d}\overline{X}(t)}{\mathrm{d}t}\cdot\nabla\wp(X,t) - \frac{1}{\tau}\frac{\partial\left[\tau\dot{y}(t)\wp(X,t)\right]}{\partial x} \\
&\quad + \frac{1}{\tau}\frac{\partial\left[\{(\tau\kappa/m)y + (|\tau|\gamma/m)\dot{y}\}\wp(X,t)\right]}{\partial\dot{x}} - \frac{\hat{\tau}\gamma k_{\mathrm{B}}T}{m^2}\frac{\partial^2\wp(X,t)}{\partial\dot{x}^2} \\
&= \frac{-\mathrm{d}\overline{X}(t)}{\mathrm{d}t}\cdot\nabla\wp(X,t) - \dot{y}\frac{\partial\wp(X,t)}{\partial x} + \left[\frac{\kappa}{m}y + \frac{\hat{\tau}\gamma}{m}\dot{y}\right]\frac{\partial\wp(X,t)}{\partial\dot{x}} \\
&\quad + \hat{\tau}\frac{\gamma}{m}\wp(X,t) - \frac{\hat{\tau}\gamma k_{\mathrm{B}}T}{m^2}\frac{\partial^2\wp(X,t)}{\partial\dot{x}^2}.
\end{aligned} \tag{3.104}$$

This gives the rate of change of an arbitrary probability distribution for the trapped Brownian particle due to the actual transition probability. Here $Y \equiv X - \overline{X}(t)$ for the most likely configuration that corresponds to the given probability distribution. In the case that this most likely configuration is the exact one, $\overline{X} = \{b(t) - \gamma\dot{b}/\kappa, \dot{b}\}$, the first term on the right-hand side becomes $-\dot{b}\partial\wp(X,t)/\partial x$.

space, here denoted X or Y, and that the system has no memory, then each choice yields a different version of the Fokker-Planck equation. Alternatively, a given Fokker-Planck equation corresponds to two different evolution equations for the stochastic variable. In the present book, and arguably in physical problems in general, the variance *never* depends on phase space (it may depend upon space and time, but not upon phase space). Also, the no memory limit is singular and must be interpreted with care. Consequently in §11.3.3 it is argued that in physical systems the stochastic calculus plays only a limited rôle, and that there is no ambiguity in the Fokker-Planck equation or in the evolution of other phase functions.

For the actual non-equilibrium probability density for the Brownian particle in the uniformly moving trap,

$$\begin{aligned} \wp(X,t) &= \frac{1}{Z(t)} e^{\beta\gamma\dot{b}^2 t} e^{-\beta\kappa[x-b(t)+\gamma\dot{b}/\kappa]^2/2} e^{-\beta m[\dot{x}-\dot{b}]^2/2} \\ &= \frac{1}{Z(t)} e^{\beta\gamma\dot{b}^2 t} e^{-\beta\kappa y^2/2} e^{-\beta m\dot{y}^2/2}, \end{aligned} \tag{3.105}$$

the Fokker-Planck equation yields

$$\begin{aligned} \frac{\partial\wp(X,t)}{\partial t} &= \left\{ \beta\kappa\dot{b}y + \beta\kappa\dot{y}y - \left[\frac{\kappa}{m}y + \frac{\hat{\tau}\gamma}{m}\dot{y}\right]\beta m\dot{y} + \hat{\tau}\frac{\gamma}{m} \right. \\ &\quad \left. + \frac{\hat{\tau}\gamma}{\beta m^2}\left[(\beta m\dot{y})^2 - \beta m\right] \right\} \wp(X,t) \\ &= \beta\kappa\dot{b}y\wp(X,t). \end{aligned} \tag{3.106}$$

This result agrees with the result that can be obtained by taking directly the partial time derivative of the explicit form for the probability distribution. This confirms the consistency of the analysis.

3.7.4 Generalised Equipartition Theorem

The equipartition theorem normally refers to an equilibrium system, and, originally, to the momentum: the average of the square of any one component of momenta is $\langle p_{i\alpha}^2 \rangle = mk_{\rm B}T$. It arises because the Maxwell-Boltzmann probability density is proportional to the exponential of the Hamiltonian, which contains the kinetic energy, which is a sum of the squares of the momenta. A slight generalisation of the equipartition theorem therefore states that for any equilibrium system, there is $k_{\rm B}T$ per quadratic degree of freedom in the Hamiltonian.

For the present case of a Brownian particle, in either a stationary or a moving trap, the Hamiltonian contains a term $\kappa y^2/2$, and since this appears in the exponent of the probability density, both equilibrium (stationary trap), and non-equilibrium (uniformly moving trap), the form of the equipartition theorem is $\langle y^2 \rangle_x = k_{\rm B}T/\kappa$. The fact that this holds in this example of a non-equilibrium system is interesting, and it is worthwhile exploring the generality of the result, particularly because it is directly relevant to the discussion of Liouville's theorem that follows.

The key to presenting the most general form of the equipartition theorem can be gleaned from Eq. (3.92), which was given in the particular context of a trapped Brownian particle. All probability densities, equilibrium and non-equilibrium, can be formally written as the exponential of the first entropy. Of the two terms on the right-hand side of Eq. (3.92) that arise from the stochastic contribution, the second derivative, $\partial^2 S/\partial\dot{y}^2$, can be interpreted as giving the coefficient of any term that appears as a quadratic in the exponent of the probability, and the square of the first derivative, $[\partial S/\partial\dot{y}]^2$, can be interpreted as giving the square of the component of such a quadratic term. Stated in this form the relationship with the original equipartition theorem is evident.

The derivation and statement of the generalised equipartition theorem is in fact rather trivial. Using the general and formally exact expression for the probability density for phase space or other microspace,[19] $\wp(X,t) = Z^{-1} e^{S_{\rm r}(X,t)/k_{\rm B}}$, one has

$$\begin{aligned} k_{\rm B}^{-2} \langle [\nabla S_{\rm r}(X,t)]\,[\nabla S_{\rm r}(X,t)] \rangle_X &= k_{\rm B}^{-2} \int {\rm d}X\, \wp(X,t)\,[\nabla S_{\rm r}(X,t)]\,[\nabla S_{\rm r}(X,t)] \\ &= k_{\rm B}^{-1} \int {\rm d}X\, [\nabla \wp(X,t)]\,[\nabla S_{\rm r}(X,t)] \\ &= -k_{\rm B}^{-1} \int {\rm d}X\, \wp(X,t) \nabla\nabla S_{\rm r}(X,t) \\ &= -k_{\rm B}^{-1} \langle \nabla\nabla S_{\rm r}(X,t) \rangle_X . \end{aligned} \tag{3.107}$$

The second equality follows from the definition of the probability density as the exponential of the reservoir entropy, and the third equality follows from an integration by parts and the vanishing of the probability for extreme fluctuations. This is the generalised equipartition theorem. For a single component it reads

$$\left\langle \left(\frac{\partial S_{\rm r}(x,t)}{k_{\rm B}\partial x} \right)^2 \right\rangle_x + \left\langle \frac{\partial^2 S_{\rm r}(x,t)}{k_{\rm B}\partial x^2} \right\rangle_x = 0. \tag{3.108}$$

The generalised equipartition theorem is equivalent to $\int {\rm d}X\, \nabla\nabla\wp(X,t) = 0$, as can be shown directly by integration and the vanishing of the probability and its gradient for extreme values. In some circumstances it may be a reasonable approximation to apply this at each point, not just on average,

$$\begin{aligned} \nabla\nabla\wp(X,t) &\equiv \left[k_{\rm B}^{-1} \nabla\nabla S_{\rm r}(X,t) + k_{\rm B}^{-2} \{\nabla S_{\rm r}(X,t)\}\{\nabla S_{\rm r}(X,t)\} \right] \wp(X,t) \\ &\approx 0. \end{aligned} \tag{3.109}$$

Although one *could* choose an X for which the bracketed term is non-negligible, the probability of such values is small. When the sum of the two terms inside the brackets is large, the probability density is small, and *vice versa.*

This second derivative arose in the total time derivative of the probability density, Eq. (3.92), and in the Fokker-Planck equation, for the partial time derivative of the probability density, Eq. (3.103). In both cases it arose directly from the stochastic part of the equations of motion. Hence, in so far as the equipartition theorem holds with negligible error at each point, then the time evolution of the probability density is given by the deterministic equations of motion. That is, the stochastic contribution can be neglected.[20]

[19] When X represents a microstate, there is no sub-system entropy associated with it, and $S_{\rm t}(X,t) = S_{\rm r}(X,t)$. When X represents a macrostate, the total entropy $S_{\rm t}(X,t) = S_{\rm s}(X,t) + S_{\rm r}(X,t)$ has to be used in place of the reservoir entropy in the derivation.

[20] The approximation of neglecting the stochastic part of the equations of motion is only valid for the evolution of the probability density itself. It is not valid for any other function of phase space. In general, deterministic non-Hamiltonian equations of motion give unphysical results.

3.7.5 Liouville's Theorem

The generic Fokker-Planck equation (3.103) may be written in the form,

$$\frac{\partial\wp(X,t)}{\partial t} = \frac{-1}{\tau}\nabla\cdot[\tau\dot{X}^0 + \overline{R}_X(X,t)\wp(X,t)] - \frac{\hat{\tau}k_\mathrm{B}}{2}\nabla\cdot[\Lambda\nabla\wp(X,t)]. \quad (3.110)$$

This follows because the variance of the fluctuations, essentially Λ, is constant with respect to X. As discussed in §3.6.3 on p. 78, this may depend upon the most likely current state, $\Lambda(\overline{X}(t))$. The fluctuations Y about this state are always relatively small and one can almost always neglect non-linear effects beyond those contained in $\overline{X}(t)$, which by definition do not depend upon X.

This form of the Fokker-Planck equation represents a conservation law for the probability, $\partial\wp(X,t)/\partial t = -\nabla\cdot J_\wp(X,t)$, with the probability flux being

$$J_\wp(X,t) = \dot{X}^\mathrm{det}\wp(X,t) + \frac{\hat{\tau}k_\mathrm{B}}{2}\Lambda\nabla\wp(X,t). \quad (3.111)$$

The deterministic velocity is $\dot{X}^\mathrm{det} \equiv \dot{X}^0 + \overline{R}_X(X,t)/\tau$, and the first term represents the flux of probability carried by the deterministic motion. The second term is obviously the flux due to the stochastic part of the equations of motion.

Liouville's theorem follows by neglecting the stochastic contribution in the Fokker-Planck equation, and rearranging it as the total time derivative,

$$\begin{aligned}\frac{\mathrm{d}\wp(X,t)}{\mathrm{d}t} &= \frac{\partial\wp(X,t)}{\partial t} + \dot{X}^\mathrm{det}(X,t)\cdot\nabla\wp(X,t) \\ &= -\left[\nabla\cdot\dot{X}^\mathrm{det}(X,t)\right]\wp(X,t).\end{aligned} \quad (3.112)$$

The deterministic velocity contains the adiabatic (Hamiltonian) velocity and other terms that come from Θ and which are independent of $\hat{\tau}$ in the velocity, and also the dissipative forces that come from Λ and which scale with $\hat{\tau}$ in the velocity. These dissipative forces are a non-Hamiltonian term that are sometimes chosen in an *ad hoc* fashion. The first equality is the formally exact definition of the convective derivative (it was denoted $\mathrm{d}^\mathrm{det}/\mathrm{d}t$ above), which is the same as the total derivative for deterministic equations of motion such as one often sees in the literature,

$$X(t+\tau) = X(t) + \tau\dot{X}^\mathrm{det}(X(t),t). \quad (3.113)$$

Here the total derivative was equated to the compressibility of the equations of motion by neglecting the stochastic contribution to the Fokker-Planck equation. The rationale for this neglect lies in the results of the preceding subsection, §3.7.4, and will be revisited shortly. First, however, the conventional justification for Liouville's theorem will be discussed.

Liouville's theorem has the appearance of a conservation law for probability, with the probability flux being $J_\wp(X,t) = \dot{X}^\mathrm{det}(X,t)\wp(X,t)$. Compared to the above expression, this neglects the stochastic term. This is consistent with the above derivation of Liouville's theorem in which the stochastic contribution to the Fokker-Planck equation was neglected. However, it must be stressed that this is *not* an exact formula for the probability flux, but an approximation that holds in those regimes where the stochastic contribution is negligible.

These particular deterministic forms for the probability flux and for Liouville's theorem follow from the ensemble interpretation of probability (see §1.3).[21] Briefly, the ensemble picture says that one has M replicas of the system of interest, and that the probability density at a point in phase space at a particular time, $\wp(X,t)$, is proportional to the number of members of the ensemble in the neighbourhood of the point Ω, $\wp(X,t) = M_\Omega(t)/\Omega M$. Assuming that ensemble members cannot be created or destroyed, then the only way for the probability density to change with time is for members of the ensemble to be carried across the boundaries of Ω by the evolution of the trajectory. Hence the divergence of the probability flux gives the rate of change per unit volume,

$$\frac{\partial \wp(X,t)}{\partial t} = -\nabla \cdot \left[\dot{X}^{\text{det}} \wp(X,t)\right]. \tag{3.114}$$

This is equivalent to the above expression for Liouville's theorem.

An alternative but related justification for Liouville's theorem has been offered, based upon the normalisation condition on the probability density

$$\int \mathrm{d}X\, \wp(X,t) = 1. \tag{3.115}$$

This has been taken to represent a conservation law on the probability density, and it has been argued that this leads to Eq. (3.114) for the rate of change of probability.[22] The normalisation condition was used above to derive Eq. (3.96), which may be seen to be an integral form of Eq. (3.114), but it is not a local condition. The reduction condition for the weight of joint states, Eq. (1.12), and the related conservation law for weight during a transition, Eq. (1.28), both of which follow from the general laws of probability, are also global conditions that hold for a sum over a complete set of states, rather than a local conservation law as assumed in Liouville's theorem.

Both arguments implicitly assume that the probability flux is $J_\wp(X,t) = \dot{X}^{\text{det}}(X,t)\wp(X,t)$, which neglects the stochastic contributions. Hence there are two issues raised by Liouville's theorem: first, whether it is valid to neglect stochastic contributions to the evolution of the probability density, and second whether it is valid to neglect stochastic contributions to the equations of motion themselves. Consequent on the second question is the validity of the deterministic evolution of an arbitrary function of the space.

---

[21] As discussed in §1.3, the ensemble interpretation of probability is an early mental image developed as one way to account for randomness in classical mechanics. It should not be mistaken for the physical basis of statistical mechanics; any dissonance between the ensemble interpretation and physical reality demands modification of the former.

[22] Zwanzig, R. (2001), *Non-equilibrium Statistical Mechanics*, (Oxford University Press, Oxford). For example, on p. 37: "Whenever a conservation law of this kind is encountered, we expect that the time derivative of the conserved quantity or density is balanced by the divergence of a flux, a velocity times that density."

First, as shown in the preceding subsection, §3.7.4, the stochastic contribution to the evolution of the probability density is generally negligible. This is because the generalised equipartition theorem holds on the likely points of the space. It therefore follows that Liouville's theorem for the evolution of the probability density is a good approximation to the Fokker-Planck equation. (This assumes that the stochastic equations of motion satisfy the fluctuation dissipation theorem.) The correct interpretation of Liouville's theorem is not that it represents the evolution of the probability density for deterministic equations of motion, but rather that it represents the evolution of the probability density for the exact dissipative, stochastic equations of motion. It is just that the stochastic contribution to the evolution of the probability density is negligible due to the generalised equipartition theorem.

Second, it is not generally valid to use deterministic, non-Hamiltonian equations of motion, and it is not valid to follow the evolution of the system or of general functions of the space of the system based on such deterministic, non-Hamiltonian equations of motion.[23] Deterministic, non-Hamiltonian equations of motion do not obey the fluctuation dissipation theorem, and as such give unphysical results. In practice the theorem is violated in two ways. First the stochastic contribution to the equations of motion is neglected. Since randomness is an inevitable consequence of the projection from the total system to the sub-system, deterministic non-Hamiltonian equations of motion cannot represent a real physical system. Second, both the magnitude and quite often the functional form of the non-Hamiltonian term are in error. Since the fluctuation dissipation theorem specifies the magnitude of the dissipative force in terms of the strength of the fluctuations, one does not have the freedom to take the zero fluctuation limit with non-zero dissipation. Nor does one have the freedom to arbitrarily choose the functional form of the non-Hamiltonian term, as is often done to obtain some or other desired time average. It is usually the case that deterministic non-Hamiltonian equations of motion do not invoke the correct functional form of the dissipation that maximises the second entropy. Deterministic non-Hamiltonian equations of motion correspond to the transition probability $\wp(X'|X,\tau) = \delta(X' - X - \tau\dot{X}^{\text{det}})$, which cannot be obtained as the zero variance limit of the exact transition probability.

That Liouville's theorem is an approximation can be seen explicitly for the Brownian particle in the moving trap. The left-hand side of Liouville's theorem in the form of Eq. (3.114) is given by Eq. (3.106),

$$\frac{\partial \wp(X,t)}{\partial t} = \beta\kappa\dot{b}y\wp(X,t). \tag{3.116}$$

The right-hand side of Eq. (3.114), using $\dot{x}^0 = \dot{x}$ and $\tau\ddot{x}^0 + \overline{R}_{\text{X}} = \tau\ddot{x}^0 + \overline{R} -$

[23]Deterministic Hamiltonian equations of motion are valid for an isolated system. They can be obtained by taking the variance to zero in the more general theory, since the dissipative force also scales with the variance. The stochastic and the dissipative forces represent the interaction with the reservoir, and the zero variance limit therefore represents an isolated system. In this case, and only in this case, Liouville's theorem is exact.

$\gamma\dot{b}\tau/m = -\tau\kappa[x - b(t)]/m - |\tau|\gamma\dot{y}/m - \gamma\dot{b}\tau/m$ (c.f. Eqs (3.66) and (3.86)) is

$$\begin{aligned}
-\nabla \cdot \left[\dot{X}^{\mathrm{det}}\wp(X,t)\right] \\
&= \frac{-\wp(X,t)}{\tau}\nabla \cdot \overline{R}_X - \left[\dot{X}^0 + \frac{\overline{R}_X}{\tau}\right] \cdot \nabla\wp(X,t) \\
&= \left\{\frac{\hat{\tau}\gamma}{m} + \beta\kappa y\dot{x} + \left[\frac{\kappa}{m}[x - b(t)] + \frac{\hat{\tau}\gamma\dot{y}}{m} + \frac{\gamma\dot{b}}{m}\right](-\beta m\dot{y})\right\}\wp(X,t) \\
&= \left\{\frac{\hat{\tau}\gamma}{m} - \hat{\tau}\beta\gamma\dot{y}^2 + \beta\kappa b y\right\}\wp(X,t).
\end{aligned} \tag{3.117}$$

Clearly, these last two equations, which are the two sides of Liouville's theorem, Eq. (3.114), are not equal to each other. However, on those points of phase space that satisfy the generalised equipartition theorem, $\dot{y}^2 \approx k_{\mathrm{B}}T/m$, or on those points that occur with negligible probability, $\wp(X,t) \approx 0$, they are approximately equal and Liouville's theorem is approximately satisfied.

Chapter 4

Heat Conduction

Steady heat flow may be regarded as the canonical non-equilibrium system, since it is the closest analogue of the canonical equilibrium system. The latter consists of a sub-system in thermal contact with a heat reservoir of a given temperature, whereas steady heat flow is accounted for by a sub-system in thermal contact with two heat reservoirs of different temperatures. As the canonical non-equilibrium system it is the archetype of all steady state flows. This chapter provides a thermodynamic formulation of the problem in terms of the second entropy, with the main aim of providing the thermodynamic justification of Fourier's law of conductive heat flow,

$$J_{\mathrm{E}}^{0} = -\lambda \nabla T, \tag{4.1}$$

where J_{E}^{0} is the conductive energy flux, λ is the thermal conductivity, and ∇T is the temperature gradient. In deriving this result, the rôle of the reservoirs in determining the not-in-equilibrium structure of the sub-system, and the relationship between the heat flow between the reservoirs and the internal relaxation of the sub-system, is explained.

4.1 Equilibrium System

For a system, isolated from its surroundings, the state is specified by the values of the extensive variables, which in the simplest case are the total energy E, number N, assuming a one-component system, and volume V. The ordinary or first entropy is a function of the state, $S(E, N, V)$. The temperature of the state, $T(E, N, V)$, is defined by[1]

$$\frac{1}{T} \equiv \frac{\partial S(E, N, V)}{\partial E}, \tag{4.2}$$

[1] Attard, P. (2002a), *Thermodynamics and Statistical Mechanics: Equilibrium by Entropy Maximisation*, (Academic Press, London).

the pressure, $p(E,N,V)$, is defined by

$$\frac{p}{T} \equiv \frac{\partial S(E,N,V)}{\partial V}, \tag{4.3}$$

and the chemical potential, $\mu(E,N,V)$, is defined by

$$\frac{\mu}{T} \equiv -\frac{\partial S(E,N,V)}{\partial N}. \tag{4.4}$$

These define the thermodynamically conjugate variables: (inverse) temperature is conjugate to energy, etc.

One can also define the entropy density, $\sigma(\varepsilon,\rho) \equiv S(E,N,V)/V$, where the energy density is $\varepsilon \equiv E/V$, and the number density is $\rho \equiv N/V$. Notice how the entropy density is a function of just two intensive variables. In terms of these densities it is straightforward to show that

$$\frac{1}{T} = \frac{\partial \sigma(\varepsilon,\rho)}{\partial \varepsilon},\ \frac{\mu}{T} = -\frac{\partial \sigma(\varepsilon,\rho)}{\partial \rho},\ \text{and } \frac{p}{T} = \sigma(\varepsilon,\rho) - \frac{\varepsilon}{T} + \frac{\mu\rho}{T}. \tag{4.5}$$

This chapter is concerned with energy flow, and henceforth the number and volume will not be shown explicitly. The heat capacity C_V can be defined from the derivative of the temperature,

$$C_\mathrm{V}^{-1} = \frac{\partial T(E)}{\partial E} = -T^2 \frac{\partial^2 S(E)}{\partial E^2}. \tag{4.6}$$

This is the heat capacity at constant volume, which is to be distinguished from the heat capacity at constant pressure. When the sub-system is connected to a thermal reservoir with which it can exchange energy, the heat capacity determines the rate of change of the sub-system energy with the reservoir temperature (the first equality), and it determines the fluctuations in the sub-system energy about the average sub-system energy (the second equality; see below).

Now consider a sub-system, subscript s, connected to a thermal reservoir, subscript r, with which it exchanges energy. The total system, sub-system plus reservoir, is isolated, and so the total energy, $E_\mathrm{total} = E_\mathrm{s} + E_\mathrm{r}$, is conserved, $\mathrm{d}E_\mathrm{total} = 0$, or, $\mathrm{d}E_\mathrm{s} = -\mathrm{d}E_\mathrm{r}$. The total entropy is

$$S_\mathrm{total}(E_\mathrm{s}|T_\mathrm{r}) = S_\mathrm{s}(E_\mathrm{s}) + S_\mathrm{r}(E_\mathrm{r}) = S_\mathrm{s}(E_\mathrm{s}) - \frac{E_\mathrm{s}}{T_\mathrm{r}}. \tag{4.7}$$

In the second equality, a Taylor expansion of the reservoir entropy has been made about E_total, and the constant part, $S_\mathrm{r}(E_\mathrm{total})$, which is independent of the sub-system, has been neglected Also neglected have been the higher-order terms in the Taylor expansion, which contain powers of the ratio of the sub-system energy to the total energy. This is called the constrained total entropy, and it holds for any value of the sub-system energy.

To find the most likely energy, $\overline{E}_\mathrm{s}(T_\mathrm{r})$, the constrained total entropy is maximised,

$$\left.\frac{\partial S_\mathrm{total}(E_\mathrm{s}|T_\mathrm{r})}{\partial E_\mathrm{s}}\right|_{E_\mathrm{s}=\overline{E}_\mathrm{s}} = 0, \text{ or } T_\mathrm{s}(\overline{E}_\mathrm{s}) = T_\mathrm{r}. \tag{4.8}$$

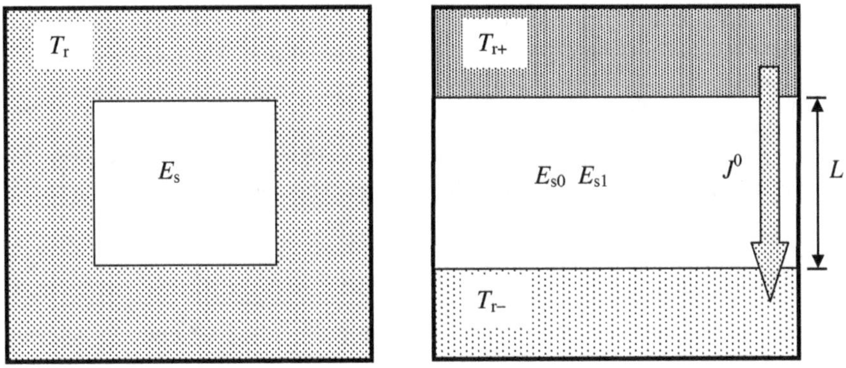

Figure 4.1: Canonical equilibrium system (left) and canonical non-equilibrium system (right).

In the equilibrium or most likely state, this says that the temperature of the sub-system is equal to that of the reservoir. The left-hand side may be denoted $\overline{T}_s$.

In most treatments of equilibrium thermodynamics, it is generally the free energy rather than the total entropy that is invoked. These are in general related to the total entropy by a factor of minus the temperature. In the present canonical case, the constrained Helmholtz free energy is

$$F(E_s|T_r) \equiv -T_r S_{\text{total}}(E_s|T_r) = E_s - T_r S_s(E_s). \tag{4.9}$$

The minimum value of the constrained Helmholtz free energy, which is often what is actually meant in textbooks, is $\overline{F}(T_r) \equiv F(\overline{E}_s|T_r)$. In terms of this, the constant volume heat capacity alluded to above is

$$C_V = \frac{-1}{T_r^2} \frac{\partial^2(\overline{F}/T_r)}{(\partial T_r^{-1})^2}. \tag{4.10}$$

As mentioned, this gives the fluctuations of the energy of the sub-system about their equilibrium value (c.f. Eqs (2.1) and (2.8)),

$$\left\langle (E_s - \overline{E}_s)^2 \right\rangle = -k_B \left[\frac{\partial^2 S_{\text{total}}}{\partial E_s^2} \right]^{-1} = k_B T_r^2 C_V. \tag{4.11}$$

4.2 First Energy Moment and First Temperature

In steady heat flow, a thermal gradient is applied to the sub-system by two reservoirs of different temperatures (see Fig. 4.1). For the present it is sufficient to consider the case of weak gradients, and to keep only the linear, leading-order term. In response, the induced energy density in the sub-system is non-uniform, and to leading-order it can be taken to have constant gradient,

$$\varepsilon(z) = \varepsilon_0 + \varepsilon_1 z. \tag{4.12}$$

Here only variations in the z-direction are considered. In this case, the total energy of the isolated sub-system, which is just the zeroth moment, is

$$E_0 = A\int_{-L/2}^{L/2} \mathrm{d}z\, \varepsilon(z) = AL\varepsilon_0. \tag{4.13}$$

The width of the sub-system is L and its area is A. The first energy moment is

$$E_1 = A\int_{-L/2}^{L/2} \mathrm{d}z\, z\varepsilon(z) = \frac{AL^3}{12}\varepsilon_1. \tag{4.14}$$

One can use these moments to define a macrostate of the sub-system, $\{E_0, E_1\}$, with entropy

$$S(E_0, E_1) = A\int_{-L/2}^{L/2} \mathrm{d}z\, \sigma(\varepsilon(z)). \tag{4.15}$$

A subscript s will be appended to these when reservoirs are added and it becomes necessary to distinguish between the energy moment of the sub-system and that of the reservoirs.

The local reciprocal temperature is

$$\begin{aligned}
\frac{\partial\sigma(\varepsilon(z))}{\partial\varepsilon(z)} &= \frac{1}{T(z)} \\
&= \frac{1}{T(0)} + z\left.\frac{\mathrm{d}T(z)^{-1}}{\mathrm{d}z}\right|_{z=0} + \mathcal{O}(T'') \\
&\equiv \frac{1}{T_0} + z\frac{1}{T_1}.
\end{aligned} \tag{4.16}$$

Here a Taylor expansion has been performed of the reciprocal temperature to linear order. This defines the zeroth temperature T_0 as the mid temperature, and it defines the first temperature T_1 as the reciprocal of the gradient of the reciprocal temperature. With a change in energy density being $\Delta\varepsilon(z) = \Delta\varepsilon_0 + z\Delta\varepsilon_1$, a change in entropy is therefore

$$\begin{aligned}
\Delta S(E_0, E_1) &= A\int_{-L/2}^{L/2} \mathrm{d}z\, \frac{\partial\sigma(\varepsilon(z))}{\partial\varepsilon(z)} \left[\Delta\varepsilon_0 + z\Delta\varepsilon_1\right] \\
&= A\int_{-L/2}^{L/2} \mathrm{d}z\, \left[\frac{1}{T_0} + z\frac{1}{T_1}\right] \left[\Delta\varepsilon_0 + z\Delta\varepsilon_1\right] \\
&= AL\frac{\Delta\varepsilon_0}{T_0} + \frac{AL^3}{12}\frac{\Delta\varepsilon_1}{T_1} \\
&= \frac{\Delta E_0}{T_0} + \frac{\Delta E_1}{T_1}.
\end{aligned} \tag{4.17}$$

One can see from this that the zeroth and first temperatures are thermodynamically conjugate to the zeroth and first energy moments, respectively,

$$\frac{\partial S(E_0, E_1)}{\partial E_0} = \frac{1}{T_0}, \text{ and } \frac{\partial S(E_0, E_1)}{\partial E_1} = \frac{1}{T_1}. \tag{4.18}$$

Conventional treatments of heat flow are in terms of temperature T and temperature gradient ∇T, rather than the present zeroth and first temperatures. One can compare the present analysis with conventional results by making the identifications, $T_0 \Rightarrow T$ and $T_1^{-1} \Rightarrow -T^{-2}\nabla T$. Both the present and conventional treatments assume small temperature gradients such that only the leading-order terms in the gradient need be retained.

4.3 Second Entropy

For steady heat flow, the primary relevant variables are the first energy moment, E_1, and the first temperature, T_1. Henceforth the zeroth energy moment and the zeroth temperature will not be shown explicitly.

The first energy moment is a non-conserved variable that fluctuates in time about its equilibrium value, $\overline{E}_1 = 0$. This is for an isolated sub-system. The equilibrium value essentially corresponds to a uniform energy profile, or to equal amounts of energy in each half of the sub-system. The first entropy in fluctuation approximation is

$$S(E_1) = \frac{1}{2} S E_1^2, \tag{4.19}$$

with the fluctuation coefficient (for this one-dimensional problem it is a scalar) being

$$S \equiv \left. \frac{\partial^2 S(E_1)}{\partial E_1^2} \right|_{E_1=0} = \frac{-k_{\mathrm{B}}}{\langle E_1(0)^2 \rangle}. \tag{4.20}$$

The averages here and below are canonical equilibrium ones. The first temperature is evidently the thermodynamic force,

$$\frac{1}{T_1} \equiv \frac{\partial S(E_1)}{\partial E_1} = S E_1. \tag{4.21}$$

The second entropy for the transition $E_1 \xrightarrow{\tau} E_1'$ is

$$S^{(2)}(E_1', E_1|\tau) = \frac{A(\tau)}{2} \left[{E_1'}^2 + {E_1}^2 \right] + B(\tau) E_1' E_1. \tag{4.22}$$

The short time expansions of the fluctuation coefficients, Eqs (2.39) and (2.40), are

$$A(\tau) = \frac{-1}{|\tau|\Lambda} + A + \mathcal{O}(\tau), \text{ and } B(\tau) = \frac{1}{|\tau|\Lambda} + B + \mathcal{O}(\tau), \tag{4.23}$$

with the reduction condition, Eq. (2.41), yielding, $A+B=S/2$. The most likely future state is

$$\begin{aligned}\overline{E}_1' &= \frac{-B(\tau)}{A(\tau)} E_1 \\ &= E_1 + \frac{|\tau|\Lambda}{2} S E_1 + \mathcal{O}(\tau^2). \end{aligned} \tag{4.24}$$

With this and the reduction condition the second entropy may be rewritten

$$\begin{aligned} S^{(2)}(E_1', E_1|\tau) &= \frac{1}{2} S E_1^2 - \frac{1}{2\Lambda|\tau|}\left[E_1' - E_1 - \frac{|\tau|\Lambda}{2} S E_1\right]^2 + \mathcal{O}(\tau^2) \\ &= \frac{1}{2} S E_1^2 - \frac{|\tau|}{2\Lambda}\left[\overset{\circ}{E}_1 - \frac{\hat{\tau}\Lambda}{2} S E_1\right]^2, \end{aligned} \tag{4.25}$$

where $\hat{\tau} = \text{sign}(\tau)$. This expansion is valid for small time intervals.

Recall that the coarse velocity is $\overset{\circ}{E}_1 \equiv [E_1' - E_1]/\tau$, and hence its most likely value is $\overline{\overset{\circ}{E}_1} = \hat{\tau}\Lambda S E_1/2$. It is now necessary to relate this rate of change of the first energy moment to the energy flux. The conductive energy flux, J_{E}^0, is the amount of energy per unit area per unit time crossing a plane by conduction. In the present geometry this is in the z-direction. The superscript 0 is appended here to make it clear that this is really the conductive heat flux, which is distinct from energy that may be transported by convection or from energy change due to work being performed. The full import of this distinction will not become clear until Ch. 5, where hydrodynamics is derived.

If the heat flux is uniform, as it is in the steady state, then no change in the energy density occurs in the interior of the system: at each plane, what flows in equals what flows out. Change only occurs at the boundaries of the sub-system, across which no energy may flow in the present isolated case. The nett effect of a uniform flux is to remove an amount of energy per unit time of AJ_{E}^0 from the boundary region at $z=-L/2$, and to add it to the boundary region at $z=L/2$. This changes the first energy moment at a rate of[2]

$$\overset{\circ}{E}_1 = \frac{-L}{2}(-AJ_{\mathrm{E}}^0) + \frac{L}{2}(AJ_{\mathrm{E}}^0) = VJ_{\mathrm{E}}^0, \tag{4.26}$$

where $V = AL$ is the volume of the sub-system. Because the flux is uniform, the energy profile of the interior of the sub-system is unchanging in time. In the steady state, the coarse velocity is the same as the instantaneous velocity.

[2] In mathematical terms, the flux is constant inside the sub-system , $J_{\mathrm{E}}^0(z) = J_{\mathrm{E}}^0$, $|z| < L/2$, and it vanishes beyond the boundaries, $J_{\mathrm{E}}^0(z) = 0$, $|z| > L/2$. Hence its derivative is a sum of δ-functions, $\mathrm{d}J_{\mathrm{E}}^0(z)/\mathrm{d}z = J_{\mathrm{E}}^0\,[\delta(z+L/2) - \delta(z-L/2)]$. The rate of change of energy density is the divergence of the flux, $\dot{\varepsilon}(z) = -\mathrm{d}J_{\mathrm{E}}^0(z)/\mathrm{d}z$. Hence the rate of change of the first energy moment is $\dot{E}_1 = A\int \mathrm{d}z\, z\dot{\varepsilon}(z) = ALJ_{\mathrm{E}}^0$.

Using this result the most likely heat flux is

$$\begin{aligned}\overline{J}_{\mathrm{E}}^{0} &= \frac{\hat{\tau}\Lambda}{2V} S E_1 \\ &= \frac{\hat{\tau}\Lambda}{2V}\frac{1}{T_1} \\ &= \frac{-\hat{\tau}\Lambda}{2VT^2}\nabla T. \end{aligned} \tag{4.27}$$

For fluxes forward in time, $\hat{\tau} = 1$, this is Fourier's law, Eq. (4.1), with the thermal conductivity identified as $\lambda \equiv \Lambda/2VT^2$.

4.4 Thermal Conductivity and Energy Correlations

The fluctuations and the regression theorem were derived in the small time limit in §2.3. In this limit the time correlation function decays linearly in time. For the present problem of steady heat flow, the thermal conductivity, $\lambda \equiv \Lambda/2VT^2$, can be obtained from the slope of the time correlation function in this small time limit. There are several notionally equivalent expressions. Multiplying both sides of the regression result, Eq. (4.24), by E_1, taking the average, and rearranging one has

$$\lambda = \frac{-1}{Vk_{\mathrm{B}}T^2|\tau|}\left\langle [E_1(t+\tau) - E_1(t)]E_1(t)\right\rangle + \mathcal{O}(\tau). \tag{4.28}$$

The right-hand side is an even function of τ. The left-hand side ought to be independent of the time interval, and so the explicit dependence on τ on the right-hand ought to cancel, which can be used as a guide to the small time regime. If one multiplies both sides by τ and differentiates with respect to τ, this may alternatively be formulated as

$$\lambda = \frac{-\hat{\tau}}{Vk_{\mathrm{B}}T^2}\left\langle \dot{E}_1(t+\tau)E_1(t)\right\rangle + \mathcal{O}(\tau), \tag{4.29}$$

where $\hat{\tau} \equiv \mathrm{sign}(\tau)$. The right-hand side is still an even function of τ.

The averages that appear here are canonical equilibrium averages. This means that one generates points in the sub-system phase space, $\mathbf{\Gamma}_0$, according to the Maxwell-Boltzmann distribution, and for each such point one computes the trajectory of the isolated sub-system according to Hamilton's equations of motion, $\mathbf{\Gamma}^0(t|\mathbf{\Gamma}_0)$, $t \in [0,\tau]$. To be specific, Eq. (4.28) can be written in Green-Kubo form,

$$\lambda = \frac{-1}{Vk_{\mathrm{B}}T^2|\tau|}\int \mathrm{d}\mathbf{\Gamma}_0\, \wp_{\mathrm{MB}}(\mathbf{\Gamma}_0|T_0)\int_0^{\tau} \mathrm{d}t\, \dot{\hat{E}}_1(\mathbf{\Gamma}^0(t|\mathbf{\Gamma}_0))\hat{E}_1(\mathbf{\Gamma}_0). \tag{4.30}$$

Obviously the shorter the time interval required for each trajectory, the more efficient will be the simulation.

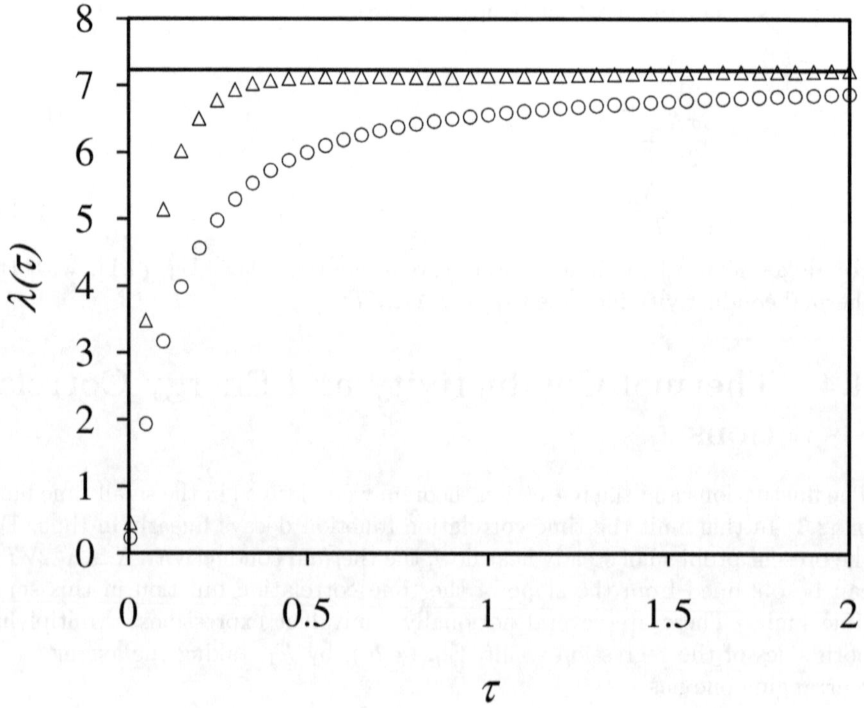

Figure 4.2: The thermal conductivity, $\lambda(\tau)$ (in units of the Lennard-Jones time and size parameter, $k_B/\tau_{LJ}\sigma_{LJ}$), at $\rho = 0.8$ and $T_0 = 2$. The symbols are simulation data, with the triangles using the instantaneous velocity, Eq. (4.29), and the circles using the coarse velocity over the interval, Eq. (4.28). The solid line is a guide to the eye.

Typical results of such a simulation are shown in Fig. 4.2. It can be seen that the terminal velocity conductivity, Eq. (4.29), reaches it short time asymptote faster than the coarse velocity conductivity, Eq. (4.28). The latter is a smoother curve, due to the fact that it integrates the former over time. However, one should not confuse smoothness with statistical error; smoothness only means that adjacent points are highly correlated. In fact, the statistical error, calculated on the basis of the fluctuations in the conductivity at a given time in independent runs, is about six times smaller for Eq. (4.29) compared with Eq. (4.28).

4.5 Reservoirs

In the preceding two sections, results for an isolated sub-system were established, including the rate of decay of spontaneous fluctuations in the energy moment. Now the treatment is extended to the case that the sub-system can exchange energy with two reservoirs of different temperatures, and it is shown that these reservoirs induce a non-zero energy moment, and that the flux of energy between the reservoirs through the sub-system is the same as the regression of a spontaneous fluctuation in an isolated system whose original magnitude is equal to the induced moment.

4.5.1 First Entropy

The boundaries between the two reservoirs and the sub-system are located at $z = \pm L/2$, and the reservoirs have respective temperatures, $T_{r\pm}$. Conduction alone is treated here (i.e. no convection), so gravity plays no rôle, and it does not matter which reservoir is the hotter. Suppose at each boundary the sub-system exchanges energy with the respective reservoir such that $\Delta E_{r\pm}$ represents the change in energy of that reservoir. In this case the change in the total reservoir entropy is

$$\begin{aligned} \Delta S_r &= \frac{\Delta E_{r+}}{T_{r+}} + \frac{\Delta E_{r-}}{T_{r-}} \\ &= \frac{\Delta E_{r0}}{T_{r0}} + \frac{\Delta E_{r1}}{T_{r1}}. \end{aligned} \tag{4.31}$$

Here has been defined the change in the zeroth and first energy moments of the reservoirs,

$$\Delta E_{r0} \equiv \Delta E_{r+} + \Delta E_{r-}, \text{ and } \Delta E_{r1} \equiv \frac{L}{2}\Delta E_{r+} - \frac{L}{2}\Delta E_{r-}, \tag{4.32}$$

and the zeroth and first reservoir temperatures,

$$\frac{1}{T_{r0}} \equiv \frac{1}{2}\left[\frac{1}{T_{r+}} + \frac{1}{T_{r-}}\right], \text{ and } \frac{1}{T_{r1}} \equiv \frac{1}{L}\left[\frac{1}{T_{r+}} - \frac{1}{T_{r-}}\right]. \tag{4.33}$$

There is an obvious analogy between these definitions of the reservoir moments and temperatures and those of the isolated sub-system given above. The definitions of the energy moments and temperatures ensure that the two forms for the change in the reservoir entropy are equivalent; no approximation has been introduced here. The first form for the change in the reservoir entropy implicitly assumes that the reservoir temperatures $T_{r\pm}$ are those at the place where the energy exchange occurs, which will be true if the boundary region is small compared with any gradient in the reservoir temperature. In practice, one generally assumes that the reservoir conductivity is infinite, so that the reservoir temperature itself is the same as the reservoir temperature at the boundary.

From energy conservation, one must have that the change in the sub-system energy at each boundary due to exchange with the reservoir must be equal and opposite to the change in the respective reservoir energy, $\Delta^r E_{s\pm} = -\Delta E_{r\pm}$. In view of the definitions above, this is equivalent to

$$\Delta^r E_{s0} = -\Delta E_{r0}, \text{ and } \Delta^r E_{s1} = -\Delta E_{r1}. \tag{4.34}$$

The reservoir-induced change in the sub-system energy moments is just

$$\Delta^{\mathrm{r}} E_{\mathrm{s}0} \equiv \Delta^{\mathrm{r}} E_{\mathrm{s}+} + \Delta^{\mathrm{r}} E_{\mathrm{s}-}, \text{ and } \Delta^{\mathrm{r}} E_{\mathrm{s}1} \equiv \frac{L}{2}\Delta^{\mathrm{r}} E_{\mathrm{s}+} - \frac{L}{2}\Delta^{\mathrm{r}} E_{\mathrm{s}-}. \tag{4.35}$$

This assumes that the boundary region in the sub-system where energy exchange with the reservoir occurs is small compared with the size of the sub-system itself.

The total change in the energy moment of the sub-system is the sum of the adiabatic change[3] (i.e. the internal change that would occur if the sub-system were isolated from the reservoirs) and the reservoir-induced change,

$$\Delta E_{\mathrm{s}0} = \Delta^0 E_{\mathrm{s}0} + \Delta^{\mathrm{r}} E_{\mathrm{s}0}, \text{ and } \Delta E_{\mathrm{s}1} = \Delta^0 E_{\mathrm{s}1} + \Delta^{\mathrm{r}} E_{\mathrm{s}1}. \tag{4.36}$$

Since the total energy is conserved in an isolated system (i.e. a system that evolves adiabatically under Hamilton's equations of motion), one has $\Delta^0 E_{\mathrm{s}0} = 0$.

From the change in the reservoir entropy given above, it is convenient to write the reservoir entropy as

$$S_{\mathrm{r}}(t) = \frac{E_{\mathrm{r}0}(t)}{T_{\mathrm{r}0}} + \frac{E_{\mathrm{r}1}(t)}{T_{\mathrm{r}1}}. \tag{4.37}$$

Here has been defined the part of the reservoir energy moment that depends upon the sub-system (i.e. the constant part of the reservoir energy that is independent of the sub-system is ignored). Although this depends upon time, due to the energy flux through the sub-system, it is assumed that the reservoirs are so large that over the time course of any experiment the reservoir temperature does not significantly change. These reservoir moments have the property that they change according to

$$\Delta E_{\mathrm{r}0}(t) = \Delta E_{\mathrm{r}0} = -\Delta^{\mathrm{r}} E_{\mathrm{s}0}, \tag{4.38}$$

and

$$\Delta E_{\mathrm{r}1}(t) = \Delta E_{\mathrm{r}1} = -\Delta^{\mathrm{r}} E_{\mathrm{s}1}. \tag{4.39}$$

With these, the total entropy of the system is

$$S_{\mathrm{total}}(E_{\mathrm{s}0}, E_{\mathrm{s}1}|T_{\mathrm{r}0}, T_{\mathrm{r}1}) = S_{\mathrm{s}}(E_{\mathrm{s}0}, E_{\mathrm{s}1}) + \frac{E_{\mathrm{r}0}(t)}{T_{\mathrm{r}0}} + \frac{E_{\mathrm{r}1}(t)}{T_{\mathrm{r}1}}. \tag{4.40}$$

The sub-system entropy may be written as a quadratic form in the first energy moment, since one is generally working in the regime of small temperature gradients.

The reservoir-induced derivatives of the total entropy are,

$$\frac{\partial S_{\mathrm{total}}}{\partial^{\mathrm{r}} E_{\mathrm{s}0}} = \frac{1}{T_{\mathrm{s}0}} - \frac{1}{T_{\mathrm{r}0}}, \tag{4.41}$$

[3]Adiabatic here and throughout means the time development of an isolated system according to Hamilton's equations of motion. This precludes thermostats, dissipative terms, artificial forces, or any other term that does not form part of the Hamiltonian of the system.

and

$$\frac{\partial S_{\text{total}}}{\partial^{\text{r}} E_{\text{s}1}} = \frac{1}{T_{\text{s}1}} - \frac{1}{T_{\text{r}1}}. \tag{4.42}$$

One sees from these that the entropy is a maximum with respect to the reservoir-induced changes when the zeroth and first temperatures of the sub-system equal those of the reservoir,

$$\overline{T}_{\text{s}0} = T_{\text{r}0}, \text{ and } \overline{T}_{\text{s}1} = T_{\text{r}1}. \tag{4.43}$$

For this non-equilibrium system this most likely state corresponds to the steady state. The result could probably have been anticipated, since it simply says that the temperature of the sub-system at each boundary is equal to that of the reservoir at that boundary.

Note that only the zeroth and first reservoir temperatures entered the analysis, and only the reservoir-induced changes in the zero and first energy moments needed to be considered. Formally, the higher-order reservoir temperatures are zero. This means that the only effect of maximising the total entropy with respect to the changes in the higher-order sub-system energy moments is to establish that the most likely values of the higher-order sub-system temperatures are zero. This means that the most likely values of the higher-order sub-system energy moments are also zero. In other words, the zeroth and first temperatures are both necessary and sufficient to characterise the steady state in the present slab geometry.

Note that one does not maximise the ordinary total entropy with respect to the adiabatic changes, Δ^0. As will be seen next, one instead performs the adiabatic maximisation of the second entropy to obtain the most likely heat flux.

4.5.2 Second Entropy

Now the second entropy of the sub-system in contact with the two thermal reservoirs will be given. The dependence on the zeroth energy moment and zeroth temperature will be suppressed, since $\Delta^0 E_0 = 0$, and also only the adiabatic change in the first moment is required for the heat flux.

In the preceding section the notation Δ^0 and Δ^{r} (and also ∂^{r}) denoted a virtual change in the constrained moment that was subsequently used to optimise the moment. In this section, the notation Δ^0_τ and Δ^{r}_τ will denote a physical change in the moment over the time interval τ. These are essentially the adiabatic and the reservoir-induced constrained fluxes, and virtual changes in them, $\partial\Delta^0_\tau$ and $\partial\Delta^{\text{r}}_\tau$, will be used to optimise them.

The second entropy for an isolated system in the small time limit, Eq. (2.50), consists of four terms: the first entropy of the sub-system , a term quadratic in the flux, a term bilinear in the flux and the thermodynamic force, and a term independent of the fluxes that enforces the reduction condition for the most likely flux. The flux referred to is the internal or adiabatic flux (i.e. for an isolated sub-system).

When reservoirs are added, the only changes to this expression are that the first entropy of the sub-system is replaced by the first entropy of the total system, and the internal rate of entropy production is replaced by the total rate of entropy production. The term quadratic in the internal flux, which represents the cost of dynamic order, remains unchanged. The term independent of the fluxes is chosen to ensure that the total second entropy for the optimum fluxes reduces to the total ordinary entropy. Making these changes gives

$$\begin{aligned} S^{(2)}_{\text{total}}&(\Delta^0_\tau E_{\text{s}1}, \Delta^{\text{r}}_\tau E_{\text{s}1}, E_{\text{s}1}|T_{\text{r}1}, \tau) \\ &= \frac{1}{2} S E_{\text{s}1}^2 + \frac{E_{\text{r}1}(t)}{T_{\text{r}1}} - \frac{1}{2\Lambda|\tau|}[\Delta^0_\tau E_{\text{s}1}]^2 + \frac{1}{2}\Delta^0_\tau E_{\text{s}1} S E_{\text{s}1} \\ &\quad + \frac{1}{2}\Delta^{\text{r}}_\tau E_{\text{s}1}\left[S E_{\text{s}1} - \frac{1}{T_{\text{r}1}}\right] + \frac{|\tau|\Lambda}{8}\left[S^2 E_{\text{s}1}^2 - \frac{2 S E_{\text{s}1}}{T_{\text{r}1}}\right]. \end{aligned} \tag{4.44}$$

Terms $\mathcal{O}(\tau^2)$ have been neglected here. The sub-system thermodynamic force is $T_{\text{s}1}^{-1} = S E_{\text{s}1}$. The first two terms on the right-hand side sum to the total ordinary entropy. The third term is the term quadratic in the internal flux. The fourth term is the entropy production due to the internal change. The fifth term is the entropy production due to exchange with the reservoir. The final term is the constant (i.e. the term independent of the fluxes) required to satisfy the reduction condition for the second entropy evaluated at the most likely fluxes.

The derivatives are

$$\frac{\partial S^{(2)}_{\text{total}}}{\partial \Delta^0_\tau E_{\text{s}1}} = \frac{-1}{\Lambda|\tau|}\Delta^0_\tau E_{\text{s}1} + \frac{1}{2} S E_{\text{s}1}, \tag{4.45}$$

$$\frac{\partial S^{(2)}_{\text{total}}}{\partial \Delta^{\text{r}}_\tau E_{\text{s}1}} = \frac{1}{2}\left[S E_{\text{s}1} - \frac{1}{T_{\text{r}1}}\right], \tag{4.46}$$

and[4]

$$\frac{\partial S^{(2)}_{\text{total}}}{\partial^{\text{r}} E_{\text{s}1}} = S E_{\text{s}1} - \frac{1}{T_{\text{r}1}} + \frac{1}{2} S\left[\Delta^0_\tau E_{\text{s}1} + \Delta^{\text{r}}_\tau E_{\text{s}1}\right] + \frac{|\tau|\Lambda}{8}\left[2S^2 E_{\text{s}1} - \frac{2S}{T_{\text{r}1}}\right]. \tag{4.47}$$

Setting the first of these to zero yields the most likely internal flux,

$$\overline{\overset{\circ}{E}{}_1^0} \equiv \frac{\overline{\Delta^0_\tau E_{\text{s}1}}}{\tau} = \frac{\hat{\tau}\Lambda}{2}\frac{1}{T_{\text{s}1}}. \tag{4.48}$$

This expression for the most likely flux is the same as for an isolated system, namely that the flux is proportional to the gradient of the sub-system temperature. Setting the second to zero shows that the most likely sub-system temperature gradient is equal to that applied by the reservoirs,

$$\overline{T_{\text{s}1}} = T_{\text{r}1}. \tag{4.49}$$

[4] In order to determine the optimum initial sub-system moment $E_{\text{s}1}$, one only takes the reservoir-induced derivative, $\partial^{\text{r}} E_{\text{s}1}$, without considering the internal or adiabatic derivative $\partial^0 E_{\text{s}1}$. This follows from the assumption that the changes in the sub-system moment are dominated by exchange with the reservoir, which is equivalent to the assumption that the conductivity of the reservoir is much greater than the conductivity of the sub-system.

Setting the third to zero shows that the internal change in the sub-system energy moment is equal and opposite to the reservoir-induced change,

$$\overline{\Delta_\tau^0 E_{s1}} = -\overline{\Delta_\tau^r E_{s1}}. \tag{4.50}$$

This result holds if, and only if, the sub-system first temperature is already in its optimum state, $\overline{T_{s1}} = T_{r1}$.

This last result means that the change in the moment due to heat exchange with the reservoir most likely cancels the relaxation of the energy moment due to internal processes, so that once the energy moment has reached its preferred value, $\overline{T_{s1}} = T_{r1}$, then the total change in moment is zero, $\overline{\Delta_\tau E_{s1}} = 0$. Whereas in the isolated sub-system the moment relaxes to zero, here this internal change is cancelled by the exchange with the reservoir, which has the nett effect of holding the sub-system moment constant.

Although the total change in energy moment is zero, it consists of two non-zero contributions: the adiabatic change and the equal and opposite reservoir-induced change. By energy conservation, the latter is itself equal and opposite to the change in the reservoir moment,

$$\overline{\Delta_\tau^0 E_{s1}} = -\overline{\Delta_\tau^r E_{s1}} = \overline{\Delta_\tau E_{r1}}. \tag{4.51}$$

This means that even though the structure of the sub-system is unchanging in time, there is nevertheless a steady heat flow between the reservoirs so that their energy distribution is steadily changing. This says that in the steady state, the heat flow between the reservoirs is equal to the adiabatic flux of an isolated sub-system with the same spontaneous temperature gradient as induced by the reservoirs. This result is one of the notions embodied by Onsager's regression hypothesis.

The second entropy evaluated for the most likely changes is

$$\begin{aligned}
& S^{(2)}_{\text{total}}(\overline{\Delta_\tau^0 E_{s1}}, \overline{\Delta_\tau^r E_{s1}}, E_{s1}|T_{r1}, \tau) \\
&= S_{\text{total}}(E_{s1}) - \frac{|\tau|\Lambda}{8}\frac{1}{T_{s1}^2} + \frac{|\tau|\Lambda}{4}\frac{1}{T_{r1}T_{s1}} + \frac{|\tau|\Lambda}{8}\left[\frac{1}{T_{s1}^2} - \frac{2}{T_{r1}T_{s1}}\right] \\
&= S_{\text{total}}(E_{s1}).
\end{aligned} \tag{4.52}$$

This is the required reduction condition.

A clarification on the notation might be useful. The internal flux was here denoted $\overset{\circ}{E}{}_1^0 \equiv \Delta^0 E_{s1}/\tau$, where the superscript 0 means internal changes (i.e. those in the absence of the reservoir). However, in the preceding section the conductive flux was introduced, which would be related to the present internal coarse velocity by $J_{\text{E}}^0 = \overset{\circ}{E}{}_1^0/V$. The superscript 0 on the left-hand side means the conductive energy flux. The notation is actually consistent in the sense that in the present analysis, number fluxes and momentum fluxes were precluded, and so the internal changes can only occur by conduction, not by convection or by the performance of work. The matter will be revisited in the next chapter, where the equations of hydrodynamics are derived.

4.6 Heat and Number Flow

In Eq. (4.27), the most likely conductive heat flux was shown to be[5]

$$\overline{J}_{\mathrm{E}}^{0} = \frac{\Lambda_{\mathrm{EE}}}{2V} \nabla \frac{1}{T}. \tag{4.53}$$

(In this and the following section, ∇T^{-1} and $\nabla(\mu/T)$ will be used instead $1/\mathbf{T}_1$ and $\mathbf{m}_1$.) From Fourier's law, Eq. (4.1), the thermal conductivity is identified as $\lambda \equiv \Lambda_{\mathrm{EE}}/2VT^2$. The temperature gradient provides the thermodynamic driving force. This result comes from maximising the second entropy, $S^{(2)}(E_1'; E_1|\tau)$.

It is of interest to discuss the combination of heat flow and number flow. The number flux, like the energy flux, is just the rate of change of the first number moment, $J_{\mathrm{N}}^0 = \dot{N}_1/V$. As will be discussed in the following chapter, due to momentum conservation the first number moment is conserved for a pure system. In order to avoid this restriction, in the present case number refers to a mobile species able to move in a fixed background lattice so that its momentum is not conserved. Hence N_1 fluctuates, and, in the language of the following chapter, J_{N}^0 is the diffusive part of the number flux, which is actually the total number flux when only one species is mobile.

The second entropy may be formulated as a quadratic form for both variables, $S^{(2)}(E_1', N_1'; E_1, N_1|\tau)$. Maximising this the most likely fluxes are

$$\begin{aligned} \overline{\mathbf{J}}_{\mathrm{E}}^{0} &= \frac{\Lambda_{\mathrm{EE}}^*}{2V} \nabla \frac{1}{T} - \frac{\Lambda_{\mathrm{EN}}^*}{2V} \nabla \frac{\mu}{T}, \\ \overline{\mathbf{J}}_{\mathrm{N}}^{0} &= \frac{\Lambda_{\mathrm{NE}}^*}{2V} \nabla \frac{1}{T} - \frac{\Lambda_{\mathrm{NN}}^*}{2V} \nabla \frac{\mu}{T}. \end{aligned} \tag{4.54}$$

Here the thermodynamic driving force for number is

$$-\mathbf{m}_1 \equiv -\nabla \frac{\mu}{T} = \frac{\partial S^{(1)}(E_1, N_1)}{\partial N_1}. \tag{4.55}$$

The cross transport coefficient, $\Lambda_{\mathrm{NE}}^* = \Lambda_{\mathrm{EN}}^*$, is related to the Soret coefficient and it gives the diffusion in response to a temperature gradient (thermodiffusion). The inverse effect, the heat flux due to a chemical potential gradient, is called the Dufour effect.

Using subscripts to denote the derivatives of the second entropy, $S_{\mathrm{EE}}^{(2)} = \lim_{\tau\to 0} |\tau| \partial^2 S^{(2)}(E_1'; E_1|\tau)/\partial E_1'^2 \big|_{E_1'=E_1=0}$, etc., one has

$$\Lambda_{\mathrm{EE}} = -[S_{\mathrm{EE}}^{(2)}]^{-1}, \tag{4.56}$$

and, from the general expression $\underline{\underline{\Lambda}}^* = -[\underline{\underline{S}}^{(2)}]^{-1}$,

$$\Lambda_{\mathrm{EE}}^* = -\left\{ \begin{pmatrix} S_{\mathrm{EE}}^{(2)} & S_{\mathrm{EN}}^{(2)} \\ S_{\mathrm{NE}}^{(2)} & S_{\mathrm{NN}}^{(2)} \end{pmatrix}^{-1} \right\}_{\mathrm{EE}} = \frac{-S_{\mathrm{NN}}^{(2)}}{S_{\mathrm{EE}}^{(2)} S_{\mathrm{NN}}^{(2)} - S_{\mathrm{EN}}^{(2)} S_{\mathrm{NE}}^{(2)}}. \tag{4.57}$$

[5]In this section and the next, it is always the most likely future flux , $\hat{\tau} = 1$.

These two transport coefficients are not equal to each other, $\Lambda_{\rm EE} \neq \Lambda^*_{\rm EE}$. Note that because $S^{(2)}(E_1'; E_1|\tau) = S^{(2)}(E_1', N_1' = 0; E_1, N_1 = 0|\tau)$, the $S^{(2)}_{\rm EE}$ that appears here is the same in both equations.

In the case that there is no number flux, $\overline{\mathbf{J}}^0_{\rm N} = 0$, Eq. (4.54) shows that the thermodynamic force for number is

$$\nabla \frac{\mu}{T} = \frac{\Lambda^*_{\rm NE}}{\Lambda^*_{\rm NN}} \nabla \frac{1}{T}. \tag{4.58}$$

In this case the energy flux is

$$\begin{aligned} \overline{\mathbf{J}}^0_{\rm E} &= \frac{\Lambda^*_{\rm EE}}{2V} \nabla \frac{1}{T} - \frac{\Lambda^*_{\rm EN}}{2V} \frac{\Lambda^*_{\rm NE}}{\Lambda^*_{\rm NN}} \nabla \frac{1}{T} \\ &= \frac{1}{2V} \left[\left\{ \left(\begin{array}{cc} \Lambda^*_{\rm EE} & \Lambda^*_{\rm EN} \\ \Lambda^*_{\rm NE} & \Lambda^*_{\rm NN} \end{array} \right)^{-1} \right\}_{\rm EE} \right]^{-1} \nabla \frac{1}{T} \\ &= \frac{-1}{2V} \left[S^{(2)}_{\rm EE} \right]^{-1} \nabla \frac{1}{T} \\ &= \frac{\Lambda_{\rm EE}}{2V} \nabla \frac{1}{T}, \; \overline{\mathbf{J}}^0_{\rm N} = \mathbf{0}. \end{aligned} \tag{4.59}$$

This is equal to the original conductive energy flux, Eq. (4.53), which was obtained when the number flux was ignored. This analysis shows that $\Lambda_{\rm EE} = \Lambda^*_{\rm EE} - \Lambda^*_{\rm EN}[\Lambda^*_{\rm NN}]^{-1}\Lambda^*_{\rm NE}$.

4.7 Heat and Current Flow

Consider the case that the mobile species treated in the preceding section has charge q, and that an electric potential $\phi(\mathbf{r}, t)$ is applied. In the most common case the potential corresponds to a constant and uniform electric field, $\phi(\mathbf{r}, t) = -E_{\rm el} z$. As discussed in the first section of the following chapter, §5.1, the internal energy density is $\varepsilon^{\rm int}(\mathbf{r}, t) = \varepsilon(\mathbf{r}, t) - qn(\mathbf{r}, t)\phi(\mathbf{r}, t)$, where $n(\mathbf{r}, t)$ is the number density, and the kinetic energy of the convective flow of the mobile species has been neglected as it is quadratic in the velocity.[6] The total energy flux is $\mathbf{J}_{\rm E} = \mathbf{J}^0_{\rm E} + \mathbf{J}^\psi_{\rm E} = \mathbf{J}^0_{\rm E} + q\phi \mathbf{J}_{\rm N}$, where the convective velocity has been neglected. The conductive energy or heat flux, $\mathbf{J}^0_{\rm E}$, was analysed above.

The rate of change of the entropy of the sub-system is derived in Eq. (5.43). Neglecting the viscous pressure tensor, chemical reaction rate, and the surface

[6]It is assumed that the mass of the fixed species is much greater than that of the mobile species, $m_0 \gg m_1$, from which it follows that the convective or barycentric velocity is much less than the velocity of the mobile species, $\mathbf{v} \ll \mathbf{v}_1$. This means that for the mobile species, the diffusive velocity is equal to the total velocity, $\tilde{\mathbf{v}}_1 = \mathbf{v}_1$, and that the diffusive and total number fluxes are the same, $\mathbf{J}^0_N = \mathbf{J}_N = n_1 \mathbf{v}_1$. In the present problem with only one mobile species, the subscript 1 will be dropped, and the velocity of the mobile species will not appear explicitly (instead $n^{-1}\mathbf{J}_N$ will be used). The formulae in Ch. 5 will be applied to the current problem by setting the convective velocity to zero, $\mathbf{v} = \mathbf{0}$, wherever it occurs in them.

integrals, for the present isolated system it is[7]

$$\begin{aligned}
\dot{S}_{\rm s} &= \int \mathrm{d}\mathbf{r} \left\{ \mathbf{J}_{\rm E}^0 \cdot \nabla \frac{1}{T} - \mathbf{J}_{\rm N} \cdot \left[\frac{q\nabla\phi}{T} + \nabla \frac{\mu}{T} \right] \right\} \\
&= \int \mathrm{d}\mathbf{r} \left\{ \left[\mathbf{J}_{\rm E}^0 - \mu \mathbf{J}_{\rm N} \right] \cdot \nabla \frac{1}{T} - \frac{1}{T} \mathbf{J}_{\rm N} \cdot \nabla [\mu + q\phi] \right\} \\
&= \int \mathrm{d}\mathbf{r} \left\{ \left[\mathbf{J}_{\rm E}^0 + q\phi \mathbf{J}_{\rm N} \right] \cdot \nabla \frac{1}{T} - \mathbf{J}_{\rm N} \cdot \nabla \frac{\mu + q\phi}{T} \right\}. \qquad (4.60)
\end{aligned}$$

These three expressions allow one to define the fluxes and conjugate thermodynamic forces as convenient.

Choosing the first equality, with fluxes $\{\mathbf{J}_{\rm E}^0, \mathbf{J}_{\rm N}\}$ and thermodynamic forces $\{\nabla T^{-1}, -qT^{-1}\nabla\phi - \nabla(\mu/T)\} = \{\mathbf{T}_1^{-1}, -qT^{-1}\nabla\phi - \mathbf{m}_1\}$, the most likely fluxes that result from maximising the second entropy are

$$\begin{aligned}
\overline{\mathbf{J}}_{\rm E}^0 &= \frac{\Lambda_{\rm EE}^*}{2V} \nabla \frac{1}{T} - \frac{\Lambda_{\rm EN}^*}{2V} \left[\frac{q\nabla\phi}{T} + \nabla \frac{\mu}{T} \right], \\
\overline{\mathbf{J}}_{\rm N} &= \frac{\Lambda_{\rm NE}^*}{2V} \nabla \frac{1}{T} - \frac{\Lambda_{\rm NN}^*}{2V} \left[\frac{q\nabla\phi}{T} + \nabla \frac{\mu}{T} \right]. \qquad (4.61)
\end{aligned}$$

These equation can either be interpreted as applying to the z-component, or else as vector equations with the Λ being a scalar times the 3×3 identity matrix. Recall the Onsager reciprocal relation, $\Lambda_{\rm NE}^* = \Lambda_{\rm EN}^*$. Evidently, since $\underline{\underline{\Lambda}}^*$ is positive definite, the most likely dissipation is positive, $\overline{\dot{S}}_{\rm s} \geq 0$.

Alternatively, from the second equality in the dissipation, Eq. (4.60), one can define $\mathbf{J}_{\rm E}^\ddagger \equiv \mathbf{J}_{\rm E}^0 - \mu \mathbf{J}_{\rm N}$. It will be shown following Eq. (4.86) below that this is closely related to the entropy flux, $\mathbf{J}_{\rm E}^\ddagger = T\mathbf{J}_{\rm S} \equiv T[(\partial\sigma/\partial\varepsilon)\mathbf{J}_{\rm E} + (\partial\sigma/\partial n)\mathbf{J}_{\rm N}]$.[8] The most likely fluxes can then be written

$$\begin{aligned}
\overline{\mathbf{J}}_{\rm E}^\ddagger &= \frac{\Lambda_{\rm EE}^\ddagger}{2V} \nabla \frac{1}{T} - \frac{\Lambda_{\rm EN}^\ddagger}{2V} \frac{\nabla [\mu + q\phi]}{T}, \\
\overline{\mathbf{J}}_{\rm N} &= \frac{\Lambda_{\rm NE}^\ddagger}{2V} \nabla \frac{1}{T} - \frac{\Lambda_{\rm NN}^\ddagger}{2V} \frac{\nabla [\mu + q\phi]}{T}, \qquad (4.62)
\end{aligned}$$

---

[7] The expression for the dissipation used here and derived in the next chapter does not agree with the expressions given by de Groot, S. R. and Mazur, P. (1984), *Non-equilibrium Thermodynamics*, (Dover, New York).

[8] The present $\mathbf{J}_{\rm E}^\ddagger$ is identical to the $\mathbf{J}_{\rm Q}^*$ given as Eq. (2B.2.1) in Pottier, N. (2010), *Non-equilibrium Statistical Physics: Linear Irreversible Processes*, (Oxford University Press, Oxford). Pottier's expressions for the entropy source, Eq. (2B.1.3) and Eq. (2B.2.3), are the same as the third and the second equalities in the present Eq. (4.60), respectively. Pottier's interpretation of the coefficient of $\mathbf{J}_{\rm N}$ in Eqs (2B.2.2) and (2B.6.6) as the entropy per particle is not strictly correct, because the entropy is not an extensive function solely of the number.

Rearranging this in the previous form one has

$$\begin{aligned}
\overline{\mathbf{J}}_{\mathrm{E}}^{0} &= \frac{1}{2V}\left[\Lambda_{\mathrm{EE}}^{\ddagger}+\mu\Lambda_{\mathrm{EN}}^{\ddagger}\right]\nabla\frac{1}{T}-\frac{\Lambda_{\mathrm{EN}}^{\ddagger}}{2V}\left[\frac{q\nabla\phi}{T}+\nabla\frac{\mu}{T}\right]+\mu\overline{\mathbf{J}}_{\mathrm{N}} \\
&= \frac{1}{2V}\left[\Lambda_{\mathrm{EE}}^{\ddagger}+2\mu\Lambda_{\mathrm{EN}}^{\ddagger}+\mu^2\Lambda_{\mathrm{NN}}^{\ddagger}\right]\nabla\frac{1}{T} \\
&\quad -\frac{1}{2V}\left[\Lambda_{\mathrm{EN}}^{\ddagger}+\mu\Lambda_{\mathrm{NN}}^{\ddagger}\right]\left[\frac{q\nabla\phi}{T}+\nabla\frac{\mu}{T}\right] \\
\overline{\mathbf{J}}_{\mathrm{N}} &= \frac{1}{2V}\left[\Lambda_{\mathrm{NE}}^{\ddagger}+\mu\Lambda_{\mathrm{NN}}^{\ddagger}\right]\nabla\frac{1}{T}-\frac{\Lambda_{\mathrm{NN}}^{\ddagger}}{2V}\left[\frac{q\nabla\phi}{T}+\nabla\frac{\mu}{T}\right].
\end{aligned} \tag{4.63}$$

Hence one can identify

$$\Lambda_{\mathrm{EE}}^{*}=\Lambda_{\mathrm{EE}}^{\ddagger}+2\mu\Lambda_{\mathrm{EN}}^{\ddagger}+\mu^2\Lambda_{\mathrm{NN}}^{\dagger},\quad \Lambda_{\mathrm{EN}}^{*}=\Lambda_{\mathrm{EN}}^{\ddagger}+\mu\Lambda_{\mathrm{NN}}^{\ddagger}, \tag{4.64}$$

and $\Lambda_{\mathrm{NN}}^{*}=\Lambda_{\mathrm{NN}}^{\ddagger}$. It is worth mentioning that these relationships can be used to show that the formula for the thermal conductivity also holds for these coefficients, $\Lambda_{\mathrm{EE}}=\Lambda_{\mathrm{EE}}^{\ddagger}-\Lambda_{\mathrm{EN}}^{\ddagger}[\Lambda_{\mathrm{NN}}^{\ddagger}]^{-1}\Lambda_{\mathrm{NE}}^{\ddagger}$.

Finally, one sees that the third equality in the dissipation, Eq. (4.60), involves the total total energy flux, $\mathbf{J}_{\mathrm{E}}\equiv\mathbf{J}_{\mathrm{E}}^{0}+q\phi\mathbf{J}_{\mathrm{N}}$. The most likely fluxes in this case are

$$\begin{aligned}
\overline{\mathbf{J}}_{\mathrm{E}} &= \frac{\Lambda_{\mathrm{EE}}^{\dagger}}{2V}\nabla\frac{1}{T}-\frac{\Lambda_{\mathrm{EN}}^{\dagger}}{2V}\nabla\frac{\mu+q\phi}{T}, \\
\overline{\mathbf{J}}_{\mathrm{N}} &= \frac{\Lambda_{\mathrm{NE}}^{\dagger}}{2V}\nabla\frac{1}{T}-\frac{\Lambda_{\mathrm{NN}}^{\dagger}}{2V}\nabla\frac{\mu+q\phi}{T}.
\end{aligned} \tag{4.65}$$

Rearranging this in the form of the original fluxes one has

$$\begin{aligned}
\overline{\mathbf{J}}_{\mathrm{E}}^{0} &= \frac{1}{2V}\left[\Lambda_{\mathrm{EE}}^{\dagger}-q\phi\Lambda_{\mathrm{EN}}^{\dagger}\right]\nabla\frac{1}{T}-\frac{\Lambda_{\mathrm{EN}}^{\dagger}}{2V}\left[\frac{q\nabla\phi}{T}+\nabla\frac{\mu}{T}\right]-q\phi\overline{\mathbf{J}}_{\mathrm{N}} \\
&= \frac{1}{2V}\left[\Lambda_{\mathrm{EE}}^{\dagger}-2q\phi\Lambda_{\mathrm{EN}}^{\dagger}+q^2\phi^2\Lambda_{\mathrm{NN}}^{\dagger}\right]\nabla\frac{1}{T} \\
&\quad -\frac{1}{2V}\left[\Lambda_{\mathrm{EN}}^{\dagger}-q\phi\Lambda_{\mathrm{NN}}^{\dagger}\right]\left[\frac{q\nabla\phi}{T}+\nabla\frac{\mu}{T}\right] \\
\overline{\mathbf{J}}_{\mathrm{N}} &= \frac{1}{2V}\left[\Lambda_{\mathrm{NE}}^{\dagger}-q\phi\Lambda_{\mathrm{NN}}^{\dagger}\right]\nabla\frac{1}{T}-\frac{\Lambda_{\mathrm{NN}}^{\dagger}}{2V}\left[\frac{q\nabla\phi}{T}+\nabla\frac{\mu}{T}\right].
\end{aligned} \tag{4.66}$$

Hence one can identify

$$\Lambda_{\mathrm{EE}}^{*}=\Lambda_{\mathrm{EE}}^{\dagger}-2q\phi\Lambda_{\mathrm{EN}}^{\dagger}+q^2\phi^2\Lambda_{\mathrm{NN}}^{\dagger},\quad \Lambda_{\mathrm{EN}}^{*}=\Lambda_{\mathrm{NE}}^{\dagger}-q\phi\Lambda_{\mathrm{NN}}^{\dagger}, \tag{4.67}$$

and $\Lambda_{\mathrm{NN}}^{*}=\Lambda_{\mathrm{NN}}^{\dagger}$.

Ohm's Law

The electric current is $q\mathbf{J}_{\mathrm{N}}$. For the case of uniform temperature, $\nabla T^{-1}=\mathbf{0}$, and chemical potential, $\nabla(\mu/T)=\mathbf{0}$, and with the electric field uniform,

$\phi(\mathbf{r},t) = -E_{\rm el}z$, the most likely current is

$$q\overline{\mathbf{J}}_{\rm N} = \frac{\Lambda^*_{\rm NN}}{2VT} q^2 \mathbf{E}_{\rm el}. \tag{4.68}$$

This is Ohm's law, $q\mathbf{J}_N = R^{-1}\mathbf{E}_{\rm el}$, with the resistivity (the reciprocal of the electrical conductivity) being $R \equiv 2VT/q^2\Lambda^*_{\rm NN}$.

Joule Effect

The heating of a conductor due to an electric current is known as the Joule effect. For uniform temperature, $\nabla T^{-1} = \mathbf{0}$ (which means that the conductor is in thermal contact with a heat reservoir of uniform temperature along its length), chemical potential, $\nabla(\mu/T) = \mathbf{0}$, internal energy, $\nabla\varepsilon^{\rm int} = \mathbf{0}$, pressure, $\nabla p = \mathbf{0}$, number density, $\nabla n = \mathbf{0}$, and electric field, $\nabla^2\phi = 0$, which mean that $\nabla \cdot \overline{\mathbf{J}}^0_{\rm E} = \nabla \cdot \overline{\mathbf{J}}_{\rm N} = 0$, the rate of energy change per unit volume is

$$\begin{aligned}
\frac{\partial \varepsilon}{\partial t} &= -\nabla \cdot \overline{\mathbf{J}}_{\rm E} \\
&= -\nabla \cdot \left[\overline{\mathbf{J}}^0_{\rm E} + q\phi\overline{\mathbf{J}}_{\rm N} \right] \\
&= q\mathbf{E}_{\rm el} \cdot \overline{\mathbf{J}}_{\rm N} \\
&= R^{-1}\mathbf{E}_{\rm el} \cdot \mathbf{E}_{\rm el} = q^2 R\overline{\mathbf{J}}_{\rm N} \cdot \overline{\mathbf{J}}_{\rm N}.
\end{aligned} \tag{4.69}$$

This is positive, as obviously also is the dissipation, $\bar{\dot{\sigma}} = T^{-1}{\rm d}\varepsilon/{\rm d}t$. (Here and below, the local dissipation is sometimes written as $\dot{\sigma} \equiv \partial\sigma(\mathbf{r},t)/\partial t$.)

Seebeck Effect

The Seebeck effect refers to the development of an electric potential between two junctions held at different temperatures and connected by two different conductors in parallel, one of which is open. Since the potential is measured in an open circuit, there is no current, $\mathbf{J}_{\rm N} = \mathbf{0}$. Using Eq. (4.61), for a single conductor in an applied potential, this condition is evidently satisfied when

$$\frac{q\nabla\phi}{T} + \nabla\frac{\mu}{T} = [\Lambda^*_{\rm NN}]^{-1}\Lambda^*_{\rm NE}\nabla T^{-1}. \tag{4.70}$$

In the absence of a temperature gradient, one can interpret this, $(q\nabla\phi/T) + \nabla(\mu/T) = \mathbf{0}$, as meaning that a gradient in the chemical potential gives rise to an internal electric potential that is equal and opposite to the externally applied potential, $\nabla(\mu/T) = -(q\nabla\phi^{\rm int}/T)$. This result means that a single conductor in a temperature gradient develops an electric potential that is equivalent to the gradient in the chemical potential.[9]

[9]Consider the following thought experiment. First, take a length of conductor and place it in a temperature gradient with no external electric field. From Eq. (4.61), charge carriers preferentially segregate at one end until $\overline{\mathbf{J}}_{\rm N} = 0$ or $\nabla(\mu/T) = [\Lambda^*_{\rm NN}]^{-1}\Lambda^*_{\rm NE}\nabla T^{-1}$. Second, cut the conductor in the middle whilst maintaining the temperature gradient so that charge cannot

In view of this, the Seebeck coefficient can be defined as the ratio of the (internal) potential gradient to the temperature gradient. With $\nabla\mu = -q\nabla\phi^{\rm int}$, and using Eq. (4.62) with $\mathbf{J}_{\rm N} = \mathbf{0}$, one obtains

$$\eta \equiv \frac{\nabla\phi^{\rm int}}{\nabla T} = \frac{-\nabla\mu}{q\nabla T} = \frac{1}{qT}\frac{\Lambda^{\ddagger}_{\rm NE}}{\Lambda^{\ddagger}_{\rm NN}}. \tag{4.71}$$

If instead one were to take $\nabla(\mu/T) = -(q\nabla\phi^{\rm int}/T)$, then Eq. (4.61) gives

$$\eta^* \equiv \frac{\nabla\phi^{\rm int}}{\nabla T} = \frac{-T\nabla(\mu/T)}{q\nabla T} = \frac{1}{qT}\frac{\Lambda^{*}_{\rm NE}}{\Lambda^{*}_{\rm NN}} = \eta + \frac{\mu}{qT}. \tag{4.72}$$

Since only the relative value of the chemical potential has physical significance, the additional term that appears here can have no physical consequences and could be set to zero, in which case $\eta^* = \eta$.

As mentioned above, the Seebeck effect is the voltage measured across a break in one of two different conductors that are connected in parallel, with the terminii held at two different temperatures. Accordingly, it is the difference in Seebeck coefficients that is important. Since the mobile charge carriers in the two conductors must be at the same chemical potential, $\mu_1 = \mu_2$,[10] one must have $\eta_1 - \eta_2 = \eta_1^* - \eta_2^*$. This reiterates the point that there is no practical difference between the two definitions. For the same reason one could also define $\eta^\dagger \equiv \Lambda^{\dagger}_{\rm NE}/qT\Lambda^{\dagger}_{\rm NN} = \eta^* + \phi/T$, with no physical consequences.

Peltier Effect

The inverse of the Seebeck effect is the Peltier effect. An electric current in a conductor at uniform temperature causes an energy flux that depends upon the transport coefficients of the material. Hence the junction of two conductors will be a source or sink of energy, depending upon the direction of current flow. The junction is thus heated or cooled at a rate that depends upon the heat capacity of the surrounding material.

In essence, the Peltier coefficient is the ratio of the energy flux to the electric current. The common definition is to use $J^{\ddagger}_{\rm E}$ for the energy flux, perhaps because of its close relationship with the entropy flux. Using Eq. (4.62) with $\nabla T = \mathbf{0}$, the Peltier coefficient is

$$\pi \equiv \left.\frac{\overline{J}^{\ddagger}_{\rm E}}{q\overline{J}_{\rm N}}\right|_{\nabla T=0} = \frac{\Lambda^{\ddagger}_{\rm EN}}{q\Lambda^{\ddagger}_{\rm NN}}. \tag{4.73}$$

rearrange. Third, bring the two pieces to equal temperature, keeping them fixed in place and disconnected. A voltmeter connected to their ends will now show a potential difference given by $q(\nabla\phi)/T = -\nabla(\mu/T)$. Since T is uniform, this is equivalent to $q\nabla\phi = -\nabla\mu$. Note that no potential difference would be measured at the end of the second stage, which suggests that the latter expression is the correct one that also holds in the presence of a temperature gradient.

[10] The subscripts 1 and 2 here and in the following subsection refer to two different conductors, *not* to the mobile and immobile charge species.

In some ways this is not entirely satisfactory since it is the total energy flux delivered at the junction that gives the temperature rise. With the total energy flux being $\mathbf{J}_{\rm E} = \mathbf{J}_{\rm E}^{\ddagger} + [\mu + q\phi]\mathbf{J}_{\rm N}$, one could instead define the Peltier coefficient as the ratio of this to the electric current at uniform temperature,

$$\tilde{\pi} \equiv \left.\frac{\overline{J}_{\rm E}}{q\overline{J}_{\rm N}}\right|_{\nabla T=0} = \left.\frac{\overline{J}_{\rm E}^{\ddagger} + [\mu + q\phi]\overline{J}_{\rm N}}{q\overline{J}_{\rm N}}\right|_{\nabla T=0} = \pi + \phi + \mu/q. \tag{4.74}$$

Since only the relative value of the electro-chemical potential has physical significance, the additional term that appears here can have no physical consequences and could be set to zero, in which case $\tilde{\pi} = \pi$.

In any case, because the potential is continuous across the barrier, $\phi_1 = \phi_2$, as is the chemical potential, $\mu_1 = \mu_2$, the difference in the Peltier coefficients is the same for either definition, $\pi_1 - \pi_2 = \tilde{\pi}_1 - \tilde{\pi}_2$. It is the difference in Peltier coefficients of the two conductors that gives the rate of heat dissipation at the junction and the consequent rate of temperature rise of the reservoir in contact with it. One can conclude that there is no practical difference between the two definitions.

The Peltier and Seebeck coefficients are related by

$$T\eta = \pi, \tag{4.75}$$

as can be confirmed by inspection. This is called the second Kelvin relation. This equality is a consequence of the symmetry of the transport matrix. This specific relationship was one of the motivating factors for Onsager's derivation and explanation of the reciprocal relations.

Thomson Effect

Energy is dissipated by a current flow and by a heat flow. The Thompson effect gives the power due to the cross coupling of current and energy flow that is in excess of that due to Joule heating and to the heat conduction.

The Joule effect discussed above gives the dissipation due to a current flow,

$$\frac{\partial \varepsilon^{\rm J}}{\partial t} = \frac{2VT}{\Lambda_{\rm NN}^{*}}\overline{\mathbf{J}}_{\rm N} \cdot \overline{\mathbf{J}}_{\rm N} = q^2 R\overline{\mathbf{J}}_{\rm N} \cdot \overline{\mathbf{J}}_{\rm N}. \tag{4.76}$$

In the derivation of the Joule effect, the current flowing in a uniform temperature and chemical potential field was used, $\overline{\mathbf{J}}_{\rm N}(0) = -q^{-1}R^{-1}\nabla\phi$. However, it is conventional in the analysis of the Thomson effect to use the total number flux in the presence of a temperature and electro-chemical gradients, $\overline{\mathbf{J}}_{\rm N} = \overline{\mathbf{J}}_{\rm N}(0) + (\Lambda_{\rm NE}^{*}/2V)\nabla T^{-1} - (\Lambda_{\rm NN}^{*}/2V)\nabla(\mu/T)$.

For heat flow in the absence of particle flow, the power is derived from Fourier's law,

$$\frac{\partial \varepsilon^{\rm F}}{\partial t} = -\nabla \cdot \overline{\mathbf{J}}_{\rm E}^{0}(\mathbf{J}_{\rm N} = \mathbf{0}) = \frac{-1}{2V}\nabla \cdot \left[\Lambda_{\rm EE}\nabla\frac{1}{T}\right], \tag{4.77}$$

where $\Lambda_{\rm EE} = \Lambda^*_{\rm EE} - \Lambda^*_{\rm EN}[\Lambda^*_{\rm NN}]^{-1}\Lambda^*_{\rm NE}$.

The Thomson effect is the power additional to these two that is dissipated when a current flows along a temperature gradient; it is the power due to the cross coupling of current and heat flow. One can imagine that the sub-system, the conductor of volume V, can exchange charge and energy with two reservoirs at either end, which provide the temperature and electro-chemical potential gradient, and that it can exchange energy with a reservoir along its length that has the same temperature gradient as the terminal reservoirs. Hence in the steady state $\nabla \cdot \overline{\mathbf{J}}_{\rm E} \neq 0$. Because charge is conserved in the conductor, in the steady state, $\nabla \cdot \overline{\mathbf{J}}_{\rm N} = 0$.

The total energy flux is $\mathbf{J}_{\rm E} = \mathbf{J}^\ddagger_{\rm E} + [\mu + q\phi]\mathbf{J}_{\rm N}$. Using Eq. (4.62),

$$\begin{aligned}
\overline{\mathbf{J}}^\ddagger_{\rm E} &= \frac{\Lambda^\ddagger_{\rm EE}}{2V}\nabla\frac{1}{T} - \frac{\Lambda^\ddagger_{\rm EN}}{2V}\frac{\nabla[\mu+q\phi]}{T}, \\
&= \frac{\Lambda^\ddagger_{\rm EE}}{2V}\nabla\frac{1}{T} - \frac{\Lambda^\ddagger_{\rm EN}}{2V}\left[\frac{-2V}{\Lambda^\ddagger_{\rm NN}}\overline{\mathbf{J}}_{\rm N} + \frac{\Lambda^\ddagger_{\rm NE}}{\Lambda^\ddagger_{\rm NN}}\nabla\frac{1}{T}\right] \\
&= \frac{1}{2V}\left[\Lambda^\ddagger_{\rm EE} - \frac{\Lambda^\ddagger_{\rm EN}\Lambda^\ddagger_{\rm NE}}{\Lambda^\ddagger_{\rm NN}}\right]\nabla\frac{1}{T} + \frac{\Lambda^\ddagger_{\rm EN}}{\Lambda^\ddagger_{\rm NN}}\overline{\mathbf{J}}_{\rm N} \\
&= \frac{\Lambda_{\rm EE}}{2V}\nabla\frac{1}{T} + qT\eta\overline{\mathbf{J}}_{\rm N}.
\end{aligned} \tag{4.78}$$

The final equality uses the definition of the Seebeck coefficient, Eq. (4.71), as well as Eqs (4.59) and (4.64), which give $\Lambda_{\rm EE} = \Lambda^\ddagger_{\rm EE} - \Lambda^\ddagger_{\rm EN}[\Lambda^\ddagger_{\rm NN}]^{-1}\Lambda^\ddagger_{\rm NE}$. The thermal conductivity is $\lambda \equiv \Lambda_{\rm EE}/2VT^2$. Also required is

$$\nabla[\mu + q\phi] = \frac{\Lambda^\ddagger_{\rm NE}}{\Lambda^\ddagger_{\rm NN}}T\nabla\frac{1}{T} - \frac{2VT}{\Lambda^\ddagger_{\rm NN}}\overline{\mathbf{J}}_{\rm N} = -q\eta\nabla T - q^2R\overline{\mathbf{J}}_{\rm N}, \tag{4.79}$$

which is a rearrangement of the second equality in Eq. (4.62).

With these, and the steady state charge conservation law, $\nabla \cdot \overline{\mathbf{J}}_{\rm N} = 0$, the power is

$$\begin{aligned}
\frac{\partial\varepsilon}{\partial t} &= -\nabla\cdot\overline{\mathbf{J}}_{\rm E} = -\nabla\cdot\left[\overline{\mathbf{J}}^\ddagger_{\rm E} + [\mu+q\phi]\overline{\mathbf{J}}_{\rm N}\right] \\
&= -\nabla\cdot\left[T^2\lambda\nabla\frac{1}{T}\right] - q\overline{\mathbf{J}}_{\rm N}\cdot\nabla(T\eta) - \overline{\mathbf{J}}_{\rm N}\cdot\nabla[\mu+q\phi] \\
&= \nabla\cdot[\lambda\nabla T] + q^2R\overline{\mathbf{J}}_{\rm N}\cdot\overline{\mathbf{J}}_{\rm N} - q\overline{\mathbf{J}}_{\rm N}\cdot\nabla(T\eta) + q\eta\overline{\mathbf{J}}_{\rm N}\cdot\nabla T \\
&= \frac{\partial\varepsilon^{\rm F}}{\partial t} + \frac{\partial\varepsilon^{\rm J}}{\partial t} - qT\frac{\partial\eta}{\partial T}\overline{\mathbf{J}}_{\rm N}\cdot\nabla T.
\end{aligned} \tag{4.80}$$

The final equality uses the fact that the local value of the Seebeck coefficient is a function of the local temperature and local density, and that the latter is relatively incompressible,

$$\nabla\eta = \frac{\partial\eta}{\partial T}\nabla T + \frac{\partial\eta}{\partial n}\nabla n \approx \frac{\partial\eta}{\partial T}\nabla T. \tag{4.81}$$

The final term in the expression for the total power is called the Thomson power,

$$\frac{\partial \varepsilon^{\rm Thomson}}{\partial t} = -qT\frac{\partial \eta}{\partial T}\overline{\mathbf{J}}_{\rm N} \cdot \nabla T. \tag{4.82}$$

The factor multiplying the current and the temperature gradient is called the Thomson coefficient, $\alpha \equiv -T\partial\eta/\partial T$.

Dissipation

The rate of change of the sub-system entropy density is derived in Eq. (5.41). For the present case (neglecting terms quadratic in the velocity, the number flux being the same as the diffusive number flux), in the general case of arbitrary fluxes it is

$$\frac{\partial \sigma(\mathbf{r},t)}{\partial t} = \frac{-1}{T}\nabla \cdot \mathbf{J}_{\rm E}^0 - \frac{q}{T}\mathbf{J}_{\rm N} \cdot \nabla\phi + \frac{1}{T}\mu\nabla \cdot \mathbf{J}_{\rm N}. \tag{4.83}$$

Recall that the convective velocity is zero in the present problem, $\mathbf{v} = \mathbf{0}$.

For a constant external potential, the most likely state is a steady state and the structure of the sub-system must be constant. Hence $\partial\sigma(\mathbf{r},t)/\partial t = 0$, $\nabla \cdot \overline{\mathbf{J}}_{\rm E}(\mathbf{r},t) = 0$, and $\nabla \cdot \overline{\mathbf{J}}_{\rm N}(\mathbf{r},t) = 0$, with the latter two identically giving the first.[11] This identity can be used to give a requirement on the conductive energy flux, namely

$$\frac{1}{T}\nabla \cdot \overline{\mathbf{J}}_{\rm E}^0 = -\frac{q}{T}\overline{\mathbf{J}}_{\rm N} \cdot \nabla\phi. \tag{4.84}$$

Similarly the number conservation requirement, $\nabla \cdot \overline{\mathbf{J}}_{\rm N}(\mathbf{r},t) = 0$, provides a condition on the temperature and chemical potential fields of the conductor, which makes them non-linear.

The rate of dissipation of the sub-system just given (both the general case and the steady state case) holds for both an isolated sub-system and for a sub-system that can exchange energy and charge with reservoirs. In the reservoir case it is shown in Eq. (5.43) that the general result can be written to show explicitly the fluxes across the boundaries,

$$\begin{aligned}\frac{\dot{S}_{\rm s}}{V} &= \mathbf{J}_{\rm E}^0 \cdot \nabla\frac{1}{T} - \mathbf{J}_{\rm N}^0 \cdot \nabla\left[\frac{q\nabla\phi}{T} + \nabla\frac{\mu}{T}\right] - \left\{\nabla\cdot\left(\frac{1}{T}\mathbf{J}_{\rm E}^0\right) - \nabla\cdot\left(\frac{\mu}{T}\mathbf{J}_{\rm N}\right)\right\} \\ &= \mathbf{J}_{\rm E}^0 \cdot \nabla\frac{1}{T} - \mathbf{J}_{\rm N}^0 \cdot \nabla\left[\frac{q\nabla\phi}{T} + \nabla\frac{\mu}{T}\right] - \frac{\dot{S}_{\rm r}}{V}.\end{aligned} \tag{4.85}$$

---

[11] The condition invoked here, $\nabla \cdot \overline{\mathbf{J}}_{\rm E}(\mathbf{r},t) = 0$, differs from the assumption made in the preceding section, $\nabla \cdot \overline{\mathbf{J}}_{\rm E}(\mathbf{r},t) \neq 0$. In that section it was assumed that the conductor could exchange energy along its length with a reservoir that had the same imposed temperature gradient. In the present analysis it is assumed that neither charge or energy can exchange with a reservoir along the conductor's length (i.e. a diathermal or adiabatic surface coating), and that such exchanges only occur at the two ends of the conductor.

The term in braces arises from the exchange with the reservoirs, and is therefore equal to the dissipation of the reservoirs, as can be confirmed explicitly,

$$\frac{\dot{S}_{\rm r}}{V} = \nabla \cdot \left[\frac{1}{T}\mathbf{J}_{\rm E}^0 - \frac{\mu}{T}\mathbf{J}_{\rm N} \right] = \nabla \cdot \left[\frac{1}{T}\mathbf{J}_{\rm E} - \frac{\mu + q\phi}{T}\mathbf{J}_{\rm N} \right]. \tag{4.86}$$

The term in brackets clearly represents the entropy flux, $\mathbf{J}_{\rm S} \equiv (\partial\sigma/\partial\varepsilon)\mathbf{J}_{\rm E} + (\partial\sigma/\partial n)\mathbf{J}_{\rm N}$. If the external potential depends upon time then one has to add a source term to the dissipation, namely $q\partial\phi/\partial t$. These expressions for the rate of change of entropy and the entropy flux hold generally.[12]

In the steady state no dissipation occurs in the conductor (the sub-system), and all of the dissipation occurs in the reservoirs at each end (see footnote 11 on p. 118). That is, in the steady state, $\overline{\dot{S}}_{\rm s} = 0$, and so the two terms above must cancel, which gives a useful expression for the reservoir dissipation. The total dissipation in the steady state is

$$\begin{aligned} \frac{\overline{\dot{S}}_{\rm total}}{V} = \frac{\overline{\dot{S}}_{\rm r}}{V} &= \overline{\mathbf{J}}_{\rm E}^0 \cdot \nabla\frac{1}{T} - \overline{\mathbf{J}}_{\rm N} \cdot \nabla \left[\frac{q\nabla\phi}{T} + \nabla\frac{\mu}{T} \right] \\ &= \underline{\underline{\Lambda}}^* : \left\{ \nabla\frac{1}{T}, \left[\frac{q\nabla\phi}{T} + \nabla\frac{\mu}{T} \right] \right\}^2 \\ &= [\underline{\underline{\Lambda}}^*]^{-1} : \left\{ \overline{\mathbf{J}}_{\rm E}^0, \overline{\mathbf{J}}_{\rm N} \right\}^2, \end{aligned} \tag{4.87}$$

where the transport coefficients that appear in Eq. (4.61) have been written in matrix form. Since the transport matrix is positive definite, this proves that the total dissipation is positive, as it must be.

This result for the dissipation was based upon setting the left-hand side of Eq. (4.85) to zero. Alternatively one can proceed from Eq. (4.86) and write the dissipation as

$$\begin{aligned} \frac{\overline{\dot{S}}_{\rm total}}{V} = \frac{\overline{\dot{S}}_{\rm r}}{V} &= \nabla \cdot \left[\frac{1}{T}\overline{\mathbf{J}}_{\rm E} - \frac{\mu + q\phi}{T}\overline{\mathbf{J}}_{\rm N} \right] \\ &= \overline{\mathbf{J}}_{\rm E} \cdot \nabla\frac{1}{T} - \overline{\mathbf{J}}_{\rm N} \cdot \nabla\frac{\mu + q\phi}{T}. \end{aligned} \tag{4.88}$$

This is readily shown to be equal to the preceding expression.

---

[12]The present author is unable to make sense of the expressions given for the rate of change of entropy in Ch. III of de Groot and Mazur, (1984). For the specific case of thermo-electric flows, their expression for the rate of change of the entropy density in the most likely state, Eq. (XIII.84), $\partial\sigma/\partial t = -\nabla \cdot \mathbf{J}_{\rm S,tot} + T\mathbf{J}_{\rm S,tot} \cdot \nabla T^{-1} - T^{-1}\mathbf{J}_{\rm N} \cdot \nabla[\mu + q\phi]$, contrasts with the present expression, $\partial\sigma/\partial t = -\nabla \cdot \mathbf{J}_{\rm S}$. Their $\mathbf{J}_{\rm S,tot}$ given by the right-hand side of their Eq. (XIII.77) can be shown to equal the present $\overline{\mathbf{J}}_{\rm S}$. It can be shown that in the most likely state (to which case all the analysis of de Groot and Mazur is restricted), the additional two terms that appear in their Eq. (XIII.84) compared to the present expression do not cancel.

The expression for the entropy flux given by Pottier, (2010), following Eq. (2B.6.6), $\mathbf{J}_{\rm S} = \mathbf{J}_{\rm Q}^*/T$, is the same as the expression given here, since $\mathbf{J}_{\rm Q}^* \equiv \mathbf{J}_{\rm E}^0 - \mu\mathbf{J}_{\rm N}$.

Chapter 5

Second Entropy for Fluctuating Hydrodynamics

The conceptual basis of hydrodynamics consists of three elements: the conservation laws, localised equilibrium thermodynamics, and the phenomenological transport laws. Respective examples of each of these are that the rate of change of density is given by the divergence in the number flux, that the change in the local energy density equals the change in the local temperature times the heat capacity, and that the conductive heat flux equals the thermal conductivity times the temperature gradient.

Although there is no doubting the practical success of hydrodynamics, there is some motivation to establish the foundations of the discipline within the broader framework of non-equilibrium thermodynamics. Firmer foundations might justify, for example, the assumed relationships of equilibrium thermodynamics in the presence of convective fluxes. They might also provide insight into extending hydrodynamics into new regimes, such as the generalisation of the phenomenological transport laws to systems rapidly varying in space and time. There is also the possibility that proceeding beyond hydrodynamics to the underlying non-equilibrium thermodynamics may reveal new insights into the causes of observed non-equilibrium phenomena, and it may offer new computational algorithms for hydrodynamics based upon the same type of variational principles that have proven useful in equilibrium thermodynamics (i.e. entropy maximisation, or, equivalently, free energy minimisation).

The connection between hydrodynamics and non-equilibrium thermodynamics occurs via the second entropy. This formulation treats the material fluxes as fluctuating quantities with a variance that is proportional to the hydrodynamic transport coefficients. Such a fluctuation approach to hydrodynamics was originally given by Landau and Lifshitz.[1] Although not well known amongst main-stream practitioners of hydrodynamics, thermodynamic researchers have

[1] Landau, L. D. and Lifshitz, E. M. (1959), *Fluid Mechanics*, (Pergammon Press, Oxford).

extended and applied the fluctuation approach to a range of different phenomena.[2] The present approach to fluctuating hydrodynamics is directly based on the second entropy, which has the advantages of making a direct connection with non-equilibrium thermodynamics, and of providing variational principles that are suitable for numerical algorithms and for a fundamental understanding of various hydrodynamic phenomena.

The chapter begins with the material conservation laws, §5.1. The entropy density for hydrodynamic flows is derived in §5.2. In §5.3 the second entropy for hydrodynamics is formulated. Here the fluxes and the fields are treated as equivalent fluctuating or constrained variables, which enables the hydrodynamic problem to be solved by variational techniques instead of by the more usual partial differential equation approach. It is shown in §5.4 that maximising the second entropy with respect to the fluxes yields the Navier-Stokes and energy equations. This shows the mathematical equivalence of the second entropy theory and the usual partial differential equation approach to hydrodynamics.[3]

A significant feature of the present approach is the reduction condition. This is invoked to explain why equilibrium thermodynamics can be invoked in hydrodynamic (non-equilibrium) problems: the solution of the hydrodynamic equations for given structural fields maximises the second entropy and reduces it to the first entropy. In practice all of hydrodynamics is based upon the assumption of local thermodynamic equilibrium (for example, the replacement of the energy density by the heat capacity times the temperature), and the provision of a sound thermodynamic foundation for local thermodynamic equilibrium in the form of the reduction condition fills a gap in the theory and provides comfort for the application of hydrodynamics in extreme regimes.

5.1 Conservation Laws

5.1.1 Densities, Velocities, and Chemical Reactions

Let $n_k(\mathbf{r}, t)$ be the number density of species k at position $\mathbf{r}$ and time t. All functions below have this dependence, but for clarity this will usually not be shown explicitly. The entire set of number densities will be denoted $\underline{n} \equiv \{n_k\}$. Using number density (i.e. per unit volume) instead of mass density (i.e. per unit mass, or specific), facilitates integration over space. The mass density is $\rho_k \equiv m_k n_k$, where m_k is the molecular mass of species k, with the entire set denoted $\underline{m} \equiv \{m_k\}$. The velocity of species k is $\mathbf{v}_k$, and its momentum density is $\mathbf{p}_k \equiv m_k n_k \mathbf{v}_k$, where bold face type signifies a three-dimensional spatial vector.

[2] Fox, R. F. and Uhlenbeck, G. E. (1970), Phys. Fluids, **13**, 1893 and 2881. Keizer, J. (1987), *Statistical Thermodynamics of Non-equilibrium Processes*, (Springer-Verlag, New York). Ortiz de Zárate, J. M. and Sengers, J. V. (2006), *Hydrodynamic Fluctuations in Fluids and Fluid Mixtures*, (Elsevier, Amsterdam).

[3] Although the two theories are formally equivalent in a mathematical sense for the optimum fluxes, the second entropy theory is in addition applicable for non-optimum fluxes, which phenomenological hydrodynamics is not. Also, there may be computational advantages in the variational approach.

The total number, mass, and momentum density are $n \equiv \sum_k n_k$, $\rho \equiv \sum_k m_k n_k = \underline{m} \cdot \underline{n}$, and $\mathbf{p} \equiv \sum_k m_k n_k \mathbf{v}_k$, respectively. In terms of these the barycentric or centre of mass velocity is

$$\mathbf{v}(\mathbf{r},t) \equiv \mathbf{p}(\mathbf{r},t)/\rho(\mathbf{r},t) = \frac{\sum_k m_k n_k(\mathbf{r},t)\mathbf{v}_k(\mathbf{r},t)}{\sum_k m_k n_k(\mathbf{r},t)}. \tag{5.1}$$

The diffusive or peculiar velocity is measured with respect to the centre of mass velocity, $\tilde{\mathbf{v}}_k(\mathbf{r},t) \equiv \mathbf{v}_k(\mathbf{r},t) - \mathbf{v}(\mathbf{r},t)$. The peculiar momentum is defined in terms of this, $\tilde{\mathbf{p}}_k(\mathbf{r},t) \equiv m_k n_k(\mathbf{r},t)\tilde{\mathbf{v}}_k(\mathbf{r},t)$, and has the property that in total it vanishes in the centre of mass frame, $\sum_k \tilde{\mathbf{p}}_k(\mathbf{r},t) = 0$.

The kinetic energy density is $\kappa = \sum_k m_k n_k v_k^2/2$. Here $v_k \equiv |\mathbf{v}_k|$, and $v_k^2 = \mathbf{v}_k \cdot \mathbf{v}_k$, and similarly below. The peculiar kinetic energy density is $\tilde{\kappa} = \sum_k m_k n_k \tilde{v}_k^2/2$, and one has $\kappa = \tilde{\kappa} + \rho v^2/2$. The external potential energy density is $n\psi \equiv \sum_k n_k \psi_k = \underline{n} \cdot \underline{\psi}$, where $\psi_k(\mathbf{r},t)$ is the external potential acting on species k, and the set of external potentials is $\underline{\psi}(\mathbf{r},t) \equiv \{\psi_k(\mathbf{r},t)\}$.

For greatest generality, the possibility of chemical reactions is included. Lower case Roman letters label the molecular species, lower case Greek letters label the reaction, and upper case Roman letters label the atomic species. Chemical reaction α is written

$$\sum_k \nu_{\alpha k} a_{kI} = 0. \tag{5.2}$$

Here $\nu_{\alpha k}$ is the stoichiometric coefficient, which is the number of molecules of type k in reaction α. This is an integer that is positive for reactants, negative for products, and zero if the molecule does not appear in the reaction. Also, a_{kI} is the number of atoms of type I in a molecule of type k. With it, the molecular mass is $m_k = \sum_I a_{kI} M_I$, where M_I is the atomic mass. Hence multiplying the reaction equation by M_I and summing over atomic species shows that $\sum_k \nu_{\alpha k} m_k = 0$, which is to say that mass is conserved in a chemical reaction. A similar conservation law holds for charge. If the external potential is due to gravity, then $\psi_k \propto m_k$, and $\sum \psi_k \nu_{\alpha k} = 0$. The same result holds if the external potential arises from an electric field.

The rate of reaction α is denoted $\dot{\xi}_\alpha$. This is defined such that the rate of change of the density of species k due to chemical reactions is

$$\dot{n}_k^{\text{react}} = -\sum_\alpha \dot{\xi}_\alpha \nu_{\alpha k}. \tag{5.3}$$

Note that, in general, the reaction rate is a function of position, $\dot{\xi}_\alpha(\mathbf{r},t)$, typically through its dependence on the local concentrations, $\dot{\xi}_\alpha(\underline{n}(\mathbf{r},t))$.

5.1.2 Number Flux

Chemical reactions provide a source in the number conservation equation. The conserved part is represented by the number convection term, which is the number flux, $\mathbf{J}_{\mathrm{N},k}(\mathbf{r},t) = n_k(\mathbf{r},t)\mathbf{v}_k(\mathbf{r},t)$. The diffusive part of this is just the flux in the centre of mass frame, $\mathbf{J}^0_{\mathrm{N},k}(\mathbf{r},t) = n_k(\mathbf{r},t)\tilde{\mathbf{v}}_k(\mathbf{r},t)$.

The number of molecules in a volume V at $\mathbf{r}$ at time t is

$$N_k(\mathbf{r},t) = \int_V \mathrm{d}\mathbf{r}'\, n_k(\mathbf{r}',t). \tag{5.4}$$

In the absence of chemical reactions, this can only change by molecules passing across the boundary. With $\hat{\mathbf{n}}$ the outward normal, and using Gauss' theorem one has

$$\begin{aligned}\frac{\partial N_k(\mathbf{r},t)}{\partial t} &= \int_V \mathrm{d}\mathbf{r}'\, \frac{\partial n_k(\mathbf{r}',t)}{\partial t} \\ &= -\oint_\Omega \mathrm{d}\mathbf{R}'\, \hat{\mathbf{n}}\cdot\mathbf{J}_{\mathrm{N},k}(\mathbf{r}',t) \\ &= -\int_V \mathrm{d}\mathbf{r}'\, \nabla\cdot\mathbf{J}_{\mathrm{N},k}(\mathbf{r}',t). \end{aligned} \tag{5.5}$$

For small enough volume this is

$$\frac{\partial n_k(\mathbf{r},t)}{\partial t} = -\nabla\cdot\mathbf{J}_{\mathrm{N},k}(\mathbf{r},t). \tag{5.6}$$

The source term when chemical reactions are present will be added to this below.

The total time derivative, also known as the convective derivative, the material derivative, or the hydrodynamic derivative, is

$$\begin{aligned}\frac{\mathrm{d}n_k(\mathbf{r},t)}{\mathrm{d}t} &\equiv \frac{\partial n_k(\mathbf{r},t)}{\partial t} + \mathbf{v}_k(\mathbf{r},t)\cdot\nabla n_k(\mathbf{r},t) \\ &= -\nabla\cdot[n_k(\mathbf{r},t)\mathbf{v}_k(\mathbf{r},t)] + \mathbf{v}_k(\mathbf{r},t)\cdot\nabla n_k(\mathbf{r},t) \\ &= -n_k(\mathbf{r},t)\nabla\cdot\mathbf{v}_k(\mathbf{r},t). \end{aligned} \tag{5.7}$$

On the left is the rate of change of the number density with time in a frame moving along with the flow. The right-hand side says that this is due to the rate of change of the co-moving volume, which is just the divergence of the velocity.

The number flux is just the momentum density, apart from a factor of the molecular weight, and this is a conserved variable. Hence for a single component, isolated system, it makes no sense to talk about fluctuations in the number flux, or to maximise the second entropy with respect to number flux. However for a multi-component system one can investigate the fluctuations in the number fluxes, subject to the constraint of the value of the total number flux. Or for a system in the presence of an external field, such as that imposed by fixed walls or by a fixed porous matrix, the fluctuations in number flux are non-zero. Another example is a solute at infinite dilution in a solvent, where one can explore the solute flux, treating the solvent as a fixed background.

Returning to the case of chemical reactions, one can simply add the source term to the divergence of the number flux. Hence the rate of change of number is

$$\frac{\partial n_k(\mathbf{r},t)}{\partial t} = \dot{n}_k^{\mathrm{react}}(\mathbf{r},t) - \nabla\cdot\mathbf{J}_{\mathrm{N},k}(\mathbf{r},t). \tag{5.8}$$

Since mass is conserved in the chemical reactions,

$$\dot{\rho}^{\mathrm{react}} \equiv \sum_k m_k \dot{n}_k^{\mathrm{react}} = \sum_{k,\alpha,I} M_I a_{kI} \dot{\xi}_\alpha \nu_{\alpha k} = 0, \tag{5.9}$$

one has that the time derivative of the mass density is

$$\frac{\partial \rho(\mathbf{r},t)}{\partial t} = -\sum_k m_k \nabla \cdot \mathbf{J}_{\mathrm{N},k}(\mathbf{r},t) = -\nabla \cdot \mathbf{p}(\mathbf{r},t). \tag{5.10}$$

5.1.3 Energy Flux

The time rate of change of the external potential is a source of energy. Apart from this, energy is conserved, so one can define an energy flux whose divergence gives the time rate of change of energy density. The energy conservation law is

$$\frac{\partial \varepsilon(\mathbf{r},t)}{\partial t} = \sum_k n_k(\mathbf{r},t) \frac{\partial \psi_k(\mathbf{r},t)}{\partial t} - \nabla \cdot \mathbf{J}_{\mathrm{E}}(\mathbf{r},t). \tag{5.11}$$

The energy density ε comprises the kinetic energy, the intermolecular potential energy, and the external potential energy (see below).

There are three types of energy flux into a volume: the heat energy conducted across the boundaries, the molecular energy convected across the boundaries, and the work performed on the volume. One can write $\mathbf{J}_E = \mathbf{J}_E^0 + \mathbf{J}_E^{\mathrm{conv}} + \mathbf{J}_E^{\psi} + \mathbf{J}_E^{\mathrm{work}}$. The two contributions that comprise the convective part have here been shown explicitly: one part is due to the mean velocity, $\mathbf{J}_E^{\mathrm{conv}}$, and the excess, $\mathbf{J}_E^{\psi}$, is the external potential carried by the diffusive flux (see next).

The thermodynamic contribution is due to conduction and occurs in the absence of molecular flow, $\mathbf{v} = \mathbf{0}$. The superscript 0 is used to denote such diffusive flow, which in this case is called the heat flow. In the steady state the heat flow is proportional to the temperature gradient, as will be seen.

The convective contribution to the energy flux is simply the transport of the energy per molecule by the total number flow,

$$\mathbf{J}_E^{\mathrm{conv}}(\mathbf{r},t) = \varepsilon(\mathbf{r},t)\mathbf{v}(\mathbf{r},t). \tag{5.12}$$

Since the energy density ε includes the external potential energy density, $n\psi \equiv \sum_k n_k \psi_k$, this term includes the potential energy carried by the mean convection, $\sum_k n_k \psi_k \mathbf{v}$. The excess is the external potential energy carried by the diffusive convective part,

$$\begin{aligned} \mathbf{J}_E^{\psi}(\mathbf{r},t) &= \sum_k n_k(\mathbf{r},t)\psi_k(\mathbf{r},t)\tilde{\mathbf{v}}_k(\mathbf{r},t) \\ &= \sum_k \psi_k(\mathbf{r},t)\mathbf{J}_{\mathrm{N},k}^0(\mathbf{r},t). \end{aligned} \tag{5.13}$$

The external field also contributes the source term, $\sum_k n_k \partial \psi_k / \partial t$.

Define the pressure tensor, $\underline{\underline{P}}(\mathbf{r},t)$, with $P_{\alpha\beta}$ being the β component of the force per unit area of the α face acting *on* the negative side of the face *by* the positive side. It can be shown that the tensor is symmetric.[4] For a fluid at rest, the pressure tensor is diagonal, with the three elements each equal to the thermodynamic pressure, p. Let the area of the α face be A_α, and let the volume be $V = A_\alpha L_\alpha$, for $\alpha = x$, y, or z (no summation convention). In general, the rate of change of energy of a molecule with velocity $\mathbf{v}$ in a force field $\mathbf{f}$ is $-\mathbf{f}\cdot\mathbf{v}$. Hence the rate of change of the energy of the volume V due to the pressure field and the motion of the molecules within it is

$$\begin{aligned}\frac{\Delta_p(V\varepsilon)}{\Delta t} &= \sum_{\alpha\beta} A_\alpha \left[P_{\alpha\beta}(\mathbf{r}-L_\alpha/2,t)v_\beta(\mathbf{r}-L_\alpha/2,t)\right.\\ &\quad \left. - P_{\alpha\beta}(\mathbf{r}+L_\alpha/2,t)v_\beta(\mathbf{r}+L_\alpha/2,t)\right]\\ &= -\sum_{\alpha\beta} A_\alpha L_\alpha \frac{\partial}{\partial x_\alpha}\left[P_{\alpha\beta}(\mathbf{r},t)v_\beta(\mathbf{r},t)\right]\\ &= -V\nabla\cdot\left[\underline{\underline{P}}(\mathbf{r},t)\cdot\mathbf{v}(\mathbf{r},t)\right]. \end{aligned} \tag{5.14}$$

Note that the difference between the two terms in the first equality arises because it is the force acting *on* the volume across each interface that is required. Hence the derivative of the energy due to the pressure work done following the flow is

$$\frac{\mathrm{d}(V\varepsilon)}{\mathrm{d}t} = -V\nabla\cdot\left[\underline{\underline{P}}\cdot\mathbf{v}\right]. \tag{5.15}$$

Using the fact that the compressibility of the flow gives the rate of change of the volume, $\mathrm{d}V/\mathrm{d}t = V\nabla\cdot\mathbf{v}$, the partial derivative of the energy density due to pressure work and to convective transport, and in the absence of an external field, is therefore

$$\begin{aligned}\frac{\partial\varepsilon}{\partial t} &= \frac{\mathrm{d}\varepsilon}{\mathrm{d}t} - \mathbf{v}\cdot\nabla\varepsilon\\ &= \frac{1}{V}\frac{\mathrm{d}(V\varepsilon)}{\mathrm{d}t} - \frac{\varepsilon}{V}\frac{\mathrm{d}V}{\mathrm{d}t} - \mathbf{v}\cdot\nabla\varepsilon\\ &= -\nabla\cdot\left[\underline{\underline{P}}\cdot\mathbf{v}\right] - \varepsilon\nabla\cdot\mathbf{v} - \mathbf{v}\cdot\nabla\varepsilon\\ &= -\nabla\cdot\left[\underline{\underline{P}}\cdot\mathbf{v}\right] - \nabla\cdot(\varepsilon\mathbf{v}). \end{aligned} \tag{5.16}$$

From this one can identify the energy flux due to the work done as

$$\mathbf{J}_E^{\mathrm{work}}(\mathbf{r},t) = \underline{\underline{P}}(\mathbf{r},t)\cdot\mathbf{v}(\mathbf{r},t). \tag{5.17}$$

In view of the above, the rate of change of energy density at a fixed position is

$$\frac{\partial\varepsilon(\mathbf{r},t)}{\partial t} = \sum_k n_k(\mathbf{r},t)\frac{\partial\psi_k(\mathbf{r},t)}{\partial t} - \nabla\cdot\mathbf{J}_{\mathrm{E}}(\mathbf{r},t), \tag{5.18}$$

[4] Faber, T. E. (1995), *Fluid Dynamics for Physicists*, (Cambridge University Press, Cambridge).

where the total energy flux is

$$\begin{aligned}\mathbf{J}_{\mathrm{E}}(\mathbf{r},t) &= \mathbf{J}^0_{\mathrm{E}}(\mathbf{r},t) + \mathbf{J}^{\mathrm{conv}}_E(\mathbf{r},t) + \mathbf{J}^{\psi}_E(\mathbf{r},t) + \mathbf{J}^{\mathrm{work}}_E(\mathbf{r},t) \\ &= \mathbf{J}^0_{\mathrm{E}}(\mathbf{r},t) + \varepsilon(\mathbf{r},t)\mathbf{v}(\mathbf{r},t) + \sum_k \psi_k(\mathbf{r},t)\mathbf{J}^0_{\mathrm{N},k}(\mathbf{r},t) \\ &\quad + \underline{\underline{P}}(\mathbf{r},t)\cdot\mathbf{v}(\mathbf{r},t). \end{aligned} \tag{5.19}$$

The pressure tensor comprises the equilibrium pressure p, and the viscous pressure tensor $\underline{\underline{\Pi}}$, $\underline{\underline{P}} = p\underline{\underline{I}} + \underline{\underline{\Pi}}$, where $\underline{\underline{I}}$ is the identity tensor. The viscous pressure tensor arises in shear flow and causes viscous heating. If it is neglected, so that $P_{\alpha\beta} = p\delta_{\alpha\beta}$, then the energy flux becomes (no external potential)

$$\mathbf{J}_E(\mathbf{r},t) = [\varepsilon(\mathbf{r},t) + p(\mathbf{r},t)]\mathbf{v}(\mathbf{r},t) + \mathbf{J}^0_E(\mathbf{r},t), \;\; \psi = \Pi = 0. \tag{5.20}$$

The bracketed term is the enthalpy density, and this form for the energy flux often appears in textbooks. The argument for neglecting the viscous terms is that the constitutive relations show that they are linear in the velocity, and hence they contribute a second-order term (i.e. $\Pi_{\alpha\beta}v_\beta \sim \mathcal{O}v^2$). This is a non-linear term, and so neglecting them is said to give a linear theory.

The internal energy density is the molecular part of the energy density, which eliminates the contribution of the external potential and the centre of mass kinetic energy density,

$$\begin{aligned}\varepsilon^{\mathrm{int}}(\mathbf{r},t) &= \varepsilon(\mathbf{r},t) - n(\mathbf{r},t)\psi(\mathbf{r},t) - \rho(\mathbf{r},t)v(\mathbf{r},t)^2/2 \\ &= \varepsilon(\mathbf{r},t) - \underline{n}(\mathbf{r},t)\cdot\underline{\psi}(\mathbf{r},t) - \frac{\mathbf{p}(\mathbf{r},t)\cdot\mathbf{p}(\mathbf{r},t)}{2\underline{m}\cdot\underline{n}(\mathbf{r},t)}. \end{aligned} \tag{5.21}$$

In terms of this, the energy flux may be written

$$\mathbf{J}_{\mathrm{E}}(\mathbf{r},t) = \mathbf{J}^0_{\mathrm{E}} + \varepsilon^{\mathrm{int}}\mathbf{v} + \frac{1}{2}\rho v^2\mathbf{v} + \sum_k \psi_k\mathbf{J}_{\mathrm{N},k} + \underline{\underline{P}}\cdot\mathbf{v}. \tag{5.22}$$

Everything on the right-hand side is a function of $\mathbf{r}$ and of t.

5.1.4 Linear Momentum

The gradient of the external potential provides a source of momentum. Hence the rate of change of momentum density is the sum of that due to the source and that due to divergence of the momentum flux,

$$\frac{\partial\mathbf{p}(\mathbf{r},t)}{\partial t} = -\sum_k n_k(\mathbf{r},t)\nabla\psi_k(\mathbf{r},t) - \nabla\cdot\underline{\underline{J}}_{\mathrm{P}}(\mathbf{r},t). \tag{5.23}$$

Recall that the momentum is $\mathbf{p} = \rho\mathbf{v}$.

The momentum flux tensor can be obtained from the rate of change of the momentum of the volume V due to the pressure field,

$$\begin{aligned}\frac{\Delta_p V p_\beta}{\Delta t} &= \sum_\alpha A_\alpha\left[P_{\alpha\beta}(\mathbf{r} - L_\alpha/2,t) - P_{\alpha\beta}(\mathbf{r} + L_\alpha/2,t)\right] \\ &= -\sum_\alpha A_\alpha L_\alpha \frac{\partial}{\partial x_\alpha}P_{\alpha\beta}(\mathbf{r},t). \end{aligned} \tag{5.24}$$

Hence the derivative of the momentum following the flow due to the pressure field is (in the absence of an external potential)

$$\rho(\mathbf{r},t)\frac{\mathrm{d}\mathbf{v}(\mathbf{r},t)}{\mathrm{d}t} = -\nabla\cdot\underline{\underline{P}}(\mathbf{r},t). \tag{5.25}$$

(The mass of the volume element is constant following the flow.) Now,

$$\begin{aligned}
\frac{\partial\mathbf{p}}{\partial t} &= \frac{\partial(\rho\mathbf{v})}{\partial t} \\
&= \frac{\mathrm{d}(\rho\mathbf{v})}{\mathrm{d}t} - \mathbf{v}\cdot\nabla(\rho\mathbf{v}) \\
&= \rho\frac{\mathrm{d}\mathbf{v}}{\mathrm{d}t} + \mathbf{v}\frac{\mathrm{d}\rho}{\mathrm{d}t} - \mathbf{v}\cdot\nabla(\rho\mathbf{v}) \\
&= \rho\frac{\mathrm{d}\mathbf{v}}{\mathrm{d}t} - \mathbf{v}\rho\nabla\cdot\mathbf{v} - \mathbf{v}\cdot\nabla(\rho\mathbf{v}) \\
&= -\nabla\cdot\underline{\underline{P}} - \nabla\cdot[\rho\mathbf{v}\mathbf{v}].
\end{aligned} \tag{5.26}$$

From this the momentum flux tensor may be identified as just the pressure tensor plus the convective momentum flux,

$$\underline{\underline{J}}_{\mathrm{P}}(\mathbf{r},t) = \underline{\underline{P}}(\mathbf{r},t) + \mathbf{p}(\mathbf{r},t)\mathbf{v}(\mathbf{r},t). \tag{5.27}$$

The diffusive part of the momentum flux is the viscous pressure tensor, $\underline{\underline{J}}_{\mathrm{P}}^{0} \equiv \underline{\underline{\Pi}}$.

5.2 Entropy Density and its Rate of Change

Now the first entropy density for the system is obtained in the presence of convecting fluxes. Then the above conservation laws are used to derive the dissipation.

The first entropy density for the actual sub-system with the external potential and fluxes is written $\sigma(\mathbf{r},t)$. The entropy density of the related sub-system without external potential or fluxes, is denoted by the subscript 0. This has energy density equal to the internal energy density of the original system, and number density equal to the original number density, and one must have

$$\sigma(\mathbf{r},t) = \sigma_0(\varepsilon^{\mathrm{int}}(\mathbf{r},t), \underline{n}(\mathbf{r},t)). \tag{5.28}$$

This follows because the entropy is essentially the number of accessible molecular configurations, and it is the internal part of the energy that is available to the molecular configurations. In view of this the entropy density may formally be written as a function of the energy, number, and momentum densities, and the external potentials, $\sigma(\varepsilon,\underline{n},\mathbf{p},\underline{\psi}) = \sigma_0(\varepsilon^{\mathrm{int}},\underline{n})$.

The energy density derivative of the entropy density gives the local temperature, $T(\mathbf{r},t)$,

$$\frac{\partial\sigma_0(\varepsilon,\underline{n})}{\partial\varepsilon} = \frac{1}{T}, \tag{5.29}$$

and the number density derivative gives the local chemical potential, $\mu_k(\mathbf{r},t)$,

$$\frac{\partial\sigma_0(\varepsilon,\underline{n})}{\partial n_k}=\frac{-\mu_k}{T},\text{ or }\frac{\partial\sigma_0(\varepsilon,\underline{n})}{\partial\underline{n}}=\frac{-\underline{\mu}}{T}. \tag{5.30}$$

It is straightforward to relate the pressure $p(\mathbf{r},t)$, to the entropy density (e.g. using the fact that the pressure is essentially the volume derivative of the entropy, or alternatively that it is the negative of the grand potential density),

$$\sigma_0(\varepsilon,\underline{n})=\left[\varepsilon+p-\underline{n}\cdot\underline{\mu}\right]/T. \tag{5.31}$$

This result will be used below. The chemical potential and the pressure are that of a stationary system without external potentials with the specified internal energy and number densities.

The above derivatives of the 'bare' entropy density, together with the internal energy given above, $\varepsilon^{\text{int}}=\varepsilon-\underline{n}\cdot\underline{\psi}-\mathbf{p}\cdot\mathbf{p}/2\underline{m}\cdot\underline{n}$, allow the derivatives of the entropy density itself to be obtained. One has

$$\left(\frac{\partial\sigma}{\partial\varepsilon}\right)_{\underline{n},\mathbf{p},\underline{\psi}}=\frac{\partial\sigma_0(\varepsilon^{\text{int}},\underline{n})}{\partial\varepsilon^{\text{int}}}=\frac{1}{T}, \tag{5.32}$$

$$\left(\frac{\partial\sigma}{\partial\underline{\psi}}\right)_{\varepsilon,\underline{n},\mathbf{p}}=\frac{\partial\sigma_0(\varepsilon^{\text{int}},\underline{n})}{\partial\varepsilon^{\text{int}}}\left(\frac{\partial\varepsilon^{\text{int}}}{\partial\underline{\psi}}\right)_{\varepsilon,\underline{n},\mathbf{p}}=\frac{-1}{T}\underline{n}, \tag{5.33}$$

$$\left(\frac{\partial\sigma}{\partial\mathbf{p}}\right)_{\varepsilon,\underline{n},\underline{\psi}}=\frac{\partial\sigma_0(\varepsilon^{\text{int}},\underline{n})}{\partial\varepsilon^{\text{int}}}\left(\frac{\partial\varepsilon^{\text{int}}}{\partial\mathbf{p}}\right)_{\varepsilon,\underline{n},\underline{\psi}}=\frac{-1}{T}\mathbf{v}, \tag{5.34}$$

and

$$\begin{aligned}\left(\frac{\partial\sigma}{\partial\underline{n}}\right)_{\varepsilon,\mathbf{p},\underline{\psi}}&=\frac{\partial\sigma_0(\varepsilon^{\text{int}},\underline{n})}{\partial\underline{n}}+\frac{\partial\sigma_0(\varepsilon^{\text{int}},\underline{n})}{\partial\varepsilon^{\text{int}}}\left(\frac{\partial\varepsilon^{\text{int}}}{\partial\underline{n}}\right)_{\varepsilon,\mathbf{p},\underline{\psi}}\\&=\frac{1}{T}\left[-\underline{\mu}-\underline{\psi}+v^2\underline{m}/2\right].\end{aligned} \tag{5.35}$$

With these thermodynamic derivatives, the rate of change of the entropy density may be obtained. One has

$$\begin{aligned}&\frac{\partial\sigma(\mathbf{r},t)}{\partial t}\\&=\frac{\partial\sigma(\varepsilon,\underline{n},\mathbf{p},\underline{\psi})}{\partial t}\\&=\frac{\partial\varepsilon}{\partial t}\left(\frac{\partial\sigma}{\partial\varepsilon}\right)_{\underline{n},\mathbf{p},\underline{\psi}}+\frac{\partial\underline{n}}{\partial t}\cdot\left(\frac{\partial\sigma}{\partial\underline{n}}\right)_{\varepsilon,\mathbf{p},\underline{\psi}}+\frac{\partial\mathbf{p}}{\partial t}\cdot\left(\frac{\partial\sigma}{\partial\mathbf{p}}\right)_{\varepsilon,\underline{n},\underline{\psi}}+\frac{\partial\underline{\psi}}{\partial t}\cdot\left(\frac{\partial\sigma}{\partial\underline{\psi}}\right)_{\varepsilon,\underline{n},\mathbf{p}}\\&=\left[\frac{\partial\varepsilon}{\partial t}-\underline{n}\cdot\frac{\partial\underline{\psi}}{\partial t}\right]\frac{1}{T}-\frac{\partial\mathbf{p}}{\partial t}\cdot\frac{\mathbf{v}}{T}-\frac{\partial\underline{n}}{\partial t}\cdot\left[\underline{\mu}+\underline{\psi}-v^2\underline{m}/2\right]\frac{1}{T}.\end{aligned} \tag{5.36}$$

Using the conservation laws to express the partial time derivatives as fluxes, this becomes

$$\begin{aligned}\frac{\partial\sigma(\mathbf{r},t)}{\partial t} &= \frac{-1}{T}\sum_k\left[\mu_k+\psi_k-m_kv^2/2\right]\left[\dot{n}_k^{\mathrm{react}}-\nabla\cdot(n_k\mathbf{v}_k)\right]\\ &\quad -\frac{1}{T}\nabla\cdot\left[\mathbf{J}_{\mathrm{E}}^0+\varepsilon^{\mathrm{int}}\mathbf{v}+\frac{\rho v^2\mathbf{v}}{2}+\sum_k\psi_k\mathbf{J}_{\mathrm{N},k}+\underline{\underline{P}}\cdot\mathbf{v}\right]\\ &\quad +\frac{\mathbf{v}}{T}\cdot\left[\sum_k n_k\nabla\psi_k+\nabla\cdot\left[\underline{\underline{P}}+\mathbf{p}\mathbf{v}\right]\right].\end{aligned} \tag{5.37}$$

There are three groups of terms of the right-hand side. The terms involving the external potential are

$$\begin{aligned}\mathrm{RHS}_\psi &= \frac{-1}{T}\sum_k\psi_k\left[\dot{n}_k^{\mathrm{react}}-\nabla\cdot(n_k\mathbf{v}_k)\right]-\frac{1}{T}\nabla\cdot\left[\sum_k\psi_kn_k\mathbf{v}_k\right]\\ &\quad +\frac{1}{T}\sum_k n_k\mathbf{v}\cdot\nabla\psi_k\\ &= -\frac{1}{T}\sum_k n_k\tilde{\mathbf{v}}_k\cdot\nabla\psi_k.\end{aligned} \tag{5.38}$$

Recall that since mass or charge is conserved in a chemical reaction, then $\sum_k\psi_k\nu_{\alpha k}=0$, which implies that $\sum_k\psi_k\dot{n}_k^{\mathrm{react}}=0$. The terms on the right-hand side involving higher powers of the velocity are

$$\begin{aligned}&\mathrm{RHS}_v\\ &= \frac{1}{T}\sum_k\frac{m_kv^2}{2}\left[\dot{n}_k^{\mathrm{react}}-\nabla\cdot(n_k\mathbf{v}_k)\right]-\frac{1}{2T}\nabla\cdot\left[\rho v^2\mathbf{v}\right]+\frac{\mathbf{v}}{T}\cdot\left[\nabla\cdot(\mathbf{p}\mathbf{v})\right]\\ &= \frac{-1}{2T}\sum_k v^2\nabla\cdot\mathbf{p}_k-\frac{1}{2T}\nabla\cdot\left[v^2\mathbf{p}\right]+\frac{\mathbf{v}}{T}\cdot\left[\nabla\cdot(\mathbf{p}\mathbf{v})\right]\\ &= \frac{-1}{2T}\sum_k v^2\nabla\cdot\mathbf{p}_k-\frac{1}{2T}\left[\mathbf{v}\cdot\mathbf{v}\nabla\cdot\mathbf{p}+2\mathbf{v}\mathbf{p}:\nabla\mathbf{v}\right]+\frac{1}{T}\left[\mathbf{v}\cdot\mathbf{v}\nabla\cdot\mathbf{p}+\mathbf{v}\mathbf{p}:\nabla\mathbf{v}\right]\\ &= \frac{-v^2}{2T}\sum_k\nabla\cdot\tilde{\mathbf{p}}_k=0.\end{aligned} \tag{5.39}$$

These manipulations may be readily confirmed by writing the scalar products as sums over the components. The terms on the right-hand side involving the pressure tensor are

$$\mathrm{RHS}_P = \frac{-1}{T}\nabla\cdot\left[\underline{\underline{P}}\cdot\mathbf{v}\right]+\frac{\mathbf{v}}{T}\cdot\left[\nabla\cdot\underline{\underline{P}}\right]=\frac{-1}{T}\left[p\nabla\cdot\mathbf{v}+\underline{\underline{\Pi}}:\nabla\mathbf{v}\right]. \tag{5.40}$$

The final bracketed term is simply $\underline{\underline{P}}:\nabla\mathbf{v}$. With these results the rate of change of entropy density becomes

$$\begin{aligned}
\frac{\partial \sigma(\mathbf{r},t)}{\partial t} \\
&= \frac{-1}{T}\sum_k \mu_k \left[\dot{n}_k^{\mathrm{react}} - \nabla\cdot(n_k\mathbf{v}_k)\right] - \frac{1}{T}\nabla\cdot\left[\mathbf{J}_{\mathrm{E}}^0 + \varepsilon^{\mathrm{int}}\mathbf{v}\right] \\
&\quad - \frac{1}{T}\sum_k n_k\tilde{\mathbf{v}}_k\cdot\nabla\psi_k - \frac{1}{T}\left[p\nabla\cdot\mathbf{v} + \underline{\underline{\Pi}} : \nabla\mathbf{v}\right] \\
&= \frac{-1}{T}\nabla\cdot\mathbf{J}_{\mathrm{E}}^0 - \frac{1}{T}\underline{\underline{\Pi}} : \nabla\mathbf{v} - \frac{1}{T}\sum_k \mu_k\dot{n}_k^{\mathrm{react}} - \frac{1}{T}\sum_k \mathbf{J}_{\mathrm{N},k}^0\cdot\nabla\psi_k \\
&\quad + \frac{1}{T}\sum_k \mu_k\nabla\cdot(n_k\tilde{\mathbf{v}}_k) + \frac{1}{T}\sum_k \mu_k\nabla\cdot(n_k\mathbf{v}) \\
&\quad - \frac{1}{T}\mathbf{v}\cdot\nabla\varepsilon^{\mathrm{int}} - \frac{1}{T}\varepsilon^{\mathrm{int}}\nabla\cdot\mathbf{v} - \frac{1}{T}p\nabla\cdot\mathbf{v} \\
&= \frac{-1}{T}\nabla\cdot\mathbf{J}_{\mathrm{E}}^0 - \frac{1}{T}\underline{\underline{\Pi}} : \nabla\mathbf{v} - \frac{1}{T}\sum_k \mu_k\dot{n}_k^{\mathrm{react}} - \frac{1}{T}\sum_k \mathbf{J}_{\mathrm{N},k}^0\cdot\nabla\psi_k \\
&\quad + \frac{1}{T}\sum_k \mu_k\nabla\cdot\mathbf{J}_{\mathrm{N},k}^0 - \mathbf{v}\cdot\nabla\sigma - \sigma\nabla\cdot\mathbf{v}.
\end{aligned} \tag{5.41}$$

This uses the expression for the entropy density, $\sigma = \left[\varepsilon^{\mathrm{int}} + p - \underline{n}\cdot\underline{\mu}\right]/T$, the result that $T\nabla\sigma = \nabla\varepsilon^{\mathrm{int}} - \sum_k \mu_k\nabla n_k$, and also the expression for the diffusive number flux, $\mathbf{J}_{\mathrm{N},k}^0 \equiv n_k\tilde{\mathbf{v}}_k$. Everything on the right-hand side is a function of $\mathbf{r}$ and t.

The source rate of change of each species is dependent upon the reaction rates, $\dot{n}_k^{\mathrm{react}} = -\sum_\alpha \dot{\xi}_\alpha\nu_{\alpha k}$. Typically, it is the reaction rates that are the independent variables. Accordingly, this contribution to the dissipation is better written in terms of the reaction rates,

$$-\frac{1}{T}\sum_k \mu_k\dot{n}_k^{\mathrm{react}} = \frac{1}{T}\sum_{k,\alpha}\dot{\xi}_\alpha\nu_{\alpha k}\mu_k = \frac{1}{T}\sum_\alpha \dot{\xi}_\alpha A_\alpha. \tag{5.42}$$

Here $A_\alpha = \sum_k \nu_{\alpha k}\mu_k$ is called the chemical affinity of the reaction. This term is a source of dissipation of the entropy density of the sub-system that arises from the chemical reactions that are occurring.

5.2.1 Sub-system Dissipation

The rate of change of the entropy density may be integrated over the volume to give the rate of change of the entropy of the sub-system, $\dot{S}_{\mathrm{s}}$. Integrating the divergences of the velocity and of the fluxes by parts, one obtains,

$$
\begin{aligned}
\dot{S}_{\mathrm{s}} &= \int \mathrm{d}\mathbf{r}\, \frac{\partial \sigma(\mathbf{r},t)}{\partial t} \\
&= \int \mathrm{d}\mathbf{r} \left\{ \frac{-1}{T} \nabla \cdot \mathbf{J}_{\mathrm{E}}^{0} - \frac{1}{T} \underline{\underline{\Pi}} : \nabla \mathbf{v} + \frac{1}{T} \sum_{\alpha} \dot{\xi}_{\alpha} A_{\alpha} \right. \\
&\quad \left. - \sum_{k} \left[\mathbf{J}_{\mathrm{N},k}^{0} \cdot \frac{\nabla \psi_k}{T} - \frac{\mu_k}{T} \nabla \cdot \mathbf{J}_{\mathrm{N},k}^{0} \right] - \nabla \cdot (\sigma \mathbf{v}) \right\} \\
&= \int \mathrm{d}\mathbf{r} \left\{ \mathbf{J}_{\mathrm{E}}^{0} \cdot \nabla \frac{1}{T} - \frac{1}{T} \underline{\underline{\Pi}} : \nabla \mathbf{v} + \frac{1}{T} \sum_{\alpha} \dot{\xi}_{\alpha} A_{\alpha} \right. \\
&\quad \left. - \sum_{k} \mathbf{J}_{\mathrm{N},k}^{0} \cdot \left[\frac{\nabla \psi_k}{T} + \nabla \frac{\mu_k}{T} \right] \right\} \\
&\quad - \oint \mathrm{d}\mathbf{r}\, \hat{\mathbf{n}} \cdot \left\{ \frac{\mathbf{J}_{\mathrm{E}}^{0}}{T} + \sigma \mathbf{v} - \sum_{k} \frac{\mu_k}{T} \mathbf{J}_{\mathrm{N},k}^{0} \right\}.
\end{aligned}
\tag{5.43}
$$

Here $\hat{\mathbf{n}}$ is the outward normal, and Stokes' theorem has been used for the partial integrations. Notice how the rate of change of entropy of the sub-system is a linear functional of the three diffusive fluxes, $\mathbf{J}_{\mathrm{E}}^{0}$, $\underline{\underline{\Pi}}$, and $\mathbf{J}_{\mathrm{N},k}^{0}$, and of the reaction rates $\dot{\xi}_{\alpha}$.

It is henceforth assumed that the boundaries are fixed and impermeable, in which case the velocities vanish on the boundaries, $\mathbf{v}_k(\mathbf{r},t) = \mathbf{0}$, $\mathbf{r} \in \mathrm{bdry}$, and only the diffusive heat flux across the boundary survives in the surface integral,

$$
\begin{aligned}
\dot{S}_{\mathrm{s}} &= \int \mathrm{d}\mathbf{r} \left\{ \mathbf{J}_{\mathrm{E}}^{0} \cdot \nabla \frac{1}{T} - \frac{1}{T} \underline{\underline{\Pi}} : \nabla \mathbf{v} + \frac{1}{T} \sum_{\alpha} \dot{\xi}_{\alpha} A_{\alpha} \right. \\
&\quad \left. - \sum_{k} \mathbf{J}_{\mathrm{N},k}^{0} \cdot \left[\frac{\nabla \psi_k}{T} + \nabla \frac{\mu_k}{T} \right] \right\} - \oint \mathrm{d}\mathbf{r}\, \hat{\mathbf{n}} \cdot \frac{\mathbf{J}_{\mathrm{E}}^{0}}{T}.
\end{aligned}
\tag{5.44}
$$

Everything in the integrand is a function of $\mathbf{r}$ and t. This simplification of fixed and impermeable boundaries is not essential since the full boundary integral will also cancel in the expression for the rate of total entropy production, as is discussed below.

This expression for the entropy production of the sub-system is quite general, and it holds for arbitrary diffusive fluxes and densities, including those with a specified time dependence. In such a constrained state, there need not be any relationship between the fluxes and the densities. There is no requirement that such a constrained state should be physically realisable; this expression for the entropy production of the sub-system holds in general.

5.2.2 Steady State

The rate of entropy production of the sub-system for constrained fluxes and fields will be used below to formulate the second entropy. But first an explicit expression for the dissipation in the steady state will be obtained. In this state there is a specific relationship between the fluxes and the densities; they cannot be chosen independently.

In the steady state, the structure of the sub-system (i.e. the densities at each location) does not change with time, and hence the entropy density is constant, $\partial\sigma(\mathbf{r},t)/\partial t = 0$, (steady state). It follows that $\dot{S}_\mathrm{s} = 0$, (steady state), or

$$\oint \mathrm{d}\mathbf{r}\,\hat{\mathbf{n}} \cdot \frac{\mathbf{J}_\mathrm{E}^0}{T} = \int \mathrm{d}\mathbf{r} \left\{ \mathbf{J}_\mathrm{E}^0 \cdot \nabla\frac{1}{T} - \frac{1}{T}\underline{\underline{\Pi}} : \nabla\mathbf{v} + \frac{1}{T}\sum_\alpha \dot{\xi}_\alpha A_\alpha - \sum_k \mathbf{J}_{\mathrm{N},k}^0 \cdot \left[\frac{\nabla\psi_k}{T} + \nabla\frac{\mu_k}{T} \right] \right\}, \text{ (steady state)}. \tag{5.45}$$

This expresses the nett heat flow across the boundary as an integral of the diffusive fluxes over the sub-system volume. This assumes that steady state fluxes exist for the specified velocity and temperature fields. In fact there is usually a limited set of such fields that are compatible with the steady state.

The reservoir entropy changes by just this heat flow across the boundary, divided by the reservoir temperature at that place on the boundary. This is true whether or not the heat flow across the boundary represents the steady state heat flow. Assuming that the sub-system temperature equals the reservoir temperature at that place on the boundary, $T(\mathbf{r},t) = T_\mathrm{r}(\mathbf{r},t)$, $\mathbf{r} \in \text{bdry}$, then the rate of change of the reservoir entropy is

$$\begin{aligned} \dot{S}_\mathrm{r} &= \oint \mathrm{d}\mathbf{r}\,\hat{\mathbf{n}} \cdot \frac{\mathbf{J}_\mathrm{E}^0}{T}, \text{ (general)} \\ &= \int \mathrm{d}\mathbf{r} \left\{ \mathbf{J}_\mathrm{E}^0 \cdot \nabla\frac{1}{T} - \frac{1}{T}\underline{\underline{\Pi}} : \nabla\mathbf{v} + \frac{1}{T}\sum_\alpha \dot{\xi}_\alpha A_\alpha \right. \\ &\qquad \left. - \sum_k \mathbf{J}_{\mathrm{N},k}^0 \cdot \left[\frac{\nabla\psi_k}{T} + \nabla\frac{\mu_k}{T} \right] \right\}, \text{ (steady state)}. \end{aligned} \tag{5.46}$$

The first equality is true in general (provided that energy is the only flux across the boundary; in the contrary case, the negative of the surface integral in Eq. (5.43) applies). The second equality holds only in the steady state. The total dissipation is the sum of the rates of entropy production of the sub-system and the reservoir, $\dot{S}_\mathrm{tot} = \dot{S}_\mathrm{r} + \dot{S}_\mathrm{s}$. Obviously in the steady state, this reduces to the entropy produced in the reservoir alone, $\dot{S}_\mathrm{tot} = \dot{S}_\mathrm{r}$, (steady state).

5.3 Second Entropy

The second entropy for a general isolated system was given in Ch. 2, Eq. (2.50). This was written in terms of fluxes, and was valid to linear order in the time step. In Ch. 4, Eq. (4.44), this was applied to the case of conductive heat flow in the presence of a reservoir. In general in the small time limit the second entropy consists of four terms: the ordinary first entropy for the structure, a term quadratic in the fluxes, a term bilinear in the fluxes and the thermodynamic forces that represents half the rate of total entropy production, and a term independent of the fluxes that ensures that the reduction condition is satisfied by the optimum fluxes.

An expression for the rate of first entropy production for the sub-system alone in the case of hydrodynamic flows was derived above, Eq. (5.43). Notice that the only fluxes that appear are the diffusive ones: the diffusive energy flux, $\mathbf{J}_{\mathrm{E}}^{0}$, which was called the conductive energy flux (or the heat flux in Ch. 4), the diffusive number fluxes, $\mathbf{J}_{\mathrm{N},k}^{0}$ or $\underline{\mathbf{J}}_{\mathrm{N}}^{0}$, the diffusive momentum flux tensor, $\underline{\underline{\Pi}}$, which is more commonly called the viscous pressure tensor, and the rates of chemical reaction, $\dot{\xi}_\alpha$ or $\underline{\dot{\xi}}$. These are essentially the molecular part of the full fluxes, and it is they that fluctuate. The convective and work parts of the fluxes are macroscopic in nature and do not directly contribute to the dissipation. This means that the second entropy for hydrodynamics is a quadratic function of the diffusive fluxes.

The dissipation of the sub-system, Eq. (5.43), contains a boundary integral that represents the heat flow into the sub-system divided by the temperature at that point. It was assumed that the boundaries were fixed and impermeable so that there were no convective fluxes across them. This is equal and opposite to the entropy production of the reservoirs for the general constrained state, the first equality in Eq. (5.46). This is true provided that the temperature of the sub-system at a point on the boundary is equal to the temperature of the reservoir at that point. This means that the boundary integral cancels, and the total dissipation is

$$\begin{aligned}\dot{S}_{\text{total}} &= \int \mathrm{d}\mathbf{r} \left\{ \mathbf{J}_{\mathrm{E}}^{0} \cdot \nabla \frac{1}{T} - \frac{1}{T} \underline{\underline{\Pi}} : \nabla \mathbf{v} + \frac{1}{T} \sum_{\alpha} \dot{\xi}_\alpha A_\alpha \right. \\ &\quad \left. - \sum_k \mathbf{J}_{\mathrm{N},k}^{0} \cdot \left[\frac{\nabla \psi_k}{T} + \nabla \frac{\mu_k}{T} \right] \right\}. \qquad (5.47)\end{aligned}$$

This is valid for arbitrary fluxes and fields. It is also valid if other fluxes are allowed to cross the boundaries of the sub-system, because also in this case, the surface integral representing the entropy production of the reservoir is equal and opposite to that representing the entropy production of the sub-system in Eq. (5.43), and the two surface integrals therefore cancel.

Now for the term in the second entropy that is quadratic in the dissipative fluxes. In principle the transport matrix has size equal to the square of the number of components of the diffusive fluxes and reaction rates. One can invoke the Curie symmetry principle to reduce this number.[5] Since the

[5]de Groot, S. R. and Mazur, P. (1984), *Non-equilibrium Thermodynamics*, (Dover, New York), §VI.2.

rate of dissipation is a scalar, for an isotropic system there can be no coupling between scalars, vectors, and traceless second rank tensors. Nor is there coupling between vector components in different directions. Accordingly, the viscous pressure tensor is decomposed into its scalar part and its traceless part, which is denoted by an asterisk, $\underline{\underline{\Pi}} = \pi \underline{I} + \underline{\underline{\Pi}}^*$. The scalar part is clearly $\pi = \mathrm{TR}[\underline{\underline{\Pi}}]/3$. The shear rate tensor may be similarly decomposed, $\nabla \mathbf{v} = [\nabla \cdot \mathbf{v}/3]\underline{I} + [\nabla \mathbf{v}]^*$. Since the viscous pressure tensor is symmetric, one has the result that $\underline{\underline{\Pi}} : \nabla \mathbf{v} = \pi \nabla \cdot \mathbf{v} + \underline{\underline{\Pi}}^* : [\nabla \mathbf{v}]^{\mathrm{sym},*}$. Because the three Cartesian coordinates are equivalent in an isotropic system, there is a single transport coefficient for the traceless part of the viscous pressure tensor, and this is related to the shear viscosity, η.

The scalar part of the viscous pressure tensor and the scalar reaction rates all couple. Let L_0 be the transport matrix for these scalars. If there are R chemical reactions, $\alpha = 1, 2, \ldots, R$, then L_0 has dimension $(R+1) \times (R+1)$. The reaction rate $\dot{\xi}_\alpha$ is the flux of a variable of even time parity, whereas the scalar part of the momentum flux tensor, π, is the flux of a variable of odd time parity. Hence L_0 is a mixed parity case. It can be written in block form, with symmetric block diagonals, and antisymmetric block adiagonals. In the event that there are no chemical reactions, the single transport coefficient is related to the bulk (or extensional) viscosity, η_b. The coupling between the scalar part of the viscous pressure tensor and the scalar reaction rates is often called the chemical viscosity, and it is generally small and often neglected.[6]

The vector heat flux and the vector diffusive fluxes all couple (in three blocks, one for each Cartesian component). If there are K species, then there are $K-1$ independent diffusive fluxes, and hence each Cartesian block has K^2 couplings and K^2 corresponding transport coefficients. An isotropic medium is assumed, and so these are the same for each block. The transport matrix $L_1 = \Lambda_1$ is symmetric, because both energy and number have even time parity. In the event that there is a single species, the single transport coefficient is related to the thermal conductivity, λ. More generally, they are related to the diffusion constants and to the Soret coefficients.

Finally, the traceless second rank tensor $\underline{\underline{\Pi}}^*$ couples to itself. The transport matrix $L_2 = \Lambda_2$ is 9×9. However for an isotropic medium it has only one independent entry, which is related to the shear viscosity, η.

These transport coefficients are possibly position dependent via their dependence on the local thermodynamic state, but this will be neglected here.

Now recall the formulae for the second entropy in the generic mixed parity case, §2.5.2. The second entropy in the small time limit, Eq. (2.92), is

$$\begin{aligned} S^{(2)}(\mathbf{x}', \mathbf{x}|\tau) &= S(\mathbf{x}) - \frac{|\tau|}{2} \Lambda^{-1} : \overset{\circ}{\mathbf{x}}{}^2 + \frac{\tau}{2} \overset{\circ}{\mathbf{x}} \cdot \mathbf{F}(\mathbf{x}; \hat{\tau}) \\ &\quad - \frac{|\tau|}{2} \Lambda^{-1} : \overline{\overset{\circ}{\mathbf{x}}}(\mathbf{x}; \hat{\tau})^2 + \mathcal{O}(\tau^2), \end{aligned} \tag{5.48}$$

[6]Kuiken, G. D. C. (1984), *Thermodynamics for Irreversible Processes*, (John Wiley and Sons, Chichester, England), pp. 133–135.

where the first entropy is $S(\mathbf{x}) = S : \mathbf{x}^2/2$, and the mixed parity force is

$$\begin{aligned}\mathbf{F}(\mathbf{x};\hat{\tau}) &\equiv 2\left[A_0 + B_0 - \hat{\tau}(A_0' + B_0')\right]\mathbf{x} \\ &= \left[I - 2\hat{\tau}\Lambda^{-1}\Theta\right] S\mathbf{x} \\ &\equiv \Lambda^{-1}L(\hat{\tau})S\mathbf{x}. \end{aligned} \tag{5.49}$$

One can see in this expression for the second entropy the four terms mentioned above: the first entropy of the structure, the term quadratic in the fluxes, the term bilinear in the fluxes and the thermodynamic forces that represents half the rate of total entropy production, and the term independent of the fluxes that ensures that the reduction condition is satisfied by the optimum fluxes. One has three matrices: the ordinary fluctuation matrix S, the flux fluctuation matrix Λ, and the transport matrix, $L(\hat{\tau})$. For fluctuations all of the same parity, $L(\hat{\tau}) = \Lambda$.

In hydrodynamics, one replaces the coarse velocities by the instantaneous fluxes. It is understood that in this limit one is dealing with future changes, so that $\tau > 0$ and $\hat{\tau} = 1$. In what follows, the first of these conventions, $|\tau| \to 0$, will be observed. However the dependence of the transport matrix and the most likely fluxes on the direction of time will be shown explicitly, with the understanding that these reduce to the conventional expression when $\hat{\tau} = 1$.

For the present hydrodynamic fluxes, in view of the couplings discussed above, one can write the second entropy as

$$S^{(2)} = S_{\rm tot} + |\tau| \left[S_0^{(2)} + S_1^{(2)} + S_2^{(2)} \right]. \tag{5.50}$$

The total first entropy is

$$S_{\rm tot} = \int \mathrm{d}\mathbf{r}\, \sigma_0(\varepsilon^{\rm int}(\mathbf{r},t), \underline{n}(\mathbf{r},t)) + \oint \mathrm{d}\mathbf{r}\, \frac{E_{\rm r}(\mathbf{r},t)}{T_{\rm r}(\mathbf{r})}. \tag{5.51}$$

The first term is evidently the sub-system entropy, and the second term is just the reservoir entropy, which is the boundary integral representation of Eq. (4.40). When optimised with respect to the internal energy density, this sets the sub-system temperature equal to the reservoir temperature on the boundary of the sub-system. In the present problem it has been assumed that only energy is exchangeable between the sub-system and the reservoir.

The second entropy for the scalar fluxes is a mixed parity case. It contains three terms: one quadratic in the flux, one linear in the flux, and one independent of the flux, One has

$$\begin{aligned} S_0^{(2)} &= \int \mathrm{d}\mathbf{r}\, \frac{-1}{2}\Lambda_0^{-1} : \{\pi, \underline{\dot{\xi}}\}^2 - \frac{1}{2}\Lambda_0^{-1} : \{\overline{\pi}, \overline{\underline{\dot{\xi}}}\}^2 \\ &\quad + \frac{\hat{\tau}}{2T}\{\pi, \underline{\dot{\xi}}\} \cdot \Lambda_0^{-1} L_0(\hat{\tau}) \{-\nabla\cdot\mathbf{v}, \underline{A}\}. \end{aligned} \tag{5.52}$$

From Eq. (5.43), one can see that the rate of entropy production due to the scalar part of the viscous pressure tensor is $(-1/T)\pi\nabla\cdot\mathbf{v}$, and that part due to chemical reactions is $(1/T)\sum_\alpha \dot{\xi}_\alpha A_\alpha$. These allow one to identify the analogue of the generic thermodynamic force $S\mathbf{x}$ in the linear flux term. The second term in the integrand depends upon the optimum value of the flux, which means that it depends upon the structure, but not upon the constrained or fluctuating value of the flux. This term makes $\overline{S}_0^{(2)} = 0$.

The second entropy for the vector fluxes is a pure parity case, which means that $L_1(\hat{\tau}) = \Lambda_1$. Again there are three terms and one has

$$\begin{aligned} S_1^{(2)} &= \int \mathrm{d}\mathbf{r} \, \frac{-1}{2}\Lambda_1^{-1} : \{\mathbf{J}_\mathrm{E}^0, \underline{\mathbf{J}}_\mathrm{N}^0\}^2 - \frac{1}{2}\Lambda_1^{-1} : \{\overline{\mathbf{J}}_\mathrm{E}^0, \overline{\underline{\mathbf{J}}}_\mathrm{N}^0\}^2 \\ &\quad + \frac{\hat{\tau}}{2}\{\mathbf{J}_\mathrm{E}^0, \underline{\mathbf{J}}_\mathrm{N}^0\} \cdot \left\{ \nabla\frac{1}{T}, \left[\frac{-1}{T}\nabla\underline{\psi} - \nabla\frac{\underline{\mu}}{T}\right]\right\}. \end{aligned} \tag{5.53}$$

In an isotropic system, there is no coupling between different Cartesian components, and so Λ_1 is composed of three identical blocks on the main diagonal.

The second entropy for the traceless second rank tensor fluxes, the traceless part of the viscous pressure tensor, is also a pure parity case, which means that $L_2(\hat{\tau}) = \Lambda_2$. One has

$$S_2^{(2)} = \int \mathrm{d}\mathbf{r} \, \frac{-1}{2}\Lambda_2^{-1} : \underline{\underline{\Pi}}^*\underline{\underline{\Pi}}^* - \frac{1}{2}\Lambda_2^{-1} : \overline{\underline{\underline{\Pi}}}^*\overline{\underline{\underline{\Pi}}}^* - \frac{\hat{\tau}}{2T}\underline{\underline{\Pi}}^* : [\nabla\mathbf{v}]^{\mathrm{sym},*}. \tag{5.54}$$

In an isotropic system, there is a single transport constant, and $\Lambda_2^{-1} : \underline{\underline{\Pi}}^*\underline{\underline{\Pi}}^* = \lambda_2^{-1}\underline{\underline{\Pi}}^* : \underline{\underline{\Pi}}^*$.

In these three expressions for the τ-dependent parts of the second entropy, the coarse velocity has been written as an instantaneous flux.

5.3.1 Variational Principle

In the next subsection and section, it will be shown that the usual partial differential equations of hydrodynamics emerge from the maximisation of the second entropy with respect to the diffusive fluxes. But it is worth pausing to point out that the expression for the second entropy, Eq. (5.50) *et seq.*, is a variational principle for the diffusive fluxes, $\dot{\underline{\xi}}$, $\mathbf{J}_\mathrm{E}^0$, $\underline{\mathbf{J}}_\mathrm{N}^0$, and $\underline{\underline{\Pi}}$, and for the thermodynamic fields, say $\varepsilon^{\mathrm{int}}$, $\underline{n}$, and $\mathbf{v}$ (since any other field can be obtained from these). Hence, maximising Eq. (5.50) with respect to all of these simultaneously, provides a novel way to solve a given hydrodynamic problem. One might expect that in many cases this would have numerical and computational advantages over the traditional partial differential equation approach.

5.3.2 Flux Optimisation

The equations of hydrodynamics apply only to the most likely fluxes. These follow by maximising the second entropy with respect to the conductive fluxes. Functional differentiation with respect to the scalar fluxes gives

$$\frac{\delta S_0^{(2)}}{\delta\{\pi(\mathbf{r}), \dot{\underline{\xi}}(\mathbf{r})\}} = -\Lambda_0^{-1}\{\pi(\mathbf{r}), \dot{\underline{\xi}}(\mathbf{r})\} + \frac{\hat{\tau}}{2T}\Lambda_0^{-1}L_0(\hat{\tau})\{-\nabla\cdot\mathbf{v}(\mathbf{r}), \underline{A}(\mathbf{r})\}. \tag{5.55}$$

This vanishes for the most likely scalar fluxes, which are therefore given by

$$\{\overline{\pi}(\mathbf{r}), \overline{\underline{\xi}}(\mathbf{r})\} = \frac{\hat{\tau}}{2T} L_0(\hat{\tau})\{-\nabla\cdot\mathbf{v}(\mathbf{r}), \underline{A}(\mathbf{r})\}. \tag{5.56}$$

Here and below, the dependence of the temperature and other field variables and of the transport matrix on the position has been neglected (unless the gradient of the field variable is required). Similarly neglected is any possible time dependence. The equations are expected to apply, provided that any time or spatial variation is sufficiently slow.

Functional differentiation with respect to the vector fluxes gives

$$\begin{aligned}\frac{\delta S_1^{(2)}}{\delta\{\mathbf{J}_\mathrm{E}^0(\mathbf{r}), \underline{\mathbf{J}}_\mathrm{N}^0(\mathbf{r})\}} &= -\Lambda_1^{-1}\{\mathbf{J}_\mathrm{E}^0(\mathbf{r}), \underline{\mathbf{J}}_\mathrm{N}^0(\mathbf{r})\} \\ &\quad + \frac{\hat{\tau}}{2}\left\{\nabla\frac{1}{T(\mathbf{r})}, \left[\frac{-1}{T(\mathbf{r})}\nabla\underline{\psi}(\mathbf{r}) - \nabla\frac{\underline{\mu}(\mathbf{r})}{T(\mathbf{r})}\right]\right\}. \end{aligned}\tag{5.57}$$

The most likely vector fluxes are therefore given by

$$\{\overline{\mathbf{J}}_\mathrm{E}^0(\mathbf{r}), \overline{\underline{\mathbf{J}}}_\mathrm{N}^0(\mathbf{r})\} = \frac{\hat{\tau}}{2}\Lambda_1\left\{\nabla\frac{1}{T(\mathbf{r})}, \left[\frac{-1}{T(\mathbf{r})}\nabla\underline{\psi}(\mathbf{r}) - \nabla\frac{\underline{\mu}(\mathbf{r})}{T(\mathbf{r})}\right]\right\}. \tag{5.58}$$

Functional differentiation with respect to the traceless momentum flux tensor gives

$$\frac{\delta S_2^{(2)}}{\delta\underline{\underline{\Pi}}^*(\mathbf{r})} = -\Lambda_2^{-1}\underline{\underline{\Pi}}^*(\mathbf{r}) - \frac{\hat{\tau}}{2T}\left[\nabla\mathbf{v}(\mathbf{r})\right]^{\mathrm{sym},*}. \tag{5.59}$$

The most likely flux is

$$\overline{\underline{\underline{\Pi}}}^*(\mathbf{r}) = -\frac{\hat{\tau}}{2T}\Lambda_2\left[\nabla\mathbf{v}(\mathbf{r})\right]^{\mathrm{sym},*}. \tag{5.60}$$

For an isotropic system this reduces to

$$\overline{\underline{\underline{\Pi}}}^*(\mathbf{r}) = -2\hat{\tau}\eta\left[\nabla\mathbf{v}(\mathbf{r})\right]^{*,\mathrm{sym}}, \tag{5.61}$$

or in component form

$$\overline{\Pi}_{\alpha\beta}(\mathbf{r}) = -\hat{\tau}\eta\left[\frac{\partial v_\alpha(\mathbf{r})}{\partial r_\beta} + \frac{\partial v_\beta(\mathbf{r})}{\partial r_\alpha} - \frac{2}{3}\delta_{\alpha\beta}\nabla\cdot\mathbf{v}(\mathbf{r})\right]. \tag{5.62}$$

Here η is called the shear viscosity.

When there are no chemical reactions, the most likely value of the scalar part of the viscous pressure tensor is

$$\overline{\pi}(\mathbf{r}) = -\hat{\tau}\eta_\mathrm{b}\nabla\cdot\mathbf{v}(\mathbf{r}), \tag{5.63}$$

where η_b is called the bulk viscosity, or the extensional viscosity.

In the absence of diffusion, the most likely value of the heat flux is

$$\overline{\mathbf{J}}_{\mathrm{E}}^{0}(\mathbf{r}) = -\hat{\tau}\lambda\nabla T(\mathbf{r}), \tag{5.64}$$

where λ is called the thermal conductivity.

These expressions for the most likely fluxes, together with the conservation laws and the equilibrium equation of state, are formally equivalent to the equations of non-linear hydrodynamics.

As mentioned in the introduction to this chapter, the present variational approach to hydrodynamics is rather similar to the theory of hydrodynamic fluctuations of Landau and Lifshitz, (1959). Fox and Uhlenbeck (1970) gave a more detailed derivation of those results that used a stochastic equation approach and Onsager's reciprocal relations. These two approaches have in common with the present second entropy theory, that they treat the fluxes as fluctuating or constrained variables that are independent of the thermodynamic forces; the linear proportionality assumed by conventional hydrodynamics emerges in the optimum state. The quadratic forms that appear in the second entropy theory can be related to time correlation functions, Ch. 2, of the same type as given by Landau and Lifshitz (1959) for their coefficients, and as given by Fox and Uhlenbeck (1970) in the exponent of their transition probability. At this quadratic level of approximation, the fluctuations in all three approaches are Gaussian in nature. A number of authors (e.g. Keizer, (1987), Ortiz de Zárate and Sengers, (2006)), have applied this flux fluctuation formulation of hydrodynamics to experimental measurement and various extensions of the theory.

5.4 Navier-Stokes and Energy Equations

This section uses the conservation laws of §5.1 to recast the hydrodynamic equations derived above by maximising the second entropy, §5.3.2, into the more familiar form of the Navier-Stokes and other convective equations. For simplicity, chemical reactions will be excluded and a single component fluid will be used.

The number conservation law is

$$\frac{\partial n(\mathbf{r},t)}{\partial t} = -\nabla \cdot \mathbf{J}_{\mathrm{N}}(\mathbf{r},t). \tag{5.65}$$

Here $n(\mathbf{r},t)$ is the number density, which is related to the mass density by $\rho(\mathbf{r},t) = mn(\mathbf{r},t)$, where m is the molecular mass, and $\mathbf{J}_{\mathrm{N}}(\mathbf{r},t) = n(\mathbf{r},t)\mathbf{v}(\mathbf{r},t)$ is the number flux, with $\mathbf{v}(\mathbf{r},t)$ being the velocity of the flow.

The energy conservation law is

$$\frac{\partial \varepsilon(\mathbf{r},t)}{\partial t} = n(\mathbf{r},t)\frac{\partial \psi(\mathbf{r},t)}{\partial t} - \nabla \cdot \mathbf{J}_{\mathrm{E}}(\mathbf{r},t). \tag{5.66}$$

Here $\psi(\mathbf{r},t)$ is the external potential, and the energy flux may be written

$$\begin{aligned}\mathbf{J}_{\mathrm{E}}(\mathbf{r},t) &= \mathbf{J}_{E}^{0}(\mathbf{r},t) + \mathbf{J}_{E}^{\mathrm{conv}}(\mathbf{r},t) + \mathbf{J}_{E}^{\mathrm{work}}(\mathbf{r},t) \\ &= \mathbf{J}_{\mathrm{E}}^{0}(\mathbf{r},t) + \varepsilon(\mathbf{r},t)\mathbf{v}(\mathbf{r},t) + \underline{\underline{P}}(\mathbf{r},t)\cdot\mathbf{v}(\mathbf{r},t).\end{aligned} \tag{5.67}$$

Here $\mathbf{J}_{\mathrm{E}}^{0}$ may be called the diffusive energy flow, or the heat flow, $\varepsilon^{\mathrm{int}} = \varepsilon - mnv^2/2 - n\psi$ is the internal energy density, and $\underline{\underline{P}} = pI + \underline{\underline{\Pi}}$ is the pressure tensor, p being the thermodynamic pressure, and $\underline{\underline{\Pi}}$ being the viscous pressure tensor, which can also be called the diffusive momentum flux. For this one-component system, the diffusive energy flow, $\mathbf{J}_{E}^{\psi}$, vanishes because $\mathbf{J}_{N}^{0} = 0$.

The momentum conservation law is

$$\frac{\partial \mathbf{p}(\mathbf{r},t)}{\partial t} = -n(\mathbf{r},t)\nabla\psi(\mathbf{r},t) - \nabla \cdot \underline{\underline{J}}_{\mathrm{P}}(\mathbf{r},t). \tag{5.68}$$

Recall that the momentum density is $\mathbf{p} = \rho\mathbf{v} = mn\mathbf{v}$. The momentum flux tensor may be identified as just the pressure tensor plus the convective momentum flux,

$$\underline{\underline{J}}_{\mathrm{P}}(\mathbf{r},t) = \underline{\underline{\Pi}}(\mathbf{r},t) + p(\mathbf{r},t)I + \mathbf{p}(\mathbf{r},t)\mathbf{v}(\mathbf{r},t). \tag{5.69}$$

The phenomenological transport laws for the most likely fluxes in this case (isotropic system, no chemical reactions, single component) are straightforward to derive. Here and below it is assumed $\hat{\tau} = 1$, which means that the equations refer to the future most likely fluxes. It is also assumed that the transport coefficients may be taken as constants, and that any time variation of the fluxes is sufficiently slow that the second entropy treatment is valid.

The most likely value of the scalar part of the viscous pressure tensor is

$$\overline{\pi}(\mathbf{r},t) = -\eta_{\mathrm{b}}\nabla\cdot\mathbf{v}(\mathbf{r},t), \tag{5.70}$$

where η_{b} is called the bulk (or volume, or extensional) viscosity. For an incompressible fluid, $\nabla\cdot\mathbf{v} = 0$, so that one can set $\overline{\pi} = 0$. For many polymer melts, an order of magnitude estimate is $\eta_{\mathrm{b}} \approx 6\eta$.[7]

In the absence of diffusion, the most likely value of the heat flux is

$$\overline{\mathbf{J}}_{\mathrm{E}}^{0}(\mathbf{r},t) = -\lambda\nabla T(\mathbf{r},t), \tag{5.71}$$

where λ is called the thermal conductivity.

The most likely value of the traceless viscous pressure tensor for an isotropic system reduces to

$$\overline{\underline{\underline{\Pi}}}^{*}(\mathbf{r},t) = -2\eta\left[\nabla\mathbf{v}(\mathbf{r},t)\right]^{*,\mathrm{sym}}, \tag{5.72}$$

or in component form

$$\overline{\Pi}_{\alpha\beta}^{*}(\mathbf{r},t) = -\eta\left[\frac{\partial v_\alpha(\mathbf{r},t)}{\partial r_\beta} + \frac{\partial v_\beta(\mathbf{r},t)}{\partial r_\alpha} - \frac{2}{3}\delta_{\alpha\beta}\nabla\cdot\mathbf{v}(\mathbf{r},t)\right]. \tag{5.73}$$

Here η is called the shear viscosity. The most likely components of the full viscous pressure tensor are

$$\overline{\Pi}_{\alpha\beta}(\mathbf{r},t) = -\eta\left[\frac{\partial v_\alpha(\mathbf{r},t)}{\partial r_\beta} + \frac{\partial v_\beta(\mathbf{r},t)}{\partial r_\alpha}\right] + \left(\frac{2\eta}{3} - \eta_{\mathrm{b}}\right)\delta_{\alpha\beta}\nabla\cdot\mathbf{v}(\mathbf{r},t). \tag{5.74}$$

[7] Chatraei, Sh. and Macosko, C. W. (1981), J. Rheology **25**, 433.

The viscous dissipation is defined as

$$\Phi(\mathbf{r},t) \equiv -\underline{\underline{\Pi}}(\mathbf{r},t) : [\nabla \mathbf{v}(\mathbf{r},t)] = -\pi \nabla \cdot \mathbf{v}(\mathbf{r},t) - \underline{\underline{\Pi}}^*(\mathbf{r},t) : [\nabla \mathbf{v}(\mathbf{r},t)]^{*,\mathrm{sym}}. \quad (5.75)$$

Its most likely value is

$$\begin{aligned} \overline{\Phi}(\mathbf{r},t) &= \eta_\mathrm{b}\theta(\mathbf{r},t)^2 + 2\eta \left([\nabla \mathbf{v}(\mathbf{r},t)]^{\mathrm{sym}} - \frac{1}{3}\theta(\mathbf{r},t) I \right)^2 \\ &= \left(\eta_\mathrm{b} - \frac{2}{3}\eta \right) \theta(\mathbf{r},t)^2 + 2\eta [\nabla \mathbf{v}(\mathbf{r},t)]^{\mathrm{sym}} : [\nabla \mathbf{v}(\mathbf{r},t)]^{\mathrm{sym}}, \end{aligned} \quad (5.76)$$

where $\theta(\mathbf{r},t) \equiv \nabla \cdot \mathbf{v}(\mathbf{r},t)$, and $[\nabla \mathbf{v}(\mathbf{r},t)]^{\mathrm{sym}} \equiv [\nabla \mathbf{v}(\mathbf{r},t) + (\nabla \mathbf{v}(\mathbf{r},t))^{\mathrm{T}}]/2$.

In these expressions, the transport coefficients, formally introduced in §5.3.2, have been replaced by their more conventional counterparts. This replacement may involve factors of temperature, the variation of which in space and time is neglected, as it is in the transport coefficients themselves.

Thermodynamics

In hydrodynamics, the traditional thermodynamic field variables are the temperature, pressure, and velocity. Hence the appropriate free energy to invoke is the Gibbs free energy, $G_0(N,p,T)$. This is for a sub-system containing N molecules that can exchange energy and volume with a reservoir of temperature T and pressure p. Here the subscript 0 emphasises that this is for a system at rest with no external potential, and hence the energy is the same as the internal energy.

The pressure derivative of the Gibbs free energy gives the volume,[8]

$$\left(\frac{\partial G_0}{\partial p} \right)_{N,T} = \overline{V}, \quad (5.77)$$

and a particular temperature derivative gives the energy, namely

$$\left(\frac{\partial (G_0/T)}{\partial T^{-1}} \right)_{N,p/T} = \overline{E}^{\mathrm{int}}. \quad (5.78)$$

The second derivatives of the Gibbs free energy give various thermodynamic coefficients that will be required. The isothermal compressibility is

$$\chi_\mathrm{T} = -\overline{V}^{-1} \left(\frac{\partial \overline{V}}{\partial p} \right)_{N,T} = -\overline{V}^{-1} \left(\frac{\partial^2 G_0}{\partial p^2} \right)_{N,T}. \quad (5.79)$$

For an incompressible fluid, $\chi_\mathrm{T} = 0$. The thermal expansivity gives the relative change in volume with temperature,

$$\alpha = \overline{V}^{-1} \left(\frac{\partial \overline{V}}{\partial T} \right)_{N,p}. \quad (5.80)$$

[8] Attard, P. (2002a), *Thermodynamics and Statistical Mechanics: Equilibrium by Entropy Maximisation*, (Academic Press, London).

The internal enthalpy per unit volume is $h^{\text{int}} = H^{\text{int}}/\overline{V} = \varepsilon^{\text{int}} + p$. The heat capacity at constant pressure is the rate of change of the enthalpy with temperature. Per unit volume this is

$$c_{\text{p}} \equiv \frac{1}{\overline{V}} \left(\frac{\partial [\overline{E}^{\text{int}} + p\overline{V}]}{\partial T} \right)_{N,p} = \left(\frac{\partial \varepsilon^{\text{int}}}{\partial T} \right)_p + \alpha h^{\text{int}}, \tag{5.81}$$

Rearranging this gives the change of internal energy density with temperature,

$$\left(\frac{\partial \varepsilon^{\text{int}}}{\partial T} \right)_p = c_{\text{p}} - \alpha h^{\text{int}}, \tag{5.82}$$

which will be required below. Also required will be its change with pressure,

$$\begin{aligned}
\left(\frac{\partial \varepsilon^{\text{int}}}{\partial p} \right)_T &= \left(\frac{\partial (\overline{E}^{\text{int}}/\overline{V})}{\partial p} \right)_{N,T} \\
&= \frac{-\varepsilon^{\text{int}}}{\overline{V}} \left(\frac{\partial \overline{V}}{\partial p} \right)_{N,T} + \frac{1}{\overline{V}} \frac{\partial}{\partial p} \left(\frac{\partial (G_0/T)}{\partial T^{-1}} \right)_{N,p/T} \\
&= \varepsilon^{\text{int}} \chi_{\text{T}} + \frac{1}{\overline{V}} \frac{\partial}{\partial p} \left[G_0 - p \left(\frac{\partial G_0}{\partial p} \right)_{N,T} + \frac{1}{T} \left(\frac{\partial G_0}{\partial T^{-1}} \right)_{N,p} \right] \\
&= \varepsilon^{\text{int}} \chi_{\text{T}} + \frac{1}{\overline{V}} \left[\overline{V} - \left(\frac{\partial G_0}{\partial p} \right)_{N,T} - p \left(\frac{\partial^2 G_0}{\partial p^2} \right)_{N,T} \right. \\
&\qquad \left. + \frac{1}{T} \frac{\partial}{\partial T^{-1}} \left(\frac{\partial G_0}{\partial p} \right)_{N,T} \right] \\
&= \chi_{\text{T}} h^{\text{int}} - \alpha T.
\end{aligned} \tag{5.83}$$

Now the convective derivatives for the density, velocity, temperature, and pressure will be derived. The convective number density derivative is

$$\frac{\mathrm{d}n}{\mathrm{d}t} = \frac{\partial n}{\partial t} + \mathbf{v} \cdot \nabla n = -n \nabla \cdot \mathbf{v}, \tag{5.84}$$

as follows from the number conservation law, Eq. (5.65). The divergence of the velocity field is just the relative rate of change of a volume element moving with the flow. Using the above thermodynamic derivatives, this may be written in terms of the rates of change of temperature and pressure,

$$\begin{aligned}
\nabla \cdot \mathbf{v} &= \frac{1}{\overline{V}} \frac{\mathrm{d}\overline{V}}{\mathrm{d}t} \\
&= \frac{1}{\overline{V}} \left(\frac{\partial \overline{V}}{\partial T} \right)_{N,p} \frac{\mathrm{d}T}{\mathrm{d}t} + \frac{1}{\overline{V}} \left(\frac{\partial \overline{V}}{\partial p} \right)_{N,T} \frac{\mathrm{d}p}{\mathrm{d}t} \\
&= \alpha \frac{\mathrm{d}T}{\mathrm{d}t} - \chi_{\text{T}} \frac{\mathrm{d}p}{\mathrm{d}t}.
\end{aligned} \tag{5.85}$$

The convective rate of change of velocity is

$$\begin{aligned} mn\frac{\mathrm{d}\mathbf{v}}{\mathrm{d}t} &= \frac{\mathrm{d}(mn\mathbf{v})}{\mathrm{d}t} - m\mathbf{v}\frac{\mathrm{d}n}{\mathrm{d}t} \\ &= \frac{\mathrm{d}\mathbf{p}}{\mathrm{d}t} + \mathbf{p}\nabla\cdot\mathbf{v} \\ &= \frac{\partial\mathbf{p}}{\partial t} + \mathbf{v}\nabla\cdot\mathbf{p} + \mathbf{p}\nabla\cdot\mathbf{v} \\ &= -n\nabla\psi - \nabla\cdot[pI + \underline{\underline{\Pi}} + \mathbf{pv}] + \nabla\cdot(\mathbf{pv}) \\ &= -n\nabla\psi - \nabla p - \nabla\cdot\underline{\underline{\Pi}}. \end{aligned} \tag{5.86}$$

This expression holds for an arbitrary viscous dissipation tensor. For the most likely viscous dissipation, this reduces to the Navier-Stokes equation,[9]

$$\begin{aligned} mn(\mathbf{r},t)\frac{\mathrm{d}\mathbf{v}(\mathbf{r},t)}{\mathrm{d}t} &= -n(\mathbf{r},t)\nabla\psi(\mathbf{r},t) - \nabla p(\mathbf{r},t) - \nabla\cdot\overline{\underline{\underline{\Pi}}}(\mathbf{r},t) \\ &= -n(\mathbf{r},t)\nabla\psi(\mathbf{r},t) - \nabla p(\mathbf{r},t) \\ &\quad + \left(\eta_{\mathrm{b}} + \frac{1}{3}\eta\right)\nabla(\nabla\cdot\mathbf{v}(\mathbf{r},t)) + \eta\nabla^2\mathbf{v}(\mathbf{r},t). \end{aligned} \tag{5.87}$$

The second equality follows from the expression for the most likely value of the viscous pressure tensor, Eqs (5.70) and (5.72). The left-hand side of the Navier-Stokes equation contains a term $mn\mathbf{v}\cdot\nabla\mathbf{v}$, which is non-linear in the velocity.

The convective derivative of the internal energy is

$$\begin{aligned} \frac{\mathrm{d}\varepsilon^{\mathrm{int}}}{\mathrm{d}t} &= \frac{\mathrm{d}}{\mathrm{d}t}[\varepsilon - n\psi - \mathbf{v}\cdot\mathbf{p}/2] \\ &= \frac{\partial\varepsilon}{\partial t} + \mathbf{v}\cdot\nabla\varepsilon - [\psi + mv^2/2]\frac{\mathrm{d}n}{\mathrm{d}t} - n\frac{\mathrm{d}\psi}{\mathrm{d}t} - mn\mathbf{v}\cdot\frac{\mathrm{d}\mathbf{v}}{\mathrm{d}t} \\ &= -\nabla\cdot\left[\mathbf{J}_{\mathrm{E}}^0 + \varepsilon\mathbf{v} + \underline{\underline{P}}\cdot\mathbf{v}\right] + \mathbf{v}\cdot\nabla\varepsilon + [\psi + mv^2/2]\,n\nabla\cdot\mathbf{v} \\ &\quad - n\frac{\partial\psi}{\partial t} - n\mathbf{v}\cdot\nabla\psi + \mathbf{v}\cdot\left[n\nabla\psi + \nabla\cdot\underline{\underline{P}}\right] \\ &= -n\frac{\partial\psi}{\partial t} - \nabla\cdot\mathbf{J}_{\mathrm{E}}^0 - \varepsilon\nabla\cdot\mathbf{v} - \underline{\underline{P}}:\nabla\mathbf{v} + [n\psi + \mathbf{v}\cdot\mathbf{p}/2]\,\nabla\cdot\mathbf{v} \\ &= -n\frac{\partial\psi}{\partial t} - \nabla\cdot\mathbf{J}_{\mathrm{E}}^0 - \left[\varepsilon^{\mathrm{int}} + p\right]\nabla\cdot\mathbf{v} - \underline{\underline{\Pi}}:\nabla\mathbf{v}. \end{aligned} \tag{5.88}$$

Now

$$\begin{aligned} \frac{\mathrm{d}\varepsilon^{\mathrm{int}}}{\mathrm{d}t} &= \left(\frac{\partial\varepsilon^{\mathrm{int}}}{\partial T}\right)_{N,p}\frac{\mathrm{d}T}{\mathrm{d}t} + \left(\frac{\partial\varepsilon^{\mathrm{int}}}{\partial p}\right)_{N,T}\frac{\mathrm{d}p}{\mathrm{d}t} \\ &= \left[c_{\mathrm{p}} - \alpha h^{\mathrm{int}}\right]\frac{\mathrm{d}T}{\mathrm{d}t} + \left[h^{\mathrm{int}}\chi_{\mathrm{T}} - \alpha T\right]\frac{\mathrm{d}p}{\mathrm{d}t}. \end{aligned} \tag{5.89}$$

[9] Yih, C.-S. (1977), *Fluid Mechanics: A Concise Introduction*, (West River Press, Ann Arbor). Drazin P. G. and Reid, W. H. (1981), *Hydrodynamic Stability*, (Cambridge University Press, Cambridge).

Hence one can rearrange the convective derivative of the internal energy density so that the left-hand side is

$$\begin{aligned}\frac{\mathrm{d}\varepsilon^{\mathrm{int}}}{\mathrm{d}t} + h^{\mathrm{int}}\nabla\cdot\mathbf{v} &= \left[c_{\mathrm{p}} - \alpha h^{\mathrm{int}}\right]\frac{\mathrm{d}T}{\mathrm{d}t} + \left[h^{\mathrm{int}}\chi_{\mathrm{T}} - \alpha T\right]\frac{\mathrm{d}p}{\mathrm{d}t} \\ &\quad + h^{\mathrm{int}}\left[\alpha\frac{\mathrm{d}T}{\mathrm{d}t} - \chi_{\mathrm{T}}\frac{\mathrm{d}p}{\mathrm{d}t}\right] \\ &= c_{\mathrm{p}}\frac{\mathrm{d}T}{\mathrm{d}t} - \alpha T\frac{\mathrm{d}p}{\mathrm{d}t}. \end{aligned} \tag{5.90}$$

Finally putting these together gives what may be called the energy equation,

$$c_{\mathrm{p}}\frac{\mathrm{d}T(\mathbf{r},t)}{\mathrm{d}t} - \alpha T(\mathbf{r},t)\frac{\mathrm{d}p(\mathbf{r},t)}{\mathrm{d}t} = -n(\mathbf{r},t)\frac{\partial\psi(\mathbf{r},t)}{\partial t} - \nabla\cdot\mathbf{J}_{\mathrm{E}}^{0}(\mathbf{r},t) + \Phi(\mathbf{r},t), \tag{5.91}$$

where Φ is the viscous dissipation defined above, Eq. (5.75). It is often the case that the external potential is not explicitly dependent upon time, $\partial\psi(\mathbf{r})/\partial t = 0$. The right-hand side can be evaluated for the most likely fluxes given above. One has three fields: temperature T, pressure p, and velocity $\mathbf{v}$, and three equations that determine them: the divergence of the velocity, Eq. (5.85), the Navier-Stokes equation, Eq. (5.87), and the energy equation, Eq. (5.91).

These three equations are equivalent to the conventional formulation of hydrodynamics. The are valid when the spatial and temporal variation of the system is negligible so that the transport and thermodynamic coefficients can be treated as constants. Most importantly, they only give the optimum fluxes for the given fields, and they are not applicable to the thermal fluctuations in the fluxes about those optimum values, nor to the transient fluxes that occur in reaching the optimum value. The second entropy formulation of fluctuation hydrodynamics is more general than the conventional phenomenological hydrodynamic equations, as it is applicable to these non-optimum fluxes. Also, it provides a thermodynamic basis for selecting non-equilibrium states that cannot be distinguished by hydrodynamics, as will be seen in the following chapter.

Chapter 6

Heat Convection and Non-Equilibrium Phase Transitions

This chapter is concerned with pattern formation in non-equilibrium systems, and with the more general phenomena of non-equilibrium phase transitions. These are characterised by the second entropy, which, after all, is the entropy of transitions. That this determines the observed outcome in such non-equilibrium phenomena is here illustrated using the hydrodynamic equations with added noise. A complementary theme of the chapter is the quantitative calculation of single time thermodynamic quantities such as the structural entropy and the dissipation, and the demonstration that they do not have any direct rôle in determining non-equilibrium pattern formation.

Rather than a generic treatment of the problem, here a particular non-equilibrium pattern forming system is studied in order to illustrate concretely the general ideas. This allows the hydrodynamic and thermodynamic calculations to be explicitly compared with experimental measurements.

The example chosen is one of the most well-established non-equilibrium phenomena, namely steady convective heat flow. This was first studied by quantitative measurement by Bénard in 1900.[1] There have been several attempts to formulate a thermodynamic theory for the occurrence of convection and the patterns that form, the most well known of which are related to the maximisation or minimisation of the dissipation, which principle was critically discussed in §1.2.2. In convection, the dissipation is given by the heat flux, and Malkus and Veronis[2] were possibly the first to assert explicitly that maximising the heat flux determines the preferred convection pattern. The broad consensus[3] from

[1]Bénard, H. (1900), Rev. Gen. Sci. Pures Appl. **11**, 1261.

[2]Malkus, W. V. R. and Veronis, G. (1958), J. Fluid Mech. **4**, 225.

[3]Getling, A. V. (1998), *Rayleigh-Bénard Convection: Structure and Dynamics*, (World Scientific, Singapore).

subsequent experimental tests is that the idea is inconsistent with the observed patterns that actually form in convection, which conclusion is quantitatively confirmed below.

As an alternative to the maximisation of the heat flux, maximisation of the entropy has been proposed as the thermodynamic principle that characterises convection. Kita suggested, at least for dilute gases, the maximisation of a type of non-equilibrium entropy based upon the Boltzmann H-function and an analysis of the collision integral, and used the approach to predict the location of the conduction-convection transition.[4] Attard proposed maximising the ordinary or structural entropy as the thermodynamic principle for convection,[5] based upon the reduction condition for the second entropy.[6] The calculations and measurements presented in §6.6 provide evidence contrary to these proposals, and the fundamental reason for the insufficiency of the entropy or free energy in non-equilibrium systems is discussed at the conclusion of the chapter.

The transition from conduction to convection is one example of a non-equilibrium phase transition, and it is treated in detail in what follows. Beyond this, it is the shape and spacing of the convective rolls that represents the non-equilibrium pattern. Hence the evolution of the patterns of convection and the transitions between different patterns are rich examples of non-equilibrium phase transitions. Specifically, this chapter focuses on the transition between straight rolls of different wavelengths since these have been measured experimentally. By calculating the heat flux and entropy, the two hypotheses described above are tested against the experimental results. Both the experiments and the hydrodynamic calculations display multiple stable patterns, from which it is concluded that in general the primary focus ought be on the transitions between patterns rather than on single time thermodynamic properties of a non-equilibrium pattern.

6.1 Hydrodynamic Equations of Convection

6.1.1 Boussinesq Approximation

Hydrodynamic calculations of convective heat flow are generally made in the so-called Boussinesq approximation. This consists in setting the compressibility to zero, $\chi_{\rm T} = 0$, neglecting the thermal expansivity everywhere except in the buoyancy force, and neglecting most of the non-linear terms in the convective equations. In this case the density equation reduces to the vanishing of the divergence of the velocity field,

$$\nabla \cdot \mathbf{v}(\mathbf{r}, t) = 0. \tag{6.1}$$

This means that the most likely value of the scalar part of the viscous pressure tensor vanishes, $\overline{\pi} = 0$.

[4] Kita, T. (2006), J. Phys. Soc. Japan **75**, 114005 and 124005.
[5] Attard, P. (2011), AIP Advances **1**, 032146.
[6] Attard, P. (2005a), J. Chem. Phys. **122**, 154101.

Retaining the thermal expansivity, the gravitational potential density is

$$n(\mathbf{r},t)\psi(\mathbf{r},t) = \{n_{00} - \alpha n_{00}[T_{\rm tot}(\mathbf{r},t) - T_{00}]\}\, mgz. \tag{6.2}$$

Here g is the acceleration due to gravity, which acts in the z-direction, and m is the molecular mass. The subscript tot is appended to the temperature and pressure to signify total; below these will be split into the conductive part, signified by the subscript 0, and the convective perturbation, such that $T_{\rm tot} = T_0 + T$ and $p_{\rm tot} = p_0 + p$. Here and below the subscript 00 denotes the reference value at the mid-point of the sub-system. Expansions to linear order in the difference from this reference point are carried out. With this and neglecting the term quadratic in the velocity, the Navier-Stokes equation becomes

$$\begin{aligned} mn_{00}\frac{\partial \mathbf{v}(\mathbf{r},t)}{\partial t} &= -\{n_{00} - \alpha n_{00}[T_{\rm tot}(\mathbf{r},t) - T_{00}]\}\, mg\hat{\mathbf{z}} - \nabla p_{\rm tot}(\mathbf{r},t) \\ & \quad + \eta\nabla^2\mathbf{v}(\mathbf{r},t). \end{aligned} \tag{6.3}$$

Neglecting the viscous dissipation, which is quadratic in the velocity, and also the thermal expansivity, and using the most likely heat flux, the energy equation becomes

$$c_{\rm p}\frac{\partial T_{\rm tot}(\mathbf{r},t)}{\partial t} + c_{\rm p}\mathbf{v}(\mathbf{r},t)\cdot\nabla T_{\rm tot}(\mathbf{r},t) = \lambda\nabla^2 T_{\rm tot}(\mathbf{r},t). \tag{6.4}$$

These three partial differential equations constitute the Boussinesq approximation that is to be solved for the temperature, pressure, and velocity fields.

6.1.2 Conduction

Although convection and conduction can occur in quite complicated geometries, here is treated the simplest case of a slab geometry with a temperature gradient imposed in the z-direction. Convective flow is treated as a perturbation from the conducting state. In conduction, the velocity field is zero, $\mathbf{v}(\mathbf{r},t) = 0$.

The boundaries of the sub-system are located at $z = \pm L_z/2$, and the temperatures of the reservoirs beyond the boundaries are $T_{\rm r\pm}$. The temperature difference is $\Delta_{\rm T} \equiv T_{\rm r+} - T_{\rm r-}$. It is assumed that the imposed temperature gradient, $\Delta_{\rm T}/L_z$, is small and that quadratic terms can be neglected. This means that it does not matter whether one deals with the difference in temperature or with the difference in inverse temperature. For convection to occur, the lower reservoir must be hotter than the upper reservoir, $\Delta_{\rm T} < 0$, at least for fluids with positive expansivity.

Since in conduction the velocity vanishes, and the temperature is steady and a function of z only, the energy equation reduces to

$$0 = \lambda\frac{{\rm d}^2 T_0(z)}{{\rm d}z^2}. \tag{6.5}$$

The subscript 0 denotes conduction. Hence the temperature field is a linear function of z that must equal the reservoirs' temperatures at the boundaries,

$$T_0(z) = T_{00} + \frac{\Delta_{\rm T}}{L_z}z,\ |z| \leq L_z/2, \tag{6.6}$$

with the mid-point temperature being $T_{00} \equiv [T_{\mathrm{r}+} + T_{\mathrm{r}-}]/2$. With this and zero velocity, the Navier-Stokes equation becomes

$$0 = -\left\{ n_{00} - \frac{\alpha n_{00}\Delta_{\mathrm{T}}}{L_z} z \right\} mg - \frac{\mathrm{d}p_0(z)}{\mathrm{d}z}. \tag{6.7}$$

Hence the conductive pressure profile is

$$p_0(z) = p_{00} - n_{00}mgz + \frac{\alpha n_{00} mg\Delta_{\mathrm{T}}}{2L_z} z^2. \tag{6.8}$$

For future use, the heat flow per unit area in conduction is

$$\overline{J^0_{\mathrm{E},0}} = -\lambda \frac{\mathrm{d}T_0(z)}{\mathrm{d}z} = \frac{-\lambda\Delta_{\mathrm{T}}}{L_z}. \tag{6.9}$$

The rate of entropy production per unit sub-system volume is

$$\dot{S}_{\mathrm{r}}/AL_z = \frac{\overline{J^0_{\mathrm{E},0}}}{L_z}\left[\frac{1}{T_{\mathrm{r}+}} - \frac{1}{T_{\mathrm{r}-}}\right] = \frac{\lambda\Delta_{\mathrm{T}}^2}{T_{00}^2 L_z^2}. \tag{6.10}$$

This is positive and independent of the sign of the temperature difference, as it ought to be.

6.1.3 Convection

Convection may be regarded as a perturbation on conduction. As mentioned above the temperature may be written as

$$T_{\mathrm{tot}}(\mathbf{r},t) = T_0(z) + T(\mathbf{r},t), \tag{6.11}$$

and similarly the pressure,

$$p_{\mathrm{tot}}(\mathbf{r},t) = p_0(z) + p(\mathbf{r},t). \tag{6.12}$$

Since the velocity is zero in conduction, the full velocity field is the same as the perturbing velocity field, $\mathbf{v}_{\mathrm{tot}}(\mathbf{r},t) = \mathbf{v}(\mathbf{r},t)$. These convective fields depend upon the Rayleigh number and, in the calculations below, the wave number of the steady state being characterised, but these will not be shown explicitly.

The full fields satisfy the density, Navier-Stokes, and energy equations. But since the conductive fields also satisfy these equations, they can be subtracted from both sides, so that one has

$$0 = \nabla \cdot \mathbf{v}(\mathbf{r}), \tag{6.13}$$

$$mn_{00}\frac{\partial \mathbf{v}(\mathbf{r},t)}{\partial t} = \alpha n_{00} T(\mathbf{r}) mg\hat{\mathbf{z}} - \nabla p(\mathbf{r}) + \eta\nabla^2\mathbf{v}(\mathbf{r}), \tag{6.14}$$

and

$$\begin{aligned} c_{\mathrm{p}}\frac{\partial T(\mathbf{r},t)}{\partial t} &= -c_{\mathrm{p}}\mathbf{v}(\mathbf{r})\cdot\nabla[T_0(z) + T(\mathbf{r})] + \lambda\nabla^2 T(\mathbf{r}) \\ &= -\frac{c_{\mathrm{p}}\Delta_{\mathrm{T}}}{L_z} v_z(\mathbf{r}) - c_{\mathrm{p}}\mathbf{v}(\mathbf{r})\cdot\nabla T(\mathbf{r}) + \lambda\nabla^2 T(\mathbf{r}). \end{aligned} \tag{6.15}$$

Since this is a steady state problem, the left-hand sides can be set to zero. Alternatively, it is sometimes convenient to solve the equations numerically by time stepping based on any non-zero value. If one regards the convective perturbation as small, then one sees that the second term on the right-hand side of the energy equation is non-linear, as it is the product of the convective temperature and the velocity. This non-linear term fixes the amplitude of the fields that give the steady state, since without it everything could be multiplied by a constant to give another solution. There are five equations (the Navier-Stokes equation is for three components) and five fields, including the three components of the velocity.

Now use L_z as the unit of length, $-\Delta_{\rm T}$ as the unit of temperature, $L_z^2 c_{\rm p}/\lambda$ as the unit of time, and $mn_{00}\lambda^2/L_z^2 c_{\rm p}^2$ as the unit of pressure. Denoting dimensionless quantities with an asterisk, one has

$$0 = \nabla^* \cdot \mathbf{v}^*, \tag{6.16}$$

$$\frac{\partial \mathbf{v}^*}{\partial t^*} = \mathcal{RP}T^*\hat{\mathbf{z}} - \nabla^* p^* + \mathcal{P}\nabla^{*2}\mathbf{v}^*, \tag{6.17}$$

and

$$\frac{\partial T^*}{\partial t^*} = v_z^* - \mathbf{v}^* \cdot \nabla^* T^* + \nabla^{*2} T^*. \tag{6.18}$$

Here the Rayleigh number is

$$\mathcal{R} \equiv -\Delta_{\rm T}\alpha g c_{\rm p} m n_{00} L_z^3/\lambda\eta, \tag{6.19}$$

and the Prandtl number is

$$\mathcal{P} \equiv \eta c_{\rm p}/m n_{00}\lambda. \tag{6.20}$$

Here and throughout, $c_{\rm p}$ is the constant pressure heat capacity per unit volume.

One can eliminate the pressure from the Navier-Stokes equations. Set the left-hand side to zero, differentiate the z-component with respect to y, the y-component with respect to z, and subtract,

$$0 = \mathcal{R}\frac{\partial \tilde{T}^*}{\partial y^*} + \nabla^{*2}\left[\frac{\partial v_z^*}{\partial y^*} - \frac{\partial v_y^*}{\partial z^*}\right]. \tag{6.21}$$

The Prandtl number has been factored out. Similarly for the x-component,

$$0 = \mathcal{R}\frac{\partial \tilde{T}^*}{\partial x^*} + \nabla^{*2}\left[\frac{\partial v_z^*}{\partial x^*} - \frac{\partial v_x^*}{\partial z^*}\right]. \tag{6.22}$$

One now has four equations (these two forms of the momentum equation, the density equation, and the energy equation), four fields (T^*, v_x^*, v_z^*, and v_z^*), and one dimensionless parameter, $\mathcal{R}$. Since everything in these equations is dimensionless and refers to the convective perturbation, the asterisk will be dropped later below.

A numerical scheme to solve these equations can be based on simple time stepping using the time rate of change of temperature given by the energy equation, and the solutions of the remaining three differential equations at each time step. This implicitly assumes rapid relaxation of the velocity at each time step, $\partial \mathbf{v}^*/\partial t^* = 0$. Such an algorithm gives the physical approach to the steady state and the transition between steady states.

6.2 Total First Entropy of Convection

Now an expression for the first or structural entropy of a convecting steady state will be obtained. More precisely, the difference between the first entropy of a convecting state and the first entropy of the conducting state at the same Rayleigh number will be obtained. Of course this can also be used to give the difference between the first entropy of convecting states of different wave numbers or of different convecting patterns, provided that the comparison is made at the same Rayleigh number.

The total entropy is the sum of the sub-system entropy and the reservoir entropy. Here the exact sub-system entropy will be given, and two forms for the reservoir contribution will be obtained. One reservoir expression is the exact change in reservoir entropy during the transition from one state to another (e.g. conduction to convection, or from one convecting state to another). The second reservoir expression is the so-called static approximation. It gives the difference in reservoir entropy between a convecting state and the conducting state, and again it can be used to obtain the difference in reservoir entropy between one convecting state and another.

The nature of the static approximation for the reservoir entropy was touched on in Ch. 3 and will be fully explained in Ch. 8 below. There the first entropy for phase space for a non-equilibrium system will be shown to consist of a static part, which is the analogue of the usual equilibrium expression, and a dynamic part, which is a correction to the static reservoir contribution that accounts for the adiabatic evolution that is unavoidably included. In the macroscopic description of this chapter, the sub-system entropy is given exactly by the usual equilibrium expression, which will be obtained explicitly. The reservoir entropy will be approximated by the static part alone, which is the usual equilibrium expression for the reservoir entropy.

The change (or difference) in entropy density of the sub-system between convection, $T_{\text{tot}}(\mathbf{r}) = T_0(z) + T(\mathbf{r})$, and conduction, $T_0(z) = T_{00} + z\Delta_T/L_z$, can be obtained by thermodynamic integration.[7] Henceforth this will simply be called the convective entropy density, the change from conduction being understood. Assuming, as in the Boussinesq approximation, that the thermal expansivity and compressibility can be neglected, the convective entropy density

[7] As mentioned above, the dependence of the temperature field (and of the change in entropy) on the Rayleigh number and on the wave number of the particular convective state is not shown explicitly. Also, dimensionless variables are *not* used in this section.

of the sub-system is

$$\begin{aligned}\sigma_\mathrm{s}(\mathbf{r}) &= \int_{\varepsilon_{\mathrm{int},0}}^{\varepsilon_{\mathrm{int},1}} \mathrm{d}\varepsilon'_{\mathrm{int}} \, \frac{\partial \sigma(\varepsilon'_{\mathrm{int}})}{\partial \varepsilon'_{\mathrm{int}}} \\ &= \int_{\varepsilon_{\mathrm{int},0}}^{\varepsilon_{\mathrm{int},1}} \mathrm{d}\varepsilon'_{\mathrm{int}} \, \frac{1}{T'} \\ &= c_\mathrm{p} \ln \left[\frac{T_0(z) + T(\mathbf{r})}{T_0(z)} \right] \\ &= c_\mathrm{p} \left[\frac{T(\mathbf{r})}{T_0(z)} - \frac{T(\mathbf{r})^2}{2T_0(z)^2} + \ldots \right] \\ &= c_\mathrm{p} \left[\frac{T(\mathbf{r})}{T_{00}} - \frac{z\Delta_T T(\mathbf{r})}{L_z T_{00}^2} - \frac{T(\mathbf{r})^2}{2T_{00}^2} + \mathcal{O}(\Delta_T^3/T_{00}^3) \right].\end{aligned} \tag{6.23}$$

This uses the fact that $\Delta\varepsilon_{\mathrm{int}} = c_\mathrm{p}\Delta T$, where c_p is the heat capacity per unit volume. Since the convective temperature is $T(\mathbf{r}) \sim \mathcal{O}(\Delta_T)$, the expansion to quadratic order in the temperature difference is justified.[8] This is the local sub-system entropy density. The global sub-system entropy density is

$$\sigma_\mathrm{s}(t) = \frac{1}{AL_z} \int \mathrm{d}\mathbf{r} \; \sigma_\mathrm{s}(\mathbf{r}, t), \tag{6.24}$$

where A is the cross-sectional area of the sub-system. For use below during convective transitions, this has been written as a function of time and invokes the instantaneous temperature, $T(\mathbf{r}, t)$.

Static Reservoir Entropy

The zeroth and first temperatures of the reservoirs are

$$\frac{1}{T_{\mathrm{r}0}} = \frac{1}{2}\left[\frac{1}{T_{\mathrm{r}+}} + \frac{1}{T_{\mathrm{r}-}}\right] = \frac{1}{T_{00}} + \mathcal{O}(\Delta_T^2/T_{00}^2), \tag{6.25}$$

and

$$\frac{1}{T_{\mathrm{r}1}} = \frac{1}{L_z}\left[\frac{1}{T_{\mathrm{r}+}} - \frac{1}{T_{\mathrm{r}-}}\right] = \frac{-\Delta_T}{L_z T_{00}^2} + \mathcal{O}(\Delta_T^3/T_{00}^3), \tag{6.26}$$

respectively. With these, the static part of the reservoir entropy associated with the sub-system is

$$S_{\mathrm{r,st}} = \frac{-\Delta E_0}{T_{\mathrm{r}0}} - \frac{\Delta E_1}{T_{\mathrm{r}1}}. \tag{6.27}$$

Again it is understood that this is the change from conduction. In view of the definition of the energy moments of the sub-system,

$$\Delta E_n = \int \mathrm{d}\mathbf{r} \; z^n \Delta\varepsilon(\mathbf{r}), \tag{6.28}$$

[8]This expansion has been checked numerically against the full logarithm and found to be accurate for ideal straight rolls over the full range of Rayleigh numbers and wave numbers.

one can instead define and invoke the static convective reservoir entropy density,

$$\sigma_{\mathrm{r,st}}(\mathbf{r}) \equiv \frac{-\Delta\varepsilon(\mathbf{r})}{T_{\mathrm{r}0}} - \frac{z\Delta\varepsilon(\mathbf{r})}{T_{\mathrm{r}1}}. \tag{6.29}$$

What appears here is the change in total energy density, which is composed of the internal energy density, the gravitational energy density, and the kinetic energy density,

$$\begin{aligned}\Delta\varepsilon(\mathbf{r}) &= \Delta\varepsilon_{\mathrm{int}}(\mathbf{r}) + \Delta\varepsilon_{\mathrm{g}}(\mathbf{r}) + \Delta\varepsilon_{\mathrm{ke}}(\mathbf{r}) \\ &= c_{\mathrm{p}}T(\mathbf{r}) - \alpha m n_{00} g z T(\mathbf{r}) + \frac{m n_{00}}{2}\mathbf{v}(\mathbf{r})\cdot\mathbf{v}(\mathbf{r}).\end{aligned} \tag{6.30}$$

Accordingly, the change in total entropy density, $\sigma_{\mathrm{tot,st}}(\mathbf{r}) = \sigma_{\mathrm{s}}(\mathbf{r}) + \sigma_{\mathrm{r,st}}(\mathbf{r})$, is composed of three analogous terms. The internal energy density contribution is

$$\begin{aligned}\sigma^{\mathrm{int}}_{\mathrm{tot,st}}(\mathbf{r}) &\equiv \sigma_{\mathrm{s}}(\mathbf{r}) - \frac{\Delta\varepsilon_{\mathrm{int}}(\mathbf{r})}{T_{\mathrm{r}0}} - \frac{z\Delta\varepsilon_{\mathrm{int}}(\mathbf{r})}{T_{\mathrm{r}1}} \\ &= c_{\mathrm{p}}\left[\frac{T(\mathbf{r})}{T_{00}} - \frac{z\Delta_T T(\mathbf{r})}{L_z T_{00}^2} - \frac{T(\mathbf{r})^2}{2T_{00}^2}\right] - \frac{c_{\mathrm{p}}T(\mathbf{r})}{T_{00}} + \frac{c_{\mathrm{p}} z\Delta_T T(\mathbf{r})}{L_z T_{00}^2} \\ &= \frac{-c_{\mathrm{p}}}{2T_{00}^2}T(\mathbf{r})^2.\end{aligned} \tag{6.31}$$

This is identical to the equilibrium fluctuation expression for the total entropy density of a sub-system in equilibrium with a reservoir of temperature T_{00} when the local fluctuation in energy is $\Delta\varepsilon_{\mathrm{int}}(\mathbf{r}) = c_{\mathrm{p}}T(\mathbf{r})$. It is what one would have expected and could have been written down directly. The fact that this is always negative means that a convecting steady state has lower total entropy than the conducting state, at least as far as the rearrangement of the internal energy in the convecting system is concerned. This latter observation is not particularly significant because the result depends upon the static approximation, and so it does not give the full change in the reservoir entropy during such a transition.

The gravitational contribution is

$$\begin{aligned}\sigma^{\mathrm{g}}_{\mathrm{tot,st}}(\mathbf{r}) &\equiv -\frac{\Delta\varepsilon_{\mathrm{g}}(\mathbf{r})}{T_{\mathrm{r}0}} - \frac{z\Delta\varepsilon_{\mathrm{g}}(\mathbf{r})}{T_{\mathrm{r}1}} \\ &= \frac{\alpha m n_{00} g z T(\mathbf{r})}{T_{00}} - \frac{\alpha m n_{00} g z^2 \Delta_T T(\mathbf{r})}{L_z T_{00}^2},\end{aligned} \tag{6.32}$$

and the kinetic energy contribution is

$$\begin{aligned}\sigma^{\mathrm{ke}}_{\mathrm{tot,st}}(\mathbf{r}) &\equiv -\frac{\Delta\varepsilon_{\mathrm{ke}}(\mathbf{r})}{T_{\mathrm{r}0}} - \frac{z\Delta\varepsilon_{\mathrm{ke}}(\mathbf{r})}{T_{\mathrm{r}1}} \\ &= -\frac{m n_{00}}{2T_{00}}\mathbf{v}(\mathbf{r})\cdot\mathbf{v}(\mathbf{r}) + \frac{m n_{00}\Delta_T z}{2L_z T_{00}^2}\mathbf{v}(\mathbf{r})\cdot\mathbf{v}(\mathbf{r}).\end{aligned} \tag{6.33}$$

Below the so-called Boussinesq symmetry will be invoked, which means that the convective temperature perturbation is anti-symmetric upon reflection through the centre of a convective roll. Effectively, the temperature is odd in z and the velocity squared is even in z. In consequence, when the entropy density is integrated over the convective volume to obtain the entropy itself, only the first term on the right-hand side of each of these last two equations contributes.

As mentioned above, density changes have been neglected everywhere except in the gravitational term, which is consistent with the Boussinesq approximation. One might expect, and it will be shown explicitly below, that the kinetic energy contribution is negligible compared to the gravitational contribution. This is consistent with neglecting the viscous dissipation in the Boussinesq approximation, which is similarly quadratic in the convective velocity.

The sum of the last three results represent the static approximation to the convective entropy density (i.e. the difference in entropy between the convective state and the conductive state). Integrating over the volume of the sub-system, the global convective entropy density is

$$\sigma_{\text{tot,st}} \equiv \frac{1}{AL_z} \int \mathrm{d}\mathbf{r} \, \left[\sigma^{\text{int}}_{\text{tot,st}}(\mathbf{r}) + \sigma^{\text{g}}_{\text{tot,st}}(\mathbf{r}) + \sigma^{\text{ke}}_{\text{tot,st}}(\mathbf{r}) \right] . \tag{6.34}$$

Change in Reservoir Entropy

In the above result for the difference in entropy between convection and conduction, it is the reservoir contribution that is approximate, since the sub-system entropy is exact. As was mentioned in the introduction to this section, one can obtain exactly the change in reservoir entropy for a transition between two non-equilibrium states. This is done by integrating over time the heat flow from one reservoir to the other through the sub-system. The rate of change of the entropy of the reservoirs is

$$\begin{aligned} \dot{S}_{\text{r}}(t) &= \int_A \mathrm{d}x\, \mathrm{d}y \left[\frac{1}{T_{\text{r}+}} \overline{J}^0_{\text{E}}(x, y, L_z/2, t) - \frac{1}{T_{\text{r}-}} \overline{J}^0_{\text{E}}(x, y, -L_z/2, t) \right] \\ &\approx \frac{-L_z \Delta_T}{T_{00}^2} \int_A \mathrm{d}x\, \mathrm{d}y\, \overline{J}^0_{\text{E}}(x, y, L_z/2, t) \\ &= \frac{-L_z \Delta_T \lambda}{T_{00}^2} \int_A \mathrm{d}x\, \mathrm{d}y\, \left. \frac{\partial T(\mathbf{r}, t)}{\partial z} \right|_{z=L_z/2} . \end{aligned} \tag{6.35}$$

The first equality is exact, whereas the second equality makes the approximation that the integrated heat flux at the two boundaries are equal. This is certainly the case in the steady state, and it is a very good approximation in the transitions between steady straight roll states that are characterised below.[9] Accordingly, the change in total entropy per unit sub-system volume during a transition over the time interval $[t_1, t_2]$ is

$$\Delta\sigma_{\text{tot}} = \sigma_{\text{s}}(t_2) - \sigma_{\text{s}}(t_1) + \frac{1}{AL_z} \int_{t_1}^{t_2} \mathrm{d}t\, \dot{S}_{\text{r}}(t), \tag{6.36}$$

where the global sub-system entropy density is given by Eq. (6.24).

[9] This approximation does not account for any nett total energy change of the sub-system during a transition. Such nett changes, which lead to an inequality of the heat flux at the two boundaries, do indeed occur in the gravitational energy and in the kinetic energy, but they are negligible compared to the total heat flux over the time interval of a transition.

6.3 Algorithm for Ideal Straight Rolls

This and the following section set out the hydrodynamic equations used for convection and describes the computer algorithms that were used to solve them. This section deals with ideal straight rolls (i.e. the rolls are considered straight and homogeneous in the x-direction), and the next section deals with the cross roll state (i.e. the combination of straight x- and y-rolls).

The applied thermal gradient and gravity are in the z-direction. The wavelength is twice the width of an individual roll, $\Lambda = 2L_y$, as they come in pairs of counter-rotating rolls. In this and the following sections, dimensionless variables are used, including for the Fourier coefficients. Hence $L_y = 1$ or $a \approx 3.1$ corresponds to a circular or square roll.

6.3.1 Hydrodynamic Equations

The hydrodynamic equations for convection were given at the end of §6.1.3. Setting the x-component of velocity and the x-derivatives to zero, the three equations for the remaining three fields are

$$0 = \frac{\partial v_y(y,z)}{\partial y} + \frac{\partial v_z(y,z)}{\partial z}, \tag{6.37}$$

$$\begin{aligned}\frac{\partial T(y,z)}{\partial t} &= v_z(y,z) - v_y(y,z)\frac{\partial T(y,z)}{\partial y} - v_z(y,z)\frac{\partial T(y,z)}{\partial z} \\ &\quad + \frac{\partial^2 T(y,z)}{\partial y^2} + \frac{\partial^2 T(y,z)}{\partial z^2},\end{aligned} \tag{6.38}$$

and

$$\begin{aligned}0 &= \mathcal{R}\frac{\partial T(y,z)}{\partial y} + \nabla^2\left[\frac{\partial v_z(y,z)}{\partial y} - \frac{\partial v_y(y,z)}{\partial z}\right] \\ &= \mathcal{R}\frac{\partial^2 T(y,z)}{\partial y^2} + \left[\frac{\partial^2}{\partial y^2} + \frac{\partial^2}{\partial z^2}\right]^2 v_z(y,z).\end{aligned} \tag{6.39}$$

The second equality follows by taking the y-derivative of the first equality and using the vanishing of the divergence of the velocity. Recall that the Rayleigh number is $\mathcal{R} \equiv -\alpha m g n_{00} c_{\rm p} \Delta_T L_z^3/\lambda\eta$.

6.3.2 Fourier Expansion

Following Busse,[10] a Galerkin method is used that invokes Fourier expansions of the fields. The temperature field is expanded as

$$T(y,z) = \sum_{l=0}^{L}\sum_{n=1}^{N} [T_{ln}^{\rm s} \sin 2n\pi z + T_{ln}^{\rm c} \cos(2n-1)\pi z] \cos lay. \tag{6.40}$$

[10] Busse, F. H. (1967), J. Math. and Phys. **46**, 140.

Here $a = \pi/L_y$ is the wave number. The form of the z-expansion is chosen to guarantee the boundary conditions, $T(y, \pm 1/2) = 0$. For the Boussinesq fluid, there is mirror plane symmetry between two rolls, $T(y,z) = T(-y,z)$, and point reflection symmetry within a roll, $T(y,z) = -T(L_y - y, -z)$. These mean that the even l coefficients of $T^{\rm c}_{ln}$ and the odd l coefficients of $T^{\rm s}_{ln}$ must vanish.

The particular solution of the differential equation for the velocity is

$$v_z^{\rm p}(y,z) = \sum_{l,n} \left[v^{\rm ps}_{z,ln} \sin 2n\pi z + v^{\rm pc}_{z,ln} \cos(2n-1)\pi z \right] \cos lay. \tag{6.41}$$

Clearly,

$$v^{\rm ps}_{z,ln} = \frac{\mathcal{R}(la)^2}{[(la)^2 + (2n\pi)^2]^2} T^{\rm s}_{ln}, \tag{6.42}$$

and

$$v^{\rm pc}_{z,ln} = \frac{\mathcal{R}(la)^2}{[(la)^2 + ((2n-1)\pi)^2]^2} T^{\rm c}_{ln}. \tag{6.43}$$

The homogeneous solution, which satisfies $\nabla^2\nabla^2 v_z^{\rm h} = 0$, is

$$\begin{aligned} v_z^{\rm h}(y,z) = & \sum_{l=1}^{L} [A^{\rm s}_l \sinh laz + B^{\rm s}_l z \cosh laz \\ & + A^{\rm c}_l \cosh laz + B^{\rm c}_l z \sinh laz] \cos lay. \end{aligned} \tag{6.44}$$

Because the system is periodic in the horizontal direction, there is a term for each expansion mode. Writing the velocity as $v_z = v_z^{\rm p} + v_z^{\rm h}$, the four boundary conditions for each mode, $v_z(y, \pm 1/2) = \partial v_z(y, \pm 1/2)/\partial z = 0$, determine the four coefficients per mode, $A^{\rm s}_l$, $B^{\rm s}_l$ $A^{\rm c}_l$, and $B^{\rm c}_l$. These coefficients have the Boussinesq symmetry discussed above. The second condition ensures the vanishing of $v_y(y, \pm 1/2)$ when the density equation is applied.

The vertical velocity field is then projected onto the Fourier grid used for the temperature field using the orthogonality of the trigonometric functions. Formally one has

$$v_z(y,z) = \sum_{l=1}^{L} \sum_{n=1}^{N} \left[v^{\rm s}_{z,ln} \sin 2n\pi z + v^{\rm c}_{z,ln} \cos(2n-1)\pi z \right] \cos lay. \tag{6.45}$$

Due to the Boussinesq symmetry, half the coefficients are zero. The horizontal velocity may be expanded as

$$v_y(y,z) = \sum_{l=1}^{L} \sum_{n=1}^{N} \left[v^{\rm c}_{y,ln} \cos 2n\pi z + v^{\rm s}_{y,ln} \sin(2n-1)\pi z \right] \sin lay. \tag{6.46}$$

The density equation gives

$$v^{\rm c}_{y,ln} = \frac{-2n\pi}{la} v^{\rm s}_{z,ln} \text{ and } v^{\rm s}_{y,ln} = \frac{(2n-1)\pi}{la} v^{\rm c}_{z,ln}. \tag{6.47}$$

Table 6.1: Measured (silicone oil, $\mathcal{P} = 930$) and computed velocity amplitudes, (μm/s), for the first three harmonics† at $a = 3.117$ and two Rayleigh numbers.

	Measured$^\sharp$	Busse$^\sharp$	Present	Measured$^\sharp$	Present
		$\mathcal{R} = 3416$		$\mathcal{R} = 11,391$	
V_y^1	132±4	133	137.6	337±10	355.1
V_y^2	5.3±0.5‡	5‡	5.1‡	13.7±1	13.0
V_y^3	1.5±0.3	-	1.2	19±1	18.3
V_z^1	145±5	138	140.6	340±10	363.0
V_z^2	0	0	0	1.7±2	0
V_z^3	4± 0.4	3.8	3.9	58±4	60.2

†The V_z are at $z^{**} = 0$, and the V_y are at $z^* = 0.28$.
‡At $z^* = 0$.
$^\sharp$Dubois, M. and Bergé, P. (1978), J. Fluid Mech. **85**, 641.

The rates of change of the temperature coefficients are obtained from the non-linear energy equation, Eq. (6.38), again using trigonometric orthogonality. The left-hand side of this equation is $\partial T/\partial t$, which is non-zero in the approach to the steady state. Hence one can update the temperature field by simple time stepping, with the new temperature coefficients obtained by adding a constant $\Delta_t \sim \mathcal{O}(10^{-4})$ times the right-hand side to the previous value.

Linear stability analysis reveals that the critical Rayleigh number is $\mathcal{R}_c = 1708$ and the critical wave number is $a_c = 3.117$.[11] For a given Rayleigh number $\mathcal{R} > \mathcal{R}_c$, there is a range of wave numbers a that yield steady state solutions. The wavelength, roll width, and wave number are related by $\Lambda = 2L_y = 2\pi/a$.

The Nusselt number (see §6.3.3) was monitored and used to halt the iterative procedure when its relative change was less than 10^{-5}. For most of the results reported below, $N = 10$ and $L = 10$. Some tests were carried out with up to $N = 16$ and $L = 16$. By comparison, Busse (1967) used up to $L + N = 12$. In general, three- or four- figure agreement was obtained between the present results and those of Busse for Rayleigh numbers up to 30,000.

The amplitudes of the first three harmonics of the velocity field in convection have been measured by Dubois and Bergé (1978) for a silicone oil ($\mathcal{P} = 930$) constrained at the critical wavelength.[12] Their results are shown in Table 6.1, together with their reports of the results of Busse's calculations, and with the

[11] Yih, C.-S. (1977), *Fluid Mechanics: A Concise Introduction*, (West River Press, Ann Arbor). Drazin P. G. and Reid, W. H. (1981), *Hydrodynamic Stability*, (Cambridge University Press, Cambridge).

[12] The physical parameters of the silicone oil are: temperature $T_{00} = 298\,$K, density $\rho_{00} = 0.960\,\mathrm{g\,cm^{-3}}$, kinematic viscosity $\nu = 1.056\,\mathrm{cm^2\,s^{-1}}$, thermal expansivity $\alpha = 0.96 \times 10^{-3}\,\mathrm{K^{-1}}$, thermal conductivity $\lambda = 3.7 \times 10^{-4}\,\mathrm{cal\,cm^{-1}\,K^{-1}\,s^{-1}}$, specific heat $c_p^\dagger = 0.337\,\mathrm{cal\,g^{-1}\,K^{-1}}$, and thermal diffusivity $\kappa = 1.14 \times 10^{-3}\,\mathrm{cm^2\,s^{-1}}$. From Table I, line 1 in Dubois, M. and Bergé, P. (1978), J. Fluid Mech. **85**, 641.

results of the present calculations, which should be equivalent to those of Busse (apart from the non-linear influence of the greater number of modes used here). There is quite good agreement between all three, which confirms both the validity of the present computational algorithm and the applicability of the hydrodynamic model to the experimental situation.

6.3.3 Nusselt Number

The Nusselt number is the ratio of the total heat flux in convection to that in conduction at a given Rayleigh number. The heat flux in conduction is just Fourier's law, $J_{\mathrm{E}}^{\mathrm{cond}} = -\lambda \Delta_T$. Since the velocity vanishes at the horizontal boundaries, the heat flux in convection is purely conductive across these boundaries. Integrating over a single convection cell, the Nusselt number is

$$\begin{aligned} \mathcal{N} &= \frac{1}{L_y J_{\mathrm{E}}^{\mathrm{cond}}} \int_{-L_y}^{0} \mathrm{d}y \, (-\lambda) \left. \frac{\partial T^{\mathrm{total}}(y,z)}{\partial z} \right|_{z=\pm 1/2} \\ &= 1 - \sum_{n=1}^{N} T_{0n}^{\mathrm{s}} 2n\pi(-1)^n. \end{aligned} \tag{6.48}$$

The temperature in the integrand is the sum of the conductive temperature field plus the convective perturbation, Eq. (6.11). The conductive part gives rise to the first term, 1, and the convective terms involving $\sin 2n\pi z$ and $l = 0$ give rise to the remainder.

As mentioned above, for a given Rayleigh number $\mathcal{R} > \mathcal{R}_{\mathrm{c}}$, there is a range of wave numbers a that yield steady state solutions. Hence the Nusselt number is a function of both, $\mathcal{N}(\mathcal{R}, a)$.

6.4 Algorithm for the Cross Roll State

6.4.1 Hydrodynamic Equations and Conditions

This section sets out the computational algorithm that is used for cross roll convection. It is supposed that the system is periodic in the x and y directions, $T(x,y,z) = T(x + m\Lambda_x, y + n\Lambda_y, z)$, $m, n = \pm 1, \pm 2, \ldots$ The wavelength is twice the width of an individual roll, as they come in counter-rotating pairs, $\Lambda_x = 2L_x$ and $\Lambda_y = 2L_y$. The wave number are $\Lambda_x = 2\pi/a_x$, and $\Lambda_y = 2\pi/a_y$.

Above the four hydrodynamic equations for the four fields, (temperature and three velocity components) were given. Explicitly they are

$$0 = \frac{\partial v_x}{\partial x} + \frac{\partial v_y}{\partial y} + \frac{\partial v_z}{\partial z}, \tag{6.49}$$

$$\frac{\partial T}{\partial t} = v_z + \frac{\partial^2 T}{\partial x^2} + \frac{\partial^2 T}{\partial y^2} + \frac{\partial^2 T}{\partial z^2} - v_x \frac{\partial T}{\partial x} - v_y \frac{\partial T}{\partial y} - v_z \frac{\partial T}{\partial z}, \tag{6.50}$$

$$0 = \mathcal{R}\frac{\partial T}{\partial x} + \nabla^2 \left[\frac{\partial v_z}{\partial x} - \frac{\partial v_x}{\partial z} \right], \tag{6.51}$$

and

$$0 = \mathcal{R}\frac{\partial T}{\partial y} + \nabla^2 \left[\frac{\partial v_z}{\partial y} - \frac{\partial v_y}{\partial z} \right]. \tag{6.52}$$

Taking the x derivative of the penultimate equation, the y derivative of the final equation, using the density equation, and adding them together gives,

$$\begin{aligned} 0 &= \mathcal{R}\left[\frac{\partial^2}{\partial x^2} + \frac{\partial^2}{\partial y^2} \right] T + \left[\frac{\partial^2}{\partial x^2} + \frac{\partial^2}{\partial y^2} + \frac{\partial^2}{\partial z^2} \right]^2 v_z \\ &\equiv \mathcal{R}\nabla_\parallel^2 T + \nabla^2 \nabla^2 v_z. \end{aligned} \tag{6.53}$$

In all these equations, the temperature and velocity fields are all functions of the position, $\mathbf{r} = \{x, y, z\}$. In the steady state they are not functions of time, but in a transition between states they are.

Boundary Conditions

The temperature that appears here is the perturbation due to convection. That is, the conductive solution has been subtracted from these equations. This means that the temperature perturbation must vanish at the upper and lower boundaries,

$$T(x, y, \pm 1/2) = 0. \tag{6.54}$$

Since no fluid can cross the boundaries one must also have

$$v_z(x, y, \pm 1/2) = 0. \tag{6.55}$$

The boundaries are solid walls at which the fluid sticks, so that one also has

$$v_x(x, y, \pm 1/2) = v_y(x, y, \pm 1/2) = 0. \tag{6.56}$$

Since this last equation implies that $\partial v_x(x, y, \pm 1/2)/\partial x = \partial v_y(x, y, \pm 1/2)/\partial y = 0$, the density equation implies that $\partial v_z(x, y, z)/\partial z|_{z=\pm 1/2} = 0$.

Symmetry

The fundamental convection cell, $-L_x \le x \le L_x$ and $-L_y \le y \le L_y$, contains two counter rotating rolls in each direction. Hence there is mirror plane symmetry at $x = 0$ and at $y = 0$. This means that

$$\begin{aligned} T(x, y, z) &= T(-x, y, z) = T(x, -y, z), \\ v_x(x, y, z) &= -v_x(-x, y, z) = v_x(x, -y, z), \\ v_y(x, y, z) &= v_y(-x, y, z) = -v_y(x, -y, z), \\ v_z(x, y, z) &= v_z(-x, y, z) = v_z(x, -y, z). \end{aligned} \tag{6.57}$$

These mean that the expansion for the temperature must be even in the lateral coordinates, and so must consist of terms like $T_{qp}(z)\cos(qa_x x)\cos(pa_y y)$, q and p being non-negative integers. A similar expansion holds for v_z. Obviously for v_x and v_y the respective cosine is replaced by a sine.

The Boussinesq symmetry refers to the reflection symmetry within a roll. For $\mathbf{r} = \{x, y, z\}$, define $\mathbf{r}^\dagger = \{L_x - x, L_y - y, -z\}$. One must have

$$T(\mathbf{r}) = -T(\mathbf{r}^\dagger), \text{ and } \mathbf{v}(\mathbf{r}) = -\mathbf{v}(\mathbf{r}^\dagger). \tag{6.58}$$

These basically say that the convective temperature perturbation at the top of an up draught must be equal and opposite to that at the bottom of a down draught. It may be confirmed that the hydrodynamic equations in Boussinesq approximation given above satisfy this symmetry.

6.4.2 Fourier Expansion

In view of the Boussinesq symmetry and the facts that $\cos(qa_x(L_x - x)) = (-1)^q\cos(qa_x x)$ and $\cos(pa_y(L_y - y)) = (-1)^p\cos(pa_y y)$, one sees that $T_{qp}(z)$ must be an odd function of z if $q+p$ is even, and it must be an even function of z if $q+p$ is odd. Since $T(x, y, \pm 1/2) = 0$, and since $\sin(2n\pi z)$ and $\cos((2n+1)\pi z)$ vanish at $z = \pm 1/2$, the expansion for the temperature is

$$T(\mathbf{r}) = \sum_{q=0}^{Q}\sum_{p=0}^{P}\sum_{n=0}^{N} T_{qpn} \cos qa_x x \cos pa_y y \begin{cases} \sin 2n\pi z, & q+p \text{ even}, \\ \cos 2n'\pi z, & q+p \text{ odd}, \end{cases} \tag{6.59}$$

where $n' \equiv (2n+1)/2$, and $T_{qp0} = 0$ if $q+p$ is even. The vertical component of velocity has a similar expansion

$$v_z(\mathbf{r}) = \sum_{q=0}^{Q}\sum_{p=0}^{P}\sum_{n=0}^{N} v^z_{qpn} \cos qa_x x \cos pa_y y \begin{cases} \sin 2n\pi z, & q+p \text{ even}, \\ \cos 2n'\pi z, & q+p \text{ odd}. \end{cases} \tag{6.60}$$

The x component of velocity has the expansion

$$v_x(\mathbf{r}) = \sum_{q=0}^{Q}\sum_{p=0}^{P}\sum_{n=0}^{N} v^x_{qpn} \sin qa_x x \cos pa_y y \begin{cases} \cos 2n'\pi z, & q+p \text{ even}, \\ \sin 2n\pi z, & q+p \text{ odd}, \end{cases} \tag{6.61}$$

with $v^x_{0pn} = 0$. Similarly

$$v_y(\mathbf{r}) = \sum_{q=0}^{Q}\sum_{p=0}^{P}\sum_{n=0}^{N} v^y_{qpn} \cos qa_x x \sin pa_y y \begin{cases} \cos 2n'\pi z, & q+p \text{ even}, \\ \sin 2n\pi z, & q+p \text{ odd}, \end{cases} \tag{6.62}$$

with $v^y_{q0n} = 0$. These latter two expansions arise from the Boussinesq symmetry and the vanishing of the velocity at $z = \pm 1/2$.

z-Component of the Velocity

Inserting the expansions for the temperature and the z-component of the velocity into Eq. (6.53) and setting the coefficients of each term to zero gives the particular solution

$$v_{qpn}^{\mathrm{zp}} = \frac{\mathcal{R}[(qa_x)^2 + (pa_y)^2]}{[(qa_x)^2 + (pa_y)^2 + \mathcal{M}_n^2]^2} T_{qpn} \tag{6.63}$$

where

$$\mathcal{M}_n \equiv \begin{cases} 2n\pi, & q+p \text{ even}, \\ (2n+1)\pi, & q+p \text{ odd}. \end{cases} \tag{6.64}$$

Note the distinction between the superscript p, for particular, and the subscript p, an integer index.

The homogeneous solution satisfies $\nabla^2\nabla^2 v_z^{\mathrm{h}}(\mathbf{r}) = 0$. It has the expansion

$$v_z^{\mathrm{h}}(\mathbf{r}) = \sum_{q=0}^{Q}\sum_{p=0}^{P} \cos qa_x x \cos pa_y y \begin{cases} f_{qp}^{\mathrm{zs}}(z), & q+p \text{ even}, \\ f_{qp}^{\mathrm{zc}}(z), & q+p \text{ odd}. \end{cases} \tag{6.65}$$

Here it may be readily verified that the odd homogeneous solution is

$$f_{qp}^{\mathrm{zs}}(z) = A_{qp}\sinh(\alpha_{qp}z) + B_{qp}z\cosh(\alpha_{qp}z), \tag{6.66}$$

and that the even homogeneous solution is

$$f_{qp}^{\mathrm{zc}}(z) = A_{qp}\cosh(\alpha_{qp}z) + B_{qp}z\sinh(\alpha_{qp}z), \tag{6.67}$$

with

$$\alpha_{qp} = \sqrt{(qa_x)^2 + (pa_y)^2}. \tag{6.68}$$

Even and odd in this context refer to the parity with respect to z.

With $v_z(\mathbf{r}) = v_z^{\mathrm{p}}(\mathbf{r}) + v_z^{\mathrm{h}}(\mathbf{r})$, the four boundary conditions, $v_z(x, y, \pm 1/2) = 0$ and $\partial v_z(x, y, z)/\partial z|_{z=\pm 1/2} = 0$, determine the coefficients A_{qp} and B_{qp} in the even and odd cases. It is straightforward to use the orthogonality properties of the trigonometric functions to obtain the Fourier coefficients, v_{qpn}^{z}.

x- and y-Components of the Velocity

The lateral components of the velocity can be written in the form

$$v_x(\mathbf{r}) = \sum_{q=0}^{Q}\sum_{p=0}^{P} f_{qp}^{x}(z) \sin qa_x x \cos pa_y y, \tag{6.69}$$

and

$$v_y(\mathbf{r}) = \sum_{q=0}^{Q}\sum_{p=0}^{P} f_{qp}^{y}(z) \cos qa_x x \sin pa_y y. \tag{6.70}$$

In view of Eqs (6.51) and (6.52), one has

$$f_{qp}^{x}(z) = qa_x f_{qp}(z), \text{ and } f_{qp}^{y}(z) = pa_y f_{qp}(z). \tag{6.71}$$

Inserting these into the density equation, $\nabla \cdot \mathbf{v}(\mathbf{r}) = 0$, and equating the lateral coefficients term by term yields

$$f_{qp}(z) = \frac{-1}{(qa_x)^2 + (pa_y)^2} \sum_{n=1}^{N} v_{qpn}^{z} \begin{cases} (2n\pi)\cos(2n\pi z), & q+p \text{ even}, \\ (-2n'\pi)\sin(2n'\pi z), & q+p \text{ odd}. \end{cases} \tag{6.72}$$

The boundary condition $\mathbf{v}(x, y, \pm 1/2) = \mathbf{0}$ is automatically satisfied, having already been invoked in the solution of the z-component of the velocity. These solutions for the horizontal velocity can now be projected onto the original z-expansions, Eq. (6.61) and Eq. (6.62).

6.5 Algorithm for Convective Transitions

The preceding two sections gave numerical procedures for the straight roll steady state and the cross roll state, respectively. For each such steady state, these allow single time quantities such as the temperature and velocity fields, the heat flow (the Nusselt number, Eq. (6.48) in the case of straight rolls, or its obvious generalisation for cross rolls), and the static part of the total entropy, Eq. (6.34), to be determined. However, since the algorithms have been formulated to give the evolution in time of the state of the system, they can also be used to characterise transitions, either from the conducting state to a convecting steady straight roll state, or from one steady straight roll state to another. This procedure is now described.

The second entropy is the entropy associated with the transition between two states over a given time interval, and hence it is the appropriate entropy that determines the non-equilibrium phase transitions just mentioned. By definition, the second entropy of two states is the logarithm of the weight of phase space points that begin in the initial macrostate and end in the final macrostate after the given time. Maximising the second entropy over the microstates in the given initial macrostate gives the most likely target macrostate. Assuming that the macrostates represent distinct phases, when the most likely target macrostate is different to the initial macrostate, a phase transition is said to have occurred.

One can avoid the explicit calculation and maximisation of the second entropy by simply choosing microstates within the initial macrostate at random and allowing the system to evolve from these for the specified time interval. (The meaning of 'at random' is discussed below.) Since the second entropy is exceedingly sharply peaked, any such random point will almost certainly evolve to the most likely state, and will remain there if it is a stable state.

Three transition algorithms were used, two based upon the ideal straight roll analysis and one based upon the cross roll analysis. In the first ideal straight roll case, a wave number a typically in the neutrally stable range $[2, 10]$ was chosen, with $L \approx N \approx 10$, and the algorithm proceeded using the equations of §6.3.[13] The initial point was chosen as a small non-zero value in some low order modes, for example $T^{\rm s}_{0,1} = T^{\rm c}_{1,1} = 10^{-3}$. No changes to the results were observed using other starting points. The algorithm converged to a final steady state that represented ideal straight roll convection parallel to the y-axis of wavelength $\Lambda = 2\pi/a$. Most of the power was in the fundamental mode a, with the next most prominent mode being $3a$. In this type of calculation the fixed wave number determines the final steady state. It is most useful for obtaining thermodynamic properties as a function of the steady state wave number, for example, the heat flux, the static part of the entropy, and the velocity fields. This calculation was used to obtain the results given in Table 6.1 on p. 156.

In the second type of straight roll calculation, a small wave number was chosen as the fundamental, $a \approx 0.2$–0.5, and a large number of modes were used, $L \approx 60$–100 and $N \approx 10$. A number of different initial states were tested including uniform distributions as well as Gaussian distributions of the temperature coefficients. It was found that the system converged to a straight roll steady state that was an odd harmonic of the small wave number, $\overline{a} = (2\overline{l}+1)a$. (The odd harmonic is demanded by the Boussinesq symmetry.) The final mode was identified from the power spectrum.

This second type of calculation modeled the conduction to straight roll convection transition. Whereas the first calculation tells the possible straight roll steady states at a given Rayleigh number, the second calculation tells the most likely straight roll steady state that results from a transition directly from the conducting state at a given Rayleigh number. It is important to note that this does not refer to the most likely state, but rather to the most likely transition from the conducting state.

This last point is worth repeating: often for non-equilibrium systems, one cannot speak of *the* most likely phase, but only of the optimum transition from the current phase.

The cross roll algorithm described in §6.4 was used to model the cross roll transition from a steady straight roll state with wave number a_y to an orthogonal straight roll state with wave number $\overline{a}_x$. As in the first form of the first algorithm, a_y was fixed typically in the range $[2, 10]$, with $P = 6$ and $N = 6$, which gave an adequate description of the straight rolls in comparison with the $L = N = 10$ calculations reported in §6.3. A small x-wave number, $a_x \approx 0.2$–0.5, and a large number of modes, $Q \approx 60$–100 were used. For the initial state, a steady straight roll state of wave number a_y generated by the first algorithm was used. In most cases, the value chosen for the initial wave number lay toward one of the extremities of the range of stable wave numbers for the Rayleigh number in the expectation of a transition to an intermediate wave number. The steady state temperature field was perturbed by adding an independent random number to each T_{qpn} (white noise). The amplitude of the noise was proportional to the square root of the total power in the initial steady state. Typically, following the addition of the perturbation, the total power in the y-modes increased by

[13] This wave number a becomes a_y in the cross roll algorithm below, and L becomes P. The converged output from this algorithm is used as the initial input to the cross roll algorithm, with noise added.

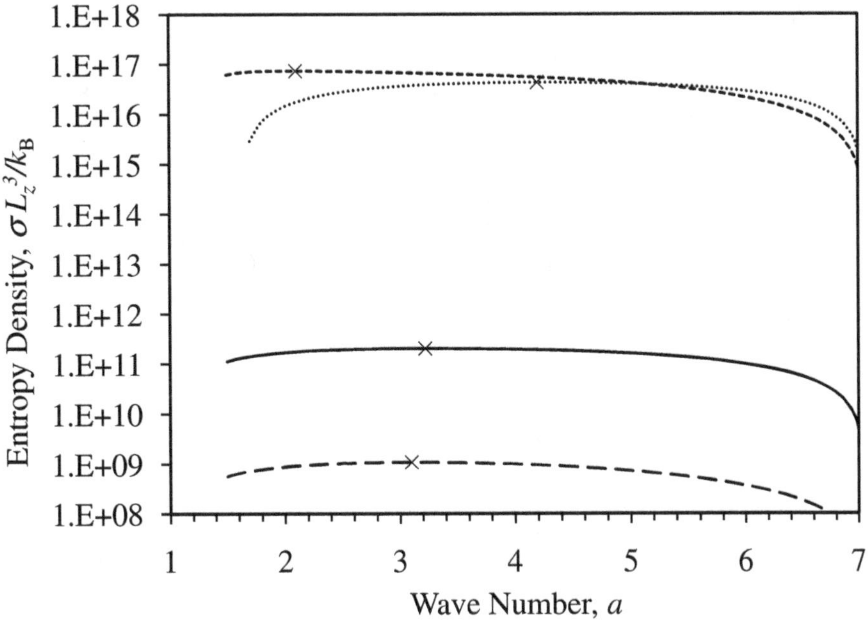

Figure 6.1: Components of the convective entropy density (log scale) at $\mathcal{R} = 5000$. The solid curve is the gravitational contribution, Eq. (6.32), the long dashed curve is the negative of the kinetic energy contribution, Eq. (6.33), and the short dashed curve is the negative of the internal energy contribution, Eq. (6.31). The dotted curve is the sub-system entropy, Eq. (6.24). The crosses mark the maximum of each curve.

about 10%, and the total power in the x-modes was twice as great as that in the y-modes, with ≈ 15 times as many x-modes as y-modes. A cross roll transition usually occurred to ideal straight rolls in the x-direction, with $\overline{a}_x = (2\overline{q}+1)a_x \approx$ 3–4.5. In some cases the computational burden was reduced by only allowing fundamental modes within a window about the likely outcome.

6.6 Convection Theory and Experiment

In this section the results of the convection calculations are presented using the material properties for a typical silicone oil (see footnote 12 on p. 156). Figure 6.1 shows the three contributions to the convective entropy density at $\mathcal{R} = 5000$. This is the difference in entropy between convection and conduction calculated using the static part of the reservoir entropy, as described in §6.2. The figure also shows the sub-system entropy density (i.e. the difference in the sub-system entropy between the convecting state and the conducting state). All of the wave numbers in this and the following figures correspond to hydrodynamic steady states, with the low wave number end of the curves signifying the limit to the

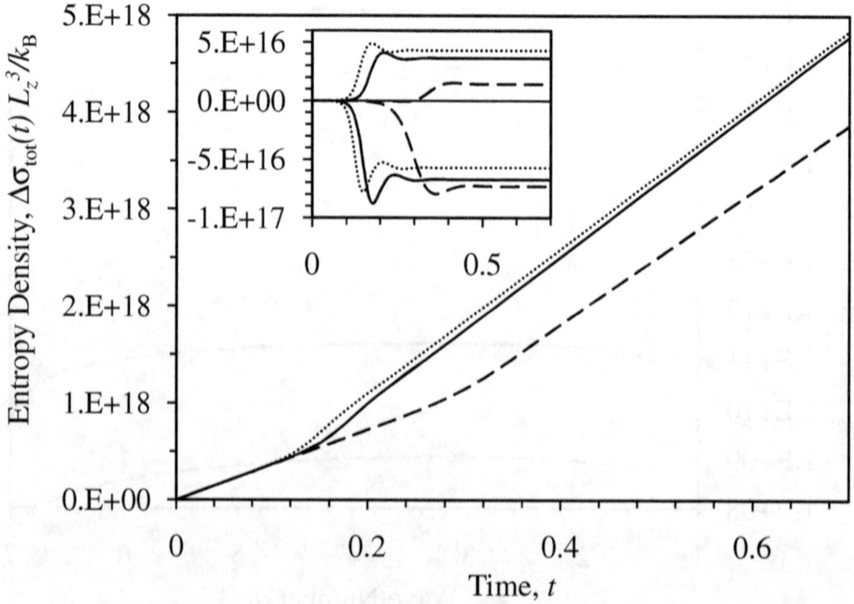

Figure 6.2: Change in entropy density during a straight roll transition from conduction at $\mathcal{R} = 5000$. The solid curves are for $a = 3$, the dashed curves are for $a = 2$, and the dotted curves are for $a = 4$. In the main figure, the curves are the change in the total entropy density, Eq. (6.36). In the inset, the lower three curves are the internal entropy part of the static convection entropy, Eq. (6.31), the upper three curves are the sub-system entropy, Eq. (6.24), and the horizontal line is a guide to the eye.

steady state ideal straight roll solutions to the Boussinesq equations. These are so-called neutrally stable states; the steady state wave numbers beyond the central region may be unstable to perturbations, such as in the case of the cross roll transitions discussed further below.

It can be seen that the entropy due to the convective temperature field itself, the internal energy contribution Eq. (6.31), is about six orders of magnitude greater than the entropy directly due to gravity, which in turn is about two orders of magnitude greater than the entropy due to the kinetic energy. These results are typical for the whole range of Rayleigh numbers. Hence due to this dominance it makes no difference whether one discusses the full static form of the convective entropy or just the internal energy contribution. The sub-system entropy itself, Eq. (6.24), is comparable in magnitude to the static convective entropy, but it is positive. It decreases in magnitude approaching the limits of the range of steady state solutions and actually becomes negative at the low wave number end in most cases.

The static convective entropy difference, which, as shown in Fig. 6.1, is dominated by the internal energy, is negative. One should *not* conclude from this

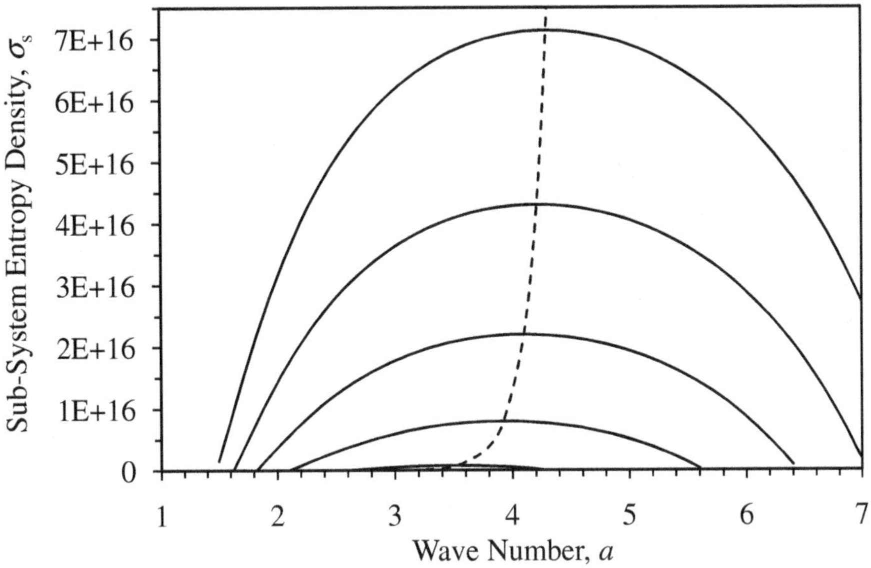

Figure 6.3: Sub-system entropy density for ideal straight roll convection for wave numbers in the region of neutral stability, for Rayleigh numbers from 2000 (bottom) to 6000 (top), in steps of 1000. The dashed curve shows the maxima.

that the convecting state is thermodynamically unfavorable, since this would contradict the hydrodynamic calculations, which show that the convecting states are stable and arise spontaneously from the conducting state with the initial perturbation. It can be seen in Fig. 6.2 that the change in the total entropy density, Eq. (6.36), during a straight roll transition is positive. This contrasts with the static part of the entropy difference, Eq. (6.31), which is negative throughout the transition. The difference in the sub-system entropy, Eq. (6.24), is mainly positive, but not during the entire transition. (The initial data for $a = 2$ in the inset, which can only just be resolved on the scale of the figure, is negative.) The entropy change of the reservoirs completely dominates the change in entropy of the total system during the transition. It is always found that the change in the total entropy density is positive at each stage of the conduction–convection transition. This means that there is indeed consistency between hydrodynamic stability and the Second Law of Thermodynamics.

Figure 6.3 shows the difference in the sub-system entropy between convection and conduction, Eq. (6.24), for ideal straight rolls as a function of wave number for several Rayleigh numbers. Steady state straight roll solutions could be obtained in the range $1708 \leq \mathcal{R} \lesssim 55,000$, with the range of neutral stability increasing with increasing Rayleigh number. It can be seen that at a given wave number, the sub-system entropy density increases with increasing Rayleigh number, and that it approaches zero toward the ends of the stable range. The linear stability analysis of the hydrodynamic equations predicts that the con-

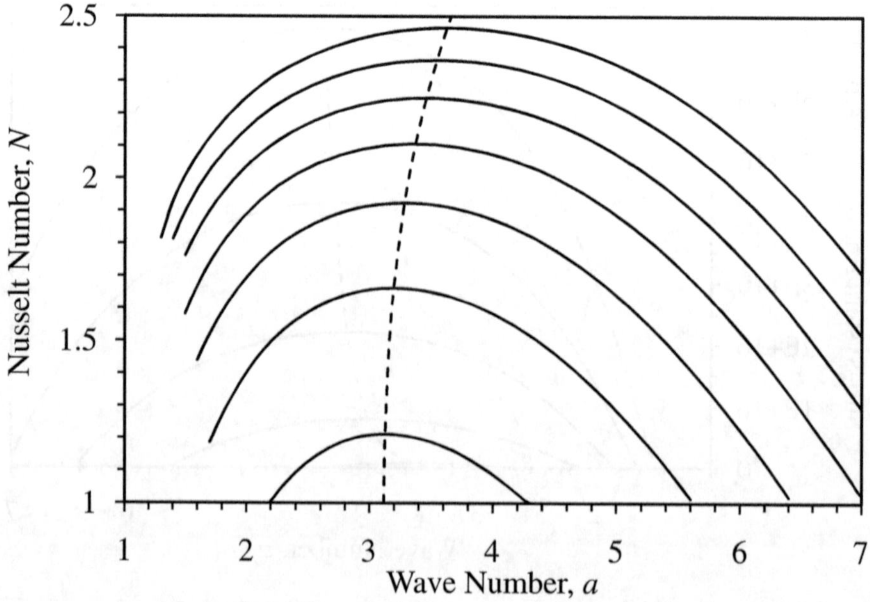

Figure 6.4: Nusselt number as a function of wave number, for Rayleigh numbers from 2000 (bottom) to 8000 (top), in steps of 1000. The maxima are shown by the dashed curve.

vective transition occurs at $\mathcal{R}_c = 1708$ and $a_c = 3.117$ (see Yih, 1977, or Drazin and Reid, 1981). As the critical Rayleigh number is approached from above along the critical wave number, the sub-system entropy density approaches zero from above. Apart from the low wave number end of the steady state range, the change in sub-system entropy density from conduction was found to be positive. As mentioned in connection with Fig. 6.2, the sub-system entropy density was not always positive during the approach to the steady state, and it was also negative in many stable states toward the low wave number end of the range.

It can be seen in Fig. 6.3 that the first and the second derivatives of the sub-system convective entropy density vanish at the critical wave number and critical Rayleigh number. This is analogous to behaviour in equilibrium systems where entropy derivatives vanish at the critical point. In convection, it is known that the hydrodynamic fluctuations diverge at the convective instability.[14]

Figure 6.4 shows the Nusselt number for various Rayleigh numbers as a function of wave number over the region of neutral stability. There is a well-defined wave number of maximum heat flux at each Rayleigh number, and this increases with increasing Rayleigh number.

Experimentally measured wave numbers for straight roll states (data from Fig. 7, Busse and Whitehead, 1971) are compared in Fig. 6.5 to the calculated

[14]See Ortiz de Zárate, J. M. and Sengers, J. V. (2001), Physica A **300**, 25, and references therein.

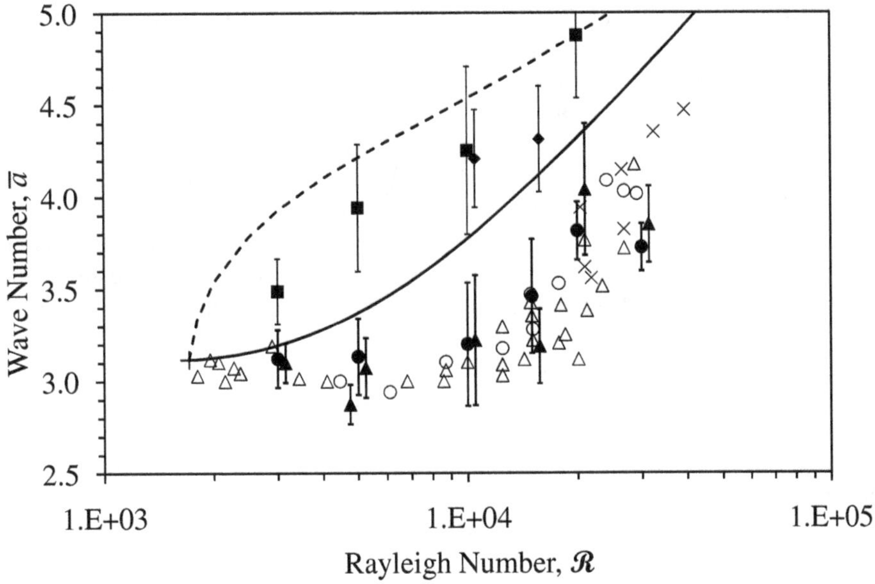

Figure 6.5: The final wave number following a spontaneous cross roll transition as a function of Rayleigh number (log scale). The open symbols are measured data [Busse F. H. and Whitehead, J. A. (1971), J. Fluid Mech. **47**, 305], and the closed circles and triangles are cross roll calculations ($P = N = 6$, Q =100–150, $a_x = 0.1$–0.2, error bars give the standard deviation, with the data shifted horizontally by $\pm 5\%$). The initial constrained state had a large wave number (triangles), a medium wave number (crosses), or a small wave number (circles). The closed squares ($L = 100$ and $a_y = 0.2$) and closed diamonds ($L = 150$ and $a_y = 0.1$) are the calculated conduction–convection transition. The solid and dashed curves give the calculated wave number of maximum Nusselt number and sub-system entropy density, Eq. (6.24), respectively.

wave numbers for maximum entropy and Nusselt number, and also to the wave number averaged over calculated cross roll transitions. The experiments were performed by initially constraining the system in a straight roll convecting state with a wave number specified by means of a periodic temperature perturbation (obtained with an intense light source and shadow mask). Upon removal of the perturbation, there often occurred a spontaneous transition via an intermediate cross roll state to an orthogonal straight roll state whose wave number is shown in the figure. Only final states that are entirely or predominantly straight rolls are analysed here. The majority of the measurements of Busse and Whitehead (1971) either did not result in a spontaneous transition, or else did not have a final straight roll state, or else had too many defects, and so could not be analysed. Almost all other measurements in the literature of the convective wavelength could not be analysed for similar reasons.

Busse and Whitehead (1971) classified the initial wave number as small, medium, or large, as indicated by the symbols in Fig. 6.5. The experimental measurements indicate that the wave number can both increase and decrease in a spontaneous transition, depending upon the initial constrained wave number and Rayleigh number. There are barriers to changing the straight roll wave number since continuous evolution of the wave number was not observed, and once the orthogonal wave number is established in the intermediate cross roll state, it remains as the wave number of the final straight roll state. Consistent with the calculations, there is a certain width or scatter in the measured final wave number at each Rayleigh number, which suggests that it depends upon the initial state or the actual destabilising perturbation.

The experimental data is compared in the figure with the calculated wave numbers that gives the maximum heat flux and the maximum sub-system convective entropy at each Rayleigh number. There is no real agreement between the observed final wave numbers and the wave number that maximises the heat flux or the sub-system entropy. There is some similarity in the observed and calculated wave numbers in that they tend to increase with increasing Rayleigh number. This similarity is no more than qualitative. These data provide evidence that neither the heat flux nor the sub-system entropy is maximised in the non-equilibrium state. Since the static part of the entropy difference is negative, one can also conclude that it does not determine the non-equilibrium state.

Some results of the cross roll transition algorithm described in §6.5 are also shown in Fig. 6.5. Both small, $a_y = 2.2$–2.4, and large, $a_y = 4.3$–6.0, wave number initial states were used, with white noise added as an initial perturbation. The number of independent trials for each initial wave number was 5–13. Each calculation was terminated when it was judged that no further transition would occur, generally on the basis that the power in the maximal x-mode was clearly dominant. The logarithm of the number of transitions observed from a given initial wave number to a given final wave number is directly related to the second entropy of the transition. It can be seen in Fig. 6.5 that there is quantitative agreement between the calculated and measured final wave numbers.

In addition to the cross roll transition, Fig. 6.5 also shows the average final wave number for the conduction–convection transition whose calculation was described in §6.5. White noise was used as a perturbation to initiate the transition. The average final wave number appeared insensitive to the wave number step ($a_y = 0.1$ or 0.2), although the standard deviation was smaller for the smaller step. It was found that in most cases the system converged to a straight roll steady state that was an odd harmonic of the small wave number, $\overline{a} = (2\overline{l}+1)a$, as demanded by the Boussinesq symmetry. Beyond $\mathcal{R} \gtrsim 20,000$, ideal straight rolls did not result, or at least there was not a single clearly dominant wave number. (Some such Bloch states are included in the averages for $\mathcal{R} = 15,000$ in Fig. 6.5.) It can be seen that the wave number resulting from the direct conduction–convection transition is larger than that resulting from a cross roll transition.

These results show that the two types of phase transition, conduction–convection and cross roll, differ significantly even though both result in the same

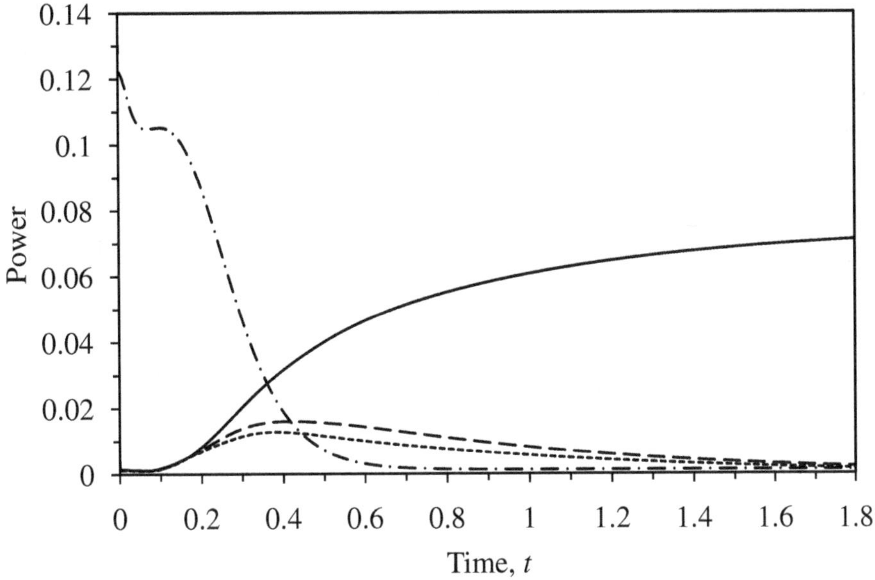

Figure 6.6: The calculated modal power during a cross roll transition ($\mathcal{R} = 5000$, $a_y = 1.7$, $\overline{a}_x = 3.41$, $a_x = 0.31$, $Q = 80$, and $P = N = 6$). The solid, dashed, and dotted curves are the three x-modes with highest power, and the dash-dotted curve is the total power in the y-modes.

type of final state, namely ideal straight rolls. At a given Rayleigh number, the ideal straight roll wave number depends upon whether it results directly from conduction, or whether it results from a cross roll transition (and, in the latter case, it depends on the initial wave number, as shown in Fig. 6.8 on p. 171). One can conclude from this that a single time variational quantity whose optimisation determines the favored non-equilibrium pattern either does not exist, or else has negligible influence compared to the barriers between the multiple possible patterns. This is likely to be a general feature of non-equilibrium systems, and instead of seeking 'the' optimum pattern or phase, one should focus on the optimum transition from a given phase.

The reason that the total first entropy does not determine the optimum convective pattern is that the non-equilibrium heat flow really represents the regression of a fluctuation of the total system. This means that one needs the conditional probability (equivalently, the conditional second entropy) for the initial constraint of reservoir temperature difference, rather than the unconditional probability (equivalently, the total first entropy) that would apply if the convective pattern represented a fluctuation from the equilibrium state of the sub-system or of the total system (i.e. no reservoir temperature difference).

The transition between two orthogonal straight roll convective states via the cross roll intermediate state, as described in §§6.4 and 6.5, can be monitored

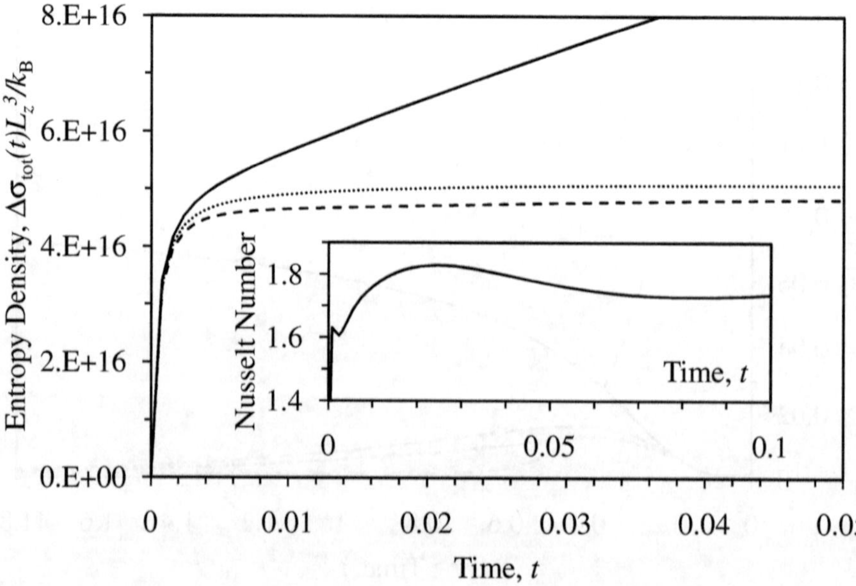

Figure 6.7: Change in entropy density during the cross roll transition of the preceding figure. The solid curve is the change in total entropy density, Eq. (6.36), the dashed curve is the change in the internal entropy part of the static convection entropy, Eq. (6.31), and the dotted curve is the change in the sub-system entropy, Eq. (6.24). The inset shows the Nusselt number.

by the evolution of the power in the various modes, as shown in Fig. 6.6. In this case the system was initially in a convecting straight roll steady state with $a_y = 1.7$, near the lower end of stable states, and at $t = 0$ white noise was added to all the modes. This increased the total power in the y-modes by about 10%, and created almost twice as much power in the x-modes as in the y-modes, spread over thirteen times as many modes. The power in the x-mode q was defined as $P_{xq} \equiv \sum_n T_{q0n}^2$, and the total power in the x-modes was defined as $P_x \equiv \sum_q P_{xq}$, and similarly for the y-modes.

The origin of the perturbations that occur in the experiments is unclear. It is possible that noise originates from mechanical vibrations or from temperature inhomogeneities and water flow in the heat baths, in which cases white noise would likely be appropriate. Although the perturbation is small on the scale of the convective temperature, it is large on molecular scales, and so treating it as a thermodynamic fluctuation (i.e. weighting it by the exponential of the entropy) would be problematic.

Figure 6.6 shows that the power in the x-modes grows over time at the expense of the initially stable y-modes. By about $t \approx 0.6$ the y-rolls have disappeared, and by about $t \approx 1.5$ a steady straight roll convecting state has been established with $\overline{a}_x = 3.41 = 11 \times 0.31$. The Nusselt number at $t = 1.88$

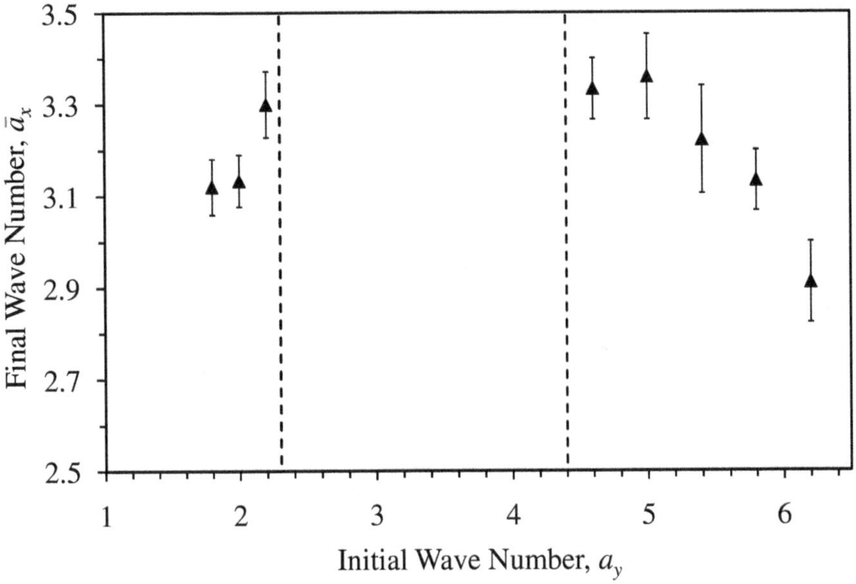

Figure 6.8: The calculated final wave number $\overline{a}_x$ as a function of the initial wave number a_y for cross roll transitions at $\mathcal{R} = 10,000$ ($Q = N = 6$, $P = 100$, and $a_x = 0.2$). The symbols signify the most likely final wave number averaged over 6–12 trials, and the vertical bars show the standard error on the mean. The vertical dashed lines bound the region of stable wave numbers for which no transition occurred.

was $\mathcal{N} = 2.110$, which compares well with $\mathcal{N} = 2.107$ calculated using the ideal straight roll algorithm of §6.3 for $\mathcal{R} = 5000$ and $a = 3.4$. It can be concluded from the figure that a cross roll transition from one straight roll state to another has occurred. It may be called spontaneous in the sense that no constraint was imposed on the final state (other than that it be an odd integer multiple of the wave number step) or indeed on whether any transition would occur at all.

Figure 6.7 shows the evolution of the total entropy during a cross roll transition. It can be seen that the total entropy monotonically increases in time. It can also be seen that the change in the static part of the convective entropy and the change in the sub-system entropy are positive during the transition. The dissipation, which is the rate of change of the reservoir entropy, is proportional to the Nusselt number. Although the Nusselt number is higher for the final state than for the initial state, it can be seen from the inset that it does not increase monotonically during the transition.

Figure 6.8 shows the final wave numbers calculated from a series of cross roll transitions for different initial wave numbers at a Rayleigh number of 10,000.[15]

[15] Since only cross roll transitions were permitted in the computations, their domain is broader here than in the experiments where other transitions can supersede them.

The most likely final wave number for any one trial is the wave number that, out of all possible wave numbers la_x, has the most power after a transition has occurred, $\overline{a}_x = (2\overline{l}+1)a_x$. This varies between trials, and it is the average of these that is plotted for each initial wave number. It can be seen that the final wave number depends upon the initial wave number. One aspect of this dependence is the existence of a region of stable initial wave numbers for which no transition was observed (6 wave numbers spanning this region were tested, with 4–12 trials in each case). A second aspect is the systematic increase in final wave number with decreasing initial wave number for large initial wave numbers. One can possibly make out the opposite trend for small initial wave numbers. This variation of the final wave number is statistically significant. It is possible that at least part of the scatter of the experimental data in Fig. 6.5 can be attributed to a similar dependence on the initial wave number.

Figure 6.8 brings the focus to the transitions rather than the states. For a given Rayleigh number and initial wave number, distinct final states (i.e. distinct transitions) occur with non-negligible probability (not shown). For example, in the case of $\mathcal{R} = 10,000$ and $a_y = 5.4$, 4 distinct transitions actually occurred in 9 trials, ($\overline{a}_x = 3.4$ occurred 4 times, $\overline{a}_x = 3.0$ occurred 3 times, $\overline{a}_x = 2.6$ occurred once, and $\overline{a}_x = 3.8$ occurred once). The existence of distinct final states from which no further transitions occur signifies that patterns with a given wave number are locally stable and that there are barriers to further transitions, which is consistent with the experimental observations.

At a given Rayleigh number, it is less meaningful to speak of the average wave number for ideal straight roll convection than it is to speak of the average wave number following a transition from a given steady state. This particular point was already made in connection with Fig. 6.5, where the distinction between the wave number resulting from a conduction–convection transition and from a cross roll transition was discussed. Even this is a severe simplification of the full convective transition phenomenon, since the final wave number of the cross roll transition depends upon the initial wave number.

Beyond heat convection, this point is true in general: for non-equilibrium phase transitions and pattern formation, it is more useful to focus on the conditional transitions between states rather than on the states themselves. By definition, patterns in non-equilibrium systems are locally stable with barriers inhibiting transitions between them. Under these circumstances any single time variational principle for the pattern structure, if it exists, would have limited utility. The same situation occurs for some difficult static problems, such as protein structure or glass formation, where the state of global entropy maximum is not related directly to the state that actually occurs. This focus on the transitions between non-equilibrium phases in the general case is consistent with the conclusion for heat convection drawn from Figs 6.5 and 6.8, namely that single time thermodynamic functions, such as the entropy or the rate of entropy production, are insufficient to characterise the observed non-equilibrium state.

Chapter 7

Equilibrium Statistical Mechanics

The equilibrium and non-equilibrium thermodynamics that have been presented in the preceding chapters were justified and derived on the basis of the statistical nature of entropy. Once the equivalence of entropy, probability, and weight has been accepted, then all of the axioms of thermodynamics follow. One can accept Boltzmann's interpretation of the statistical nature of entropy in a qualitative sense, without having to quantify exactly the number or weight of molecular configurations in a macrostate. The cost of this is that thermodynamic parameters such as the heat capacity or the thermal conductivity can only be obtained by experimental measurement, and not from any first principles calculation.

The task of statistical mechanics is to provide a way of quantifying the entropy from the nature of the interactions of the molecules that comprise the system. This is equivalent to obtaining the probability distribution of the molecules. Since all thermodynamic parameters can be written as functions of the molecular configuration, the average value of the thermodynamic parameters of the system can then be obtained.

This chapter gives an account of classical equilibrium statistical mechanics, and the following chapter extends the analysis to the non-equilibrium case. First, phase space is introduced as representing the microstates of the system, and Hamilton's equations of motion are given. A very important result is then established, namely that the phase space probability density for an isolated system is uniform on an energy hypersurface. From this is derived the Maxwell-Boltzmann distribution for a sub-system of a heat reservoir, which is the canonical equilibrium system. This presentation of relatively well-known results for canonical equilibrium occupies §§7.1–7.3.

In §7.4 and the following sections, time is introduced into the canonical equilibrium system in the form of the transition probability and the evolution equations. The nature of the dissipation in the stochastic equations of motion of the sub-system is derived from the second entropy, and it is shown that the fluctuation dissipation theorem ensures that the Maxwell-Boltzmann distribution is stationary. The Fokker-Planck equation, §7.4.5, and Liouville's theorem, §7.5.3, are discussed for this canonical equilibrium case, as well as the trajectory probability and time correlation functions, §7.7.

7.1 Hamilton's Equations of Motion

The fundamental definition of entropy given in §1.4 was based upon set theory, in which the notions of macrostates and microstates were defined. In order to quantify the entropy, the first task is to identify the appropriate microstates for a system composed of molecules.

The microstates must have the property that they form a complete set (at any time the system is in a microstate), that they are disjoint (at any time the system is in only one microstate), and that they are indivisible. The last property of indivisibility is to be interpreted as meaning that the microstates are the finest level of description that is going to be used to characterise the system. Whilst internal sub-division of the microstates into smaller states may be possible physically, one has taken the decision not to do so explicitly, but to rather incorporate such internal configurations into the internal entropy or weight of the microstates themselves.

For a system of N molecules, the microstates are generally taken to be phase space, $\mathbf{\Gamma} \equiv \{\mathbf{q}^N, \mathbf{p}^N\}$, where $\mathbf{q}_i$ is the position of molecule i, and $\mathbf{p}_i$ is its momentum. At a given time t the system has a specific molecular configuration, $\mathbf{\Gamma}(t)$, and so phase space satisfies the first two requirements for a set of microstates. Evidently, the rotational configurations of non-spherical molecules have been ignored here, as well as the configurations of the individual electrons or nucleons of each atom, as well as any structural rearrangements in the case of a macromolecule.[1] Obviously one should have reason to believe that it is not necessary for the problem at hand to take into account these internal configurations explicitly. As well, and as will become clearer below, it is often the case that $\mathbf{\Gamma}$ represents the phase space of a sub-system of a larger system, and the molecular configurations of the reservoir external to the sub-system are also not represented explicitly. All of these 'internal' and 'external' configurations that comprise a phase space microstate can be incorporated into the internal entropy or weight of each phase space point.

Phase space is not the only possible choice for the microstates of the system. One could for example choose instead the space of molecular positions and velocities. However, it turns out that the phase space representation of an isolated system has certain properties that simplify the probability distribution.

[1] Most of the following analysis is readily generalised to the case that $\mathbf{\Gamma}$ includes some or other of these internal coordinates. The present choice of position and momentum coordinates represents a compromise that develops statistical mechanics at the molecular level, whilst keeping the notation and concepts simple enough that the basic principles are not obscured. Experience shows that this level of description is sufficient for quantitatively accurate results in a diverse range of applications.

The time development of an isolated system is governed by Hamilton's equations. These are equivalent to Newton's equations of motion, but they are more convenient because they are first order equations, and they offer a relatively simple and transparent path for the derivation of a number of results. The Hamiltonian $\mathcal{H}(\mathbf{\Gamma})$ is the total energy of the system, and is generally the sum of the kinetic energy of the molecules and their potential energy. In the non-equilibrium case the potential energy may be explicitly time dependent, but this will not be considered at this stage.

The derivatives of the Hamiltonian give the rate of change of the configuration,

$$\dot{\mathbf{q}}_i^0 = \frac{\partial \mathcal{H}(\mathbf{\Gamma})}{\partial \mathbf{p}_i}, \text{ and } \dot{\mathbf{p}}_i^0 = \frac{-\partial \mathcal{H}(\mathbf{\Gamma})}{\partial \mathbf{q}_i}. \tag{7.1}$$

Here and throughout, a vector derivative represents an equation for each component of the vector. The superscript 0 has been used to emphasise the adiabatic motion. The trajectory through phase space may be denoted $\mathbf{\Gamma}^0(t|\mathbf{\Gamma}_0, 0)$, or more simply $\mathbf{\Gamma}^0(t|\mathbf{\Gamma}_0)$, which is the configuration of the system at time t given that it was at $\mathbf{\Gamma}_0$ at time $t = 0$.

An isolated system has evolution fixed by Hamilton's equations of motion. There are no non-Hamiltonian terms, and there are no stochastic terms. The time evolution of any function of phase space is similarly determined by the Hamiltonian, and only by the Hamiltonian.

An isolated system, which at present is the main concern, has fixed energy. This means that energy must be conserved during the Hamiltonian evolution of the system. One can confirm that this is indeed true since one has

$$\begin{aligned}
\frac{\mathrm{d}\mathcal{H}(\mathbf{\Gamma})}{\mathrm{d}t} &= \dot{\mathbf{\Gamma}}^0 \cdot \frac{\partial \mathcal{H}(\mathbf{\Gamma})}{\partial \mathbf{\Gamma}} \\
&= \sum_{i=1}^{N} \left[\dot{\mathbf{p}}_i^0 \cdot \frac{\partial \mathcal{H}(\mathbf{\Gamma})}{\partial \mathbf{p}_i} + \dot{\mathbf{q}}_i^0 \cdot \frac{\partial \mathcal{H}(\mathbf{\Gamma})}{\partial \mathbf{q}_i} \right] \\
&= \sum_{i=1}^{N} \left[\dot{\mathbf{p}}_i^0 \cdot \dot{\mathbf{q}}_i^0 - \dot{\mathbf{q}}_i^0 \cdot \dot{\mathbf{p}}_i^0 \right] \\
&= 0.
\end{aligned} \tag{7.2}$$

In §§ 3.7.3 and 3.7.5 the Fokker-Planck equation and Liouville's theorem were derived, and both involved the compressibility of the deterministic part of the equations of motion, $\nabla \cdot \dot{\mathbf{\Gamma}}^{\mathrm{det}}$. (The derivation given at the time was applied to a Brownian particle and was given in terms of a variable x, but it was mentioned that the results held with x interpreted quite generally.) In the present case of an isolated system, the deterministic velocity is given by Hamilton's equations, $\dot{\mathbf{\Gamma}}^{\mathrm{det}} = \dot{\mathbf{\Gamma}}^0$. A Hamiltonian trajectory is incompressible, which is readily shown by invoking the divergence operator in phase space and the

Hamiltonian expressions for the rates of change of the positions and momenta,

$$\begin{aligned}\nabla \cdot \dot{\mathbf{\Gamma}}^0 &= \sum_{i=1}^{N} \sum_{\alpha=x,y,z} \left[\frac{\partial \dot{p}_{i\alpha}^0}{\partial p_{i\alpha}} + \frac{\partial \dot{q}_{i\alpha}^0}{\partial q_{i\alpha}} \right] \\ &= \sum_{i=1}^{N} \sum_{\alpha=x,y,z} \left[\frac{-\partial^2 \mathcal{H}(\mathbf{\Gamma})}{\partial p_{i\alpha} \partial q_{i\alpha}} + \frac{\partial^2 \mathcal{H}(\mathbf{\Gamma})}{\partial q_{i\alpha} \partial p_{i\alpha}} \right] \\ &= 0. \end{aligned} \tag{7.3}$$

7.1.1 Classical versus Quantum Statistical Mechanics

A small digression is in order here to address the choice of classical over quantum mechanics. Although the ultimate theory of matter and motion is quantum, when there is a sufficiently large superposition of wave functions the behaviour tends to the classical limit. This means that larger atoms composed of many electrons and nucleons behave classically. The cut-off in practical terms is around about the argon atom, where the quantum correction to classical statistical mechanics for the liquid state is on the order of a few per cent.[2] Smaller objects than argon, such as hydrogen, helium, electrons, or photons, require quantum statistical mechanics. For larger atoms or molecules, classical statistical mechanics suffices.

The interaction potential between atoms, whether intermolecular or intramolecular, is predominantly due to their electrons, and this should be obtained by quantum calculations or by experimental measurement or fit. However, once the interaction potential has been obtained, it can be used in the classical equations of motion.

The two most common intermolecular potentials for atoms are the hard sphere and the Lennard-Jones pair potentials. Both of these have a short-range repulsion, which models the Fermi exclusion of the outermost electrons as the atoms come together. The Lennard-Jones potential has in addition a long range r^{-6} attractive tail, which arises from the correlated fluctuations of the electrons of each atom (induced dipole–induced dipole), and whose magnitude is proportional to the square of their polarisability, which can be obtained quantum mechanically or by fitting. Some molecules have in addition a nett charge or a permanent dipole moment, and the electrostatic potential due to these is purely classical.

7.2 Probability Density of an Isolated System

There are essentially three ways of establishing the probability density for phase space of an isolated system.[3] The first is to simply postulate it, which is the traditional approach to the subject, and this is called the ergodic hypothesis.

[2] Barker J. A. and Klein, M. L. (1973), Phys. Rev. B **7**, 4707.

[3] Attard, P. (2002a), *Thermodynamics and Statistical Mechanics: Equilibrium by Entropy Maximisation*, (Academic Press, London).

The second is to derive it by relating a phase space average to a time average. And the third way is to derive it by analysing the evolution of the probability density on a trajectory. Each of these three approaches are now successively described.

7.2.1 Ergodic Hypothesis

As just mentioned, the traditional way of presenting statistical mechanics begins with the *Ergodic Hypothesis.* This consists of two postulates:

- all points of the same energy in the phase space of an isolated system are connected by a single trajectory,
- points in phase space of equal energy are equally likely.

If phase space comprised isolated regions, contrary to the first postulate, then a time average over a trajectory would yield different answers depending upon which basin it began in. Non-ergodic systems, which include glasses, annealed systems, or biological and other macromolecules, are problematic if these disconnected equal energy regions of phase space have measurably different properties. In this work it is assumed that the region of phase space covered by a trajectory in the relevant time interval, is representative of all regions of phase space with the same energy. In this sense a time integral over this accessible part of phase space is the same as an integral over all regions of phase space with the same energy. With the possible exception of the non-ergodic systems just mentioned, experience shows that this is a realistic practical assumption for actual physical systems.

The second postulate of equal *a priori* energy states is in many ways more important because it determines the fundamental weight that is to be given to phase space. The probability distribution and the entropy, not only of an isolated system but also for sub-systems of reservoirs, are quantitatively dependent upon this assumption. Because this is so important, in the following two sections a proof of the postulate is offered that depends upon a deeper axiom, namely that the probability density is uniform in time.

7.2.2 Time, Volume, and Surface Averages

The macrostates of classical statistical mechanics represent the values of experimental observables, and these in general are functions of phase space, $f(\mathbf{\Gamma})$. These could be the energy, the pressure, the number of molecules, the pair distribution function, etc. Generally it is the average value of the phase function that is needed, and this requires the probability density.

Confining attention to an isolated system, there are three different ways of obtaining the average. The first and possibly most well known is the temporal average. This is the average of the phase space function along a trajectory in time, and it is written as

$$\begin{aligned}\langle f\rangle_s &= \int_0^s \mathrm{d}t\, \wp(t) f(\mathbf{\Gamma}^0(t|\mathbf{\Gamma}_0)) \\ &= \frac{1}{s}\int_0^s \mathrm{d}t\, f(\mathbf{\Gamma}^0(t|\mathbf{\Gamma}_0)).\end{aligned} \tag{7.4}$$

Here s is some time interval that is long compared to molecular time scales.[4] The trajectory $\mathbf{\Gamma}^0(t|\mathbf{\Gamma}_0)$ gives the position of the system at time t starting from $\mathbf{\Gamma}_0$ at time $t = 0$. The superscript zero signifies that this is the Hamiltonian or adiabatic trajectory that is applicable to an isolated system. In this case the average is independent of the starting position of the trajectory, although obviously it does depend upon the energy of the isolated system, $E_0 \equiv \mathcal{H}(\mathbf{\Gamma}_0)$.

The second equality follows assuming a uniform probability in time, $\wp(t) = 1/s$. This is a fundamental axiom: the weight of a state is linearly proportional to the time the system spends in the state. In consequence a time average is an unweighted average.

The second average is to take the phase space average of the function. This requires the conditional probability density, $\wp(\mathbf{\Gamma}|E_0)$. It depends upon the fixed energy E_0, since the isolated system must have energy within $\mathrm{d}E$ of E_0. This density gives the probability that the system will be found within $\mathrm{d}\mathbf{\Gamma}$ of $\mathbf{\Gamma}$. The average of the phase function is just the integral over phase space weighted by the probability density,

$$\langle f\rangle_{\mathbf{\Gamma}} = \int \mathrm{d}\mathbf{\Gamma}\, \wp(\mathbf{\Gamma}|E_0) f(\mathbf{\Gamma}). \tag{7.5}$$

The probability density can be defined as the temporal average of a Dirac δ-function over the trajectory,

$$\wp(\mathbf{\Gamma}_1|E_0) = \frac{1}{s}\int_0^s \mathrm{d}t\, \delta(\mathbf{\Gamma}^0(t|\mathbf{\Gamma}_0) - \mathbf{\Gamma}_1). \tag{7.6}$$

Direct substitution and interchange of the order of integration shows that phase and time averages are equivalent, $\langle f\rangle_{\mathbf{\Gamma}} = \langle f\rangle_s$.

The third average invokes a surface average. Because of the fixed energy of an isolated system, the trajectory does not cover all of phase space but is confined to the $(6N-1)$-dimensional hypersurface $\mathcal{H}(\mathbf{\Gamma}) = E_0$. In view of this one can define a coordinate system for the surface, $\gamma(\mathbf{\Gamma})$. The relationship is one-to-one, so that one also has $\mathbf{\Gamma}(\gamma, E_0)$. The trajectory can be written $\gamma^0(t|\gamma_0) = \gamma(\mathbf{\Gamma}^0(t|\mathbf{\Gamma}_0))$. The surface probability density is again defined as the temporal average of a δ-function,

$$\wp^{(E)}(\gamma_1) = \frac{1}{s}\int_0^s \mathrm{d}t\, \delta(\gamma^0(t|\gamma_0) - \gamma_1). \tag{7.7}$$

[4]See the discussion above of the first postulate of the ergodic hyopothesis, namely that the time interval has to be long enough that the trajectory covers a representative portion of phase space.

In terms of this the surface average is

$$\langle f \rangle_\gamma = \oint_{E_0} \mathrm{d}\gamma\, \wp^{(E)}(\gamma) f(\mathbf{\Gamma}(\gamma, E_0)), \tag{7.8}$$

and again one may confirm the equivalence of surface and time averages, $\langle f \rangle_\gamma = \langle f \rangle_s$. Here the integral is over the energy hypersurface of thickness $\mathrm{d}E$ to which the isolated system is constrained, and $\mathrm{d}\gamma$ is the element of area.

The element of phase space volume is related to the element of hypersurface area by

$$\mathrm{d}\mathbf{\Gamma} = \mathrm{d}\gamma\, \mathrm{d}n = \mathrm{d}\gamma\, \mathrm{d}E/|\nabla \mathcal{H}(\mathbf{\Gamma})|, \tag{7.9}$$

where $\mathrm{d}n$ is the differential length of the normal to the energy hypersurface. The generalised gradient is $\nabla \equiv \partial/\partial\mathbf{\Gamma}$. Where the energy gradient is steep, the volume of phase space associated with the element of area is small, because adjacent hypersurfaces crowd together. From this equation one sees that the distance between two energy hypersurfaces is not constant in phase space.

The phase space probability density is related to the surface probability density by the usual transformation of probability densities,

$$\begin{aligned} \wp(\mathbf{\Gamma}|E_0) &= \wp^{(E)}(\gamma) \left| \frac{\mathrm{d}\gamma}{\mathrm{d}\mathbf{\Gamma}} \right|, \ |\mathcal{H}(\mathbf{\Gamma}) - E_0| < \mathrm{d}E \\ &= \wp^{(E)}(\gamma) \frac{|\nabla \mathcal{H}(\mathbf{\Gamma})|}{\mathrm{d}E}, \ |\mathcal{H}(\mathbf{\Gamma}) - E_0| < \mathrm{d}E \\ &= \wp^{(E)}(\gamma)\, |\nabla \mathcal{H}(\mathbf{\Gamma})|\, \delta(\mathcal{H}(\mathbf{\Gamma}) - E_0), \end{aligned} \tag{7.10}$$

where $\gamma = \gamma(\mathbf{\Gamma})$. In the final equality, the limit $\mathrm{d}E \to 0$ has been taken, which gives the δ-function.

It is straightforward to verify the equality of surface and phase space averages,

$$\begin{aligned} \langle f \rangle_{\mathbf{\Gamma}} &= \int \mathrm{d}\mathbf{\Gamma}\, \wp(\mathbf{\Gamma}|E) f(\mathbf{\Gamma}) \\ &= \int \frac{\mathrm{d}\gamma\, \mathrm{d}E}{|\nabla \mathcal{H}(\mathbf{\Gamma})|} |\nabla \mathcal{H}(\mathbf{\Gamma})|\, \delta(\mathcal{H}(\mathbf{\Gamma}) - E) \wp^{(E)}(\gamma) f(\mathbf{\Gamma}) \\ &= \oint_{E_0} \mathrm{d}\gamma\, \wp^{(E)}(\gamma) f(\mathbf{\Gamma}(\gamma, E)) \\ &= \langle f \rangle_\gamma . \end{aligned} \tag{7.11}$$

Hence one can see that temporal, phase space, and hypersurface averages are equivalent, and hence that this formulation of classical statistical mechanics is internally consistent.

The characteristic function may be defined from the ratio of the surface and the phase space probability densities given above,

$$\chi(\mathbf{\Gamma}, E_0) \equiv |\nabla \mathcal{H}(\mathbf{\Gamma})|\, \delta(\mathcal{H}(\mathbf{\Gamma}) - E_0). \tag{7.12}$$

This is the area density of the hypersurface in phase space. Hence the area of the energy hypersurface is

$$\begin{aligned} A(E_0) &= \oint_{E_0} \mathrm{d}\gamma \\ &= \int \mathrm{d}\mathbf{\Gamma}\, |\nabla \mathcal{H}(\mathbf{\Gamma})|\, \delta(\mathcal{H}(\mathbf{\Gamma}) - E_0) = \int \mathrm{d}\mathbf{\Gamma}\, \chi(\mathbf{\Gamma}, E_0). \end{aligned} \tag{7.13}$$

The equality of these follows by direct substitution of the volume and area elements given above.

The above results have been cast in what might be called the natural representation, namely in terms of the energy hypersurface, based upon the constancy of the energy. However, any function of the energy, say $g(E)$, is also a constant of the motion, and there is no real reason to single out the energy hypersurface as the preferred surface.[5] This means that the results should be invariant to the representation. One can verify that this is so from the characteristic function,

$$\begin{aligned} \chi(\mathbf{\Gamma}, E_0) &= |\nabla \mathcal{H}(\mathbf{\Gamma})|\, \delta(\mathcal{H}(\mathbf{\Gamma}) - E_0) \\ &= |\nabla g_H(\mathbf{\Gamma})|\, \delta(g_H(\mathbf{\Gamma}) - G_0). \end{aligned} \tag{7.14}$$

Here $g_H(\mathbf{\Gamma}) \equiv g(\mathcal{H}(\mathbf{\Gamma}))$ and $G_0 \equiv g(E_0)$. The final equality demonstrates that the characteristic function is invariant with respect to the representation of the hypersurface. On physical grounds this is an essential requirement of the formulation of the theory. The following results are given in the 'natural' energy representation, but they are unchanged in any other representation.

7.2.3 Energy Uniformity

The probability density on the hypersurface is now obtained explicitly from the uniform temporal probability density, $\wp(t) = 1/s$. This is then transformed to the probability density in phase space using the characteristic function defined above.

The probability density on the hypersurface follows from the usual rules for the transformation of probability densities,

$$\wp^{(E)}(\gamma) = \wp(t) \left| \frac{\mathrm{d}t}{\mathrm{d}\gamma} \right| \propto \frac{1}{|\dot{\gamma}^0|}. \tag{7.15}$$

This assumes that the experimental time scale s is short compared to the system recurrence time, so that the trajectory only passes within $\mathrm{d}\gamma$ of γ once. It also assumes that s is long enough so that it covers a representative selection of points on the hypersurface. The physical interpretation follows from the temporal representation, Eq. (7.7). The integral over the δ-function gives the total time that is spent in the neighbourhood of γ, namely $t_\gamma = \mathrm{d}\gamma/|\dot{\gamma}^0|$. The probability of being in the area element $\mathrm{d}\gamma$ is proportional to this. Dividing by

[5] Arguments based upon linear additivity do give energy a preferred rôle in statistical mechanics, but these are not relevant at this stage of the development of the theory.

the area element gives the above expression for the probability density. Since the velocity of the trajectory is not a constant of the motion, one can immediately conclude from this result that the surface probability density is *not* uniform on the surface.

In general terms it is more convenient to operate with the phase space probability density rather than the hypersurface probability density. Using the above result for the latter, Eq. (7.15), and the conversion rule Eq. (7.10), the phase space probability density is

$$\begin{aligned} \wp(\mathbf{\Gamma}|E) &\propto \frac{|\nabla\mathcal{H}(\mathbf{\Gamma})|}{|\dot{\gamma}^0|}\frac{1}{\mathrm{d}E}, \quad |\mathcal{H}(\mathbf{\Gamma}) - E| < \mathrm{d}E \\ &\propto \frac{|\nabla\mathcal{H}(\mathbf{\Gamma})|}{|\dot{\mathbf{\Gamma}}^0|}\delta(\mathcal{H}(\mathbf{\Gamma}) - E). \end{aligned} \tag{7.16}$$

(Here and henceforth the initial energy E_0 has been more simply written E.) This has the physical interpretation that the probability density is proportional to the time that the system spends in an elementary cell, which remains t_γ, as given above, (since $|\dot{\mathbf{\Gamma}}^0| = |\dot{\gamma}^0|$ because $|\gamma_1 - \gamma_2| = |\mathbf{\Gamma}(\gamma_1, E) - \mathbf{\Gamma}(\gamma_2, E)| = |\mathbf{\Gamma}_1 - \mathbf{\Gamma}_2|$), and it is inversely proportional to the volume element $\mathrm{d}\mathbf{\Gamma} = \mathrm{d}\gamma\mathrm{d}E/|\nabla\mathcal{H}(\mathbf{\Gamma})|$.[6]

For the present physical problem of an isolated system that evolves according to Hamilton's equations of motion, the phase space probability density takes on a remarkably simple form. First one can introduce a normalising factor, the partition function or total weight, $Z'(E)$, so that the proportionality in the above expression may be replaced by an equality,

$$\wp(\mathbf{\Gamma}|E) = \frac{|\nabla\mathcal{H}(\mathbf{\Gamma})|}{|\dot{\mathbf{\Gamma}}^0|Z'(E)}\delta(\mathcal{H}(\mathbf{\Gamma}) - E). \tag{7.17}$$

Now in the space of canonical positions and momenta, one has for a Hamiltonian system,[7]

$$\begin{aligned} |\nabla\mathcal{H}(\mathbf{\Gamma})|^2 &= \sum_{i=1}^{N} \frac{\partial\mathcal{H}}{\partial\mathbf{q}_i}\cdot\frac{\partial\mathcal{H}}{\partial\mathbf{q}_i} + \frac{\partial\mathcal{H}}{\partial\mathbf{p}_i}\cdot\frac{\partial\mathcal{H}}{\partial\mathbf{p}_i} \\ &= \sum_{i=1}^{N} \dot{\mathbf{p}}_i^0\cdot\dot{\mathbf{p}}_i^0 + \dot{\mathbf{q}}_i^0\cdot\dot{\mathbf{q}}_i^0 \\ &= |\dot{\mathbf{\Gamma}}^0|^2. \end{aligned} \tag{7.18}$$

[6]Hamilton's equations of motion have not been assumed in deriving this result, although it has been assumed that $\mathcal{H}(\mathbf{\Gamma})$ is a constant of the motion. One can possibly imagine some system that evolves under arbitrary equations of motion, with $\mathcal{H}(\mathbf{\Gamma})$ representing some conserved variable. In such a case, these two desiderata could be taken as the self-evident axioms that determine the probability density of such a non-Hamiltonian system.

[7]On physical grounds these measures of length are improper because they add quantities with different dimensions. Hence a metric element really ought to appear in these sums. This will change the value of the length of the gradient and of the speed of the trajectory, but it will make no difference to the final conclusion that the two are equal.

The equality of these is a remarkable coincidence, and is a unique feature of Hamilton's equations of motion.

With this result, the numerator and the denominator cancel leaving a particularly simple expression for the probability density in phase space,

$$\wp(\mathbf{\Gamma}|E, N, V) = \frac{\delta(\mathcal{H}(\mathbf{\Gamma}) - E)}{Z'(E, N, V)}. \tag{7.19}$$

For completeness, the volume V and number of molecules N of the system have also been signified here.

One can conclude from this result that phase space points with the same energy are equally likely. This is not the same as saying that the probability density of an isolated system is uniform in phase space, since the δ-function is highly nonuniform (but see §7.2.6 below for phase space itself). One might interpret this as saying that the phase space probability density is uniform on the energy hypersurface, provided that one keeps in mind that the surface probability density itself is non-uniform on the energy hypersurface, due to the fact that it varies in inverse proportion to the variation in the trajectory speed.

Using the explicit results for the probability densities, the average of a function of phase space for an isolated Hamiltonian system of energy E is

$$\begin{aligned} \langle f(\mathbf{\Gamma}) \rangle &= \frac{1}{Z'(E, N, V)} \int \mathrm{d}\mathbf{\Gamma}\, f(\mathbf{\Gamma}) \delta(\mathcal{H}(\mathbf{\Gamma}) - E) \\ &= \frac{1}{Z'(E, N, V)} \oint_E \frac{\mathrm{d}\gamma}{|\dot{\gamma}|} f(\mathbf{\Gamma}(\gamma, E)). \end{aligned} \tag{7.20}$$

This version of the partition function is (but see also below),

$$Z'(E, N, V) = \int \mathrm{d}\mathbf{\Gamma}\; \delta(\mathcal{H}(\mathbf{\Gamma}) - E) = \oint_E \frac{\mathrm{d}\gamma}{|\dot{\gamma}^0|}. \tag{7.21}$$

7.2.4 Trajectory Uniformity

In Ch. 1, Eq. (1.28), a conservation law for weight was given, $\sum_j \omega(j, i|\tau) = \omega(i)$, which says that the microstate transition weight distributes the original weight of microstate i amongst all the target microstates j. This conservation law will now be used to give an alternative demonstration that phase space points of equal energy have equal probability.

Let $\mathbf{\Gamma}^0(t|\mathbf{\Gamma}_0)$ be the position in phase space at time t of a system that was at $\mathbf{\Gamma}_0$ at time $t = 0$. The system is isolated and this trajectory is determined by Hamilton's equations of motion, as signified by the superscript 0. Let $\wp_0(\mathbf{\Gamma})$ be the initial phase space probability density at $t = 0$, and let $\wp(\mathbf{\Gamma}, t)$ be the phase space probability density at time t that evolves from this initial density. The first task is to show that $\wp(\mathbf{\Gamma}|t)$ is constant following the trajectory, and the second task is to show that this implies that the probability density is uniform on the trajectory.

The total rate of change of the probability density is

$$\frac{\mathrm{d}\wp(\mathbf{\Gamma},t)}{\mathrm{d}t} = \frac{\partial\wp(\mathbf{\Gamma},t)}{\partial t} + \dot{\mathbf{\Gamma}}^0 \cdot \nabla\wp(\mathbf{\Gamma},t). \tag{7.22}$$

This is just the definition of the total derivative. Note that in contrast to the analysis in §3.7, there are no stochastic terms here, because the equations of motion of an isolated system are completely determined by Hamilton's equations.

Because of the deterministic nature of the Hamiltonian trajectory, the conditional transition probability is formally $\wp(\mathbf{\Gamma}_2|\mathbf{\Gamma}_1,t_{21}) = \delta(\mathbf{\Gamma}_2 - \mathbf{\Gamma}^0(t_{21}|\mathbf{\Gamma}_1))$. This means that

$$\begin{aligned}\int \mathrm{d}\mathbf{\Gamma}_2\, \wp(\mathbf{\Gamma}_2,t_2;\mathbf{\Gamma}_1,t_1) &= \int \mathrm{d}\mathbf{\Gamma}_2\, \delta(\mathbf{\Gamma}_2 - \mathbf{\Gamma}^0(t_{21}|\mathbf{\Gamma}_1))\wp(\mathbf{\Gamma}_1,t_1) \\ &= \wp(\mathbf{\Gamma}_1,t_1).\end{aligned} \tag{7.23}$$

This is the just the conservation law for probability, Eq. (1.28). The conservation law implies that there exists a probability flux, $\dot{\mathbf{\Gamma}}^0\wp(\mathbf{\Gamma},t)$, which is just the adiabatic limit of the probability flux discussed in §3.7.5. With this the rate of change of probability density at a fixed position is given by the negative divergence of this flux,

$$\frac{\partial\wp(\mathbf{\Gamma},t)}{\partial t} = -\nabla\cdot[\dot{\mathbf{\Gamma}}^0\nabla\wp(\mathbf{\Gamma},t)]. \tag{7.24}$$

Since phase space is incompressible under Hamilton's equations of motion, $\nabla\cdot\dot{\mathbf{\Gamma}}^0 = 0$, (§7.1), one therefore has

$$\frac{\partial\wp(\mathbf{\Gamma},t)}{\partial t} = -\dot{\mathbf{\Gamma}}^0\cdot\nabla\wp(\mathbf{\Gamma},t), \tag{7.25}$$

from which it follows that the total time derivative of a probability density in an isolated system vanishes,

$$\frac{\mathrm{d}\wp(\mathbf{\Gamma},t)}{\mathrm{d}t} = 0. \tag{7.26}$$

This says that the phase space probability density is constant, moving along with a trajectory.

The physical interpretation of this is as follows. The incompressibility of phase space under Hamilton's equations of motion implies that volume is preserved during its evolution. Hamiltonian trajectories do not cross, and so the evolution of a volume is well defined: trajectories initially inside the boundary remain inside, and those outside remain outside. The conservation law for probability implies therefore that the weight inside the volume remains inside the volume, and since the volume itself is constant, this means that the probability density is constant following a trajectory.

An equivalent argument was originally given by Boltzmann. Consider two regions of phase space the second of which evolved from the first after a time

Figure 7.1: A volume of phase space, its evolution, and a single trajectory passing through them.

τ (Fig. 7.1). The regions have different shapes but the same volume. Consider a trajectory passing through these volumes, now regarded as fixed regions of phase space. Let t_A and t_B be the times it entered and left the original volume, and let t_C and t_D be the times it entered and left the evolved volume. Since the boundary of the second region evolves from the boundary of the first region after time τ, one must have $t_C - t_A = t_D - t_B = \tau$. This implies that the trajectory spent an equal amount of time in the two regions, $t_B - t_A = t_D - t_C$. Since the probability of a region of phase space is linearly proportional to the time a system spends in it, it follows that evolved volumes of phase space lying on a single trajectory are equally likely.

These arguments show that the probability density is constant following a trajectory, $\wp(\mathbf{\Gamma}^0(t_1|\mathbf{\Gamma}_0), t_1) = \wp(\mathbf{\Gamma}^0(t_2|\mathbf{\Gamma}_0), t_2)$. To prove from this that the probability is uniform on the trajectory, $\wp(\mathbf{\Gamma}^0(t_1|\mathbf{\Gamma}_0), t) = \wp(\mathbf{\Gamma}^0(t_2|\mathbf{\Gamma}_0), t)$, one simply notes that for a system in equilibrium, the partial temporal derivative of the probability density must vanish,

$$\frac{\partial \wp(\mathbf{\Gamma}, t)}{\partial t} = 0. \tag{7.27}$$

Since the probability density also moves along the trajectory, this implies that the probability density must be uniform along the trajectory,

$$\wp(\mathbf{\Gamma}^0(t|\mathbf{\Gamma}_0), t) = \wp_0(\mathbf{\Gamma}_0), \text{ all } t. \tag{7.28}$$

Taking into account the first postulate that forms the ergodic hypothesis—all points with the same energy lie on a single trajectory—this is equivalent to Eq. (7.19): phase space points with the same energy are equally likely.

7.2.5 Partition Function and Entropy

The partition function, $Z'(E, N, V)$, which normalises the probability density, represents the total weight of the isolated system, and its logarithm is the total entropy. The mathematical formulation of entropy discussed in §1.4 invokes sums over distinct states. In the present phase space, the particles are indistinguishable, which means that if two particles are swapped, then the system remains in the same microstate. For example, with $\gamma \equiv \{\mathbf{q}, \mathbf{p}\}$, the microstate $\mathbf{\Gamma}_1 \equiv \{\gamma', \gamma'', \gamma_3, \ldots, \gamma_N\}$ and the microstate $\mathbf{\Gamma}_2 \equiv \{\gamma'', \gamma', \gamma_3, \ldots, \gamma_N\}$ are physically identical and cannot be counted as distinct microstates. Hence integrals over phase space, including the partition function, should be reduced by a factor of $N!$ to count only distinct states.[8] This becomes important for systems with variable numbers of particles and for the number derivative of the entropy.

The probability density should have the dimensions of inverse phase space volume. Also, the partition function should be dimensionless so that its logarithm may be properly taken to give the total entropy. Planck's constant, $h = 6.63 \times 10^{-34}$ J s, has the dimensions of momentum times length, and one can define a convenient measure of phase space volume as $\Delta \equiv h^{3N}$. One also needs to cancel the dimensions of inverse energy of the δ-function by introducing an energy measure Δ_E.

These two considerations lead to the final expression for the phase space probability density of an isolated system,

$$\wp(\mathbf{\Gamma}|E, V, N) = \frac{\Delta_\mathrm{E}\delta(\mathcal{H}(\mathbf{\Gamma}) - E)}{h^{3N} N!\, Z(E, V, N)}. \tag{7.29}$$

Normalisation of this is ensured by the dimensionless partition function,

$$Z(E, V, N) = \frac{\Delta_\mathrm{E}}{h^{3N} N!} \int \mathrm{d}\mathbf{\Gamma}\, \delta(\mathcal{H}(\mathbf{\Gamma}) - E)\,. \tag{7.30}$$

The energy constant Δ_E represents the width of the energy hypersurface. The δ-function is really a coarse grained top hat function. The numerical value of Δ_E has no physical consequences. It turns out that the partition function is an exponentially varying function of the energy, and so Δ_E could even vary slowly with energy without measurable consequence. It is a convenience rather than a necessity to introduce the energy constant explicitly.

Using Planck's constant to give the correct dimensions of phase space might suggest a quantum mechanical justification for it. However, this is not the case; there is nothing in classical mechanics that demands the existence of quantum mechanics, and the present development of statistical mechanics is purely classical. The value chosen for the constant has no physical consequences, which means that one could proceed by setting it to unity if one wanted.[9]

The entropy of the isolated system is the logarithm of its total weight,

$$\begin{aligned} S(E, N, V) &= k_\mathrm{B} \ln Z(E, N, V) \\ &= k_\mathrm{B} \ln \frac{\Delta_E}{h^{3N} N!} \int \mathrm{d}\mathbf{\Gamma}\, \delta(\mathcal{H}(\mathbf{\Gamma}) - E). \end{aligned} \tag{7.31}$$

[8] For example, for three particles on a line, and $f(x_1, x_2, x_3)$ unchanged by permutation of the arguments, $\int \mathrm{d}x_1 \int_{x_2 > x_1} \mathrm{d}x_2 \int_{x_3 > x_2} \mathrm{d}x_3\, f(x_1, x_2, x_3) = \int \mathrm{d}x^3\, f(x_1, x_2, x_3)/3!$, where the first integral is over distinct states and the second integral is over all states.

[9] Below, sub-systems that can exchange a conserved variable (e.g. energy, number, or volume) with a reservoir will be treated. For such an exchangeable parameter A, one can always define the probability density or partition function with an arbitrary factor of $\Delta_\mathrm{A} \exp \lambda_\mathrm{A} A$, with no physical consequences.

The derivatives of this give the various thermodynamic properties. For example, the temperature of the isolated system is given by $\partial S(E, N, V)/\partial E = 1/T(E, N, V)$.

7.2.6 Internal Entropy of Phase Space Points

In the next section the Maxwell-Boltzmann distribution is derived, which is the probability density for the phase space of a sub-system of a thermal reservoir. In order to obtain the correct result, the internal entropy of points in phase space is required.

The energy weight density, which is the total weight per unit energy, is $\omega(E, N, V) = Z(E, N, V)/\Delta_E$. This may also be written as the integral over the weight density of phase space, $\omega(\mathbf{\Gamma})$, namely

$$\omega(E, N, V) = \int \mathrm{d}\mathbf{\Gamma}\, \omega(\mathbf{\Gamma})\delta(\mathcal{H}(\mathbf{\Gamma}) - E). \tag{7.32}$$

From these two and the expression for the partition function, one concludes that unconditional phase space has uniform weight density,

$$\omega(\mathbf{\Gamma}) = \frac{1}{h^{3N} N!}. \tag{7.33}$$

This weight density enables the definition of the entropy of a point in phase space,

$$S(\mathbf{\Gamma}) = k_{\mathrm{B}} \ln \left[\omega(\mathbf{\Gamma})\Delta_{\mathbf{\Gamma}}\right]. \tag{7.34}$$

In terms of these the unconditional phase space probability density is

$$\wp(\mathbf{\Gamma}) = \frac{\omega(\mathbf{\Gamma})}{Z''} = \frac{e^{S(\mathbf{\Gamma})/k_{\mathrm{B}}}}{\Delta_{\mathbf{\Gamma}} Z''}. \tag{7.35}$$

The arbitrary volume element $\Delta_{\mathbf{\Gamma}}$ again has no physical consequence, even if it is taken to vary with $\mathbf{\Gamma}$. Where it appears explicitly in the probability density cancels its implicit appearance in the definition of the entropy, as the first equality shows. It is included here so that the logarithm is taken of a dimensionless number and so that the probability density has explicitly the correct dimensions. Also, it ensures that the entropy is invariant with respect to transformations of phase space. For convenience, one could take it to be constant, or even set it equal to $N!h^{3N}$.

7.3 Canonical Equilibrium System

7.3.1 Maxwell-Boltzmann Distribution

In this section the total isolated system will be taken to consist of a sub-system labelled s and a thermal reservoir labelled r. The total energy $E_{\mathrm{t}} = E_{\mathrm{s}} + E_{\mathrm{r}}$ is fixed, but the individual energy fluctuates due to exchange across the thermal boundary between the two systems. The total Hamiltonian is the sum of the Hamiltonian of each system, $\mathcal{H}_t = \mathcal{H}_s + \mathcal{H}_r$, which assumes that the interactions across the boundary are relatively negligible. The particles are confined to their respective volumes, so that the phase integrals permute particles within each system but not between each system. This means that the correction for counting distinct states is $N_s!N_r!$ rather than $(N_s + N_r)!$.

The total weight or partition function when sub-system s has energy E_s is

$$\begin{aligned} & Z(E_s|E, V_s, N_s, V_r, N_r) \\ &= \frac{\Delta_E^2}{h^{3N_s}h^{3N_r}} \int \frac{d\mathbf{\Gamma}_s}{N_s!} \int \frac{d\mathbf{\Gamma}_r}{N_r!} \delta(E_s - \mathcal{H}_s)\, \delta(E_t - \mathcal{H}_t) \\ &= \frac{\Delta_E^2}{h^{3N_s}h^{3N_r}} \int \frac{d\mathbf{\Gamma}_s}{N_s!} \delta(E_s - \mathcal{H}_s) \int \frac{d\mathbf{\Gamma}_r}{N_r!} \delta(E_t - E_s - \mathcal{H}_r) \\ &= Z(E_s, N_s, V_s)\, Z(E_t - E_s, N_r, V_r). \end{aligned} \tag{7.36}$$

The total weight is the product of the weights of the two systems in their respective states. Taking the logarithm of this gives the total entropy as the sum of the individual entropies,

$$\begin{aligned} S(E_s|E_t, V_s, N_s, V_r, N_r) &= S(E_s, N_s, V_s) + S(E_t - E_s, N_r, V_r) \\ &= S(E_s, N_s, V_s) - \frac{E_s}{T_r} + \text{const.} \end{aligned} \tag{7.37}$$

In the second equality, the second system has been taken to be a reservoir infinitely larger than the sub-system, and a Taylor expansion has been performed about the total energy E_t, neglecting higher-order terms that scale as powers of E_s/E_t. The constant $S(E_t, N_r, V_r)$ is independent of the sub-system s and will henceforth be neglected. The quantity $-E_s/T_r$ will be called the reservoir entropy, although it would be more precise to call it the sub-system dependent part of the reservoir entropy. The second equality uses the definition of temperature, namely that its inverse is the energy derivative of the entropy. Henceforth the subscripts s and r can be dropped because the reservoir only enters through its temperature, and all other quantities refer to the sub-system.

The probability for the sub-system to have a particular energy is the exponential of the constrained total entropy,

$$\wp(E|N, V, T) = \frac{1}{\Delta_E Z(N, V, T)} e^{S(E,N,V)/k_B} e^{-E/k_B T}. \tag{7.38}$$

The final factor is called the Boltzmann factor. This signifies the fact that an isothermal system, which is a system that can exchange energy with a heat reservoir, has probability proportional to the exponential of the negative of its energy divided by temperature. This exponent is just the state-dependent reservoir entropy. Obviously there is a second factor here, the exponential of the internal entropy of the sub-system in the state (see below). The normalising, dimensionless partition function is[10]

$$\begin{aligned} Z(N,V,T) &= \frac{1}{\Delta_E}\int \mathrm{d}E\, e^{S(E,N,V)/k_\mathrm{B}} e^{-E/k_\mathrm{B}T} \\ &= \frac{1}{N!h^{3N}}\int \mathrm{d}E\, e^{-E/k_\mathrm{B}T}\int \mathrm{d}\mathbf{\Gamma}\, \delta(\mathcal{H}(\mathbf{\Gamma})-E) \\ &= \frac{1}{N!h^{3N}}\int \mathrm{d}\mathbf{\Gamma}\, e^{-\mathcal{H}(\mathbf{\Gamma})/k_\mathrm{B}T}, \end{aligned} \tag{7.39}$$

where the final line follows upon interchanging the order of integration.

The most important probability density for an isothermal system is that for the sub-system phase space. This can be obtained from the energy probability density,

$$\begin{aligned} \wp(\mathbf{\Gamma}|N,V,T) &= \int \mathrm{d}E\, \wp(\mathbf{\Gamma}|E,N,V)\wp(E|N,V,T) \\ &= \int \mathrm{d}E\, \frac{\Delta_\mathrm{E}\delta(\mathcal{H}(\mathbf{\Gamma})-E)}{N!h^{3N}Z(E,N,V)}\, \frac{e^{S(E,N,V)/k_\mathrm{B}}e^{-E/k_\mathrm{B}T}}{\Delta_\mathrm{E}Z(N,V,T)} \\ &= \frac{e^{-\mathcal{H}(\mathbf{\Gamma})/k_\mathrm{B}T}}{N!h^{3N}Z(N,V,T)}. \end{aligned} \tag{7.40}$$

The first equality is a standard application of the laws of probability;[11] the first probability is for an isolated sub-system with energy E, which is independent of the reservoir temperature. The partition function in the denominator and the entropy in the exponent cancel because $S(E,N,V) = k_\mathrm{B}\ln Z(E,N,V)$. Finally, the Dirac δ is used to undo the integration. This phase space probability density has the desired dimensions and is correctly normalised.

This result is the most common form of the Maxwell-Boltzmann distribution. The lack of internal entropy in this form compared to Eq. (7.38) results from the fact that points in phase space have equal weight density, Eq. (7.33). The exponent is the sub-system part of the reservoir entropy, and this Maxwell-Boltzmann distribution can be simply interpreted as the exponential of the total entropy of a sub-system phase point for a sub-system in contact with a thermal reservoir.

7.3.2 Helmholtz Free Energy

In general there are two contributions to the total entropy (that of the reservoir and that of the sub-system), and there are three types of total entropy, which correspond to statistical mechanics, fluctuation theory, and thermodynamics, respectively. These, and the corresponding Helmholtz free energies, are now

[10] Note that $Z(E,N,V)$ and $Z(N,V,T)$ are two completely different functions; here and throughout the form of the arguments is used to indicate the type of the function.

[11] One has $\wp(a|c) = \sum_b \wp(ab|c) = \sum_b \wp(a|bc)\wp(b|c) = \sum_b \wp(a|b)\wp(b|c)$, the final equality holding if and only if $b \cap c = b$. In this case the temperature of the reservoir T (the c) is irrelevant when the sub-system energy E_1 (the b) is specified.

derived for the canonical equilibrium system of a sub-system and a thermal reservoir.

The total entropy is the sum of that of the reservoir and that of the sub-system,

$$S_{\rm t}(N,V,T) = S_{\rm r}(N,V,T) + S_{\rm s}(N,V,T). \tag{7.41}$$

This may be evaluated explicitly by writing

$$\begin{aligned} S_{\rm t}(N,V,T) &= k_{\rm B} \ln Z(N,V,T) \\ &= k_{\rm B} \int \mathrm{d}\mathbf{\Gamma}\, \wp(\mathbf{\Gamma}|N,V,T) \ln Z(N,V,T) \\ &= k_{\rm B} \int \mathrm{d}\mathbf{\Gamma}\, \wp(\mathbf{\Gamma}|N,V,T) \left[\ln \frac{e^{-\beta\mathcal{H}(\mathbf{\Gamma})}}{N!h^{3N}} - \ln \wp(\mathbf{\Gamma}|N,V,T) \right] \\ &= \frac{-\langle \mathcal{H} \rangle_T}{T} - k_{\rm B} \int \mathrm{d}\mathbf{\Gamma}\, \wp(\mathbf{\Gamma}|N,V,T) \ln \left[N!h^{3N} \wp(\mathbf{\Gamma}|N,V,T) \right]. \end{aligned} \tag{7.42}$$

This is the unconstrained total entropy, by which is meant it is the sum total of the weights over all the states. The first term on the right-hand side of the final equality is evidently the average reservoir entropy, (c.f. Eq. (7.37) above),

$$S_{\rm r}(N,V,T) = \frac{-\langle \mathcal{H} \rangle_T}{T}. \tag{7.43}$$

Hence the remaining term is the sub-system entropy,

$$S_{\rm s}(N,V,T) = -k_{\rm B} \int \mathrm{d}\mathbf{\Gamma}\, \wp(\mathbf{\Gamma}|N,V,T) \ln \left[N!h^{3N} \wp(\mathbf{\Gamma}|N,V,T) \right]. \tag{7.44}$$

In Ch. 1, Eq. (1.22), the total entropy was written as the average internal entropy of the macrostate minus the average logarithm of the probability. In the present case the phase space microstates of the isolated sub-system are all equally likely, and hence they have no internal entropy as such. However, when the sub-system is in contact with the heat reservoir there is an entropy associated with each microstate due to the loss of energy from the reservoir.

The general definition of a free energy is that it is minus the temperature times the total entropy.[12] For an isothermal system it is called the Helmholtz free energy and in statistical mechanics this is

$$\begin{aligned} F_{\rm SM}(N,V,T) &= -TS_{\rm t}(N,V,T) \\ &= -k_{\rm B}T \ln Z(N,V,T) \\ &= \langle \mathcal{H} \rangle_T - TS_{\rm s}(N,V,T). \end{aligned} \tag{7.45}$$

This is the unconstrained free energy. This is a characteristically statistical mechanical result, with the averages of phase functions appearing.

[12] This definition—the free energy is minus the temperature times the total entropy—is completely general and applies to all systems, not just isothermal systems, and not just equilibrium systems. Hence maximising the total entropy is mathematically equivalent to minimising the free energy. Generally it is the constrained quantities that are extremised.

The second type of entropy is the constrained entropy. This is the total entropy when the sub-system is constrained to have energy E,

$$S_{\rm t}(E|N,V,T) = \frac{-E}{T} + S_{\rm s}(E,N,V). \tag{7.46}$$

The first term on the right-hand side is the sub-system dependent reservoir entropy, and the second term is the sub-system entropy, both evaluated in the macrostate E (i.e. when the sub-system has energy E). The corresponding free energy is the constrained Helmholtz free energy, which is just minus the temperature times this,

$$F(E|N,V,T) = E - TS_{\rm s}(E,N,V). \tag{7.47}$$

The constrained free energy, like the constrained total entropy, has four independent variables as arguments. These constrained functions are appropriate for fluctuation theory, since the probability of the sub-system to have energy E is proportional to their exponential.

The final type of entropy is the maximal constrained entropy. The most likely value of the sub-system energy, $\overline{E}(N,V,T)$, maximises the total constrained entropy,

$$\left.\frac{\partial S_{\rm t}(E|N,V,T)}{\partial E}\right|_{E=\overline{E}} = 0. \tag{7.48}$$

Equivalently, it minimises the constrained Helmholtz free energy. Direct substitution shows that this optimal energy is the one that makes the sub-system temperature equal to the reservoir temperature, $T_{\rm s}(\overline{E},N,V) = T$. The maximal constrained total entropy is the constrained total entropy evaluated at this most likely energy,

$$S_{\rm t}(\overline{E}|N,V,T) = \frac{-\overline{E}}{T} + S_{\rm s}(\overline{E},N,V). \tag{7.49}$$

The corresponding free energy is the Helmholtz free energy that one finds in thermodynamics texts,

$$F_{\rm TD}(N,V,T) = \overline{E} - TS_{\rm s}(\overline{E},N,V). \tag{7.50}$$

In general the average energy of the sub-system can be replaced by the most likely energy,

$$\langle\mathcal{H}\rangle_T \approx \overline{E}(N,V,T). \tag{7.51}$$

The three types of total entropy and the corresponding Helmholtz free energies are summarised in Table 7.1. In these and all other cases, the free energy is minus the temperature times the total entropy. As a general rule, thermodynamics always refers to the most likely state, which is approximately the same as the average state, whereas statistical mechanics refers to the sum over states.

Table 7.1: The canonical total entropies and Helmholtz free energies

	Entropy	Free Energy	Formula
unconstrained	$S_{\rm t}(N,V,T)$	$F_{\rm SM}(N,V,T)$	$= -k_{\rm B}T\ln Z(N,V,T)$
constrained	$S_{\rm t}(E\vert N,V,T)$	$F(E\vert N,V,T)$	$= E - TS_{\rm s}(E,N,V)$
max. constrained	$S_{\rm t}(\overline{E}\vert N,V,T)$	$F_{\rm TD}(N,V,T)$	$= \overline{E} - TS_{\rm s}(\overline{E},N,V)$

The three total entropies are ordered such that

$$S_{\rm t}(N,V,T) > S_{\rm t}(\overline{E}|N,V,T) \geq S_{\rm t}(E|N,V,T). \tag{7.52}$$

The first inequality follows because the unconstrained total entropy is a sum of the weights of all the states, including the weight of the maximal state, which is the middle term. If fluctuations are negligible, which they are in the thermodynamic limit, then $S_{\rm t}(N,V,T) \approx S_{\rm t}(\overline{E}|N,V,T)$, which justifies using the same notation, $F(N,V,T)$, for both the statistical mechanical and for the thermodynamic Helmholtz free energy. This equivalence of the thermodynamic and statistical mechanical pictures may be shown explicitly by evaluating the total unconstrained entropy in fluctuation approximation

$$\begin{aligned}
&-TS_{\rm t}(N,V,T) \\
&\quad= -k_{\rm B}T\ln Z(N,V,T) \\
&\quad= -k_{\rm B}T\ln \int {\rm d}E\, \frac{1}{\Delta_E} e^{-E/k_{\rm B}T} e^{S_{\rm s}(E,N,V)/k_{\rm B}} \\
&\quad\approx -k_{\rm B}T\ln \left[e^{-\overline{E}/k_{\rm B}T} e^{S(\overline{E},N,V)/k_{\rm B}} \int \frac{{\rm d}E}{\Delta_E} e^{S_{EE}(E-\overline{E})^2/2k_{\rm B}} \right] \\
&\quad= \overline{E} - TS(\overline{E},N,V) + \frac{k_{\rm B}T}{2} \ln \frac{-2\pi k_{\rm B}}{\Delta_E^2 S_{EE}}.
\end{aligned} \tag{7.53}$$

The second derivative of the sub-system entropy, $S_{EE}(\overline{E},N,V)$, is related to the heat capacity. The right-hand side is the thermodynamic Helmholtz free energy with a logarithmic correction that for macroscopic systems may be neglected. That is, in the thermodynamic limit, $V \to \infty$, $E/V \to$ const., $N/V \to$ const., the logarithmic correction is relatively negligible compared to the first two terms, which scale with the volume. One concludes from this explicit result that $F_{\rm SM}(N,V,T) = F_{\rm TD}(N,V,T)$ with negligible error in the thermodynamic limit.

The temperature derivative of the total unconstrained entropy is

$$\begin{aligned}\frac{\partial S_{\mathrm{t}}(N,V,T)}{\partial T} &= k_{\mathrm{B}}\frac{\partial \ln Z(N,V,T)}{\partial T}\\ &= \frac{k_{\mathrm{B}}}{Z(N,V,T)}\frac{\partial}{\partial T}\int \mathrm{d}E\,\frac{1}{\Delta_E}e^{-E/k_{\mathrm{B}}T}e^{S_{\mathrm{s}}(E,N,V)/k_{\mathrm{B}}}\\ &= \frac{k_{\mathrm{B}}}{Z(N,V,T)}\frac{1}{k_{\mathrm{B}}T^2}\int \mathrm{d}E\,\frac{E}{\Delta_E}e^{-E/k_{\mathrm{B}}T}e^{S_{\mathrm{s}}(E,N,V)/k_{\mathrm{B}}}\\ &= \frac{1}{T^2}\langle E\rangle_T\,. \end{aligned} \tag{7.54}$$

This is an example of a general result: derivatives of the total unconstrained entropy (equivalently, the logarithm of the partition function) yield average values. The corresponding thermodynamic quantity is

$$\begin{aligned}\frac{\partial[-F_{\mathrm{TD}}(N,V,T)/T]}{\partial T} &= \frac{\partial}{\partial T}\left[\frac{-\overline{E}}{T}+S_{\mathrm{s}}(\overline{E},N,V)\right]_{N,V}\\ &= -\overline{E}\frac{\partial T^{-1}}{\partial T}+\left(\frac{\partial \overline{E}}{\partial T}\right)_{N,V}\left.\frac{\partial S_{\mathrm{t}}(E|N,V,T)}{\partial E}\right|_{E=\overline{E}}\\ &= \frac{\overline{E}}{T^2}. \end{aligned} \tag{7.55}$$

Notice how the variational nature of the constrained total entropy made the second term of the right-hand side of the second equality vanish. This is a significant advantage of the present variational formulation that holds for all systems, not just the present canonical isothermal one. The second general result that is embodied here is that derivatives of the thermodynamic free energy yield most likely values. The equality of average values and most likely values confirms the equivalence of the statistical mechanical and the thermodynamic pictures, $\langle E\rangle_T \approx \overline{E}$.

7.3.3 Probability Distribution for Other Systems

Table 7.2 summarises the thermodynamic potentials, sometimes called free energies, of the most common thermodynamic systems. The partial derivatives of the free energies are given by the coefficients in the total differential. In the table, p is pressure and μ is chemical potential. The over-line on the free energies indicates the minimal or thermodynamic value, for example $\overline{F} \equiv F(\overline{E}|N,V,T) = F_{\mathrm{TD}}(N,V,T)$.

The free energies in the table can be used to obtain the reservoir entropy and hence the probability density for phase space for the common thermodynamic systems. One simply multiplies the free energy by $-1/T$ to obtain the total entropy, subtracts the sub-system entropy $S(E,N,V)$ to obtain the reservoir entropy, and replaces the energy by the Hamiltonian, $E \Rightarrow \mathcal{H}(\mathbf{\Gamma})$.

For the case of an isobaric system, this gives for the reservoir entropy

$$S_{\mathrm{r}}(\mathbf{\Gamma},V|N,p,T) = \frac{-\mathcal{H}(\mathbf{\Gamma})}{T} - pVT, \tag{7.56}$$

Table 7.2: Summary of the common thermodynamic systems†

System	Constrained potential Parameters	Equilibrium potential Differential
Isolated	– N, V, E	Entropy, S $T\mathrm{d}S = \mathrm{d}E + p\mathrm{d}V - \mu\mathrm{d}N$
Isothermal	$F = E - TS(E, V, N)$ $E\|N, V, T$	Helmholtz, $\overline{F}$ $\mathrm{d}\overline{F} = -\overline{S}\mathrm{d}T - \overline{p}\mathrm{d}V + \overline{\mu}\mathrm{d}N$
Open	$\Omega = E - TS(E, V, N) - \mu N$ $N, E\|\mu, V, T$	Grand, $\overline{\Omega}$ $\mathrm{d}\overline{\Omega} = -\overline{S}\mathrm{d}T - \overline{p}\mathrm{d}V - \overline{N}\mathrm{d}\mu$
Isobaric	$G = E - TS(E, V, N) + pV$ $E, V\|N, p, T$	Gibbs, $\overline{G}$ $\mathrm{d}\overline{G} = -\overline{S}\mathrm{d}T + \overline{V}\mathrm{d}p + \overline{\mu}\mathrm{d}N$
Isenthalpic	$\Phi = -TS(H - pV, V, N)$ $V\|N, H, p; T$	$\overline{\Phi}$ $\overline{T}\mathrm{d}\overline{\Phi}/T = -\mathrm{d}H + \overline{V}\mathrm{d}p + \overline{\mu}\mathrm{d}N$ $+\overline{\Phi}\,\overline{T}\mathrm{d}T/T^2$
Isentropic	$H = E(S, V, N) + pV$ $V\|N, S, p$	Enthalpy, $\overline{H}$ $\mathrm{d}\overline{H} = \overline{V}\mathrm{d}p + \overline{\mu}\mathrm{d}N + \overline{T}\mathrm{d}S$

†From Attard (2002a), Ch. 3.

and for the probability density

$$\wp(\mathbf{\Gamma}, V|N, p, T) = \frac{e^{-\beta\mathcal{H}(\mathbf{\Gamma})}e^{-\beta pV}}{N!h^{3N}\Delta_{\mathrm{V}}Z(N, p, T)}, \tag{7.57}$$

where Δ_{V} is an arbitrary constant with the dimensions of volume. This is the probability density for a sub-system that can exchange energy and volume with a reservoir of constant pressure and temperature. The Hamiltonian is generally a function of the volume, since it usually includes potential energy terms that represent the walls of the sub-system, and this must be taken into account in assessing the volume probability.

Another common system is the grand system, which can exchange particles and energy with a reservoir of constant temperature and chemical potential. For a multi-component system, $\underline{N} \equiv \{N_1, N_2, \ldots, N_M\}$, with reservoir chemical potentials $\underline{\mu} \equiv \{\mu_1, \mu_2, \ldots, \mu_M\}$, the reservoir entropy is

$$S_{\mathrm{r}}(\mathbf{\Gamma}, \underline{N}|\underline{\mu}, V, T) = \frac{-\mathcal{H}(\mathbf{\Gamma})}{T} + \frac{\underline{N}\cdot\underline{\mu}}{T}, \tag{7.58}$$

and the probability density is

$$\wp(\mathbf{\Gamma}, \underline{N}|\underline{\mu}, V, T) = \frac{e^{-\beta \mathcal{H}(\mathbf{\Gamma})} e^{\beta \underline{N} \cdot \underline{\mu}}}{\prod_\alpha \left[N_\alpha ! h^{3N_\alpha} \right] Z(\underline{\mu}, V, T)}. \tag{7.59}$$

Again the Hamiltonian is a function of number, and its change as particles are exchanged with the reservoir must be taken into account for the number probability density.

7.3.4 Equipartition Theorem

The Hamiltonian is the total energy of the isolated system, and can be written as the sum of the kinetic energy and the potential energy, $\mathcal{H}(\mathbf{\Gamma}) = \mathcal{K}(\mathbf{p}^N) + U(\mathbf{q}^N)$. In the simplest case, the kinetic energy is the usual quadratic form of the momenta,

$$\mathcal{K}(\mathbf{p}^N) = \frac{1}{2m} \sum_{i=1}^{N} \mathbf{p}_i \cdot \mathbf{p}_i. \tag{7.60}$$

This assumes that all of the molecules have the same mass m.

The separability of the momentum and position coordinates, and also the quadratic form for the kinetic energy enables a factorisation of the partition function integrals, from which several simple analytic results can be obtained. One example is the equipartition theorem, which follows from the average kinetic energy,

$$\begin{aligned} \langle \mathcal{K} \rangle_T &= \int \mathrm{d}\mathbf{\Gamma}\, \wp(\mathbf{\Gamma}|N, V, T) \sum_i p_i^2/2m \\ &= \frac{1}{2m} \sum_i \frac{\int \mathrm{d}\mathbf{p}^N \int \mathrm{d}\mathbf{q}^N\, e^{-\mathcal{K}(\mathbf{p}^N)/k_\mathrm{B}T} e^{-U(\mathbf{q}^N)/k_\mathrm{B}T} p_i^2}{\int \mathrm{d}\mathbf{p}^N \int \mathrm{d}\mathbf{q}^N\, e^{-\mathcal{K}(\mathbf{p}^N)/k_\mathrm{B}T} e^{-U(\mathbf{q}^N)/k_\mathrm{B}T}} \\ &= \frac{1}{2m} \sum_i \frac{\int \mathrm{d}\mathbf{p}_i\, e^{-\mathbf{p}_i \cdot \mathbf{p}_i/2mk_\mathrm{B}T} p_i^2}{\int \mathrm{d}\mathbf{p}_i\, e^{-\mathbf{p}_i \cdot \mathbf{p}_i/2mk_\mathrm{B}T}} \\ &= \frac{1}{2m} \sum_i 3mk_\mathrm{B}T = \frac{3Nk_\mathrm{B}T}{2}. \end{aligned} \tag{7.61}$$

Colloquially, there is said to be half a $k_\mathrm{B}T$ per quadratic term in the Hamiltonian. This is called the equipartition theorem. The average kinetic energy is often used to measure the temperature of a sub-system, including the local temperature in a non-equilibrium system.[13]

A generalised equipartition theorem was discussed in §3.7.4. In the present canonical equilibrium case, the entropy is $S(\mathbf{\Gamma}) = -\mathcal{H}(\mathbf{\Gamma})/T$. With $x \equiv p_{i\alpha}$,

[13]There are other phase functions whose average gives the temperature. Rugh, H. H. (1997), Phys. Rev. Lett. **78**, 772. Attard, (2002a), §§7.2.5–6.

Eq. (3.108) becomes

$$0 = \left\langle \left(\frac{\partial S(\mathbf{\Gamma})}{k_\mathrm{B} \partial p_{i\alpha}} \right)^2 \right\rangle + \left\langle \frac{\partial^2 S(\mathbf{\Gamma})}{k_\mathrm{B} \partial p_{i\alpha}^2} \right\rangle = \left\langle (-\beta p_{i\alpha}/m)^2 \right\rangle_{\mathbf{\Gamma}} - \beta/m. \tag{7.62}$$

That is, $\langle p_{i\alpha}^2 \rangle = m k_\mathrm{B} T$, in agreement with the equipartition theorem as conventionally derived here.

7.4 Transition Probability

7.4.1 Stochastic Equations of Motion

The aim of this section is to obtain an explicit expression for the transition probability for the canonical equilibrium system. In order to abbreviate the notation, number N and volume V will not be shown explicitly, and use will be made of the inverse temperature $\beta \equiv 1/k_\mathrm{B}T$. Also the factor $N!h^{3N}$ will be incorporated into the partition function, so that the Maxwell-Boltzmann probability distribution reads

$$\wp_\mathrm{MB}(\mathbf{\Gamma}|\beta) = \frac{e^{-\beta \mathcal{H}(\mathbf{\Gamma})}}{Z(\beta)} = \frac{e^{S_\mathrm{r}(\mathbf{\Gamma})/k_\mathrm{B}}}{Z(\beta)}. \tag{7.63}$$

Here the exponent has been written as the reservoir entropy, which in this canonical equilibrium case is of course $S_\mathrm{r}(\mathbf{\Gamma}) = -\mathcal{H}(\mathbf{\Gamma})/T$. This is done to foreshadow the general non-equilibrium results that will be given in the following chapter. The partition function is the total weight (modulo the scale factors), and it normalises the probability distribution,

$$Z(\beta) = \int \mathrm{d}\mathbf{\Gamma}\, e^{-\beta \mathcal{H}(\mathbf{\Gamma})}. \tag{7.64}$$

Finally, a point in the sub-system phase space is denoted $\mathbf{\Gamma} \equiv \{\mathbf{q}^N, \mathbf{p}^N\} \equiv \{\mathbf{\Gamma}_\mathrm{q}, \mathbf{\Gamma}_\mathrm{p}\}$. The momentum components, $\mathbf{\Gamma}_\mathrm{p}$, will play a central rôle in the second entropy for the transition probability.

The evolution of the probability distribution is described by the unconditional transition probability, $\wp(\mathbf{\Gamma}_2, t_2; \mathbf{\Gamma}_1, t_1) = \wp(\mathbf{\Gamma}_2, \mathbf{\Gamma}_1)$, where the time arguments are suppressed in the second form if there is no ambiguity. This gives the probability that the sub-system is in the state $\mathbf{\Gamma}_2$ at t_2 and $\mathbf{\Gamma}_1$ at t_1. Accordingly, the order of the arguments is unimportant, $\wp(\mathbf{\Gamma}_2, t_2; \mathbf{\Gamma}_1, t_1) = \wp(\mathbf{\Gamma}_1, t_1; \mathbf{\Gamma}_2, t_2)$. This says nothing about the reversibility of the transition probability, which will be discussed in §7.6 below. The relationship between the conditional and unconditional transition probabilities is of course $\wp(\mathbf{\Gamma}_2, t_2; \mathbf{\Gamma}_1, t_1) = \wp(\mathbf{\Gamma}_2, t_2|\mathbf{\Gamma}_1, t_1)\wp(\mathbf{\Gamma}_1, t_1)$. Again for brevity, the conditional transition probability may be written $\wp(\mathbf{\Gamma}_2|\mathbf{\Gamma}_1, t_{21})$. The normalisation gives the evolution of the probability density,

$$\begin{aligned}\wp(\mathbf{\Gamma}_2, t_2) &= \int \mathrm{d}\mathbf{\Gamma}_1\, \wp(\mathbf{\Gamma}_2, t_2; \mathbf{\Gamma}_1, t_1) \\ &= \int \mathrm{d}\mathbf{\Gamma}_1\, \wp(\mathbf{\Gamma}_2, t_2|\mathbf{\Gamma}_1, t_1)\wp(\mathbf{\Gamma}_1, t_1).\end{aligned} \tag{7.65}$$

The transition probability for the canonical equilibrium system must be such that the Maxwell-Boltzmann distribution is stationary under its action,

$$\wp_{\mathrm{MB}}(\mathbf{\Gamma}_2|\beta) = \int \mathrm{d}\mathbf{\Gamma}_1\, \wp(\mathbf{\Gamma}_2|\mathbf{\Gamma}_1, t_{21})\wp_{\mathrm{MB}}(\mathbf{\Gamma}_1|\beta). \tag{7.66}$$

This provides an important constraint and test of the transition probability.

The stochastic dissipative equations of motion give the trajectory of the sub-system through its phase space, and also the evolution of phase functions such as the probability density. In an infinitesimal time step Δ_t, the change in the location in phase space is given by Hamilton's equations for the isolated sub-system, which is the adiabatic part, plus a stochastic perturbation that is due to the thermal reservoir. (See §7.4.3 below for a more formal derivation of the equations of motion from mixed parity fluctuation theory.) For the transition $\mathbf{\Gamma}_1 \stackrel{\Delta_t}{\rightarrow} \mathbf{\Gamma}_2$, $\Delta_t \equiv t_2 - t_1$, one has

$$\mathbf{\Gamma}_2 = \mathbf{\Gamma}_1 + \Delta_t \dot{\mathbf{\Gamma}}^0 + \mathbf{R},\ \Delta_t = t_2 - t_1, \tag{7.67}$$

where $\dot{\mathbf{\Gamma}}^0$ is the adiabatic velocity of the isolated sub-system, which is given by Hamilton's equations of motion, Eq. (7.1). The reservoir force, $\mathbf{R}$, is stochastic in nature, and is drawn from the probability distribution $\wp(\mathbf{R}|\mathbf{\Gamma}_1, \Delta_t)$, which is in essence the same as the conditional transition probability, $\wp(\mathbf{\Gamma}_2|\mathbf{\Gamma}_1, \Delta_t)$. (Strictly speaking, it is $\mathbf{R}/\Delta_t$ that is the force, but this semantic point is not observed here.) It will turn out that the probability distribution for the random force will depend upon the magnitude of the time step via the variance, $\sigma \propto |\Delta_t|$, but not upon its direction. The random force is taken to apply to $\mathbf{\Gamma}_1$ at the beginning of the time step.

It is important to note that the equations of motion represent a conditional transition. These equations can be either forward or backward in time. The forward equations give where the system will probably go to in the future. The backward equations give where the system probably came from in the past; they do not imply that time physically runs backwards. If $\Delta_t > 0$, then Eq. (7.67) represents a forward transition. The backward transition is $\mathbf{\Gamma}_2 \stackrel{-\Delta_t}{\rightarrow} \mathbf{\Gamma}_1$, $\Delta_t \equiv t_2 - t_1 > 0$, and the stochastic equations of motion are

$$\mathbf{\Gamma}_1 = \mathbf{\Gamma}_2 - \Delta_t \dot{\mathbf{\Gamma}}^0 + \mathbf{R},\ \Delta_t = t_2 - t_1. \tag{7.68}$$

The reservoir force is drawn from the probability distribution $\wp(\mathbf{R}|\mathbf{\Gamma}_2, -\Delta_t)$. Notice how the time step changed signs between the two equations, but the stochastic force did not.[14] This means that if one wants to rearrange Eq. (7.67) so that $\mathbf{\Gamma}_1$ appears on the left-hand side, one has to make a decision whether the $\mathbf{R}$ that appears has a specific fixed value, in which case it changes sign when taken to the other side, or whether it is a random variable chosen from the appropriate probability distribution and so it appears with unchanged sign, as in Eq. (7.68). The probability distribution for the random force will be given in terms of the second entropy below. It will turn out that there is a single most likely value $\overline{\mathbf{R}}$ that does not depend upon the sign of the time step. This makes the stochastic dissipative equations of motion irreversible (see §7.4.4).

In this book the notion of reversibility will be discussed in the context either of the conjugate system with velocities reversed, or of calculating the trajectory leading up to the current point, as in the backwards version of the equations of motion. Phrases such as 'reversing time' or 'time running backwards' will be avoided, as they tend to confuse the two aspects of reversibility, and in any case they imply behaviour that is at odds with the actual progress of time in the physical universe.

The reservoir perturbation is stochastic in nature and comprises a dissipative term and a fluctuation term. They may be written $\mathbf{R} = \overline{\mathbf{R}} + \tilde{\mathbf{R}}$, where the over-line indicates the most likely perturbation (the dissipation), and the tilde indicates a random variable of mean zero (the fluctuation). In this book, the force of zero mean, $\tilde{\mathbf{R}}$, will usually be referred to as 'the' random or stochastic force, even though strictly speaking the full reservoir force, $\mathbf{R}$, is random. The deterministic part of the equations of motion comprises the adiabatic term plus the dissipative term, which is the most likely reservoir perturbation. It is denoted $\Delta\mathbf{\Gamma}^{\rm det} \equiv \Delta_t\dot{\mathbf{\Gamma}}^0 + \overline{\mathbf{R}}$. The transition $\mathbf{\Gamma}_1 \stackrel{\Delta_t}{\rightarrow} \mathbf{\Gamma}_2$, can be equivalently written

$$\mathbf{\Gamma}_2 = \mathbf{\Gamma}_1 + \Delta\mathbf{\Gamma}^{\rm det} + \tilde{\mathbf{R}}. \tag{7.69}$$

The reservoir perturbation $\mathbf{R}$ is a real force, by which is meant that it acts on momentum components only. The reason for this is that the position is essentially the time integral of the momentum, and since the reservoir force acting on the momentum is $\mathcal{O}(\Delta_t)$, its influence on the position is $\mathcal{O}(\Delta_t^2)$, which can be neglected for an infinitesimal time step. This will be justified more formally in the mixed parity derivation of the stochastic equations of motion that follows in §7.4.3. The fact that the reservoir perturbation is a real force means that only the momentum components of phase space are directly affected by the stochastic forces, and the forward transition can be equivalently written

$$\begin{aligned} \mathbf{\Gamma}_{\rm q2} &= \mathbf{\Gamma}_{\rm q1} + \Delta_t\dot{\mathbf{\Gamma}}^0_{\rm q}, \\ \mathbf{\Gamma}_{\rm p2} &= \mathbf{\Gamma}_{\rm p1} + \Delta_t\dot{\mathbf{\Gamma}}^0_{\rm p} + \mathbf{R}. \end{aligned} \tag{7.70}$$

[14]It is possible to define a separate forward and backward random force, $\mathbf{R}_\pm$, that formally enters the equations of motion with different signs for the forward and reverse cases, and that has separate probability distributions. It is mathematically equivalent to the present approach and it only changes the appearance rather than the substance of the results.

In component form, with $i \in [1, N]$ and $\alpha \in \{x, y, z\}$, these are

$$\begin{aligned} q_{2,i\alpha} &= q_{1,i\alpha} + \Delta_t \dot{q}^0_{i\alpha}, \\ p_{2,i\alpha} &= p_{2,i\alpha} + \Delta_t \dot{p}^0_{i\alpha} + R_{i\alpha}. \end{aligned} \tag{7.71}$$

7.4.2 Second Entropy

The conditional probabilities that give the distribution of the random reservoir force $\mathbf{R}$ will now be determined from the second entropy. (See §7.4.3 below for a more formal derivation of the second entropy from mixed parity fluctuation theory.) Maximising this will give the functional form of the dissipative force, and will relate its magnitude to the variance of the fluctuations; this is the molecular origin of the fluctuation dissipation theorem.

The transition in the position $\mathbf{\Gamma}_{\rm q1} \stackrel{\Delta_t}{\to} \mathbf{\Gamma}_{\rm q2}$, is determined by Hamilton's equations and is independent of the reservoir and hence of the second entropy.[15] It is the transition in the momentum states $\mathbf{\Gamma}_{\rm p1} \stackrel{\Delta_t}{\to} \mathbf{\Gamma}_{\rm p2}$, that is determined by the second entropy. This is a case of a pure parity transition, since $\mathbf{\Gamma}_{\rm p}$ is the collection of momentum macrostates, each of which has odd parity.

In the pure parity case, the thermodynamic driving force is the gradient of the total entropy, and the second entropy contains a term that is equal to half the rate of total entropy production (c.f. §2.4.3, Eq. (2.50)). In the present case of phase space microstates, the total entropy is the same as the reservoir entropy, because the weight of phase space points is uniform, Eq. (7.33).

In the most general case one would have a $3N \times 3N$ transport matrix that determined the variance of the stochastic forces. However, one can assume that the forces are uncorrelated with each other,

$$\left\langle \tilde{R}_{i\alpha} \right\rangle = 0, \text{ and } \left\langle \tilde{R}_{i\alpha} \tilde{R}_{j\gamma} \right\rangle = \sigma \delta_{ij} \delta_{\alpha\gamma}. \tag{7.72}$$

These hold for both the forward and the backward forces.[16] Here the variance σ has been chosen to be the same for each molecule and for each component, which makes the transport matrix proportional to the identity matrix. This simplification is not essential to the theory, as is discussed further below. One can also assume that the random forces are uncorrelated at successive times.

In general terms, the second entropy consists of four terms: the first entropy, a quadratic term in the fluxes, a linear term in the fluxes that equals half the rate of total first entropy production, and a term constant with respect to the fluxes that ensures that the reduction condition is satisfied (c.f. §2.4.3, Eq. (2.50)). In

[15]If one used a non-infinitesimal time step, then $\mathbf{\Gamma}_{\rm q2}$ would not be determined solely by the Hamiltonian velocity $\dot{\mathbf{\Gamma}}^0_{\rm q}(\mathbf{\Gamma})$, but by an integral of the trajectory over the interval that includes the stochastic contributions from the reservoir. In this case one would need the second entropy for the full phase space $\mathbf{\Gamma}$, which is not a pure parity case. It may also be that this is a non-Markov case.

[16]One must have $\sigma \propto |\Delta_t|$, as is clearer in the mixed parity analysis, §7.4.3, where $\sigma = |\Delta_t| k_{\rm B} \Lambda_{\rm pp}$, with Λ independent of the time step because it is the coefficient of $|\Delta_t|^{-1}$ in the small time expansion of the fluctuation matrices for the second entropy.

the current presentation, the flux is the pure parity $\mathbf{R}$ (momentum components only); see §7.4.3 below for a more rigorous mixed parity derivation. Also, the total entropy is the same as the reservoir entropy (apart from the immaterial uniform sub-system entropy). The second entropy, like the transition probability, is symmetric in the two arguments. But as mentioned above, the forward and the backward reservoir forces are drawn from the respective conditional transition probabilities, and it remains to show that these do not depend upon the sign of the time step.

First consider the transition $\mathbf{\Gamma}_1 \stackrel{\Delta_t}{\rightarrow} \mathbf{\Gamma}_2$. This will be called the original transition, in order to distinguish it from the reverse transition that is treated next. It is not necessary to specify the sign of the time step. The conditional second entropy for the original transition is

$$\begin{aligned} & S^{(2)}(\mathbf{\Gamma}_2|\mathbf{\Gamma}_1,\Delta_t) \\ = \ & \frac{-k_\mathrm{B}}{2\sigma}\mathbf{R}\cdot\mathbf{R}+\frac{1}{2}\left[S_\mathrm{r}(\mathbf{\Gamma}_2)-S_\mathrm{r}(\mathbf{\Gamma}_1)\right]+C(\mathbf{\Gamma}_1)+\mathcal{D}(\mathbf{\Gamma}_{\mathrm{q}2},\mathbf{\Gamma}_{\mathrm{q}1},\Delta_t) \\ = \ & \frac{-k_\mathrm{B}}{2\sigma}\mathbf{R}\cdot\mathbf{R}+\frac{1}{2}\mathbf{R}\cdot\nabla_\mathrm{p}S_\mathrm{r}(\mathbf{\Gamma}_1) \\ & +\frac{k_\mathrm{B}}{2\sigma}\overline{\mathbf{R}}\cdot\overline{\mathbf{R}}-\frac{1}{2}\overline{\mathbf{R}}\cdot\nabla_\mathrm{p}S_\mathrm{r}(\mathbf{\Gamma}_1)+\mathcal{D}(\mathbf{\Gamma}_{\mathrm{q}2},\mathbf{\Gamma}_{\mathrm{q}1},\Delta_t) \\ = \ & \frac{-k_\mathrm{B}}{2\sigma}\tilde{\mathbf{R}}\cdot\tilde{\mathbf{R}}+\mathcal{D}(\mathbf{\Gamma}_{\mathrm{q}2},\mathbf{\Gamma}_{\mathrm{q}1},\Delta_t). \end{aligned} \tag{7.73}$$

The function $\mathcal{D}(\mathbf{\Gamma}_{\mathrm{q}2},\mathbf{\Gamma}_{\mathrm{q}1},\Delta_t) \equiv -(k_\mathrm{B}/2\alpha)\left[\mathbf{\Gamma}_{\mathrm{q}2}-\mathbf{\Gamma}_{\mathrm{q}1}-\Delta_t\dot{\mathbf{\Gamma}}^0_\mathrm{q}\right]^2$, $\alpha\rightarrow 0$, gives a δ-function for the adiabatic development of the position coordinates. The notation ∇_p means the momentum components of the gradient operator (i.e. the $\partial/\partial p_{i\alpha}$), which arises because the stochastic force only has momentum components. The term $\mathbf{R}\cdot\nabla_\mathrm{p}S_\mathrm{r}(\mathbf{\Gamma})/2$ is half the total entropy produced in the time step since the adiabatic evolution does not produce any entropy (points on an adiabatic trajectory are equally likely). The function $C(\mathbf{\Gamma}_1)$, which is constant with respect to $\mathbf{\Gamma}_2$, is required to make the conditional second entropy vanish at the most likely value, $\overline{\mathbf{\Gamma}}_2(\mathbf{\Gamma}_1)$, equivalently, $\overline{\mathbf{R}}(\mathbf{\Gamma}_1,\Delta_t)$. This vanishing is equivalent to the reduction condition, as is now shown.

The unconditional second entropy is the conditional second entropy plus the entropy of the conditioning state,

$$S^{(2)}(\mathbf{\Gamma}_2,\mathbf{\Gamma}_1|\Delta_t)=S^{(2)}(\mathbf{\Gamma}_2|\mathbf{\Gamma}_1,\Delta_t)+S_\mathrm{r}(\mathbf{\Gamma}_1). \tag{7.74}$$

The vanishing of the conditional second entropy in the most likely state means that the unconditional second entropy satisfies the reduction condition, namely that the value of the second entropy in the most likely future state equals the entropy of the current state,

$$S^{(2)}(\overline{\mathbf{\Gamma}}_2,\mathbf{\Gamma}_1|\Delta_t)=S_\mathrm{r}(\mathbf{\Gamma}_1). \tag{7.75}$$

The probability distribution for the stochastic forward reservoir force is proportional to the exponential of the conditional second entropy, $\wp(\mathbf{R}|\mathbf{\Gamma}_1, \Delta_t) \propto \exp S^{(2)}(\mathbf{\Gamma}_2|\mathbf{\Gamma}_1, \Delta_t)/k_\mathrm{B}$. Using the fact that at constant $\mathbf{\Gamma}_1$, $\mathrm{d}\mathbf{\Gamma}_{\mathrm{p}2} = \mathrm{d}\mathbf{R}$, the derivative of either the conditional or unconditional second entropy with respect to $\mathbf{\Gamma}_{\mathrm{p}2}$ is

$$\frac{\partial S^{(2)}(\mathbf{\Gamma}_2, \mathbf{\Gamma}_1|\Delta_t)}{\partial \mathbf{\Gamma}_{\mathrm{p}2}} = \frac{-k_\mathrm{B}}{\sigma}\mathbf{R} + \frac{1}{2}\nabla_\mathrm{p} S_\mathrm{r}(\mathbf{\Gamma}_1). \tag{7.76}$$

This vanishes at the most likely force,

$$\overline{\mathbf{R}}(\mathbf{\Gamma}) = \frac{\sigma}{2k_\mathrm{B}}\nabla_\mathrm{p} S_\mathrm{r}(\mathbf{\Gamma}). \tag{7.77}$$

To leading-order, it does not matter whether the gradient of the reservoir entropy is evaluated at $\mathbf{\Gamma}_1$ or $\mathbf{\Gamma}_2$. Note that this depends upon the magnitude of the time step (via the variance, $\sigma \propto |\Delta_t|$), but not upon its sign.

A similar analysis can be performed for the reverse transition, $\mathbf{\Gamma}_2 \stackrel{-\Delta_t}{\rightarrow} \mathbf{\Gamma}_1$. The second entropy conditioned upon what is now the initial point of the transition is

$$\begin{aligned} S^{(2)}&(\mathbf{\Gamma}_1|\mathbf{\Gamma}_2, -\Delta_t) \\ &= \frac{-k_\mathrm{B}}{2\sigma}\mathbf{R}\cdot\mathbf{R} + \frac{1}{2}\left[S_\mathrm{r}(\mathbf{\Gamma}_1) - S_\mathrm{r}(\mathbf{\Gamma}_2)\right] + C'(\mathbf{\Gamma}_2) + \mathcal{D}(\mathbf{\Gamma}_{\mathrm{q}1}, \mathbf{\Gamma}_{\mathrm{q}2}, -\Delta_t) \\ &= \frac{-k_\mathrm{B}}{2\sigma}\mathbf{R}\cdot\mathbf{R} + \frac{1}{2}\mathbf{R}\cdot\nabla_\mathrm{p} S_\mathrm{r}(\mathbf{\Gamma}_2) \\ &\quad + \frac{k_\mathrm{B}}{2\sigma}\overline{\mathbf{R}}\cdot\overline{\mathbf{R}} - \frac{1}{2}\overline{\mathbf{R}}\cdot\nabla_\mathrm{p} S_\mathrm{r}(\mathbf{\Gamma}_2) + \mathcal{D}(\mathbf{\Gamma}_{\mathrm{q}1}, \mathbf{\Gamma}_{\mathrm{q}2}, -\Delta_t) \\ &= \frac{-k_\mathrm{B}}{2\sigma}\tilde{\mathbf{R}}\cdot\tilde{\mathbf{R}} + \mathcal{D}(\mathbf{\Gamma}_{\mathrm{q}1}, \mathbf{\Gamma}_{\mathrm{q}2}, -\Delta_t). \end{aligned} \tag{7.78}$$

Note that it is half the change in first entropy for this reverse transition that appears here (final state minus initial state, irrespective of time order). The probability distribution for the reservoir force is again proportional to the exponential of the conditional second entropy, $\wp(\mathbf{R}|\mathbf{\Gamma}_2, -\Delta_t) \propto \exp S^{(2)}(\mathbf{\Gamma}_1|\mathbf{\Gamma}_2, -\Delta_t)/k_\mathrm{B}$. Using the fact that at constant $\mathbf{\Gamma}_2$, $\mathrm{d}\mathbf{\Gamma}_{\mathrm{p}1} = \mathrm{d}\mathbf{R}$, the derivative of this conditional second entropy is

$$\frac{\partial S^{(2)}(\mathbf{\Gamma}_1|\mathbf{\Gamma}_2, -\Delta_t)}{\partial \mathbf{\Gamma}_{\mathrm{p}1}} = \frac{-k_\mathrm{B}}{\sigma}\mathbf{R} + \frac{1}{2}\nabla_\mathrm{p} S_\mathrm{r}(\mathbf{\Gamma}_2). \tag{7.79}$$

This vanishes at the most likely backward force,

$$\overline{\mathbf{R}}(\mathbf{\Gamma}) = \frac{\sigma}{2k_\mathrm{B}}\nabla_\mathrm{p} S_\mathrm{r}(\mathbf{\Gamma}). \tag{7.80}$$

Again, it does not matter to leading-order whether this is evaluated at the initial or final point of the transition. This is the same as the most likely force for the original transition, Eq. (7.77). One can conclude that the probability distribution for the random forces is independent of the sign of the time step. This gives rise to irreversibility in the equations of motion, as is discussed below.

As for the original transition, the unconditional second entropy is related to the conditional second entropy for the reverse transition by $S^{(2)}(\mathbf{\Gamma}_2, \mathbf{\Gamma}_1|\Delta_t) = S^{(2)}(\mathbf{\Gamma}_1|\mathbf{\Gamma}_2, -\Delta_t) + S_\mathrm{r}(\mathbf{\Gamma}_2)$, and it satisfies the reduction condition on the most likely initial state,

$$S^{(2)}(\mathbf{\Gamma}_2, \overline{\mathbf{\Gamma}}_1|\Delta_t) = S_\mathrm{r}(\mathbf{\Gamma}_2). \tag{7.81}$$

In view of these results the unconditional second entropy can be written in a form symmetric in the two states involved in the transition. In both transitions, the stochastic equations of motion show that half the change in entropy plus the entropy of the initial state is equal to the average entropy of the two states,[17]

$$\begin{aligned} \frac{1}{2}\left[S_\mathrm{r}(\mathbf{\Gamma}_1) + S_\mathrm{r}(\mathbf{\Gamma}_2)\right] &= \frac{1}{2}\mathbf{R}\cdot\nabla_\mathrm{p} S_\mathrm{r}(\mathbf{\Gamma}_1) + S_\mathrm{r}(\mathbf{\Gamma}_1) + \mathcal{O}(\Delta_t^2) \\ &= \frac{1}{2}\mathbf{R}\cdot\nabla_\mathrm{p} S_\mathrm{r}(\mathbf{\Gamma}_2) + S_\mathrm{r}(\mathbf{\Gamma}_2) + \mathcal{O}(\Delta_t^2). \end{aligned} \tag{7.82}$$

Hence the unconditional second entropy may be written

$$\begin{aligned} S^{(2)}(\mathbf{\Gamma}_2, \mathbf{\Gamma}_1|\Delta_t) &= \frac{-k_\mathrm{B}}{2\sigma}\mathbf{R}\cdot\mathbf{R} + \frac{1}{2}\left[S_\mathrm{r}(\mathbf{\Gamma}_1) + S_\mathrm{r}(\mathbf{\Gamma}_2)\right] + \mathcal{D}(\mathbf{\Gamma}_{\mathrm{q}2}, \mathbf{\Gamma}_{\mathrm{q}1}, \Delta_t) \\ &\quad + \frac{k_\mathrm{B}}{2\sigma}\overline{\mathbf{R}}(\mathbf{\Gamma})\cdot\overline{\mathbf{R}}(\mathbf{\Gamma}) - \frac{1}{2}\overline{\mathbf{R}}(\mathbf{\Gamma})\cdot\nabla_\mathrm{p} S_\mathrm{r}(\mathbf{\Gamma}) \\ &= \frac{-k_\mathrm{B}}{2\sigma}\mathbf{R}\cdot\mathbf{R} + \frac{1}{2}\left[S_\mathrm{r}(\mathbf{\Gamma}_1) + S_\mathrm{r}(\mathbf{\Gamma}_2)\right] + \mathcal{D}(\mathbf{\Gamma}_{\mathrm{q}2}, \mathbf{\Gamma}_{\mathrm{q}1}, \Delta_t) \\ &\quad - \frac{k_\mathrm{B}}{2\sigma}\overline{\mathbf{R}}(\mathbf{\Gamma})\cdot\overline{\mathbf{R}}(\mathbf{\Gamma}). \end{aligned} \tag{7.83}$$

See §7.4.3 below for an alternative derivation of this result. Maximising this with respect to $\mathbf{\Gamma}_1$ or $\mathbf{\Gamma}_2$ yields the most likely values given above. The terms involving $\overline{\mathbf{R}}(\mathbf{\Gamma})$ and $\nabla_\mathrm{p} S_\mathrm{r}(\mathbf{\Gamma})$ can be held constant during the differentiation as they contribute terms $\mathcal{O}(\Delta_t)$ that are negligible compared to the remaining terms that are $\mathcal{O}(\Delta_t^0)$.

The reservoir entropy for the canonical equilibrium system is $S_\mathrm{r}(\mathbf{\Gamma}) = -\mathcal{H}(\mathbf{\Gamma})/T$. Hence the most likely forces in this case are

$$\overline{\mathbf{R}}(\mathbf{\Gamma}) = \frac{\sigma}{2k_\mathrm{B}}\nabla_\mathrm{p} S_\mathrm{r}(\mathbf{\Gamma}) = \frac{-\sigma}{2k_\mathrm{B}T}\nabla_\mathrm{p}\mathcal{H}(\mathbf{\Gamma}) = \frac{-\sigma}{2mk_\mathrm{B}T}\mathbf{\Gamma}_\mathrm{p}. \tag{7.84}$$

In component form this is

$$\overline{R}_{i\alpha} = \frac{-\sigma}{2k_\mathrm{B}T}\frac{\partial\mathcal{H}(\mathbf{\Gamma})}{\partial p_{i\alpha}} = \frac{-\sigma}{2mk_\mathrm{B}T}p_{i\alpha}. \tag{7.85}$$

This has the form of a friction or drag force, $f = -\mu_\pm v$ with coefficient $\mu_\pm = \pm\sigma/2k_\mathrm{B}T|\Delta_t|$. (It is $\overline{R}_{i\alpha}/\Delta_t$ that is the force; μ_+ refers to the case $\Delta_t > 0$ and

[17] The quadratic term in this Taylor expansion, $\mathbf{RR} : \nabla_\mathrm{p}\nabla_\mathrm{p} S_\mathrm{r}/2$, is negligible compared to the quadratic term in the second entropy, $(-k_\mathrm{B}/2\sigma)\mathbf{R}\cdot\mathbf{R}$, because $\sigma \propto |\Delta_t|$.

μ_- refers to the case $\Delta_t < 0$). Note that for the most likely backward force, which gives the most likely previous state, $\Delta_t < 0$, the coefficient μ is negative. This is correct and will be discussed further below under irreversibility.

With this result the stochastic, dissipative equations of motion in a canonical equilibrium system for the transition $\mathbf{\Gamma}_1 \stackrel{\Delta_t}{\to} \mathbf{\Gamma}_2$ become

$$\begin{aligned} q_{2,i\alpha} &= q_{1,i\alpha} + \Delta_t \dot{q}^0_{i\alpha}, \\ p_{2,i\alpha} &= p_{1,i\alpha} + \Delta_t \dot{p}^0_{i\alpha} - \frac{\sigma}{2mk_{\mathrm{B}}T} p_{i\alpha} + \tilde{R}_{i\alpha}, \end{aligned} \tag{7.86}$$

with $\tilde{R}_{i\alpha}$ a random variable of mean zero and variance σ. This holds for positive or negative values of the time step. Positive values give the future evolution of the system, and negative values allow the past history of the system to be obtained, both in a probabilistic sense. It is important to note that the dissipative force does not depend upon the sign of the time step.

The stochastic, dissipative equations of motion are just Langevin's equation for Brownian motion (c.f. Ch. 3), but applied to molecules. The term $-\sigma p_{i\alpha}/2mk_{\mathrm{B}}T$ represents the dissipation, and the term $\tilde{R}_{i\alpha}$ represents the fluctuations. The magnitude of both are determined by the variance. One sees therefore that this is the form of the fluctuation dissipation theorem that is appropriate for the canonical equilibrium system.

Both terms are essential for the correct treatment of the canonical equilibrium system; the common procedure that sets the fluctuation term to zero and that treats the dissipative term as a thermostat that determines the temperature via the equipartition theorem violates the fluctuation dissipation theorem. One cannot have dissipation without fluctuation.

Just as important is the origin of the functional form of the dissipative force, $f = -\mu v$. Originally, in the context of Brownian motion, Langevin postulated the dissipative force with this functional form as arising from Stokes' hydrodynamic drag force. The present derivation based upon the second entropy shows that the result is much more general, since it arises from the gradient of the first entropy. The notion that hydrodynamics and Stokes' drag operate at the present molecular level is unrealistic. The real origin of the dissipative term in the stochastic equations of motion (Brownian, molecular, and even hydrodynamics itself) lies in the second entropy, and the present statistical mechanical derivation of these stochastic equations of motion for molecules is significant in this regard. One concludes that dissipation is a general statistical phenomenon rather than a particular hydrodynamic result.

Finally, the variance above was taken to be the same for each molecule and for each component of direction. There is nothing that determines the magnitude of the variance *per se*, except that it should scale with the time step. (For a mathematical derivation of this, see Eq. (7.91) *et seq.*, where the reservoir terms comprise the time step times a transport matrix that is independent of the time step; the variance is $\sigma = |\Delta_t| k_{\mathrm{B}} \Lambda_{\mathrm{pp}}$. For a numerical exploration of this point, see the computer simulation of heat flow discussed in Chs 9 and 11.) One could have individual variances, provided of course that the fluctuation dissipation theorem was satisfied in each case, $\overline{R}_{i\alpha} = -\sigma_{i\alpha} p_{i\alpha}/2mk_{\mathrm{B}}T$. Optimising the variance for each component of a mixture may have computational advantages.

7.4.3 Mixed Parity Derivation of the Second Entropy and the Equations of Motion

It is worthwhile giving an alternative, more formal, derivation of the second entropy and the stochastic equations of motion based upon the results for fluctuations of mixed parity, §2.5.1, as exemplified in the analysis of driven Brownian motion, §3.6.3. Let $\mathbf{\Gamma}_1 \equiv \mathbf{\Gamma}(t)$ denote the phase space point at time t, and denote the adjacent point by $\mathbf{\Gamma}_2 \equiv \mathbf{\Gamma}(t+\Delta_t)$, with conditional most likely value $\overline{\mathbf{\Gamma}}_2 \equiv \overline{\mathbf{\Gamma}}(t+\Delta_t|\mathbf{\Gamma}_1, t)$. With the coarse velocity $\stackrel{\circ}{\mathbf{\Gamma}} \equiv [\mathbf{\Gamma}_2 - \mathbf{\Gamma}_1]/\Delta_t$, the second entropy for mixed parity fluctuations, Eq. (2.92), is

$$\begin{aligned} S^{(2)}(\mathbf{\Gamma}_2, \mathbf{\Gamma}_1|\Delta_t) &= S_{\mathrm{r}}(\mathbf{\Gamma}_1) - \frac{1}{2|\Delta_t|}\Lambda^{-1} : [\mathbf{\Gamma}_2 - \mathbf{\Gamma}_1]^2 - \frac{1}{2|\Delta_t|}\Lambda^{-1} : [\overline{\mathbf{\Gamma}}_2 - \mathbf{\Gamma}_1]^2 \\ &\quad + [\mathbf{\Gamma}_2 - \mathbf{\Gamma}_1] \cdot \left[\frac{1}{2} - \hat{t}\Lambda^{-1}\Theta\right] \nabla S_{\mathrm{r}}(\mathbf{\Gamma}_1) + \mathcal{O}(\Delta_t^2), \end{aligned} \tag{7.87}$$

where $\hat{t} \equiv \mathrm{sign}(\Delta_t)$. Compared to Eq. (2.92), the fluctuation expression for the entropy, $S : \mathbf{x}^2/2$ has been replaced by the entropy itself, $S_{\mathrm{r}}(\mathbf{\Gamma})$, and the related thermodynamic force, $S\mathbf{x}$, has been replaced by the entropy gradient, $\nabla S_{\mathrm{r}}(\mathbf{\Gamma})$. Such terms arose from the results that $A_0 + B_0 = S/2$ and $A_0' + B_0' = \Lambda^{-1}\Theta S$. Also, one can identify the symmetric transport matrix with the variance by making the replacement $|\Delta_t|\Lambda \Rightarrow k_{\mathrm{B}}^{-1}\sigma\mathrm{I}$. In the case of phase space the total entropy is the same as the reservoir entropy, because phase space points are microstates and have no internal entropy.

Now group together the same parity components, $\mathbf{\Gamma} = \{\mathbf{\Gamma}_{\mathrm{q}}, \mathbf{\Gamma}_{\mathrm{p}}\}$. Since the transport matrix Λ, like the entropy matrix S, only couples variables of the same parity, in this grouped representation it is block diagonal, with non-zero sub-matrices Λ_{qq}, and Λ_{pp}, which are symmetric. The antisymmetric transport matrix Θ only couples variables of opposite parity, and so in this grouped representation it is block adiagonal, with non-zero sub-matrices $\Theta_{\mathrm{qp}} = -\Theta_{\mathrm{pq}}^{\mathrm{T}}$. Accordingly the second entropy may be written more explicitly as

$$\begin{aligned} & S^{(2)}(\mathbf{\Gamma}_2, \mathbf{\Gamma}_1|\Delta_t) \\ &= S_{\mathrm{r}}(\mathbf{\Gamma}_1) - \frac{1}{2|\Delta_t|}\Lambda_{\mathrm{qq}}^{-1} : [\mathbf{\Gamma}_{2\mathrm{q}} - \mathbf{\Gamma}_{1\mathrm{q}}]^2 - \frac{1}{2|\Delta_t|}\Lambda_{\mathrm{pp}}^{-1} : [\mathbf{\Gamma}_{2\mathrm{p}} - \mathbf{\Gamma}_{1\mathrm{p}}]^2 \\ &\quad - \frac{1}{2|\Delta_t|}\Lambda_{\mathrm{qq}}^{-1} : [\overline{\mathbf{\Gamma}}_{2\mathrm{q}} - \mathbf{\Gamma}_{1\mathrm{q}}]^2 - \frac{1}{2|\Delta_t|}\Lambda_{\mathrm{pp}}^{-1} : [\overline{\mathbf{\Gamma}}_{2\mathrm{p}} - \mathbf{\Gamma}_{1\mathrm{p}}]^2 \\ &\quad + \frac{1}{2}[\mathbf{\Gamma}_{2\mathrm{q}} - \mathbf{\Gamma}_{1\mathrm{q}}] \cdot \nabla_{\mathrm{q}} S_{\mathrm{r}}(\mathbf{\Gamma}_1) + \frac{1}{2}[\mathbf{\Gamma}_{2\mathrm{p}} - \mathbf{\Gamma}_{1\mathrm{p}}] \cdot \nabla_{\mathrm{p}} S_{\mathrm{r}}(\mathbf{\Gamma}_1) \\ &\quad - \hat{t}[\mathbf{\Gamma}_{2\mathrm{q}} - \mathbf{\Gamma}_{1\mathrm{q}}] \cdot \Lambda_{\mathrm{qq}}^{-1}\Theta_{\mathrm{qp}}\nabla_{\mathrm{p}} S_{\mathrm{r}}(\mathbf{\Gamma}_1) - \hat{t}[\mathbf{\Gamma}_{2\mathrm{p}} - \mathbf{\Gamma}_{1\mathrm{p}}] \cdot \Lambda_{\mathrm{pp}}^{-1}\Theta_{\mathrm{pq}}\nabla_{\mathrm{q}} S_{\mathrm{r}}(\mathbf{\Gamma}_1). \end{aligned} \tag{7.88}$$

Maximising this, the conditional most likely adjacent configuration is

$$\begin{aligned}\overline{\mathbf{\Gamma}}_{2\mathrm{q}} &= \mathbf{\Gamma}_{1\mathrm{q}} - \Delta_t \Theta_{\mathrm{qp}} \nabla_{\mathrm{p}} S_{\mathrm{r}}(\mathbf{\Gamma}) + \frac{|\Delta_t|}{2} \Lambda_{\mathrm{qq}} \nabla_{\mathrm{q}} S_{\mathrm{r}}(\mathbf{\Gamma}) + \mathcal{O}(\Delta_t^2), \\ \overline{\mathbf{\Gamma}}_{2\mathrm{p}} &= \mathbf{\Gamma}_{1\mathrm{p}} - \Delta_t \Theta_{\mathrm{pq}} \nabla_{\mathrm{q}} S_{\mathrm{r}}(\mathbf{\Gamma}) + \frac{|\Delta_t|}{2} \Lambda_{\mathrm{pp}} \nabla_{\mathrm{p}} S_{\mathrm{r}}(\mathbf{\Gamma}) + \mathcal{O}(\Delta_t^2). \end{aligned} \tag{7.89}$$

Here and below $\mathbf{\Gamma}$ can be replaced by either $\mathbf{\Gamma}_1$ or by $\mathbf{\Gamma}_2$ since this does not change the results to the exhibited order. In view of the analysis of §3.6.3, the terms proportional to Δ_t contain the adiabatic development, and the terms proportional to $|\Delta_t|$ are the most likely reservoir force (c.f. Eq. (3.66) and the associated discussion). In view of the stochastic equations of motion given above, Eq. (7.67), one may suppose that the adiabatic development dominates the terms proportional to Δ_t, and that any other contribution to this term is negligible. Since $\dot{\mathbf{\Gamma}}^0_{\mathrm{q}} = \partial \mathcal{H}(\mathbf{\Gamma})/\partial \mathbf{\Gamma}_{\mathrm{p}}$ and $\dot{\mathbf{\Gamma}}^0_{\mathrm{p}} = -\partial \mathcal{H}(\mathbf{\Gamma})/\partial \mathbf{\Gamma}_{\mathrm{q}}$, and since $S_{\mathrm{r}}(\mathbf{\Gamma}) = -\mathcal{H}(\mathbf{\Gamma})/T$, these mean that

$$\Theta_{\mathrm{qp}} = T\mathrm{I}, \text{ and } \Theta_{\mathrm{pq}} = -T\mathrm{I}, \tag{7.90}$$

where I is the identity matrix. These obviously satisfy the requirement that $\Theta_{\mathrm{qp}} = -\Theta_{\mathrm{pq}}^{\mathrm{T}}$. With these, the conditional most likely next configuration is

$$\begin{aligned}\overline{\mathbf{\Gamma}}_{2\mathrm{q}} &= \mathbf{\Gamma}_{1\mathrm{q}} + \Delta_t \dot{\mathbf{\Gamma}}^0_{\mathrm{q}} + \frac{|\Delta_t|}{2} \Lambda_{\mathrm{qq}} \nabla_{\mathrm{q}} S_{\mathrm{r}}(\mathbf{\Gamma}), \\ \overline{\mathbf{\Gamma}}_{2\mathrm{p}} &= \mathbf{\Gamma}_{1\mathrm{p}} + \Delta_t \dot{\mathbf{\Gamma}}^0_{\mathrm{p}} + \frac{|\Delta_t|}{2} \Lambda_{\mathrm{pp}} \nabla_{\mathrm{p}} S_{\mathrm{r}}(\mathbf{\Gamma}). \end{aligned} \tag{7.91}$$

To put this in the form of Eq. (7.70), one notes that the reservoir contribution to the position evolution is negligible, $|\Delta_t| \Lambda_{\mathrm{qq}} \nabla_{\mathrm{q}} S_{\mathrm{r}}(\mathbf{\Gamma})/2 \sim \mathcal{O}(\Delta_t^2)$. This must be the case because the reservoir contribution to the evolution of the position coordinates must come from the integral over the time interval of the reservoir contribution to the evolution of the momentum coordinates, which is itself $\mathcal{O}(\Delta_t)$. This is consistent with what was found for driven Brownian motion, where the reservoir contribution to the position evolution was explicitly shown to be $\mathcal{O}(\Delta_t^2)$, Eq. (3.65). The remaining term can be identified as the most likely reservoir force $\overline{R} \equiv |\Delta_t| \Lambda_{\mathrm{pp}} \nabla_{\mathrm{p}} S_{\mathrm{r}}(\mathbf{\Gamma})/2$. Finally, adding the stochastic contribution to this, $R = \overline{R} + \tilde{R}$, one recovers Eq. (7.70).

With these identifications, the second entropy is now

$$\begin{aligned} & S^{(2)}(\mathbf{\Gamma}_2, \mathbf{\Gamma}_1 | \Delta_t) \\ &= S_{\mathrm{r}}(\mathbf{\Gamma}_1) - \frac{1}{2|\Delta_t|} \Lambda_{\mathrm{qq}}^{-1} : [\mathbf{\Gamma}_{2\mathrm{q}} - \mathbf{\Gamma}_{1\mathrm{q}}]^2 - \frac{1}{2|\Delta_t|} \Lambda_{\mathrm{pp}}^{-1} : [\mathbf{\Gamma}_{2\mathrm{p}} - \mathbf{\Gamma}_{1\mathrm{p}}]^2 \\ & \quad - \frac{1}{2|\Delta_t|} \Lambda_{\mathrm{qq}}^{-1} : [\overline{\mathbf{\Gamma}}_{2\mathrm{q}} - \mathbf{\Gamma}_{1\mathrm{q}}]^2 - \frac{1}{2|\Delta_t|} \Lambda_{\mathrm{pp}}^{-1} : [\overline{\mathbf{\Gamma}}_{2\mathrm{p}} - \mathbf{\Gamma}_{1\mathrm{p}}]^2 \\ & \quad + \frac{1}{2} [\mathbf{\Gamma}_{2\mathrm{q}} - \mathbf{\Gamma}_{1\mathrm{q}}] \cdot \nabla_{\mathrm{q}} S_{\mathrm{r}}(\mathbf{\Gamma}) + \frac{1}{2} [\mathbf{\Gamma}_{2\mathrm{p}} - \mathbf{\Gamma}_{1\mathrm{p}}] \cdot \nabla_{\mathrm{p}} S_{\mathrm{r}}(\mathbf{\Gamma}) \\ & \quad + \hat{t} [\mathbf{\Gamma}_{2\mathrm{q}} - \mathbf{\Gamma}_{1\mathrm{q}}] \cdot \Lambda_{\mathrm{qq}}^{-1} \dot{\mathbf{\Gamma}}^0_{\mathrm{q}} + \hat{t} [\mathbf{\Gamma}_{2\mathrm{p}} - \mathbf{\Gamma}_{1\mathrm{p}}] \cdot \Lambda_{\mathrm{pp}}^{-1} \dot{\mathbf{\Gamma}}^0_{\mathrm{p}} + \mathcal{O}(\Delta_t^2). \end{aligned} \tag{7.92}$$

Since $\Lambda_{\rm qq}\nabla_{\rm q}S_{\rm r}(\mathbf{\Gamma}) \sim \mathcal{O}(\Delta_t)$, in the limit of an infinitesimal time step, then $\Lambda_{\rm qq}^{-1} \sim \mathcal{O}(\Delta_t^{-1})$. This means that the terms involving $\Lambda_{\rm qq}^{-1}$ must make a δ-function for the adiabatic development of the position coordinates, which is the $\mathcal{D}(\mathbf{\Gamma}_{1\rm q}, \mathbf{\Gamma}_{2\rm q}, \Delta_t)$ of Eq. (7.73) and Eq. (7.83). Also, the term involving $\overline{\mathbf{\Gamma}}_{2\rm p}$ is just a constant, $C(\mathbf{\Gamma}_1)$, which has an analogue in each of those equations, that ensures that the reduction condition is satisfied. Hence one has

$$\begin{aligned}
S^{(2)}(\mathbf{\Gamma}_2, \mathbf{\Gamma}_1|\Delta_t) &= S_{\rm r}(\mathbf{\Gamma}_1) - \frac{1}{2|\Delta_t|}\Lambda_{\rm pp}^{-1} : [\mathbf{\Gamma}_{2\rm p} - \mathbf{\Gamma}_{1\rm p}]^2 \\
&\quad + \frac{1}{2}[\mathbf{\Gamma}_{2\rm q} - \mathbf{\Gamma}_{1\rm q}] \cdot \nabla_{\rm q}S_{\rm r}(\mathbf{\Gamma}) + \frac{1}{2}[\mathbf{\Gamma}_{2\rm p} - \mathbf{\Gamma}_{1\rm p}] \cdot \nabla_{\rm p}S_{\rm r}(\mathbf{\Gamma}) \\
&\quad + \hat{t}[\mathbf{\Gamma}_{2\rm p} - \mathbf{\Gamma}_{1\rm p}] \cdot \Lambda_{\rm pp}^{-1}\dot{\mathbf{\Gamma}}_{\rm p}^0 + \mathcal{D}(\mathbf{\Gamma}_{2\rm q}, \mathbf{\Gamma}_{1\rm q}, \Delta_t) + C'(\mathbf{\Gamma}_1) \\
&= S_{\rm r}(\mathbf{\Gamma}_1) - \frac{1}{2|\Delta_t|}\Lambda_{\rm pp}^{-1} : [\mathbf{\Gamma}_{2\rm p} - \mathbf{\Gamma}_{1\rm p} - \Delta_t\dot{\mathbf{\Gamma}}_{\rm p}^0]^2 \\
&\quad + \frac{1}{2}[S_{\rm r}(\mathbf{\Gamma}_2) - S_{\rm r}(\mathbf{\Gamma}_1)] + \mathcal{D}(\mathbf{\Gamma}_{2\rm q}, \mathbf{\Gamma}_{1\rm q}, \Delta_t) + C(\mathbf{\Gamma}_1) \\
&= \frac{-1}{2|\Delta_t|}\Lambda_{\rm pp}^{-1} : R^2 + \frac{1}{2}[S_{\rm r}(\mathbf{\Gamma}_2) + S_{\rm r}(\mathbf{\Gamma}_1)] + \mathcal{D}(\mathbf{\Gamma}_{2\rm q}, \mathbf{\Gamma}_{1\rm q}, \Delta_t) \\
&\quad - \frac{1}{2|\Delta_t|}\Lambda_{\rm pp}^{-1} : \overline{R}^2 + \mathcal{O}(\Delta_t^2). \qquad (7.93)
\end{aligned}$$

The second equality uses the fact that the terms involving the gradient of the entropy represent the change in the total entropy over the transition. Also the two terms involving $\Lambda_{\rm pp}^{-1}$ in the first equality have been combined in the second equality and the resultant extra term involving $|\Delta_t|\Lambda_{\rm pp}^{-1} : \dot{\mathbf{\Gamma}}_{\rm p}^0\dot{\mathbf{\Gamma}}_{\rm p}^0/2$ has been incorporated into the constant $C(\mathbf{\Gamma}_1)$. In the final equality, $R = \mathbf{\Gamma}_{2\rm p} - \mathbf{\Gamma}_{1\rm p} - \Delta_t\dot{\mathbf{\Gamma}}_{\rm p}^0$. With $|\Delta_t|^{-1}\Lambda_{\rm pp}^{-1} = (k_{\rm B}/\sigma)\mathrm{I}$, one can see that this final expression is identical to Eq. (7.83).

Note that

$$\frac{1}{2}[S_{\rm r}(\mathbf{\Gamma}_2) - S_{\rm r}(\mathbf{\Gamma}_1)] = \frac{1}{2}R \cdot \nabla_{\rm p}S_{\rm r}(\mathbf{\Gamma}), \qquad (7.94)$$

because in this case Hamilton's equations of motion cause the adiabatic evolution to vanish, $\tau\dot{\mathbf{\Gamma}}^0 \cdot \nabla S_{\rm r}(\mathbf{\Gamma}) = (-\tau/T)\dot{\mathbf{\Gamma}}^0 \cdot \nabla\mathcal{H}(\mathbf{\Gamma}) = 0$. This result is in fact more general than the present canonical equilibrium case or the identification of the adiabatic terms in the equations of motion. Replacing the adiabatic velocity by the more fundamental antisymmetric transport matrix, $\dot{\mathbf{\Gamma}}^0 \Rightarrow -\Theta\nabla S_{\rm r}(\mathbf{\Gamma})$, one has

$$[\nabla S_{\rm r}(\mathbf{\Gamma})] \cdot \dot{\mathbf{\Gamma}}^0 = -[\nabla S_{\rm r}(\mathbf{\Gamma})] \cdot \Theta\nabla S_{\rm r}(\mathbf{\Gamma}) = 0. \qquad (7.95)$$

(A symmetric double scalar product of an antisymmetric matrix always vanishes, as can be seen by taking the transpose.)

7.4.4 Irreversibility and Dissipation

The fact that the reservoir force is independent of the sign of the time step means that the most likely trajectory is irreversible. To show this, consider $\mathbf{\Gamma}_2$ as the future point of two distinct transitions that have $\mathbf{\Gamma}_1$ and $\mathbf{\Gamma}_3$ as the respective previous points. Let $\mathbf{\Gamma}_1$ be the initial point for which $\mathbf{\Gamma}_2$ is the most likely future point; this is the forward transition $\mathbf{\Gamma}_1 \stackrel{\Delta_t}{\rightarrow} \mathbf{\Gamma}_2$, or $\mathbf{\Gamma}_2 = \mathbf{\Gamma}_1 + \Delta_t\dot{\mathbf{\Gamma}}^0 + \overline{\mathbf{R}}$. And let $\mathbf{\Gamma}_3$ be the most likely prior point of $\mathbf{\Gamma}_2$; this is the backward transition $\mathbf{\Gamma}_2 \stackrel{-\Delta_t}{\rightarrow} \mathbf{\Gamma}_3$, or $\mathbf{\Gamma}_3 = \mathbf{\Gamma}_2 - \Delta_t\dot{\mathbf{\Gamma}}^0 + \overline{\mathbf{R}}$. Since the most likely forces are the same in both transitions, one can conclude that $\mathbf{\Gamma}_2$ is the most likely destination of $\mathbf{\Gamma}_1$, but $\mathbf{\Gamma}_3$, not $\mathbf{\Gamma}_1$, is the most likely prior point of $\mathbf{\Gamma}_2$. The difference between these two prior states is

$$\begin{aligned} \mathbf{\Gamma}_3 - \mathbf{\Gamma}_1 &= \left[\mathbf{\Gamma}_2 - \Delta_t\dot{\mathbf{\Gamma}}^0 + \overline{\mathbf{R}}\right] - \left[\mathbf{\Gamma}_2 - \Delta_t\dot{\mathbf{\Gamma}}^0 - \overline{\mathbf{R}}\right] \\ &= 2\overline{\mathbf{R}} = \sigma\nabla_{\mathrm{p}}S_{\mathrm{r}}(\mathbf{\Gamma})/k_{\mathrm{B}}. \end{aligned} \tag{7.96}$$

Notice that in rearranging the expression for $\mathbf{\Gamma}_1$, the most likely force $\overline{\mathbf{R}}$ is treated as a specific value rather than as a stochastic variable. Under Hamilton's equations, there would be a unique initial point and the right-hand side would be zero. That this is non-zero indicates that the trajectory is irreversible. The irreversibility is proportional to the magnitude of the gradient in the reservoir entropy. This means that the closer the system is to the equilibrium state, the more reversible is the trajectory, and *vice versa.* This matter will be discussed further in §7.6.2 below.

For the fixed values of $\mathbf{\Gamma}_3$, $\mathbf{\Gamma}_1$, and $\mathbf{R}^* \equiv \overline{\mathbf{R}}$, the backward transition $\mathbf{\Gamma}_2 \stackrel{-\Delta_t}{\rightarrow} \mathbf{\Gamma}_3$ is much more likely than the forward transition $\mathbf{\Gamma}_3 \stackrel{\Delta_t}{\rightarrow} \mathbf{\Gamma}_2$. The value of the random force in the forward transition is obviously $\mathbf{R} = -\mathbf{R}^*$, which is equal and opposite to the value that gives the most likely forward transition. Similarly, the forward transition $\mathbf{\Gamma}_1 \stackrel{\Delta_t}{\rightarrow} \mathbf{\Gamma}_2$, with random force again $\mathbf{R}^* \equiv \overline{\mathbf{R}}$, is much more likely than the backward transition $\mathbf{\Gamma}_2 \stackrel{-\Delta_t}{\rightarrow} \mathbf{\Gamma}_1$, which again requires a random force in the opposite direction to the most likely one, $\mathbf{R} = -\mathbf{R}^*$. One must have $\wp(\mathbf{\Gamma}_3|\mathbf{\Gamma}_2, -\Delta_t) \gg \wp(\mathbf{\Gamma}_2|\mathbf{\Gamma}_3, \Delta_t)$, and $\wp(\mathbf{\Gamma}_2|\mathbf{\Gamma}_1, \Delta_t) \gg \wp(\mathbf{\Gamma}_1|\mathbf{\Gamma}_2, -\Delta_t)$. The two likely transitions have about the same probability, and the two unlikely transitions are also approximately equally probable, so that conditioned on the future point $\mathbf{\Gamma}_2$, one has $\wp(\mathbf{\Gamma}_3|\mathbf{\Gamma}_2, -\Delta_t) \gg \wp(\mathbf{\Gamma}_1|\mathbf{\Gamma}_2, -\Delta_t)$. Given $\mathbf{\Gamma}_2$ as the current point, this says that it is much more likely that the system came from $\mathbf{\Gamma}_3$ than from $\mathbf{\Gamma}_1$, even though $\mathbf{\Gamma}_2$ is the most likely destination of $\mathbf{\Gamma}_1$. The relative likelihoods depend upon the magnitude of $\nabla_{\mathrm{p}}S_{\mathrm{r}}(\mathbf{\Gamma})$. A physical analogy of the situation is depicted in the following chapter in Fig. 8.1, on p. 267. A sketch of typical phase trajectories is given in Fig. 7.4 on p. 222.

Now the change in entropy for the two transitions forward in time is obtained (again fixing the values of all the points and the forces). For the most likely forward transition, $\mathbf{\Gamma}_1 \stackrel{\Delta_t}{\rightarrow} \mathbf{\Gamma}_2$, the change in entropy in the transition *to* the most likely state is

$$\begin{aligned} S_{\mathrm{r}}(\mathbf{\Gamma}_2) - S_{\mathrm{r}}(\mathbf{\Gamma}_1) &= \overline{\mathbf{R}}(\mathbf{\Gamma}, t) \cdot \nabla_{\mathrm{p}}S_{\mathrm{r}}(\mathbf{\Gamma}) \\ &= \frac{\sigma}{2k_{\mathrm{B}}}\nabla_{\mathrm{p}}S_{\mathrm{r}}(\mathbf{\Gamma}) \cdot \nabla_{\mathrm{p}}S_{\mathrm{r}}(\mathbf{\Gamma}). \end{aligned} \tag{7.97}$$

This is positive or zero, which means that the reservoir entropy increases in the most likely forward transition.

For the most likely backward transition, $\mathbf{\Gamma}_2 \stackrel{-\Delta_t}{\rightarrow} \mathbf{\Gamma}_3$, the change in entropy in the transition *from* the most likely state is

$$\begin{aligned} S_{\rm r}(\mathbf{\Gamma}_2) - S_{\rm r}(\mathbf{\Gamma}_3) &= -\overline{\mathbf{R}}(\mathbf{\Gamma}) \cdot \nabla_{\rm p} S_{\rm r}(\mathbf{\Gamma}) \\ &= \frac{-\sigma}{2k_{\rm B}} \nabla_{\rm p} S_{\rm r}(\mathbf{\Gamma}) \cdot \nabla_{\rm p} S_{\rm r}(\mathbf{\Gamma}). \end{aligned} \tag{7.98}$$

This is negative or zero, which says that the entropy decreases on the transition from the most likely prior point. These make sense because the most likely state for a transition will usually have the highest entropy of all states in the neighbourhood. These two results are consistent with what was deduced from fluctuation theory in §2.4.6.

7.4.5 The Fokker-Planck Equation and Stationarity of the Equilibrium Probability

For the transition $\mathbf{\Gamma}_1 \stackrel{\Delta_t}{\rightarrow} \mathbf{\Gamma}_2$, with corresponding equations of motion $\mathbf{\Gamma}_2 = \mathbf{\Gamma}_1 + \Delta_t \dot{\mathbf{\Gamma}}^0 + \mathbf{R}$, the conditional transition probability is

$$\wp(\mathbf{\Gamma}_2|\mathbf{\Gamma}_1, \Delta_t) = \frac{1}{Z^{(2)}(\mathbf{\Gamma}_1)} e^{S^{(2)}(\mathbf{\Gamma}_2|\mathbf{\Gamma}_1,\Delta_t)}. \tag{7.99}$$

The conditional second entropy for this transition is given by Eq. (7.73). With this, the normalising partition function is

$$\begin{aligned} Z^{(2)}(\mathbf{\Gamma}_1) &= \int \mathrm{d}\mathbf{\Gamma}_{\rm p2}\, \mathrm{d}\mathbf{\Gamma}_{\rm q2}\, e^{(-1/2\sigma)\tilde{\mathbf{R}}\cdot\tilde{\mathbf{R}} + \mathcal{D}(\mathbf{\Gamma}_{\rm q2},\mathbf{\Gamma}_{\rm q1},\Delta_t)/k_{\rm B}} \\ &= (2\pi\alpha)^{3N/2} (2\pi\sigma)^{3N/2}, \end{aligned} \tag{7.100}$$

since $\tilde{\mathbf{R}} \equiv \mathbf{\Gamma}_{\rm p2} - \mathbf{\Gamma}_{\rm p1} - \Delta_t \dot{\mathbf{\Gamma}}^0_{\rm p}(\mathbf{\Gamma}_1) - \overline{\mathbf{R}}(\mathbf{\Gamma}_1)$. The partition function is independent of the initial point of the transition, $\mathbf{\Gamma}_1$.

Suppose that the system is in some transient state corresponding to an arbitrary probability distribution $\wp(\mathbf{\Gamma}, t)$. With the transition as above, $\mathbf{\Gamma}_2 = \mathbf{\Gamma}_1 + \Delta_t \dot{\mathbf{\Gamma}}^0 + \mathbf{R}$, and defining $\overline{\mathbf{\Gamma}}_2(\mathbf{\Gamma}_1) \equiv \mathbf{\Gamma}_1 + \Delta_t \dot{\mathbf{\Gamma}}^0(\mathbf{\Gamma}_1) + \overline{\mathbf{R}}(\mathbf{\Gamma}_1)$, so that $\mathrm{d}\overline{\mathbf{\Gamma}}_2 = \mathrm{d}\mathbf{\Gamma}_1 \left[1 + \nabla_{\rm p} \cdot \overline{\mathbf{R}}\right]$, the evolution of $\wp(\mathbf{\Gamma}, t)$ under the action of the conditional transition probability for the canonical equilibrium system given above is

$$\begin{aligned}
\wp(\mathbf{\Gamma}_2, t_2) &= \int d\mathbf{\Gamma}_1\, \wp(\mathbf{\Gamma}_2|\mathbf{\Gamma}_1, \Delta_t)\wp(\mathbf{\Gamma}_1, t_1) \\
&= \frac{1}{Z^{(2)}} \int d\mathbf{\Gamma}_{p1}\, d\mathbf{\Gamma}_{q1}\, e^{-[\mathbf{\Gamma}_2 - \overline{\mathbf{\Gamma}}_2(\mathbf{\Gamma}_1)]^2/2\sigma} e^{\mathcal{D}(\mathbf{\Gamma}_{q2}, \mathbf{\Gamma}_{q1}, \Delta_t)} \wp(\mathbf{\Gamma}_1, t_1) \\
&= \frac{1}{(2\pi\sigma)^{3N/2}} \int d\overline{\mathbf{\Gamma}}_2\, [1 - \nabla_{\rm p} \cdot \overline{\mathbf{R}}] e^{-[\mathbf{\Gamma}_2 - \overline{\mathbf{\Gamma}}_2(\mathbf{\Gamma}_1)]^2/2\sigma} \\
&\quad \times \wp(\overline{\mathbf{\Gamma}}_2 - \Delta_t \dot{\mathbf{\Gamma}}^0 - \overline{\mathbf{R}}, t_1) \\
&= \frac{1}{(2\pi\sigma)^{3N/2}} \int d\overline{\mathbf{\Gamma}}_2\, [1 - \nabla_{\rm p} \cdot \overline{\mathbf{R}}] e^{-[\mathbf{\Gamma}_2 - \overline{\mathbf{\Gamma}}_2(\mathbf{\Gamma}_1)]^2/2\sigma} \\
&\quad \times \Big\{ \wp(\mathbf{\Gamma}_2, t_1) + \left[\overline{\mathbf{\Gamma}}_2 - \mathbf{\Gamma}_2 - \Delta_t \dot{\mathbf{\Gamma}}^0 - \overline{\mathbf{R}}\right] \cdot \nabla \wp(\mathbf{\Gamma}_2, t_1) \\
&\quad + \frac{1}{2} \left[\overline{\mathbf{\Gamma}}_2 - \mathbf{\Gamma}_2 - \Delta_t \dot{\mathbf{\Gamma}}^0 - \overline{\mathbf{R}}\right]^2 : \nabla\nabla \wp(\mathbf{\Gamma}_2, t_1) \Big\} \\
&= [1 - \nabla_{\rm p} \cdot \overline{\mathbf{R}}(\mathbf{\Gamma}_2)] \Big\{ \wp(\mathbf{\Gamma}_2, t_1) - \left[\Delta_t \dot{\mathbf{\Gamma}}^0 + \overline{\mathbf{R}}\right] \cdot \nabla \wp(\mathbf{\Gamma}_2, t_1) \\
&\quad + \frac{\sigma}{2} \nabla_{\rm p}^2 \wp(\mathbf{\Gamma}_2, t_1) \Big\} + \mathcal{O}(\Delta_t^2). \qquad (7.101)
\end{aligned}$$

The integral has been evaluated by expanding the probability distribution about the peak of the Gaussian. Discarding terms quadratic in the time step, and rearranging gives the partial time derivative of the probability density,

$$\begin{aligned}
\frac{\partial \wp(\mathbf{\Gamma}, t)}{\partial t} &= \frac{-1}{\Delta_t} \left[\nabla_{\rm p} \cdot \overline{\mathbf{R}}(\mathbf{\Gamma})\right] \wp(\mathbf{\Gamma}, t) - \left[\dot{\mathbf{\Gamma}}^0 + \frac{1}{\Delta_t} \overline{\mathbf{R}}\right] \cdot \nabla \wp(\mathbf{\Gamma}, t) \\
&\quad + \frac{\sigma}{2\Delta_t} \nabla_{\rm p}^2 \wp(\mathbf{\Gamma}, t). \qquad (7.102)
\end{aligned}$$

This is a general expression for the evolution of any phase space density under the stochastic equations of motion. It is in fact the Fokker-Planck equation. This equation was already derived in the specific case of Brownian notion, §3.7.3, Eq. (3.103). Note that the most likely force $\overline{\mathbf{R}}$ and the variance σ depend upon the magnitude of the time step but not upon its sign. Hence the right-hand side differs for the forward ($\Delta_t > 0$) and backward ($\Delta_t < 0$) derivatives. This is a general property of stochastic equations of motion and differs from an ordinary derivative which is insensitive to the direction in which it is taken.

The canonical equilibrium probability density, the Maxwell-Boltzmann probability distribution, is stationary under the action of this transition probability (equivalently, these stochastic equations of motion). Using

$$\wp_{\rm MB}(\mathbf{\Gamma}) = \frac{1}{Z(\beta)} e^{S_{\rm r}(\mathbf{\Gamma})/k_{\rm B}} = \frac{1}{Z(\beta)} e^{-\beta\mathcal{H}(\mathbf{\Gamma})}, \qquad (7.103)$$

and

$$\overline{\mathbf{R}} = \frac{\sigma}{2k_{\rm B}} \nabla_{\rm p} S_{\rm r}(\mathbf{\Gamma}) = \frac{-\beta\sigma}{2m} \mathbf{\Gamma}_{\rm p}, \qquad (7.104)$$

the three terms on the right-hand side of Eq. (7.102) are

$$\begin{aligned}\frac{-1}{\Delta_t}\left[\nabla_{\rm p}\cdot\overline{\mathbf{R}}(\mathbf{\Gamma})\right]\wp_{\rm MB}(\mathbf{\Gamma}) &= \frac{\beta\sigma}{2m\Delta_t}[\nabla_{\rm p}\cdot\mathbf{\Gamma}_{\rm p}]\wp_{\rm MB}(\mathbf{\Gamma}) \\ &= \frac{3N\beta\sigma}{2m\Delta_t}\wp_{\rm MB}(\mathbf{\Gamma}),\end{aligned} \tag{7.105}$$

$$\begin{aligned}&-\left[\dot{\mathbf{\Gamma}}^0+\frac{1}{\Delta_t}\overline{\mathbf{R}}\right]\cdot\nabla\wp_{\rm MB}(\mathbf{\Gamma}) \\ &= \beta[\dot{\mathbf{\Gamma}}^0\cdot\nabla\mathcal{H}(\mathbf{\Gamma})]\wp_{\rm MB}(\mathbf{\Gamma})+\frac{\beta}{\Delta_t}[\overline{\mathbf{R}}\cdot\nabla_{\rm p}\mathcal{H}(\mathbf{\Gamma})]\wp_{\rm MB}(\mathbf{\Gamma}) \\ &= 0-\frac{\beta^2\sigma}{2m^2\Delta_t}\mathbf{\Gamma}_{\rm p}\cdot\mathbf{\Gamma}_{\rm p}\wp_{\rm MB}(\mathbf{\Gamma}) \\ &= \frac{-\beta^2\sigma}{2m^2\Delta_t}\sum_{i,\alpha}p_{i\alpha}^2\,\wp_{\rm MB}(\mathbf{\Gamma}),\end{aligned} \tag{7.106}$$

and

$$\begin{aligned}&\frac{\sigma}{2\Delta_t}\nabla_{\rm p}^2\wp_{\rm MB}(\mathbf{\Gamma}) \\ &= \frac{-\beta\sigma}{2\Delta_t}\nabla_{\rm p}^2\mathcal{H}(\mathbf{\Gamma})\,\wp_{\rm MB}(\mathbf{\Gamma})+\frac{\beta^2\sigma}{2\Delta_t}[\nabla_{\rm p}\mathcal{H}(\mathbf{\Gamma})]\cdot[\nabla_{\rm p}\mathcal{H}(\mathbf{\Gamma})]\,\wp_{\rm MB}(\mathbf{\Gamma}) \\ &= \frac{-3N\beta\sigma}{2m\Delta_t}\wp_{\rm MB}(\mathbf{\Gamma})+\frac{\beta^2\sigma}{2m^2\Delta_t}\sum_{i,\alpha}p_{i\alpha}^2\,\wp_{\rm MB}(\mathbf{\Gamma}).\end{aligned} \tag{7.107}$$

Adding these together, Eq. (7.102) shows that the partial time derivative vanishes exactly

$$\frac{\partial\wp_{\rm MB}(\mathbf{\Gamma})}{\partial t}=0. \tag{7.108}$$

Obviously one could have obtained this result by inspection, since the Maxwell-Boltzmann probability density does not depend explicitly upon time. The real point of the exercise is to show the consistency between the second entropy transition probability and the Maxwell-Boltzmann probability density. It also confirms the Fokker-Planck equation, Eq. (7.102).

It is important to note that all terms were required to obtain the exact cancellation that made the Maxwell-Boltzmann distribution stationary. However, it is a reasonable approximation to neglect the stochastic term, $(\sigma/2\Delta_t)\nabla_{\rm p}^2\wp(\mathbf{\Gamma},t)$. This was discussed in Ch. 3 in the specific case of Brownian motion, for example the equality of the full and deterministic derivatives of the probability, Eq. (3.94), the generalised equipartition theorem, §3.7.4, and the justification for Liouville's theorem, §3.7.5. The equipartition argument is quite general, and in the present canonical equilibrium case one has explicitly

$$\begin{aligned}\int \mathrm{d}\mathbf{\Gamma}\,\frac{\sigma}{2\Delta_t}\nabla_{\rm p}^2\wp_{\rm MB}(\mathbf{\Gamma}) &= \int \mathrm{d}\mathbf{\Gamma}\,\wp_{\rm MB}(\mathbf{\Gamma})\left\{\frac{-3N\beta\sigma}{2m\Delta_t}+\frac{\beta^2\sigma}{2m^2\Delta_t}\sum_{i,\alpha}p_{i\alpha}^2\right\} \\ &= 0.\end{aligned} \tag{7.109}$$

If one ignores fluctuations, then in the thermodynamic limit the equipartition theorem may be taken to apply at each point in phase space (c.f. §§3.7.4 and 3.7.5). More precisely, this is true on the likely points of phase space so that weighted by the probability distribution to a reasonable approximation one may take

$$\sum_{i,\alpha} p_{i\alpha}^2 \wp_{\mathrm{MB}}(\mathbf{\Gamma}) \approx 3Nmk_{\mathrm{B}}T\wp_{\mathrm{MB}}(\mathbf{\Gamma}). \tag{7.110}$$

Hence to this level of approximation one may neglect the stochastic term for any probability distribution yielding the equipartition theorem, $\nabla_{\mathrm{p}}^2\wp(\mathbf{\Gamma},t) \approx 0$. Under these circumstances, the deterministic equations of motion that keep $\overline{\mathbf{R}} = (-\beta\sigma/2m)\mathbf{\Gamma}_{\mathrm{p}}$ and neglect the stochastic term $\tilde{\mathbf{R}} = 0$, will preserve the Maxwell-Boltzmann distribution, at least on those points of phase space that satisfy the equipartition theorem.

Finally, the convective or deterministic time derivative of the Maxwell-Boltzmann distribution is non-zero,

$$\begin{aligned} \frac{\mathrm{d}^{\mathrm{det}}\wp_{\mathrm{MB}}(\mathbf{\Gamma})}{\mathrm{d}t} &\equiv \frac{\partial\wp_{\mathrm{MB}}(\mathbf{\Gamma})}{\partial t} + \left[\dot{\mathbf{\Gamma}}^0 + \frac{1}{\Delta_t}\overline{\mathbf{R}}\right] \cdot \nabla\wp_{\mathrm{MB}}(\mathbf{\Gamma}) \\ &= \frac{\beta^2\sigma}{2m^2\Delta_t}\sum_{i,\alpha} p_{i\alpha}^2 \wp_{\mathrm{MB}}(\mathbf{\Gamma}) \\ &\approx \frac{3N\beta\sigma}{2m\Delta_t}\wp_{\mathrm{MB}}(\mathbf{\Gamma}). \end{aligned} \tag{7.111}$$

This monotonic increase of the probability density (for the forward time derivative) is, in the expression for the partial derivative, cancelled by the decrease due to the expansion of the phase space volume (assuming the validity of the equipartition theorem at each point).

7.5 Evolution in Phase Space

7.5.1 Various Phase Functions

The rate of change of a phase function is obtained from the change during the transition $\mathbf{\Gamma}_1 \stackrel{\Delta_t}{\rightarrow} \mathbf{\Gamma}_2$, $\mathbf{\Gamma}_2 = \mathbf{\Gamma}_1 + \Delta_t\dot{\mathbf{\Gamma}}^0 + \mathbf{R}$, $\Delta_t \equiv t_2 - t_1$,

$$\begin{aligned} \frac{\mathrm{d}f(\mathbf{\Gamma},t)}{\mathrm{d}t} &= \frac{1}{\Delta_t}\left[f(\mathbf{\Gamma}_1 + \Delta_t\dot{\mathbf{\Gamma}}^0 + \mathbf{R}, t_1 + \Delta_t) - f(\mathbf{\Gamma}_1, t_1)\right] \\ &= \dot{f}^0(\mathbf{\Gamma},t) + \frac{1}{\Delta_t}\mathbf{R}\cdot\nabla_{\mathrm{p}}f(\mathbf{\Gamma},t) + \frac{1}{2\Delta_t}\mathbf{R}\mathbf{R} : \nabla_{\mathrm{p}}\nabla_{\mathrm{p}}f(\mathbf{\Gamma},t). \end{aligned} \tag{7.112}$$

Since the probability distribution for the random force depends upon the magnitude but not the sign of the time step, the rate of change of a function is different for the forward derivative, $\Delta_t > 0$, and the backward derivative $\Delta_t < 0$; see the discussion of reversibility, §7.4.4 above and §7.6.2 below. In this the adiabatic derivative is

$$\dot{f}^0(\mathbf{\Gamma},t) = \frac{\partial f(\mathbf{\Gamma},t)}{\partial t} + \dot{\mathbf{\Gamma}}^0 \cdot \nabla f(\mathbf{\Gamma},t). \tag{7.113}$$

Averaged over the stochastic forces the total time derivative is

$$\begin{aligned}
&\left\langle \frac{\mathrm{d}f(\mathbf{\Gamma},t)}{\mathrm{d}t} \right\rangle_{\tilde{R}} \\
&= \dot{f}^0(\mathbf{\Gamma},t) + \frac{1}{\Delta_t}\overline{\mathbf{R}} \cdot \nabla_{\mathrm{p}} f(\mathbf{\Gamma},t) + \frac{\sigma}{2\Delta_t}\nabla_{\mathrm{p}}^2 f(\mathbf{\Gamma},t) \\
&= \dot{f}^0(\mathbf{\Gamma},t) + \frac{\sigma}{2k_{\mathrm{B}}\Delta_t}\nabla_{\mathrm{p}} S_{\mathrm{r}}(\mathbf{\Gamma}) \cdot \nabla_{\mathrm{p}} f(\mathbf{\Gamma},t) + \frac{\sigma}{2\Delta_t}\nabla_{\mathrm{p}}^2 f(\mathbf{\Gamma},t) \\
&= \dot{f}^0(\mathbf{\Gamma},t) - \frac{\beta\sigma}{2m\Delta_t}\mathbf{\Gamma}_{\mathrm{p}} \cdot \nabla_{\mathrm{p}} f(\mathbf{\Gamma},t) + \frac{\sigma}{2\Delta_t}\nabla_{\mathrm{p}}^2 f(\mathbf{\Gamma},t).
\end{aligned} \tag{7.114}$$

Note that $\sigma/|\Delta_t| \sim \mathcal{O}(1)$ and that $\beta = 1/k_{\mathrm{B}}T$. The average over the stochastic forces is often assumed implicitly and the angled brackets are not shown. Also, this total time derivative is also called the convective derivative, the material derivative, or the hydrodynamic derivative, and may also be denoted $\dot{f}(\mathbf{\Gamma},t)$.

One can define the deterministic derivative, as the rate of change on the most likely trajectory in the absence of stochastic forces. It is

$$\dot{f}^{\mathrm{det}}(\mathbf{\Gamma},t) \equiv \dot{f}^0(\mathbf{\Gamma},t) + \frac{1}{\Delta_t}\overline{\mathbf{R}} \cdot \nabla_{\mathrm{p}} f(\mathbf{\Gamma},t). \tag{7.115}$$

Again because the most likely random force depends upon the magnitude but not the sign of the time step, the forward and backward deterministic derivatives have different values. This is equal to the total time derivative if, and only if, the term arising from the stochastic forces, $\nabla_{\mathrm{p}}^2 f$, is negligible.

The evolution of several useful phase space functions may be obtained from this. For a position coordinate, $f(\mathbf{\Gamma},t) \equiv q_{i\alpha}$, one obtains

$$\left\langle \frac{\mathrm{d}q_{i\alpha}}{\mathrm{d}t} \right\rangle_{\tilde{R}} = \dot{q}_{i\alpha}^0 \tag{7.116}$$

and for a momentum coordinate, $f(\mathbf{\Gamma},t) \equiv p_{i\alpha}$, one obtains

$$\left\langle \frac{\mathrm{d}p_{i\alpha}}{\mathrm{d}t} \right\rangle_{\tilde{R}} = \dot{p}_{i\alpha}^0 - \frac{\beta\sigma}{2m\Delta_t}p_{i\alpha}. \tag{7.117}$$

These of course could have been obtained directly by averaging the stochastic equations of motion.

Taking $f(\mathbf{\Gamma},t) \equiv S_{\mathrm{r}}(\mathbf{\Gamma}) = -\mathcal{H}(\mathbf{\Gamma})/T$, the average of the total rate of change of the reservoir entropy is

$$\begin{aligned}
\left\langle \frac{\mathrm{d}S_{\mathrm{r}}(\mathbf{\Gamma})}{\mathrm{d}t} \right\rangle_{\tilde{R}} &= \frac{\beta\sigma}{2m^2\Delta_t T}\sum_{i,\alpha} p_{i\alpha}^2 - \frac{\sigma}{2\Delta_t T}\sum_{i,\alpha} m^{-1} \\
&= \frac{\sigma}{2m\Delta_t T}\left[\frac{\beta}{m}\sum_{i,\alpha} p_{i\alpha}^2 - 3N\right].
\end{aligned} \tag{7.118}$$

The adiabatic term vanishes because energy is conserved on a Hamiltonian trajectory, $\dot{\mathbf{\Gamma}}^0 \cdot \nabla \mathcal{H}(\mathbf{\Gamma}) = 0$, and hence only the dissipative term and the fluctuation term contribute to this. In view of the equipartition theorem, the average over phase space vanishes,

$$\int \mathrm{d}\mathbf{\Gamma}\, \wp_{\mathrm{MB}}(\mathbf{\Gamma}|\beta) \left\langle \frac{\mathrm{d}S_{\mathrm{r}}(\mathbf{\Gamma})}{\mathrm{d}t} \right\rangle_{\tilde{R}} = 0. \tag{7.119}$$

This result is what one would expect, because in an equilibrium system there should be no nett change of the reservoir entropy with time, even though the reservoir entropy may increase or decrease at individual points in the sub-system phase space. In fact, in the thermodynamic limit, $N \to \infty$, fluctuations are relatively negligible, so that the instantaneous value of the kinetic energy can be taken to be the same as the average value, $\sum_{i,\alpha} p_{i\alpha}^2/2m = 3Nk_{\mathrm{B}}T/2$. That is, the bracketed term can be set to zero with negligible error, and one can conclude that the reservoir entropy is a constant of the motion at every point in phase space that has non-negligible probability.

For an arbitrary or transient probability distribution, $f(\mathbf{\Gamma}, t) \equiv \wp(\mathbf{\Gamma}, t)$, and using the result for the partial time derivative obtained with the canonical equilibrium conditional transition probability, Eq. (7.102), the total rate of change averaged over the stochastic forces, Eq. (7.114), becomes

$$\begin{aligned} \left\langle \frac{\mathrm{d}\wp(\mathbf{\Gamma},t)}{\mathrm{d}t} \right\rangle_{\tilde{R}} &= \frac{\partial \wp(\mathbf{\Gamma},t)}{\partial t} + \left[\dot{\mathbf{\Gamma}}^0 + \frac{1}{\Delta_t} \overline{\mathbf{R}} \right] \cdot \nabla \wp(\mathbf{\Gamma},t) + \frac{\sigma}{2\Delta_t} \nabla_{\mathrm{p}}^2 \wp(\mathbf{\Gamma},t) \\ &= \frac{-1}{\Delta_t} \left[\nabla_{\mathrm{p}} \cdot \overline{\mathbf{R}}(\mathbf{\Gamma}) \right] \wp(\mathbf{\Gamma},t) + \frac{\sigma}{\Delta_t} \nabla_{\mathrm{p}}^2 \wp(\mathbf{\Gamma},t). \end{aligned} \tag{7.120}$$

Notice how the stochastic term on the right-hand side contributes in addition to the compressibility. Applying this to the Maxwell-Boltzmann distribution, $\wp_{\mathrm{MB}}(\mathbf{\Gamma}|\beta) = Z^{-1} \exp[-\beta \mathcal{H}(\mathbf{\Gamma})]$, and using the result for the compressibility, Eq. (7.105), this becomes

$$\begin{aligned} \left\langle \frac{\mathrm{d}\wp_{\mathrm{MB}}(\mathbf{\Gamma}|\beta)}{\mathrm{d}t} \right\rangle_{\tilde{R}} &= \frac{3N\beta\sigma}{2m\Delta_t} \wp_{\mathrm{MB}}(\mathbf{\Gamma}|\beta) + \frac{\sigma}{\Delta_t} \left[\beta^2 \nabla_{\mathrm{p}} \mathcal{H}(\mathbf{\Gamma}) \cdot \nabla_{\mathrm{p}} \mathcal{H}(\mathbf{\Gamma}) \right. \\ &\quad \left. - \beta \nabla_{\mathrm{p}}^2 \mathcal{H}(\mathbf{\Gamma}) \right] \wp_{\mathrm{MB}}(\mathbf{\Gamma}|\beta) \\ &= \frac{3N\beta\sigma}{2m\Delta_t} \wp_{\mathrm{MB}}(\mathbf{\Gamma}|\beta) + \left[\frac{\beta^2 \sigma}{m^2 \Delta_t} \sum_{i,\alpha} p_{i\alpha}^2 - \frac{3N\beta\sigma}{m\Delta_t} \right] \wp_{\mathrm{MB}}(\mathbf{\Gamma}|\beta) \\ &\approx \frac{3N\beta\sigma}{2m\Delta_t} \wp_{\mathrm{MB}}(\mathbf{\Gamma}|\beta). \end{aligned} \tag{7.121}$$

As discussed above, the last approximation holds in the thermodynamic limit when fluctuations are negligible and the equipartition theorem can be taken to hold at each point in phase space of non-negligible probability. In this case, the stochastic contribution to the evolution is negligible and the total time derivative of the probability density is the same as the deterministic derivative, Eq. (7.111). This cancellation of the stochastic term was discussed in the context of the generalised equipartition theorem, §3.7.4, and of Liouville's theorem in §3.7.5. Irrespective of the approximation, one has the exact result for the integral of the total time derivative of the probability density over phase space,

$$\int \mathrm{d}\mathbf{\Gamma} \left\langle \frac{\mathrm{d}\wp_{\mathrm{MB}}(\mathbf{\Gamma}|\beta)}{\mathrm{d}t} \right\rangle_{\tilde{R}} = \frac{3N\beta\sigma}{2m\Delta_t}. \tag{7.122}$$

In this case the cancellation of the stochastic term is exact, and the non-zero result stems entirely from the dissipative contribution to the evolution of the Maxwell-Boltzmann distribution.

It is worth comparing this result to the result for the reservoir entropy, Eq. (7.119), the vanishing in which case resulted from the cancellation of the stochastic with the deterministic terms. The deterministic derivative alone is

$$\int \mathrm{d}\mathbf{\Gamma}\, \wp_{\mathrm{MB}}(\mathbf{\Gamma}|\beta)\dot{S}_{\mathrm{r}}^{\mathrm{det}}(\mathbf{\Gamma})/k_{\mathrm{B}} = \int \mathrm{d}\mathbf{\Gamma}\, \dot{\wp}_{\mathrm{MB}}^{\mathrm{det}}(\mathbf{\Gamma}|\beta) = \frac{3N\beta\sigma}{2m\Delta_t}, \tag{7.123}$$

whereas

$$\int \mathrm{d}\mathbf{\Gamma}\, \wp_{\mathrm{MB}}(\mathbf{\Gamma}|\beta) \left\langle \frac{\mathrm{d}S_{\mathrm{r}}(\mathbf{\Gamma})}{k_{\mathrm{B}}\mathrm{d}t} \right\rangle_{\tilde{R}} \neq \int \mathrm{d}\mathbf{\Gamma} \left\langle \frac{\mathrm{d}\wp_{\mathrm{MB}}(\mathbf{\Gamma}|\beta)}{\mathrm{d}t} \right\rangle_{\tilde{R}}. \tag{7.124}$$

This is an example of the general principle that the chain rule of differentiation holds for the deterministic time derivative, but not for the total time derivative that includes the stochastic contribution because the latter arises from a second derivative.

Now the compressibility of the equations of motion is discussed. Using the result for the compressibility, Eq. (7.105), and assuming the validity of the equipartition theorem at each point, $\beta \sum_{i,\alpha} p_{i\alpha}^2/m \approx 3N$, the total time derivative of the Maxwell-Boltzmann distribution, Eq. (7.121), can be written

$$\left\langle \frac{\mathrm{d}\wp_{\mathrm{MB}}(\mathbf{\Gamma}|\beta)}{\mathrm{d}t} \right\rangle_{\tilde{R}} \approx \frac{3N\beta\sigma}{2m\Delta_t}\wp_{\mathrm{MB}}(\mathbf{\Gamma}|\beta) = \frac{-1}{\Delta_t}\left[\nabla_{\mathrm{p}} \cdot \overline{\mathbf{R}}\right] \wp_{\mathrm{MB}}(\mathbf{\Gamma}|\beta). \tag{7.125}$$

This says that the total rate of change of the probability density is related to the negative of the divergence of the deterministic part of the equations of motion.

A related result may be derived in an alternative fashion based upon the normalisation of the probability. Let $\mathbf{\Gamma}' = \mathbf{\Gamma} + \Delta\mathbf{\Gamma}^{\mathrm{det}} = \mathbf{\Gamma} + \Delta_t\dot{\mathbf{\Gamma}}^0 + \overline{\mathbf{R}}$ be the deterministic evolution. For an arbitrary normalised probability density, one has

$$\begin{aligned} 1 &= \int \mathrm{d}\mathbf{\Gamma}'\, \wp(\mathbf{\Gamma}', t) \\ &= \int \mathrm{d}\mathbf{\Gamma}\, \left[1 + \nabla_{\mathrm{p}} \cdot \overline{\mathbf{R}}\right] \left\{\wp(\mathbf{\Gamma}, t) + \left[\Delta_t\dot{\mathbf{\Gamma}}^0 + \overline{\mathbf{R}}\right] \cdot \nabla\wp(\mathbf{\Gamma}, t)\right\}. \end{aligned} \tag{7.126}$$

Clearly the two terms linear in the time step must cancel, which gives

$$\int \mathrm{d}\mathbf{\Gamma}\, \left[\Delta_t\dot{\mathbf{\Gamma}}^0 + \overline{\mathbf{R}}\right] \cdot \nabla\wp(\mathbf{\Gamma}, t) = -\int \mathrm{d}\mathbf{\Gamma}\, \wp(\mathbf{\Gamma}, t)\, \nabla_{\mathrm{p}} \cdot \overline{\mathbf{R}}. \tag{7.127}$$

Inserting the Maxwell-Boltzmann distribution, one sees that this is an identity equivalent to the equipartition theorem. Since the adiabatic derivative of the Maxwell-Boltzmann distribution vanishes, $\dot{\wp}^0_{\mathrm{MB}}(\mathbf{\Gamma}|\beta) = 0$, the integrand on the left-hand side is just its deterministic derivative. Assuming that fluctuations are negligible and that this holds at each point in phase space, this says

$$\dot{\wp}^{\mathrm{det}}_{\mathrm{MB}}(\mathbf{\Gamma}|\beta) \approx \left\langle \frac{\mathrm{d}\wp_{\mathrm{MB}}(\mathbf{\Gamma}|\beta)}{\mathrm{d}t} \right\rangle_{\tilde{R}} \approx \frac{-1}{\Delta_t} \wp_{\mathrm{MB}}(\mathbf{\Gamma}|\beta) \nabla_{\mathrm{p}} \cdot \overline{\mathbf{R}}(\mathbf{\Gamma}). \tag{7.128}$$

This is the same as Eq. (7.125). This says that the deterministic derivative of the Maxwell-Boltzmann distribution on the most likely trajectory neglecting stochastic forces is equal to the full derivative averaged over the stochastic forces. This result, which is more general than the present canonical equilibrium system, will be compared to Liouville's theorem for non-Hamiltonian deterministic equations of motion below.

7.5.2 Compressibility

The compressibility of the equations of motion does not have a straightforward physical interpretation, as distinct from the mathematical interpretation of the rate of change of a volume element. It is clear that

$$\nabla \cdot \Delta\mathbf{\Gamma}^{\mathrm{det}} = \nabla_{\mathrm{p}} \cdot \overline{\mathbf{R}} = \frac{\sigma}{2k_{\mathrm{B}}} \nabla^2_{\mathrm{p}} S_{\mathrm{r}}(\mathbf{\Gamma}). \tag{7.129}$$

The first equality is general; the second equality assumes that the deterministic non-Hamiltonian force is derived from the reservoir entropy, as mandated by maximisation of the second entropy. On the left-hand side is the divergence of the deterministic part of the equations of motion, which have the interpretation of the most likely reservoir perturbation, the dissipative force. From the second entropy, this divergence of the dissipation is the Laplacian of the reservoir entropy, which is related to the fluctuation part of the reservoir perturbation. That these two are so linked is another manifestation of the fluctuation dissipation theorem.

One sometimes sees in the literature an attempt to interpret the compressibility in terms of either the entropy production of the sub-system or else the entropy production of the reservoir. Neither of these is correct and it is instructive to examine the arguments to see why they are flawed. One argument begins by asserting that the sub-system entropy of a point in its phase space is the logarithm of a volume associated with that point, $S_{\mathrm{s}}(\mathbf{\Gamma}, t) = k_{\mathrm{B}} \ln v_{\mathrm{s}}(\mathbf{\Gamma}, t)$. The argument continues by asserting that therefore the rate of change of the sub-system entropy is the logarithmic derivative of the sub-system volume, which is just the compressibility,

$$\dot{S}_{\mathrm{s}}(\mathbf{\Gamma}, t)/k_{\mathrm{B}} = \frac{\dot{v}_{\mathrm{s}}(\mathbf{\Gamma}, t)}{v_{\mathrm{s}}(\mathbf{\Gamma}, t)} = \frac{1}{\Delta_t} \nabla \cdot \mathbf{\Gamma}^{\mathrm{det}}, \text{ (err.)} \tag{7.130}$$

The second argument says that the total volume associated with a point in the sub-system phase space, $\mathbf{\Gamma}$, is the product of the sub-system volume and the reservoir volume, $v_{\rm tot}(\mathbf{\Gamma},t) = v_{\rm s}(\mathbf{\Gamma},t)v_{\rm r}(\mathbf{\Gamma},t)$. Since the total system is isolated and evolves adiabatically according to Hamilton's equations, the total volume is a constant of the motion, $\dot{v}_{\rm tot}(\mathbf{\Gamma},t) = 0$. Taking the reservoir entropy to be the logarithm of its volume, $S_{\rm r}(\mathbf{\Gamma},t) = k_{\rm B} \ln v_{\rm r}(\mathbf{\Gamma},t)$, this would imply that the rate of entropy production of the reservoir is

$$\dot{S}_{\rm r}(\mathbf{\Gamma},t)/k_{\rm B} = \frac{\dot{v}_{\rm r}(\mathbf{\Gamma},t)}{v_{\rm r}(\mathbf{\Gamma},t)} = \frac{-\dot{v}_{\rm s}(\mathbf{\Gamma},t)}{v_{\rm s}(\mathbf{\Gamma},t)} = \frac{-1}{\Delta_t}\nabla\cdot\mathbf{\Gamma}^{\rm det}, \text{ (err.)} \tag{7.131}$$

Although it is certainly correct that the compressibility is equal to the relative rate of change of a sub-system volume, it is not correct that it equals either the rate of sub-system entropy production, or the negative of the rate of reservoir entropy production. The problem with the first argument, Eq. (7.130), is the assumption that entropy is proportional to the logarithm of the volume associated with each point in the isolated systems phase space. Equation (7.33) shows that the weight density is constant and uniform in phase space, and hence the entropy of a volume of phase space is simply that volume times the logarithm of the weight density. It is true that the argument of the logarithm includes an arbitrary volume scale (c.f. Δ_Γ in Eq. (7.34)), but there would of course be no point in choosing a scale that changed with time or position in phase space. One might speculate that the motivation for asserting that the entropy is the logarithm of an evolving volume comes from using the $\wp \ln \wp$ expression for part of the entropy discussed in Ch. 1, Eq. (1.23), together with a conservation law for ensemble members in a volume of the sort discussed in Ch. 3, §3.7.5.

The problems with the second argument are similar. There is no fundamental justification for asserting that the entropy is the logarithm of a volume in the total phase space. Moreover, whilst it is true that the volume of the total phase space does not change during the adiabatic (Hamiltonian) evolution of the isolated total system, it is not necessarily true that the postulated deterministic equations of motion for the projected sub-system phase space preserve this property. In fact, the concept of the evolution of a sub-system volume is not well defined, at least not in the physically realistic case of stochastic dissipative equations of motion.

The notion of the evolution of a projected volume requires careful thought. A projected volume can break up, and conversely it can merge with other projected volumes, as is sketched in Fig. 7.2. This occurs because the principle that trajectories cannot cross that results from Hamilton's equations for an isolated system does not hold in the case of projection onto a sub-system. Some effects of the projection operation are sketched in Fig. 7.3. In the projected sub-space, trajectories can cross, be created, and be destroyed.[18] These crossing, separation, and union points are a manifestation of the stochastic part

[18]Points of creation and destruction of trajectories in the projected sub-space occur in pairs. If one follows a trajectory from a unification point, one will eventually reach a bifurcation point from which the two original trajectories emerge. The single trajectory between these two points in the sub-space corresponds to the coincidence of two separate trajectories in the

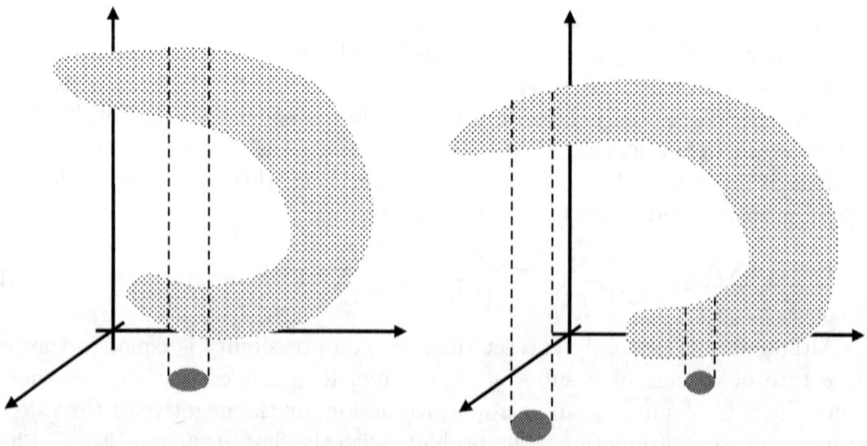

Figure 7.2: The energy hypersurface in the phase space of the total system and the projection of parts of it onto the sub-system. The left-hand sketch shows two disjoint volumes of the total system that project onto a single volume of the sub-system. The right-hand sketch shows that over time the projected volume is not constant and becomes disconnected.

of the sub-space trajectory, since they preclude the one-to-one determinism of Hamilton's equations for the full phase space. The concept of the evolution of a volume of the projected phase space is meaningless, because trajectories that were originally in the interior of the volume can cross the evolving boundary and end up exterior to the volume. Because trajectories can be created and destroyed at different points, the change in the number of trajectories interior to a fixed volume is not given by the flux in the number across the fixed boundary to that volume. In other words, the probability flux is not equal to the velocity times the probability density, $\mathbf{J}_\wp \neq \dot{\mathbf{\Gamma}}^{\mathrm{det}} \wp(\mathbf{\Gamma}, t)$, but rather one has to include the additional contribution to the flux due to the stochastic terms that result from the projection operation (c.f. Eq. (3.111)).

It is emphasised that this is the real behaviour of a physical sub-system. It would of course be possible to develop artificial deterministic equations of motion in which the trajectories do not cross, and are neither created nor destroyed. However, this is not what happens in the real world, which suggests that such equations would likely have other unphysical consequences.

7.5.3 Liouville's Theorem

Liouville's theorem was given as the deterministic limit of the Fokker-Planck equation in §3.7.5. An alternative but ultimately equivalent derivation can be given for phase space. The deterministic Hamiltonian or non-Hamiltonian

total space. Hence there is a global conservation law for trajectories in the sub-space that is reflected in the global normalisation of the probability density.

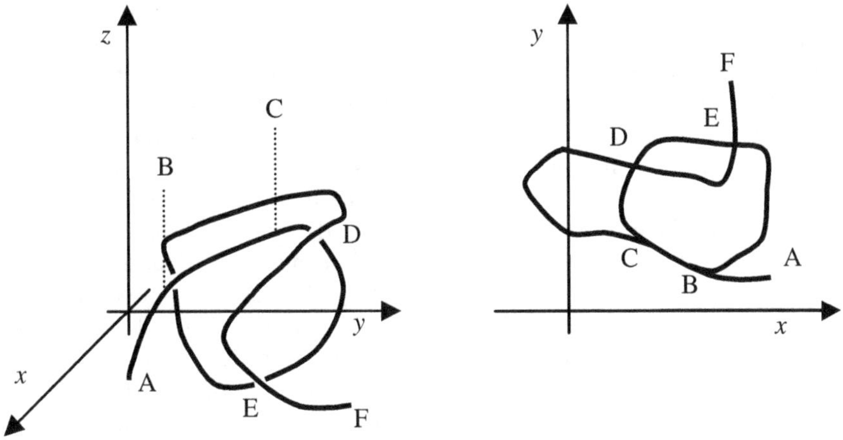

Figure 7.3: A trajectory in the full three-dimensional space (left), and its projection onto a two-dimensional sub-space (right). Note that the full trajectory does not cross itself, and is not created or destroyed (not counting the beginning (A) and end (F)), whereas the projection shows two points where trajectories cross (D and E), one point where two trajectories unite to become one trajectory (B), and one point where a single trajectory bifurcates into two trajectories (C).

equations of motion, with the original sub-system phase space point at time t_1 being $\mathbf{\Gamma}_1$, give the next point on the trajectory at time $t_2 = t_1 + \Delta_t$, this being $\mathbf{\Gamma}_2 = \mathbf{\Gamma}_1 + \Delta\mathbf{\Gamma}^{\text{det}}$. The deterministic force is the most likely force, $\Delta\mathbf{\Gamma}^{\text{det}} = \Delta_t \dot{\mathbf{\Gamma}}^0(\mathbf{\Gamma}_1) + \overline{\mathbf{R}}(\mathbf{\Gamma}_1)$. The corresponding conditional transition probability is $\delta(\mathbf{\Gamma}_2 - \mathbf{\Gamma}_1 - \Delta\mathbf{\Gamma}^{\text{det}})$, and the next probability density is

$$\begin{aligned}
\wp(\mathbf{\Gamma}_2, t_2) &= \int \mathrm{d}\mathbf{\Gamma}_1\, \delta(\mathbf{\Gamma}_2 - \mathbf{\Gamma}_1 - \Delta\mathbf{\Gamma}^{\text{det}}) \wp(\mathbf{\Gamma}_1, t_1) \\
&= \wp(\mathbf{\Gamma}_2 - \Delta\mathbf{\Gamma}^{\text{det}}, t_1) \left| \frac{\mathrm{d}\mathbf{\Gamma}_1}{\mathrm{d}\left[\mathbf{\Gamma}_1 + \Delta\mathbf{\Gamma}^{\text{det}}\right]} \right|_{\mathbf{\Gamma}_2} \\
&= \wp(\mathbf{\Gamma}_2, t_1) - \Delta\mathbf{\Gamma}^{\text{det}} \cdot \nabla \wp(\mathbf{\Gamma}_2, t_1) \\
&\quad - \nabla \cdot \Delta\mathbf{\Gamma}^{\text{det}} \wp(\mathbf{\Gamma}_2, t_1) + \mathcal{O}(\Delta_t^2).
\end{aligned} \tag{7.132}$$

This gives the deterministic derivative as

$$\begin{aligned}
\dot{\wp}^{\text{det}}(\mathbf{\Gamma}, t) &\equiv \frac{\partial \wp(\mathbf{\Gamma}, t)}{\partial t} + \frac{1}{\Delta_t} \Delta\mathbf{\Gamma}^{\text{det}} \cdot \nabla \wp(\mathbf{\Gamma}, t) \\
&= \frac{-1}{\Delta_t} \left[\nabla \cdot \Delta\mathbf{\Gamma}^{\text{det}}\right] \wp(\mathbf{\Gamma}, t) \\
&= \frac{-1}{\Delta_t} \left[\nabla \cdot \overline{\mathbf{R}}\right] \wp(\mathbf{\Gamma}, t).
\end{aligned} \tag{7.133}$$

The final equality follows from the incompressibility of Hamilton's equations, $\nabla \cdot \dot{\mathbf{\Gamma}}^0 = 0$. This is the non-Hamiltonian form of Liouville's theorem, in which the deterministic time derivative is interpreted as the total time derivative. For a Hamiltonian trajectory of an isolated system, $\overline{\mathbf{R}} = \mathbf{0}$, the equation is formally correct. More generally, it is the same as the Fokker-Planck equation (7.102) with the stochastic term neglected. As discussed in the general case in the context of the generalised equipartition theorem, §3.7.4, and more particularly above for the canonical equilibrium system with the Maxwell-Boltzmann distribution, Eq. (7.109) and following, it is justifiable to neglect the stochastic contribution to the evolution of the probability density. This, however, does not justify the use of deterministic non-Hamiltonian equations of motion in general.

The conditional transition probability, $\delta(\mathbf{\Gamma}_2-\mathbf{\Gamma}_1-\Delta\mathbf{\Gamma}^{\mathrm{det}})$, based upon deterministic non-Hamiltonian equations of motion, says in essence that trajectories are conserved, and that they do not cross each other. This is compatible with, and has the same status as, the ensemble interpretation of probability in which the number of members is conserved during the evolution of the ensemble.[19] In general neither are compatible with stochastic equations of motion, in which trajectories may cross, be created, or be destroyed, as was discussed in connection with Figs 7.2 and 7.3. It is only in the case of the evolution of the probability density that deterministic non-Hamiltonian equations of motion can possibly be valid as an approximation, as in Eqs (7.125) and (7.128), and even then the non-Hamiltonian term must have a specific form. More generally, the stochastic terms do not cancel for the evolution of any other phase function, and it is likely that adding arbitrary non-Hamiltonian terms to the equations of motion and discarding the stochastic terms will have unintended, unphysical consequences.

7.6 Reversibility

The concept of time reversibility is fundamental to mechanics and to non-equilibrium theory. In statistical mechanics three types of reversibility occur: there is a probabilistic reversibility, which is rather trivial, there is a mechanical reversibility that involves the conjugate trajectory, and there is a mathematical reversibility that involves retracing a trajectory. The latter two are quite important in equilibrium and non-equilibrium theory.

The first of these, probabilistic, is manifest in the joint probability function, where the order of the arguments is irrelevant,

$$\wp(\mathbf{\Gamma}_2, t_2; \mathbf{\Gamma}_2, t_1) = \wp(\mathbf{\Gamma}_2, t_1; \mathbf{\Gamma}_2, t_2). \tag{7.134}$$

In words, this is the probability that the sub-system is in the microstate $\mathbf{\Gamma}_2$ at time t_2 *and* that it is in the microstate $\mathbf{\Gamma}_1$ at time t_1. The conjunction 'and' obviously makes the order of the states in the argument of the probability irrelevant.[20] This symmetry rule does not depend upon the physical nature of

[19] The conservation law for weight during a transition, Eq. (1.28), and the related reduction condition for the weight of joint states, Eq. (1.12), both of which follow from the general laws of probability, are global conditions that hold for a sum over a complete set of states. They do not represent a local conservation law, as assumed in Liouville's theorem.

[20] A state here is the combination of the phase space point and the time, $\{\mathbf{\Gamma}, t\}$.

a trajectory. One can define the signed time interval $\Delta_t \equiv t_2 - t_1$ and write $\wp(\mathbf{\Gamma}_2, t_2; \mathbf{\Gamma}_2, t_1) \equiv \wp(\mathbf{\Gamma}_2|\mathbf{\Gamma}_1, \Delta_t)\wp(\mathbf{\Gamma}_1, t_1)$. With this conditional probability the symmetry rule reads

$$\wp(\mathbf{\Gamma}_2|\mathbf{\Gamma}_1, \Delta_t)\wp(\mathbf{\Gamma}_1, t_1) = \wp(\mathbf{\Gamma}_1|\mathbf{\Gamma}_2, -\Delta_t)\wp(\mathbf{\Gamma}_2, t_2). \tag{7.135}$$

Again, this is just a consequence of the definitions of probability that give two equivalent ways of writing the probability that $\mathbf{\Gamma}_2$ occurs a time Δ_t after $\mathbf{\Gamma}_1$. One should not confuse $-\Delta_t$ with the unphysical concept of time running backwards.

7.6.1 Isolated System

The remaining two types of reversibility are related to the mechanical or physical reversibility that comes from Hamilton's equations. In common parlance, the Hamiltonian trajectory is said to retrace its course if time is reversed. This can be a rather confusing picture, because it involves two distinct concepts. There is the concept of conjugacy, the second type of reversibility mentioned above, in which all the momenta are reversed but time is calculated forward. And there is the third type of reversibility mentioned above, namely that of retracing the trajectory, where the momenta remain the same but time is calculated backwards, as was briefly discussed in §7.4.1 on p. 197.

The conjugate phase space point has all its momenta reversed: with $\mathbf{\Gamma} = \{q^{3N}, p^{3N}\}$, then the conjugate point is $\mathbf{\Gamma}^\dagger = \{q^{3N}, (-p)^{3N}\}$. Now Hamilton's equations show that the velocity at the conjugate phase point is the negative of the conjugate of the velocity of the original phase point,

$$\dot{\mathbf{\Gamma}}^0(\mathbf{\Gamma}^\dagger) = -\left[\dot{\mathbf{\Gamma}}^0(\mathbf{\Gamma})\right]^\dagger. \tag{7.136}$$

This can be shown directly using the fact that the Hamiltonian is an even function of the momenta, $\mathcal{H}(\mathbf{\Gamma}^\dagger) = \mathcal{H}(\mathbf{\Gamma})$. Hence

$$\dot{q}^0_{i\alpha}(\mathbf{\Gamma}^\dagger) = \frac{\partial \mathcal{H}(\mathbf{\Gamma}^\dagger)}{\partial p^\dagger_{i\alpha}} = -\frac{\partial \mathcal{H}(\mathbf{\Gamma})}{\partial p_{i\alpha}} = -\dot{q}^0_{i\alpha}(\mathbf{\Gamma}) = -[\dot{q}^0_{i\alpha}(\mathbf{\Gamma})]^\dagger, \tag{7.137}$$

and

$$\dot{p}^0_{i\alpha}(\mathbf{\Gamma}^\dagger) = -\frac{\partial \mathcal{H}(\mathbf{\Gamma}^\dagger)}{\partial q^\dagger_{i\alpha}} = -\frac{\partial \mathcal{H}(\mathbf{\Gamma})}{\partial q_{i\alpha}} = \dot{p}^0_{i\alpha}(\mathbf{\Gamma}) = -[\dot{p}^0_{i\alpha}(\mathbf{\Gamma})]^\dagger. \tag{7.138}$$

Note that the phase space velocity $\dot{\mathbf{\Gamma}}$ obeys the same conjugacy rule as phase space itself, namely if $\dot{\mathbf{\Gamma}} = \{\dot{\mathbf{\Gamma}}_\mathrm{q}, \dot{\mathbf{\Gamma}}_\mathrm{p}\}$, then $[\dot{\mathbf{\Gamma}}]^\dagger = \{\dot{\mathbf{\Gamma}}_\mathrm{q}, -\dot{\mathbf{\Gamma}}_\mathrm{p}\}$.

The nature of an isolated system embodied by this property of Hamilton's equations, $\dot{\mathbf{\Gamma}}^0(\mathbf{\Gamma}^\dagger) = -\left[\dot{\mathbf{\Gamma}}^0(\mathbf{\Gamma})\right]^\dagger$, means that a Hamiltonian transition is reversible in the sense that if

$$\mathbf{\Gamma}_2 = \mathbf{\Gamma}_1 + \Delta_t \dot{\mathbf{\Gamma}}^0(\mathbf{\Gamma}), \tag{7.139}$$

then

$$\mathbf{\Gamma}_1^\dagger = \mathbf{\Gamma}_2^\dagger + \Delta_t \dot{\mathbf{\Gamma}}^0(\mathbf{\Gamma}^\dagger). \tag{7.140}$$

This says that if $\mathbf{\Gamma}_2$ is the transition point of $\mathbf{\Gamma}_1$ going forward in time, then $\mathbf{\Gamma}_1^\dagger$ is the transition point of $\mathbf{\Gamma}_2^\dagger$, also going forward in time. (The argument of the velocity can be evaluated at either phase space point to the exhibited order in the time step.) Integrating this along a trajectory gives

$$\mathbf{\Gamma}_2 = \mathbf{\Gamma}^0(t_2|\mathbf{\Gamma}_1, t_1) \Leftrightarrow \mathbf{\Gamma}_1^\dagger = \mathbf{\Gamma}^0(t_2|\mathbf{\Gamma}_2^\dagger, t_1). \tag{7.141}$$

This is the sense in which a Hamiltonian system is time reversible.

The conditional transition probability for the isolated system is just

$$\wp_0(\mathbf{\Gamma}_2|\mathbf{\Gamma}_1, \Delta_t) = \delta(\mathbf{\Gamma}_2 - \mathbf{\Gamma}_1 - \Delta_t \dot{\mathbf{\Gamma}}_1^0). \tag{7.142}$$

Since there is a one-to-one relationship between the forward and backward transition, one must have $\delta(\mathbf{\Gamma}_2 - \mathbf{\Gamma}_1 - \Delta_t \dot{\mathbf{\Gamma}}^0(\mathbf{\Gamma})) = \delta(\mathbf{\Gamma}_1^\dagger - \mathbf{\Gamma}_2^\dagger - \Delta_t \dot{\mathbf{\Gamma}}^0(\mathbf{\Gamma}^\dagger))$, and hence

$$\wp_0(\mathbf{\Gamma}_2|\mathbf{\Gamma}_1, \Delta_t) = \wp_0(\mathbf{\Gamma}_1^\dagger|\mathbf{\Gamma}_2^\dagger, \Delta_t). \tag{7.143}$$

This says that reversing the velocities gives the reverse transition going forward in time. A corollary is that Hamilton's equations show that

$$\wp_0(\mathbf{\Gamma}_2|\mathbf{\Gamma}_1, \Delta_t) \neq \wp_0(\mathbf{\Gamma}_1|\mathbf{\Gamma}_2, \Delta_t). \tag{7.144}$$

The third type of reversibility, namely reversing time but leaving the momenta unchanged, holds rather trivially for Hamilton's equations simply by swapping terms on the two sides of the original form of Hamilton's equations, Eq. (7.139),

$$\mathbf{\Gamma}_1 = \mathbf{\Gamma}_2 - \Delta_t \dot{\mathbf{\Gamma}}^0(\mathbf{\Gamma}). \tag{7.145}$$

This has the form of Hamilton's equations but with a negative time step. It enables the trajectory to be retraced in time. Hamilton's equations of motion are reversible in this third sense in that $\mathbf{\Gamma}' = \mathbf{\Gamma}^0(t|\mathbf{\Gamma}) \Leftrightarrow \mathbf{\Gamma} = \mathbf{\Gamma}^0(-t|\mathbf{\Gamma}')$. In terms of conditional probability this corresponds to

$$\wp_0(\mathbf{\Gamma}_2|\mathbf{\Gamma}_1, \Delta_t) = \wp_0(\mathbf{\Gamma}_1|\mathbf{\Gamma}_2, -\Delta_t), \tag{7.146}$$

which is essentially the same as is demanded by the laws of probability in the case that the probability density is unchanged during adiabatic evolution.

7.6.2 Canonical Equilibrium System

Now focus on the canonical equilibrium system. Since the Hamiltonian has even parity, the Maxwell-Boltzmann distribution is insensitive to the direction of the momenta,

$$\wp_{\mathrm{MB}}(\mathbf{\Gamma}^\dagger) = \wp_{\mathrm{MB}}(\mathbf{\Gamma}). \tag{7.147}$$

In the first instance, take the conditional transition probability to be the adiabatic one given above, $\wp_0(\mathbf{\Gamma}_2|\mathbf{\Gamma}_1, \Delta_t)$.[21] Because energy is conserved in an adiabatic transition, one has $\mathcal{H}(\mathbf{\Gamma}_1) = \mathcal{H}(\mathbf{\Gamma}_2) = \mathcal{H}(\mathbf{\Gamma}_2^\dagger)$, and hence $\wp_{\rm MB}(\mathbf{\Gamma}_1) = \wp_{\rm MB}(\mathbf{\Gamma}_2) = \wp_{\rm MB}(\mathbf{\Gamma}_2^\dagger)$. Accordingly the unconditional transition probability satisfies

$$\wp_0(\mathbf{\Gamma}_2|\mathbf{\Gamma}_1, \Delta_t)\wp_{\rm MB}(\mathbf{\Gamma}_1) = \wp_0(\mathbf{\Gamma}_1^\dagger|\mathbf{\Gamma}_2^\dagger, \Delta_t)\wp_{\rm MB}(\mathbf{\Gamma}_2^\dagger). \tag{7.148}$$

This says that for the canonical equilibrium system, the unconditional forward transition is equally likely to occur as the reverse conjugate transition (in the case that the reservoir perturbations are so negligible as to justify using the adiabatic transition probability).

In contrast to the Hamiltonian reversibility, Eq. (7.136), the most likely stochastic perturbation of the equations of motion is not reversible. With the most likely reservoir forces being,

$$\overline{\mathbf{R}}_{\rm q} \equiv 0, \text{ and } \overline{\mathbf{R}}_{\rm p} = \frac{\sigma \nabla_{\rm p} S_{\rm r}(\mathbf{\Gamma})}{2k_{\rm B}} = \frac{-\sigma\beta}{2m}\mathbf{\Gamma}_{\rm p}, \tag{7.149}$$

it follows that

$$\overline{\mathbf{R}}(\mathbf{\Gamma}^\dagger) = -\overline{\mathbf{R}}(\mathbf{\Gamma}) = \left[\overline{\mathbf{R}}(\mathbf{\Gamma})\right]^\dagger. \tag{7.150}$$

This is exactly opposite to the behaviour of the Hamiltonian contribution to the deterministic trajectory. This means that in the presence of perturbations from the reservoir, the trajectory passing through a conjugate point is no longer the conjugate of the original trajectory. That is if

$$\mathbf{\Gamma}_2 = \mathbf{\Gamma}_1 + \Delta_t \dot{\mathbf{\Gamma}}^0(\mathbf{\Gamma}) + \overline{R}(\mathbf{\Gamma}), \tag{7.151}$$

then

$$\mathbf{\Gamma}_1^\dagger \neq \mathbf{\Gamma}_2^\dagger + \Delta_t \dot{\mathbf{\Gamma}}^0(\mathbf{\Gamma}^\dagger) + \overline{R}(\mathbf{\Gamma}^\dagger). \tag{7.152}$$

This is one sense (violating the second type of reversibility) that thermodynamic systems are irreversible.

For the third type of reversibility, the time step is negated, $\Delta_t \Rightarrow -\Delta_t$. As mentioned briefly in §7.4.1 on p. 197, and in more detail in §7.4.4, the most likely pre-transition point of $\mathbf{\Gamma}_2$,

$$\mathbf{\Gamma}_3 = \mathbf{\Gamma}_2 - \Delta_t \dot{\mathbf{\Gamma}}^0(\mathbf{\Gamma}) + \overline{R}(\mathbf{\Gamma}), \ \Delta_t > 0, \tag{7.153}$$

[21] The correct conditional transition probability is the second entropy form given earlier in the chapter, from which is derived the stochastic dissipative equations of motion. The adiabatic transition probability based on Hamilton's equations of motion is equivalent to setting the variance to zero, $\sigma = 0$, so that both the stochastic terms and the dissipative terms vanish leaving only the adiabatic part of the equations of motion. Since the Maxwell-Boltzmann distribution is stationary under the stochastic dissipative equations of motion, it remains stationary in the adiabatic limit, as may be confirmed explicitly from the constancy of the Hamiltonian.

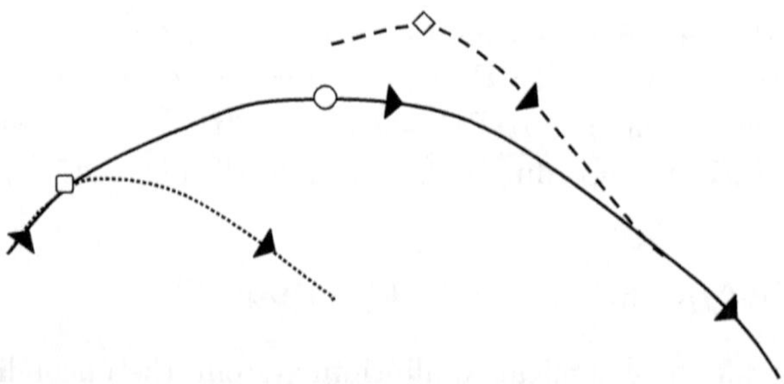

Figure 7.4: The most likely trajectory $\overline{\mathbf{\Gamma}}(t'|\mathbf{\Gamma},t)$ (solid curve), passing through $\mathbf{\Gamma}$ (circle) at time t, the most likely trajectory, $\overline{\mathbf{\Gamma}}(t'|\mathbf{\Gamma}_1,t_1)$ (dotted curve), passing through $\mathbf{\Gamma}_1$ (square) at the earlier time $t_1 < t$, and the most likely trajectory, $\overline{\mathbf{\Gamma}}(t'|\mathbf{\Gamma}_2,t_2)$ (dashed curve), passing through $\mathbf{\Gamma}_2$ (diamond) at the later time $t_2 > t$.

does not equal $\mathbf{\Gamma}_1$. As shown in Eq. (7.96), to linear order in the time step, the difference between these is

$$\mathbf{\Gamma}_3 - \mathbf{\Gamma}_1 = 2\overline{R}(\mathbf{\Gamma}). \tag{7.154}$$

This irreversibility is a direct consequence of the time symmetry of the regression of fluctuations: most likely the system came *from* equilibrium and most likely it is going *to* equilibrium. In terms of the most likely trajectory, the fact that $\mathbf{\Gamma}_2$ lies on the trajectory originating at $\mathbf{\Gamma}_1$, $\mathbf{\Gamma}_2 = \overline{\mathbf{\Gamma}}(t_2|\mathbf{\Gamma}_1,t_1)$, does *not* imply that $\mathbf{\Gamma}_1$ lies on the trajectory originating at $\mathbf{\Gamma}_2$, $\mathbf{\Gamma}_1 \neq \overline{\mathbf{\Gamma}}(t_1|\mathbf{\Gamma}_2,t_2)$ (going backwards in time). In other words, one cannot speak of *the* most likely trajectory; one can only speak of the most likely trajectory that originates at a given point at a given time. Figure 7.4 sketches the situation that can occur for projected trajectories.

Although the stochastic, dissipative trajectory is not individually reversible under the conjugation operation, or under reversing the time step, it is reversible when both of these are combined. Consider the most likely next point for the transition $\mathbf{\Gamma}_1 \stackrel{\Delta_t}{\rightarrow} \mathbf{\Gamma}_2$,

$$\mathbf{\Gamma}_2 = \mathbf{\Gamma}_1 + \Delta_t\dot{\mathbf{\Gamma}}^0(\mathbf{\Gamma}) + \overline{\mathbf{R}}(\mathbf{\Gamma}), \tag{7.155}$$

where to linear order in the time step, it is immaterial whether the argument $\mathbf{\Gamma}$ is set equal to $\mathbf{\Gamma}_1$ or to $\mathbf{\Gamma}_2$. Taking the conjugate of this yields

$$\begin{aligned}\mathbf{\Gamma}_2^\dagger &= \mathbf{\Gamma}_1^\dagger + \Delta_t\left[\dot{\mathbf{\Gamma}}^0(\mathbf{\Gamma})\right]^\dagger + \left[\overline{\mathbf{R}}(\mathbf{\Gamma})\right]^\dagger \\ &= \mathbf{\Gamma}_1^\dagger - \Delta_t\dot{\mathbf{\Gamma}}^0(\mathbf{\Gamma}^\dagger) + \overline{\mathbf{R}}(\mathbf{\Gamma}^\dagger).\end{aligned} \tag{7.156}$$

The last equality is just the stochastic dissipative equations of motion going backward in time, $\mathbf{\Gamma}_1^\dagger \stackrel{-\Delta_t}{\rightarrow} \mathbf{\Gamma}_2^\dagger$. That is, if $\mathbf{\Gamma}_2$ is the most likely post-transition point of $\mathbf{\Gamma}_1$, then $\mathbf{\Gamma}_2^\dagger$ is the most likely pre-transition point of $\mathbf{\Gamma}_1^\dagger$. Integrating

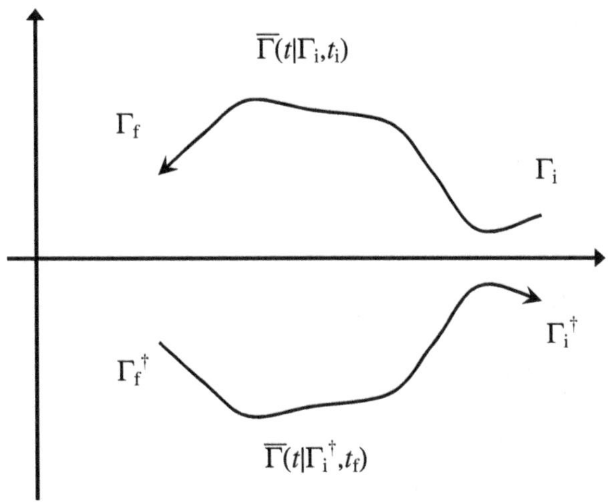

Figure 7.5: A forward most likely trajectory, (top), and its backward conjugate, (bottom), for $t_i \leq t \leq t_f$. The arrow in each case signifies the positive time direction.

this one sees that the reverse of a trajectory is its conjugate, namely if $\mathbf{\Gamma}_f = \overline{\mathbf{\Gamma}}(t_f|\mathbf{\Gamma}_i, t_i)$, then $\mathbf{\Gamma}_f^\dagger = \overline{\mathbf{\Gamma}}(t_i|\mathbf{\Gamma}_i^\dagger, t_f)$. For $t_f \leq t_i$, the first is the end point of a backward trajectory from the initial point $\mathbf{\Gamma}_i$, and the second is the end point of a forward trajectory from the initial point $\mathbf{\Gamma}_i^\dagger$. This reversibility is sketched in Fig. 7.5. Note that due to irreversibility, $\mathbf{\Gamma}_i \neq \overline{\mathbf{\Gamma}}(t_i|\mathbf{\Gamma}_f, t_f)$ (forward trajectory), and $\mathbf{\Gamma}_i^\dagger \neq \overline{\mathbf{\Gamma}}(t_f|\mathbf{\Gamma}_f^\dagger, t_i)$ (backward trajectory). Reversibility only holds on the conjugate trajectory travelling in the opposite time direction.

It remains to prove that for the canonical equilibrium system, it is equally likely to observe the unconditional forward transition as it is to observe the unconditional backward conjugate transition. This was proved above using the adiabatic transition probability, Eq. (7.148), but now the full stochastic dissipative transition probability is used.

The two preceding numbered equations give the most likely transition (forward), and its conjugate backward transition. The corresponding stochastic forward transition is

$$\mathbf{\Gamma}_2 = \mathbf{\Gamma}_1 + \Delta_t \dot{\mathbf{\Gamma}}^0(\mathbf{\Gamma}) + \mathbf{R}(\mathbf{\Gamma}), \tag{7.157}$$

which has conjugate

$$\begin{aligned} \mathbf{\Gamma}_2^\dagger &= \mathbf{\Gamma}_1^\dagger + \Delta_t \left[\dot{\mathbf{\Gamma}}^0(\mathbf{\Gamma})\right]^\dagger + [\mathbf{R}(\mathbf{\Gamma})]^\dagger \\ &= \mathbf{\Gamma}_1^\dagger - \Delta_t \dot{\mathbf{\Gamma}}^0(\mathbf{\Gamma}^\dagger) + \mathbf{R}(\mathbf{\Gamma}^\dagger). \end{aligned} \tag{7.158}$$

The latter represents the conjugate backward stochastic transition. The second entropies that correspond to these must be equal

$$S^{(2)}(\mathbf{\Gamma}_2, t_2; \mathbf{\Gamma}_1, t_1) = S^{(2)}(\mathbf{\Gamma}_2^\dagger, t_1; \mathbf{\Gamma}_1^\dagger, t_2), \tag{7.159}$$

with $\Delta_t \equiv t_2 - t_1$. Explicitly this follows from the expression for the unconditional second entropy, Eq. (7.83),

$$\begin{aligned}
& S^{(2)}(\mathbf{\Gamma}_2, \mathbf{\Gamma}_1|\Delta_t) \\
&= \frac{-k_\mathrm{B}}{2\sigma}\mathbf{R}\cdot\mathbf{R} + \frac{1}{2}\left[S_\mathrm{r}(\mathbf{\Gamma}_1) + S_\mathrm{r}(\mathbf{\Gamma}_2)\right] + \mathcal{D}(\mathbf{\Gamma}_{\mathrm{q}2}, \mathbf{\Gamma}_{\mathrm{q}1}, \Delta_t) \\
&\quad + \frac{k_\mathrm{B}}{2\sigma}\overline{\mathbf{R}}(\mathbf{\Gamma})\cdot\overline{\mathbf{R}}(\mathbf{\Gamma}) - \frac{1}{2}\overline{\mathbf{R}}(\mathbf{\Gamma})\cdot\nabla_\mathrm{p}S_\mathrm{r}(\mathbf{\Gamma}) \\
&= \frac{-k_\mathrm{B}}{2\sigma}\mathbf{R}\cdot\mathbf{R} + \frac{1}{2}\left[S_\mathrm{r}(\mathbf{\Gamma}_1^\dagger) + S_\mathrm{r}(\mathbf{\Gamma}_2^\dagger)\right] + \mathcal{D}(\mathbf{\Gamma}_{\mathrm{q}1}^\dagger, \mathbf{\Gamma}_{\mathrm{q}2}^\dagger, \Delta_t) \\
&\quad + \frac{k_\mathrm{B}}{2\sigma}\overline{\mathbf{R}}(\mathbf{\Gamma}^\dagger)\cdot\overline{\mathbf{R}}(\mathbf{\Gamma}^\dagger) - \frac{1}{2}\overline{\mathbf{R}}(\mathbf{\Gamma}^\dagger)\cdot\nabla_\mathrm{p}S_\mathrm{r}(\mathbf{\Gamma}^\dagger) \\
&= S^{(2)}(\mathbf{\Gamma}_1^\dagger, \mathbf{\Gamma}_2^\dagger|\Delta_t).
\end{aligned} \tag{7.160}$$

In the first and subsequent equalities, terms that are $\mathcal{O}(\Delta_t)$ can be equally well evaluated at $\mathbf{\Gamma}_1$ or $\mathbf{\Gamma}_2$, which is simply signified by $\mathbf{\Gamma}$. In the second equality, the facts have been used that the reservoir entropy has even parity, $S_\mathrm{r}(\mathbf{\Gamma}^\dagger) = S_\mathrm{r}(\mathbf{\Gamma})$, that the most likely reservoir force has odd parity, $\overline{\mathbf{R}}(\mathbf{\Gamma}^\dagger) = -\overline{\mathbf{R}}(\mathbf{\Gamma})$, and that the momentum gradient of the entropy has odd parity, $\nabla_\mathrm{p}S_\mathrm{r}(\mathbf{\Gamma}^\dagger) = -\nabla_\mathrm{p}S_\mathrm{r}(\mathbf{\Gamma})$.

Since the second entropy gives the unconditional transition probability, this is equivalent to

$$\wp(\mathbf{\Gamma}_2, \mathbf{\Gamma}_1|\Delta_t) = \wp(\mathbf{\Gamma}_1^\dagger, \mathbf{\Gamma}_2^\dagger|\Delta_t). \tag{7.161}$$

In more verbose notation, this is $\wp(\mathbf{\Gamma}_2, t_2; \mathbf{\Gamma}_1, t_1) = \wp(\mathbf{\Gamma}_2^\dagger, t_1; \mathbf{\Gamma}_1^\dagger, t_2)$. This completes the proof that in the canonical equilibrium system, the forward transition is as unconditionally likely to occur as the conjugate backward transition. This generalises the reversibility shown in Fig. 7.5 from a most likely trajectory to an arbitrary trajectory.

With $\Delta_t = t_2 - t_1$, the left-hand side may be written as a conditional probability,

$$\wp(\mathbf{\Gamma}_2, \mathbf{\Gamma}_1|\Delta_t) = \wp(\mathbf{\Gamma}_2|\mathbf{\Gamma}_1, \Delta_t)\wp_\mathrm{MB}(\mathbf{\Gamma}_1) = \wp(\mathbf{\Gamma}_1|\mathbf{\Gamma}_2, -\Delta_t)\wp_\mathrm{MB}(\mathbf{\Gamma}_2). \tag{7.162}$$

And similarly for the right-hand side,

$$\wp(\mathbf{\Gamma}_1^\dagger, \mathbf{\Gamma}_2^\dagger|\Delta_t) = \wp(\mathbf{\Gamma}_1^\dagger|\mathbf{\Gamma}_2^\dagger, \Delta_t)\wp_\mathrm{MB}(\mathbf{\Gamma}_2^\dagger) = \wp(\mathbf{\Gamma}_2^\dagger|\mathbf{\Gamma}_1^\dagger, -\Delta_t)\wp_\mathrm{MB}(\mathbf{\Gamma}_1^\dagger). \tag{7.163}$$

Equating the forward and backward transitions conditioned on $\mathbf{\Gamma}_1$ and $\mathbf{\Gamma}_1^\dagger$, and using the fact that $\wp_\mathrm{MB}(\mathbf{\Gamma}_1) = \wp_\mathrm{MB}(\mathbf{\Gamma}_1^\dagger)$, these imply

$$\wp(\mathbf{\Gamma}_2|\mathbf{\Gamma}_1, \Delta_t) = \wp(\mathbf{\Gamma}_2^\dagger|\mathbf{\Gamma}_1^\dagger, -\Delta_t). \tag{7.164}$$

This says that the forward transition for one state to another is as likely as the backward transition from that state with the velocities reversed.

The two forward transitions may also be equated to each other. Rearranging both sides shows that the conditional transition probabilities are in inverse

Table 7.3: Possible transitions between two neighbouring points, Γ_1 and Γ_2

$\Gamma_1 \stackrel{\Delta_t}{\rightarrow} \Gamma_2$	$\Gamma_2 = \Gamma_1 + \Delta_t \dot{\Gamma}^0(\Gamma) + \mathbf{R}(\Gamma)$	$\wp(\Gamma_2 \vert \Gamma_1, \Delta_t)$
$\Gamma_1^\dagger \stackrel{-\Delta_t}{\rightarrow} \Gamma_2^\dagger$	$\Gamma_2^\dagger = \Gamma_1^\dagger - \Delta_t \dot{\Gamma}^0(\Gamma^\dagger) + \mathbf{R}(\Gamma^\dagger)$	$= \wp(\Gamma_2^\dagger \vert \Gamma_1^\dagger, -\Delta_t)$
$\Gamma_2 \stackrel{-\Delta_t}{\rightarrow} \Gamma_1$	$\Gamma_1 = \Gamma_2 - \Delta_t \dot{\Gamma}^0(\Gamma) + \mathbf{R}(\Gamma)$	$\wp(\Gamma_1 \vert \Gamma_2, -\Delta_t)$
$\Gamma_2^\dagger \stackrel{\Delta_t}{\rightarrow} \Gamma_1^\dagger$	$\Gamma_1^\dagger = \Gamma_2^\dagger + \Delta_t \dot{\Gamma}^0(\Gamma^\dagger) + \mathbf{R}(\Gamma^\dagger)$	$= \wp(\Gamma_1^\dagger \vert \Gamma_2^\dagger, \Delta_t)$
$\Gamma_1 \stackrel{-\Delta_t}{\rightarrow} \Gamma_2$	$\Gamma_2 = \Gamma_1 - \Delta_t \dot{\Gamma}^0(\Gamma) + \mathbf{R}(\Gamma)$	$\wp(\Gamma_2 \vert \Gamma_1, -\Delta_t)$
$\Gamma_1^\dagger \stackrel{\Delta_t}{\rightarrow} \Gamma_2^\dagger$	$\Gamma_2^\dagger = \Gamma_1^\dagger + \Delta_t \dot{\Gamma}^0(\Gamma^\dagger) + \mathbf{R}(\Gamma^\dagger)$	$= \wp(\Gamma_2^\dagger \vert \Gamma_1^\dagger, \Delta_t)$
$\Gamma_2 \stackrel{\Delta_t}{\rightarrow} \Gamma_1$	$\Gamma_1 = \Gamma_2 + \Delta_t \dot{\Gamma}^0(\Gamma) + \mathbf{R}(\Gamma)$	$\wp(\Gamma_1 \vert \Gamma_2, \Delta_t)$
$\Gamma_2^\dagger \stackrel{-\Delta_t}{\rightarrow} \Gamma_1^\dagger$	$\Gamma_1^\dagger = \Gamma_2^\dagger - \Delta_t \dot{\Gamma}^0(\Gamma^\dagger) + \mathbf{R}(\Gamma^\dagger)$	$= \wp(\Gamma_1^\dagger \vert \Gamma_2^\dagger, -\Delta_t)$

proportion to the probability of the initial state,

$$\frac{\wp_{\mathrm{MB}}(\mathbf{\Gamma}_1)}{\wp_{\mathrm{MB}}(\mathbf{\Gamma}_2^\dagger)} = \frac{\wp_{\mathrm{MB}}(\mathbf{\Gamma}_1)}{\wp_{\mathrm{MB}}(\mathbf{\Gamma}_2)} = \frac{\wp(\mathbf{\Gamma}_1^\dagger|\mathbf{\Gamma}_2^\dagger, \Delta_t)}{\wp(\mathbf{\Gamma}_2|\mathbf{\Gamma}_1, \Delta_t)}. \tag{7.165}$$

The interpretation is that conditional transitions to a more probable state are more likely than conditional transitions to a less probable state. The generic version of this result based on the axioms of probability was given as Eq. (1.37). This is essentially the phase space version of the relationship between Boltzmann's statistical entropy and Clausius' Second Law of Thermodynamics, as was discussed in §§1.1 and 1.4.4.

Table 7.3 shows all the possible transitions between two neighbouring points in phase space. There are eight transitions that stem from three binary choices: choose either Γ_1 or else Γ_2 as the initial point of the transition, choose either a forward transition or else a backward transition, and choose either the original points or else their conjugates. Of these there are two distinct sets of equivalent adiabatic transitions, $\mathbf{R} = \mathbf{0}$, as grouped in the upper and lower halves of the table. There are four distinct sets of stochastic transitions, and these are grouped into pairs of equivalent transitions, as shown by the equality of the conditional probabilities, Eq. (7.164).

This completes the discussion of reversibility in equilibrium systems. Whilst time does not run backwards in the physical universe, it is still legitimate to enquire from whence the system came. This corresponds to the mathematical calculation of the prior trajectory, which is formulated as a conditional transition probability with a negative time step.

7.7 Trajectory Probability and Time Correlation Functions

7.7.1 Trajectory Probability

This section obtains the probability of a trajectory in phase space for a canonical equilibrium system, and analyses its properties and consequences. Consider a sub-system and a thermal reservoir of temperature T, and let $\mathbf{\Gamma}(t)$ denote the position in the sub-system phase space at time t. Consider a time interval $t \in [0, t_f]$ that is discretised into $f+1$ uniformly spaced nodes, $\Delta_t \equiv t_f/f$, so that $t_n = n\Delta_t$, $n = 0, 1, \ldots, f$. Denote a particular trajectory by an $(f+1)$-component vector, $\underline{\mathbf{\Gamma}} \equiv \{\mathbf{\Gamma}_0, \mathbf{\Gamma}_1, \ldots, \mathbf{\Gamma}_f\}$, where the value of the trajectory at the nodes is written $\mathbf{\Gamma}_n \equiv \mathbf{\Gamma}(t_n)$.

The probability of a particular trajectory is

$$\wp(\underline{\mathbf{\Gamma}}) = \wp_{\mathrm{MB}}(\mathbf{\Gamma}_0) \prod_{n=1}^{f} \wp(\mathbf{\Gamma}_n|\mathbf{\Gamma}_{n-1}, \Delta_t), \tag{7.166}$$

where the conditional transition probability is

$$\wp(\mathbf{\Gamma}_n|\mathbf{\Gamma}_{n-1}, \Delta_t) = \frac{1}{\xi} e^{-\tilde{\mathbf{R}}_{n-1}\cdot\tilde{\mathbf{R}}_{n-1}/2\sigma} \delta\Big(\mathbf{\Gamma}_{\mathrm{q},n} - \mathbf{\Gamma}_{\mathrm{q},n-1} - \Delta_t \dot{\mathbf{\Gamma}}^0_{\mathrm{q}}(\mathbf{\Gamma}_{n-1})\Big), \tag{7.167}$$

with the stochastic part of the force being

$$\tilde{\mathbf{R}}_{n-1} \equiv \mathbf{\Gamma}_{\mathrm{p},n} - \mathbf{\Gamma}_{\mathrm{p},n-1} - \Delta_t \dot{\mathbf{\Gamma}}^0_{\mathrm{p}}(\mathbf{\Gamma}_{n-1}) - \overline{\mathbf{R}}(\mathbf{\Gamma}_{n-1}). \tag{7.168}$$

Note the Markovian nature of the trajectory probability, which holds for a small enough time step Δ_t.

Although this is an explicit expression for the trajectory probability, there is an alternative way of proceeding. One can generate a trajectory by the stochastic, dissipative equations of motion, Eq. (7.70) or Eq. (7.86). That is

$$\begin{aligned} \mathbf{\Gamma}_{\mathrm{q},n+1} &= \mathbf{\Gamma}_{\mathrm{q},n} + \Delta_t \dot{\mathbf{\Gamma}}^0_{\mathrm{q}}(\mathbf{\Gamma}_n), \\ \mathbf{\Gamma}_{\mathrm{p},n+1} &= \mathbf{\Gamma}_{\mathrm{p},n} + \Delta_t \dot{\mathbf{\Gamma}}^0_{\mathrm{p}}(\mathbf{\Gamma}_n) + \mathbf{R}_n, \quad n = 0, 1, \ldots f-1. \end{aligned} \tag{7.169}$$

Here the random reservoir force at t_n is $\mathbf{R}_n \equiv \overline{\mathbf{R}}(\mathbf{\Gamma}_n) + \tilde{\mathbf{R}}_n$, with the dissipative or most likely force being $\overline{\mathbf{R}}(\mathbf{\Gamma}) = \sigma \nabla_{\mathrm{p}} S_{\mathrm{r}}(\mathbf{\Gamma})/2k_{\mathrm{B}} = -\beta\sigma\mathbf{\Gamma}_{\mathrm{p}}/2m$, and the fluctuation force $\tilde{\mathbf{R}}_n$ being a $3N$-component vector of independent Gaussian random variables of mean zero and variance σ. The initial point $\mathbf{\Gamma}_0 \equiv \mathbf{\Gamma}(t_0)$ is normally chosen from the Maxwell-Boltzmann distribution. Suppose that one generates a large number of trajectories, say M, in this way, $\underline{\mathbf{\Gamma}}_m$, $m = 1, 2, \ldots, M$. Hence $\mathbf{\Gamma}_{mn} \equiv \mathbf{\Gamma}_m(t_n)$ is the nth phase space point on the mth trajectory. With this the trajectory probability is

$$\wp(\underline{\mathbf{\Gamma}}) = \frac{1}{M} \sum_{m=1}^{M} \delta(\underline{\mathbf{\Gamma}} - \underline{\mathbf{\Gamma}}_m) = \frac{1}{M} \sum_{m=1}^{M} \prod_{n=0}^{f} \delta(\mathbf{\Gamma}(t_n) - \mathbf{\Gamma}_m(t_n)). \tag{7.170}$$

7.7.2 Equilibrium Averages

The canonical average of a phase function, $g(\mathbf{\Gamma})$, is usually written as an integral over phase space weighted by the Maxwell-Boltzmann distribution

$$\langle g(\mathbf{\Gamma})\rangle = \int \mathrm{d}\mathbf{\Gamma}\, \wp_{\mathrm{MB}}(\mathbf{\Gamma}) g(\mathbf{\Gamma}). \tag{7.171}$$

However it can also be written as an unweighted average over one of the trajectories that has been generated by the stochastic, dissipative equations of motion,

$$\langle g(\mathbf{\Gamma})\rangle = \frac{1}{f+1}\sum_{n=0}^{f} g(\mathbf{\Gamma}_m(t_n)). \tag{7.172}$$

This is a time average. This can further be averaged over all M trajectories, in which case it is readily shown that this is formally equivalent to an average over the trajectory probability

$$\begin{aligned}
\langle g(\mathbf{\Gamma})\rangle &= \int \mathrm{d}\mathbf{\Gamma}_0 \ldots \mathrm{d}\mathbf{\Gamma}_f\, \wp(\underline{\mathbf{\Gamma}}) \frac{1}{f+1}\sum_{n=0}^{f} g(\mathbf{\Gamma}_n) \\
&= \frac{1}{f+1}\sum_{n=0}^{f} \int \mathrm{d}\mathbf{\Gamma}_0 \ldots \mathrm{d}\mathbf{\Gamma}_f\, g(\mathbf{\Gamma}_n) \frac{1}{M}\sum_{m=1}^{M} \delta(\underline{\mathbf{\Gamma}} - \underline{\mathbf{\Gamma}}_m) \\
&= \frac{1}{M(f+1)}\sum_{m=1}^{M}\sum_{n=0}^{f} g(\mathbf{\Gamma}_m(t_n)).
\end{aligned} \tag{7.173}$$

7.7.3 Time Correlation Functions

The time correlation function for the canonical equilibrium system can be obtained as an average over the trajectory probability. For the simplest case of two phase functions, $g(\mathbf{\Gamma})$ and $h(\mathbf{\Gamma})$, their correlation function is

$$\begin{aligned}
C_{gh}(t_f) &\equiv \langle g(t_f) h(0)\rangle \\
&= \frac{1}{M}\sum_{m=1}^{M} g(\mathbf{\Gamma}_{mf}) h(\mathbf{\Gamma}_{m0}) \\
&= \frac{1}{M}\sum_{m=1}^{M} \int \mathrm{d}\mathbf{\Gamma}_0\, \wp_{\mathrm{MB}}(\mathbf{\Gamma}_0) g(\mathbf{\Gamma}_{mf}) h(\mathbf{\Gamma}_0),
\end{aligned} \tag{7.174}$$

where $t_0 = 0$. The right-hand side of the first equality is a common but rather loose notation for the time correlation function. In the second equality, $\mathbf{\Gamma}_{mf} \equiv \mathbf{\Gamma}_m(t_f|\mathbf{\Gamma}_{m0}, 0)$ is the end point of one of the M trajectories generated by the stochastic dissipative equations of motion, starting from $\mathbf{\Gamma}_{m0} \equiv \mathbf{\Gamma}_m(t_0)$ that has been chosen independently for each m from the Maxwell-Boltzmann distribution. In the third equality, all these stochastic, dissipative trajectories start from the same $\mathbf{\Gamma}_0$, and then a weighted average over phase space is performed.

The importance of an average over the M independent trajectories must be emphasised.[22] Ordinarily one cannot obtain the time correlation function from a single trajectory because of the stochastic nature of the equations of motion (but see Eq. (7.178) below). There is no such thing as 'the' trajectory. In general it is not a good approximation to evaluate the time correlation function solely on the most likely trajectory, which is the same as the trajectory averaged over the stochastic forces, because this would be equivalent to asserting that the average of a function of the trajectory is equal to the function of the average trajectory.

One has the symmetry

$$C_{gh}(t_f) = C_{hg}(-t_f). \tag{7.175}$$

This follows from the time homogeneity of an equilibrium system,

$$\langle g(t_f)h(0)\rangle = \langle g(0)h(-t_f)\rangle = \langle h(-t_f)g(0)\rangle . \tag{7.176}$$

The time correlation function can also be written as an average over the trajectory probability,

$$C_{gh}(t_f) = \int \mathrm{d}\mathbf{\Gamma}_0 \ldots \mathrm{d}\mathbf{\Gamma}_f \, \wp(\underline{\mathbf{\Gamma}}) g(\mathbf{\Gamma}_f) h(\mathbf{\Gamma}_0). \tag{7.177}$$

This can be confirmed by substitution of the δ-function representation. It is straightforward to extend these expressions for the time correlation function beyond the pair level. Because the equations of motion generate points according to the trajectory probability, the time correlation function can also be obtained as a simple average over the points of a single stochastic dissipative trajectory, $\mathbf{\Gamma}(t)$,

$$C_{gh}(t_n) = \frac{1}{L}\sum_{i=1}^{L} g(\mathbf{\Gamma}(t_i + t_n))\, h(\mathbf{\Gamma}(t_i)), \tag{7.178}$$

where the t_i are a set of nodes, most simply equally spaced. This is just the time average given above, Eq. (7.172).

van Hove Function

The number density has phase space representation

$$\rho(\mathbf{r}, t) = \sum_{i=1}^{N} \delta(\mathbf{r} - \mathbf{r}_i(t)), \tag{7.179}$$

[22]The averages referred to in this paragraph are over the stochastic forces for the subsequent trajectory for fixed initial point $\mathbf{\Gamma}_0$ or $\mathbf{\Gamma}_{m0}$. One cannot obtain the time correlation function from a single trajectory always starting from this initial point. However, if one uses the time homogeneity of an equilibrium system, one can perform an average by moving the time interval over the single trajectory, Eq. (7.178).

where $\mathbf{r}_i$ is the position of molecule i. The integral of this is $\int \mathrm{d}\mathbf{r}\, \rho(\mathbf{r}, t) = N$. Hence for a homogenous system, $\rho \equiv N/V$.

The density-density correlation function, may be obtained from the time correlation formalism above by taking $g \equiv \rho(\mathbf{r}' + \mathbf{r}, t)$ and $f \equiv \rho(\mathbf{r}', 0)$. This is usually written,

$$G(\mathbf{r}, \mathbf{r}', t) \equiv N^{-1} \left\langle \rho(\mathbf{r}' + \mathbf{r}, t)\rho(\mathbf{r}', 0) \right\rangle . \tag{7.180}$$

This is the probability of finding a molecule at $\mathbf{r}'$ at time $t = 0$ and a molecule displaced from $\mathbf{r}'$ by $\mathbf{r}$ at a time t later. The second molecule may be the same as or different to the first.

For a homogenous system only the displacement is relevant, and this may be integrated over the volume to obtain the van Hove function,

$$\begin{aligned} G(\mathbf{r}, t) &= \int \mathrm{d}\mathbf{r}'\, G(\mathbf{r}, \mathbf{r}', t) \\ &= \int \mathrm{d}\mathbf{r}' \frac{1}{N} \left\langle \sum_{i=1}^{N} \sum_{j=1}^{N} \delta(\mathbf{r}' + \mathbf{r} - \mathbf{r}_i(t))\delta(\mathbf{r}' - \mathbf{r}_j(0)) \right\rangle \\ &= \frac{1}{N} \sum_{i=1}^{N} \sum_{j=1}^{N} \left\langle \delta(\mathbf{r} + \mathbf{r}_j(0) - \mathbf{r}_i(t)) \right\rangle . \end{aligned} \tag{7.181}$$

This is equivalent to $G(\mathbf{r}, t) = VG(\mathbf{r}, \mathbf{0}, t) = \rho^{-1} \left\langle \rho(\mathbf{r}, t)\rho(\mathbf{0}, 0) \right\rangle$.

The self part of this gives the correlation or movement of the same molecule,

$$G_{\mathrm{s}}(\mathbf{r}, t) = \frac{1}{N} \sum_{i=1}^{N} \left\langle \delta(\mathbf{r} + \mathbf{r}_i(0) - \mathbf{r}_i(t)) \right\rangle , \tag{7.182}$$

which is essentially the probability that a molecule moves $\mathbf{r}$ in time t. The distinct part gives the correlation of different molecules

$$G_{\mathrm{d}}(\mathbf{r}, t) = \frac{1}{N} \sum_{i=1}^{N} \sum_{j=1}^{N}{}^{(i \neq j)} \left\langle \delta(\mathbf{r} + \mathbf{r}_i(0) - \mathbf{r}_i(t)) \right\rangle . \tag{7.183}$$

At large distances and at large times the molecules are uncorrelated, so that one has $\lim_{\mathbf{r}\to\infty} G_{\mathrm{d}}(\mathbf{r}, t) = \lim_{t\to\infty} G_{\mathrm{d}}(\mathbf{r}, t) = \rho$.

7.7.4 Reversibility

In §7.6.2, the reversibility of the most likely trajectory was discussed (see Fig. 7.5 on p. 223). Unlike an adiabatic trajectory, the most likely stochastic dissipative trajectory is irreversible under the individual operations of velocity conjugation and time reversal. It is however reversible under their combination. A similar situation holds more generally for the stochastic dissipative trajectory.

Suppose one has a specific trajectory $\mathbf{\Gamma}(t)$, on a time interval of length $|t_{\mathrm{i}} - t_{\mathrm{f}}|$, with end points $\mathbf{\Gamma}_{\mathrm{i}}$ and $\mathbf{\Gamma}_{\mathrm{f}}$. To be definite, take $t_{\mathrm{f}} \geq t_{\mathrm{i}}$, in which case $t_{\mathrm{i}} \to t_{\mathrm{f}}$ is a forward transition, and $t_\mathrm{f} \to t_\mathrm{i}$ is a backward transition. In the adiabatic case, the reversibility of the equations of motion gives four equivalent transitions,

$$\begin{aligned} &\mathbf{\Gamma}_\mathrm{i} \to \mathbf{\Gamma}_\mathrm{f},\ t_\mathrm{i} \to t_\mathrm{f}, \qquad &&\mathbf{\Gamma}_\mathrm{f} \to \mathbf{\Gamma}_\mathrm{i},\ t_\mathrm{f} \to t_\mathrm{i}, \\ &\mathbf{\Gamma}_\mathrm{f}^\dagger \to \mathbf{\Gamma}_\mathrm{i}^\dagger,\ t_\mathrm{i} \to t_\mathrm{f}, \qquad &&\mathbf{\Gamma}_\mathrm{i}^\dagger \to \mathbf{\Gamma}_\mathrm{f}^\dagger,\ t_\mathrm{f} \to t_\mathrm{i}. \end{aligned} \tag{7.184}$$

For the case of stochastic dissipative equations of motion, reversibility gives only two equivalent transitions, namely

$$\mathbf{\Gamma}_\mathrm{i} \to \mathbf{\Gamma}_\mathrm{f},\ t_\mathrm{i} \to t_\mathrm{f}, \quad \text{and} \quad \mathbf{\Gamma}_\mathrm{i}^\dagger \to \mathbf{\Gamma}_\mathrm{f}^\dagger,\ t_\mathrm{f} \to t_\mathrm{i}. \tag{7.185}$$

The first is the forward transition, and the second is the backward transition of the conjugate points. These equivalences were enumerated for an infinitesimal time step in Table 7.3 on p. 225.

In terms of notation, on the conjugate trajectory, the subscript of the phase space point is not the same as the subscript of the time point. In view of this now a trajectory is denoted by $\{\underline{\mathbf{\Gamma}}, \underline{t}\} \equiv \{\mathbf{\Gamma}_0, t_0; \mathbf{\Gamma}_1, t_1; \ldots; \mathbf{\Gamma}_f, t_f\}$. The conjugate of a phase point on the trajectory is $\mathbf{\Gamma}_n^\dagger = \{\mathbf{q}^N, (-\mathbf{p})^N\}$, and the conjugate time point is defined as $t_n^\dagger = t_0 + t_f - t_n = t_{f-n}$. With these the conjugate backward trajectory is $\{\underline{\mathbf{\Gamma}}^\dagger, \underline{t}^\dagger\} \equiv \{\mathbf{\Gamma}_0^\dagger, t_f; \mathbf{\Gamma}_1^\dagger, t_{f-1}; \ldots; \mathbf{\Gamma}_f^\dagger, t_0\}$.

In this discrete vector notation, the order of the elements is important, since whether a trajectory is forward or backward is signified by whether the time arguments increase or decrease. Furthermore, the first position argument signifies the initial or known point of the trajectory, the others being determined from this by the stochastic dissipative equations of motion.

In the continuous case, the original trajectory may be written $\mathbf{\Gamma}(t|\mathbf{\Gamma}_0, t_0)$, $t \in [t_0, t_f]$. Often, $t_0 = 0$. The backward conjugate trajectory in the continuous case may be denoted $\mathbf{\Gamma}^\ddagger(t|\mathbf{\Gamma}_0, t_0) \equiv \mathbf{\Gamma}(t|\mathbf{\Gamma}_0^\dagger, t_f) = [\mathbf{\Gamma}(t_0 + t_f - t|\mathbf{\Gamma}_0, t_0)]^\dagger$, $t \in [t_0, t_f]$. For brevity, these two will sometimes be written simply $\mathbf{\Gamma}(t)$ and $\mathbf{\Gamma}^\ddagger(t)$, the initial point $\{\mathbf{\Gamma}_0, t_0\}$ being understood.

Reversibility of the stochastic dissipative equations of motion implies that

$$\wp(\mathbf{\Gamma}_\mathrm{f}|\mathbf{\Gamma}_\mathrm{i}, \Delta_t) = \wp(\mathbf{\Gamma}_\mathrm{f}^\dagger|\mathbf{\Gamma}_\mathrm{i}^\dagger, -\Delta_t). \tag{7.186}$$

This is just Eq. (7.164). For adjacent points on a trajectory,

$$\wp(\mathbf{\Gamma}_n, t_n|\mathbf{\Gamma}_{n-1}, t_{n-1}) = \wp(\mathbf{\Gamma}_n|\mathbf{\Gamma}_{n-1}, \Delta_t) = \wp(\mathbf{\Gamma}_n^\dagger|\mathbf{\Gamma}_{n-1}^\dagger, -\Delta_t). \tag{7.187}$$

This assumes equally spaced nodes and time homogeneity of an equilibrium system.

Using this, the probability for the conjugate backward trajectory to occur in the canonical equilibrium system is

$$\begin{aligned}
\wp(\{\underline{\Gamma}^\dagger,\underline{t}^\dagger\}) &= \wp_{\rm MB}(\Gamma_0^\dagger)\prod_{n=1}^{f}\wp(\Gamma_n^\dagger,t_n^\dagger|\Gamma_{n-1}^\dagger,t_{n-1}^\dagger)\\
&= \wp_{\rm MB}(\Gamma_0^\dagger)\prod_{n=1}^{f}\wp(\Gamma_n^\dagger|\Gamma_{n-1}^\dagger,-\Delta_t)\\
&= \wp_{\rm MB}(\Gamma_0)\prod_{n=1}^{f}\wp(\Gamma_n|\Gamma_{n-1},\Delta_t)\\
&= \wp(\{\underline{\Gamma},\underline{t}\}). \qquad (7.188)
\end{aligned}$$

Here the nodes have been taken to be equally spaced, $\Delta_t = t_n - t_{n-1}$. This says that, for the canonical equilibrium system, one is just as likely to observe a forward trajectory as its backwards twin, where backwards means the conjugate trajectory with the momenta reversed. This is the generalisation of the single time step result given above as Eq. (7.161).

Suppose that the phase functions are of pure parity, $g(\Gamma^\dagger) = \epsilon_g g(\Gamma)$ and $h(\Gamma^\dagger) = \epsilon_h h(\Gamma)$. The parity signature is $\epsilon = 1$ for an even parity function, and $\epsilon = -1$ for an odd parity function. Then the time correlation function has the property

$$\begin{aligned}
C_{gh}(t_f) &= \int \mathrm{d}\Gamma_0 \ldots \mathrm{d}\Gamma_f\, \wp(\{\underline{\Gamma},\underline{t}\}) g(\Gamma_f) h(\Gamma_0)\\
&= \int \mathrm{d}\Gamma_0^\dagger \ldots \mathrm{d}\Gamma_f^\dagger\, \wp(\{\underline{\Gamma}^\dagger,\underline{t}^\dagger\}) \epsilon_g g(\Gamma_f^\dagger)\epsilon_h h(\Gamma_0^\dagger)\\
&= \epsilon_g\epsilon_h C_{gh}(-t_f). \qquad (7.189)
\end{aligned}$$

The second equality uses the facts that $\mathrm{d}\Gamma = \mathrm{d}\Gamma^\dagger$, and $\wp(\{\underline{\Gamma},\underline{t}\}) = \wp(\{\underline{\Gamma}^\dagger,\underline{t}^\dagger\})$, and the final equality uses the fact that the $\Gamma_n^\dagger$ are dummy variables of integration, and the fact that time is reversed on the conjugate backward trajectory. Whereas the symmetry Eq. (7.175) arises from time homogeneity of an equilibrium system, the present symmetry arises from microscopic conjugate reversibility in the canonical equilibrium system.

For the canonical equilibrium system, there is no preferred direction in time. One way of seeing this is to examine the change in entropy on a trajectory, which is formally

$$\Delta S_{\rm r}(\underline{\Gamma}) = S_{\rm r}(\Gamma_f) - S_{\rm r}(\Gamma_0). \qquad (7.190)$$

The change in entropy is defined as $\Delta S_{\rm r} \equiv S_{\rm r}(t_f) - S_{\rm r}(t_0)$, so that by definition, the probability for a change over an interval obeys the symmetry $\wp(\Delta S_{\rm r}|t_f - t_0) = \wp(-\Delta S_{\rm r}|t_0 - t_f)$. But from the properties of the conjugate reverse trajectory one has

$$\begin{aligned}
\wp(\Delta S_{\mathrm{r}}|t_f - t_0) &= \int \mathrm{d}\mathbf{\Gamma}_0 \ldots \mathrm{d}\mathbf{\Gamma}_f \, \wp(\{\underline{\mathbf{\Gamma}}, \underline{t}\}) \delta(\Delta S_{\mathrm{r}} - S_{\mathrm{r}}(\mathbf{\Gamma}_f) + S_{\mathrm{r}}(\mathbf{\Gamma}_0)) \\
&= \int \mathrm{d}\mathbf{\Gamma}_0^{\dagger} \ldots \mathrm{d}\mathbf{\Gamma}_f^{\dagger} \, \wp(\{\underline{\mathbf{\Gamma}}^{\dagger}, \underline{t}^{\dagger}\}) \delta\Big(\Delta S_{\mathrm{r}} - S_{\mathrm{r}}(\mathbf{\Gamma}_f^{\dagger}) + S_{\mathrm{r}}(\mathbf{\Gamma}_0^{\dagger})\Big) \\
&= \wp(\Delta S_{\mathrm{r}}|t_0 - t_f) \\
&= \wp(-\Delta S_{\mathrm{r}}|t_f - t_0). \qquad (7.191)
\end{aligned}$$

Recall that $\mathbf{\Gamma}^{\ddagger}(t|\mathbf{\Gamma}_0, t_0) \equiv \mathbf{\Gamma}(t|\mathbf{\Gamma}_0^{\dagger}, t_f)$. This says that one is just as likely to see the entropy increase over an interval as to decrease. This obviously must be the case for an equilibrium system that undergoes fluctuations.

Chapter 8

Non-Equilibrium Statistical Mechanics

This chapter sets out the theory of non-equilibrium statistical mechanics. It extends the classical phase space formulation of the preceding chapter for the canonical equilibrium case to the case of time dependent systems. The goal is to derive an explicit expression for the reservoir entropy associated with each point in the sub-system phase space, $S_{\mathrm{r}}(\mathbf{\Gamma}, t)$, and hence the probability density, $\wp(\mathbf{\Gamma}, t)$, and to elucidate their properties.

There are two generic non-equilibrium systems: a sub-system with an energy or material flux due to a thermodynamic gradient imposed by spatially separated reservoirs, and mechanical motion due to a time varying external potential. Respective examples of these are the steady heat flow treated at the thermodynamic level in Chs 4 and 6, and the driven Brownian motion treated at the fluctuation level in Ch. 3. The generic molecular-level results of the present chapter will be applied to these two cases in the two following chapters. In this chapter the theory of non-equilibrium statistical mechanics is developed in a general fashion that largely avoids the details of specific systems, although initially the approach is motivated and illustrated by using a time dependent Hamiltonian for a mechanical non-equilibrium system.

8.1 General Considerations

It was shown in the preceding chapter that points in the phase space of an isolated system have equal weight (energy unspecified). This sub-system weight is a property of the sub-system itself, and is unaffected by either adding a time varying external potential to the Hamiltonian, or an external thermodynamic gradient. Such external terms do affect the probability density via their effect on the reservoir entropy, and so formally one may write for the non-equilibrium probability density

$$\wp(\mathbf{\Gamma}, t) = \frac{e^{S_r(\mathbf{\Gamma},t)/k_B}}{h^{3N} N! Z'(t)} = \frac{e^{S_r(\mathbf{\Gamma},t)/k_B}}{Z(t)}. \tag{8.1}$$

Here $\mathbf{\Gamma} \equiv \{\mathbf{\Gamma}_q, \mathbf{\Gamma}_p\} \equiv \{\mathbf{q}^N, \mathbf{p}^N\}$ denotes a point in the phase space of the sub-system. In the second equality, the $N!$ factor that accounts for distinct states, and the h^{3N} factor for the uniform weight of the sub-system phase space, have been incorporated into the partition function Z. The non-trivial task is to obtain an explicit expression for the reservoir entropy for the non-equilibrium system, $S_r(\mathbf{\Gamma}, t)$. A rather general formulation of the reservoir entropy can be obtained before having to particularise to specific examples of non-equilibrium systems in the following chapters.

The single defining characteristic of a non-equilibrium system is that there is a preferred direction of time, namely that of entropy increase. Reversing the direction of time is equivalent to reversing the molecular velocities, the conjugation operation defined earlier, $\mathbf{\Gamma}^\dagger \equiv \{\mathbf{\Gamma}_q, -\mathbf{\Gamma}_p\} \equiv \{\mathbf{q}^N, (-\mathbf{p})^N\}$. From this one can immediately conclude that for the non-equilibrium system the probability density is *not* invariant with respect to conjugation

$$\wp(\mathbf{\Gamma}, t) \neq \wp(\mathbf{\Gamma}^\dagger, t). \tag{8.2}$$

This has far reaching consequences, an important one of which can be established immediately. Consider a sub-system that is acted upon by a time varying external potential and that is in thermal contact with a reservoir of temperature T. The Hamiltonian of the sub-system is

$$\mathcal{H}(\mathbf{\Gamma}, t) = \mathcal{H}^{\text{bare}}(\mathbf{\Gamma}) + U^{\text{ext}}(\mathbf{\Gamma}_q, t). \tag{8.3}$$

The kinetic energy of the bare system is quadratic in the velocities, and the external potential acts on the position coordinates and is independent of the velocities. These mean that the full sub-system Hamiltonian is unchanged upon velocity reversal,

$$\mathcal{H}(\mathbf{\Gamma}, t) = \mathcal{H}(\mathbf{\Gamma}^\dagger, t). \tag{8.4}$$

One can immediately conclude from this and Eq. (8.2) that the non-equilibrium probability distribution cannot have Maxwell-Boltzmann form,

$$\wp(\mathbf{\Gamma}, t) \neq \frac{1}{Z(t)} e^{-\beta \mathcal{H}(\mathbf{\Gamma}, t)}. \tag{8.5}$$

From this example, it is clear that the sub-system dependent part of the reservoir entropy cannot simply be written as the negative of the current sub-system energy divided by the reservoir temperature. Nevertheless the concept of the reservoir entropy is quite well defined. In fact in this example, using $\underline{\mathbf{\Gamma}} \equiv \{\mathbf{\Gamma}(t')\}$, $t \geq t' \geq 0$, to denote the actual trajectory followed by the sub-system up to the present time, the change in the total energy is just the work done on the trajectory,

$$W(\underline{\Gamma}) \equiv \Delta E_{\mathrm{tot}}(\underline{\Gamma}) = \int_0^t \mathrm{d}t' \, \frac{\partial U^{\mathrm{ext}}(\mathbf{\Gamma}_{\mathrm{q}}(t'), t')}{\partial t'}. \tag{8.6}$$

This follows because $\dot{\mathbf{\Gamma}}_{\mathrm{total}} \cdot \nabla_{\mathrm{total}} \mathcal{H}_{\mathrm{total}} = 0$, the total system being isolated and therefore evolving via Hamilton's equations of motion. With this the reservoir entropy is[1]

$$S_{\mathrm{r}}(\underline{\Gamma}) = \frac{E_{\mathrm{r}}(\underline{\Gamma})}{T} = \frac{W(\underline{\Gamma}) - E_{\mathrm{s}}(\underline{\Gamma})}{T}. \tag{8.7}$$

The reservoir entropy therefore depends upon the work done on the trajectory leading up to the current point in phase space. This in turn is sensitive to the molecular velocities: the work done up to the present point is different to the work done on a trajectory leading up to the conjugate point.

This formal expression for the reservoir entropy is a candidate for the non-equilibrium probability, and one is tempted to write

$$\wp(\mathbf{\Gamma}, t) = \frac{1}{Z(t)} e^{S_{\mathrm{r}}(\underline{\Gamma})/k_{\mathrm{B}}}, \text{ (err.)} \tag{8.8}$$

But this is not satisfactory because on the left-hand side appears just a single point in the sub-system phase space, and on the right-hand side appears a specific trajectory leading up to this point. This raises two questions: how to determine the trajectories leading to a given point at a given time, and from these, how to determine the unique trajectory and hence the unique reservoir entropy associated with each point, $S_{\mathrm{r}}(\mathbf{\Gamma}, t)$.

8.2 Reservoir Entropy

8.2.1 Trajectory Entropy

In the canonical equilibrium case, the total energy, $E_{\mathrm{tot}} = \mathcal{H}(\mathbf{\Gamma}) + E_{\mathrm{r}}(\mathbf{\Gamma}, t)$, was constant with respect to both the sub-system phase space point $\mathbf{\Gamma}$ and time t. Hence the reservoir entropy could be written

$$S_{\mathrm{r}}(\mathbf{\Gamma}, t) = S_{\mathrm{r}}(E_{\mathrm{r}}(\mathbf{\Gamma}, t)) = S_{\mathrm{r}}(E_{\mathrm{tot}} - \mathcal{H}(\mathbf{\Gamma})) = S_{\mathrm{r}}(E_{\mathrm{tot}}) - \frac{\mathcal{H}(\mathbf{\Gamma})}{T}, \text{ (equil.)} \tag{8.9}$$

Since the first term on the right-hand side of the final equality is constant with respect to points in phase space it can be neglected (or, more precisely,

[1] Strictly speaking these are the change in the reservoir entropy, the change in the reservoir energy, and the change in the sub-system energy from the initial point on the trajectory. However, here and below it is assumed that the time interval is long enough such that $\mathbf{\Gamma}(0)$ is uncorrelated with $\mathbf{\Gamma}(t)$. This means that any functions of the initial point are constant with respect to the current point and time and can be incorporated into the normalising partition function.

incorporated into the partition function). Exponentiating this, one immediately obtains the Maxwell-Boltzmann distribution.

For a mechanical non-equilibrium system with time dependent external potential of the form given in Eq. (8.3), $\mathcal{H}(\mathbf{\Gamma},t) = \mathcal{H}^{\rm bare}(\mathbf{\Gamma}) + U^{\rm ext}(\mathbf{\Gamma}_{\rm q},t)$, the total energy is no longer fixed, but depends upon the work done, which in turn depends upon the specific trajectory leading up to the present point, $\Delta E_{\rm tot}(\underline{\mathbf{\Gamma}}) \equiv W(\underline{\mathbf{\Gamma}})$, Eq. (8.6). The integrand of that equation is just the adiabatic rate of change of the sub-system energy,

$$\dot{\mathcal{H}}^0(\mathbf{\Gamma},t) \equiv \frac{\partial \mathcal{H}(\mathbf{\Gamma},t)}{\partial t} + \dot{\mathbf{\Gamma}}^0 \cdot \nabla \mathcal{H}(\mathbf{\Gamma},t) = \frac{\partial U^{\rm ext}(\mathbf{\Gamma}_{\rm q},t)}{\partial t}. \tag{8.10}$$

With the sub-system energy of the trajectory equal to the energy of the final point of the trajectory, $E_{\rm s}(\underline{\mathbf{\Gamma}}) = \mathcal{H}(\mathbf{\Gamma},t)$, one can rewrite the change in reservoir entropy on a particular trajectory as

$$\begin{aligned} S_{\rm r}(\underline{\mathbf{\Gamma}}) &= \frac{W(\underline{\mathbf{\Gamma}}) - E_{\rm s}(\underline{\mathbf{\Gamma}})}{T} \\ &= \frac{-\mathcal{H}(\mathbf{\Gamma},t)}{T} + \frac{1}{T}\int_0^t \mathrm{d}t'\, \dot{\mathcal{H}}^0(\mathbf{\Gamma}(t'),t') \\ &\equiv S_{\rm r,st}(\mathbf{\Gamma},t) + S_{\rm r,dyn}(\underline{\mathbf{\Gamma}}). \end{aligned} \tag{8.11}$$

Again these are strictly the change from the initial time, but the initial values, including the initial sub-system energy $\mathcal{H}(\mathbf{\Gamma}(0),0)$, are not shown (see footnote 1 on p. 235). In the final equality the static reservoir entropy has been defined, which could also be called the instantaneous reservoir entropy, or the equilibrium expression for the reservoir entropy,

$$S_{\rm r,st}(\mathbf{\Gamma},t) = \frac{-\mathcal{H}(\mathbf{\Gamma},t)}{T}. \tag{8.12}$$

This is defined as the entropy given by the usual equilibrium formula for the entropy. The specifically non-equilibrium part of the reservoir entropy, which could be called the dynamic part of the reservoir entropy, is defined as

$$S_{\rm r,dyn}(\underline{\mathbf{\Gamma}}) \equiv -\int_0^t \mathrm{d}t'\, \dot{S}^0_{\rm r,st}(\mathbf{\Gamma}(t'),t') = \frac{1}{T}\int_0^t \mathrm{d}t'\, \dot{\mathcal{H}}^0(\mathbf{\Gamma}(t'),t'). \tag{8.13}$$

This represents what remains after invoking the instantaneous reservoir entropy, and it should not be interpreted as having the physical meaning of some non-equilibrium entropy. To be sure, this term vanishes for an equilibrium system, but in fact it represents no more than a correction for a double counting in the expression for the static entropy when it is applied to a non-equilibrium system.

To see this, $S_{\rm r,st}(\mathbf{\Gamma},t)$ is based upon the total change in the sub-system energy from the initial time, and this contains contributions from the adiabatic motion of the sub-system and contributions from the reservoir perturbation. The reservoir entropy changes only when the reservoir energy changes, and, by energy conservation, these are the reservoir induced changes in the sub-system energy. Hence adding the term $S_{\mathrm{r,dyn}}(\underline{\boldsymbol{\Gamma}})$ corresponds to removing the adiabatic change from the total change in sub-system energy leaving only the reservoir induced change.

The notation $\dot{S}^{0}_{\mathrm{r,st}}(\boldsymbol{\Gamma}, t)$, represents the adiabatic rate of change of the static part of the reservoir entropy. It should be clear that the reservoir entropy *itself* is unaffected by the adiabatic evolution of the sub-system. Nevertheless, the static part of the reservoir entropy generally involves sub-system quantities that do evolve adiabatically, and their effects have to be subtracted.

Although the present result has been derived for a mechanical non-equilibrium system, the concepts and indeed notation carry over for thermodynamic non-equilibrium systems. The remarks in the preceding paragraph provide the interpretation and mathematical definition of the change in the dynamic part of the reservoir entropy in such cases. In most of what follows in this chapter, including the following section, the reservoir entropy is to be interpreted in a generic sense.

8.2.2 Reduction to the Point Entropy

The reduction condition on the second entropy was established in Ch. 1 (e.g. Eq. (1.46)), and was used in the presentation of fluctuation theory in Ch. 2 (e.g. Eqs (2.19) and (2.41)). The reduction condition was used to derive the non-equilibrium probability distribution for driven Brownian motion in §§3.6 and 3.6.3. In this section the reduction condition is used to derive the non-equilibrium probability distribution for the general case.

The reduction condition states that the value of the second entropy at the most likely value of one of the points gives the first entropy of the remaining point. Equivalently, the maximum value of the second entropy with respect to one of the points equals the first entropy of the remaining point. The generalisation of this to the third and higher-order entropies is obvious: each maximisation reduces the order of the entropy by one. For the path entropy, maximisation with respect to all points but one, reduces it to the first entropy of the remaining point. For the present problem, since the most likely trajectory maximises the trajectory entropy, the reduction condition yields the entropy of the final point,

$$S_{\mathrm{r}}(\boldsymbol{\Gamma}, t) = S_{\mathrm{r}}(\underline{\overline{\boldsymbol{\Gamma}}}),\ \underline{\overline{\boldsymbol{\Gamma}}} \equiv \left\{\overline{\boldsymbol{\Gamma}}(t'|\boldsymbol{\Gamma}, t)\right\},\ t' \leq t. \tag{8.14}$$

Because one is dealing with the sub-system phase space, the reservoir entropy is the same as the total entropy.

In almost all cases below, the most likely trajectory means the most likely backward trajectory passing through $\boldsymbol{\Gamma}$ at time t, $\underline{\overline{\boldsymbol{\Gamma}}} \equiv \{\overline{\boldsymbol{\Gamma}}(t'|\boldsymbol{\Gamma}, t)\}$, $t' \leq t$. It is necessary to distinguish the forward and backward trajectories due to the irreversibility of the stochastic dissipative equations of motion, as will be discussed below (see also §§7.4.4 and 7.6.2 above).

Putting these together, the reservoir entropy for a point in the sub-system phase space is formally

$$\begin{aligned} S_\mathrm{r}(\mathbf{\Gamma},t) &= S_\mathrm{r,st}(\mathbf{\Gamma},t)+S_\mathrm{r,dyn}(\mathbf{\Gamma},t) \\ &\equiv S_\mathrm{r,st}(\mathbf{\Gamma},t)-\int_0^t \mathrm{d}t'\,\dot{S}^0_\mathrm{r,st}(\overline{\mathbf{\Gamma}}(t'|\mathbf{\Gamma},t),t'). \end{aligned} \tag{8.15}$$

The constant term $-S_\mathrm{r,st}(\overline{\mathbf{\Gamma}}(0|\mathbf{\Gamma},t),0)$ has been neglected (see footnote 1 on p. 235). This result for the reservoir entropy holds for both mechanical non-equilibrium systems and for thermodynamic non-equilibrium systems. The recipe for the most likely trajectory will be derived below.

This result can be rewritten by once more including the neglected constant term, and noting that the difference between the final and the initial value of the static part of the reservoir entropy is just the integral of its total time derivative, so the change in reservoir entropy over the time interval is

$$\begin{aligned} S_\mathrm{r}(\mathbf{\Gamma},t)-S_\mathrm{r}(\overline{\mathbf{\Gamma}}(0),0) &= S_\mathrm{r,st}(\mathbf{\Gamma},t)-S_\mathrm{r,st}(\overline{\mathbf{\Gamma}}(0),0)-\int_0^t \mathrm{d}t'\,\dot{S}^0_\mathrm{r,st}(\overline{\mathbf{\Gamma}}(t'|\mathbf{\Gamma},t),t') \\ &= \int_0^t \mathrm{d}t'\Big[\dot{S}^-_\mathrm{r,st}(\overline{\mathbf{\Gamma}}(t'|\mathbf{\Gamma},t),t')-\dot{S}^0_\mathrm{r,st}(\overline{\mathbf{\Gamma}}(t'|\mathbf{\Gamma},t),t')\Big]. \end{aligned} \tag{8.16}$$

Here $\dot{S}^-_\mathrm{r,st}$ is the total time derivative of the static part of the reservoir entropy; due to irreversibility, it is necessary to signify it as the backward derivative. In this form one can see explicitly that subtracting the adiabatic contribution isolates the change in entropy due solely to the reservoir.

The integrand of the dynamic part of the reservoir entropy, Eq. (8.15), the adiabatic rate of change of the static part of the reservoir entropy, has asymptote,[2]

$$\dot{S}^0_\mathrm{r,st}(\overline{\mathbf{\Gamma}}(t'|\mathbf{\Gamma},t),t)\rightarrow\overline{\dot{S}^0_\mathrm{r,st}}(t'),\ |t'-t|\rightarrow\infty. \tag{8.17}$$

The concept behind this result is generally applicable: in the future or in the past, with overwhelming probability, the system lies closer to the most likely value of any phase function than it does at the present time,

$$\frac{\left|f(\overline{\mathbf{\Gamma}}(t'|\mathbf{\Gamma},t),t)-\overline{f}(t')\right|}{\left|f(\mathbf{\Gamma},t)-\overline{f}(t)\right|}\rightarrow 0,\ |t'-t|\rightarrow\infty. \tag{8.18}$$

This asymptote justifies the phrase 'reservoir entropy' rather than 'change in reservoir entropy', since they manifest the fact that the integrand of the dynamic part of the reservoir entropy tends to a value independent of the current point in phase space at the lower limit of the integral.

[2]Here and below $\overline{f}(t)\equiv f(\overline{\mathbf{\Gamma}}(t),t)$ is used to denote the most likely value of a given phase function at time t.

8.2.3 Fluctuation Form for the Reservoir Entropy

A complementary expression for the reservoir entropy associated with each point in the sub-system phase space may be obtained from fluctuation theory. As discussed in §3.6.3 on driven Brownian motion, the entropy of fluctuations about the most likely non-equilibrium state is the same as the entropy of fluctuations in the equilibrium system. Accordingly, let $\overline{\mathbf{\Gamma}}(t) = \langle \mathbf{\Gamma}(t) \rangle$ be the most likely configuration of the sub-system at time t. This depends upon the initial configuration, and will be taken to be unique. Let $\gamma \equiv \mathbf{\Gamma} - \overline{\mathbf{\Gamma}}(t)$ be the fluctuation or the departure from the most likely point.

The static part of the reservoir entropy for the system of interest will be assumed known. For example, for the canonical mechanical non-equilibrium system it is

$$S_{\mathrm{r,st}}(\mathbf{\Gamma}, t) = -\mathcal{H}(\mathbf{\Gamma}, t)/T, \tag{8.19}$$

where $\mathcal{H}(\mathbf{\Gamma}, t)$ is the Hamiltonian or total energy of the sub-system, and T is the temperature of the reservoir. For a steady heat flow, which is the canonical thermodynamic non-equilibrium system, it is

$$S_{\mathrm{r,st}}(\mathbf{\Gamma}, t) = \frac{-E_0(\mathbf{\Gamma}, t)}{T_0} - \frac{E_1(\mathbf{\Gamma}, t)}{T_1}, \tag{8.20}$$

where $E_n(\mathbf{\Gamma}, t)$ is the n-th moment of the sub-system energy, and T_n is the n-th temperature of the reservoirs (see Ch. 4).

The static part of the reservoir entropy may be characterised by its properties at the most likely configuration. Define

$$\begin{aligned} \overline{S}(t) &\equiv S_{\mathrm{r,st}}(\overline{\mathbf{\Gamma}}(t), t), \\ \overline{S'}(t) &\equiv \nabla S_{\mathrm{r,st}}(\overline{\mathbf{\Gamma}}(t), t), \text{ and} \\ \overline{S''}(t) &\equiv \nabla\nabla S_{\mathrm{r,st}}(\overline{\mathbf{\Gamma}}(t), t). \end{aligned} \tag{8.21}$$

For brevity, the time dependence of these will usually not be exhibited. The fact that these depend upon $\overline{\mathbf{\Gamma}}(t)$ makes the theory non-linear (see the discussion in §3.6.3 on p. 78). In terms of these, the value of the static part of the reservoir entropy for the current configuration is[3]

$$S_{\mathrm{r,st}}(\mathbf{\Gamma}, t) = \overline{S}(t) + \overline{S'}(t) \cdot \gamma + \frac{1}{2}\overline{S''}(t) : \gamma^2 + \mathcal{O}(\gamma^3). \tag{8.22}$$

Also, because $\mathrm{d}\mathbf{\Gamma} = \mathrm{d}\gamma$, its gradient is obviously

$$\nabla S_{\mathrm{r,st}}(\mathbf{\Gamma}, t) = \overline{S'}(t) + \overline{S''}(t)\gamma + \mathcal{O}(\gamma^2). \tag{8.23}$$

The non-trivial axiom is that the fluctuation matrix is the same in the equilibrium system as in the non-equilibrium system. This axiom was used to derive the results for driven Brownian motion in §3.6.3. This means that the

[3] Recall the convention used throughout that the square of a vector, a^2, denotes either the dyadic product aa, or else the scalar product $a \cdot a$, depending on the context, and similarly for the product of vectors ab.

non-equilibrium entropy can be expressed in terms of the fluctuations,

$$S_{\rm r}(\mathbf{\Gamma},t) = \overline{S}_{\rm r}(t) + \frac{1}{2}\overline{S''}(t) : \gamma^2 + \mathcal{O}(\gamma^3). \tag{8.24}$$

Here $\overline{S}_{\rm r}(t)$ is a time dependent constant in phase space that will be shown in Eq. (8.127) below to be equal to the reservoir entropy most likely produced to date. This constant term could be neglected (i.e. incorporated into the partition function) as far as the probability distribution is concerned, but it is better to retain it explicitly as it has a quite definite physical interpretation, and it does affect the various time derivatives and their interpretation. The simplicity of this form for the reservoir entropy is a little misleading since it cannot be evaluated without knowing the most likely configuration $\overline{\mathbf{\Gamma}}(t)$.

Expanding the formal expression obtained in the preceding section in powers of the fluctuation, one has

$$\begin{aligned} S_{\rm r}(\mathbf{\Gamma},t) &= S_{\rm r,st}(\mathbf{\Gamma},t) + S_{\rm r,dyn}(\mathbf{\Gamma},t) \\ &= \overline{S}(t) + \overline{S'}(t)\cdot\gamma + \frac{1}{2}\overline{S''}(t) : \gamma^2 + \mathcal{O}(\gamma^3) + S_{\rm r,dyn}(\mathbf{\Gamma},t). \end{aligned} \tag{8.25}$$

Equating this with the fluctuation expression, one concludes that the dynamic part of the reservoir entropy has the expansion

$$S_{\rm r,dyn}(\mathbf{\Gamma},t) = \overline{S}_{\rm r}(t) - \overline{S}(t) - \overline{S'}(t)\cdot\gamma + \mathcal{O}(\gamma^3). \tag{8.26}$$

This means that $\nabla S_{\rm r,dyn}(\mathbf{\Gamma},t) = -\nabla S_{\rm r,st}(\overline{\mathbf{\Gamma}}(t),t) + \mathcal{O}(\gamma^2)$, which is a time dependent constant in phase space. Also $\overline{S}_{\rm r}(t) - \overline{S}(t) = \overline{S}_{\rm r,dyn}(t)$, which makes sense.

8.3 Transitions and Motion in Phase Space

8.3.1 Foundations for Time Dependent Weight

In the Prologue, §§1.4.2–1.4.4, the laws of probability and the relationship between probability, weight, and entropy were established in a generic fashion based on set theory. That analysis invoked weights that were independent of time, and so was most appropriate for an equilibrium system (although the material on transitions is a valid approach to treating steady state non-equilibrium systems). With some relatively minor modifications, that generic analysis can be given for time dependent weights, which is most important for transitions and the second entropy in the non-equilibrium case that is treated in this chapter.

In the time dependent case, denote the microstate weight by $\omega(i,t)$, the macrostate weight by $\omega(\alpha,t) = \sum_{i\in\alpha}\omega(i,t)$, and the total weight by $W(t) = \sum_i \omega(i,t) = \sum_\alpha \omega(\alpha,t)$. The entropies are as usual $S(i,t) = k_{\rm B}\ln\omega(i,t)$, $S(\alpha,t) = k_{\rm B}\ln\omega(\alpha,t)$, and $S(t) = k_{\rm B}\ln W(t)$. Again as usual the probabilities are $\wp(i,t) = \omega(i,t)/W(t) = \exp[S(i,t)/k_{\rm B}]/W(t)$ and $\wp(\alpha,t) = \omega(\alpha,t)/W(t) = \exp[S(\alpha,t)/k_{\rm B}]/W(t)$. So far these are formally identical to the results for the time independent case.

Now consider the macrostate transition $\{\alpha, t\} \to \{\beta, t'\}$, and let its weight be $\omega(\beta, t'; \alpha, t)$. (The following results also hold for microstates, which case will not be dealt with separately.) This is a physical transition and so $t' > t$. This time order carries all the information about the actual transition, since from statistical symmetry the order of the arguments is unimportant, $\omega(\beta, t'; \alpha, t) = \omega(\alpha, t; \beta, t')$. This is the weight attached to the system being in the macrostate α at time t *and* in the macrostate β at time t'.[4]

In the time independent case, two related reduction conditions were established, namely that for the weight of the system being simultaneously in the two macrostates, Eq. (1.12), $\sum_\beta \omega(\alpha, \beta) = \omega(\alpha)$, and that for the transition between two macrostates, Eq. (1.30), $\sum_\beta \omega(\alpha, \beta|\tau) = \omega(\alpha)$, and $\sum_\alpha \omega(\alpha, \beta|\tau) = \omega(\beta)$. The reduction condition for the transition can be interpreted as a conservation law for weight. Since these rules stem from the disjoint complete nature of the macrostates, a similar rule must hold for the time dependent case, with a modification due to the fact that the total weight is not conserved but now changes with time, $W(t)$. In this case the reduction law incorporates a scaling based upon the time dependence of the total weight,[5]

$$\sum_\beta \omega(\beta, t'; \alpha, t) = \sqrt{\frac{W(t')}{W(t)}} \omega(\alpha, t). \tag{8.27}$$

With this the total weight of the transition is

$$\sum_{\alpha\beta} \omega(\beta, t'; \alpha, t) = \sqrt{W(t')W(t)}. \tag{8.28}$$

One sees that the total weight of the transition is the geometric mean of the total weights at the two terminal times. Compared to the time independent case of §1.4.4, there is now a modified conservation law for weight, namely that the weight of the initial state α is now scaled and distributed amongst all the target states β. This scaling has to occur at each transition if the weights of the states and the total weight change with time. This reduction condition given above for the sum over the target macrostates also holds for the sum over initial states,

$$\sum_\alpha \omega(\beta, t'; \alpha, t) = \sqrt{\frac{W(t)}{W(t')}} \omega(\beta, t'). \tag{8.29}$$

---

[4]The unconditional weight $\omega(\beta, t'; \alpha, t)$ or probability $\wp(\beta, t'; \alpha, t)$ always refers to the physical transition: $\alpha \to \beta$ if $t' > t$, or $\beta \to \alpha$ if $t > t'$. The conditional probability $\wp(\beta, t'|\alpha, t)$ refers to the mathematical transition $\alpha \mapsto \beta$; this is a forward transition if $t' > t$ and it is a backward transition if $t > t'$. See also the discussion in §8.4 on p. 262.

[5]The most general case is to write $\sum_\beta \omega(\beta, t'; \alpha, t) = f(t', t)\omega(\alpha, t)$ and consequently $\sum_{\alpha,\beta} \omega(\beta, t'; \alpha, t) = f(t', t)W(t)$. Demanding that this second result be symmetric in the two times leads to $f(t', t) = W(t')^\mu W(t)^{\mu-1}$. Finally, demanding in the time independent case that the right-hand side of the second result be W fixes $\mu = 1/2$, which gives $f(t', t) = \sqrt{W(t')/W(t)}$, as used in the text.

This may be interpreted as saying that all the weight in β must come from somewhere, after scaling. It is emphasised that these two reduction laws are required for the consistency of the laws of probability and lead directly to Bayes' theorem.

Using the total transition weight as the normalising factor, the unconditional transition probability is

$$\wp(\beta,t';\alpha,t) = \frac{\omega(\beta,t';\alpha,t)}{\sqrt{W(t')W(t)}}. \tag{8.30}$$

With this the conditional transition probability is

$$\begin{aligned} \wp(\beta,t'|\alpha,t) &= \frac{\wp(\beta,t';\alpha,t)}{\wp(\alpha,t)} \\ &= \frac{\omega(\beta,t';\alpha,t)}{\omega(\alpha,t)}\sqrt{\frac{W(t)}{W(t')}} \\ &= \frac{\omega(\beta,t';\alpha,t)}{\sum_\beta \omega(\beta,t';\alpha,t)}. \end{aligned} \tag{8.31}$$

The second entropy is defined as the logarithm of the transition weight, $S^{(2)}(\beta,t';\alpha,t) = k_\mathrm{B} \ln \omega(\beta,t';\alpha,t)$, and with it the transition probability is

$$\wp(\beta,t';\alpha,t) = \frac{e^{S^{(2)}(\beta,t';\alpha,t)/k_\mathrm{B}}}{\sqrt{W(t')W(t)}}. \tag{8.32}$$

Accordingly the conditional transition probability is

$$\wp(\beta,t'|\alpha,t) = \sqrt{\frac{W(t)}{W(t')}}e^{\left[S^{(2)}(\beta,t';\alpha,t)-S(\alpha,t)\right]/k_\mathrm{B}} \equiv e^{S^{(2)}(\beta,t'|\alpha,t)/k_\mathrm{B}}. \tag{8.33}$$

The definition of the conditional second entropy,

$$S^{(2)}(\beta,t'|\alpha,t) \equiv S^{(2)}(\beta,t';\alpha,t) - S(\alpha,t) + [S(t) - S(t')]/2, \tag{8.34}$$

is rather arbitrary, and the final constant term could just as easily be retained explicitly as a normalisation factor for the conditional probability. Even though these results refer to the physical transition $\alpha \to \beta$, $t' > t$, they remain valid for the so-called reverse transition, namely the conditional probability that prior to the current state β, the system was in the state α,

$$\begin{aligned} \wp(\alpha,t|\beta,t') &= \frac{\wp(\beta,t';\alpha,t)}{\wp(\beta,t')} \\ &= \sqrt{\frac{W(t')}{W(t)}}e^{\left[S^{(2)}(\beta,t';\alpha,t)-S(\beta,t')\right]/k_\mathrm{B}} \\ &= e^{S^{(2)}(\alpha,t|\beta,t')/k_\mathrm{B}}. \end{aligned} \tag{8.35}$$

Note that this is the conditional probability that the system *was* in the state α at time t, given that it is currently in the state β at time $t' > t$. This is often still called a conditional transition probability, but it should be kept in mind that in this case the transition is conditioned on the future state, not the past state.

The important reduction condition is modified slightly from the time independent case of §1.4.4. The most likely target state $\overline{\beta} \equiv \overline{\beta}(t'|\alpha, t)$ is the state that maximises the second entropy,

$$\left. \frac{\partial S^{(2)}(\beta, t'; \alpha, t)}{\partial \beta} \right|_{\beta = \overline{\beta}} = 0. \tag{8.36}$$

The entropy of the initial state α is related to the maximal value of the second entropy via the reduction condition

$$\begin{aligned} S(\alpha, t) &= k_{\mathrm{B}} \ln \omega(\alpha, t) \\ &= k_{\mathrm{B}} \ln \left[\sqrt{\frac{W(t)}{W(t')}} \sum_{\beta} \omega(\beta, t'; \alpha, t) \right] \\ &\approx k_{\mathrm{B}} \ln \left[\sqrt{\frac{W(t)}{W(t')}} \omega(\overline{\beta}, t'; \alpha, t) \right] \\ &= S^{(2)}(\overline{\beta}, t'; \alpha, t) + [S(t) - S(t')]/2. \end{aligned} \tag{8.37}$$

The third approximation follows because in physical systems the probability distributions are sharply peaked, and hence the logarithm of a sum over states is approximately equal to the logarithm of the largest term in the sum. This means that this reduction condition for the most likely state is valid when fluctuations are relatively negligible. This is more conveniently written

$$S^{(2)}(\overline{\beta}, t'; \alpha, t) = S(\alpha, t) + [S(t') - S(t)]/2. \tag{8.38}$$

A similar reduction condition holds for the most likely prior state,

$$S^{(2)}(\beta, t'; \overline{\alpha}, t) = S(\beta, t') + [S(t) - S(t')]/2, \tag{8.39}$$

where $\overline{\alpha} \equiv \overline{\alpha}(t|\beta, t')$. Recall that due to statistical symmetry the order of the arguments in the second entropy is irrelevant. Compared to the time independent case of §1.4.4, where the maximal value of the second entropy reduces simply to the first or ordinary entropy, Eq. (1.46), one sees that for the present time varying non-equilibrium system, the maximal value of the second entropy reduces to the first entropy plus half the difference in the total entropy of the two states.

8.3.2 Fluctuation Form of the Second Entropy

The most general quadratic form for the second entropy for the fluctuations is

$$\begin{aligned} S^{(2)}(\mathbf{\Gamma}_2, t_2; \mathbf{\Gamma}_1, t_1) &= \frac{1}{2} A : \gamma_2\gamma_2 + B : \gamma_2\gamma_1 + \frac{1}{2} C : \gamma_1\gamma_1 \\ &\quad + \frac{1}{2}\left[\overline{S}_{\mathrm{r}}(t_2) + \overline{S}_{\mathrm{r}}(t_1)\right]. \end{aligned} \tag{8.40}$$

Here the coefficients are a function of the two times, and will be written $A(t_{21}, t)$, $B(t_{21}, t)$, and $C(t_{21}, t)$, where $t \equiv (t_2 + t_1)/2$ and $t_{21} \equiv t_2 - t_1 = -t_{12}$. Usually t will not be shown explicitly. The final time dependent constant term arises from Eq. (8.28), where it was shown that for time dependent weights, the transition weight is normalised to the geometric mean of the two terminal states.[6] Because the coefficients are second derivatives, A and C are symmetric matrices. Because $S^{(2)}(\mathbf{\Gamma}_2, t_2; \mathbf{\Gamma}_1, t_1) = S^{(2)}(\mathbf{\Gamma}_1, t_1; \mathbf{\Gamma}_2, t_2)$,[7] one must have

$$A(t_{21}) = C(t_{12}) \text{ and } B(t_{21}) = B(t_{12})^{\mathrm{T}}. \tag{8.41}$$

This may be termed the statistical symmetry requirement. It also holds for equilibrium fluctuations, where it is equivalent to the requirement of time homogeneity, Eq. (2.64). With this result almost all of the analysis for mixed parity equilibrium fluctuations given in §2.5 holds, the exception being those results that depend upon macroscopic reversibility, which cannot be invoked in the present non-equilibrium case.

Specifically, the small time expansions given in §2.5.2, Eqs (2.83) and (2.84), rely only upon statistical symmetry and therefore they also hold in the present case,

$$A(\tau, t) = \frac{-1}{|\tau|}\Lambda(t)^{-1} + A_0(t) + \hat{\tau}A_0'(t) + \mathcal{O}(\tau) \tag{8.42}$$

and

$$B(\tau, t) = \frac{1}{|\tau|}\Lambda(t)^{-1} + B_0(t) + \hat{\tau}B_0'(t) + \mathcal{O}(\tau), \tag{8.43}$$

with $\Lambda(t)$, $A_0(t)$, $A_0'(t)$, and $B_0(t)$ being symmetric, and $B_0'(t)$ being antisymmetric.

---

[6] Here and throughout, the entropy at time t is taken to equal the most likely value, $\overline{S}_{\mathrm{r}}(t)$. This neglects the contribution to the entropy from the fluctuations about the most likely value, which is the integral over the quadratic terms.

[7] In the general theory of probability, the joint probability is symmetric in its arguments: $\wp(a, b) = \wp(b, a)$. In non-equilibrium statistical mechanics, the unconditional transition probability preserves this symmetry property, $\wp(\mathbf{\Gamma}_2, t_2; \mathbf{\Gamma}_1, t_1) = \wp(\mathbf{\Gamma}_1, t_1; \mathbf{\Gamma}_2, t_2)$. The probability of the system being in the state $\mathbf{\Gamma}_1$ at t_1 *and* in the state $\mathbf{\Gamma}_2$ at t_2 corresponds unambiguously to the unconditional probability of the transition from the earlier state to the later state. This physical transition forward in time nevertheless leads to two possible conditional transitions: where the system is likely to go *to*, and where the system has likely come *from*. See the distinction between the physical transition $\rightarrow$ and the mathematical transition $\mapsto$ discussed in §8.4 on p. 262 and in footnotes 4 and 14 on pp 241 and 266.

Maximising the second entropy with respect to γ_2, one obtains the conditional most likely state as

$$\begin{aligned}\overline{\gamma}_2 &= -A(t_{21})^{-1}B(t_{21})\gamma_1 \\ &= \gamma_1 + t_{21}\Lambda[A_0' + B_0']\gamma_1 + |t_{21}|\Lambda[A_0 + B_0]\gamma_1 + \mathcal{O}(t_{21}^2). \end{aligned} \tag{8.44}$$

In view of the generic result, Eq. (8.38), the reduction condition is

$$\begin{aligned} S^{(2)}(\overline{\boldsymbol{\Gamma}}_2, t_2; \boldsymbol{\Gamma}_1, t_1) &= S(\boldsymbol{\Gamma}_1, t_1) + \frac{1}{2}\left[\overline{S}_{\rm r}(t_2) - \overline{S}_{\rm r}(t_1)\right] \\ &= \frac{1}{2}\overline{S''}(t_1) : \gamma_1\gamma_1 + \frac{1}{2}\left[\overline{S}_{\rm r}(t_1) + \overline{S}_{\rm r}(t_2)\right]. \end{aligned} \tag{8.45}$$

It is clear that the final constant term here is equal to that in Eq. (8.40). From the coefficient of the quadratic term one deduces that

$$C(t_{21}, t) - B(t_{21}, t)^{\rm T} A(t_{21}, t)^{-1} B(t_{21}, t) = \overline{S''}(t_1). \tag{8.46}$$

These may be compared to the most likely state conditioned on γ_2,

$$\begin{aligned}\overline{\gamma}_1 &= -C(t_{21})^{-1}B(t_{21})^{\rm T}\gamma_2 \\ &= \gamma_2 - t_{21}\Lambda[A_0' + B_0']\gamma_2 + |t_{21}|\Lambda[A_0 + B_0]\gamma_2 + \mathcal{O}(t_{21}^2), \end{aligned} \tag{8.47}$$

where the symmetry requirements, Eq. (8.41), have been used. The reduction condition, $S^{(2)}(\boldsymbol{\Gamma}_2, t_2; \overline{\boldsymbol{\Gamma}}_1, t_1) = S(\boldsymbol{\Gamma}_2, t_2) + \left[\overline{S}(t_1) - \overline{S}(t_2)\right]/2$, in this case yields

$$A(t_{21}, t) - B(t_{21}, t) C(t_{21}, t)^{-1} B(t_{21}, t)^{\rm T} = \overline{S''}(t_2). \tag{8.48}$$

It should be noted that this equation is entirely equivalent to the previous reduction condition, Eq. (8.46), and it can be obtained directly by interchanging t_2 and t_1 and using the symmetry rules, Eq. (8.41).

From the reduction condition an important relationship between some of the second entropy fluctuation coefficients and the first entropy fluctuation coefficient can be obtained. Expanding the left-hand side of Eq. (8.46) yields

$$\begin{aligned} \text{LHS} &= \frac{-1}{|t_{21}|}\Lambda^{-1} + A_0 - \hat{t}A_0' + \left[\frac{1}{|t_{21}|}\Lambda^{-1} + B_0 - \hat{t}B_0'\right] \\ &\quad \times \{\mathrm{I} + t_{21}\Lambda[A_0' + B_0'] + |t_{21}|\Lambda[A_0 + B_0]\} + \mathcal{O}(t_{21}) \\ &= A_0 - \hat{t}A_0' + \hat{t}[A_0' + B_0'] + [A_0 + B_0] + B_0 - \hat{t}B_0' + \mathcal{O}(t_{21}) \\ &= 2[A_0(t) + B_0(t)] + \mathcal{O}(t_{21}). \end{aligned} \tag{8.49}$$

The right-hand side is

$$\overline{S''}(t_1) = \overline{S''}(t) - \frac{t_{21}}{2}\frac{\mathrm{d}\overline{S''}(t)}{\mathrm{d}t}. \tag{8.50}$$

Hence

$$A_0(t) + B_0(t) = \frac{1}{2}\overline{S''}(t). \tag{8.51}$$

With this the reservoir part of the most likely change in state in Eq. (8.44) that will be used frequently below is

$$\begin{aligned}\overline{\mathbf{R}}_\gamma(\mathbf{\Gamma},t,t_{21}) &\equiv |t_{21}|\Lambda(t)\,[A_0(t)+B_0(t)]\,\gamma \\ &= \frac{|t_{21}|}{2}\Lambda(t)\overline{S''}(t)\gamma \\ &= \frac{|t_{21}|}{2}\Lambda(t)\nabla S_{\rm r}(\mathbf{\Gamma},t). \end{aligned} \tag{8.52}$$

It is convenient to rewrite the second entropy into several equivalent forms using the small time expansions for the coefficients. For this the preceding result will be used, as well as a result to be derived below in Eqs (8.63), (8.73), and (8.84), namely

$$[A_0'+B_0']\gamma = -\Lambda^{-1}\Phi\overline{S''}\gamma \Rightarrow -\Lambda^{-1}\Phi^0\overline{S''}\gamma = \Lambda^{-1}\dot{\gamma}^0. \tag{8.53}$$

All of these coefficients are possibly functions of time, $\overline{S''}(t)=\overline{S''}(\overline{\mathbf{\Gamma}}(t),t)$, etc., and $\dot{\gamma}^0$ is the adiabatic velocity of the fluctuation. With these, the second entropy, Eq. (8.40), may be rewritten as in Eq. (2.92),

$$\begin{aligned} & S^{(2)}(\mathbf{\Gamma}_2,t_2;\mathbf{\Gamma}_1,t_1) \\ &= \frac{-1}{2|t_{21}|}\Lambda^{-1}:[\gamma_2-\gamma_1]^2+[\gamma_2-\gamma_1]\cdot\left[\frac{1}{2}\overline{S''}-\hat{t}_{21}\Lambda^{-1}\Phi^0\overline{S''}\right]\gamma_1 \\ &\quad +\frac{1}{2}\overline{S''}:\gamma_1\gamma_1+\frac{1}{2}\left[\overline{S}_{\rm r}(t_1)+\overline{S}_{\rm r}(t_2)\right]-\frac{1}{2|t_{21}|}\Lambda^{-1}:[\overline{\gamma_2}-\gamma_1]^2 \\ &= \frac{-\Lambda^{-1}}{2|t_{21}|}:[\gamma_2-\gamma_1]^2+\frac{\gamma_2-\gamma_1}{2}\cdot\nabla S_{\rm r}(\mathbf{\Gamma}_1,t_1)+\hat{t}_{21}\,[\gamma_2-\gamma_1]\cdot\Lambda^{-1}\dot{\gamma}_1^0 \\ &\quad +S_{\rm r}(\mathbf{\Gamma}_1,t_1)+\frac{1}{2}\left[\overline{S}_{\rm r}(t_2)-\overline{S}_{\rm r}(t_1)\right]-\frac{1}{2|t_{21}|}\Lambda^{-1}:[\overline{\gamma_2}-\gamma_1]^2 \\ &= \frac{-\Lambda^{-1}}{2|t_{21}|}:[\gamma_2-\overline{\gamma_2}]^2+S_{\rm r}(\mathbf{\Gamma}_1,t_1)+\frac{1}{2}\left[\overline{S}_{\rm r}(t_2)-\overline{S}_{\rm r}(t_1)\right]+\mathcal{O}(\gamma^2 t_{21}). \end{aligned} \tag{8.54}$$

Because terms linear in t_{21} have been neglected in the expansions of the coefficients, this expression for the second entropy neglects terms $\mathcal{O}(\gamma^2 t_{21})$. As mentioned in footnote 6 on p. 244, it also neglects terms that arise from the normalisation of the Gaussian fluctuations; in view of Eq. (8.28), strictly speaking one should subtract from this expression for the second entropy a term $[k_{\rm B}/2]\ln\left(|2\pi k_{\rm B}|t_{21}|\Lambda(t)||2\pi k_{\rm B}\overline{S''}(t)^{-1}|\right)$, but this is negligible in comparison with $\overline{S}_{\rm r}(t)$. This is written in the form of a transition from $\mathbf{\Gamma}_1$, so that it can be maximised with respect to $\mathbf{\Gamma}_2$, (but see next). In Eq. (8.84) below it will be shown that the only part of the transport matrix that needs to be retained is the block $\Lambda_{\rm pp}$.

The original form of the second entropy is symmetric in the two termini of the transition. It can be written in the form of the reverse transition by interchanging the subscripts 1 and 2. Even though the above form is written from the point of view as the transition from $\mathbf{\Gamma}_1$, it is equally valid as representing the transition from $\mathbf{\Gamma}_2$ to the exhibited order in the time interval. One way to see this is to maximise it with respect to $\mathbf{\Gamma}_1$. The derivative of the second equality in Eq. (8.54) is

$$\begin{aligned}\frac{\partial S^{(2)}(\mathbf{\Gamma}_2,t_2;\mathbf{\Gamma}_1,t_1)}{\partial \mathbf{\Gamma}_1} &= \frac{1}{|t_{21}|}\Lambda^{-1}:[\gamma_2-\gamma_1]-\frac{1}{2}\nabla S_{\rm r}(\mathbf{\Gamma}_1,t_1)\\ &\quad -\hat{t}_{21}\Lambda^{-1}\dot{\gamma}_1^0+\nabla S_{\rm r}(\mathbf{\Gamma}_1,t)+\mathcal{O}(t_{21}), \qquad (8.55)\end{aligned}$$

where the derivatives of $\nabla S_{\rm r}(\mathbf{\Gamma}_1,t_1)$, $\dot{\gamma}^0(\mathbf{\Gamma}_1,t_1)$, and $\overline{\gamma}_2(t_{21}|\mathbf{\Gamma}_1)$ have been neglected because they occur in terms that are $\mathcal{O}(t_{21})$. Setting this to zero gives an expression for $\overline{\gamma_1}$ that is the same Eq. (8.47).

The symmetry can also be seen directly since one has

$$\begin{aligned}
& S^{(2)}(\mathbf{\Gamma}_2,t_2;\mathbf{\Gamma}_1,t_1)\\
&= \frac{-1}{2|t_{21}|}\Lambda^{-1}:[\gamma_2-\overline{\gamma_2}]^2+S_{\rm r}(\mathbf{\Gamma}_1,t_1)+\frac{1}{2}\left[\overline{S}_{\rm r}(t_2)-\overline{S}_{\rm r}(t_1)\right]+\mathcal{O}(\gamma^2 t_{21})\\
&= \frac{-1}{2|t_{21}|}\Lambda^{-1}:\left[\overline{\gamma}_1-\gamma_1-|t_{21}|\Lambda\overline{S''}\{\gamma_1+\gamma_2\}/2\right]^2\\
&\quad +\frac{1}{2}\overline{S''}:\gamma_1^2+\frac{1}{2}\left[\overline{S}_{\rm r}(t_2)+\overline{S}_{\rm r}(t_1)\right]+\mathcal{O}(\gamma^2 t_{21})\\
&= \frac{-1}{2|t_{21}|}\Lambda^{-1}:[\overline{\gamma}_1-\gamma_1]^2+\frac{1}{2}[\overline{\gamma}_1-\gamma_1]\cdot\overline{S''}\{\gamma_1+\gamma_2\}\\
&\quad +\frac{1}{2}\overline{S''}:\gamma_1^2+\frac{1}{2}\left[\overline{S}_{\rm r}(t_2)+\overline{S}_{\rm r}(t_1)\right]+\mathcal{O}(\gamma^2 t_{21})\\
&= \frac{-1}{2|t_{21}|}\Lambda^{-1}:[\overline{\gamma}_1-\gamma_1]^2+S_{\rm r}(\mathbf{\Gamma}_2,t_2)+\frac{1}{2}\left[\overline{S}_{\rm r}(t_1)-\overline{S}_{\rm r}(t_2)\right]+\mathcal{O}(\gamma^2 t_{21})\\
&= S^{(2)}(\mathbf{\Gamma}_1,t_1;\mathbf{\Gamma}_2,t_2). \qquad (8.56)
\end{aligned}$$

The second equality neglects the difference in the adiabatic evolution, $t_{21}\{\dot{\gamma}_2^0-\dot{\gamma}_1^0\}$ as this contributes terms $\mathcal{O}(\gamma^2 t_{21})$. The penultimate equality uses the fact that $\overline{\gamma}_1-\gamma_1=\gamma_2-\gamma_1+\mathcal{O}(\gamma^2 t_{21})$.

8.3.3 Time Correlation Function

The time correlation function for the fluctuations is

$$\begin{aligned}Q(t_2,t_1) &\equiv k_{\rm B}^{-1}\langle\gamma_2\,\gamma_1\rangle_{t_2,t_1}\\ &= k_{\rm B}^{-1}\int {\rm d}\mathbf{\Gamma}_2\,{\rm d}\mathbf{\Gamma}_1\,\wp^{(2)}(\mathbf{\Gamma}_2,t_2;\mathbf{\Gamma}_1,t_1)\,\gamma_2\,\gamma_1. \qquad (8.57)\end{aligned}$$

The subscript t_i means that the average is taken over ${\rm d}\mathbf{\Gamma}_i$. From statistical symmetry, one must have

$$Q(t_2,t_1)=Q(t_1,t_2)^{\rm T}. \qquad (8.58)$$

The quadratic form of the second entropy makes the transition probability Gaussian, and so one can reduce the two time average to a single time average by replacing the mean by the mode. This can be done in two ways, both of which must yield the same result, as is now shown. First one can invoke the most likely state conditioned on γ_1,

$$\begin{aligned} Q(t_2,t_1) &= k_{\mathrm{B}}^{-1}\left\langle [(\gamma_2-\overline{\gamma}_2)+\overline{\gamma}_2]\,\gamma_1\right\rangle_{t_2,t_1} \\ &= k_{\mathrm{B}}^{-1}\left\langle \overline{\gamma}_2\,\gamma_1\right\rangle_{t_2,t_1} \\ &= k_{\mathrm{B}}^{-1}\left\langle \left[-A^{-1}B\gamma_1\right]\gamma_1\right\rangle_{t_1} \\ &= A(t_{21})^{-1}B(t_{21})\overline{S''}(t_1)^{-1}. \end{aligned} \tag{8.59}$$

The second equality follows because $\langle\gamma_2-\overline{\gamma}_2\rangle_{t_2}=0$. A similar argument that invokes the most likely state conditioned on γ_2 gives

$$\begin{aligned} Q(t_2,t_1) &= k_{\mathrm{B}}^{-1}\left\langle \gamma_2\,\overline{\gamma}_1\right\rangle_{t_2,t_1} \\ &= \overline{S''}(t_2)^{-1}B(t_{21})C(t_{21})^{-1}. \end{aligned} \tag{8.60}$$

Interchanging t_1 and t_2 in this gives

$$\begin{aligned} Q(t_1,t_2) &= \overline{S''}(t_1)^{-1}B(t_{12})C(t_{12})^{-1} \\ &= \overline{S''}(t_1)^{-1}B(t_{21})^{\mathrm{T}}A(t_{21})^{-1} \\ &= Q(t_2,t_1)^{\mathrm{T}}, \end{aligned} \tag{8.61}$$

which is consistent with the above symmetry requirement, Eq. (8.58).

These expressions for the time correlation function are formally exact for any time interval, but they do invoke the fluctuation approximation for $S^{(2)}$ to quadratic order. Inserting the small time expansion into Eq. (8.59) one obtains

$$\begin{aligned} Q(t_2,t_1) &= -\left[\mathrm{I}+t_{21}\Lambda[A_0'+B_0']+\frac{|t_{21}|}{2}\Lambda\overline{S''}+\mathcal{O}(t_{21}^2)\right]\overline{S''}(t_1)^{-1} \\ &= -\overline{S''}(t)^{-1}+\frac{t_{21}}{2}\frac{\mathrm{d}\overline{S''}(t)^{-1}}{\mathrm{d}t}-t_{21}\Lambda[A_0'+B_0']\overline{S''}(t)^{-1} \\ &\quad -\frac{|t_{21}|}{2}\Lambda+\mathcal{O}(t_{21}^2). \end{aligned} \tag{8.62}$$

One can define the coefficient of t_{21} as $\Theta(t)\equiv\Phi(t)+\theta(t)$, with

$$\Phi(t)\equiv-\Lambda(t)[A_0'(t)+B_0'(t)]\overline{S''}(t)^{-1}, \tag{8.63}$$

which has transpose $\Phi(t)^{\mathrm{T}}=-\overline{S''}(t)^{-1}[A_0'(t)-B_0'(t)]\Lambda(t)$, and

$$\theta(t)\equiv\frac{\mathrm{d}\overline{S''}(t)^{-1}}{2\mathrm{d}t}=\frac{-1}{2}\overline{S''}(t)^{-1}\dot{\overline{S''}}(t)\overline{S''}(t)^{-1}, \tag{8.64}$$

which is symmetric. The condition that $Q(t_2,t_1)=Q(t_1,t_2)^{\mathrm{T}}$ means that $\Theta(t)$ must be an antisymmetric matrix, $\Theta(t)=-\Theta(t)^{\mathrm{T}}$. Hence the symmetric part of $\Phi(t)$ must cancel with $\theta(t)$,

$$\frac{1}{2}\left[\Phi(t)+\Phi(t)^{\mathrm{T}}\right]=-\theta(t)=-\frac{\mathrm{d}\overline{S''}(t)^{-1}}{2\mathrm{d}t}. \tag{8.65}$$

Evidently the symmetric component of Φ is half the rate of change of the inverse of $\overline{S''}(t)$. For many systems this matrix is independent of time, in which case Φ itself is an antisymmetric matrix. In such systems Φ and Θ are the same matrix, and are the analogue of the antisymmetric matrix Θ used in the mixed parity fluctuation analysis of §2.5.2. That analysis was for an equilibrium system, which is of course homogeneous in time, and so in that case $\theta = 0$. Other examples of non-equilibrium systems that are homogeneous in time are the particular model of driven Brownian motion of Ch. 3, and the steady state system of Ch. 4.

With this notation the time correlation function becomes

$$Q(t_{21};t) = -\overline{S''}(t)^{-1} + t_{21}\Theta(t) - \frac{|t_{21}|}{2}\Lambda(t) + \mathcal{O}(t_{21}^2). \tag{8.66}$$

Further properties of the coefficient Θ are derived in the following section.

8.3.4 Stochastic, Dissipative Equations of Motion

The stochastic equation corresponding to the expansion of the coefficients of the second entropy is

$$\begin{aligned} \gamma_2 &= \overline{\gamma}_2 + \tilde{\mathbf{R}} \\ &= \gamma_1 + t_{21}\Lambda[A_0' + B_0']\gamma_1 + |t_{21}|\Lambda[A_0 + B_0]\gamma_1 + \tilde{\mathbf{R}} + \mathcal{O}(t_{21}^2) \\ &= \gamma_1 - t_{21}\Phi\overline{S''}\gamma_1 + \frac{1}{2}|t_{21}|\Lambda\overline{S''}\gamma_1 + \tilde{\mathbf{R}} + \mathcal{O}(t_{21}^2), \end{aligned} \tag{8.67}$$

with the most likely part having been given in Eq. (8.52), and the variance of the random force being

$$\left\langle \tilde{\mathbf{R}}\tilde{\mathbf{R}} \right\rangle_t = |t_{21}| k_{\mathrm{B}} \Lambda(t). \tag{8.68}$$

This follows because when the second entropy is rewritten in terms of the departure from the most likely state, Eq. (8.54), the leading-order of the coefficient matrix is $-\Lambda^{-1}/2|t_{21}|$.

One sees immediately two consequences of the fluctuation dissipation theorem for the non-equilibrium system. First, that the magnitude of the fluctuations is linearly proportional to the magnitude of the dissipation, both being determined by Λ, the leading-order coefficient of the second entropy expansion. Second, that it is the gradient in the entropy that drives the evolution, since $\overline{S''}(t)\gamma = \nabla S_{\mathrm{r}}(\mathbf{\Gamma}, t) + \mathcal{O}(\gamma^2)$.

One also sees that one can write the most likely value of the stochastic equations of motion as

$$\overline{\gamma}_2 = -Q(t_{21}, t)\nabla S_{\mathrm{r}}(\mathbf{\Gamma}_1, t_1). \tag{8.69}$$

Even though the expansion of $\overline{\gamma}_2$ and $Q(t_{21}, t)$ differ by $\theta(t)$ due to the difference between $\Theta(t)$ and $\Phi(t)$, this difference is cancelled by transforming the gradient from t_1 to t. To see this, take the right-hand side of the above across to the left-hand side and expand both terms

$$\begin{aligned}
\overline{\gamma}_2 + Q(t_{21},t)\nabla S_{\rm r}(\boldsymbol{\Gamma}_1,t_1) &= \left\{\gamma_1 - t_{21}\Phi\overline{S''}\gamma_1 + \frac{1}{2}|t_{21}|\Lambda\overline{S''}\gamma_1\right\} \\
&\quad + \left\{-\overline{S''}(t)^{-1} + t_{21}\Theta - \frac{|t_{21}|}{2}\Lambda\right\}\overline{S''}(t_1)\gamma_1 \\
&= t_{21}\theta\overline{S''}(t)\gamma_1 - \frac{t_{12}}{2}\overline{S''}(t)^{-1}\frac{\mathrm{d}\overline{S''}(t)}{\mathrm{d}t}\gamma_1 \\
&= 0 + \mathcal{O}(t_{21}^2). \qquad (8.70)
\end{aligned}$$

Adiabatic Velocity

The adiabatic velocity of the fluctuation is defined as $\dot{\gamma}^0 \equiv \dot{\boldsymbol{\Gamma}}^0(\boldsymbol{\Gamma},t) - \overline{\dot{\boldsymbol{\Gamma}}^0}(t)$, where $\overline{\dot{\boldsymbol{\Gamma}}^0}(t) \equiv \dot{\boldsymbol{\Gamma}}^0(\overline{\boldsymbol{\Gamma}}(t),t)$. This must be included in the above stochastic equations of motion via the term $\Phi\nabla S_{\rm r}$, since this is the only term proportional to t_{21}. To elucidate the relationship between the two, one requires the Taylor expansion of the adiabatic velocity,

$$\dot{\gamma}^0 \equiv \dot{\boldsymbol{\Gamma}}^0(\boldsymbol{\Gamma},t) - \dot{\boldsymbol{\Gamma}}^0(\overline{\boldsymbol{\Gamma}}(t),t) = \gamma\cdot\nabla\overline{\dot{\boldsymbol{\Gamma}}^0}(t) + \mathcal{O}(\gamma^2). \qquad (8.71)$$

The most likely gradient of the adiabatic velocity that appears here is the dyadic matrix,

$$\nabla\overline{\dot{\boldsymbol{\Gamma}}^0}(t) = \left(\begin{array}{cc} \partial\dot{\boldsymbol{\Gamma}}^0_{\rm q}/\partial\boldsymbol{\Gamma}_{\rm q} & \partial\dot{\boldsymbol{\Gamma}}^0_{\rm p}/\partial\boldsymbol{\Gamma}_{\rm q} \\ \partial\dot{\boldsymbol{\Gamma}}^0_{\rm q}/\partial\boldsymbol{\Gamma}_{\rm p} & \partial\dot{\boldsymbol{\Gamma}}^0_{\rm p}/\partial\boldsymbol{\Gamma}_{\rm p} \end{array}\right)_{\overline{\boldsymbol{\Gamma}}(t)} = \left(\begin{array}{cc} 0 & -\overline{\mathcal{H}''_{\rm qq}}(t) \\ \overline{\mathcal{H}''_{\rm pp}}(t) & 0 \end{array}\right). \qquad (8.72)$$

Splitting the coefficient of t_{21} into an adiabatic part and a reservoir part, $\Phi \equiv \Phi^0 + \Phi^{\rm r}$, the adiabatic velocity must be given by

$$\dot{\gamma}^0 = -\Phi^0\overline{S''}\gamma + \mathcal{O}(\gamma^2) = -\Phi^0\nabla S_{\rm r}(\boldsymbol{\Gamma},t) + \mathcal{O}(\gamma^2). \qquad (8.73)$$

Like the Taylor expansion, this is a linear function of the fluctuation. Multiplying on the right by γ and taking the average one must have

$$\Phi^0 = k_{\rm B}^{-1}\left\langle\dot{\gamma}^0\gamma\right\rangle_t. \qquad (8.74)$$

From either this or directly, one can see that $\Phi^0 = -[\nabla\overline{\dot{\boldsymbol{\Gamma}}^0}(t)]^{\rm T}\overline{S''}(t)^{-1}$. Since $\overline{S''}$ is block diagonal and $\nabla\overline{\dot{\boldsymbol{\Gamma}}^0}$ is block adiagonal, then Φ^0 must be block adiagonal,

$$\Phi^0 = \left(\begin{array}{cc} 0 & \Phi^0_{\rm qp} \\ \Phi^0_{\rm pq} & 0 \end{array}\right), \qquad (8.75)$$

with

$$\begin{aligned} \left\{\Phi^0_{\rm qp}\right\}_{ij} &= k_{\rm B}^{-1} \langle \dot{\gamma}^0_{{\rm q}i}\, \gamma_{{\rm p}j} \rangle_{\mathbf{\Gamma}} \\ &= k_{\rm B}^{-1} \sum_k \left. \frac{\partial^2 \mathcal{H}(\mathbf{\Gamma},t)}{\partial p_i \partial p_k} \right|_{\overline{\mathbf{\Gamma}}(t)} \langle \gamma_{{\rm p}k}\, \gamma_{{\rm p}j} \rangle_{\mathbf{\Gamma}} \\ &= -\sum_k \left\{ \overline{\mathcal{H}''}(t) \right\}_{{\rm p}i,{\rm p}k} \left\{ \overline{S''}(t)^{-1} \right\}_{{\rm p}k,{\rm p}j}, \end{aligned} \tag{8.76}$$

and

$$\begin{aligned} \left\{\Phi^0_{\rm pq}\right\}_{ij} &= k_{\rm B}^{-1} \langle \dot{\gamma}^0_{{\rm p}i}\, \gamma_{{\rm q}j} \rangle_{\mathbf{\Gamma}} \\ &= -k_{\rm B}^{-1} \sum_k \left. \frac{\partial^2 \mathcal{H}(\mathbf{\Gamma},t)}{\partial q_i \partial q_k} \right|_{\overline{\mathbf{\Gamma}}(t)} \langle \gamma_{{\rm q}k}\, \gamma_{{\rm q}j} \rangle_{\mathbf{\Gamma}} \\ &= \sum_k \left\{ \overline{\mathcal{H}''}(t) \right\}_{{\rm q}i,{\rm q}k} \left\{ \overline{S''}(t)^{-1} \right\}_{{\rm q}k,{\rm q}j}. \end{aligned} \tag{8.77}$$

In summary, one has

$$\Phi^0_{\rm qq} = \Phi^0_{\rm pp} = 0, \tag{8.78}$$

and

$$\Phi^0_{\rm qp} = -\overline{\mathcal{H}''_{\rm pp}}\, \overline{S''_{\rm pp}}^{-1}, \text{ and } \Phi^0_{\rm pq} = \overline{\mathcal{H}''_{\rm qq}}\, \overline{S''_{\rm qq}}^{-1}. \tag{8.79}$$

In general it is $\Theta(t)$, not $\Phi(t)$, that is antisymmetric. However in the case that $\theta(t) = 0$, which occurs for systems homogeneous in time, then these are identical and $\Phi(t)$ is antisymmetric. In such a case one has $\Phi^0_{\rm qp} = -\Phi^{0,{\rm T}}_{\rm pq}$, and one must have the condition

$$\overline{S''_{\rm qq}}^{-1} \overline{\mathcal{H}''_{\rm qq}} = \overline{\mathcal{H}''_{\rm pp}}\, \overline{S''_{\rm pp}}^{-1}. \tag{8.80}$$

It is of interest to show explicitly that this result holds identically in certain systems.

For mechanical non-equilibrium systems, $\overline{S''}(t) = -\overline{\mathcal{H}''}(t)/T$. Hence one can immediately conclude that in this case the equality does in fact hold, and one has the explicit result that $\Phi^0_{\rm qp}(t) = T\mathrm{I}$. The case of driven Brownian motion treated in Ch. 3 was homogeneous in time in the sense that $\dot{\overline{S''}}(t) = 0$. However if, for example, either the mass of the Brownian particle or the curvature of the harmonic potential were time dependent, then $\dot{\overline{S''}}(t) \neq 0$, and $\Phi(t)$ would not be antisymmetric. Nevertheless even in this case Eq. (8.80) would still hold and $\Phi^0(t)$ would be antisymmetric.

Equation (8.80) also appears to hold for thermodynamic non-equilibrium systems, as is demonstrated by the following argument for the case of steady heat flow that was treated in Ch. 4. In this steady state case $\dot{\overline{S''}}(t) = 0$ and $\Phi(t)$ must be antisymmetric, and from the above analysis Eq. (8.80) must be true. To show this directly one notes that the only non-zero entries in $\overline{S''}$ and $\overline{\mathcal{H}''}$

are those representing molecules that are close together. In the case that the applied thermodynamic gradient from the reservoirs is a temperature gradient, this means that both molecules feel the same local applied temperature. For two such interacting molecules, say i and j, this component of the entropy is $\overline{S''_{ij}}/k_{\mathrm{B}} = -\nabla_i\nabla_j[\beta_0 E_{0;ij} + \beta_1 E_{1;ij}] = -\nabla_i\nabla_j[\beta(z_i)\mathcal{H}] = -\beta(\overline{z}_i)\nabla_i\nabla_j\mathcal{H} = -\beta(\overline{z}_j)\nabla_i\nabla_j\mathcal{H}$. In bringing the temperature outside the gradient operators, second-order terms, $\nabla\nabla T$ and $(\nabla T)^2$, have been neglected.[8] Hence one can write $\overline{S''}(t)/k_{\mathrm{B}} = -\overline{\beta}(t)\overline{\mathcal{H}''}(t) = -\overline{\mathcal{H}''}(t)\overline{\beta}(t)$, where the $\overline{\beta}$ matrix is diagonal with the inverse temperature down the diagonal. This shows explicitly that Eq. (8.80) holds in the case of an applied temperature gradient, and that in this case one has $\Phi^0_{\mathrm{qp}}(t) = k_{\mathrm{B}}^{-1}\overline{\beta}^{-1}$.

Thus Eq. (8.80) has just been shown explicitly to hold in general for mechanical non-equilibrium systems, and also for the particular thermodynamic non-equilibrium system that consists of an applied temperature gradient. Reasoning by induction, one might conclude that it is generally valid. If this is indeed the case, then $\Phi^0(t)$ is an antisymmetric matrix. This means that if $\theta(t)$ is non-zero, then it must be part of $\Theta^{\mathrm{r}}(t)$, and it must cancel with the symmetric part of $\Phi^{\mathrm{r}}(t)$.

Before returning to the equations of motion, there is an important result that one can draw from this representation of the adiabatic velocity and the antisymmetric nature of Φ^0, namely

$$\dot{\gamma}^0 \cdot \nabla S_{\mathrm{r}}(\mathbf{\Gamma}, t) = -\left[\Phi^0\nabla S_{\mathrm{r}}(\mathbf{\Gamma}, t) + \mathcal{O}(\gamma^2)\right] \cdot \nabla S_{\mathrm{r}}(\mathbf{\Gamma}, t) = 0 + \mathcal{O}(\gamma^3). \tag{8.81}$$

This follows from the general property of an antisymmetric matrix, namely that the symmetric double scalar product must vanish.

Position and Momentum Form of the Equations of Motion

In view of the above, the stochastic equation for the evolution of the fluctuation is

$$\gamma_2 = \gamma_1 + t_{21}\dot{\gamma}^0 - t_{21}\Phi^{\mathrm{r}}\nabla S_{\mathrm{r}} + \frac{1}{2}|t_{21}|\Lambda\nabla S_{\mathrm{r}} + \tilde{\mathbf{R}} + \mathcal{O}(t_{21}^2). \tag{8.82}$$

Here, to linear order in the time step, all terms that are multiplied by the time step, including the adiabatic velocity, $\dot{\gamma}^0(\mathbf{\Gamma}, t)$, the gradient in the entropy, $\nabla S_{\mathrm{r}}(\mathbf{\Gamma}, t) = \overline{S''}(t)\gamma$, and the two coefficients of the reservoir perturbation, $\Phi^{\mathrm{r}}(t)$, and $\Lambda(t)$, can be evaluated at either terminus of the transition.

This may be written in terms of the position and momentum components,

$$\begin{aligned}
\gamma_{2\mathrm{q}} &= \gamma_{1\mathrm{q}} + t_{21}\dot{\gamma}^0_{\mathrm{q}} + \mathcal{O}(t_{21}^2), \\
\gamma_{2\mathrm{p}} &= \gamma_{1\mathrm{p}} + t_{21}\dot{\gamma}^0_{\mathrm{p}} - t_{21}\Phi^{\mathrm{r}}_{\mathrm{pq}}\overline{S''_{\mathrm{qq}}}\gamma_{\mathrm{q}} - t_{21}\Phi^{\mathrm{r}}_{\mathrm{pp}}\overline{S''_{\mathrm{pp}}}\gamma_{1\mathrm{p}} + \frac{1}{2}|t_{21}|\Lambda_{\mathrm{pp}}\overline{S''_{\mathrm{pp}}}\gamma_{\mathrm{p}} \\
&\quad + \tilde{\mathbf{R}}_{\mathrm{p}} + \mathcal{O}(t_{21}^2).
\end{aligned} \tag{8.83}$$

[8] With $\overline{f}(t) \equiv f(\overline{\mathbf{\Gamma}}(t), t)$, one has $\overline{S''_{\mathrm{r}}}/k_{\mathrm{B}} = -\overline{(\beta\mathcal{H})''} = -\overline{\beta''}\,\overline{\mathcal{H}} - 2\overline{\beta'}\,\overline{\mathcal{H}'} - \overline{\beta}\,\overline{\mathcal{H}''} = -\overline{\beta}\,\overline{\mathcal{H}''} + \mathcal{O}(\overline{\beta''}, \overline{\beta'}^2)$. This uses the fact that $\overline{\mathcal{H}'} = \mathcal{O}(\overline{\beta'})$, since $\overline{\mathcal{H}'} = \overline{(\beta^{-1}\beta\mathcal{H})'} = -\overline{\beta}^{-1}\overline{\beta'}\,\overline{\mathcal{H}} - \overline{\beta}^{-1}\overline{S'_{\mathrm{r}}}/k_{\mathrm{B}}$, the second contribution vanishing since $\overline{S'_{\mathrm{r}}} = 0$ by the definition of $\overline{\mathbf{\Gamma}}(t)$.

There is no reservoir perturbation to the position evolution at the linear order in the time step. This is because the position is essentially the integral of the momentum over the time step, and so the reservoir perturbation to the position evolution comes from integrating the reservoir perturbation to the momentum evolution, which gives terms $\mathcal{O}(\Delta_t^2)$ that can be neglected. Since Λ is symmetric, the fact that $|t_{21}|\Lambda_{\rm qp} \sim \mathcal{O}(t_{21}^2)$, means that $|t_{21}|\Lambda_{\rm pq} \sim \mathcal{O}(t_{21}^2)$, and so it may also be neglected. (If $\Phi(t)$ were antisymmetric due to $\theta(t) = 0$, one could neglect $\Phi^{\rm r}_{\rm pq}(t)$ for the same reason.) This is why only the adiabatic term contributes to the evolution of the position coordinates in the equations of motion, and why the stochastic part of the reservoir force, $\tilde{\mathbf{R}}_{\rm p}$, has only momentum components.

The equations of motion contain two types of terms, namely the reversible terms that are proportional to t_{21}, and the irreversible terms that are proportional to $|t_{21}|$. The transition is dominated by the adiabatic term, which is the term due directly to the interactions between the molecules of the sub-system. The forces from the reservoir are a perturbation on the adiabatic evolution. One can argue that of the two contributions to the reversible evolution, $\Phi^{\rm r}(t)$ can be neglected compared with the adiabatic contribution. One cannot make a similar argument to neglect $\Lambda(t)$ because irreversibility is essential for the correct statistical properties of the non-equilibrium system, and this is the only irreversible term.

A second argument for neglecting $\Phi^{\rm r}(t)$ but retaining $\Lambda(t)$ is that both terms represent the thermodynamic driving force down the entropy gradient to the optimum state. Since to leading-order a fluctuation is as likely to be returning to the optimum state as it is to be coming from the optimum state, then the thermodynamic driving force must be dominated by the time symmetric irreversible term, $|t_{21}|\Lambda(t)$. This second argument only holds for infinitesimal time steps, since on longer time scales the regression need not be symmetric.

A third argument for neglecting $\Phi^{\rm r}(t)$ is based upon the fact that the position variables are essentially 'slave' variables, in the sense that changes in them are fully determined by the integral of the momentum variables over the time step, $\gamma'_{\rm q} - \gamma_{\rm q} = \Delta_t \gamma_{\rm p}/m$, where m is the mass in the simplest case. (This is also equal to $\Delta_t \dot{\gamma}^0_{\rm q}$, but that is not directly relevant for what follows.) Since once the deterministic adiabatic contributions have been accounted for, only the statistical fluctuations of the momentum variables remain, and it is the pure parity fluctuation theory of §2.3 that is more relevant to the present results than the mixed parity theory of §2.5. Because of the symmetry of the second entropy in the pure parity case, the reservoir force is linearly proportional to $|t_{21}|$, and there are no reservoir terms that depend upon t_{21}.

Of course over longer time scales the reversible non-adiabatic evolution must be non-negligible, due to the cumulative effects of the reversible adiabatic and the irreversible reservoir terms on the evolution of the sub-system. It is only to linear order in the time step that $\Phi^{\rm r}(t)$ is arguably negligible.[9]

[9] Stochastic equations of motion equivalent to $\Phi^{\rm r}(t) = 0$, and $\Lambda(t) \neq 0$ have been successfully tested with molecular dynamics computer simulations for the cases of equilibrium systems, driven Brownian motion, and steady heat flow, [Attard, P. (2009b), Annu. Rep. Prog. Chem., Sect. C **105**, 63]. It may or may not be significant that these three systems have in

Accepting the above arguments for neglecting $\Phi^{\rm r}(t)$, the stochastic dissipative equations of motion written in terms of the position and momentum components, are

$$\begin{aligned}\gamma_{2\rm q} &= \gamma_{1\rm q} + t_{21}\dot{\gamma}^0_{\rm q} + \mathcal{O}(t_{21}^2),\\ \gamma_{2\rm p} &= \gamma_{1\rm p} + t_{21}\dot{\gamma}^0_{\rm p} + \overline{\mathbf{R}}_{\gamma\rm p} + \tilde{\mathbf{R}}_{\rm p} + \mathcal{O}(t_{21}^2).\end{aligned} \tag{8.84}$$

Recall that $\gamma \equiv \mathbf{\Gamma} - \overline{\mathbf{\Gamma}}(t)$, $\dot{\gamma}^0 \equiv \dot{\mathbf{\Gamma}}^0 - \overline{\dot{\mathbf{\Gamma}}^0}(t)$, and $\overline{\dot{\mathbf{\Gamma}}^0}(t) \equiv \dot{\mathbf{\Gamma}}^0(\overline{\mathbf{\Gamma}}(t), t)$. The total reservoir force for the fluctuation will be written $\mathbf{R}_\gamma = \overline{\mathbf{R}}_\gamma + \tilde{\mathbf{R}}$. As in Eq. (8.52) the most likely part of this is

$$\overline{\mathbf{R}}_\gamma(\mathbf{\Gamma}, t, \Delta_t) \equiv \frac{1}{2}|\Delta_t|\Lambda_{\rm pp}(t)\overline{S''_{\rm pp}}(t)\gamma_{\rm p} = \frac{1}{2}|\Delta_t|\Lambda_{\rm pp}(t)\nabla_{\rm p}S_{\rm r}(\mathbf{\Gamma}, t), \tag{8.85}$$

where the time step is $\Delta_t = t_{21}$ in this particular case. In the fluctuation approximation, contributions to this that are $\mathcal{O}(\gamma^2)$ have been neglected.

Transport Matrix

The entropy and transport matrices depend upon time, $\Lambda(t) \equiv \Lambda(\overline{\mathbf{\Gamma}}(t), t)$ etc. This is where non-linearity enters the theory (see the discussion in §3.6.3 on p. 78). For brevity, the time dependence of the matrices will often be suppressed. As just discussed, only the momentum components occur, and in most cases one can take the symmetric transport matrix to be diagonal,

$$\Lambda_{\rm pp}(t) = \frac{\sigma(t)}{k_{\rm B}|\Delta_t|}\mathrm{I}_{\rm pp}, \tag{8.86}$$

with the scalar σ being the variance of the reservoir force, and $\mathrm{I}_{\rm pp}$ being the momentum block of the identity matrix. This is the simplest and most important case. It is straightforward to generalise this, for example, to a multi-component system with a different variance σ_α for each species. To obtain the most general case from the results below, simply make the replacements $\sigma(t)a \Rightarrow k_{\rm B}|\Delta_t|\Lambda_{\rm pp}(t)a$, and $\sigma(t)a \cdot b \Rightarrow k_{\rm B}|\Delta_t|\Lambda_{\rm pp}(t) : ab$, where a and b are the momentum components of phase space vectors. Although it is possible to choose a variance dependent upon time, $\sigma(t) \equiv \sigma(\overline{\mathbf{\Gamma}}(t), t)$, generally this will not be shown.

The variance σ (equivalently, Λ) represents the strength of the reservoir perturbation in the equations of motion. Since the reservoir is an idealisation of reality, to a large extent the numerical value of the variance can be freely chosen.[10] In particular, it is not a molecular property and it is not directly determined by the Hamiltonian or thermodynamic state of the sub-system. Provided

---

common $\theta(t) = 0$.

[10]Since Λ is the coefficient matrix in the expansion in powers of the time step of the time correlation function (equivalently, the second entropy fluctuation matrices), it must be independent of the time step. Hence the above definition of the variance means that it can be freely chosen provided that it is linearly proportional to the magnitude of the time step, $\sigma \propto |\Delta_t|$.

that the fluctuation dissipation theorem is satisfied, reasonable results can be obtained for quite a wide range of values of σ. Reasonable means that the transport, temperature, and other thermodynamic properties of the sub-system are insensitive to the precise choice of σ.

With this simplest form for the variance, the most likely reservoir force for the fluctuation is given by

$$\begin{aligned}\overline{\mathbf{R}}_\gamma(\boldsymbol{\Gamma},t,\Delta_t) &= \frac{|\Delta_t|}{2}\Lambda_{\rm pp}(t)\nabla_{\rm p}S_{\rm r}(\boldsymbol{\Gamma},t)\\ &\Rightarrow \frac{\sigma}{2k_{\rm B}}\nabla_{\rm p}S_{\rm r}(\boldsymbol{\Gamma},t)\\ &= \frac{\sigma}{2k_{\rm B}}\overline{S''_{\rm pp}}(t)\gamma_{\rm p}\\ &= \frac{\sigma}{2k_{\rm B}}\left[\nabla_{\rm p}S_{\rm r,st}(\boldsymbol{\Gamma},t)-\overline{S'_{\rm p}}(t)\right].\end{aligned} \tag{8.87}$$

The final equality follows from Eq. (8.23). This says that the most likely reservoir force for the fluctuations depends upon the magnitude but not the sign of the time step. As mentioned following the presentation of the second entropy above, the fluctuations have variance $\left\langle \tilde{\mathbf{R}}\tilde{\mathbf{R}}\right\rangle_{\tilde{\mathbf{R}}} = k_{\rm B}|\Delta_t|\Lambda_{\rm pp}(t) \Rightarrow \sigma \mathrm{I}_{\rm pp}$. This is the molecular-level, non-equilibrium version of the fluctuation dissipation theorem.

Configuration Evolution

Transforming the stochastic dissipative equations of motion for fluctuations to those for the actual configuration yields

$$\begin{aligned}\boldsymbol{\Gamma}_{2\rm q} &= \boldsymbol{\Gamma}_{1\rm q}+t_{21}\dot{\boldsymbol{\Gamma}}^0_{\rm q}(\boldsymbol{\Gamma},t)+\mathcal{O}(t_{21}^2),\\ \boldsymbol{\Gamma}_{2\rm p} &= \boldsymbol{\Gamma}_{1\rm p}+t_{21}\dot{\boldsymbol{\Gamma}}^0_{\rm p}(\boldsymbol{\Gamma},t)+\mathbf{R}(\boldsymbol{\Gamma},t,t_{21})+\mathcal{O}(t_{21}^2).\end{aligned} \tag{8.88}$$

The relation between the actual reservoir force, $\mathbf{R}$, which appears here, and the reservoir force for the fluctuation, $\mathbf{R}_\gamma$, which was used above, will be given shortly. For brevity, the equations of motion will often be written

$$\boldsymbol{\Gamma}_2=\boldsymbol{\Gamma}_1+t_{21}\dot{\boldsymbol{\Gamma}}^0(\boldsymbol{\Gamma},t)+\mathbf{R}(\boldsymbol{\Gamma},t,t_{21})+\mathcal{O}(t_{21}^2), \tag{8.89}$$

with the understanding that the reservoir perturbation has momentum components only. The reverse transition is

$$\boldsymbol{\Gamma}_1=\boldsymbol{\Gamma}_2+t_{12}\dot{\boldsymbol{\Gamma}}^0(\boldsymbol{\Gamma},t)+\mathbf{R}(\boldsymbol{\Gamma},t,t_{12})+\mathcal{O}(t_{21}^2). \tag{8.90}$$

The apparent algebraic inconsistency between these last two equations (the latter is *not* related to the former by transferring terms from one side to the other, $\mathbf{R}(\boldsymbol{\Gamma},t,t_{21}) \neq -\mathbf{R}(\boldsymbol{\Gamma},t,t_{12})$) arises because $\mathbf{R}$ is a stochastic variable.[11]

[11] Alternatively, one could explicitly signify the stochastic nature of the variable by writing $\boldsymbol{\Gamma}_2=\boldsymbol{\Gamma}_1+t_{21}\dot{\boldsymbol{\Gamma}}^0+\mathcal{R}$, with the reservoir perturbation distributed according to $\wp(\mathcal{R}|\boldsymbol{\Gamma},t,t_{21})$. With this notation it is clear that one does not change the sign of the stochastic variable when writing the reverse transition: $\boldsymbol{\Gamma}_1=\boldsymbol{\Gamma}_2+t_{12}\dot{\boldsymbol{\Gamma}}^0+\mathcal{R}$, $\wp(\mathcal{R}|\boldsymbol{\Gamma},t,t_{12})$. In particular, $\wp(\mathcal{R}|\boldsymbol{\Gamma},t,t_{21}) \neq \wp(-\mathcal{R}|\boldsymbol{\Gamma},t,t_{12})$.

It reflects the irreversibility of the equations of motion (see §8.4.2 below). The reservoir force for the actual configuration, $\mathbf{R}(\mathbf{\Gamma}, t, \Delta_t)$, is a linear function of both Δ_t and $|\Delta_t|$, as will now be shown. This force is characterised by a probability distribution, $\wp(\mathbf{R}|\mathbf{\Gamma}, t, \Delta_t)$, and can be written as the sum of its most likely value and its stochastic value, $\mathbf{R} = \overline{\mathbf{R}} + \tilde{\mathbf{R}}$. (The reservoir force has momentum components only.)

The stochastic part of the force $\tilde{\mathbf{R}}$ is the same for the actual coordinates and for the fluctuations $\tilde{\mathbf{R}} \equiv \tilde{\mathbf{R}}_\gamma$. It is linearly proportional to $|\Delta_t|$ and is related to the most likely force by the fluctuation dissipation theorem, as was shown above.

One has the reservoir force for fluctuations and the reservoir force acting on the actual coordinates. From the equations of motion, the most likely values of these are related by

$$\begin{aligned}
\overline{\mathbf{R}}(\mathbf{\Gamma}, t, \Delta_t) &\equiv \overline{\mathbf{\Gamma}}(t + \Delta_t|\mathbf{\Gamma}, t) - \mathbf{\Gamma} - \Delta_t \dot{\mathbf{\Gamma}}^0(\mathbf{\Gamma}, t) \\
&= \overline{\gamma}(t + \Delta_t|\mathbf{\Gamma}, t) + \overline{\mathbf{\Gamma}}(t + \Delta_t) - \gamma - \overline{\mathbf{\Gamma}}(t) \\
&\quad - \Delta_t \left[\dot{\gamma}^0(\mathbf{\Gamma}, t) + \dot{\overline{\mathbf{\Gamma}}^0}(t) \right] \\
&= \overline{\mathbf{R}}_\gamma(\mathbf{\Gamma}, t, \Delta_t) + \Delta_t \left[\dot{\overline{\mathbf{\Gamma}}}(t) - \dot{\overline{\mathbf{\Gamma}}^0}(t) \right]. \qquad (8.91)
\end{aligned}$$

Here one sees explicitly that the actual reservoir force, $\overline{\mathbf{R}}$, depends upon the sign of the time step, unlike the reservoir force for the fluctuation, $\overline{\mathbf{R}}_\gamma$, which is proportional to $|\Delta_t|$ (when $\Phi^{\mathrm{r}}(t)$ is neglected). As mentioned in the preceding paragraph, the stochastic part of the reservoir force, $\tilde{\mathbf{R}}$, is the same for the evolution both of the actual configuration and of the fluctuation.

The most likely trajectory $\overline{\mathbf{\Gamma}}(t)$ is calculated forward in time. Hence

$$\begin{aligned}
\dot{\overline{\mathbf{\Gamma}}}(t) - \dot{\overline{\mathbf{\Gamma}}^0}(t) &= \frac{1}{\Delta_t} \overline{\mathbf{R}}(\overline{\mathbf{\Gamma}}(t), t, \Delta_t), \quad \Delta_t > 0 \\
&\equiv \frac{1}{|\Delta_t|} \overline{\overline{\mathbf{R}}}(t, |\Delta_t|). \qquad (8.92)
\end{aligned}$$

An explicit expression for $\overline{\overline{\mathbf{R}}}(t, |\Delta_t|)$ will be derived in Eq. (8.126) below. Only the momentum components of this are non-zero because, as discussed above, the reservoir perturbation does not affect the evolution of the position to linear order in the time step. In particular, $\dot{\overline{\mathbf{\Gamma}}}_{\mathrm{q}}(t) = \dot{\overline{\mathbf{\Gamma}}}{}^0_{\mathrm{q}}(t)$. With this Eq. (8.91) is more simply written,

$$\overline{\mathbf{R}}(\mathbf{\Gamma}, t, \Delta_t) = \overline{\mathbf{R}}_\gamma(\mathbf{\Gamma}, t, |\Delta_t|) + \hat{t}\, \overline{\overline{\mathbf{R}}}(t, |\Delta_t|), \qquad (8.93)$$

where $\hat{t} \equiv \mathrm{sign}(\Delta_t)$. If the system is in the optimum state (i.e. the fluctuation is zero), then the most likely value of the reservoir force on the fluctuation is also zero, $\overline{\mathbf{R}}_\gamma(\overline{\mathbf{\Gamma}}(t), t, |\Delta_t|) = 0$. Hence one sees that $\overline{\mathbf{R}}(\overline{\mathbf{\Gamma}}(t), t, \Delta_t)$ is linearly proportional to Δ_t (whereas $\overline{\overline{\mathbf{R}}}(t, |\Delta_t|)$ is defined such that it is linearly proportional to $|\Delta_t|$).

Using Eq. (8.91) and the final equality in Eq. (8.87), the most likely actual reservoir force is

$$\begin{aligned}\overline{\mathbf{R}}(\mathbf{\Gamma},t,\Delta_t) &= \overline{\mathbf{R}}_\gamma(\mathbf{\Gamma},t,\Delta_t)+\Delta_t\left[\dot{\overline{\mathbf{\Gamma}}}_{\rm p}(t)-\overline{\dot{\mathbf{\Gamma}}^0_{\rm p}(t)}\right]\\ &= \frac{\sigma}{2k_{\rm B}}\nabla_{\rm p}S_{\rm r,st}(\mathbf{\Gamma},t)-\frac{\sigma}{2k_{\rm B}}\overline{S'_{\rm p}}(t)+\hat{t}\overline{\overline{\mathbf{\Gamma}}}_{\rm p}(t)\\ &= \frac{\sigma}{2k_{\rm B}}\nabla_{\rm p}S_{\rm r,st}(\mathbf{\Gamma},t)+\frac{\sigma}{2k_{\rm B}}\left(\hat{t}-1\right)\overline{S'_{\rm p}}(t),\end{aligned} \tag{8.94}$$

where $\hat{t}\equiv \text{sign}(\Delta_t)$, $\overline{\dot{\mathbf{\Gamma}}^0(t)}\equiv\dot{\mathbf{\Gamma}}^0(\overline{\mathbf{\Gamma}}(t),t)$, and only momentum components appear (because $\dot{\overline{\mathbf{\Gamma}}}_{\rm q}(t)=\overline{\dot{\mathbf{\Gamma}}^0_{\rm q}(t)}$). It is not until Eq. (8.126) below that the final equality will be justified. This contains a phase space dependent term, which is proportional to the gradient of the static part of the reservoir entropy, and a constant term (i.e. independent of phase space) that depends upon time and upon the direction of the time step.

The important qualitative difference between the most likely value of the actual reservoir force, $\overline{\mathbf{R}}$, and the most likely value of the reservoir force acting on the fluctuation, $\overline{\mathbf{R}}_\gamma$, is that the latter does not depend upon the sign of the time step and is linearly proportional to $|\Delta_t|$ only, $\mathbf{R}_\gamma(\mathbf{\Gamma},t,\Delta_t)\propto|\Delta_t|$. As argued in the justification for the neglect of $\Phi^{\rm r}(t)$ leading to Eq. (8.84), the reason for this is the time symmetry of the fluctuation: the regression of a fluctuation *to* the optimum state is statistically identical to its progression *from* the optimum state.

There is a pleasing symmetry in the reservoir forces. The most likely reservoir force on the fluctuation is

$$\overline{\mathbf{R}}_\gamma=\frac{\sigma}{2k_{\rm B}}\nabla_{\rm p}S_{\rm r}(\mathbf{\Gamma},t). \tag{8.95}$$

In comparison, the most likely reservoir force on the configuration going forward in time is

$$\overline{\mathbf{R}}=\frac{\sigma}{2k_{\rm B}}\nabla_{\rm p}S_{\rm r,st}(\mathbf{\Gamma},t),\ \Delta_t>0. \tag{8.96}$$

Since in general an explicit expression for $S_{\rm r,st}(\mathbf{\Gamma},t)$ is available from purely equilibrium considerations, and since it can be evaluated at the current point in phase space with no reference to the prior history of the system, this is a powerful result that allows the trajectory of the sub-system in terms of its actual phase space coordinates to be calculated going forward in time. In essence this is a molecular dissipative force, and it is of the usual drag or friction form such as used in the Langevin equation. The result has been justified purely on statistical grounds without reference to any macroscopic hydrodynamic or friction considerations.

8.3.5 Transition Probability and Fokker-Planck Equation

The transition probability is just the exponential of the second entropy. Using the expansion, Eq. (8.54), it is explicitly

$$\begin{aligned}\wp(\mathbf{\Gamma}_2,t_2;\mathbf{\Gamma}_1,t_1) &= \frac{1}{Z^{(2)}(t_2,t_1)} e^{S^{(2)}(\mathbf{\Gamma}_2,t_2;\mathbf{\Gamma}_1,t_1)/k_\mathrm{B}} \\ &= \frac{e^{-\Lambda^{-1}:[\gamma_2-\overline{\gamma}_2]^2/2k_\mathrm{B}|t_{21}|}}{|2\pi k_\mathrm{B} t_{21}\Lambda|^{1/2}} \frac{e^{[S_\mathrm{r}(\mathbf{\Gamma}_1,t_1)+\{\overline{S}_\mathrm{r}(t_2)-\overline{S}_\mathrm{r}(t_1)\}/2]/k_\mathrm{B}}}{Z^{(2)\prime}(t_2,t_1)} \\ &= \frac{e^{-\Lambda_\mathrm{pp}^{-1}:[\gamma_{2\mathrm{p}}-\overline{\gamma}_{2\mathrm{p}}]^2/2k_\mathrm{B}|t_{21}|}}{|2\pi k_\mathrm{B} t_{21}\Lambda_\mathrm{pp}|^{1/2}} \delta(\mathbf{\Gamma}_{2\mathrm{q}}-\mathbf{\Gamma}_{\mathrm{q}1}-t_{21}\dot{\mathbf{\Gamma}}_\mathrm{q}^0)\,\wp(\mathbf{\Gamma}_1,t_1),\end{aligned} \tag{8.97}$$

where $Z^{(2)\prime}(t_2,t_1) \equiv Z(t_1)e^{\{\overline{S}_\mathrm{r}(t_2)-\overline{S}_\mathrm{r}(t_1)\}/2k_\mathrm{B}}$. The determinant of the transport matrix normalises the Gaussian. In the final equality, the limit $\Lambda_\mathrm{qq} \to 0$ has been taken, which gives a δ-function for the adiabatic evolution of the position coordinates. As above, one often makes the replacement $|t_{21}|\Lambda_\mathrm{pp} \Rightarrow k_\mathrm{B}^{-1}\sigma \mathrm{I}_\mathrm{pp}$.

It is evident from the final form of the unconditional transition probability that it obeys the standard reduction condition,

$$\int \mathrm{d}\mathbf{\Gamma}_2\, \wp(\mathbf{\Gamma}_2,t_2;\mathbf{\Gamma}_1,t_1) = \wp(\mathbf{\Gamma}_1,t_1). \tag{8.98}$$

From the symmetry of the second entropy, one has an analogous result for the integral over $\mathbf{\Gamma}_1$,

$$\int \mathrm{d}\mathbf{\Gamma}_1\, \wp(\mathbf{\Gamma}_2,t_2;\mathbf{\Gamma}_1,t_1) = \wp(\mathbf{\Gamma}_2,t_2). \tag{8.99}$$

One can conclude from these that the transition probability based upon the second entropy preserves the non-equilibrium probability density.

The Fokker-Planck equation was derived in the canonical equilibrium case as Eq. (7.102). It was also derived in the specific case of Brownian motion, Eq. (3.103). In the present non-equilibrium case, it can be written in terms of the actual coordinate,

$$\begin{aligned}\frac{\partial \wp(\mathbf{\Gamma},t)}{\partial t} &= \frac{-1}{\Delta_t}\left[\nabla_\mathrm{p}\cdot\overline{\mathbf{R}}(\mathbf{\Gamma},t)\right]\wp(\mathbf{\Gamma},t) - \left[\dot{\mathbf{\Gamma}}^0 + \frac{1}{\Delta_t}\overline{\mathbf{R}}\right]\cdot\nabla\wp(\mathbf{\Gamma},t) \\ &\quad + \frac{\sigma}{2\Delta_t}\nabla_\mathrm{p}^2\wp(\mathbf{\Gamma},t),\end{aligned} \tag{8.100}$$

or, equivalently, in terms of the fluctuation,

$$\begin{aligned}\frac{\partial \wp(\gamma,t)}{\partial t} &= \frac{-1}{\Delta_t}\left[\nabla_\mathrm{p}\cdot\overline{\mathbf{R}}_\gamma(\gamma,t)\right]\wp(\gamma,t) - \left[\dot{\gamma}^0 + \frac{1}{\Delta_t}\overline{\mathbf{R}}_\gamma\right]\cdot\nabla\wp(\gamma,t) \\ &\quad + \frac{\sigma}{2\Delta_t}\nabla_\mathrm{p}^2\wp(\gamma,t).\end{aligned} \tag{8.101}$$

Using the fluctuation form, the first term on the right-hand side yields

$$\begin{aligned}
\frac{-1}{\Delta_t} \nabla_{\rm p} \cdot \overline{\mathbf{R}}_\gamma(\gamma, t) &= \frac{-\hat{t}}{2} \mathrm{TR}\{\Lambda_{\rm pp}(t) \overline{S''_{\rm pp}}(t)\} \\
&\Rightarrow \frac{-\sigma}{2k_{\rm B}\Delta_t} \mathrm{TR}\{\overline{S''_{\rm pp}}(t)\} \\
&= \frac{-\sigma}{2k_{\rm B}\Delta_t} \nabla_{\rm p}^2 S_{\rm r}(\boldsymbol{\Gamma}, t). \qquad (8.102)
\end{aligned}$$

The second term on the right-hand side yields

$$\frac{-1}{k_{\rm B}} \left[\dot{\gamma}^0 + \frac{1}{\Delta_t} \overline{\mathbf{R}}_\gamma \right] \cdot \nabla S_{\rm r}(\boldsymbol{\Gamma}, t) = \frac{-\sigma}{2k_{\rm B}^2 \Delta_t} \nabla_{\rm p} S_{\rm r}(\boldsymbol{\Gamma}, t) \cdot \nabla_{\rm p} S_{\rm r}(\boldsymbol{\Gamma}, t), \qquad (8.103)$$

since $\dot{\gamma}^0 \cdot \nabla S_{\rm r}(\boldsymbol{\Gamma}, t) = 0$, Eq. (8.81). The final term on the right-hand side yields

$$\frac{\sigma(t)}{2k_{\rm B}\Delta_t} \left\{ \nabla_{\rm p}^2 S_{\rm r}(\boldsymbol{\Gamma}, t) + k_{\rm B}^{-1} [\nabla_{\rm p} S_{\rm r}(\boldsymbol{\Gamma}, t)] \cdot [\nabla_{\rm p} S_{\rm r}(\boldsymbol{\Gamma}, t)] \right\}. \qquad (8.104)$$

Adding these three together the right-hand side is

$$\begin{aligned}
\mathrm{RHS} &= \wp(\gamma, t) \left[\frac{-\sigma}{2k_{\rm B}\Delta_t} \nabla_{\rm p}^2 S_{\rm r}(\boldsymbol{\Gamma}, t) - \frac{\sigma}{2k_{\rm B}^2\Delta_t} \nabla_{\rm p} S_{\rm r}(\boldsymbol{\Gamma}, t) \cdot \nabla_{\rm p} S_{\rm r}(\boldsymbol{\Gamma}, t) \right. \\
&\quad \left. + \frac{\sigma(t)}{2k_{\rm B}\Delta_t} \left\{ \nabla_{\rm p}^2 S_{\rm r}(\boldsymbol{\Gamma}, t) + k_{\rm B}^{-1} [\nabla_{\rm p} S_{\rm r}(\boldsymbol{\Gamma}, t)] \cdot [\nabla_{\rm p} S_{\rm r}(\boldsymbol{\Gamma}, t)] \right\} \right] \\
&= 0. \qquad (8.105)
\end{aligned}$$

The left-hand side is

$$\begin{aligned}
\frac{\partial \wp(\gamma, t)}{\partial t} &= \left\{ \frac{-\dot{Z}(t)}{Z(t)} + \frac{\partial S_{\rm r}(\gamma, t)}{k_{\rm B} \partial t} \right\} \wp(\gamma, t) \\
&= \left\{ \frac{-\dot{Z}(t)}{Z(t)} + k_{\rm B}^{-1} \dot{\overline{S}}_{\rm r}(t) \right\} \wp(\gamma, t) \\
&= 0. \qquad (8.106)
\end{aligned}$$

This follows because the factor in the numerator of $\exp \overline{S}_{\rm r}(t)/k_{\rm B}$ cancels with the same factor in the partition function (see also the first equality in Eq. (8.139) below).[12] That these agree confirms the Fokker-Planck equation for the exact non-equilibrium probability density in fluctuation form.

Using the Fokker-Planck equation written in terms of the actual coordinates, because $\overline{\mathbf{R}}$ only differs from $\overline{\mathbf{R}}_\gamma$ by a constant, the first term on the right-hand

[12] In deriving that result the time dependence of $\overline{S''}(t)$ was neglected. If this were included one would have $[\dot{\overline{S''}} : \gamma^2/2k_{\rm B}]\wp + (1/2)\wp \mathrm{TR}(\dot{\overline{S''}}\,\overline{S''}^{-1})$. Invoking the equipartition theorem, $\gamma^2 \Rightarrow \langle \gamma^2 \rangle = -k_{\rm B}\overline{S''}^{-1}$, this is on average zero. Even without the equipartition theorem, the individual terms here are of zeroth order in the volume, compared with the other terms in the Fokker-Planck equation that are linear in the volume, and hence this contribution is relatively negligible in the thermodynamic limit.

side is unchanged. The third term on the right-hand side is the same in both representations. The second term on the right-hand side yields

$$\begin{aligned}
&\frac{-1}{k_{\mathrm{B}}}\left[\dot{\mathbf{\Gamma}}^0 + \frac{1}{\Delta_t}\overline{\mathbf{R}}\right] \cdot \nabla S_{\mathrm{r}}(\mathbf{\Gamma}, t) \\
&= \frac{-1}{k_{\mathrm{B}}}\left[\overline{\dot{\mathbf{\Gamma}}^0}(t) + \frac{1}{\Delta_t}\overline{\mathbf{R}}_\gamma + \frac{\mathrm{d}\overline{\mathbf{\Gamma}}(t)}{\mathrm{d}t} - \overline{\dot{\mathbf{\Gamma}}^0}(t)\right] \cdot \nabla S_{\mathrm{r}}(\mathbf{\Gamma}, t) \\
&= \frac{-\sigma}{2k_{\mathrm{B}}^2\Delta_t} \nabla_{\mathrm{p}} S_{\mathrm{r}}(\mathbf{\Gamma}, t) \cdot \nabla_{\mathrm{p}} S_{\mathrm{r}}(\mathbf{\Gamma}, t) - \frac{\mathrm{d}\overline{\mathbf{\Gamma}}(t)}{k_{\mathrm{B}}\mathrm{d}t} \cdot \nabla S_{\mathrm{r}}(\mathbf{\Gamma}, t).
\end{aligned} \tag{8.107}$$

The first term in the brackets on the right-hand side of the first equality follows from Eq. (8.81), and the remaining terms follow from Eq. (8.94). Compared with the second term on the right-hand side of the fluctuation form, the final term is new. Now since $(\mathrm{d}\gamma)_{\mathbf{\Gamma}} = -(\mathrm{d}\overline{\mathbf{\Gamma}}(t)/\mathrm{d}t)\mathrm{d}t$, the left-hand side is

$$\begin{aligned}
\frac{\partial \wp(\mathbf{\Gamma}, t)}{\partial t} &= \frac{\partial \wp(\gamma, t)}{\partial t} - \frac{\mathrm{d}\overline{\mathbf{\Gamma}}(t)}{\mathrm{d}t} \cdot \nabla \wp(\mathbf{\Gamma}, t) \\
&= \frac{-\mathrm{d}\overline{\mathbf{\Gamma}}(t)}{k_{\mathrm{B}}\mathrm{d}t} \cdot [\nabla S_{\mathrm{r}}(\mathbf{\Gamma}, t)]\wp(\mathbf{\Gamma}, t).
\end{aligned} \tag{8.108}$$

This differs from the left-hand side of the fluctuation form of the Fokker-Planck equation by exactly the same amount as the right-hand side differs. Hence one can conclude that the Fokker-Planck equation is also consistent with the exact non-equilibrium probability density in configuration form.

8.3.6 Most Likely Force with Constraints

The reservoir formalism is an abstraction and idealisation of reality. It provides a convenient way of focusing in molecular detail on the part of the system of primary interest, the sub-system, and of incorporating the more distant parts, the reservoir, in a probabilistic sense that requires only the values of a few macroscopic thermodynamic parameters.

In the case of a thermodynamic non-equilibrium system, the reservoir provides a thermodynamic gradient acting on the sub-system, and a source and sink for the material flux through the sub-system. This may be called boundary driven flow, since the direct interaction between the sub-system and the reservoirs only occurs over a region of molecular width at the boundaries. Similarly for a mechanical non-equilibrium system, the reservoir provides a sink for the heat dissipated by the work done on the sub-system, and this heat is transferred at the boundaries.

In the idealisation that is the reservoir formulation of non-equilibrium statistical mechanics, the direct interaction between the sub-system and the reservoir is represented by the stochastic reservoir force $\mathbf{R}$, which is applied equally to all the molecules in the sub-system, not just those at the boundaries. It obviously simplifies the equations considerably not to have to apply a differential force that depends upon the location of the molecules, the recipe for which is likely to be *ad hoc* and specific to each system. However, this does raise the question of whether artifacts might be introduced by this invocation of bulk rather than boundary stochastic forces.

One quantity that often occurs in thermodynamic non-equilibrium systems is the adiabatic rate of change of the static part of the reservoir entropy, $\dot{S}^0_{\rm r,st}(\mathbf{\Gamma}, t)$. As will be shown in later chapters, the autocorrelation function of this gives the hydrodynamic transport coefficients via the Green-Kubo formulae. The time correlation function represents the adiabatic relaxation of the static part of the reservoir entropy. This adiabatic relaxation represents an ordered flux, and there is reason to believe that this order can be significantly perturbed by the stochastic forces from the reservoir when these are applied throughout the sub-system in the idealised reservoir formalism, as distinct from their application only at the boundaries in a real system.

One way of circumventing this problem is to minimise the perturbation of the adiabatic flux by the stochastic reservoir forces. This can be done by constraining the latter to be orthogonal to the gradient of the adiabatic flux,

$$\overline{\mathbf{R}} \cdot \nabla_{\rm p} \dot{S}^0_{\rm r,st}(\mathbf{\Gamma}, t) = 0. \tag{8.109}$$

The Lagrange multiplier technique can be used to maximise the second entropy subject to this constraint. The result is

$$\overline{\mathbf{R}} = \frac{\sigma}{2k_{\rm B}} \left[\nabla_{\rm p} S_{\rm r,st}(\mathbf{\Gamma}, t) + (\hat{t} - 1)\overline{S'_{\rm p}}(t) \right] + \frac{\mu\sigma}{2k_{\rm B}} \nabla_{\rm p} \dot{S}^0_{\rm r,st}(\mathbf{\Gamma}, t). \tag{8.110}$$

The Lagrange multiplier is evidently given by

$$\mu(\mathbf{\Gamma}, t) = \frac{-\left[\nabla_{\rm p} S_{\rm r,st} + (\hat{t} - 1)\overline{S'_{\rm p}} \right] \cdot \nabla_{\rm p} \dot{S}^0_{\rm r,st}}{\nabla_{\rm p} \dot{S}^0_{\rm r,st} \cdot \nabla_{\rm p} \dot{S}^0_{\rm r,st}}. \tag{8.111}$$

Except for $\overline{S'_{\rm p}}(t) \equiv \nabla_{\rm p} S_{\rm r,st}(\overline{\mathbf{\Gamma}}(t), t)$, everything on the right-hand side is evaluated at $(\mathbf{\Gamma}, t)$.

Most Likely Force for Mechanical Work

For the case of mechanical work, an explicit expression for the most likely reservoir force is readily obtained. As was shown above in §8.2.1, the equilibrium or static part of the reservoir entropy was proportional to the Hamiltonian, $S_{\rm r,st}(\mathbf{\Gamma}, t) = -\beta \mathcal{H}(\mathbf{\Gamma}, t)$, where $\beta = 1/k_{\rm B}T$ and $\mathbf{\Gamma}$ is a point in the sub-system phase space. It was also shown, in §8.1, that the mechanical work term arose from a time dependent external potential that depended only on the position coordinates, so that the sub-system Hamiltonian could be written as $\mathcal{H}(\mathbf{\Gamma}, t) = \mathcal{H}^{\rm bare}(\mathbf{\Gamma}) + U^{\rm ext}(\mathbf{\Gamma}_{\rm q}, t)$. Hence the adiabatic derivative of the Hamiltonian was independent of the momenta, $\dot{\mathcal{H}}^0(\mathbf{\Gamma}, t) = \partial U^{\rm ext}(\mathbf{\Gamma}_{\rm q}, t)/\partial t$. From these results it follows that the momentum gradient of the adiabatic flux vanishes,

$$\nabla_{\rm p} \dot{S}^0_{\rm r,st}(\mathbf{\Gamma}, t) = 0. \tag{8.112}$$

This means that the constraint discussed above is automatically satisfied, $\overline{\mathbf{R}} \cdot \nabla_\mathrm{p} \dot{S}^0_{\mathrm{r,st}}(\mathbf{\Gamma}, t) = 0$.

The most likely reservoir force in the case of mechanical work is explicitly $\overline{\mathbf{R}} = (\sigma/2k_\mathrm{B}) \nabla_\mathrm{p} S_{\mathrm{r,st}}(\mathbf{\Gamma}, t) = (-\beta\sigma/2) \nabla_\mathrm{p} \mathcal{H}(\mathbf{\Gamma}, t)$, $\Delta_t > 0$. In component form this is

$$\overline{R}_{i\alpha} = \frac{-\beta\sigma}{2} \frac{\partial \mathcal{H}(\mathbf{\Gamma}, t)}{\partial p_{i\alpha}} = \frac{-\beta\sigma}{2m_i} p_{i\alpha}, \ \Delta_t > 0. \tag{8.113}$$

This has the same functional form as in the equilibrium case, namely it appears as a friction or drag force, with the coefficient related to the strength of the stochastic perturbations by the generalised fluctuation dissipation theorem.

8.4 Changes in Entropy and Time Derivatives

There is a difference between the meanings of the words 'transition' and 'change' as occurs in the physical universe and as used in mathematics. A physical transition only occurs forward in time, and physical change means the future value minus the past value. A mathematical transition can be either forward or backward in time, and mathematical change means the final state minus the initial state.

To be clear on this point, consider two states x_1 and x_2 at the respective times $t_1 < t_2$. A physical transition is always from the past to the future, $x_1 \to x_2$, whereas a mathematical transition can be either forwards, $x_1 \mapsto x_2$, or backwards, $x_2 \mapsto x_1$, in time. The use of the symbol $\mapsto$ to denote a mathematical transition reflects the conditional probabilities $\wp(x_2|x_1)$ and $\wp(x_1|x_2)$ where they most often occur. In this example the physical change is $\Delta f = f(x_2, t_2) - f(x_1, t_1)$, irrespective of the type of mathematical transition, whereas mathematical change is $\Delta f = f(x_2, t_2) - f(x_1, t_1)$ for the forward transition $x_1 \mapsto x_2$ and it is $\Delta f = f(x_1, t_1) - f(x_2, t_2)$ for the backward transition $x_2 \mapsto x_1$. Hence for a forward transition, physical change and mathematical change are the same, whereas for a backward transition they are equal and opposite.

It ought to be clear that there is nothing unphysical about the mathematical concept of a backward transition. The question 'where did the system come from?' is perfectly legitimate. The most likely answer to that question is the backward mathematical transition, $\overline{x}_1 \equiv x(t_1|x_2, t_2)$. This is used in answering the question 'what is the most likely physical change in reaching the current state?', namely $\Delta f = f(x_2, t_2) - f(\overline{x}_1, t_1)$.

8.4.1 Change in Entropy

One has to distinguish between the change in entropy due to a transition, $\vec{\Delta} S_\mathrm{r}$, and the difference in entropy between the end points of a transition, ΔS_r. In general these are not equal, but for one particular case they are equal, as detailed below. Both change and difference in this subsection is defined as the physical quantity, namely the future minus the past.

For the physical transition $\mathbf{\Gamma}_1 \stackrel{\Delta_t}{\to} \mathbf{\Gamma}_2$, $\Delta_t > 0$, the end points are related by $\mathbf{\Gamma}_2 = \mathbf{\Gamma}_1 + \Delta_t \dot{\mathbf{\Gamma}}^0 + \mathbf{R}$, and the *change* in entropy is

$$\begin{aligned} \vec{\Delta} S_{\rm r} &= S_{\rm r}(\mathbf{\Gamma}_2, t_2 | \mathbf{\Gamma}_1, t_1) - S_{\rm r}(\mathbf{\Gamma}_1, t_1) \\ &= \mathbf{R} \cdot \nabla S_{\rm r,st}(\mathbf{\Gamma}, t), \end{aligned} \tag{8.114}$$

with $\mathbf{R} = \overline{\mathbf{R}} + \tilde{\mathbf{R}}$. The justification for this result is that the reservoir entropy can only change via exchange of a conserved variable with the sub-system. The reservoir force in the stochastic dissipative equations of motion is the mechanism by which this exchange occurs.[13] In general, the static part of the reservoir entropy is defined from such conservation laws.

For the same general physical transition, $\mathbf{\Gamma}_1 \stackrel{\Delta_t}{\to} \mathbf{\Gamma}_2$, $\Delta_t \equiv t_{21} > 0$, one can again write $\mathbf{\Gamma}_2 = \mathbf{\Gamma}_1 + \Delta_t \dot{\mathbf{\Gamma}}^0 + \mathbf{R}(\Delta_t)$. Going forward in time, the Taylor expansion of the expression for the reservoir entropy derived in the preceding sections gives the *difference* in entropy,

$$\begin{aligned} \Delta S_{\rm r} &\equiv S_{\rm r}(\mathbf{\Gamma}_2, t_2) - S_{\rm r}(\mathbf{\Gamma}_1, t_1) \\ &= \Delta_t \dot{S}_{\rm r}^0(\mathbf{\Gamma}, t) + \mathbf{R}(\Delta_t) \cdot \nabla_{\rm p} S_{\rm r}(\mathbf{\Gamma}, t) \qquad (8.115) \\ &= \Delta_t \left[\dot{S}^0_{\rm r,st}(\mathbf{\Gamma}, t) + \dot{S}^0_{\rm r,dyn}(\mathbf{\Gamma}, t) \right] + \mathbf{R}(\Delta_t) \cdot \nabla_{\rm p} \left[S_{\rm r,st}(\mathbf{\Gamma}, t) + S_{\rm r,dyn}(\mathbf{\Gamma}, t) \right]. \end{aligned}$$

The quadratic term is not required here because the primary interest is in the entropy difference for a particular reservoir force, rather than the average entropy difference. The adiabatic evolution here and below is always

$$\begin{aligned} \dot{S}_{\rm r}^0(\mathbf{\Gamma}, t) &\equiv \left(\frac{\partial S_{\rm r}(\mathbf{\Gamma}, t)}{\partial t} \right)_{\mathbf{\Gamma}} + \dot{\mathbf{\Gamma}}^0 \cdot \nabla S_{\rm r}(\mathbf{\Gamma}, t) \\ &= \left(\frac{\partial S_{\rm r}(\mathbf{\Gamma}, t)}{\partial t} \right)_{\gamma} + \dot{\gamma}^0 \cdot \nabla S_{\rm r}(\mathbf{\Gamma}, t) - \left[\frac{{\rm d}\overline{\mathbf{\Gamma}}(t)}{{\rm d}t} - \overline{\dot{\mathbf{\Gamma}}^0}(t) \right] \cdot \nabla S_{\rm r}(\mathbf{\Gamma}, t) \\ &= \left(\frac{\partial S_{\rm r}(\mathbf{\Gamma}, t)}{\partial t} \right)_{\gamma} - \frac{1}{|\Delta_t|} \overline{\overline{\mathbf{R}}}(t, |\Delta_t|) \cdot \nabla S_{\rm r}(\mathbf{\Gamma}, t) + \mathcal{O}(\gamma^3). \qquad (8.116) \end{aligned}$$

The final equality uses Eqs (8.81) and (8.92) (see also Eq. (8.126) below).

In general, for an arbitrary physical transition $\mathbf{\Gamma}_1 \stackrel{\Delta_t}{\to} \mathbf{\Gamma}_2$, $\Delta_t > 0$, the change in entropy is not equal to the difference in entropy,

$$\vec{\Delta} S_{\rm r} \neq \Delta S_{\rm r}, \quad \mathbf{R} = \mathbf{\Gamma}_2 - \mathbf{\Gamma}_1 - \Delta_t \dot{\mathbf{\Gamma}}^0 \text{ (arbitrary)}. \tag{8.117}$$

However, for the particular case that the initial point of the transition is the most likely prior point of $\mathbf{\Gamma}_2$,

$$\begin{aligned} \mathbf{\Gamma}_1 &= \overline{\mathbf{\Gamma}}(t_1 | \mathbf{\Gamma}_2, t_2) \\ &= \mathbf{\Gamma}_2 - \Delta_t \dot{\mathbf{\Gamma}}^0 + \overline{\mathbf{R}}(-\Delta_t), \quad \Delta_t \equiv t_2 - t_1 > 0, \end{aligned} \tag{8.118}$$

[13] This is in essence Newton's third law, which in turn is the basis of the First Law of Thermodynamics.

the difference in entropy is by design equal to the physical change in entropy for the transition. Formally the latter is given by Eq. (8.114),

$$\vec{\Delta} S_{\rm r} = S_{\rm r}(\mathbf{\Gamma}_2, t_2|\overline{\mathbf{\Gamma}}_1, t_1) - S_{\rm r}(\overline{\mathbf{\Gamma}}_1, t_1) = \overline{\mathbf{R}}(-\Delta_t) \cdot \nabla S_{\rm r,st}(\mathbf{\Gamma}, t). \tag{8.119}$$

For the change in entropy, one evaluates the Taylor expansion from the formal expression given above, $S_{\rm r}(\mathbf{\Gamma}, t) = S_{\rm r,st}(\mathbf{\Gamma}, t) + S_{\rm r,dyn}(\mathbf{\Gamma}, t)$. Because the dynamic part of the reservoir entropy, Eq. (8.15), is an integral over the backward most likely trajectory, by the fundamental theorem of calculus

$$\overline{\Delta} S_{\rm r,dyn} \equiv S_{\rm r,dyn}(\mathbf{\Gamma}_2, t_2) - S_{\rm r,dyn}(\overline{\mathbf{\Gamma}}_1, t_1) = -\Delta_t \dot{S}^0_{\rm r,st}(\mathbf{\Gamma}, t). \tag{8.120}$$

With this

$$\begin{aligned}
\overline{\Delta} S_{\rm r} &\equiv S_{\rm r}(\mathbf{\Gamma}_2, t_1) - S_{\rm r}(\overline{\mathbf{\Gamma}}_1, t_1) \\
&= \overline{\Delta} S_{\rm r,st} - \overline{\Delta} S_{\rm r,dyn} \\
&= \Delta_t \dot{S}^0_{\rm r,st}(\mathbf{\Gamma}, t) + \overline{\mathbf{R}}(-\Delta_t) \cdot \nabla_{\rm p} S_{\rm r,st}(\mathbf{\Gamma}, t) - \Delta_t \dot{S}^0_{\rm r,st}(\mathbf{\Gamma}, t) \\
&= \overline{\mathbf{R}}(-\Delta_t) \cdot \nabla_{\rm p} S_{\rm r,st}(\mathbf{\Gamma}, t) \\
&= \vec{\Delta} S_{\rm r}, \ \ \overline{\mathbf{\Gamma}}_1 \mapsto \mathbf{\Gamma}_2, \ \ \Delta_t \equiv t_2 - t_1 > 0.
\end{aligned} \tag{8.121}$$

This equality of the change in entropy and the difference in entropy on the most likely backward transition will play a central rôle in a number of the results below.

The Taylor expansion of the change in the dynamic part of the reservoir entropy on the most likely backward transition is

$$\overline{\Delta} S_{\rm r,dyn} = \Delta_t \dot{S}^0_{\rm r,dyn}(\mathbf{\Gamma}, t) + \overline{\mathbf{R}}(-\Delta_t) \cdot \nabla S_{\rm r,dyn}(\mathbf{\Gamma}, t) + \mathcal{O}(\Delta_t^2), \ \ \Delta_t > 0. \tag{8.122}$$

Equating this to Eq. (8.120) and adding together the two adiabatic derivatives, one deduces that the adiabatic rate of change of the reservoir entropy is

$$\begin{aligned}
\dot{S}^0_{\rm r}(\mathbf{\Gamma}, t) &= \frac{-1}{\Delta_t} \overline{\mathbf{R}}(-\Delta_t) \cdot \nabla S_{\rm r,dyn}(\mathbf{\Gamma}, t), \ \Delta_t > 0, \\
&= \frac{-\sigma}{2 k_{\rm B} |\Delta_t|} \left[\nabla_{\rm p} S_{\rm r,st}(\mathbf{\Gamma}, t) - 2 \overline{S'_{\rm p}}(t) \right] \cdot \overline{S'_{\rm p}}(t).
\end{aligned} \tag{8.123}$$

The second equality invokes Eqs (8.26) and (8.94) above, the latter of which invokes Eq. (8.126) below. The first equality was deduced for $\Delta_t > 0$, but the final result is independent of the sign of the time step. In fact, because $\sigma \propto |\Delta_t|$, one can see that the final result is independent of both the sign and the magnitude of the time step. It ought to be stressed that this is the mathematical expression for the adiabatic rate of change of the most likely estimate of the reservoir entropy, $S_{\rm r}(\mathbf{\Gamma}, t) = S_{\rm r,st}(\mathbf{\Gamma}, t) + S_{\rm r,dyn}(\mathbf{\Gamma}, t)$. It is not simply related to the rate of change of the actual reservoir entropy on a trajectory; it is more closely related to the difference rather than the change in reservoir entropy, as distinguished above.

It remains to obtain an explicit expression for two time dependent constants used above, namely $\overline{\overline{\mathbf{R}}}(t)$ and $c(t)$. The quantity $\overline{\overline{\mathbf{R}}}(t)$, which was introduced just before Eq. (8.93), is the most likely reservoir force acting on the actual coordinates on the most likely trajectory going forward in time. The quantity $c(t)$ is the configuration independent part of the fluctuation form for the reservoir entropy, $S_{\rm r}(\mathbf{\Gamma},t) = c(t) + \overline{S''}(t)\gamma^2/2$, which first appeared back in Eq. (8.24). This was assumed without proof in that equation and in all subsequent equations to represent the most likely value of the reservoir entropy, and it was denoted $\overline{S}_{\rm r}(t)$. Here it will finally be proven that $c(t) = \overline{S}_{\rm r}(t)$.

The desired results follow from the condition $\overline{\Delta} S_{\rm r}(\mathbf{\Gamma},t) = \overline{\vec{\Delta}} S_{\rm r}(\mathbf{\Gamma},t)$, $\Delta_t < 0$, which is Eq. (8.121). Using the fluctuation form for the reservoir entropy, the left-hand side of Eq. (8.121) is

$$\begin{aligned} \overline{\Delta} S_{\rm r}(\mathbf{\Gamma},t) &= -|\Delta_t| \left(\frac{\partial S_{\rm r}(\mathbf{\Gamma},t)}{\partial t}\right)_\gamma - |\Delta_t|\dot{\gamma}^0 \cdot \nabla S_{\rm r}(\mathbf{\Gamma},t) + \overline{\mathbf{R}}_\gamma \cdot \nabla S_{\rm r}(\mathbf{\Gamma},t) \\ &= -|\Delta_t|\dot{c}(t) - \mathcal{O}(\gamma^3) + \frac{|\Delta_t|}{2}\gamma_{\rm p} \cdot \overline{S''_{\rm pp}}\Lambda_{\rm pp}\overline{S''}\gamma_{\rm p} + \mathcal{O}(\gamma^3) \\ &= -|\Delta_t|\dot{c}(t) + \mathcal{O}(\gamma^2). \end{aligned} \tag{8.124}$$

This uses $S_{\rm r}(\mathbf{\Gamma},t) = c(t) + \overline{S''}(t)\gamma^2/2$, which is Eq. (8.24), $\dot{\gamma}^0 \cdot \nabla S_{\rm r}(\mathbf{\Gamma},t) = \mathcal{O}(\gamma^3)$, which is Eq. (8.81), and $\overline{\mathbf{R}}_\gamma = (|\Delta_t|/2)\Lambda_{\rm pp}(t)\overline{S''_{\rm pp}}(t)\gamma + \mathcal{O}(\gamma^2)$, which is Eq. (8.87). The right-hand side of Eq. (8.121) is

$$\begin{aligned} &\overline{\vec{\Delta}} S_{\rm r}(\mathbf{\Gamma},t) \\ &= \overline{\mathbf{R}} \cdot \nabla S_{\rm r,st}(\mathbf{\Gamma},t),\ \Delta_t < 0 \\ &= \left[\frac{|\Delta_t|}{2}\Lambda_{\rm pp}\overline{S''_{\rm pp}}\gamma_{\rm p} + \mathcal{O}(\gamma^2) - \overline{\overline{\mathbf{R}}}\right] \cdot \left[\overline{S'_{\rm p}} + \overline{S''_{\rm pp}}\gamma_{\rm p} + \mathcal{O}(\gamma^2)\right] \\ &= -\overline{\overline{\mathbf{R}}} \cdot \overline{S'_{\rm p}} + \frac{|\Delta_t|}{2}\overline{S'_{\rm p}} \cdot \Lambda_{\rm pp}\overline{S''_{\rm pp}}\gamma_{\rm p} - \overline{\overline{\mathbf{R}}} \cdot \overline{S''_{\rm pp}}\gamma_{\rm p} + \mathcal{O}(\gamma^2). \end{aligned} \tag{8.125}$$

This uses $\overline{\mathbf{R}} = \overline{\mathbf{R}}_\gamma + \hat{t}\overline{\overline{\mathbf{R}}}$, which is Eq. (8.93), and $\nabla S_{\rm r,st}(\mathbf{\Gamma},t) = \overline{S'}(t) + \overline{S''}(t)\gamma + \mathcal{O}(\gamma^2)$, which is Eq. (8.23).

Equating these two term by term, the coefficients of γ^1 yield

$$\overline{\overline{\mathbf{R}}}(t,|\Delta_t|) = \frac{|\Delta_t|}{2}\Lambda_{\rm pp}(t)\overline{S'_{\rm p}}(t). \tag{8.126}$$

This result was used to give an explicit expression for the reservoir force acting on the actual coordinates, Eq. (8.94) above. Equating the coefficients of γ^0 yields

$$\begin{aligned} \dot{c}(t) &= \frac{1}{|\Delta_t|}\overline{\overline{\mathbf{R}}}(t,|\Delta_t|) \cdot \overline{S'_{\rm p}}(t) \\ &= \frac{1}{2}\overline{S'_{\rm p}}(t) \cdot \Lambda_{\rm pp}(t)\overline{S'_{\rm p}}(t). \end{aligned} \tag{8.127}$$

The right-hand side of the first equality is the definition of the forward rate of reservoir entropy production on the most likely trajectory, $\dot{\overline{S}}_{\rm r}(t)$. This finally justifies the identification $c(t) \equiv \overline{S}_{\rm r}(t)$, which was first introduced in Eq. (8.24).

The second equality, which follows from the preceding expression for $\overline{\overline{\mathbf{R}}}$, shows that the most likely forward rate of reservoir entropy production is positive (because Λ must be positive definite in order for the second entropy to be negative and for the transition probability to be well defined). No information can be extracted from the second-order terms, γ^2, because these have not been carried through on the right-hand side.

8.4.2 Irreversibility and Dissipation

As in the equilibrium case, §§7.4.4 and 7.6.2, in a non-equilibrium system the most likely trajectory is irreversible. Let $\mathbf{\Gamma}_1$ at t_1 be a terminal state. There are two future states of interest at $t_2 > t_1$: the forward transition, $(\mathbf{\Gamma}_1, t_1) \mapsto (\mathbf{\Gamma}_2, t_2)$, and the backward transition, $(\mathbf{\Gamma}_2', t_2) \mapsto (\mathbf{\Gamma}_1, t_1)$.[14] For the forward transition, the most likely future state is $\mathbf{\Gamma}_2 = \overline{\mathbf{\Gamma}}(t_2|\mathbf{\Gamma}_1, t_1) = \mathbf{\Gamma}_1 + \Delta_t \dot{\mathbf{\Gamma}}^0 + \overline{\mathbf{R}}(\Delta_t)$, $\Delta_t \equiv t_2 - t_1 > 0$. For the the backward transition, there is the state $\mathbf{\Gamma}_2'$ of which $\mathbf{\Gamma}_1$ is the most likely preceding state, $\mathbf{\Gamma}_1 = \overline{\mathbf{\Gamma}}(t_1|\mathbf{\Gamma}_2', t_2)$, or, equivalently, $\mathbf{\Gamma}_1 = \mathbf{\Gamma}_2' - \Delta_t \dot{\mathbf{\Gamma}}^0 + \overline{\mathbf{R}}(-\Delta_t)$, $\Delta_t \equiv t_2 - t_1 > 0$. Note that it is the same most likely force that appears in both cases, except that part of it depends upon the sign of the time step. The difference between these two future states is

$$\begin{aligned} \mathbf{\Gamma}_2 - \mathbf{\Gamma}_2' &= \mathbf{\Gamma}_1 + \Delta_t \dot{\mathbf{\Gamma}}^0 + \overline{\mathbf{R}}(\Delta_t) - \left[\mathbf{\Gamma}_1 + \Delta_t \dot{\mathbf{\Gamma}}^0 - \overline{\mathbf{R}}(-\Delta_t) \right] \\ &= \overline{\mathbf{R}}_\gamma + \Delta_t \frac{\mathrm{d}\overline{\mathbf{\Gamma}}(t)}{dt} - \Delta_t \overline{\dot{\mathbf{\Gamma}}^0}(t) + \left[\overline{\mathbf{R}}_\gamma - \Delta_t \frac{\mathrm{d}\overline{\mathbf{\Gamma}}(t)}{dt} + \Delta_t \overline{\dot{\mathbf{\Gamma}}^0}(t) \right] \\ &= 2\overline{\mathbf{R}}_\gamma \\ &= \sigma k_{\mathrm{B}}^{-1} \nabla_{\mathrm{p}} S_{\mathrm{r}}(\mathbf{\Gamma}, t) + \mathcal{O}(\Delta_t^2). \end{aligned} \tag{8.128}$$

The irreversibility of the most likely trajectory is manifest in the fact that $\mathbf{\Gamma}_2 \neq \mathbf{\Gamma}_2'$. The irreversibility is proportional to the magnitude of the gradient of the reservoir entropy. This means that the closer the system is to the optimum state, the more reversible is the trajectory, and *vice versa*.

The mathematical transition $(\mathbf{\Gamma}_1, t_1) \mapsto (\mathbf{\Gamma}_2, t_2)$ is much more likely than its reverse $(\mathbf{\Gamma}_2, t_2) \mapsto (\mathbf{\Gamma}_1, t_1)$, because the former uses the most likely force, $\overline{\mathbf{R}}(\Delta_t)$, whereas the latter uses its opposite, $-\overline{\mathbf{R}}(\Delta_t)$, which by definition must be quite unlikely. Hence $\wp(\mathbf{\Gamma}_2, t_2|\mathbf{\Gamma}_1, t_1) \gg \wp(\mathbf{\Gamma}_1, t_1|\mathbf{\Gamma}_2, t_2)$. For the same reason $\wp(\mathbf{\Gamma}_1, t_1|\mathbf{\Gamma}_2', t_2) \gg \wp(\mathbf{\Gamma}_2', t_2|\mathbf{\Gamma}_1, t_1)$. The two most likely transitions have about the same likelihood, and the two unlikely transitions are about equally unlikely. Hence conditioned on the current point $\mathbf{\Gamma}_1$, the probabilities of the two future points are related such that $\wp(\mathbf{\Gamma}_2, t_2|\mathbf{\Gamma}_1, t_1) \gg \wp(\mathbf{\Gamma}_2', t_2|\mathbf{\Gamma}_1, t_1)$. This says that given the current point $\mathbf{\Gamma}_1$, it is much more likely that the system will go to $\mathbf{\Gamma}_2$ than it will go to $\mathbf{\Gamma}_2'$ even though $\mathbf{\Gamma}_1$ is the most likely prior state of $\mathbf{\Gamma}_2'$. A

[14] As mentioned above, in footnote 4 on p. 241 and in §8.4 on p. 262, the notation $a \mapsto b$ is used to denote the mathematical transition conditioned on the state a at time t_a, corresponding to $\wp(b, t_b|a, t_a)$, irrespective of which is the past state and which is the future state. If $t_b > t_a$ it is called a forward transition, and if $t_b < t_a$ it is called a backward or reverse transition. The notation $a \rightarrow b$ is used to denote the physical transition for $t_b > t_a$.

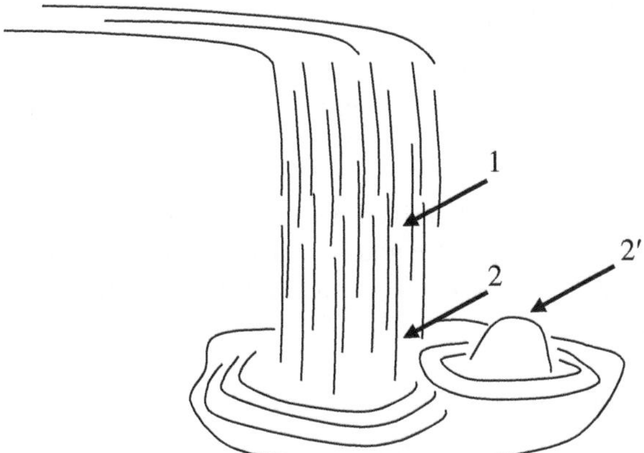

Figure 8.1: Idyllic waterfall. A rock 2′ is splashed by a small amount of water that most likely came from 1. Most of the water from 1 goes to 2.

physical analogy of the situation is depicted in Fig. 8.1. A sketch of the phase trajectories is given in Fig. 7.4 on p. 222.

Now the change in entropy for the most likely next state, $(\mathbf{\Gamma}_1, t_1) \to (\mathbf{\Gamma}_2, t_2)$, will be compared to that for the transition from the most likely previous state, $(\mathbf{\Gamma}_1, t_1) \to (\mathbf{\Gamma}'_2, t_2)$. (Note these are both physical transitions. Note also that change is defined as future minus past.) Using $\mathbf{\Gamma}_2 = \mathbf{\Gamma}_1 + \Delta_t \dot{\mathbf{\Gamma}}^0 + \overline{\mathbf{R}}(\Delta_t)$, the change in entropy in the transition *to* the most likely state is

$$\begin{aligned} \vec{\Delta} S_{\rm r}(\mathbf{\Gamma}_1 \to \mathbf{\Gamma}_2) &= \overline{\mathbf{R}}(\mathbf{\Gamma}, t, \Delta_t) \cdot \nabla_{\rm p} S_{\rm r,st}(\mathbf{\Gamma}, t) \\ &= \frac{\sigma}{2k_{\rm B}} \nabla_{\rm p} S_{\rm r,st}(\mathbf{\Gamma}, t) \cdot \nabla_{\rm p} S_{\rm r,st}(\mathbf{\Gamma}, t). \end{aligned} \tag{8.129}$$

This uses Eq. (8.94), $\overline{\mathbf{R}} = (\sigma/2k_{\rm B})[\nabla_{\rm p} S_{\rm r,st}(\mathbf{\Gamma}, t) + (\hat{t}-1)\overline{S'_{\rm p}}(t)]$, the second term vanishing in this case since $\Delta_t > 0$. The change in entropy is non-negative, which means that the entropy increases in the most likely forward transition.

Using $\mathbf{\Gamma}_1 = \mathbf{\Gamma}'_2 - \Delta_t \dot{\mathbf{\Gamma}}^0 + \overline{\mathbf{R}}(-\Delta_t)$, the change in entropy in the transition *from* the most likely prior state is

$$\begin{aligned} \vec{\Delta} S_{\rm r}(\mathbf{\Gamma}_1 \to \mathbf{\Gamma}'_2) &= -\overline{\mathbf{R}}(\mathbf{\Gamma}, t, -\Delta_t) \cdot \nabla_{\rm p} S_{\rm r,st}(\mathbf{\Gamma}, t) \\ &= \frac{-\sigma}{2k_{\rm B}} \left[\nabla_{\rm p} S_{\rm r,st}(\mathbf{\Gamma}, t) - 2\overline{S'_{\rm p}}(t)\right] \cdot \nabla_{\rm p} S_{\rm r,st}(\mathbf{\Gamma}, t). \end{aligned} \tag{8.130}$$

Most likely this is positive (since $\nabla_{\rm p} S_{\rm r,st}(\overline{\mathbf{\Gamma}}(t), t) = \overline{S'_{\rm p}}(t)$), but it can be negative depending upon the fluctuation from the optimum state. In the most likely state, $\mathbf{\Gamma}_1 = \overline{\mathbf{\Gamma}}(t)$, both expressions are equal to $\vec{\overline{\Delta}} S_{\rm r} = \sigma \overline{S'_{\rm p}}(t) \cdot \overline{S'_{\rm p}}(t)/2k_{\rm B}$, which is positive. (This in fact is the maximum value of $\vec{\Delta} S_{\rm r}(\mathbf{\Gamma}_1 \to \mathbf{\Gamma}'_2)$.)

These two results for the non-equilibrium system are the same as in the equilibrium case, §7.4.4, and are consistent with what was deduced from fluctuation theory in §2.4.6. In those cases there was no term $\overline{S}_{\rm r}(t)$.

8.4.3 Various Time Derivatives

In this section, various results for the time derivatives of the entropy, probability density, and partition function are given. This is done to illustrate various techniques for obtaining such derivatives, and also to show the physical content of the theory.

In general the expressions derived below will hold for both positive and negative time steps, unless explicitly stated otherwise. Hence change will generally be used in the mathematical sense (final state minus initial state), as discussed in §8.4 on p. 262.

General Expressions

The rate of change of a phase function, $f(\mathbf{\Gamma}, t)$, for a transition $(\mathbf{\Gamma}_1, t_1) \mapsto (\mathbf{\Gamma}_2, t_2)$ characterised by the stochastic equations of motion, $\mathbf{\Gamma}_2 = \mathbf{\Gamma}_1 + \Delta_t \dot{\mathbf{\Gamma}}^0 + \mathbf{R}$, $\Delta_t \equiv t_2 - t_1$, is the same as in the equilibrium case, Eq. (7.112),

$$\frac{\mathrm{d}f(\mathbf{\Gamma}, t)}{\mathrm{d}t} = \dot{f}^0(\mathbf{\Gamma}, t) + \frac{1}{\Delta_t}\mathbf{R} \cdot \nabla_\mathrm{p} f(\mathbf{\Gamma}, t) + \frac{1}{2\Delta_t}\mathbf{R}\mathbf{R} : \nabla_\mathrm{p}\nabla_\mathrm{p} f(\mathbf{\Gamma}, t). \tag{8.131}$$

The superscript 0 denotes the total adiabatic change,

$$\dot{f}^0(\mathbf{\Gamma}, t) = \frac{\partial f(\mathbf{\Gamma}, t)}{\partial t} + \dot{\mathbf{\Gamma}}^0 \cdot \nabla f(\mathbf{\Gamma}, t). \tag{8.132}$$

Averaged over the stochastic forces the total derivative is[15]

$$\left\langle \frac{\mathrm{d}f(\mathbf{\Gamma}, t)}{\mathrm{d}t} \right\rangle_{\tilde{\mathbf{R}}} = \dot{f}^0(\mathbf{\Gamma}, t) + \frac{1}{\Delta_t}\overline{\mathbf{R}} \cdot \nabla_\mathrm{p} f(\mathbf{\Gamma}, t) + \frac{\sigma}{2\Delta_t}\nabla_\mathrm{p}^2 f(\mathbf{\Gamma}, t). \tag{8.133}$$

The rate of change on a most likely transition is

$$\dot{\overline{f}}(\mathbf{\Gamma}, t) = \dot{f}^0(\mathbf{\Gamma}, t) + \frac{1}{\Delta_t}\overline{\mathbf{R}} \cdot \nabla_\mathrm{p} f(\mathbf{\Gamma}, t), \tag{8.134}$$

since $\overline{\mathbf{R}}\,\overline{\mathbf{R}} \sim \mathcal{O}(\Delta_t^2)$. Note that this is *not* the most likely rate of change.

Partition Function

The probability density is simply the exponential of the reservoir entropy

$$\wp(\mathbf{\Gamma}, t) = \frac{e^{S_\mathrm{r}(\mathbf{\Gamma}, t)/k_\mathrm{B}}}{Z(t)}, \tag{8.135}$$

which is normalised to unity by the partition function

$$Z(t) = \int \mathrm{d}\mathbf{\Gamma}\, e^{S_\mathrm{r}(\mathbf{\Gamma}, t)/k_\mathrm{B}}. \tag{8.136}$$

[15]Here and below an average over the stochastic forces is denoted by the subscript $\tilde{\mathbf{R}}$. Averages over phase space, which occur most commonly in what follows, have no subscript.

Taking the time derivative of this shows that

$$\frac{\dot{Z}(t)}{Z(t)} = \int \mathrm{d}\boldsymbol{\Gamma}\, \wp(\boldsymbol{\Gamma},t) \frac{\partial S_\mathrm{r}(\boldsymbol{\Gamma},t)}{k_\mathrm{B}\partial t} \equiv \left\langle \frac{\partial S_\mathrm{r}(\boldsymbol{\Gamma},t)}{k_\mathrm{B}\partial t} \right\rangle . \tag{8.137}$$

Since, $S_\mathrm{total}(t) = k_\mathrm{B} \ln Z(t)$, this is essentially the rate of change of the unconstrained total entropy.

Using the fluctuation form for the reservoir entropy, Eq. (8.24), one also has

$$\begin{aligned} Z(t) &= \int \mathrm{d}\gamma\, e^{[\overline{S}_\mathrm{r}(t) + \overline{S''}(t):\gamma\gamma/2]/k_\mathrm{B}} \\ &= e^{\overline{S}_\mathrm{r}(t)/k_\mathrm{B}} \sqrt{\left|2\pi k_\mathrm{B} \overline{S''}(t)^{-1}\right|}. \end{aligned} \tag{8.138}$$

The exponential term dominates this. According to Eq. (8.127), $\overline{S}_\mathrm{r}(t)$ represents the total entropy produced in the reservoir to date. Hence the logarithmic derivative of the partition function is (neglecting the variation of the second factor),

$$\frac{\dot{Z}(t)}{Z(t)} = \frac{\dot{\overline{S}}_\mathrm{r}(t)}{k_\mathrm{B}} = \frac{\sigma}{2k_\mathrm{B}^2|\Delta_t|} \overline{S'_\mathrm{p}}(t) \cdot \overline{S'_\mathrm{p}}(t). \tag{8.139}$$

Invoking the adiabatic development, $\boldsymbol{\Gamma}_0 \equiv \boldsymbol{\Gamma} + \Delta_t \dot{\boldsymbol{\Gamma}}^0$, one also has

$$\begin{aligned} Z(t) &= \int \mathrm{d}\boldsymbol{\Gamma}\, e^{S_\mathrm{r}(\boldsymbol{\Gamma},t)/k_\mathrm{B}} \\ &= \int \mathrm{d}\boldsymbol{\Gamma}_0\, e^{\left[S_\mathrm{r}(\boldsymbol{\Gamma}_0,t) - \Delta_t \dot{\boldsymbol{\Gamma}}^0 \cdot \nabla S_\mathrm{r}(\boldsymbol{\Gamma}_0,t)\right]/k_\mathrm{B}} \\ &= \int \mathrm{d}\boldsymbol{\Gamma}_0\, e^{S_\mathrm{r}(\boldsymbol{\Gamma}_0,t)/k_\mathrm{B}} \left[1 - \Delta_t \dot{\boldsymbol{\Gamma}}^0 \cdot \nabla S_\mathrm{r}(\boldsymbol{\Gamma}_0,t)/k_\mathrm{B} + \mathcal{O}(\Delta_t^2)\right]. \end{aligned} \tag{8.140}$$

From the definition of the partition function the excess must vanish, which shows that

$$\left\langle \dot{\boldsymbol{\Gamma}}^0(\boldsymbol{\Gamma},t) \cdot \nabla S_\mathrm{r}(\boldsymbol{\Gamma},t) \right\rangle = 0. \tag{8.141}$$

This together with Eq. (8.137) shows that $\left\langle \dot{S}_\mathrm{r}^0(\boldsymbol{\Gamma},t) \right\rangle = k_\mathrm{B}\dot{Z}(t)/Z(t)$.

Writing the left-hand side of Eq. (8.141) as $\left\langle [\dot{\gamma}^0(\boldsymbol{\Gamma},t) + \overline{\dot{\boldsymbol{\Gamma}}^0}(t)] \cdot \nabla S_\mathrm{r}(\boldsymbol{\Gamma},t) \right\rangle$, and noting that $\langle \nabla S_\mathrm{r}(\boldsymbol{\Gamma},t) \rangle = \nabla \overline{S}_\mathrm{r}(t) = 0$, one sees that it is equivalent to

$$\langle \dot{\gamma}^0(\boldsymbol{\Gamma},t) \cdot \nabla S_\mathrm{r}(\boldsymbol{\Gamma},t) \rangle = 0. \tag{8.142}$$

This is consistent with, but weaker than, Eq. (8.81). Using $\langle \dot{\gamma}^0(\boldsymbol{\Gamma},t) \rangle = \overline{\dot{\gamma}^0}(t) \propto \overline{\gamma}(t) = 0$, and $\nabla S_\mathrm{r}(\boldsymbol{\Gamma},t) = \nabla S_\mathrm{r,st}(\boldsymbol{\Gamma},t) - \overline{S'}(t)$, this is also equivalent to

$$\left\langle \dot{\boldsymbol{\Gamma}}^0(\boldsymbol{\Gamma},t) \cdot \nabla S_\mathrm{r,st}(\boldsymbol{\Gamma},t) \right\rangle = \overline{\dot{\boldsymbol{\Gamma}}^0}(t) \cdot \overline{S'}(t). \tag{8.143}$$

Probability Density

Using the result established above that the average of the adiabatic evolution of the reservoir entropy is equal to the logarithmic derivative of the partition function, $\left\langle \dot{S}_{\rm r}^{0}(\mathbf{\Gamma},t)\right\rangle = k_{\rm B}\dot{Z}(t)/Z(t)$, one sees that the adiabatic derivative of the non-equilibrium probability density vanishes when integrated over phase space,

$$\int {\rm d}\mathbf{\Gamma}\, \dot{\wp}^{0}(\mathbf{\Gamma},t) = \int {\rm d}\mathbf{\Gamma} \left[\frac{-\dot{Z}(t)}{Z(t)} + \frac{\dot{S}_{\rm r}^{0}(\mathbf{\Gamma},t)}{k_{\rm B}} \right] \wp(\mathbf{\Gamma},t) = 0. \tag{8.144}$$

The total rate of change of the non-equilibrium probability density on a transition is

$$\begin{aligned} \frac{{\rm d}\wp(\mathbf{\Gamma},t)}{{\rm d}t} &= \dot{\wp}^{0}(\mathbf{\Gamma},t) + \frac{\wp(\mathbf{\Gamma},t)}{k_{\rm B}\Delta_t} \Big\{ \mathbf{R}\cdot\nabla_{\rm p}S_{\rm r}(\mathbf{\Gamma},t) + \frac{1}{2}\mathbf{R}\mathbf{R}:\nabla_{\rm p}\nabla_{\rm p}S_{\rm r}(\mathbf{\Gamma},t) \\ &\quad + \frac{1}{2k_{\rm B}}\left[\mathbf{R}\cdot\nabla_{\rm p}S_{\rm r}(\mathbf{\Gamma},t)\right]^2 \Big\}. \end{aligned} \tag{8.145}$$

The derivative for the most likely backward transition is

$$\begin{aligned} \dot{\overline{\wp}}(\mathbf{\Gamma},t) &= \frac{-\dot{Z}(t)}{Z(t)}\wp(\mathbf{\Gamma},t) + \frac{{\rm d}^{\rm det}S_{\rm r}(\mathbf{\Gamma},t)}{k_{\rm B}{\rm d}t}\wp(\mathbf{\Gamma},t) \\ &= \frac{-\dot{Z}(t)}{Z(t)}\wp(\mathbf{\Gamma},t) + \frac{1}{k_{\rm B}\Delta_t}\overline{\mathbf{R}}\cdot\left[\nabla_{\rm p}S_{\rm r,st}(\mathbf{\Gamma},t)\right]\wp(\mathbf{\Gamma},t), \quad \Delta_t < 0 \\ &= \frac{-\sigma}{2k_{\rm B}^2|\Delta_t|}\overline{S'_{\rm p}}(t)\cdot\overline{S'_{\rm p}}(t)\wp(\mathbf{\Gamma},t) \\ &\quad - \frac{\sigma}{2k_{\rm B}^2|\Delta_t|}\left[\nabla_{\rm p}S_{\rm r,st}(\mathbf{\Gamma},t) - 2\overline{S'_{\rm p}}(t)\right]\cdot\left[\nabla_{\rm p}S_{\rm r,st}(\mathbf{\Gamma},t)\right]\wp(\mathbf{\Gamma},t) \\ &= -\frac{\sigma}{2k_{\rm B}^2|\Delta_t|}\gamma_{\rm p}\cdot\overline{S''_{\rm pp}}\,\overline{S''_{\rm pp}}\gamma_{\rm p}. \end{aligned} \tag{8.146}$$

Actually, since the final equality uses the fluctuation expression, Eq. (8.23), $\nabla S_{\rm r,st}(\mathbf{\Gamma},t) = \overline{S'}(t) + \overline{S''}(t)\gamma + \mathcal{O}(\gamma^2)$, which has neglected terms $\mathcal{O}(\gamma^2)$, this is technically zero.

Averaged over the stochastic forces the full derivative is

$$\begin{aligned} \left\langle \frac{{\rm d}\wp(\mathbf{\Gamma},t)}{{\rm d}t} \right\rangle_{\tilde{\mathbf{R}}} &= \frac{-\dot{Z}(t)}{Z(t)}\wp(\mathbf{\Gamma},t) + \frac{{\rm d}^{\rm det}S_{\rm r}(\mathbf{\Gamma},t)}{k_{\rm B}{\rm d}t}\wp(\mathbf{\Gamma},t) + \frac{\wp(\mathbf{\Gamma},t)}{k_{\rm B}^2\Delta_t} \\ &\quad \times\left\{ \frac{\sigma}{2}\nabla_{\rm p}^2 S_{\rm r}(\mathbf{\Gamma},t) + \frac{\sigma}{2k_{\rm B}}\left[\nabla_{\rm p}S_{\rm r}(\mathbf{\Gamma},t)\right]^2 \right\} \\ &= \frac{-\dot{Z}(t)}{Z(t)}\wp(\mathbf{\Gamma},t) + \frac{1}{k_{\rm B}\Delta_t}\overline{\mathbf{R}}\cdot\left[\nabla_{\rm p}S_{\rm r,st}(\mathbf{\Gamma},t)\right]\wp(\mathbf{\Gamma},t) \\ &\quad + \frac{\wp(\mathbf{\Gamma},t)}{k_{\rm B}^2\Delta_t}\left\{ \frac{\sigma}{2}\nabla_{\rm p}^2 S_{\rm r}(\mathbf{\Gamma},t) + \frac{\sigma}{2k_{\rm B}}\left[\nabla_{\rm p}S_{\rm r}(\mathbf{\Gamma},t)\right]^2 \right\}, \\ &\quad \Delta_t < 0. \end{aligned} \tag{8.147}$$

The second equality holds for $\Delta_t < 0$. Recall that the transport matrix has been taken as $\Lambda_{\rm pp} = (\sigma/k_{\rm B}|\Delta_t|)\mathrm{I}_{\rm pp}$. The rate of change on the most likely trajectory neglects the embraced term. Integrating the derivative over phase space yields

$$\begin{aligned}
-|\Delta_t| \int \mathrm{d}\mathbf{\Gamma} & \left\langle \frac{\mathrm{d}\wp(\mathbf{\Gamma},t)}{\mathrm{d}t} \right\rangle_{\tilde{\mathbf{R}}} \\
&= \frac{|\Delta_t|\dot{Z}(t)}{Z(t)} + k_{\rm B}^{-1} \left\langle \overline{\mathbf{R}}(\mathbf{\Gamma},t,-|\Delta_t|) \cdot \nabla_{\rm p} S_{\rm r,st}(\mathbf{\Gamma},t) \right\rangle \\
&\quad + \frac{\sigma}{2k_{\rm B}} \left\langle \nabla_{\rm p}^2 S_{\rm r}(\mathbf{\Gamma},t) \right\rangle + \frac{\sigma}{2k_{\rm B}^2} \left\langle [\nabla_{\rm p} S_{\rm r}(\mathbf{\Gamma},t)] \cdot [\nabla_{\rm p} S_{\rm r}(\mathbf{\Gamma},t)] \right\rangle \\
&= \frac{|\Delta_t|\dot{Z}(t)}{Z(t)} + k_{\rm B}^{-1} \left\langle \overline{\mathbf{R}}(\mathbf{\Gamma},t,-|\Delta_t|) \cdot \nabla_{\rm p} S_{\rm r,st}(\mathbf{\Gamma},t) \right\rangle \\
&= \frac{\sigma}{2k_{\rm B}^2} \left\langle \nabla_{\rm p} S_{\rm r,st}(\mathbf{\Gamma},t) \cdot \nabla_{\rm p} S_{\rm r,st}(\mathbf{\Gamma},t) \right\rangle - \frac{\sigma}{2k_{\rm B}^2} \overline{S'_{\rm p}}(t)^2,
\end{aligned} \tag{8.148}$$

since $\Delta_t < 0$. Using the fluctuation expression, Eq. (8.23), $\nabla S_{\rm r,st}(\mathbf{\Gamma},t) = \overline{S'}(t) + \overline{S''}(t)\gamma + \mathcal{O}(\gamma^2)$, one has $\langle \nabla_{\rm p} S_{\rm r,st}(\mathbf{\Gamma},t) \cdot \nabla_{\rm p} S_{\rm r,st}(\mathbf{\Gamma},t) \rangle_{\mathbf{\Gamma}} = \overline{S'}(t) \cdot \overline{S'}(t) + \mathcal{O}(\overline{S''}(t))$, so that to leading-order this vanishes. The second equality, in which the stochastic terms cancel, follows from the generalised equipartition theorem, (see §3.7.4, Eq. (3.108)),

$$\left\langle \nabla_{\rm p}^2 S_{\rm r}(\mathbf{\Gamma},t) \right\rangle = -k_{\rm B}^{-1} \left\langle [\nabla_{\rm p} S_{\rm r}(\mathbf{\Gamma},t)] \cdot [\nabla_{\rm p} S_{\rm r}(\mathbf{\Gamma},t)] \right\rangle . \tag{8.149}$$

Actually the more general form of this, which follows from $\int \mathrm{d}\mathbf{\Gamma} \nabla_{\rm p} \nabla_{\rm p} \wp(\mathbf{\Gamma},t) = 0$, is

$$\left\langle \nabla_{\rm p} \nabla_{\rm p} S_{\rm r}(\mathbf{\Gamma},t) \right\rangle = -k_{\rm B}^{-1} \left\langle [\nabla_{\rm p} S_{\rm r}(\mathbf{\Gamma},t)][\nabla_{\rm p} S_{\rm r}(\mathbf{\Gamma},t)] \right\rangle . \tag{8.150}$$

This means that the cancellation in the above result occurs also for the general $\Lambda_{\rm pp}(t)$, not just when there is a single variance.

Denoting the previous, $\Delta_t < 0$, most likely point by a prime, $\mathbf{\Gamma}' = \mathbf{\Gamma} + \Delta_t \dot{\mathbf{\Gamma}}^0 + \overline{\mathbf{R}}$, the normalisation condition yields

$$\begin{aligned}
1 &= \int \mathrm{d}\mathbf{\Gamma}' \, \wp(\mathbf{\Gamma}', t+\Delta_t) \\
&= \int \mathrm{d}\mathbf{\Gamma} \, \left[1 + \nabla_{\rm p} \cdot \overline{\mathbf{R}}(\mathbf{\Gamma},t)\right] \\
&\quad \times \left[\wp(\mathbf{\Gamma},t) - \frac{\Delta_t \dot{Z}(t)}{Z(t)} \wp(\mathbf{\Gamma},t) + \frac{\wp(\mathbf{\Gamma},t)}{k_{\rm B}} \overline{\mathbf{R}} \cdot \nabla_{\rm p} S_{\rm r,st}(\mathbf{\Gamma},t) \right] .
\end{aligned} \tag{8.151}$$

The excess terms must cancel each other. In general one must have

$$\left\langle \nabla_{\rm p} \cdot \overline{\mathbf{R}}(\mathbf{\Gamma},t) \right\rangle = \left\langle \nabla_{\rm p} \cdot \overline{\mathbf{R}}_\gamma(\mathbf{\Gamma},t) \right\rangle = \frac{\Delta_t \dot{Z}(t)}{Z(t)} - \Delta_t \left\langle \frac{\mathrm{d}^{\rm det} S_{\rm r}(\mathbf{\Gamma},t)}{k_{\rm B} \mathrm{d}t} \right\rangle , \tag{8.152}$$

where the first equality follows because the difference between $\overline{\mathbf{R}}$ and $\overline{\mathbf{R}}_\gamma$ depends upon time only.

Table 8.1: Various changes in entropy and other selected results

$\nabla S_{\mathrm{r,st}}(\boldsymbol{\Gamma},t)$	$=$	$\overline{S'}(t)+\overline{S''}(t)\gamma+\mathcal{O}(\gamma^2)$	(8.23)
$\nabla S_{\mathrm{r,dyn}}(\boldsymbol{\Gamma},t)$	$=$	$-\overline{S'}(t)+\mathcal{O}(\gamma^2)$	(8.26)
$\vec{\Delta} S_{\mathrm{r}}$	$=$	$\mathbf{R}\cdot\nabla S_{\mathrm{r,st}}(\boldsymbol{\Gamma},t)$	(8.114)
ΔS_{r}	$=$	$\Delta_t\dot{S}^0_{\mathrm{r}}(\boldsymbol{\Gamma},t)+\mathbf{R}\cdot\nabla_{\mathrm{p}}S_{\mathrm{r}}(\boldsymbol{\Gamma},t)$	(8.115)
	$=$	$\overline{\vec{\Delta}} S_{\mathrm{r}},\ \mathbf{R}=\overline{\mathbf{R}}(-\Delta_t),\ \Delta_t>0$	(8.121)
$\dot{S}^0_{\mathrm{r}}(\boldsymbol{\Gamma},t)$	$=$	$-\sigma\left[\nabla_{\mathrm{p}}S_{\mathrm{r,st}}(\boldsymbol{\Gamma},t)-2\overline{S'_{\mathrm{p}}}(t)\right]\cdot\overline{S'_{\mathrm{p}}}(t)/2k_{\mathrm{B}}\lvert\Delta_t\rvert$	(8.123)
$\dot{\overline{S}}_{\mathrm{r}}(t)$	$=$	$\sigma\overline{S'_{\mathrm{p}}}(t)\cdot\overline{S'_{\mathrm{p}}}(t)/2k_{\mathrm{B}}\lvert\Delta_t\rvert$	(8.127)
$\overline{\mathbf{R}}_\gamma(\boldsymbol{\Gamma},t)$	$=$	$\sigma\nabla_{\mathrm{p}}S_{\mathrm{r}}(\boldsymbol{\Gamma},t)/2k_{\mathrm{B}}$	(8.87)
$\overline{\mathbf{R}}(\boldsymbol{\Gamma},t)$	$=$	$\sigma\left[\nabla_{\mathrm{p}}S_{\mathrm{r,st}}(\boldsymbol{\Gamma},t)+(\hat{t}-1)\,\overline{S'_{\mathrm{p}}}(t)\right]/2k_{\mathrm{B}}$	(8.94)
$\dot{Z}(t)/Z(t)$	$=$	$\langle\partial S_{\mathrm{r}}(\boldsymbol{\Gamma},t)/k_{\mathrm{B}}\partial t\rangle$	(8.137)
	$=$	$k_{\mathrm{B}}^{-1}\dot{\overline{S}}_{\mathrm{r}}(t)$	(8.139)

Since $\overline{\mathbf{R}}_\gamma(\boldsymbol{\Gamma},t)=\sigma\overline{S''_{\mathrm{pp}}}(t)\gamma_{\mathrm{p}}/2k_{\mathrm{B}}$, the left-hand side is

$$\nabla_{\mathrm{p}}\cdot\overline{\mathbf{R}}(\boldsymbol{\Gamma},t)=\nabla_{\mathrm{p}}\cdot\overline{\mathbf{R}}_\gamma(\boldsymbol{\Gamma},t)=\frac{\sigma}{2k_{\mathrm{B}}}\mathrm{TR}(\overline{S''_{\mathrm{pp}}}(t)),\tag{8.153}$$

where TR means trace.

For the backward derivative, the right-hand side of the penultimate equation is

$$\begin{aligned}
\mathrm{RHS} &= \frac{\Delta_t\dot{Z}(t)}{Z(t)}-k_{\mathrm{B}}^{-1}\left\langle\overline{\mathbf{R}}(\boldsymbol{\Gamma},t)\cdot\nabla_{\mathrm{p}}S_{\mathrm{r,st}}(\boldsymbol{\Gamma},t)\right\rangle,\Delta_t<0.\\
&= \frac{-\lvert\Delta_t\rvert\dot{Z}(t)}{Z(t)}-\frac{\sigma}{2k_{\mathrm{B}}^2}\left\langle\nabla_{\mathrm{p}}S_{\mathrm{r,st}}(\boldsymbol{\Gamma},t)\cdot\nabla_{\mathrm{p}}S_{\mathrm{r,st}}(\boldsymbol{\Gamma},t)\right\rangle\\
&\quad+\frac{\sigma}{k_{\mathrm{B}}^2}\left\langle\overline{S'_{\mathrm{p}}}(t)\cdot\nabla_{\mathrm{p}}S_{\mathrm{r,st}}(\boldsymbol{\Gamma},t)\right\rangle\\
&= \frac{\sigma}{2k_{\mathrm{B}}^2}\overline{S'_{\mathrm{p}}}(t)^2-\frac{\sigma}{2k_{\mathrm{B}}^2}\left\langle\nabla_{\mathrm{p}}S_{\mathrm{r,st}}(\boldsymbol{\Gamma},t)\cdot\nabla_{\mathrm{p}}S_{\mathrm{r,st}}(\boldsymbol{\Gamma},t)\right\rangle\\
&= \frac{\sigma}{2k_{\mathrm{B}}}\mathrm{TR}(\overline{S''_{\mathrm{pp}}}(t)).
\end{aligned}\tag{8.154}$$

This agrees with the direct result for the left-hand side. This result can be used to rearrange Eq. (8.148).

The different changes in entropy that have been derived above are shown in Table 8.1.

8.4.4 Steady State System

Non-equilibrium systems can be sub-divided into two main types: mechanical and thermodynamic. The most common example of the first type is that of mechanical work being performed on a sub-system in contact with a thermal reservoir. In such a case the sub-system Hamiltonian has an explicit time dependence, Eq. (8.3). Most generally the arbitrary time variation of the sub-system Hamiltonian gives rise to a transient non-equilibrium system.

The second type of non-equilibrium system is that in which a constant thermodynamic gradient is applied to the sub-system by reservoirs on either side. The reservoirs have different values of a field variable, and they can exchange the conjugate extensive variable with the sub-system. Typically the reservoirs are so large that the values of their field variables are constant in time. As a result, a steady flux flows between the reservoirs and through the sub-system, while the sub-system itself is macroscopically constant in time. This is a thermodynamic steady state system. Mechanical non-equilibrium systems can also be steady state systems. Conversely, it is also possible to create a transient thermodynamic non-equilibrium system.

Since the sub-system is macroscopically constant, a steady state system can be defined as a system in which the probability density is explicitly independent of time,

$$\frac{\partial \wp_{\rm ss}(\mathbf{\Gamma},t)}{\partial t} = 0. \tag{8.155}$$

Here, as above, $\mathbf{\Gamma}$ denotes a point in the sub-system phase space. In view of the fact that this is independent of time, one can write $\wp_{\rm ss}(\mathbf{\Gamma})$. This in essence defines a steady state system as one that is homogeneous in time. The time scales must be small enough that the accumulated flux between the reservoirs does not measurably change their field variables.

The case of a Brownian particle trapped by an external potential in uniform motion, §3.6.3, is not a steady state system by this criterion. However, a variable transformation to the frame of reference of the trap, $\tilde{x} = \{x - b(t), \dot{x}\}$, can be used as the basis of a steady state formulation. In the fluctuation variable, the non-equilibrium probability density, Eq. (3.90), has vanishing partial time derivative.

From the usual definition, $\wp_{\rm ss}(\mathbf{\Gamma},t) = e^{S_{\rm r}(\mathbf{\Gamma},t)/k_{\rm B}}/Z_{\rm ss}(t)$, it follows that for a steady state system,

$$\frac{\partial S_{\rm r}(\mathbf{\Gamma},t)}{\partial t} = \frac{k_{\rm B}\dot{Z}(t)}{Z(t)} = \dot{\overline{S}}_{\rm r}(t). \tag{8.156}$$

This is consistent with, but much stronger than, the result in the general non-equilibrium case, Eq. (8.137), since it says that the partial derivative is constant throughout phase space.

The preceding result implies that

$$S_{\rm r,st}(\mathbf{\Gamma},t) = S_{\rm r,st}(\mathbf{\Gamma}) + t\dot{c}_{\rm st}, \text{ and } S_{\rm r,dyn}(\mathbf{\Gamma},t) = S_{\rm r,dyn}(\mathbf{\Gamma}) + t\dot{c}_{\rm dyn}, \tag{8.157}$$

with $\dot{c}_{\text{st}} + \dot{c}_{\text{dyn}} = k_{\text{B}}\dot{Z}(t)/Z(t) = \dot{\overline{S}}_{\text{r}}(t)$. The form of $S_{\text{r,st}}(\mathbf{\Gamma}, t)$ is fixed by the fact that the structure of a steady state system is constant in time, and so any time dependence of the static part of the reservoir entropy is uniform in phase space.

In view of the time independence of a steady state system, the most likely adiabatic rate of change of the equilibrium part of the reservoir entropy must be a constant in time, and hence the asymptote of the integrand of the dynamic part of the reservoir entropy, is

$$\dot{S}^0_{\text{r,st}}(\overline{\mathbf{\Gamma}}(t'|\mathbf{\Gamma}, t), t') \to \dot{\overline{S}}^0_{\text{r,st}}, \quad |t - t'| \to \infty. \tag{8.158}$$

Since a steady state system is macroscopically constant in time, it must be homogeneous in time. In particular the most likely trajectory must satisfy

$$\overline{\mathbf{\Gamma}}(t'|\mathbf{\Gamma}, t) = \overline{\mathbf{\Gamma}}(t' + t''|\mathbf{\Gamma}, t + t''), \tag{8.159}$$

for an arbitrary time shift t''. In view of this the dynamic part of the reservoir entropy at the adjacent time step is related to that at the current time step by

$$\begin{aligned}
& S_{\text{r,dyn}}(\mathbf{\Gamma}, t + \Delta_t) \\
&= -\int_0^{t+\Delta_t} \mathrm{d}t' \, \dot{S}^0_{\text{r,st}}(\overline{\mathbf{\Gamma}}(t'|\mathbf{\Gamma}, t + \Delta_t), t') \\
&= -\int_0^{t+\Delta_t} \mathrm{d}t' \, \left[\dot{S}^0_{\text{r,st}}(\overline{\mathbf{\Gamma}}(t' - \Delta_t|\mathbf{\Gamma}, t)) + \dot{c}_{\text{st}}\right] \\
&= -(t + \Delta_t)\dot{c}_{\text{st}} - \int_{-\Delta_t}^{t} \mathrm{d}t'' \, \dot{S}^0_{\text{r,st}}(\overline{\mathbf{\Gamma}}(t''|\mathbf{\Gamma}, t)) \\
&= -(t + \Delta_t)\dot{c}_{\text{st}} + [S_{\text{r,dyn}}(\mathbf{\Gamma}, t) + t\dot{c}_{\text{st}}] - \Delta_t \dot{S}^0_{\text{r,st}}(\overline{\mathbf{\Gamma}}(0|\mathbf{\Gamma}, t)) \\
&= -(t + \Delta_t)\dot{c}_{\text{st}} + [S_{\text{r,dyn}}(\mathbf{\Gamma}, t) + t\dot{c}_{\text{st}}] - \Delta_t \left[\dot{\overline{S}}^0_{\text{r,st}} - \dot{c}_{\text{st}}\right] \\
&= S_{\text{r,dyn}}(\mathbf{\Gamma}, t) - \Delta_t \dot{\overline{S}}^0_{\text{r,st}}.
\end{aligned} \tag{8.160}$$

This means that for a steady state system,

$$\frac{\partial S_{\text{r,dyn}}(\mathbf{\Gamma}, t)}{\partial t} = -\dot{\overline{S}}^0_{\text{r,st}}. \tag{8.161}$$

Hence $\dot{c}_{\text{dyn}} = -\dot{\overline{S}}^0_{\text{r,st}}$, and $\dot{c}_{\text{st}} = \dot{\overline{S}}^0_{\text{r,st}} + \dot{\overline{S}}_{\text{r}}(t)$.

The most likely backward rate of change of the dynamic part of the reservoir entropy is $\dot{\overline{S}}^-_{\text{r,dyn}}(\mathbf{\Gamma}, t) = -\dot{S}^0_{\text{r,st}}(\mathbf{\Gamma}, t)$. Hence

$$\left[\dot{\mathbf{\Gamma}}^0 + \frac{\overline{\mathbf{R}}(\Delta_t)}{\Delta_t}\right] \cdot \nabla S_{\text{r,dyn}}(\mathbf{\Gamma}, t) = \dot{\overline{S}}^0_{\text{r,st}} - \dot{S}^0_{\text{r,st}}(\mathbf{\Gamma}, t), \; \Delta_t < 0. \tag{8.162}$$

8.5 Odd Projection of the Dynamic Reservoir Entropy

In this section the odd projection of the dynamic part of the reservoir entropy is analysed for the general non-equilibrium case. There are three reasons for focusing on the odd projection. First, in many important cases only the odd projection of the dynamic part of the reservoir entropy is required. Examples include the average of various fluxes that are typically required to obtain the linear transport coefficients via the Green-Kubo relations (e.g. Eq. (9.90) below), and the ratio of the non-equilibrium probability density to its conjugate. Second, there are computational advantages in evaluating the odd projection, which include the fact that the integrand is short-ranged, and the fact that adiabatic trajectories rather than stochastic dissipative trajectories can be used. Third, there is reason to believe that the even projection of the dynamic part of the reservoir entropy is relatively negligible, and that the odd projection is the only part of the reservoir entropy that is essential to the dynamic part of the probability density. An explicit calculation involving an odd projection for steady heat flow is given in §9.5.3 below.

As discussed at the beginning of this chapter, §8.1, a non-equilibrium system is characterised by a probability distribution that is sensitive to the direction of time, which is to say that it must depend upon the molecular velocities, not just their speeds. For a sub-system phase space point $\mathbf{\Gamma} = \{\mathbf{q}^N, \mathbf{p}^N\}$, the conjugate phase space point is the one with the velocities reversed, $\mathbf{\Gamma}^\dagger = \{\mathbf{q}^N, (-\mathbf{p})^N\}$. Accordingly one must have

$$\wp(\mathbf{\Gamma}, t) \neq \wp(\mathbf{\Gamma}^\dagger, t). \tag{8.163}$$

The static part of the reservoir entropy necessarily has even parity, $S_{\mathrm{r,st}}(\mathbf{\Gamma}, t) = S_{\mathrm{r,st}}(\mathbf{\Gamma}^\dagger, t)$, which implies that

$$S_{\mathrm{r,dyn}}(\mathbf{\Gamma}, t) \neq S_{\mathrm{r,dyn}}(\mathbf{\Gamma}^\dagger, t). \tag{8.164}$$

The argument that the odd projection is dominant goes like this. One can always formally separate a function into even and odd parity projections. Hence in the case of the dynamic part of the reservoir entropy, one can write

$$S_{\mathrm{r,dyn}}(\mathbf{\Gamma}, t) = S^{\mathrm{even}}_{\mathrm{r,dyn}}(\mathbf{\Gamma}, t) + S^{\mathrm{odd}}_{\mathrm{r,dyn}}(\mathbf{\Gamma}, t), \tag{8.165}$$

where the even projection is

$$S^{\mathrm{even}}_{\mathrm{r,dyn}}(\mathbf{\Gamma}, t) \equiv \frac{1}{2}\left[S_{\mathrm{r,dyn}}(\mathbf{\Gamma}, t) + S_{\mathrm{r,dyn}}(\mathbf{\Gamma}^\dagger, t)\right], \tag{8.166}$$

and the odd projection is

$$S^{\mathrm{odd}}_{\mathrm{r,dyn}}(\mathbf{\Gamma}, t) \equiv \frac{1}{2}\left[S_{\mathrm{r,dyn}}(\mathbf{\Gamma}, t) - S_{\mathrm{r,dyn}}(\mathbf{\Gamma}^\dagger, t)\right]. \tag{8.167}$$

One expects that in general it is the equilibrium part of the reservoir entropy that dominates the non-equilibrium probability distribution. This is because this part contains the molecular interactions of the sub-system via $\mathcal{H}(\mathbf{\Gamma},t)$, which in turn determines the instantaneous structure. The specifically non-equilibrium parts of the system are generally a perturbation about the equilibrium structure. Hence one expects that the equilibrium part of the reservoir entropy dominates the even projection of the dynamic part of the reservoir entropy,

$$S_{\rm r,st}(\mathbf{\Gamma},t) \gg S^{\rm even}_{\rm r,dyn}(\mathbf{\Gamma},t). \tag{8.168}$$

One cannot say the same for the odd parity part because the static part of the reservoir entropy has zero odd parity projection. The arrow of time is crucial for a non-equilibrium system, and the odd parity projection of the dynamic part of the reservoir entropy is the only part of the expression for the non-equilibrium probability density that is sensitive to the direction of time. One concludes that one need only retain, and that one must retain, the odd parity projection of the dynamic part of the reservoir entropy. Hence it is at least a good approximation, and it may even be exact in the thermodynamic limit, to write

$$S_{\rm r}(\mathbf{\Gamma},t) \approx S_{\rm r,st}(\mathbf{\Gamma},t) + S^{\rm odd}_{\rm r,dyn}(\mathbf{\Gamma},t). \tag{8.169}$$

Of course when the time dependent part of the system is truly a perturbation, this argument is formally exact.

Now the properties of the odd projection are analysed. Recall that the dynamic part of the reservoir entropy was defined in Eq. (8.15),

$$S_{\rm r,dyn}(\mathbf{\Gamma},t) = -\int_0^t {\rm d}t'\, \dot{S}^0_{\rm r,st}(\overline{\mathbf{\Gamma}}(t'|\mathbf{\Gamma},t),t'), \tag{8.170}$$

where $\dot{S}^0_{\rm r,st}(\mathbf{\Gamma},t) = \partial S_{\rm r,st}(\mathbf{\Gamma},t)/\partial t + \dot{\mathbf{\Gamma}}^0 \cdot \nabla S_{\rm r,st}(\mathbf{\Gamma},t)$ is the adiabatic rate of change of the static part of the reservoir entropy, and $\overline{\mathbf{\Gamma}}(t'|\mathbf{\Gamma},t)$ is the most likely backward trajectory from the point $\mathbf{\Gamma}$ at time $t > 0$. The time t is assumed large enough so that there is no correlation between the current point $\mathbf{\Gamma}$ and the initial point $\overline{\mathbf{\Gamma}}(0|\mathbf{\Gamma},t)$, which is to say that the integrand is assumed to reach its asymptotic value before the integral reaches its lower limit.

The odd projection of the dynamic part of the reservoir entropy is

$$\begin{aligned} S^{\rm odd}_{\rm r,dyn}(\mathbf{\Gamma},t) &\equiv \frac{1}{2}\left[S^{\rm odd}_{\rm r,dyn}(\mathbf{\Gamma},t) - S^{\rm odd}_{\rm r,dyn}(\mathbf{\Gamma}^\dagger,t)\right] \\ &= \frac{-1}{2}\int_0^t {\rm d}t' \left[\dot{S}^0_{\rm r,st}(\overline{\mathbf{\Gamma}}(t'|\mathbf{\Gamma},t),t') - \dot{S}^0_{\rm r,st}(\overline{\mathbf{\Gamma}}(t'|\mathbf{\Gamma}^\dagger,t),t')\right] \\ &\approx \frac{-1}{2}\int_0^t {\rm d}t' \left[\dot{S}^{0,\rm odd}_{\rm r,st}(\overline{\mathbf{\Gamma}}(t'|\mathbf{\Gamma},t),t') - \dot{S}^{0,\rm odd}_{\rm r,st}(\overline{\mathbf{\Gamma}}(t'|\mathbf{\Gamma}^\dagger,t),t')\right]. \end{aligned} \tag{8.171}$$

The finally approximation invokes the projections of the adiabatic rate of change of the static part of the reservoir entropy,

$$\dot{S}^{0,\rm even}_{\rm r,st}(\mathbf{\Gamma},t) \equiv \frac{\partial S_{\rm r,st}(\mathbf{\Gamma},t)}{\partial t}, \text{ and } \dot{S}^{0,\rm odd}_{\rm r,st}(\mathbf{\Gamma},t) \equiv \dot{\mathbf{\Gamma}}^0 \cdot \nabla S_{\rm r,st}(\mathbf{\Gamma},t). \tag{8.172}$$

These follow because $S_{\rm r,st}(\mathbf{\Gamma},t)$ is of even parity. For a steady state system, there is no explicit time dependence, so that one has $S_{\rm r,st}(\mathbf{\Gamma})$, and hence $\dot S^{0,\rm even}_{\rm r,st}(\mathbf{\Gamma}) = 0$. Hence the final approximation in Eq. (8.171) is exact for a steady state system. It is also a good approximation for a mechanical non-equilibrium system with a time dependent Hamiltonian. This is because the trajectories starting from $\mathbf{\Gamma}$ and from $\mathbf{\Gamma}^\dagger$ tend to their respective starting points as $t' \to t$, and because $\dot S^{0,\rm even}_{\rm r,st}$ tends to an asymptote independent of the starting point as $t' \to 0$, as will be discussed shortly. Hence the even contribution to the integrand of $S^{\rm odd}_{\rm r,dyn}(\mathbf{\Gamma},t)$ vanishes at the two limits of the integral,

$$\begin{aligned} &\dot S^{0,\rm even}_{\rm r,st}(\overline{\mathbf{\Gamma}}(t'|\mathbf{\Gamma},t),t') - \dot S^{0,\rm even}_{\rm r,st}(\overline{\mathbf{\Gamma}}(t'|\mathbf{\Gamma}^\dagger,t),t') \\ &\left\{ \begin{array}{lllll} \to & \dot S^{0,\rm even}_{\rm r,st}(\mathbf{\Gamma},t') - \dot S^{0,\rm even}_{\rm r,st}(\mathbf{\Gamma}^\dagger,t') & = & 0, & t' \to t, \\ \to & \overline{\dot S^{0,\rm even}_{\rm r,st}}(t') - \overline{\dot S^{0,\rm even}_{\rm r,st}}(t') & = & 0, & t' \to 0. \end{array} \right. \end{aligned} \tag{8.173}$$

Since this vanishes at the two limits of the integral, it is reasonable to neglect it entirely compared to the integration of the odd projection. This point will be invoked in obtaining the final result.

In view of the definition of the odd projection of the dynamic part of the reservoir entropy, one is motivated to analyse also the forward trajectory, as well as the trajectories from the conjugate phase space point. Hence one defines

$$S_\pm(\mathbf{\Gamma},t) \equiv \mp \int_t^{t\pm t} \mathrm{d}t'\ \dot S^0_{\rm r,st}(\overline{\mathbf{\Gamma}}(t'|\mathbf{\Gamma},t),t'), \tag{8.174}$$

and also

$$S^\dagger_\pm(\mathbf{\Gamma},t) \equiv S_\pm(\mathbf{\Gamma}^\dagger,t) = \mp \int_t^{t\pm t} \mathrm{d}t'\ \dot S^0_{\rm r,st}(\overline{\mathbf{\Gamma}}(t'|\mathbf{\Gamma}^\dagger,t),t'). \tag{8.175}$$

In terms of these, the odd projection of the dynamic part of the reservoir entropy is $S^{\rm odd}_{\rm r,dyn}(\mathbf{\Gamma},t) = [S_-(\mathbf{\Gamma},t) - S^\dagger_-(\mathbf{\Gamma},t)]/2$.

The above definitions invoke the future $t' > t$. In the general non-equilibrium case of a time dependent Hamiltonian, one has to extend the past definition into the future by defining an extended system with Hamiltonian $\tilde{\mathcal{H}}(\mathbf{\Gamma},t';t) = \mathcal{H}(\mathbf{\Gamma},2t-t')$ for $t' > t$. This extended Hamiltonian is an even function of time about the current time t. In many cases, including an equilibrium system, a steady state system, and a harmonic system, the extended system is identical to the original system. In what follows below all future quantities $t' > t$ refer to the extended system.

Figure 8.2 sketches the time evolution of the adiabatic rate of change of the equilibrium part of the reservoir entropy along both actual and adiabatic trajectories. The limit on the integrals should be chosen to be sufficiently large for the integrand to reach its asymptotic value,

$$\dot S^0_{\rm r,st}(\overline{\mathbf{\Gamma}}(t'|\mathbf{\Gamma},t),t') \to \overline{\dot S^0_{\rm r,st}}(t';t), \quad |t'-t| \to \infty. \tag{8.176}$$

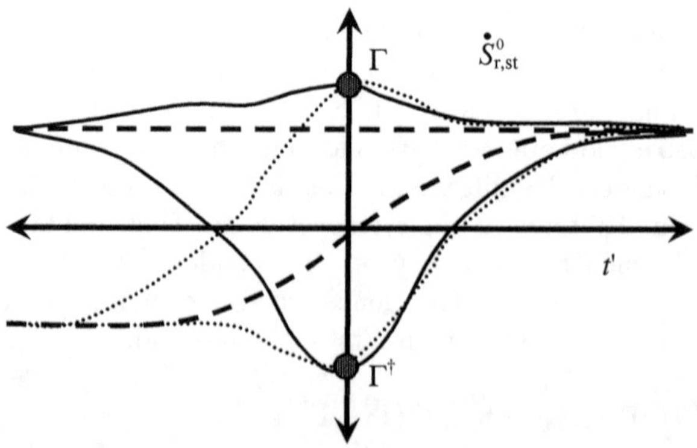

Figure 8.2: Sketch of $\dot{S}^0_{\mathrm{r,st}}$, moving forward and backward in time from a phase space point and from its conjugate. The solid curves are actual trajectories, the dotted curves are adiabatic trajectories, and the dashed line and curve are the respective asymptotes for the two trajectories. The constancy of the asymptotes means that a steady state system has been implicitly assumed, and that the time scales are long enough to reach the asymptote, but short enough such that $|t'\dot{S}^0_{\mathrm{r,st}}| \ll |S_{\mathrm{r,st}}|$.

This asymptote arises from the fact that with overwhelming probability the system came from its most likely value in the past (and will return there in the future), independent of the current phase space point of the sub-system. The dependence of the asymptote on the current time only occurs for the future, $t' > t$, and arises because the extended Hamiltonian is an even function of time about the current time. For a steady state system, the asymptote is independent of t', and for a mechanical non-equilibrium system with time dependent Hamiltonian, it is approximately an even function of t' about t.

The asymptotic behaviour on the adiabatic trajectory is somewhat different from that on the most likely stochastic, dissipative trajectory. One has

$$\dot{S}^0_{\mathrm{r,st}}(\mathbf{\Gamma}^0(t'|\mathbf{\Gamma},t),t') \;\sim\; \mathrm{sign}(t'-t)\,\overline{\dot{S}^0_{\mathrm{r,st}}}(t';t),\;\; |t'-t| \sim \tau_{\mathrm{relax}}. \tag{8.177}$$

Here τ_{relax} is a relaxation time that is long enough for the system to reach its asymptote, but not so long that the structure has changed significantly, $|\tau_{\mathrm{relax}}\dot{S}^0_{\mathrm{r,st}}| \ll |S_{\mathrm{r,st}}|$. (One does not need to impose this condition for the stochastic, dissipative trajectory, because the interactions with the reservoir maintain the structure of the sub-system.) For an isolated system, the structure represents a fluctuation, and $\dot{S}^0_{\mathrm{r,st}}$ represents its regression, which must be an odd function of time, at least for a steady state system. For $t' > 0$, the adiabatic asymptote and the actual asymptote approximately coincide, which is just Onsager's regression hypothesis.

The integrand of the dynamic part of the reservoir entropy, is approximately an even function of $t' - t$ (at least it is on long enough time scales). This follows because a fluctuation in the most likely flux is as equally likely to be returning *to* the optimum state as it is to be coming *from* the optimum state. It therefore follows that

$$S_+(\mathbf{\Gamma}, t) \approx S_-(\mathbf{\Gamma}, t), \text{ and } S_+^\dagger(\mathbf{\Gamma}, t) \approx S_-^\dagger(\mathbf{\Gamma}, t). \tag{8.178}$$

For $t' > t$, the most likely trajectory in the presence of the reservoir approximately coincides with the adiabatic trajectory, $\overline{\mathbf{\Gamma}}(t'|\mathbf{\Gamma}, t) \approx \mathbf{\Gamma}^0(t'|\mathbf{\Gamma}, t)$, $t' > t$. (This is not true for $t' < t$.) Hence it follows that

$$S_+(\mathbf{\Gamma}, t) \approx S_+^0(\mathbf{\Gamma}, t), \text{ and } S_+^\dagger(\mathbf{\Gamma}, t) \approx S_+^{\dagger,0}(\mathbf{\Gamma}, t). \tag{8.179}$$

This is essentially Onsager's regression hypothesis. Here the dynamic part of the reservoir entropy calculated on an adiabatic trajectory is denoted by a superscript 0.

The adiabatic past trajectories may be converted to future trajectories. Because of the time reversibility of Hamilton's equations of motion and the extended system Hamiltonian, one has the exact result[16]

$$\mathbf{\Gamma}^0(t'|\mathbf{\Gamma}^\dagger, t) = \mathbf{\Gamma}^0(2t - t'|\mathbf{\Gamma}, t)^\dagger. \tag{8.180}$$

The simplification that results from these three results and Eq. (8.171) is that, to a good approximation, the odd projection of the dynamic part of the reservoir entropy depends only upon the adiabatic trajectories,

$$\begin{aligned} S_{\mathrm{r,dyn}}^{\mathrm{odd}}(\mathbf{\Gamma}, t) &= \frac{-1}{2} \int_0^t \mathrm{d}t' \left[\dot{S}_{\mathrm{r,st}}^0(\overline{\mathbf{\Gamma}}(t'|\mathbf{\Gamma}, t), t') - \dot{S}_{\mathrm{r,st}}^0(\overline{\mathbf{\Gamma}}(t'|\mathbf{\Gamma}^\dagger, t), t') \right] \\ &\approx \frac{-1}{2} \int_t^{2t} \mathrm{d}t' \left[\dot{S}_{\mathrm{r,st}}^0(\overline{\mathbf{\Gamma}}(t'|\mathbf{\Gamma}, t), t') - \dot{S}_{\mathrm{r,st}}^0(\overline{\mathbf{\Gamma}}(t'|\mathbf{\Gamma}^\dagger, t), t') \right] \\ &\approx \frac{-1}{2} \int_t^{2t} \mathrm{d}t' \left[\dot{S}_{\mathrm{r,st}}^0(\mathbf{\Gamma}^0(t'|\mathbf{\Gamma}, t), t') - \dot{S}_{\mathrm{r,st}}^0(\mathbf{\Gamma}^0(t'|\mathbf{\Gamma}^\dagger, t), t') \right] \\ &= \frac{-1}{2} \int_0^t \mathrm{d}t' \left[\dot{S}_{\mathrm{r,st}}^0(\mathbf{\Gamma}^0(t'|\mathbf{\Gamma}^\dagger, t)^\dagger, t') - \dot{S}_{\mathrm{r,st}}^0(\mathbf{\Gamma}^0(t'|\mathbf{\Gamma}, t)^\dagger, t') \right] \\ &\approx \frac{-1}{2} \int_0^t \mathrm{d}t' \left[-\dot{S}_{\mathrm{r,st}}^0(\mathbf{\Gamma}^0(t'|\mathbf{\Gamma}^\dagger, t), t') + \dot{S}_{\mathrm{r,st}}^0(\mathbf{\Gamma}^0(t'|\mathbf{\Gamma}, t), t') \right] \\ &\equiv S_{\mathrm{r,dyn}}^{\mathrm{odd};0}(\mathbf{\Gamma}, t). \end{aligned} \tag{8.181}$$

The first equality is the definition. The second equality invokes the fact that the most likely trajectory is approximately even. The third equality invokes the fact that the most likely trajectory and the adiabatic trajectory approximate each other in the future (the regression hypothesis). The fourth equality invokes the reversibility of the adiabatic trajectory in the extended system. The fifth

[16] One does not have an equivalent result for the stochastic, dissipative trajectory, $\overline{\mathbf{\Gamma}}(t'|\mathbf{\Gamma}^\dagger, t) \neq \overline{\mathbf{\Gamma}}(2t - t'|\mathbf{\Gamma}, t)^\dagger$, because the dissipative force depends upon the current value of the static part of the reservoir entropy, and this differs on the two trajectories after large enough times. This is in contrast with the transitions in Table 7.3 on p. 225, which hold only for an infinitesimal time interval.

equality invokes the fact that the even parity projection of the integrand is negligible, as discussed in deriving Eq. (8.171). (That derivation also holds for adiabatic trajectories.) These various equalities are sketched in Fig. 8.2. The penultimate equality can also be rewritten as an integral from 0 to $2t$, although the present final expression has the advantage that it does not invoke the extended system. The final expression can also be integrated exactly, at least formally, but so far this has not appeared to be useful. One could subtract the asymptotes explicitly in the integrand (either because they are constant in phase space, or else because they integrate to zero), which shows explicitly that the integrand is short-ranged and that therefore $S^{\text{odd}}_{\text{r,dyn}}$ does not scale with t. Amongst other things this means that the lower limit of the integral could be replaced by $t-\tau$, for some fixed time interval. This is an important and useful property when it comes to numerically computing the dynamic part of the reservoir entropy.

This completes the full argument. It says that in some circumstances the odd projection of the dynamic part of the reservoir entropy is either dominant or is all that is required. Further it says that the odd projection of the dynamic part of the reservoir entropy may be evaluated on the past adiabatic trajectories. It should be emphasised that this adiabatic replacement is essential in order to derive the Green-Kubo theory from the present theory for non-equilibrium statistical mechanics. An explicit example of this is given for the thermal conductivity in §9.5 below, and it is shown in §9.5.3 that an odd projection of the dynamic part of the reservoir entropy gives the leading-order contribution for steady heat flow. The odd projection is also discussed for driven Brownian motion in §11.2.3.

8.6 Path Entropy and Transitions

8.6.1 Path Entropy

The change in entropy for the transition $\mathbf{\Gamma}_1 \to \mathbf{\Gamma}_2$ was given above as Eq. (8.114),

$$\begin{aligned}
\vec{\Delta} S_{\text{r}}(\mathbf{\Gamma}_2, t_2; \mathbf{\Gamma}_1, t_1) &\equiv S_{\text{r}}(\mathbf{\Gamma}_2, t_2 | \mathbf{\Gamma}_1, t_1) - S_{\text{r}}(\mathbf{\Gamma}_1, t_1) \\
&= \mathbf{R}_2 \cdot \nabla S_{\text{r,st}}(\mathbf{\Gamma}, t) \\
&= \left[\mathbf{\Gamma}_2 - \mathbf{\Gamma}_1 - t_{21} \dot{\mathbf{\Gamma}}^0 \right] \cdot \nabla S_{\text{r,st}}(\mathbf{\Gamma}, t) \qquad (8.182) \\
&= \left[S_{\text{r,st}}(\mathbf{\Gamma}_2, t_2) - S_{\text{r,st}}(\mathbf{\Gamma}_1, t_1) - t_{21} \dot{S}^0_{\text{r,st}}(\mathbf{\Gamma}, t) \right].
\end{aligned}$$

To obtain the final equality, the partial time derivative, $\partial S_{\text{r,st}}(\mathbf{\Gamma}, t)/\partial t$, has been added and subtracted. Here and below this will be taken to be the physical or forward transition, $t_{21} \equiv t_2 - t_1 > 0$. This change in entropy was distinguished from the entropy difference, $\Delta S_{\text{r}} \equiv S_{\text{r}}(\mathbf{\Gamma}_2, t_2) - S_{\text{r}}(\mathbf{\Gamma}_1, t_1)$. The two are equal on the most likely backward transition, but this fact is not required here. The reservoir force that appears here is the sum of the most likely force (the deterministic part), and the stochastic or random force, $\mathbf{R} = \overline{\mathbf{R}} + \tilde{\mathbf{R}}$, and it gives the departure from the adiabatic evolution, $\mathbf{R}_2 = \mathbf{\Gamma}_2 - \mathbf{\Gamma}_1 - t_{21}\dot{\mathbf{\Gamma}}^0$. The most likely force is given by Eq. (8.94),

$$\overline{\mathbf{R}}_2 \equiv \overline{\mathbf{R}}(\mathbf{\Gamma}, t, t_{21}) = \frac{|t_{21}|}{2}\Lambda_{\rm pp}\left[\nabla_{\rm p}S_{\rm r,st}(\mathbf{\Gamma}, t) + \left[\hat{t}_{21} - 1\right]\overline{S'_{\rm p}}(t)\right], \tag{8.183}$$

where only momentum components appear (because $\Lambda_{\rm qq} \sim \mathcal{O}(t_{21})$). Recall that one interchange $\Lambda_{\rm pp} \Leftrightarrow \sigma \mathrm{I}_{\rm pp}/k_{\rm B}|\Delta_t|$. For the present case, $\hat{t}_{21} \equiv \mathrm{sign}(t_{21}) = 1$, and so only the first term contributes, $\overline{\mathbf{R}}_2 = |t_{21}|\Lambda_{\rm pp}\nabla_{\rm p}S_{\rm r,st}(\mathbf{\Gamma}, t)/2$.

The second entropy for this transition was given above in Eq. (8.54) as a quadratic in the stochastic force,

$$S^{(2)}(\mathbf{\Gamma}_2, t_2; \mathbf{\Gamma}_1, t_1) = \frac{-1}{2|t_{21}|}\Lambda_{\rm pp}^{-1} : \tilde{\mathbf{R}}_2\tilde{\mathbf{R}}_2 + S_{\rm r}(\mathbf{\Gamma}_1, t_1) + \frac{1}{2}\left[\overline{S}_{\rm r}(t_2) - \overline{S}_{\rm r}(t_1)\right]. \tag{8.184}$$

As mentioned in footnote 6 on p. 244, compared to Eq. (8.28) this neglects terms that arise from the normalisation of the Gaussian fluctuations. Writing the random part of the force as $\tilde{\mathbf{R}}_2 = \mathbf{R}_2 - \overline{\mathbf{R}}_2$, the conditional second entropy as defined in Eq. (8.34) is

$$\begin{aligned} & S^{(2)}(\mathbf{\Gamma}_2, t_2|\mathbf{\Gamma}_1, t_1) \\ & \quad \equiv S^{(2)}(\mathbf{\Gamma}_2, t_2; \mathbf{\Gamma}_1, t_1) - S_{\rm r}(\mathbf{\Gamma}_1, t_1) - \frac{1}{2}\left[\overline{S}_{\rm r}(t_2) - \overline{S}_{\rm r}(t_1)\right] \\ & \quad = \frac{-1}{2|t_{21}|}\Lambda_{\rm pp}^{-1} : \tilde{\mathbf{R}}_2\tilde{\mathbf{R}}_2 \\ & \quad = \frac{-1}{2|t_{21}|}\Lambda_{\rm pp}^{-1} : \left[\mathbf{R}_2 - \overline{\mathbf{R}}_2\right]^2 \\ & \quad = \frac{-1}{2|t_{21}|}\Lambda_{\rm pp}^{-1} : \left[\mathbf{R}_2^2 + \overline{\mathbf{R}}_2^2\right] + \frac{1}{|t_{21}|}\mathbf{R}_2 \cdot \Lambda_{\rm pp}^{-1}\overline{\mathbf{R}}_2 \\ & \quad = \frac{-1}{2|t_{21}|}\Lambda_{\rm pp}^{-1} : \left[\mathbf{R}_2^2 + \overline{\mathbf{R}}_2^2\right] + \frac{1}{2}\mathbf{R}_2 \cdot \nabla_{\rm p}S_{\rm r,st}(\mathbf{\Gamma}, t) \\ & \quad = \frac{-1}{2|t_{21}|}\Lambda_{\rm pp}^{-1} : \left[\mathbf{R}_2^2 + \overline{\mathbf{R}}_2^2\right] + \frac{1}{2}\vec{\Delta}S_{\rm r}(\mathbf{\Gamma}_2, t_2; \mathbf{\Gamma}_1, t_1). \end{aligned} \tag{8.185}$$

It can be seen that the conditional second entropy scales with the size of the time step. The final equality utilises the change in entropy on a transition, Eq. (8.182). Note the distinction between this the conditional second entropy and the conditional reservoir entropy used in Eq. (8.182). The reason for transforming from the second equality to the final form is to exhibit explicitly the even and odd terms with respect to the conjugacy operation (see below).

The conditional transition probability is

$$\wp(\mathbf{\Gamma}_2, t_2|\mathbf{\Gamma}_1, t_1) \equiv \frac{1}{Z(t_2, t_1)}e^{S^{(2)}(\mathbf{\Gamma}_2, t_2|\mathbf{\Gamma}_1, t_1)/k_{\rm B}}, \tag{8.186}$$

with the normalising partition function being

$$\begin{aligned} Z(t_2,t_1) &= \int \mathrm{d}\mathbf{\Gamma}_2\, e^{S^{(2)}(\mathbf{\Gamma}_2,t_2|\mathbf{\Gamma}_1,t_1)/k_\mathrm{B}} \\ &= \int \mathrm{d}\mathbf{\Gamma}_2\, e^{-\Lambda_\mathrm{pp}^{-1}:\tilde{\mathbf{R}}_2\tilde{\mathbf{R}}_2/2|t_{21}|k_\mathrm{B}} \\ &= \left|2\pi|t_{21}|k_\mathrm{B}\Lambda_\mathrm{pp}(t)\right|^{1/2}. \end{aligned} \tag{8.187}$$

The reason that one has to explicitly normalise this expression for the conditional probability compared to the generic expression, Eq. (8.28), is because here the fluctuation contributions have been neglected in using the most likely value of the entropy (c.f. footnote 6 on p. 244). The unconditional transition probability is $\wp(\mathbf{\Gamma}_2,t_2;\mathbf{\Gamma}_1,t_1) = \wp(\mathbf{\Gamma}_2,t_2|\mathbf{\Gamma}_1,t_1)\,\wp(\mathbf{\Gamma}_1,t_1)$.

Now consider a particular trajectory of the sub-system, $\underline{\mathbf{\Gamma}}$, on the interval $t \in [t_0, t_f]$. As in §8.1, use the notation $\underline{\mathbf{\Gamma}}$ to denote this particular trajectory, in order to distinguish it from $\mathbf{\Gamma}(t)$, which generally means a particular point in phase space at a particular time. The notation $f(\underline{\mathbf{\Gamma}})$ denotes a functional of the whole trajectory. Discretise the interval, $t_n = t_0 + n\Delta_t$, $n \in [0, f]$, with $\mathbf{\Gamma}_n \equiv \mathbf{\Gamma}(t_n)$, and $\Delta_t \equiv [t_f - t_0]/f > 0$. In the discrete representation, $\underline{\mathbf{\Gamma}}$ is an $(f+1)$-component vector.

The change in entropy for this trajectory is

$$\begin{aligned} \vec{\Delta} S_\mathrm{r}(\underline{\mathbf{\Gamma}}) &= \sum_{n=1}^{f} [S_\mathrm{r}(\mathbf{\Gamma}_n,t_n|\mathbf{\Gamma}_{n-1},t_{n-1}) - S_\mathrm{r}(\mathbf{\Gamma}_{n-1},t_{n-1})] \\ &= \sum_{n=1}^{f} \left[S_\mathrm{r,st}(\mathbf{\Gamma}_n,t_n) - S_\mathrm{r,st}(\mathbf{\Gamma}_{n-1},t_{n-1}) - \Delta_t \dot{S}^0_\mathrm{r,st}(\mathbf{\Gamma}_{n-1},t_{n-1})\right] \\ &= S_\mathrm{r,st}(\mathbf{\Gamma}_f,t_f) - S_\mathrm{r,st}(\mathbf{\Gamma}_0,t_0) - \int_{t_0}^{t_f} \mathrm{d}t'\, \dot{S}^0_\mathrm{r,st}(\mathbf{\Gamma}(t'),t') \\ &\equiv S_\mathrm{r,st}(\mathbf{\Gamma}_f,t_f) - S_\mathrm{r,st}(\mathbf{\Gamma}_0,t_0) + S_\mathrm{r,dyn}(\underline{\mathbf{\Gamma}}). \end{aligned} \tag{8.188}$$

The final equality defines the dynamic part of the reservoir entropy for the actual trajectory.

The Markov-type nature of this expression for the change in entropy implies that the reservoir has no memory at this level. This is entirely appropriate for these molecular-level transitions because Hamilton's equations themselves are Markovian. The time scales of the reservoir and the sub-system are the same at the level of phase space. It is only when one integrates out fast modes, in, for example Brownian dynamics, that one has to include non-Markovian memory effects that correspond to the time integration of the reservoir and solvent phase space coordinates. (See, for example, Ch. 10 and §11.3.)

For a sufficiently long trajectory,

$$\vec{\Delta} S_\mathrm{r}(\underline{\mathbf{\Gamma}}) \approx \left[\overline{S}_\mathrm{r}(t_f) - \overline{S}_\mathrm{r}(t_0)\right] + S_\mathrm{r,dyn}(\underline{\mathbf{\Gamma}}). \tag{8.189}$$

The part of the difference in the reservoir entropies of the end points that scales with $|t_f - t_0|$ is the most likely part. Hence both terms on the right-hand side scale with the length of the trajectory.

Note that this, the change in entropy for the trajectory, is not equal to the difference in the reservoir entropy of the end points of the total transition,

$$\begin{aligned} \Delta S_{\rm r} &= S_{\rm r}(\mathbf{\Gamma}_f, t_f) - S_{\rm r}(\mathbf{\Gamma}_0, t_0) \\ &= S_{\rm r,st}(\mathbf{\Gamma}_f, t_f) - S_{\rm r,st}(\mathbf{\Gamma}_0, t_0) + S_{\rm r,dyn}(\mathbf{\Gamma}_f, t_f) - S_{\rm r,dyn}(\mathbf{\Gamma}_0, t_0) \\ &= S_{\rm r,st}(\mathbf{\Gamma}_f, t_f) - S_{\rm r,st}(\mathbf{\Gamma}_0, t_0) - \int_{t_0}^{t_f} {\rm d}t'\, \dot{S}^0_{\rm r,st}(\overline{\mathbf{\Gamma}}(t'|\mathbf{\Gamma}_f, t_f), t'). \end{aligned} \tag{8.190}$$

The distinction is that in the former it is the actual trajectory rather than the most likely trajectory that appears in the integrand, $\Delta S_{\rm r} \neq \vec{\Delta} S_{\rm r}[\mathbf{\Gamma}]$.

The trajectory entropy conditioned upon the starting point is just the sum of the conditional second entropy, Eq. (8.185), for each transition,

$$\begin{aligned} & S(\underline{\mathbf{\Gamma}}|\mathbf{\Gamma}_0, t_0) \\ &= \sum_{n=1}^{f} S^{(2)}(\mathbf{\Gamma}_n, t_n|\mathbf{\Gamma}_{n-1}, t_{n-1}) \\ &= \sum_{n=1}^{f} \left\{ \frac{-1}{2|\Delta_t|} \Lambda_{\rm pp}^{-1} : \left[\mathbf{R}_n^2 + \overline{\mathbf{R}}_n^2 \right] + \frac{1}{2} \vec{\Delta} S_{\rm r}(\mathbf{\Gamma}_n, t_n; \mathbf{\Gamma}_{n-1}, t_{n-1}) \right\} \\ &= \frac{1}{2} \vec{\Delta} S_{\rm r}(\underline{\mathbf{\Gamma}}) - \frac{1}{2|\Delta_t|} \sum_{n=1}^{f} \Lambda_{\rm pp}^{-1} : \left[\mathbf{R}_n^2 + \overline{\mathbf{R}}_n^2 \right]. \end{aligned} \tag{8.191}$$

Recall that $\mathbf{R}_n \equiv \mathbf{\Gamma}_n - \mathbf{\Gamma}_{n-1} - \Delta_t \dot{\mathbf{\Gamma}}^0$ is the total reservoir perturbation for that transition, with the adiabatic velocity being evaluated at either end of the interval. Note that there is a certain arbitrariness in whether constant terms are included here or else in the normalising partition function.

The probability of the trajectory given the starting point $\mathbf{\Gamma}_0$ is just the product of the conditional pair transition probabilities. Writing this in terms of the fluctuation force it is

$$\begin{aligned} \wp(\underline{\mathbf{\Gamma}}|\mathbf{\Gamma}_0, t_0) &= \prod_{n=1}^{f} \wp(\mathbf{\Gamma}_n, t_n|\mathbf{\Gamma}_{n-1}, t_{n-1}) \\ &= \prod_{n=1}^{f} \frac{e^{S^{(2)}(\mathbf{\Gamma}_n, t_n|\mathbf{\Gamma}_{n-1}, t_{n-1})/k_{\rm B}}}{Z(t_n, t_{n-1})} \\ &= \prod_{n=1}^{f} \frac{\Theta_\sigma(\tilde{\mathbf{R}}_n)}{|2\pi\Delta_t k_{\rm B} \Lambda_{\rm pp}(t_n)|^{1/2}}, \end{aligned} \tag{8.192}$$

where the Gaussian probability distribution without normalisation is

$$\Theta_\sigma(\tilde{\mathbf{R}}_n) \equiv e^{-\Lambda_{\rm pp}(t_n)^{-1} : \tilde{\mathbf{R}}_n^2 / 2\Delta_t k_{\rm B}}. \tag{8.193}$$

(The subscript σ is used to remind the reader that $\Lambda_{\rm pp}$ is usually replaced by $\sigma \mathrm{I}_{\rm pp}/k_{\rm B}|\Delta_t|$.) This can also be written as the exponential of the trajectory entropy (as can be seen from the second equality),

$$\begin{aligned} \wp(\underline{\mathbf{\Gamma}}|\mathbf{\Gamma}_0, t_0) &= \frac{1}{Z[t]} e^{S(\underline{\mathbf{\Gamma}}|\mathbf{\Gamma}_0,t_0)/k_{\rm B}} \\ &= e^{[S_{\rm r,st}(\mathbf{\Gamma}_f,t_f) - S_{\rm r,st}(\mathbf{\Gamma}_0,t_0)]/2k_{\rm B}} e^{S_{\rm r,dyn}(\underline{\mathbf{\Gamma}})/2k_{\rm B}} \\ &\quad \times \prod_{n=1}^{f} \frac{\Theta_\sigma(\mathbf{R}_n)\Theta_\sigma(\overline{\mathbf{R}}_n)}{|2\pi\Delta_t k_{\rm B}\Lambda_{\rm pp}(t_n)|^{1/2}}. \end{aligned} \tag{8.194}$$

Note that the arguments of the Gaussian distributions are the actual reservoir force, $\mathbf{R}_n = \mathbf{\Gamma}_n - \mathbf{\Gamma}_{n-1} - \Delta_t \dot{\mathbf{\Gamma}}^0$, and the most like value of the actual reservoir force, $\overline{\mathbf{R}}_n = \Delta_t \Lambda_{\rm pp}(t_n) \nabla_{\rm p} S_{\rm st}(\mathbf{\Gamma}_n, t_n)$, with $\Delta_t > 0$, and the right-hand side being evaluated at either terminal of the time step. The above two expressions for $\wp(\underline{\mathbf{\Gamma}}|\mathbf{\Gamma}_0, t_0)$ are exactly equal to each other.

Usually the initial state is distributed according to the non-equilibrium probability distribution found above, and the unconditional trajectory probability is just $\wp(\underline{\mathbf{\Gamma}}) \equiv \wp(\underline{\mathbf{\Gamma}}|\mathbf{\Gamma}_0, t_0)\wp(\mathbf{\Gamma}_0, t_0)$. It is sometimes convenient to assume instead that the initial state is distributed according to the static probability, $\wp_{\rm st}(\mathbf{\Gamma}_0, t_0) = e^{S_{\rm r,st}(\mathbf{\Gamma}_0,t_0)/k_{\rm B}}/Z_{\rm st}(t_0)$. In this case the unconditional trajectory probability is

$$\begin{aligned} \wp_{\rm st}(\underline{\mathbf{\Gamma}}) &\equiv \wp(\underline{\mathbf{\Gamma}}|\mathbf{\Gamma}_0, t_0)\wp_{\rm st}(\mathbf{\Gamma}_0, t_0) \\ &= \frac{1}{Z_{\rm st}(t_0)} e^{[S_{\rm r,st}(\mathbf{\Gamma}_f,t_f) + S_{\rm r,st}(\mathbf{\Gamma}_0,t_0)]/2k_{\rm B}} e^{S_{\rm r,dyn}(\underline{\mathbf{\Gamma}})/2k_{\rm B}} \\ &\quad \times \prod_{n=1}^{f} \frac{\Theta_\sigma(\mathbf{R}_n)\Theta_\sigma(\overline{\mathbf{R}}_n)}{|2\pi\Delta_t k_{\rm B}\Lambda_{\rm pp}(t_n)|^{1/2}}. \end{aligned} \tag{8.195}$$

For a long enough trajectory the actual distribution of the initial state makes negligible difference to the trajectory probability.

Conjugate Trajectory

Now consider the conjugate system that is characterised by the Hamiltonian $\tilde{\mathcal{H}}(\mathbf{\Gamma}, t) = \mathcal{H}(\mathbf{\Gamma}, t_0 + t_f - t)$, $t \in [t_0, t_f]$, where $\mathcal{H}(\mathbf{\Gamma}, t)$ is the possibly time dependent Hamiltonian of the original system. Consider the conjugate trajectory in this conjugate system, $\tilde{\underline{\mathbf{\Gamma}}}$, which is defined such that $\tilde{\mathbf{\Gamma}}(t) \equiv \mathbf{\Gamma}(t_0 + t_f - t)^\dagger$, $t \in [t_0, t_f]$, or, equivalently, $\tilde{\mathbf{\Gamma}}_n = \mathbf{\Gamma}^\dagger_{f-n}$. Recall that if $\mathbf{\Gamma}(t) = \{\mathbf{q}(t)^N, \mathbf{p}(t)^N\}$, then $\mathbf{\Gamma}(t)^\dagger = \{\mathbf{q}(t)^N, (-\mathbf{p}(t))^N\}$. It is important to note that time proceeds in the positive direction for the conjugate system, $\Delta_t > 0$. The initial point of the conjugate trajectory is the conjugate of the final point of the original trajectory, $\tilde{\mathbf{\Gamma}}_0 = \mathbf{\Gamma}^\dagger_f$, and the final point of the conjugate trajectory is the conjugate of the initial point of the original trajectory, $\tilde{\mathbf{\Gamma}}_f = \mathbf{\Gamma}^\dagger_0$.

Because the static entropy has even parity, it has the same value at the corresponding points on the two trajectories, $\tilde{S}_{\rm r,st}(\tilde{\mathbf{\Gamma}}(t), t) = S_{\rm r,st}(\mathbf{\Gamma}(t_0 + t_f -$

$t)^\dagger, t_0 + t_f - t) = S_{\mathrm{r,st}}(\mathbf{\Gamma}(t_0 + t_f - t), t_0 + t_f - t)$. It is necessary to designate the entropy of the conjugate system with a tilde because it is a function of $\tilde{\mathcal{H}}$. The adiabatic rate of change of the static part of the entropy in the conjugate system is the negative of that at the corresponding point in the original system,

$$\dot{\tilde{S}}^0_{\mathrm{r,st}}(\tilde{\mathbf{\Gamma}}(t), t) = -\dot{S}^0_{\mathrm{r,st}}(\mathbf{\Gamma}(t_0 + t_f - t), t_0 + t_f - t). \tag{8.196}$$

This follows because the partial time derivative of the static part of the entropy is proportional to the partial time derivative of the Hamiltonian, and this is equal and opposite at the respective points on the two trajectories by definition of the Hamiltonian of the conjugate system. Also the operator $\dot{\mathbf{\Gamma}}^0 \cdot \nabla$ is an odd parity operator, and since $S_{\mathrm{r,st}}$ has even parity, this term is also equal and opposite at the respective points. It important to note that invoking a conjugate system with Hamiltonian defined as above was essential to this result. The specific results which follow only hold for such a conjugate system.

It should be mentioned that in some cases the conjugate system is the same as the original. The most obvious such case is when the Hamiltonian does not explicitly depend upon time, such as for an equilibrium system, or for a thermodynamic steady state system. Another example is when the Hamiltonian is harmonic with time dependence of the form $\cos 2\pi l t/(t_0+t_f)$, with l an integer.

The conditional probability of the conjugate trajectory given the starting point $\tilde{\mathbf{\Gamma}}_0 = \mathbf{\Gamma}_f^\dagger$ is

$$\begin{aligned}
&\wp(\tilde{\underline{\mathbf{\Gamma}}}|\tilde{\mathbf{\Gamma}}_0;\tilde{\mathcal{H}}) \\
&= e^{[\tilde{S}_{\mathrm{r,st}}(\tilde{\mathbf{\Gamma}}_f,t_f)-\tilde{S}_{\mathrm{r,st}}(\tilde{\mathbf{\Gamma}}_0,t_0)]/2k_{\mathrm{B}}} e^{\tilde{S}_{\mathrm{r,dyn}}(\tilde{\underline{\mathbf{\Gamma}}})/2k_{\mathrm{B}}} \prod_{n=1}^{f} \frac{\Theta_\sigma(\mathbf{R}_n^{\mathrm{conj}})\Theta_\sigma(\overline{\mathbf{R}}_n^{\mathrm{conj}})}{|2\pi\Delta_t k_{\mathrm{B}}\tilde{\Lambda}_{\mathrm{pp}}(t_n)|^{1/2}} \\
&= e^{[S_{\mathrm{r,st}}(\mathbf{\Gamma}_0,t_0)-S_{\mathrm{r,st}}(\mathbf{\Gamma}_f,t_f)]/2k_{\mathrm{B}}} e^{-S_{\mathrm{r,dyn}}(\underline{\mathbf{\Gamma}})/2k_{\mathrm{B}}} \prod_{n=1}^{f} \frac{\Theta_\sigma(\mathbf{R}_n)\Theta_\sigma(\overline{\mathbf{R}}_n)}{|2\pi\Delta_t k_{\mathrm{B}}\Lambda_{\mathrm{pp}}(t_n)|^{1/2}}.
\end{aligned} \tag{8.197}$$

The final equality follows from three results. First, since the adiabatic rate of change of the static part of the entropy satisfies $\dot{\tilde{S}}^0_{\mathrm{r,st}}(\tilde{\mathbf{\Gamma}}(t), t) = -\dot{S}^0_{\mathrm{r,st}}(\mathbf{\Gamma}(t_0 + t_f - t), t_0 + t_f - t)$, the dynamic part of the entropy on the conjugate trajectory is the negative of that on the original trajectory,

$$\begin{aligned}
\tilde{S}_{\mathrm{r,dyn}}(\tilde{\underline{\mathbf{\Gamma}}}) &= -\int_{t_0}^{t_f} \mathrm{d}t'\, \dot{\tilde{S}}^0_{\mathrm{r,st}}(\tilde{\mathbf{\Gamma}}(t'), t') \\
&= \int_{t_0}^{t_f} \mathrm{d}t'\, \dot{S}^0_{\mathrm{r,st}}(\mathbf{\Gamma}(t_0 + t_f - t'), t_0 + t_f - t') \\
&= \int_{t_0}^{t_f} \mathrm{d}t''\, \dot{S}^0_{\mathrm{r,st}}(\mathbf{\Gamma}(t''), t'') \\
&= -S_{\mathrm{r,dyn}}(\underline{\mathbf{\Gamma}}).
\end{aligned} \tag{8.198}$$

Second, the actual reservoir force on the conjugate trajectory is essentially the negative conjugate of the actual reservoir force on the original trajectory,

$$\begin{aligned}
\mathbf{R}_n^{\text{conj}} &= \tilde{\mathbf{\Gamma}}_n - \tilde{\mathbf{\Gamma}}_{n-1} - \Delta_t \dot{\tilde{\mathbf{\Gamma}}}^0(\tilde{\mathbf{\Gamma}}_{n-1}, t_{n-1}) \\
&= \mathbf{\Gamma}_{f-n}^{\dagger} - \mathbf{\Gamma}_{f-n+1}^{\dagger} - \Delta_t \dot{\mathbf{\Gamma}}^0(\mathbf{\Gamma}_{f-n+1}^{\dagger}, t_{f-n+1}) \\
&= \mathbf{\Gamma}_{f-n}^{\dagger} - \mathbf{\Gamma}_{f-n+1}^{\dagger} + \Delta_t \dot{\mathbf{\Gamma}}^0(\mathbf{\Gamma}_{f-n+1}, t_{f-n+1})^{\dagger} \\
&= -\mathbf{R}_{f-n+1}^{\dagger}.
\end{aligned} \tag{8.199}$$

Hence the Gaussian of it is unchanged, $\Lambda_{\text{pp}} : \left[-\mathbf{R}_{f-n+1}^{\dagger}\right]^2 = \Lambda_{\text{pp}} : [\mathbf{R}_{f-n+1}]^2$. Third, the most likely reservoir force has a similar symmetry,

$$\begin{aligned}
\overline{\mathbf{R}}_n^{\text{conj}} &= \frac{|\Delta_t|}{2} \Lambda_{\text{pp}} \nabla_{\text{p}} \tilde{S}_{\text{r,st}}(\tilde{\mathbf{\Gamma}}_n, t_n) \\
&= \frac{-|\Delta_t|}{2} \Lambda_{\text{pp}} \nabla_{\text{p}} S_{\text{r,st}}(\mathbf{\Gamma}_{f-n}, t_{f-n}),
\end{aligned} \tag{8.200}$$

and so the Gaussian factor of it is also unchanged. These last two results are used in conjunction with the fact that n is a dummy index for the product symbol that can be replaced by $f - n$. It has also been assumed that the transport matrix is either independent of time, or else satisfies $\tilde{\Lambda}(t_n) = \Lambda(t_f + t_0 - t_n)$. Again it is mentioned that the various functions can be evaluated at either terminus of each time step without affecting the results to linear order in the time step.

The ratio of the conditional probabilities of the original and of the conjugate trajectory is exactly given by

$$\begin{aligned}
\frac{\wp(\underline{\mathbf{\Gamma}}|\mathbf{\Gamma}_0)}{\wp(\tilde{\underline{\mathbf{\Gamma}}}|\tilde{\mathbf{\Gamma}}_0; \tilde{\mathcal{H}})} &= e^{[S_{\text{r,st}}(\mathbf{\Gamma}_f, t_f) - S_{\text{r,st}}(\mathbf{\Gamma}_0, t_0)]/k_{\text{B}}} e^{S_{\text{r,dyn}}(\underline{\mathbf{\Gamma}})/k_{\text{B}}} \\
&= e^{\vec{\Delta} S_{\text{r}} \underline{\mathbf{\Gamma}}/k_{\text{B}}}.
\end{aligned} \tag{8.201}$$

The exponent is the change in the reservoir entropy on the trajectory of the original system. Since $\vec{\Delta} S_{\text{r}}(\underline{\mathbf{\Gamma}}) = -\vec{\Delta} \tilde{S}_{\text{r}}(\tilde{\underline{\mathbf{\Gamma}}})$, this obeys exactly the conjugation operation (i.e. taking the conjugate of the left-hand side is equivalent to swapping the numerator and the denominator of the fraction). For a long trajectory the initial conditions ought to have negligible influence, so one expects that this result for the ratio of conditional probabilities to also dominate the ratio of unconditional probabilities.

Assuming that the system was in the non-equilibrium state prior to the start of the original and of the conjugate trajectory, the ratio of the unconditional probabilities of the two trajectories is

$$\begin{aligned}
\frac{\wp(\underline{\mathbf{\Gamma}}|\mathbf{\Gamma}_0)\wp(\mathbf{\Gamma}_0, t_0)}{\wp(\tilde{\underline{\mathbf{\Gamma}}}|\tilde{\mathbf{\Gamma}}_0; \tilde{\mathcal{H}})\wp(\tilde{\mathbf{\Gamma}}_0, t_0; \tilde{\mathcal{H}})} &= \frac{Z(t_0; \tilde{\mathcal{H}})}{Z(t_0)} \frac{e^{S_{\text{r,dyn}}(\mathbf{\Gamma}_0, t_0)/k_{\text{B}}}}{e^{S_{\text{r,dyn}}(\tilde{\mathbf{\Gamma}}_0, t_0; \tilde{\mathcal{H}})/k_{\text{B}}}} e^{S_{\text{r,dyn}}(\underline{\mathbf{\Gamma}})/k_{\text{B}}} \\
&\approx e^{S_{\text{r,dyn}}(\underline{\mathbf{\Gamma}})/k_{\text{B}}}.
\end{aligned} \tag{8.202}$$

Neither of the two factors involving $S_{\rm r,dyn}$ depend upon the length of the time interval, and so they can be neglected compared to $S_{\rm r,dyn}(\underline{\mathbf{\Gamma}})$ for a long enough trajectory. For a harmonic or cyclic Hamiltonian with period $t_f - t_0$, the two partition functions are equal and so they cancel. The only case when the final approximation may not be accurate is for the case when the time dependence of the Hamiltonian is monotonic over the interval.

Alternatively, one can specify that the system was in the static state prior to the start of the original and of the conjugate trajectory. In this case the ratio of the unconditional probabilities of the two trajectories is

$$\begin{aligned} \frac{\wp(\underline{\mathbf{\Gamma}}|\mathbf{\Gamma}_0)\wp_{\rm st}(\mathbf{\Gamma}_0,t_0)}{\wp(\tilde{\underline{\mathbf{\Gamma}}}|\tilde{\mathbf{\Gamma}}_0;\tilde{\mathcal{H}})\wp_{\rm st}(\tilde{\mathbf{\Gamma}}_0,t_0;\tilde{\mathcal{H}})} &= \frac{Z_{\rm st}(t_f)}{Z_{\rm st}(t_0)} e^{S_{\rm r,dyn}(\underline{\mathbf{\Gamma}})/k_{\rm B}} \\ &\approx e^{S_{\rm r,dyn}(\underline{\mathbf{\Gamma}})/k_{\rm B}}. \end{aligned} \tag{8.203}$$

In the general case the logarithm of the partition function is a free energy, such as the Helmholtz free energy or one of its analogues, $F_{\rm st}(t) = -k_{\rm B}T \ln Z_{\rm st}(t)$. Hence the ratio that appears in the first equality corresponds to the exponential of the change in the static free energy. (The static free energy is the same as an equilibrium free energy in the case of mechanical work. It is a generalisation of an equilibrium free energy in the case of an applied thermodynamic gradient.) For a cyclic trajectory, $\mathcal{H}(\mathbf{\Gamma},t_f) = \mathcal{H}(\mathbf{\Gamma},t_0)$, which includes a harmonic trajectory but is more general, the final approximation is exact as $Z_{\rm st}(t_f) = Z_{\rm st}(t_0)$. Again the final approximation may not be accurate when the explicitly time dependent part of the Hamiltonian varies monotonically with time over the trajectory.

These results show that the state of the system is immaterial outside of the time interval of interest for a large enough time interval. In this regime this ratio of probabilities of the forward and conjugate trajectories is just the exponential of the change in the reservoir entropy over the interval.

8.6.2 Fluctuation and Work Theorem

In a non-equilibrium system, on average the entropy of the reservoir increases over a time interval. It is of interest to characterise the fluctuations in this entropy change about its most likely value. The probability of a particular entropy change may be obtained from the expressions given above for the trajectory probability. More precisely, the probability of observing the entropy of the reservoir change by $\vec{\Delta}S_{\rm r}$ over the given time interval is related to the probability of observing the opposite change by

$$\begin{aligned} \wp(\vec{\Delta}S_{\rm r}) &= \int {\rm d}\underline{\mathbf{\Gamma}}\, \delta(\vec{\Delta}S_{\rm r} - \vec{\Delta}S_{\rm r}(\underline{\mathbf{\Gamma}}))\wp(\underline{\mathbf{\Gamma}}) \\ &= \int {\rm d}\tilde{\underline{\mathbf{\Gamma}}}\, \delta(\vec{\Delta}S_{\rm r} - \vec{\Delta}S_{\rm r}(\underline{\mathbf{\Gamma}}))\wp(\tilde{\underline{\mathbf{\Gamma}}};\tilde{\mathcal{H}}) e^{\vec{\Delta}S_{\rm r}(\underline{\mathbf{\Gamma}})/k_{\rm B}} \\ &= e^{\vec{\Delta}S_{\rm r}/k_{\rm B}} \int {\rm d}\tilde{\underline{\mathbf{\Gamma}}}\, \delta(\vec{\Delta}S_{\rm r} + \vec{\Delta}S_{\rm r}(\tilde{\underline{\mathbf{\Gamma}}}))\wp(\tilde{\underline{\mathbf{\Gamma}}};\tilde{\mathcal{H}}) \\ &= e^{\vec{\Delta}S_{\rm r}/k_{\rm B}} \wp(-\vec{\Delta}S_{\rm r};\tilde{\mathcal{H}}). \end{aligned} \tag{8.204}$$

Note the appearance of the probability of the conjugate system on the right-hand side of the second and subsequent equalities. It is essential to the result that the conjugate system be invoked. Recall that in some cases the conjugate system is the same as the original system. Examples where this can be done exactly include equilibrium systems, non-equilibrium mechanical systems with harmonic Hamiltonian, systems with a steady applied thermodynamic gradient, and systems with an harmonic applied thermodynamic gradient. An approximate example is a cyclic system.

This result is exactly true if the conditional trajectory probability is used (i.e. the beginning and end of the trajectory is fixed). Otherwise it holds approximately for a sufficiently long trajectory where end effects can be neglected. This result says in essence that the probability of a positive increase in entropy is exponentially greater than the probability of a decrease in entropy over a time interval. In essence this is the fluctuation theorem that was first derived by Bochkov and Kuzovlev,[17] and subsequently by Evans *et al.*[18] The original versions of the theorem relied upon either an adiabatic trajectory, or else a deterministic trajectory with an *ad hoc* thermostat, and they were restricted to mechanical work. The present version of the theorem[19] has two key generalisations: it holds not only for non-equilibrium mechanical systems but also for non-equilibrium thermodynamic systems (e.g. heat flow, driven diffusion, and shear), and it explicitly accounts for the stochastic and dissipative interactions with the reservoir along the trajectory.

Closely related to the fluctuation theorem is a class of results that have come to be called the work theorem. Consider the average over the trajectory of the exponential of the negative of the entropy change,

$$\begin{aligned} \left\langle e^{-\vec{\Delta} S_{\rm r}(\underline{\mathbf{\Gamma}})/k_{\rm B}} \right\rangle &= \int \mathrm{d}\underline{\mathbf{\Gamma}}\, \wp(\underline{\mathbf{\Gamma}}) e^{-\vec{\Delta} S_{\rm r}(\underline{\mathbf{\Gamma}})/k_{\rm B}} \\ &= \int \mathrm{d}\underline{\tilde{\mathbf{\Gamma}}}\, \wp(\underline{\tilde{\mathbf{\Gamma}}}; \tilde{\mathcal{H}}) \\ &= 1. \end{aligned} \tag{8.205}$$

Note that the exponent on the left-hand side is the negative of the change in reservoir entropy over the trajectory, neglecting end effects. Similarly, in going from the first to the second equality, it has been assumed that the trajectory is long enough that the end effects may be neglected. Although the averand on the left-hand side scales exponentially with time, this result shows that this particular average is not extensive in time (i.e. it does not depend on $|t_f - t_0|$).

An illustration of this result is given by the average of the exponential of the negative of the non-equilibrium part of the reservoir entropy (or thermodynamic

[17] Bochkov, G. N. and Kuzovlev, Yu. E. (1981), Physica, **106A**, 443.

[18] Evans, D. J., Cohen, E. G. D., and Morriss, G. P. (1993), Phys. Rev. Lett. **71**, 2401. Evans, D. J. (2003), Molec. Phys. **101**, 1551.

[19] Attard, P. (2006), J. Chem. Phys. **124**, 024109.

work), assuming static equilibrium at the interval termini,

$$\begin{aligned}
&\left\langle e^{-S_{\mathrm{r,dyn}}(\underline{\Gamma})/k_{\mathrm{B}}}\right\rangle \\
&= \int \mathrm{d}\underline{\Gamma}\, \wp(\underline{\Gamma}|\Gamma_0)\wp_{\mathrm{st}}(\Gamma_0,t_0)e^{-S_{\mathrm{r,dyn}}(\underline{\Gamma})/k_{\mathrm{B}}} \\
&= \int \mathrm{d}\tilde{\underline{\Gamma}}\, \wp(\tilde{\underline{\Gamma}}|\tilde{\Gamma}_0;\tilde{\mathcal{H}})\wp_{\mathrm{st}}(\tilde{\Gamma}_0,t_0;\tilde{\mathcal{H}})\frac{\wp_{\mathrm{st}}(\Gamma_0,t_0)}{\wp_{\mathrm{st}}(\tilde{\Gamma}_0,t_0;\tilde{\mathcal{H}})} \\
&\quad \times e^{-S_{\mathrm{r,st}}(\Gamma_0,t_0)/k_{\mathrm{B}}}e^{S_{\mathrm{r,st}}(\Gamma_f,t_f)/k_{\mathrm{B}}} \\
&= \frac{Z_{\mathrm{st}}(t_f)}{Z_{\mathrm{st}}(t_0)} \\
&= e^{-\beta\Delta F_{\mathrm{st}}}. \qquad (8.206)
\end{aligned}$$

The penultimate equality uses the fact that $Z_{\mathrm{st}}(t_0;\tilde{\mathcal{H}}) = Z_{\mathrm{st}}(t_f)$, and the final equality uses the fact that for a static system the free energy is $-k_{\mathrm{B}}T$ times the logarithm of the partition function. One can interpret the change in entropy on a trajectory as the thermodynamic work, in which case one sees that the averand on the left-hand side is the exponential of the negative of this. The exponent on the right-hand side is the negative of the difference in the static free energy divided by $k_{\mathrm{B}}T$.

As in the case of the fluctuation theorem, the work theorem was originally restricted to mechanical work. It was first given by Bochkov and Kuzovlev, (1981), who assumed a long cyclic trajectory, and it was later given by Jarzynski,[20] who also treated a mechanical non-equilibrium system and assumed a Boltzmann distribution at the beginning and end of the trajectory. The present result is valid both for applied thermodynamic gradients and for external time dependent mechanical fields.

8.7 Path Entropy for Mechanical Work

8.7.1 Evolution of the Reservoir Entropy and Transitions

This section analyses a sub-system that can exchange energy with a thermal reservoir of temperature T and that is characterised by an explicitly time dependent Hamiltonian $\mathcal{H}(\Gamma,t)$, where $\Gamma \equiv \{\mathbf{q}^{3N},\mathbf{p}^{3N}\}$ is a point in the phase space of the sub-system. The sub-system Hamiltonian may be separated into potential and kinetic energy terms, $\mathcal{H}(\Gamma,t) = U(\mathbf{q}^{3N},t) + \mathcal{K}(\mathbf{p}^{3N})$, and with little loss of generality the kinetic energy may be taken to be $\mathcal{K}(\mathbf{p}^N) = \sum_{i\alpha} p_{i\alpha}^2/2m$, where i indexes the atoms and α indexes the axes.

This particular system is the archetypal mechanical non-equilibrium system, and one of the goals of this section is to characterise the mechanical work done on the sub-system by the external potential over time.

For the present mechanical non-equilibrium case, the static part of the reservoir entropy is the usual Boltzmann factor, $S_{\mathrm{r,st}}(\Gamma,t) = -\mathcal{H}(\Gamma,t)/T$. With this,

[20] Jarzynski, C. (1997), Phys. Rev. Lett. **78**, 2690.

the change in entropy on a transition, $\mathbf{\Gamma}_2 \stackrel{\Delta_t}{\to} \mathbf{\Gamma}_1$, $\Delta_t \equiv t_2 - t_1$, is

$$\vec{\Delta} S_{\rm r}(\mathbf{\Gamma}_2, t_2; \mathbf{\Gamma}_1, t_1) = \mathbf{R} \cdot \nabla_{\rm p} S_{\rm r,st}(\mathbf{\Gamma}, t) = \frac{-1}{mT} \sum_{i\alpha} R_{i\alpha}\, p_{i\alpha}. \tag{8.207}$$

Here the total reservoir force is the excess over the adiabatic evolution, $\mathbf{R} = \mathbf{\Gamma}_2 - \mathbf{\Gamma}_1 - \Delta_t \dot{\mathbf{\Gamma}}^0$, and it may be decomposed into a dissipative part and a stochastic part, $\mathbf{R} = \overline{\mathbf{R}} + \tilde{\mathbf{R}}$. The variance of the stochastic part of the force is

$$\left\langle \tilde{\mathbf{R}} \tilde{\mathbf{R}} \right\rangle = \sigma \mathrm{I}_{\rm pp}, \tag{8.208}$$

and the most likely reservoir force is

$$\begin{aligned} \overline{\mathbf{R}} &= \frac{\sigma}{2k_{\rm B}} \left[\nabla_{\rm p} S_{\rm r,st}(\mathbf{\Gamma}, t) + \{\hat{t} - 1\} \overline{S'_{\rm p}}(t) \right] \\ &= \frac{-\beta\sigma}{2m} \left[\mathbf{\Gamma}_{\rm p} + \{\hat{t} - 1\} \overline{\mathbf{\Gamma}}_{\rm p}(t) \right]. \end{aligned} \tag{8.209}$$

Here the simplest transport matrix has been assumed, $\Lambda_{\rm pp} = \sigma \mathrm{I}_{\rm pp} / k_{\rm B} |\Delta_t|$, with the variance, $\sigma \propto |\Delta_t|$, the inverse temperature, $\beta \equiv 1/k_{\rm B}T$, and $\hat{t} \equiv \mathrm{sign}(\Delta_t)$. The variance represents the strength of the influence of the reservoir on the sub-system and it can be chosen arbitrarily within quite a wide range. In most of the results below the time step will be taken to be positive, in which case the pre-factor of the most likely momenta vanishes, $\{\hat{t} - 1\} = 0$.

The stochastic dissipative equations of motion for this non-equilibrium mechanical system are

$$\begin{aligned} q_{i\alpha}(t + \Delta_t) &= q_{i\alpha}(t) + \Delta_t \dot{q}^0_{i\alpha}(t), \\ p_{i\alpha}(t + \Delta_t) &= p_{i\alpha}(t) + \Delta_t \dot{p}^0_{i\alpha}(t) - \frac{\beta\sigma}{2m} \left[p_{i\alpha} + \{\hat{t} - 1\} \overline{p}_{i\alpha}(t) \right] + \tilde{R}_{i\alpha}. \end{aligned} \tag{8.210}$$

Here the adiabatic velocities are

$$\dot{q}^0_{i\alpha}(t) = \frac{\partial \mathcal{H}(\mathbf{\Gamma}, t)}{\partial p_{i\alpha}} = \frac{p_{i\alpha}}{m} \text{ and } \dot{p}^0_{i\alpha}(t) = \frac{-\partial \mathcal{H}(\mathbf{\Gamma}, t)}{\partial q_{i\alpha}}. \tag{8.211}$$

The stochastic forces are chosen independently at each time step. The dissipative force, $-(\beta\sigma/2m) p_{i\alpha}$, has the form of a friction or drag force, and this term corresponds to a so-called Hoover thermostat. The additional term that appears here, $-(\beta\sigma/2m)\{\hat{t} - 1\}\overline{p}_{i\alpha}(t)$, is only non-zero going backward in time. The relationship between the friction coefficient and the variance of the stochastic force is called the fluctuation dissipation theorem. The form of the equations of motion, with a stochastic term and a dissipative term, is the same form as the Langevin equation for Brownian motion.

The conditional second entropy for the transition is

$$\begin{aligned} S^{(2)}(\mathbf{\Gamma}_2, t_2|\mathbf{\Gamma}_1, t_1) \\ &= \frac{-k_\mathrm{B}}{2\sigma}\tilde{\mathbf{R}}\cdot\tilde{\mathbf{R}} \\ &= \frac{-k_\mathrm{B}}{2\sigma}\left[\mathbf{R}\cdot\mathbf{R}+\overline{\mathbf{R}}\cdot\overline{\mathbf{R}}\right]+\frac{1}{2}\vec{\Delta}S_\mathrm{r}(\mathbf{\Gamma}_2, t_2;\mathbf{\Gamma}_1, t_1) \\ &= \frac{-k_\mathrm{B}}{2\sigma}\sum_{i\alpha}\left[R_{i\alpha}^2+\frac{\beta^2\sigma^2}{4m^2}p_{i\alpha}^2\right]-\frac{1}{2mT}\sum_{i\alpha}R_{i\alpha}\,p_{i\alpha}. \end{aligned} \tag{8.212}$$

The second and third equalities only hold for $\Delta_t > 0$, which restriction is henceforth assumed. Strictly speaking one should add to this a term of the form $-\sum_{i\alpha}[q_{2;i\alpha} - q_{1;i\alpha} - \Delta_t p_{1;i\alpha}/m]^2/2\sigma'$, which comes from $\Lambda_{\mathrm{qq}} \to 0$, and which represents a Dirac-δ for the adiabatic evolution of the position coordinates (c.f. Eq. (8.97)). The conditional transition entropy is

$$\begin{aligned} \wp(\mathbf{\Gamma}_2, t_2|\mathbf{\Gamma}_1, t_1) &= \frac{1}{Z(t_2,t_1)}e^{S^{(2)}(\mathbf{\Gamma}_2,t_2|\mathbf{\Gamma}_1,t_1)/k_\mathrm{B}} \\ &= \frac{1}{Z(t_2,t_1)}e^{-\tilde{\mathbf{R}}\cdot\tilde{\mathbf{R}}/2\sigma}\delta\left(\mathbf{\Gamma}_{\mathrm{q}2}-\mathbf{\Gamma}_{\mathrm{q}1}-\Delta_t\dot{\mathbf{\Gamma}}_\mathrm{q}^0\right) \\ &= \frac{1}{Z(t_2,t_1)}\delta\left(\mathbf{\Gamma}_{\mathrm{q}2}-\mathbf{\Gamma}_{\mathrm{q}1}-\Delta_t\dot{\mathbf{\Gamma}}_\mathrm{q}^0\right)\prod_{i\alpha}e^{-\left[R_{i\alpha}^2+\beta^2\sigma^2p_{i\alpha}^2/4m^2\right]/2\sigma} \\ &\quad\times\prod_{i\alpha}e^{-\beta R_{i\alpha}\,p_{i\alpha}/2m}, \end{aligned} \tag{8.213}$$

with the normalisation factor being $Z(t_2, t_1) = (2\pi\sigma)^{3N/2}$.

The exponent in the final factor, $-\beta\sum_{i\alpha} R_{i\alpha}\,p_{i\alpha}/2m = -\beta\mathbf{R}\cdot\nabla_\mathrm{p}\mathcal{H}/2$, is half the change in entropy for the transition (divided by Boltzmann's constant). The factor of $\beta/2$ emerges in the present analysis of the transition probability from the second entropy approach. It is worth mentioning that this term has previously appeared as the basis of Glauber or Kawasaki dynamics, which are essentially a stochastic approach to dynamical systems that neglect adiabatic evolution.[21] This form for the conditional stochastic transition probability satisfies detailed balance for an equilibrium Boltzmann distribution, and has previously been used with the adiabatic development in the context of a stochastic molecular dynamics algorithm for *equilibrium* systems.[22] In the non-equilibrium context, it has been used for the development of a *non-equilibrium* molecular dynamics algorithm for steady heat flow and for driven Brownian motion.[23]

---

[21] Glauber, R. J. (1963), J. Math. Phys. **4**, 294. Kawasaki, K. (1966), Phys. Rev. **145**, 145. Langer, J. S. (1969), Annals Phys. **54**, 258. Metiu, H., Kitahara, K., and Ross, J., (1975), J. Chem. Phys. **63**, 5116.

[22] Attard, P. (2002b), J. Chem. Phys. **116**, 9616. Boinepalli, S. and Attard, P. (2003), J. Chem. Phys. **119**, 12769.

[23] Attard, P. (2006), J. Chem. Phys. **124**, 024109. Attard, P. (2009a), J. Chem. Phys. **130**, 194113.

8.7.2 Transition Theorems

For the trajectory $\underline{\mathbf{\Gamma}}$, discretised as discussed above in §8.6, the change in reservoir entropy for the present mechanical non-equilibrium system is

$$\begin{aligned}\vec{\Delta} S_{\mathrm{r}}(\underline{\mathbf{\Gamma}}) &= S_{\mathrm{r,st}}(\mathbf{\Gamma}_f, t_f) - S_{\mathrm{r,st}}(\mathbf{\Gamma}_0, t_0) - \int_{t_0}^{t_f} \mathrm{d}t' \, \dot{S}^0_{\mathrm{r,st}}(\mathbf{\Gamma}(t'), t') \\ &= \frac{1}{T}\left[\mathcal{H}(\mathbf{\Gamma}_0, t_0) - \mathcal{H}(\mathbf{\Gamma}_f, t_f)\right] + \frac{1}{T}\int_{t_0}^{t_f} \mathrm{d}t' \, \frac{\partial U(\mathbf{\Gamma}', t')}{\partial t'}. \end{aligned} \tag{8.214}$$

where $\mathbf{\Gamma}' \equiv \mathbf{\Gamma}(t')$. The final term here, the dynamic part of the change in entropy over the trajectory, is the actual work done over the trajectory divided by temperature, $S_{\mathrm{ne}}(\underline{\mathbf{\Gamma}}) = W(\underline{\mathbf{\Gamma}})/T$, with the work done being

$$W(\underline{\mathbf{\Gamma}}) = \int_{t_0}^{t_f} \mathrm{d}t' \, \dot{\mathcal{H}}^0(\mathbf{\Gamma}(t'), t') = \Delta_t \sum_{n=0}^{f-1} \dot{\mathcal{H}}^0(\mathbf{\Gamma}_n, t_n), \tag{8.215}$$

where the adiabatic rate of change of the energy is just the partial time derivative of the potential, $\dot{\mathcal{H}}^0(\mathbf{\Gamma}, t) = \partial U(\mathbf{\Gamma}, t)/\partial t$. There are of course more refined ways of transforming the integral to a sum, but they make no difference to leading-order. In particular, it matters not whether the sum is from 0 to $f-1$ or from 1 to f, or whether the summand is taken as the average of the termini of each time step. In the expression for the change in entropy, $\vec{\Delta} S_{\mathrm{r}}(\underline{\mathbf{\Gamma}})$, for a sufficiently long trajectory the contributions of the initial and final states are negligible compared to the work term, which is the only term that grows with the trajectory (unless the time dependent part of the potential also scales with the time interval).

The conditional trajectory entropy is

$$\begin{aligned} S(\underline{\mathbf{\Gamma}}|\mathbf{\Gamma}_0, t_0) &= \frac{1}{2}\vec{\Delta} S_{\mathrm{r}}(\underline{\mathbf{\Gamma}}) - \frac{k_{\mathrm{B}}}{2\sigma}\sum_{n=1}^{f}\sum_{i\alpha}\left[R^2_{n;i\alpha} + \frac{\beta^2\sigma^2}{4m^2}p^2_{n;i\alpha}\right] \\ &= \frac{1}{2T}\left[\mathcal{H}(\mathbf{\Gamma}_0, t_0) - \mathcal{H}(\mathbf{\Gamma}_f, t_f)\right] + \frac{W(\underline{\mathbf{\Gamma}})}{T} \\ &\quad - \frac{k_{\mathrm{B}}}{2\sigma}\sum_{n=1}^{f}\sum_{i\alpha}\left[R^2_{n;i\alpha} + \frac{\beta^2\sigma^2}{4m^2}p^2_{n;i\alpha}\right]. \end{aligned} \tag{8.216}$$

Recall that $\mathbf{R}_n \equiv \mathbf{\Gamma}_{\mathrm{p},n} - \mathbf{\Gamma}_{\mathrm{p},n-1} - \Delta_t \dot{\mathbf{\Gamma}}^0_{\mathrm{p}}$ is the total reservoir perturbation for that transition, with the adiabatic velocity being evaluated at either terminus, $\dot{\mathbf{\Gamma}}^0_{\mathrm{p}}(\mathbf{\Gamma}_n, t_n)$, or $\dot{\mathbf{\Gamma}}^0_{\mathrm{p}}(\mathbf{\Gamma}_{n-1}, t_{n-1})$, or the average of the two. The conditional probability of the trajectory is essentially the exponential of this,

$$
\begin{aligned}
\wp(\underline{\Gamma}|\Gamma_0,t_0) &= \prod_{n=1}^{f} \frac{\Theta_\sigma(\tilde{\mathbf{R}}_n)}{(2\pi\sigma)^{3N/2}} \\
&= e^{\vec{\Delta}S_{\rm r}(\underline{\Gamma})/2k_{\rm B}} \prod_{n=1}^{f} \frac{\Theta_\sigma(\mathbf{R}_n)\Theta_\sigma(\overline{\mathbf{R}}_n)}{(2\pi\sigma)^{3N/2}} \\
&= e^{-\beta[\mathcal{H}(\Gamma_f,t_f)-\mathcal{H}(\Gamma_0,t_0)]/2} e^{\beta W(\underline{\Gamma})/2} \prod_{n=1}^{f} \frac{\Theta_\sigma(\mathbf{R}_n)\Theta_\sigma(\overline{\mathbf{R}}_n)}{(2\pi\sigma)^{3N/2}},
\end{aligned} \tag{8.217}
$$

where the Gaussian probability distribution without normalisation is

$$
\Theta_\sigma(\mathbf{R}_n) \equiv e^{-\mathbf{R}_n\cdot\mathbf{R}_n/2\sigma}. \tag{8.218}
$$

As in the preceding section, §8.6, define a conjugate system, $\tilde{\mathcal{H}}(\Gamma,t) = \mathcal{H}(\Gamma,t_f+t_0-t)$, $t\in[t_0,t_f]$, and a conjugate trajectory, $\tilde{\underline{\Gamma}}$, such that $\tilde{\Gamma}(t) = \Gamma(t_f+t_0-t)^\dagger$. The ratio of the conditional probabilities is

$$
\frac{\wp(\underline{\Gamma}|\Gamma_0)}{\wp(\tilde{\underline{\Gamma}}|\tilde{\Gamma}_0;\tilde{\mathcal{H}})} = e^{\vec{\Delta}S_{\rm r}(\underline{\Gamma})/k_{\rm B}} = e^{-\beta[\mathcal{H}(\Gamma_f,t_f)-\mathcal{H}(\Gamma_0,t_0)]} e^{\beta W(\underline{\Gamma})}. \tag{8.219}
$$

Assuming that the system was in the non-equilibrium state prior to the start of the original and of the conjugate trajectory, the ratio of the unconditional probabilities of the two trajectories is

$$
\frac{\wp(\underline{\Gamma}|\Gamma_0)\wp(\Gamma_0,t_0)}{\wp(\tilde{\underline{\Gamma}}|\tilde{\Gamma}_0;\tilde{\mathcal{H}})\wp(\tilde{\Gamma}_0,t_0;\tilde{\mathcal{H}})} = \frac{Z(t_0;\tilde{\mathcal{H}})}{Z(t_0)} \frac{e^{\beta\overline{W}(\Gamma_0,t_0)}}{e^{\beta\overline{W}(\tilde{\Gamma}_0,t_0;\tilde{\mathcal{H}})}} e^{\beta W(\underline{\Gamma})}. \tag{8.220}
$$

Here the non-equilibrium probability density is $\wp(\Gamma,t) = \exp[-\beta\mathcal{H}(\Gamma,t) + \beta\overline{W}(\Gamma,t)]/Z(t)$, where the most likely work done is essentially the dynamic part of the reservoir entropy,

$$
\begin{aligned}
\overline{W}(\Gamma,t) &\equiv TS_{\rm r,dyn}(\Gamma,t) \\
&= -T\int_0^t {\rm d}t'\, \dot{S}^0_{\rm r}(\overline{\Gamma}(t'|\Gamma,t),t') \\
&= \int_0^t {\rm d}t'\, \frac{U(\Gamma',t')}{\partial t'}, \quad \Gamma' \equiv \overline{\Gamma}(t'|\Gamma,t),
\end{aligned} \tag{8.221}
$$

and the partition function is

$$
Z(t) = \int {\rm d}\Gamma\, e^{-\beta\mathcal{H}(\Gamma,t)} e^{\beta\overline{W}(\Gamma,t)}. \tag{8.222}
$$

In the case of a cyclic system, $\mathcal{H}(\Gamma,t_0) = \mathcal{H}(\Gamma,t_f)$, with the definition of the conjugate system extended such that $\tilde{\mathcal{H}}(\Gamma,t) = \mathcal{H}(\Gamma,t)$, $t\le t_0$, then $S_{\rm r,dyn}(\Gamma,t;\tilde{\mathcal{H}}) = S_{\rm r,dyn}(\Gamma,t)$ and it follows that $Z(t_0;\tilde{\mathcal{H}}) = Z(t_0)$. In any case, for a cyclic system it is clear that neither $\ln[Z(t_0;\tilde{\mathcal{H}})/Z(t_0)]$ nor $\overline{W}(\mathbf{\Gamma}_0,t_0) - \overline{W}(\tilde{\mathbf{\Gamma}}_0,t_0;\tilde{\mathcal{H}})$ scales with the size of the time interval.

A similar result holds if the system is dynamically disordered prior to and after the trajectories. This corresponds to the Boltzmann distribution,

$$\wp_{\mathrm{B}}(\mathbf{\Gamma},t) \equiv \frac{1}{Z_{\mathrm{B}}(t)} e^{-\beta\mathcal{H}(\mathbf{\Gamma},t)}, \quad t = t_0, \text{ and } t = t_f, \tag{8.223}$$

in which case the ratio of the unconditional probabilities is given exactly by,

$$\begin{aligned} \frac{\wp(\underline{\mathbf{\Gamma}}|\mathbf{\Gamma}_0)\wp_{\mathrm{B}}(\mathbf{\Gamma}_0,t_0)}{\wp(\underline{\tilde{\mathbf{\Gamma}}}|\tilde{\mathbf{\Gamma}}_0;\tilde{\mathcal{H}})\wp_{\mathrm{B}}(\tilde{\mathbf{\Gamma}}_0,t_0;\tilde{\mathcal{H}})} &= \frac{Z_{\mathrm{B}}(t_0;\tilde{\mathcal{H}})}{Z_{\mathrm{B}}(t_0)} e^{\beta W(\underline{\mathbf{\Gamma}})} \\ &= e^{-\beta[F_{\mathrm{st}}(t_f)-F_{\mathrm{st}}(t_0)]} e^{\beta W(\underline{\mathbf{\Gamma}})}, \end{aligned} \tag{8.224}$$

where the static or instantaneous Helmholtz free energy is simply the logarithm of the partition function,

$$F_{\mathrm{st}}(t) = -T\ln Z_{\mathrm{B}}(t) = -T\ln\int \mathrm{d}\mathbf{\Gamma}\; e^{-\beta\mathcal{H}(\mathbf{\Gamma},t)}. \tag{8.225}$$

For a cyclic system, $\mathcal{H}(\mathbf{\Gamma},t_0) = \mathcal{H}(\mathbf{\Gamma},t_f)$, the static partition functions are exactly equal to each other, and there is no change in Helmholtz free energy over the time interval.

The fluctuation and work theorems were given above for the general non-equilibrium case, Eqs (8.204), and (8.205). Those results also hold for mechanical work, but it may be worth repeating Eq. (8.206) explicitly. Assuming thermal equilibrium at the beginning and at the end of the interval, the average Boltzmann factor of the work done is

$$\begin{aligned} \left\langle e^{-\beta W(\underline{\mathbf{\Gamma}})} \right\rangle &= \int \mathrm{d}\underline{\mathbf{\Gamma}}\; \wp(\underline{\mathbf{\Gamma}}|\mathbf{\Gamma}_0)\wp_{\mathrm{B}}(\mathbf{\Gamma}_0,t_0)e^{-\beta W(\underline{\mathbf{\Gamma}})} \\ &= \int \mathrm{d}\underline{\tilde{\mathbf{\Gamma}}}\; \tilde{\wp}(\underline{\tilde{\mathbf{\Gamma}}}|\tilde{\mathbf{\Gamma}}_0)\tilde{\wp}_{\mathrm{B}}(\tilde{\mathbf{\Gamma}}_0,t_0)e^{-\beta[F_{\mathrm{st}}(t_f)-F_{\mathrm{st}}(t_0)]} \\ &= e^{-\beta[F_{\mathrm{st}}(t_f)-F_{\mathrm{st}}(t_0)]}. \end{aligned} \tag{8.226}$$

The averand on the left-hand side is the exponential of the negative of the mechanical work actually done over the time interval. The exponent on the right-hand side is the negative of the difference in the Helmholtz free energy divided by $k_{\mathrm{B}}T$. The assumption that the system is in an equilibrium state at the termini of the trajectory is valid if the time dependence of the Hamiltonian is negligible near the termini of the trajectory: $\mathcal{H}(\mathbf{\Gamma},t) \approx \mathcal{H}(\mathbf{\Gamma},t_0)$, for $t \in [t_0, t_0-\tau]$, $\mathcal{H}(\mathbf{\Gamma},t) \approx \mathcal{H}(\mathbf{\Gamma},t_f)$, for $t \in [t_f, t_f+\tau]$, where τ is some relaxation time.

As has been mentioned, the mechanical work theorem was first given by Bochkov and Kuzovlev, (1981) (for the case of a long cyclic trajectory), and was later given by Jarzynski, (1997). The present derivation explicitly accounts for the exchange of heat between the sub-system and the reservoir during the performance of the work.

Chapter 9

Statistical Mechanics of Steady Flow: Heat and Shear

This chapter returns to the subject of steady heat flow, previously treated in Ch. 4 at the level of macroscopic non-equilibrium thermodynamics, which material is reprised in §9.1. The bulk of the present chapter is focused on the molecular-level description of steady heat flow. The non-equilibrium probability density for phase space is given in §9.2. The stochastic dissipative equations of motion, the fluctuation dissipation theorem, and the most likely reservoir force are given in §9.3. A version of the equipartition theorem appropriate for steady heat flow is given in §9.4. The Green-Kubo time correlation expression for the thermal conductivity is given in §9.5, where the relationship with the odd projection of the dynamic part of the reservoir entropy is discussed. The chapter concludes with a summary of the statistical mechanics of shear flow in §9.6.

9.1 Thermodynamics of Steady Heat Flow

9.1.1 Canonical Equilibrium System

The canonical equilibrium system consists of a sub-system that can exchange energy with a reservoir of fixed temperature. Temperature and energy are conjugate variables, and temperature is defined as the energy derivative of the entropy,

$$\frac{1}{T} \equiv \frac{\partial S(E, N, V)}{\partial E}. \tag{9.1}$$

Here N is the number of molecules and V is the volume of the relevant system. In what follows, T is the temperature of the reservoir and T_s is the temperature of the sub-system, which is not necessarily equal to that of the reservoir except in the equilibrium or optimum state.

The relevant thermodynamic potential for the canonical equilibrium system is the Helmholtz free energy, $F(N,V,T)$. This is actually related to the total entropy of the total system, $F(N,V,T) = -TS_{\rm tot}(N,V,T) = -k_{\rm B}TZ(N,V,T)$, where Z is the partition function (see below). More generally, the constrained Helmholtz free energy is (see §4.1),

$$\begin{aligned} F(E|N,V,T) &\equiv -TS_{\rm tot}(E|N,V,T) \\ &= E - TS_{\rm s}(E,N,V). \end{aligned} \tag{9.2}$$

This expression is valid for arbitrary values of the energy of the sub-system, E, and hence it characterises fluctuations and constrained states. The most likely sub-system energy, $\overline{E}(N,V,T)$, minimises the constrained Helmholtz free energy, or, equivalently, maximises the constrained total entropy. The ordinary thermodynamic Helmholtz free energy is approximately equal to the minimum value of the constrained Helmholtz free energy $F(N,V,T) \approx F(\overline{E}|N,V,T)$. The approximation is exact in the thermodynamic limit, $N \to \infty$, $V \to \infty$, $N/V =$ const. Note the difference between the entropy of the total system, $S_{\rm tot}$, and that of the sub-system alone, $S_{\rm s}$.

The temperature of the reservoir is $T_{\rm s}^{-1} \equiv \partial S_{\rm s}(E,N,V)/\partial E$. Setting the energy derivative of the constrained Helmholtz free energy to zero, one sees that it is minimised when the temperature of the sub-system is equal to that of the reservoir,

$$\left.\frac{\partial F(E|N,V,T)}{\partial E}\right|_{E=\overline{E}} = 0, \text{ or } \overline{T}_{\rm s} = T, \tag{9.3}$$

where $\overline{T}_{\rm s} \equiv T_{\rm s}(\overline{E}(N,V,T),V,T) \equiv (\partial S_{\rm s}(E,N,V)/\partial E|_{E=\overline{E}})^{-1}$. This is of course equivalent to maximising the total system entropy.

The probability distribution for the canonical equilibrium system is the Maxwell-Boltzmann distribution,

$$\wp_{\rm MB}(\mathbf{\Gamma}|N,V,T) = \frac{1}{Z(N,V,T)} e^{-\beta\mathcal{H}(\mathbf{\Gamma})}, \tag{9.4}$$

where $\beta \equiv 1/k_{\rm B}T$, $\mathcal{H}$ is the Hamiltonian of the sub-system, and $\mathbf{\Gamma} = \{\mathbf{q}^N, \mathbf{p}^N\}$ is a point in the sub-system phase space. This represents a statement that the reservoir entropy is $S_{\rm r}(\mathbf{\Gamma}) = -\beta\mathcal{H}(\mathbf{\Gamma})$, which includes the assumption that phase space points in the sub-system have equal weight (i.e. have no internal entropy). The partition function normalises the probability, $Z(N,V,T) = \int {\rm d}\mathbf{\Gamma}\, e^{-\beta\mathcal{H}(\mathbf{\Gamma})}$, and Boltzmann's constant times its logarithm is the total entropy of the total system. Henceforth number and volume are not shown explicitly.

9.1.2 Fourier's Law of Heat Conduction

The obvious non-equilibrium analogue of the canonical equilibrium system is for two heat reservoirs of different temperatures. In this, the canonical non-equilibrium system, the sub-system that separates the two reservoirs mediates the energy flow between them, (see Fig. 4.1 on p. 99). For fixed reservoir temperatures the optimum state is a steady state. It will be shown that in this optimum state the local energy density at each point in the sub-system will be such that there is a linear temperature profile between the two reservoirs. The constrained state is the state with non-optimum heat flux (and non-optimum energy profile, which will shortly be shown to be equivalent to a non-optimum temperature gradient). The second entropy increases during spontaneous changes in the heat flux, and it reaches a maximum in the steady state.

Let the boundaries between the sub-system and the reservoirs be located at $z = \pm L/2$ and let the reservoir temperatures be $T_\pm$. Define the zeroth and first temperatures as[1]

$$\frac{1}{T_0} \equiv \frac{1}{2}\left[\frac{1}{T_+} + \frac{1}{T_-}\right], \text{ and } \frac{1}{T_1} \equiv \frac{1}{L}\left[\frac{1}{T_+} - \frac{1}{T_-}\right]. \tag{9.5}$$

The zeroth temperature is essentially the average or mid temperature, and the reciprocal of the first temperature is essentially the temperature gradient (more precisely, the gradient of the reciprocal temperature). It is straightforward to show (see §4.2) that the zeroth and first temperatures are thermodynamically conjugate to the zeroth and first energy moments, respectively

$$\frac{1}{T_{s0}} \equiv \frac{\partial S_s(E_0, E_1)}{\partial E_0}, \text{ and } \frac{1}{T_{s1}} \equiv \frac{\partial S_s(E_0, E_1)}{\partial E_1}. \tag{9.6}$$

Here the energy moments are used to label a constrained sub-system macrostate; the other state variables are not shown explicitly. In the present case the system is taken to be homogenous in the x- and y-directions, and the energy moments are measured in the z-direction,

$$E_n \equiv A \int_{-L/2}^{L/2} \mathrm{d}z\, \varepsilon(z) z^n, \; n = 0, 1, \tag{9.7}$$

where A is the cross-sectional area of the sub-system and ε is the local energy density.

The energy flux J_E^0 is the energy per unit area per unit time crossing a plane perpendicular to the z-axis. As shown by Onsager[2] (see also Ch. 4), this is related to the adiabatic rate of change of the first energy moment.[3] This follows from energy conservation: the rate of change of the energy density is just the negative divergence of the flux, which in the present one-dimensional case is its gradient, $\dot{\varepsilon}^0(z) = -\mathrm{d}J_E^0(z)/\mathrm{d}z$. Hence considering the sub-system as isolated (so that $J_E^0(\pm L/2) = 0$, and $\dot{E}_1 = \dot{E}_1^0$), the adiabatic rate of change of the first

[1] Attard, P. (2005a), J. Chem. Phys. **122**, 154101.

[2] Onsager, L. (1931), Phys. Rev. **37**, 405, and **38**, 2265.

[3] This relationship has to be modified in the case of convection to account for the non-uniformity of the sub-system.

energy moment is (see also footnote 2 on p. 102),

$$\begin{aligned}\dot{E}_1^0 &= A\int_{-L/2}^{L/2} \mathrm{d}z\, z\dot{\varepsilon}^0(z) \\ &= A\int_{-L/2}^{L/2} \mathrm{d}z\, z\frac{-\mathrm{d}J_{\mathrm{E}}^0(z)}{\mathrm{d}z} \\ &= -AzJ_{\mathrm{E}}^0(z)\Big|_{-L/2}^{L/2} + A\int_{-L/2}^{L/2} \mathrm{d}z\, J_{\mathrm{E}}^0(z) \\ &= VJ_{\mathrm{E,av}}^0, \end{aligned} \tag{9.8}$$

where $V = AL$ is the volume of the sub-system. In the final equality, $J_{\mathrm{E,av}}^0 \equiv L^{-1}\int_{-L/2}^{L/2} \mathrm{d}z\, J_{\mathrm{E}}^0(z)$ is the flux averaged over the sub-system. In the steady state, the flux must be uniform (otherwise the energy density would change with time anywhere where the gradient of the flux was non-zero), which means that the spatially averaged flux is the same as the flux itself. An isolated system can be in a steady state everywhere except near the boundaries. Here and throughout, adiabatic means the evolution of the isolated sub-system according to Hamilton's equations of motion. This expression is formally exact and holds in non-optimum constrained states as well as in the optimum steady state.

Fourier's law states that the conductive energy flux is minus the thermal conductivity times the temperature gradient,

$$J_{\mathrm{E}}^0 = -\lambda\nabla T. \tag{9.9}$$

In typical hydrodynamic and other non-equilibrium applications, Fourier's law is taken to apply locally. The present canonical non-equilibrium case of a sub-system between two thermal reservoirs, will be sufficient to derive Fourier's law for the relationship between flux and temperature gradient, and this may be extended piece-wise for an arbitrary temperature profile $T(z)$. In the present case one expects (and one can show) that in the optimum state, the inverse temperature of the sub-system linearly interpolates between that of the reservoirs, $\overline{T}_{\mathrm{s}}(z)^{-1} = T_0^{-1} + zT_1^{-1}$.[4] This has constant gradient, $\nabla\left[\overline{T}_{\mathrm{s}}(z)^{-1}\right] = T_1^{-1}$. Alternatively, $\overline{T}_{\mathrm{s}}(z) = T_0 - zT_0^2/T_1 + \mathcal{O}(T_1^{-2})$. Hence in the linear regime of small temperature gradients, $\nabla T = -T_0^2/T_1$, and Fourier's law is equivalent to

$$\overline{\dot{E}_1^0} = \lambda V T_0^2 \frac{1}{T_1}. \tag{9.10}$$

In fact this is the most fundamental form of Fourier's and its regime of validity is greater than that of the original expression. The over-line *must* be exhibited on the left-hand side because this law only holds for the most likely flux. More generally, the flux is a constrained variable and so can be given any desired value independent of the temperature gradient. The task is now to derive Fourier's law from the second entropy theory.

[4] In general in thermodynamics, inverse temperature is a more natural physical quantity than temperature, and many thermodynamic relationships are simpler when expressed in terms of inverse temperature.

9.1.3 Second Entropy for Heat Flow

Isolated System

For an isolated sub-system, the first entropy may be written as a quadratic form in the first energy moment,

$$S_s(E_1) = \frac{1}{2} S E_1^2. \tag{9.11}$$

For this isolated system, the moment fluctuates about zero, $\langle E_1 \rangle = \overline{E}_1 = 0$. The fluctuation coefficient is $S = -k_B / \langle E_1^2 \rangle$. Since the heat flux is related to the rate of change of the first energy moment, the zeroth energy moment is largely irrelevant to the present analysis and is not shown here or below. The first temperature is the gradient of the first entropy, which is the thermodynamic force,

$$\frac{1}{T_{s1}(E_1)} = S E_1. \tag{9.12}$$

The first energy moment has even parity in phase space, $E_1(\mathbf{\Gamma}^\dagger) = E_1(\mathbf{\Gamma})$, and hence this is a case of pure parity fluctuations. The second entropy for the isolated sub-system is (see §4.3),

$$S_s^{(2)}(E_1', E_1|\tau) = \frac{1}{2} A(\tau)[E_1'^2 + E_1^2] + B(\tau) E_1' E_1, \tag{9.13}$$

where E_1 is at time t, and E_1' is at time $t + \tau$. This is maximised with respect to E_1' by

$$\overline{E_1'} \equiv \overline{E_1'}(E_1, \tau) = -A(\tau)^{-1} B(\tau) E_1. \tag{9.14}$$

The short time expansions of the fluctuation coefficients in the pure parity case, Eqs (2.39) and (2.40), are

$$A(\tau) = \frac{-1}{|\tau|\Lambda} + A + \mathcal{O}(\tau), \text{ and } B(\tau) = \frac{1}{|\tau|\Lambda} + B + \mathcal{O}(\tau), \tag{9.15}$$

with the reduction condition, Eq. (2.41), yielding, $A + B = S/2$. Short time means small on macroscopic time scales, but long on molecular time scales (see Fig. 2.1 on p. 40 and Fig. 4.2 on p. 104). With this the most likely future state in the short time limit is

$$\overline{E}_1' = E_1 + \frac{|\tau|\Lambda}{2} S E_1 + \mathcal{O}(\tau^2). \tag{9.16}$$

With this and the reduction condition the second entropy may be rewritten

$$\begin{aligned} S^{(2)}(E_1', E_1|\tau) &= \frac{1}{2} S E_1^2 - \frac{1}{2\Lambda|\tau|} \left[E_1' - E_1 - \frac{|\tau|\Lambda}{2} S E_1 \right]^2 + \mathcal{O}(\tau^2) \\ &= \frac{1}{2} S E_1^2 - \frac{|\tau|}{2\Lambda} \left[\overset{\circ}{E}_1 - \frac{\hat{\tau}\Lambda}{2} S E_1 \right]^2, \end{aligned} \tag{9.17}$$

which is valid for small time intervals. Here $\hat{\tau} \equiv \text{sign}(\tau)$.

The most likely rate of change of the first energy moment for a given fluctuation is

$$\overline{\overset{\circ}{E}{}_1^0} \equiv \frac{\overline{E}_1' - E_1}{\tau} = \frac{\hat{\tau}\Lambda}{2} S E_1 = \frac{\hat{\tau}\Lambda}{2} \frac{1}{T_{s1}}. \tag{9.18}$$

For this isolated system, the rate of change is an adiabatic quantity, which explains the superscript 0 on the left-hand side. In view of Fourier's law above, one can identify the thermal conductivity as

$$\lambda = \frac{\Lambda}{2VT_{s0}^2}. \tag{9.19}$$

Note that Fourier's law only applies forward in time, $\hat{\tau} = 1$.

The time correlation function is (see Eq. 2.22),

$$\begin{aligned} Q(\tau) &= k_B^{-1} \langle E_1(t+\tau) E_1(t) \rangle_0 = k_B^{-1} \left\langle \overline{E_1'} E_1 \right\rangle_0 \\ &= -k_B^{-1} A(\tau)^{-1} B(\tau) \langle E_1 E_1 \rangle_0 = A(\tau)^{-1} B(\tau) S^{-1} \\ &\sim -\left[1 + \frac{|\tau|\Lambda}{2} S \right] S^{-1} + \mathcal{O}(\tau^2) \\ &= -S^{-1} - |\tau| V T_{s0}^2 \lambda. \end{aligned} \tag{9.20}$$

The correlations in the first energy moment are Markovian, and hence the time correlation function decays exponentially on longer time scales, (see Fig. 2.1 on p. 40). The time correlation function and the averages are for a sub-system that evolves adiabatically and that is initially distributed either according to the Maxwell-Boltzmann distribution, or else uniformly in phase space on the energy hypersurface $E_0 = \langle \mathcal{H} \rangle_{N,V,T}$; since the fluctuations in $\mathcal{H}$ in the Maxwell-Boltzmann case are relatively negligible in the thermodynamic limit, the two are approximately equal. If one imagines that this truncated expansion formally holds at all times such that this defines a time dependent thermal conductivity $\lambda(\tau)$, then one can rewrite this as

$$\begin{aligned} \lambda(\tau) &= \frac{-1}{VT_{s0}^2|\tau|} [Q(\tau) - Q(0)] \\ &= \frac{-1}{Vk_B T_{s0}^2|\tau|} \langle [E_1(t+\tau) - E_1(t)] \, E_1(t) \rangle_0 \\ &= \frac{-1}{Vk_B T_{s0}^2|\tau|} \int_0^\tau \mathrm{d}t' \left\langle \dot{E}_1^0(t'+t) E_1(t) \right\rangle_0 \\ &= \frac{-1}{Vk_B T_{s0}^2|\tau|} \int_0^\tau \mathrm{d}t' \left\langle \dot{E}_1^0(\mathbf{\Gamma}^0(t'|\mathbf{\Gamma},0)) E_1(\mathbf{\Gamma}) \right\rangle_0. \end{aligned} \tag{9.21}$$

Since the system is isolated, it is the adiabatic trajectory that appears here. The right-hand side becomes independent of τ for τ in the short time regime (larger than molecular time scales, but smaller than macroscopic time scales;

see Fig. 4.2 on p. 104). It is permissible to invoke the limit $|\tau| \to \infty$ *after* the thermodynamic limit $L \to \infty$, $A \to \infty$, $N \to \infty$, $N/AL = \text{const}$. This is the Green-Kubo expression for the thermal conductivity.[5]

Reservoirs

Now two reservoirs of different temperatures are added to the isolated system that was just treated. This is the canonical non-equilibrium system; the analysis here is an abbreviated version of §4.5. Again suppressing the dependence on the zeroth energy moment, the total first entropy is

$$S_{\rm tot}(E_1|T_1) = S_{\rm s}(E_1) - \frac{E_1^{\rm (r)}(t)}{T_1}. \tag{9.22}$$

The first temperature T_1 is that of the reservoirs, and E_1 remains the first energy moment of the sub-system. The quantity $E_1^{\rm (r)}(t)$ is the total reservoir contribution to the first moment of the sub-system energy to the present time; it is equal to $-E_{\rm r1}(t)$, which was used in §4.5. Since the first moment is not a conserved variable, it can change by adiabatic means, or by exchange with the reservoirs, $\Delta E_1 = \Delta^0 E_1 + \Delta^{\rm r} E_1$. Here $E_1^{\rm (r)}(t) = \int_0^t \mathrm{d}^{\rm r} E_1$, so that $\mathrm{d}E_1^{\rm (r)}/\mathrm{d}^0 E_1 = 0$, and $\mathrm{d}E_1^{\rm (r)}/\mathrm{d}^{\rm r} E_1 = 1$. Of course one also has the results $\mathrm{d}E_1/\mathrm{d}^0 E_1 = 1$, and $\mathrm{d}E_1/\mathrm{d}^{\rm r} E_1 = 1$.

Maximisation of the total first entropy with respect to $\Delta^{\rm r} E_1$ gives

$$\frac{1}{T_{\rm s1}(\overline{E}_1)} = \frac{1}{T_1}, \tag{9.23}$$

which is to say that in the steady state the sub-system first temperature is equal to the first temperature of the reservoirs.

As just mentioned, the first energy moment is not a conserved variable. It can change by internal processes, $\Delta^0 E_1 = \tau \dot{E}_1^0$, and by energy exchange with the reservoir, $\Delta^{\rm r} E_1 = -\Delta E_{\rm r,1}$. Hence define the transition as $E_1'' = E_1 + \Delta^0 E_1 + \Delta^{\rm r} E_1 \equiv E_1' + \Delta^{\rm r} E_1$. The total second entropy is the sum of that due to the sub-system and that due to the reservoir. The sub-system second entropy given above for the isolated system may be rewritten

$$\begin{aligned} & S_{\rm s}^{(2)}(E_1', E_1|\tau) \\ &= \frac{1}{2} A(\tau)[E_1' - E_1]^2 + [A(\tau) + B(\tau)] E_1' E_1 \\ &= \frac{-1}{2\Lambda|\tau|}[\Delta^0 E_1]^2 + \frac{1}{2} S E_1 \Delta^0 E_1 + \frac{1}{2} S E_1^2. \end{aligned} \tag{9.24}$$

[5] Green, M. S. (1954), J. Chem. Phys. **22**, 398. Kubo, R. (1966), Rep. Progr. Phys. **29**, 255. Kubo, R., Toda, M., and Hashitsume, N. (1978), *Statistical Physics II. Non-equilibrium Statistical Mechanics*, (Springer-Verlag, Berlin). Hansen, J.-P. and McDonald, I. R. (1986), *Theory of Simple Liquids*, (Academic Press, London). Zwanzig, R. (2001), *Non-equilibrium Statistical Mechanics*, (Oxford University Press, Oxford).

The second equality follows for small τ. Recall that the thermal conductivity is $\lambda = \Lambda/2VT_{s0}^2$.

This has the following physical interpretation. The first term is always negative and hence it is unfavourable. It represents the entropic cost of ordering the system to create the flux. It is quadratic in the flux as it must be from symmetry arguments. This first term limits the flux. The second term is positive when the flux opposes the first moment, which is to say that it reduces it. This is half the entropy production, since it is a flux times a thermodynamic force. This term is linear in the flux, as it must be from symmetry, and it is what drives the flux. The final term is independent of the flux and represents the first entropy cost of the non-zero first moment.

In the present case that the sub-system can exchange energy with two encompassing reservoirs of temperature $T_\pm$, this expression for the second entropy has to be modified to include the exchange $\Delta^{\mathrm{r}}E_1$. The term quadratic in the flux is unchanged, as this is purely internal to the sub-system. To the linear term is added half the first entropy production gained by the sub-system, and half that gained by the reservoir,

$$\frac{1}{2}SE_1\Delta^{\mathrm{r}}E_1 - \frac{1}{2T_1}\Delta^{\mathrm{r}}E_1. \tag{9.25}$$

The negative sign for the second term arises because $\Delta^{\mathrm{r}}E_1 = -\Delta E_{\mathrm{r},1}$, where the left-hand side is the reservoir induced change in the sub-system energy moment, and the right-hand side is the negative of the change in the energy moment of the reservoirs. To the final term is simply added the sub-system dependent part of the first entropy of the reservoir, $-E_1^{(\mathrm{r})}(t)/T_1$. Hence the total second entropy is

$$S_{\text{total}}^{(2)}(\Delta^0 E_1, \Delta^{\mathrm{r}}E_1, E_1|T_1, \tau) = \\ \frac{-(\Delta^0 E_1)^2}{2\Lambda|\tau|} + \frac{SE_1}{2}[\Delta^0 E_1 + \Delta^{\mathrm{r}}E_1] - \frac{\Delta^{\mathrm{r}}E_1}{2T_1} + \frac{1}{2}SE_1^2 - \frac{E_1^{(\mathrm{r})}(t)}{T_1}. \tag{9.26}$$

The derivatives are

$$\frac{\partial S_{\text{total}}^{(2)}}{\partial \Delta^0 E_1} = \frac{-1}{\Lambda|\tau|}\Delta^0 E_1 + \frac{1}{2}SE_1, \tag{9.27}$$

$$\frac{\partial S_{\text{total}}^{(2)}}{\partial \Delta^{\mathrm{r}} E_1} = \frac{1}{2}SE_1 - \frac{1}{2T_1}, \tag{9.28}$$

and

$$\frac{\partial S_{\text{total}}^{(2)}}{\partial^{\mathrm{r}} E_1} = \frac{1}{2}S[\Delta^0 E_1 + \Delta^{\mathrm{r}}E_1] + SE_1 - \frac{1}{T_1}. \tag{9.29}$$

(Note that $\Delta^{\mathrm{r}}E_1$ refers to the transition in time τ, whereas $\partial^{\mathrm{r}}E_1$ refers to the prior reservoir induced change in E_1. As discussed in footnote 4 on p. 108, one assumes that equilibration with the reservoirs is much faster than the internal relaxation of the sub-system and does not consider the derivative $\partial^0 E_1$.) These vanish when

$$\overline{\Delta^0} E_1 = \frac{|\tau|\Lambda}{2} S E_1, \tag{9.30}$$

which is unchanged from the isolated system,

$$\frac{1}{\overline{T}_{s1}} \equiv S\overline{E}_1 = \frac{1}{T_1}, \tag{9.31}$$

which is the expected moment induced by the reservoir, and

$$\overline{\Delta^0} E_1 = -\overline{\Delta^{\mathrm{r}}} E_1. \tag{9.32}$$

This last result means that the stochastic change in the moment due to heat exchange with the reservoir most likely cancels the deterministic relaxation of the energy moment due to internal processes. Whereas in the isolated sub-system the moment relaxes to zero, the cancelling exchange with the reservoir holds the sub-system moment constant. Its non-zero value is such that the induced temperature gradient of the sub-system is equal to that imposed by the reservoir. The induced gradient is essentially the first temperature T_1, and so this is the same as saying that in the steady state the internal thermodynamic force is equal to the external thermodynamic force. This is the content of Onsager's regression hypothesis (Onsager, 1931).

9.2 Phase Space Probability Density

In what follows of this chapter, the statistical mechanics of steady heat flow is presented, beginning with this section on the phase space probability distribution. The general expression given in Ch. 8 will be applied to the present problem.

9.2.1 Explicit Hamiltonian and First Energy Moment

Suppose that the Hamiltonian of the sub-system does not depend explicitly on time, and that it consists of the kinetic energy and the potential energy. The kinetic energy is taken to be the usual quadratic form in the momenta, and the potential energy consists of one-, two-, three-, ... body potentials,

$$\begin{aligned}
\mathcal{H}_{\rm s}(\mathbf{\Gamma}) &= \mathcal{K}(\mathbf{\Gamma}) + U^{(1)}(\mathbf{\Gamma}) + U^{(2)}(\mathbf{\Gamma}) + U^{(3)}(\mathbf{\Gamma}) + \ldots \\
&= \frac{1}{2m}\sum_{i=1}^{N} p_i^2 + \sum_{i=1}^{N} u^{(1)}(\mathbf{q}_i) + \sum_{i<j}^{N} u^{(2)}(\mathbf{q}_i, \mathbf{q}_j) \\
&\quad + \sum_{i<j<k}^{N} u^{(3)}(\mathbf{q}_i, \mathbf{q}_j, \mathbf{q}_k) + \ldots \\
&= \frac{1}{2m}\sum_{i=1}^{N} p_i^2 + \sum_{i=1}^{N} u^{(1)}(\mathbf{q}_i) + \frac{1}{2!}\sum_{i,j}^{N}{}' u^{(2)}(\mathbf{q}_i, \mathbf{q}_j) \\
&\quad + \frac{1}{3!}\sum_{i,j,k}^{N}{}' u^{(3)}(\mathbf{q}_i, \mathbf{q}_j, \mathbf{q}_k) + \ldots \qquad (9.33)
\end{aligned}$$

Here the simplest form for the kinetic energy has been assumed, although sometimes below, the possibility that the mass m_i differs for different atoms will be allowed for. The prime on the summation means that terms where any of the indeces have the same value are excluded. The prefactors of the summations in the final form correct for multiple counting of the energy of each set of interacting atoms. The one body potential often represents an external field, such as gravity, or else the walls of a confined system.

The zeroth energy moment is the Hamiltonian itself, $E_0(\mathbf{\Gamma}) = \mathcal{H}_{\rm s}(\mathbf{\Gamma})$, which is the total energy of the system. The first energy moment in the z-direction is explicitly

$$\begin{aligned}
E_1(\mathbf{\Gamma}) &= \frac{1}{2m}\sum_{i=1}^{N} z_i p_i^2 + \sum_{i=1}^{N} z_i u^{(1)}(\mathbf{q}_i) + \sum_{i<j}^{N} \frac{z_i + z_j}{2} u^{(2)}(\mathbf{q}_i, \mathbf{q}_j) \\
&\quad + \sum_{i<j<k}^{N} \frac{z_i + z_j + z_k}{3} u^{(3)}(\mathbf{q}_i, \mathbf{q}_j, \mathbf{q}_k) + \ldots \\
&= \frac{1}{2m}\sum_{i=1}^{N} z_i p_i^2 + \sum_{i=1}^{N} z_i u^{(1)}(\mathbf{q}_i) + \frac{1}{2!}\sum_{i,j}^{N}{}' \frac{z_i + z_j}{2} u^{(2)}(\mathbf{q}_i, \mathbf{q}_j) \\
&\quad + \frac{1}{3!}\sum_{i,j,k}^{N}{}' \frac{z_i + z_j + z_k}{3} u^{(3)}(\mathbf{q}_i, \mathbf{q}_j, \mathbf{q}_k) + \ldots \qquad (9.34)
\end{aligned}$$

Here the z position of the ith atom is $z_i \equiv q_{iz}$. Notice how each atom in a set of interacting atoms contributes equally to the moment, which is to say that their average z-position appears, such as $[z_i + z_j]/2$ for the pair term, $[z_i + z_j + z_k]/3$ for the triplet term etc.

In what follows, the adiabatic rate of change of the first energy moment will be required. Recall that the heat flux is this divided by the sub-system volume, $J_{\rm E}^0 = \dot{E}_1^0/V$. In terms of the kinetic energy and the many body potential energies, the adiabatic rate of change of the first energy moment is

$$\begin{aligned}
\dot{E}_1^0(\mathbf{\Gamma}) &= \dot{\mathbf{\Gamma}}^0 \cdot \nabla E_1(\mathbf{\Gamma}) \\
&= \sum_{l=1}^{N} \dot{\mathbf{p}}_l^0 \cdot \frac{\partial E_1(\mathbf{\Gamma})}{\partial \mathbf{p}_l} + \sum_{l=1}^{N} \dot{\mathbf{q}}_l^0 \cdot \frac{\partial E_1(\mathbf{\Gamma})}{\partial \mathbf{q}_l} \\
&= \sum_{l=1}^{N} z_l\, \dot{\mathbf{p}}_l^0 \cdot \mathbf{p}_l/m_l + \sum_{l=1}^{N} \left[z_l\, \dot{\mathbf{q}}_l^0 \cdot \frac{\partial u^{(1)}(\mathbf{q}_l)}{\partial \mathbf{q}_l} + \dot{z}_l^0 u^{(1)}(\mathbf{q}_l) \right] \\
&\quad + \frac{1}{2!} \sum_{l,j}^{N}{}' \left[\{z_l + z_j\}\, \dot{\mathbf{q}}_l^0 \cdot \frac{\partial u^{(2)}(\mathbf{q}_l, \mathbf{q}_j)}{\partial \mathbf{q}_l} + \dot{z}_l^0 u^{(2)}(\mathbf{q}_l, \mathbf{q}_j) \right] \\
&\quad + \frac{1}{3!} \sum_{l,j,k}^{N}{}' \left[\{z_l + z_j + z_k\}\, \dot{\mathbf{q}}_l^0 \cdot \frac{\partial u^{(3)}(\mathbf{q}_l, \mathbf{q}_j, \mathbf{q}_k)}{\partial \mathbf{q}_l} \right. \\
&\quad \left. + \dot{z}_l^0 u^{(3)}(\mathbf{q}_l, \mathbf{q}_j, \mathbf{q}_k) \right] + \ldots \qquad (9.35)
\end{aligned}$$

This follows because the potentials are symmetric with respect to interchange of their arguments, and the summation indeces can be relabelled. One has of course $\dot{\mathbf{q}}_l^0 = \mathbf{p}_l/m_l$.

For the particular case that the single potential represents a wall or slit pore, $u^{(1)}(\mathbf{q}) = w(z)$, that the pair potential is central, $u^{(2)}(\mathbf{q}_i, \mathbf{q}_j) = u(q_{ij})$, $q_{ij} \equiv |\mathbf{q}_i - \mathbf{q}_j|$, and that there are no triplet or higher-order potentials, one can write this as[6]

$$\dot{E}_1^0(\mathbf{\Gamma}) = \sum_{i=1}^{N} \sum_{\alpha=x,y,z} \kappa_{i\alpha} p_{i\alpha}, \qquad (9.36)$$

with

$$\kappa_{i\alpha} \equiv \frac{\epsilon_i}{m_i} \delta_{\alpha,z} - \sum_{j=1}^{N}{}^{j\neq i} u'(q_{ij}) \frac{q_{iz} - q_{jz}}{2 m_i q_{ij}} [q_{i\alpha} - q_{j\alpha}], \qquad (9.37)$$

where $\delta_{\alpha,z}$ is a Kronecker delta, and ϵ_i is the energy of the ith atom,

$$\epsilon_i = \frac{p_i^2}{2m_i} + w(q_{iz}) + \frac{1}{2} \sum_{j=1}^{N}{}^{j\neq i} u(q_{ij}). \qquad (9.38)$$

If one wants to treat a bulk fluid, set $w(z) = 0$. Notice that κ_i does not depend upon the momentum except in the kinetic energy contribution to the energy per molecule, which fact will prove useful below.

---

[6] Attard, P. (2006), J. Chem. Phys. **124**, 024109.

9.2.2 Reservoir Entropy and Probability Density

The phase space probability distribution is essentially the exponential of the reservoir entropy. The reservoir entropy is formally given by Eq. (8.15),

$$S_{\rm r}(\mathbf{\Gamma},t) = S_{\rm r,st}(\mathbf{\Gamma}) + S_{\rm r,dyn}(\mathbf{\Gamma},t). \tag{9.39}$$

In the present case, the static (or equilibrium, or reversible) part of the reservoir entropy is

$$S_{\rm st}(\mathbf{\Gamma}) = \frac{-1}{T_0}E_0(\mathbf{\Gamma}) - \frac{1}{T_1}E_1(\mathbf{\Gamma}). \tag{9.40}$$

This is just the analogue of the Boltzmann factor, which is the reservoir entropy in the canonical equilibrium case. The dynamic (or non-equilibrium, or irreversible) part of the reservoir entropy in the present case is

$$\begin{aligned} S_{\rm dyn}(\mathbf{\Gamma},t) &\equiv -\int_0^t {\rm d}t'\, \dot S^0_{\rm r,st}(\overline{\mathbf{\Gamma}}(t'|\mathbf{\Gamma},t),t') \\ &= \frac{1}{T_1}\int_0^t {\rm d}t'\, \dot E^0_1(\overline{\mathbf{\Gamma}}(t'|\mathbf{\Gamma},t)). \end{aligned} \tag{9.41}$$

This only depends upon the first energy moment because the adiabatic rate of change of the zeroth energy moment vanishes. The dynamic part of the reservoir entropy has an explicit time dependence that reflects the entropy produced in the reservoirs up to the present time due to the net heat flow between them.

Because this is a steady state system, the most likely value of the heat flux is independent of time, $\overline{\dot E^0_1}$. Also, the most likely trajectory is invariant with respect to translations in time, $\overline{\mathbf{\Gamma}}(t'|\mathbf{\Gamma},t) = \overline{\mathbf{\Gamma}}(t'+\tau|\mathbf{\Gamma},t+\tau)$. Finally, in the long time limit, on the most likely trajectory the heat flux must tend to its most likely value, whatever its current value, $\dot E^0_1(\overline{\mathbf{\Gamma}}(t'|\mathbf{\Gamma},t)) \to \overline{\dot E^0_1}$, $|t'-t| \to \infty$. These three facts mean that the partial time derivative of the dynamic part of the reservoir entropy is

$$\begin{aligned} &\frac{\partial S_{\rm dyn}(\mathbf{\Gamma},t)}{\partial t} \\ &= \frac{1}{T_1\Delta_t}\left\{\int_0^{t+\Delta_t} {\rm d}t'\, \dot E^0_1(\overline{\mathbf{\Gamma}}(t'|\mathbf{\Gamma},t+\Delta_t)) - \int_0^t {\rm d}t'\, \dot E^0_1(\overline{\mathbf{\Gamma}}(t'|\mathbf{\Gamma},t))\right\} \\ &= \frac{1}{T_1\Delta_t}\left\{\int_{-\Delta_t}^t {\rm d}t''\, \dot E^0_1(\overline{\mathbf{\Gamma}}(t''|\mathbf{\Gamma},t)) - \int_0^t {\rm d}t'\, \dot E^0_1(\overline{\mathbf{\Gamma}}(t'|\mathbf{\Gamma},t))\right\} \\ &= \frac{\overline{\dot E^0_1}}{T_1}. \end{aligned} \tag{9.42}$$

This is in agreement with the general steady state result, Eq. (8.161).

The phase space probability for the canonical non-equilibrium problem

$$\begin{aligned} \wp(\mathbf{\Gamma}|T_0,T_1,t) &= \frac{1}{Z(T_0,T_1,t)} e^{S_{\rm r}(\mathbf{\Gamma},t)/k_{\rm B}} \\ &= \frac{1}{Z(T_0,T_1,t)} e^{-\beta_0 E_0(\mathbf{\Gamma})-\beta_1 E_1(\mathbf{\Gamma})} e^{S_{\rm r,dyn}(\mathbf{\Gamma},t)/k_{\rm B}}, \end{aligned} \tag{9.43}$$

where $\beta_n \equiv 1/k_B T_n$, $n = 0, 1$. In order to simplify the notation, in most of this chapter the temperature arguments will not be shown explicitly. The partition function normalises this to unity,

$$Z(t) = \int \mathrm{d}\mathbf{\Gamma}\, e^{S_\mathrm{r}(\mathbf{\Gamma},t)/k_\mathrm{B}}. \tag{9.44}$$

Because this is a steady state system one must have $\partial\wp(\mathbf{\Gamma}, t)/\partial t = 0$, which implies

$$\frac{1}{Z(t)}\frac{\mathrm{d}Z(t)}{\mathrm{d}t} = \frac{\partial S_{\mathrm{r,dyn}}(\mathbf{\Gamma}, t)}{k_\mathrm{B}\partial t} = \beta_1 \overline{\dot{E}_1^0}. \tag{9.45}$$

Hence $Z(t) = \tilde{Z} \exp \beta_1 \overline{\dot{E}_1^0} t$.

In view of this time dependence and the asymptotic behaviour of the integrand, it is possible to cancel the exponential time dependence of the partition function by making the integrand of the dynamic part of the reservoir entropy short-ranged. That is one can write

$$\wp(\mathbf{\Gamma}|T_0, T_1) = \frac{1}{\tilde{Z}(T_0, T_1)} e^{-\beta_0 E_0(\mathbf{\Gamma}) - \beta_1 E_1(\mathbf{\Gamma})} e^{\tilde{S}_{\mathrm{r,dyn}}(\mathbf{\Gamma})/k_\mathrm{B}}, \tag{9.46}$$

with

$$\tilde{S}_{\mathrm{dyn}}(\mathbf{\Gamma}) = \frac{1}{T_1}\int_{-\tau}^{0} \mathrm{d}t' \left\{ \dot{E}_1^0(\overline{\mathbf{\Gamma}}(t'|\mathbf{\Gamma}, 0)) - \overline{\dot{E}_1^0} \right\}. \tag{9.47}$$

The integrand is sufficiently short-ranged so that the integral is independent of the lower limit for τ larger than molecular times, but smaller than macroscopic times. Although this version may have certain numerical and computational advantages, the full formulation is retained here because of the physical interpretation of the time dependent parts.

The logarithm of the partition function times Boltzmann's constant is the total entropy of the total system (or, more precisely, the sub-system dependent part thereof), $S_{\mathrm{tot}}(T_0, T_1, t) = k_\mathrm{B} \ln Z(T_0, T_1, t)$. This is related to the maximum value of the total entropy used in the thermodynamic analysis of the preceding section, Eq. (9.22) with the zeroth temperature added,

$$S_{\mathrm{tot}}(T_0, T_1, t) > S_{\mathrm{tot}}(\overline{E}_0, \overline{E}_1|T_0, T_1, t) \equiv S_\mathrm{s}(\overline{E}_0, \overline{E}_1) - \frac{\overline{E}_0}{T_0} - \frac{\overline{E}_1^{(\mathrm{r})}(t)}{T_1}. \tag{9.48}$$

In the thermodynamic limit, this strict upper bound becomes an approximate equality (because the far left-hand side includes the entropy of the fluctuations about the most likely state, and, since these scale as the square root of the system size, they are relatively negligible in the thermodynamic limit). The thermodynamic potential or free energy for steady heat flow, which is the analogue of the Helmholtz free energy, is

$$\begin{aligned} F(T_0, T_1, t) &= -T_0 S_{\mathrm{tot}}(\overline{E}_0, \overline{E}_1|T_0, T_1, t) \\ &\approx -T_0 S_{\mathrm{tot}}(T_0, T_1, t) \\ &= -k_\mathrm{B} T_0 \ln Z(T_0, T_1, t). \end{aligned} \tag{9.49}$$

The time derivative of the thermodynamic potential is proportional to the logarithmic derivative of the partition function, which is proportional to that of the total entropy. From the above one has

$$\frac{\partial S_{\rm tot}(T_0, T_1, t)}{\partial t} = \frac{\overline{\dot{E}_1^0}}{T_1}, \text{ and } \frac{\partial F(T_0, T_1, t)}{\partial t} = \frac{-T_0}{T_1}\overline{\dot{E}_1^0}. \tag{9.50}$$

This implies $\mathrm{d}\overline{E}_1^{(\mathrm{r})}(t)/\mathrm{d}t = -\overline{\dot{E}_1^0}$, which is to say that most likely the rate of relaxation of the first energy moment of the sub-system due to the reservoirs is equal and opposite to the rate of its change due to internal influences.

9.3 Most Likely Trajectory

The general stochastic dissipative equations of motion are given by Eq. (8.89)

$$\mathbf{\Gamma}_2 = \mathbf{\Gamma}_1 + t_{21}\dot{\mathbf{\Gamma}}^0(\mathbf{\Gamma}) + \mathbf{R}(\mathbf{\Gamma}, t_{21}) + \mathcal{O}(t_{21}^2), \tag{9.51}$$

where the random force $\mathbf{R}$ has only momentum components. The adiabatic velocities are

$$\dot{q}_{i\alpha}^0 = \frac{\partial \mathcal{H}(\mathbf{\Gamma})}{\partial p_{i\alpha}} = q_{i\alpha}/m_i, \tag{9.52}$$

and

$$\dot{p}_{i\alpha}^0 = \frac{-\partial \mathcal{H}(\mathbf{\Gamma})}{\partial q_{i\alpha}} = -w'(q_{iz})\delta_{\alpha,z} - \sum_{j=1}^{N}{}^{j\neq i} u'(q_{ij})\frac{q_{i\alpha} - q_{j\alpha}}{q_{ij}}. \tag{9.53}$$

Here the atoms are labelled by Roman letters, $i = 1, 2, \ldots, N$, and the coordinates are labelled by Greek letters, $\alpha = x, y, z,$. The Hamiltonian has been assumed to consist of only kinetic energy, an external wall potential, and a central pair potential between the atoms (c.f. Eq. (9.38)).

The random force, which only has momentum components, is the sum of a stochastic, $\tilde{\mathbf{R}}$, and a dissipative, $\overline{\mathbf{R}}$, force. The stochastic forces have variance

$$\left\langle \tilde{R}_{i\alpha}\tilde{R}_{j\gamma} \right\rangle = \sigma\delta_{ij}\delta_{\alpha\gamma}. \tag{9.54}$$

The variance represents the stochastic influence of the heat reservoirs. In practical terms its value can be chosen arbitrarily within a relatively wide range, provided that it is of the same order as the time step, $\sigma \sim |t_{21}|$.[7] By this is meant that the change in momenta at each time step due to the reservoirs

[7] The relationship between the variance and the time step independent transport coefficient Λ shows that $\sigma \propto |t_{21}|$. This means that in comparing the results of simulations for equations of motion with different time steps, strictly speaking the variances ought to be in the ratio of the time steps. However since Λ is arbitrary within wide limits, this strict requirement in practice need not be honored. See, for example, the molecular dynamics results shown in Figs 11.4–11.6 on pp. 406–408.

should be comparable to the adiabatic change. The variance cannot be too small because the energy exchange with the reservoirs has to be large enough to counteract the adiabatic energy flux. Conversely, the variance cannot be too large, because the stochastic dissipative equations of motion represent an expansion of the second entropy that assumes that quadratic terms in the time step and in the variance can be neglected. Also, too large a variance unnaturally disorders the adiabatic flux in a way that does not occur in the actual boundary driven flows that exist in the real world. (This latter effect can be mitigated by the constraint that is implemented on the dissipative force below.) In practice one can explore several values of the variance and check that the relevant physical properties of the system are insensitive to the precise value chosen.

The dissipative force is given by Eq. (8.94),

$$\begin{aligned}
\overline{R}_{i\alpha}(\mathbf{\Gamma}, t_{21}) &= \frac{\sigma}{2k_\mathrm{B}} \frac{\partial S_{\mathrm{r,st}}(\mathbf{\Gamma})}{\partial p_{i\alpha}} + \frac{\sigma}{2k_\mathrm{B}} (\hat{t}-1)\, \overline{S'_{pi\alpha}} \\
&= \frac{-\sigma}{2k_\mathrm{B}} \left[\frac{1}{T_0} \frac{\partial \mathcal{H}(\mathbf{\Gamma})}{\partial p_{i\alpha}} + \frac{1}{T_1} \frac{\partial E_1(\mathbf{\Gamma})}{\partial p_{i\alpha}} \right] + \frac{\sigma}{2k_\mathrm{B}} (\hat{t}-1)\, \overline{\frac{\partial S_{\mathrm{r,st}}(\mathbf{\Gamma})}{\partial p_{i\alpha}}} \\
&= \frac{-\sigma}{2k_\mathrm{B}} \left[\frac{p_{i\alpha}}{T_0 m_i} + \frac{z_i p_{i\alpha}}{T_1 m_i} \right] - \frac{\sigma}{2k_\mathrm{B}} (\hat{t}-1) \left[\frac{\overline{p_{i\alpha}}}{T_0 m_i} + \frac{\overline{z_i p_{i\alpha}}}{T_1 m_i} \right] \\
&= \frac{-\sigma}{2m_i} \beta(z_i) p_{i\alpha} - \frac{\sigma}{2m_i} (\hat{t}-1)\, \overline{\beta(z_i) p_{i\alpha}}. \qquad (9.55)
\end{aligned}$$

The local inverse temperature has been defined as $\beta(z) \equiv k_\mathrm{B}^{-1}[1/T_0 + z/T_1]$, which is just the linear interpolation between the inverse temperatures of the two reservoirs.

Here $\hat{t} \equiv \mathrm{sign}(t_{21})$, and so the final term is only non-zero for a backward time step. For this final term, it has been assumed that the gradient of the static part of the reservoir entropy at the most likely point can be replaced by the most likely gradient of the static part of the reservoir entropy. There is an argument to say that this term is zero, namely that in the conductive regime the number flux at each point z is zero.

The first term represents a drag or friction force. It is of the same force as the Langevin dissipation at the local temperature of the atom. It of course satisfies the fluctuation dissipation theorem, because its magnitude is determined in essence by the variance σ. It is the satisfaction of this theorem that allows the wide choice for the variance without affecting the physical properties of the system.

Constraints

As discussed in §8.3.6, one can impose constraints on the reservoir perturbation so as to minimise the influence on the physical properties of interest. In the case of heat flow possibly the most important property of the system is the heat flux, which is embodied in the adiabatic rate of change of the first energy moment, $J_\mathrm{E}^0 = \dot{E}_1^0/V$. In order to disturb this as little as possible, one can make the dissipative force orthogonal to its gradient,[8]

$$\overline{\mathbf{R}}(\mathbf{\Gamma}, t_{21}) \cdot \nabla_{\mathrm{p}} \dot{E}_1^0(\mathbf{\Gamma}) = 0. \tag{9.56}$$

Denoting the unconstrained or free force given above by $\overline{\mathbf{R}}^{\mathrm{f}}(\mathbf{\Gamma}, t_{21})$ and introducing a Lagrange multiplier, $\mu(\mathbf{\Gamma})$, the constrained most likely force becomes

$$\overline{\mathbf{R}}(\mathbf{\Gamma}, t_{21}) = \overline{\mathbf{R}}^{\mathrm{f}}(\mathbf{\Gamma}, t_{21}) + \mu(\mathbf{\Gamma}) \nabla_{\mathrm{p}} \dot{E}_1^0(\mathbf{\Gamma}), \tag{9.57}$$

with the Lagrange multiplier being

$$\mu(\mathbf{\Gamma}) = \frac{-\overline{\mathbf{R}}^{\mathrm{f}}(\mathbf{\Gamma}, t_{21}) \cdot \nabla_{\mathrm{p}} \dot{E}_1^0(\mathbf{\Gamma})}{\nabla_{\mathrm{p}} \dot{E}_1^0(\mathbf{\Gamma}) \cdot \nabla_{\mathrm{p}} \dot{E}_1^0(\mathbf{\Gamma})}. \tag{9.58}$$

Using the expression for the adiabatic rate of change of first energy moment given in Eq. (9.36), and bearing in mind that the momentum dependence of κ comes only via the kinetic energy per particle, one has explicitly

$$\frac{\partial \dot{E}_1^0(\mathbf{\Gamma})}{\partial p_{i\alpha}} = \kappa_{i\alpha} + \frac{p_{i\alpha} p_{iz}}{m_i^2}. \tag{9.59}$$

Hence the most likely force orthogonal to the heat flux in component form is

$$\overline{\mathbf{R}}_{i\alpha} = \frac{-\sigma}{2m_i} \beta(z_i) p_{i\alpha} - \frac{\sigma}{2m_i} (\hat{t} - 1) \overline{\beta(z_i) p_{i\alpha}} + \mu \left[\kappa_{i\alpha} + \frac{p_{i\alpha} p_{iz}}{m_i^2} \right], \tag{9.60}$$

with

$$\mu = \frac{\sigma \sum_{i\alpha} \left[\beta(z_i) p_{i\alpha} + (\hat{t} - 1) \overline{\beta(z_i) p_{i\alpha}} \right] \left[\kappa_{i\alpha} + p_{i\alpha} p_{iz} / m_i^2 \right] / 2m_i}{\sum_{i\alpha} \left[\kappa_{i\alpha} + p_{i\alpha} p_{iz} / m_i^2 \right]^2}. \tag{9.61}$$

9.4 Equipartition Theorem for Heat Flow

Global Equipartition Theorem

The generalised equipartition theorem, Eq. (8.150) (see also §3.7.4),

$$\langle \nabla_{\mathrm{p}} \nabla_{\mathrm{p}} S_{\mathrm{r}}(\mathbf{\Gamma}, t) \rangle_{\mathrm{ne}} = -k_{\mathrm{B}}^{-1} \langle [\nabla_{\mathrm{p}} S_{\mathrm{r}}(\mathbf{\Gamma}, t)][\nabla_{\mathrm{p}} S_{\mathrm{r}}(\mathbf{\Gamma}, t)] \rangle_{\mathrm{ne}}, \tag{9.62}$$

in component form is

$$\left\langle \frac{\partial^2 S_{\mathrm{r}}(\mathbf{\Gamma}, t)}{k_{\mathrm{B}} \partial p_{i\alpha} \partial p_{j\gamma}} \right\rangle_{\mathrm{ne}} = - \left\langle \frac{\partial S_{\mathrm{r}}(\mathbf{\Gamma}, t)}{k_{\mathrm{B}} \partial p_{i\alpha}} \frac{\partial S_{\mathrm{r}}(\mathbf{\Gamma}, t)}{k_{\mathrm{B}} \partial p_{j\gamma}} \right\rangle_{\mathrm{ne}}. \tag{9.63}$$

---

[8] In §8.3.6, the gradient of the adiabatic rate of change of the static part of the reservoir entropy, $\dot{S}_{\mathrm{r}}^0 = -\dot{E}_1^0/T_1$, was used instead. That expression equates to the present expression when both sides of it are multiplied by $-T_1$.

These are averages in the non-equilibrium system. In the present case, with $S_{\rm r}(\mathbf{\Gamma},t) = S_{\rm r,st}(\mathbf{\Gamma}) + \tilde{S}_{\rm r,dyn}(\mathbf{\Gamma})$, since the time dependent contribution to $S_{\rm r,dyn}$ is a constant that does not contribute to the gradient, this becomes

$$\left\langle \frac{-\beta(z_i)}{m_i}\delta_{ij}\delta_{\alpha\gamma} + \frac{\partial^2 \tilde{S}_{\rm r,dyn}(\mathbf{\Gamma})}{k_{\rm B}\partial p_{i\alpha}\partial p_{j\gamma}} \right\rangle_{\rm ne} = - \left\langle \left[\frac{-\beta(z_i)p_{i\alpha}}{m_i} + \frac{\partial \tilde{S}_{\rm r,dyn}(\mathbf{\Gamma})}{k_{\rm B}\partial p_{i\alpha}} \right] \left[\frac{-\beta(z_j)p_{j\gamma}}{m_j} + \frac{\partial \tilde{S}_{\rm r,dyn}(\mathbf{\Gamma})}{k_{\rm B}\partial p_{j\gamma}} \right] \right\rangle_{\rm ne} . \tag{9.64}$$

Recall that $\beta(z) \equiv \beta_0 + \beta_1 z = [1/k_{\rm B}T_0] + [z/k_{\rm B}T_1]$.

Now this is expanded for small temperature gradients, $\beta_1 \to 0$, recalling that $S_{\rm r,dyn} \sim \beta_1$. It will also be assumed that the dynamic part of the reservoir entropy is dominated by the odd projection, $\tilde{S}_{\rm r,dyn}(\mathbf{\Gamma}) \approx \tilde{S}^{\rm odd}_{\rm r,dyn}(\mathbf{\Gamma})$. (Symmetry arguments outlined in footnote 9 on p. 316 show that $\tilde{S}^{\rm even}_{\rm r,dyn} \sim \mathcal{O}(\beta_1^2)$.) The non-equilibrium averages will be expanded as

$$\begin{aligned}
\langle f(\mathbf{\Gamma})\rangle_{\rm ne} &= \int \mathrm{d}\mathbf{\Gamma}\, \frac{1}{Z_{\rm ne}} e^{S_{\rm r,st}(\mathbf{\Gamma})/k_{\rm B}} e^{\tilde{S}_{\rm r,dyn}(\mathbf{\Gamma})/k_{\rm B}} f(\mathbf{\Gamma}) \\
&= \int \mathrm{d}\mathbf{\Gamma}\, \frac{e^{S_{\rm r,st}(\mathbf{\Gamma})/k_{\rm B}}}{Z_{\rm st} + \mathcal{O}(\beta_1^2)} \left[1 + \tilde{S}^{\rm odd}_{\rm r,dyn}(\mathbf{\Gamma})/k_{\rm B} + \mathcal{O}(\beta_1^2) \right] f(\mathbf{\Gamma}) \\
&= \left\langle f^{\rm even}(\mathbf{\Gamma}) + k_{\rm B}^{-1}\tilde{S}^{\rm odd}_{\rm r,dyn}(\mathbf{\Gamma}) f^{\rm odd}(\mathbf{\Gamma}) \right\rangle_{\rm st} + \mathcal{O}(\beta_1^2).
\end{aligned} \tag{9.65}$$

With this, the left-hand side of the generalised equipartition theorem becomes

$$\begin{aligned}
\text{LHS} &= \frac{-\delta_{ij}\delta_{\alpha\gamma}}{m_i} \langle \beta(z_i)\rangle_{\rm st} + k_{\rm B}^{-2} \left\langle \frac{\partial^2 \tilde{S}^{\rm odd}_{\rm r,dyn}(\mathbf{\Gamma})}{\partial p_{i\alpha}\partial p_{j\gamma}} \tilde{S}^{\rm odd}_{\rm r,dyn}(\mathbf{\Gamma}) \right\rangle_{\rm st} \\
&= \frac{-\delta_{ij}\delta_{\alpha\gamma}\beta_0}{m_i} + \mathcal{O}(\beta_1^2).
\end{aligned} \tag{9.66}$$

The right-hand side is

$$\begin{aligned}
\text{RHS} &= -\left\langle \frac{\beta(z_i)\beta(z_j)p_{i\alpha}p_{j\gamma}}{m_i m_j} \right\rangle_{\rm st} + \mathcal{O}(\beta_1^2) \\
&= \frac{-\beta_0^2}{m_i m_j} \langle p_{i\alpha}p_{j\gamma}\rangle_{\rm MB} + \mathcal{O}(\beta_1^2).
\end{aligned} \tag{9.67}$$

Equating these two gives the usual equipartition theorem for the canonical equilibrium system for the mid-temperature. (The temperature gradient drops out of this result because the average is taken over the whole system.)

Local Equipartition Theorem

This section obtains the average of the square of the momenta at a fixed position in the sub-system. This gives a local form of the equipartition theorem. For heat flow, the non-equilibrium probability density is

$$\wp_{\rm ne}(\mathbf{\Gamma}) = \frac{e^{S_{\rm r,st}(\mathbf{\Gamma})/k_{\rm B}} e^{\tilde{S}_{\rm r,dyn}(\mathbf{\Gamma})/k_{\rm B}}}{\tilde{Z}_{\rm ne}} = \frac{e^{S_{\rm r,st}(\mathbf{\Gamma})/k_{\rm B}}}{\tilde{Z}_{\rm ne}} e^{\beta_1 W_1(\mathbf{\Gamma})}. \tag{9.68}$$

Use will be made of the fact that the dynamic part of the reservoir entropy is linear in the temperature gradient, and so in order to facilitate a small gradient expansion here has been defined $\tilde{S}_{\mathrm{r,dyn}}(\mathbf{\Gamma})/k_{\mathrm{B}} \equiv \beta_1 W_1(\mathbf{\Gamma})$. In what follows use will be made of the static probability distribution,

$$\wp_{\mathrm{st}}(\mathbf{\Gamma}) = \frac{e^{S_{\mathrm{r,st}}(\mathbf{\Gamma})/k_{\mathrm{B}}}}{Z_{\mathrm{st}}} = \frac{e^{-\beta_0 E_0(\mathbf{\Gamma})-\beta_1 E_1(\mathbf{\Gamma})}}{Z_{\mathrm{st}}}. \tag{9.69}$$

The average of a function of the sub-system phase space over all the coordinates except z_i is

$$\begin{aligned}\langle f\rangle_{\mathrm{ne}/i} &\equiv \int \mathrm{d}\mathbf{\Gamma}/z_i\, \wp_{\mathrm{ne}}(\mathbf{\Gamma}) f(\mathbf{\Gamma}) \\ &= \frac{\langle f + \beta_1 W_1 f + \beta_1^2 W_1^2 f^2/2\rangle_{\mathrm{st}/i}}{\langle 1 + \beta_1 W_1 + \beta_1^2 W_1^2/2\rangle_{\mathrm{st}/i}} + \mathcal{O}\beta_1^3 \\ &= \langle f\rangle_{\mathrm{st}/i} + \beta_1 \left\langle \left[f - \langle f\rangle_{\mathrm{st}/i}\right]\left[W_1 - \langle W_1\rangle_{\mathrm{st};i}\right]\right\rangle_{\mathrm{st}/i} \\ &\quad + \frac{\beta_1^2}{2}\left\langle \left[f - \langle f\rangle_{\mathrm{st}/i}\right]\left[W_1 - \langle W_1\rangle_{\mathrm{st}/i}\right]^2\right\rangle_{\mathrm{st}/i} + \mathcal{O}(\beta_1^3). \end{aligned} \tag{9.70}$$

Now the dynamic part of the reservoir entropy will be assumed to be dominated by its odd projection, $\tilde{S}_{\mathrm{r,dyn}}(\mathbf{\Gamma}) \approx \tilde{S}^{\mathrm{odd}}_{\mathrm{r,dyn}}(\mathbf{\Gamma})$, or, equivalently, $W_1(\mathbf{\Gamma}) \approx W_1^{\mathrm{odd}}(\mathbf{\Gamma})$. (Note that $\tilde{S}^{\mathrm{even}}_{\mathrm{r,dyn}} \sim \mathcal{O}(\beta_1^2)$; see footnote 9 on p. 316.) Because the static probability distribution has even parity, $\left\langle W_1^{\mathrm{odd}}\right\rangle_{\mathrm{st}/i} = 0$. For the case of an even parity function, then $\left\langle f^{\mathrm{even}} W_1^{\mathrm{odd}}\right\rangle_{\mathrm{st}/i} = 0$, and the average simplifies to

$$\langle f^{\mathrm{even}}\rangle_{\mathrm{ne}/i} = \langle f^{\mathrm{even}}\rangle_{\mathrm{st}/i} + \frac{\beta_1^2}{2}\left\langle \left[f^{\mathrm{even}} - \langle f^{\mathrm{even}}\rangle_{\mathrm{st}/i}\right]\left(W_1^{\mathrm{odd}}\right)^2\right\rangle_{\mathrm{st}/i} + \mathcal{O}(\beta_1^4). \tag{9.71}$$

To obtain the equipartition theorem for steady heat flow, the non-equilibrium average of the square of the momenta is required. In view of the above general result, the static average needs to be evaluated, and invoking the quadratic form of the kinetic energy this is

$$\begin{aligned}\langle p_{i\alpha}^2\rangle_{\mathrm{st}/i} &= \frac{\int \mathrm{d}\mathbf{q}^N/z_i\, e^{-\beta_0 E_0(\mathbf{q}^N)-\beta_1 E_1(\mathbf{q}^N)} \int \mathrm{d}\mathbf{p}^N e^{-\beta_0 K_0(\mathbf{p}^N)-\beta_1 K_1(\mathbf{p}^N)} p_{i\alpha}^2}{\int \mathrm{d}\mathbf{q}^N/z_i\, e^{-\beta_0 E_0(\mathbf{q}^N)-\beta_1 E_1(\mathbf{q}^N)} \int \mathrm{d}\mathbf{p}^N e^{-\beta_0 K_0(\mathbf{p}^N)-\beta_1 K_1(\mathbf{p}^N)}} \\ &= \frac{\int_{-\infty}^{\infty} \mathrm{d}p_{i\alpha}\, e^{-\beta(z_i)p_{i\alpha}^2/2m_i} p_{i\alpha}^2}{\int_{-\infty}^{\infty} \mathrm{d}p_{i\alpha}\, e^{-\beta(z_i)p_{i\alpha}^2/2m_i}} \\ &= \frac{m_i}{\beta(z_i)}. \end{aligned} \tag{9.72}$$

Here $\beta(z) \equiv \beta_0 + \beta_1 z$. Using this one has

$$\begin{aligned} \left\langle p_{i\alpha}^2 \right\rangle_{\mathrm{ne}/i} &= \frac{m_i}{\beta(z_i)} + \frac{\beta_1^2}{2} \left\langle \left[p_{i\alpha}^2 - \frac{m_i}{\beta(z_i)} \right] \left(W_1^{\mathrm{odd}}\right)^2 \right\rangle_{\mathrm{st}/i} + \mathcal{O}(\beta_1^4) \\ &= \frac{m_i}{\beta(z_i)} + \mathcal{O}(\beta_1^2). \end{aligned} \tag{9.73}$$

The second term in the first equality, which is of quadratic order in the temperature gradient, depends upon the departure of the local kinetic energy from its static, average value, and so may be expected to be small but non-zero. The leading term is what one would get for the equipartition theorem in an equilibrium system at the local temperature imposed by the reservoirs. The second term is a dynamic or non-equilibrium effect. The difference between this local equipartition theorem and the generalised equipartition theorem given above, Eqs. (9.66) and (9.67), is that the generalized result utilises an integral over the system volume, which cancels the contributions linear in the gradient and so reduces to the canonical equilibrium result for the mid-temperature.

9.5 Green-Kubo Expressions for the Thermal Conductivity

9.5.1 Isolated System

For an isolated system the thermal conductivity was shown in Eq. (9.21) to be

$$\begin{aligned} \lambda(\tau) &= \frac{-1}{V k_{\mathrm{B}} T_0^2 |\tau|} \left[Q(\tau) - Q(0) \right] \\ &= \frac{-1}{V k_{\mathrm{B}} T_0^2 |\tau|} \left\langle \left[E_1(t+\tau) - E_1(t) \right] E_1(t) \right\rangle_0 \\ &= \frac{-1}{V k_{\mathrm{B}} T_0^2 |\tau|} \int_0^\tau \mathrm{d}t' \left\langle \dot{E}_1^0(t'+t) E_1(t) \right\rangle_0 \\ &= \frac{-1}{V k_{\mathrm{B}} T_0^2 |\tau|} \int_0^\tau \mathrm{d}t' \left\langle \dot{E}_1^0(\mathbf{\Gamma}^0(t'|\mathbf{\Gamma},0)) E_1(\mathbf{\Gamma}) \right\rangle_0 \\ &= \frac{-1}{V k_{\mathrm{B}} T_0^2 |\tau|} \int_0^\tau \mathrm{d}t' \int \mathrm{d}\mathbf{\Gamma} \, \frac{\delta(\mathcal{H}(\mathbf{\Gamma}) - \overline{E}_0)}{Z(N,V,\overline{E}_0)} \, \dot{E}_1^0(\mathbf{\Gamma}^0(t'|\mathbf{\Gamma},0)) E_1(\mathbf{\Gamma}). \end{aligned} \tag{9.74}$$

As the average is for an isolated system, the distribution of the initial point is uniform in phase space over the energy hypersurface of energy $\overline{E}_0(N, V, T_0)$. This is independent of τ for τ large on molecular time scales, but small on macroscopic time scales.

Using the microscopic reversibility of Hamilton's equations, $\mathbf{\Gamma}^0(t'|\mathbf{\Gamma},0) = \mathbf{\Gamma}^0(-t'|\mathbf{\Gamma}^\dagger,0)^\dagger$, and the fact that E_1 has even phase space parity, and that $\dot{E}_1^0$

has odd parity, this can be written in various equivalent forms,

$$
\begin{aligned}
\lambda(\tau) &= \frac{-1}{Vk_{\rm B}T_0^2|\tau|}\int_0^\tau \mathrm{d}t' \left\langle \dot{E}_1^0(\mathbf{\Gamma}^0(t'|\mathbf{\Gamma},0))E_1(\mathbf{\Gamma})\right\rangle_0 \\
&= \frac{-1}{Vk_{\rm B}T_0^2|\tau|}\int_{-\tau}^0 \mathrm{d}t' \left\langle \dot{E}_1^0(\mathbf{\Gamma}^0(-t'|\mathbf{\Gamma},0))E_1(\mathbf{\Gamma})\right\rangle_0 \\
&= \frac{1}{Vk_{\rm B}T_0^2|\tau|}\int_{-\tau}^0 \mathrm{d}t' \left\langle \dot{E}_1^0(\mathbf{\Gamma}^0(t'|\mathbf{\Gamma}^\dagger,0))E_1(\mathbf{\Gamma}^\dagger)\right\rangle_0 \\
&= \frac{-1}{2Vk_{\rm B}T_0^2\tau}\int_{-\tau}^\tau \mathrm{d}t' \,\mathrm{sign}(t') \left\langle \dot{E}_1^0(\mathbf{\Gamma}^0(t'|\mathbf{\Gamma},0))\, E_1(\mathbf{\Gamma})\right\rangle_0 . \qquad (9.75)
\end{aligned}
$$

The expression for the thermal conductivity may also be rewritten using the fact that it must be independent of τ (for τ larger than some relaxation time). Taking τ over to the other side and differentiating with respect to τ, the first expression for the thermal conductivity is therefore equivalent to

$$
\begin{aligned}
\lambda &= \frac{-\hat{\tau}}{Vk_{\rm B}T_0^2}\left\langle \dot{E}_1^0(\mathbf{\Gamma}^0(\tau|\mathbf{\Gamma},0))E_1(\mathbf{\Gamma})\right\rangle_0 \\
&= \frac{\hat{\tau}}{Vk_{\rm B}T_0^2}\left\langle \dot{E}_1^0(\mathbf{\Gamma}^0(\tau|\mathbf{\Gamma},0))\left[E_1(\mathbf{\Gamma}^0(\tau|\mathbf{\Gamma},0)) - E_1(\mathbf{\Gamma})\right]\right\rangle_0 \\
&= \frac{\hat{\tau}}{Vk_{\rm B}T_0^2}\int_0^\tau \mathrm{d}t' \left\langle \dot{E}_1^0(\mathbf{\Gamma}^0(\tau|\mathbf{\Gamma},0))\dot{E}_1^0(\mathbf{\Gamma}^0(t'|\mathbf{\Gamma},0))\right\rangle_0 \\
&= \frac{\hat{\tau}}{Vk_{\rm B}T_0^2}\int_0^\tau \mathrm{d}t' \left\langle \dot{E}_1^0(\mathbf{\Gamma}'')\dot{E}_1^0(\mathbf{\Gamma}^0(t'|\mathbf{\Gamma}'',\tau))\right\rangle_0, \quad \mathbf{\Gamma}'' \equiv \mathbf{\Gamma}^0(\tau|\mathbf{\Gamma},0) \\
&= \frac{\hat{\tau}}{Vk_{\rm B}T_0^2}\int_{-\tau}^0 \mathrm{d}t'' \left\langle \dot{E}_1^0(\mathbf{\Gamma}'')\dot{E}_1^0(\mathbf{\Gamma}^0(t''|\mathbf{\Gamma}'',0))\right\rangle_0, \quad t'' \equiv t' - \tau \\
&= \frac{\hat{\tau}}{Vk_{\rm B}T_0^2}\int_0^\tau \mathrm{d}t \left\langle \dot{E}_1^0(\mathbf{\Gamma}'')\dot{E}_1^0(\mathbf{\Gamma}^0(t|\mathbf{\Gamma}''^\dagger,0)^\dagger)\right\rangle_0, \quad t \equiv -t'' \\
&= \frac{\hat{\tau}}{Vk_{\rm B}T_0^2}\int_0^\tau \mathrm{d}t \left\langle \dot{E}_1^0(\mathbf{\Gamma})\dot{E}_1^0(\mathbf{\Gamma}^0(t|\mathbf{\Gamma},0))\right\rangle_0, \quad \mathbf{\Gamma} \equiv \mathbf{\Gamma}''^\dagger \\
&= \frac{1}{2Vk_{\rm B}T_0^2}\int_{-\infty}^\infty \mathrm{d}t \left\langle \dot{E}_1^0(\mathbf{\Gamma})\dot{E}_1^0(\mathbf{\Gamma}^0(t|\mathbf{\Gamma},0))\right\rangle_0 . \qquad (9.76)
\end{aligned}
$$

This uses the fact that there is no instantaneous correlation between the flux and the moment, $\left\langle \dot{E}_1^0(\mathbf{\Gamma})E_1(\mathbf{\Gamma})\right\rangle_0 = 0$, which follows because the flux is odd, $\dot{E}_1^0(\mathbf{\Gamma}^\dagger) = -\dot{E}_1^0(\mathbf{\Gamma})$, and the moment is even, $E_1(\mathbf{\Gamma}^\dagger) = E_1(\mathbf{\Gamma})$. It also uses the fact that $\mathbf{\Gamma}$ is a dummy variable for the average, and the fact that any point on a Hamiltonian trajectory can be used as the starting point. In the final equality, $|\tau|$ has been sent to infinity, on the grounds that the integrand is short-ranged.

For future reference, the penultimate equality may be rearranged as

$$
\lambda = \frac{1}{2Vk_{\rm B}T_0^2}\int_{-\infty}^0 \mathrm{d}t \left\langle \dot{E}_1^0(\mathbf{\Gamma})\left[\dot{E}_1^0(\mathbf{\Gamma}^0(t|\mathbf{\Gamma},0)) - \dot{E}_1^0(\mathbf{\Gamma}^0(t|\mathbf{\Gamma}^\dagger,0))\right]\right\rangle_0 , \qquad (9.77)
$$

as is readily proven.

The isolated system averages that appear here may be replaced by the canonical equilibrium system average with negligible error in the thermodynamic limit. These various results are the Green-Kubo expression for the thermal conductivity, (see footnote 5 on p. 301). Generally in the literature the conductive energy flux appears instead of the rate of change of the first energy moment, $J_{\rm E}^0 \equiv \dot{E}_1^0/V$.

9.5.2 Heat Reservoirs

In the presence of the heat reservoirs, §9.1.3, one has exactly the same relationship between heat flux and thermodynamic gradient as for the isolated system (compare Eqs (9.18) and (9.30)). Hence multiplying both sides of Eq. (9.30) by E_1 and taking the average, the thermal conductivity is given by the same expression as above, with the isolated system average replaced by the non-equilibrium average,

$$\begin{aligned}
\lambda(\tau) &= \frac{-1}{V k_{\rm B} T_0^2 |\tau|} \left\langle \left[E_1^0(t+\tau) - E_1(t) \right] E_1(t) \right\rangle_{\rm ne} \\
&= \frac{-1}{V k_{\rm B} T_0^2 |\tau|} \int_t^{t+\tau} {\rm d}t' \left\langle \dot{E}_1^0(t') E_1(t) \right\rangle_{\rm ne} \\
&= \frac{-1}{V k_{\rm B} T_0^2 |\tau|} \int_t^{t+\tau} {\rm d}t' \int {\rm d}\underline{\boldsymbol{\Gamma}}\, \wp(\underline{\boldsymbol{\Gamma}})\, \dot{E}_1^0(\boldsymbol{\Gamma}(t')) E_1(\boldsymbol{\Gamma}) \\
&\approx \frac{-1}{V k_{\rm B} T_0^2 |\tau|} \int_t^{t+\tau} {\rm d}t' \int {\rm d}\boldsymbol{\Gamma}\, \wp_{\rm ne}(\boldsymbol{\Gamma}, t)\, \dot{E}_1^0(\overline{\boldsymbol{\Gamma}}(t'|\boldsymbol{\Gamma}, t)) E_1(\boldsymbol{\Gamma}). \qquad (9.78)
\end{aligned}$$

This expression holds for both a forward trajectory, $(\tau > 0)$, and a backward trajectory, $(\tau < 0)$, and is independent of τ for τ large on molecular scales but small on macroscopic scales. In obtaining the final equality, the average over all trajectories has been replaced by the most likely trajectory averaged over the initial point.

The linear relationship between flux and temperature gradient embodied in Fourier's law is valid for small gradients, $\beta_1 \to 0$, or, equivalently, $T_1 \to \infty$. (The present formulation of non-equilibrium statistical mechanics is valid for arbitrary gradients.) Accordingly the expression for the thermal conductivity based upon the non-equilibrium probability ought be expanded with respect to β_1, and only the leading-order term ought to be retained. Hence the non-equilibrium probability may be replaced by the Maxwell-Boltzmann probability distribution,

$$\wp_{\rm ne}(\boldsymbol{\Gamma}, t) \sim \wp_{\rm MB}(\boldsymbol{\Gamma}) + \mathcal{O}(\beta_1), \qquad (9.79)$$

and one has

$$\lambda(\tau) = \frac{-1}{V k_{\rm B} T_0^2 |\tau|} \int_t^{t+\tau} {\rm d}t' \left\langle \dot{E}_1^0(\overline{\boldsymbol{\Gamma}}(t'|\boldsymbol{\Gamma}, t)) E_1(\boldsymbol{\Gamma}) \right\rangle_{\rm MB} + \mathcal{O}(\beta_1). \qquad (9.80)$$

This expresses the thermal conductivity as a canonical equilibrium average, albeit over the most likely non-equilibrium trajectory.

For $\tau < 0$ one can write

$$\begin{aligned}\int_t^{t+\tau} \mathrm{d}t' \, \dot{E}_1^0(\overline{\mathbf{\Gamma}}(t'|\mathbf{\Gamma},t)) &= -\int_{t-|\tau|}^{t} \mathrm{d}t' \, \dot{E}_1^0(\overline{\mathbf{\Gamma}}(t'|\mathbf{\Gamma},t)) \\ &= -\int_0^{t} \mathrm{d}t' \, \dot{E}_1^0(\overline{\mathbf{\Gamma}}(t'|\mathbf{\Gamma},t)) - \int_{t-|\tau|}^{0} \mathrm{d}t' \, \overline{\dot{E}_1^0} \\ &= -T_1 S_{\mathrm{r,dyn}}(\mathbf{\Gamma},t) + (t-|\tau|)\overline{\dot{E}_1^0} \\ &= -T_1 \tilde{S}_{\mathrm{r,dyn}}(\mathbf{\Gamma}) - |\tau|\overline{\dot{E}_1^0}. \end{aligned} \tag{9.81}$$

Recall that $\tilde{S}_{\mathrm{r,dyn}}(\mathbf{\Gamma})$ is equal to $S_{\mathrm{r,dyn}}(\mathbf{\Gamma},t)$ with the asymptote of the integrand subtracted, Eq. (9.47). Inserting this into the preceding expression for the thermal conductivity one obtains

$$\lambda(\tau) = \frac{-1}{V k_{\mathrm{B}} T_0^2 |\tau|} \left\langle \left[-T_1 \tilde{S}_{\mathrm{r,dyn}}(\mathbf{\Gamma}) - |\tau| \overline{\dot{E}_1^0} \right] E_1(\mathbf{\Gamma}) \right\rangle_{\mathrm{MB}}, \quad \tau < 0. \tag{9.82}$$

As mentioned above, the right-hand side has to be independent of τ. This means that

$$\left\langle \tilde{S}_{\mathrm{r,dyn}}(\mathbf{\Gamma}) E_1(\mathbf{\Gamma}) \right\rangle_{\mathrm{MB}} = 0. \tag{9.83}$$

Since both the Maxwell-Boltzmann distribution and the first energy moment have even phase space parity, one way to ensure this is if the even projection of the dynamic part of the reservoir entropy is negligible, $\tilde{S}^{\mathrm{even}}_{\mathrm{r,dyn}}(\mathbf{\Gamma}) \approx 0$. In fact it appears most likely that $\tilde{S}^{\mathrm{even}}_{\mathrm{r,dyn}}(\mathbf{\Gamma}) = \mathcal{O}(\beta_1^2)$.[9]

Defining the projections as

$$S^{\mathrm{even/odd}}_{\mathrm{r,dyn}}(\mathbf{\Gamma},t) \equiv \frac{1}{2}\left[S_{\mathrm{r,dyn}}(\mathbf{\Gamma},t) \pm S_{\mathrm{r,dyn}}(\mathbf{\Gamma}^{\dagger},t) \right], \tag{9.84}$$

this condition is equivalent to

$$S^{\mathrm{odd}}_{\mathrm{r,dyn}}(\mathbf{\Gamma},t) \approx \tilde{S}_{\mathrm{r,dyn}}(\mathbf{\Gamma}), \text{ and } S^{\mathrm{even}}_{\mathrm{r,dyn}}(\mathbf{\Gamma},t) \approx t\overline{\dot{E}_1^0}/T_1. \tag{9.85}$$

It should be stressed that this is a sufficient, not necessary, condition for the thermal conductivity to be independent of τ; it is stronger than is strictly required.

[9] Fluxes in the z-direction are determined by the temperature gradient in the same direction. Hence the average of functions that have odd temporal parity should be odd functions of the temperature gradient. Conversely, the averages of even parity functions ought be even functions of the temperature gradient. The result that $S^{\mathrm{even}}_{\mathrm{r,dyn}}(\mathbf{\Gamma},t) \approx -t\overline{\dot{E}_1^0}/T_1$ obeys this rule.

What remains of the last expression for the thermal conductivity is just an identity,

$$\begin{aligned}\lambda(\tau) &= \frac{1}{Vk_{\rm B}T_0^2}\left\langle \overline{\dot{E}_1^0} E_1(\mathbf{\Gamma})\right\rangle_{\rm MB}, \quad \tau<0\\ &= \frac{1}{Vk_{\rm B}T_0^2}\frac{\hat{\tau}\Lambda}{2}S\left\langle E_1(\mathbf{\Gamma})E_1(\mathbf{\Gamma})\right\rangle_{\rm MB}, \quad \tau<0\\ &= \frac{-1}{Vk_{\rm B}T_0^2}\frac{\Lambda}{2}S\left[-k_{\rm B}S^{-1}\right]\\ &= \lambda. \end{aligned} \tag{9.86}$$

This uses Eq. (9.18), with the most likely coarse velocity replaced by the instantaneous velocity, and also the fact that the fluctuation matrix may be obtained in the canonical equilibrium system.

An alternative, more explicitly non-equilibrium expression is obtained as follows. One has from Eq. (9.30) (taking the coarse forward velocity in the non-equilibrium system, $\hat{\tau}=1$, to be the same as the instantaneous velocity),

$$\left\langle \dot{E}_1^0(\mathbf{\Gamma})\right\rangle_{\rm ne} = VT_0^2\lambda S\left\langle E_1(\mathbf{\Gamma})\right\rangle_{\rm ne}. \tag{9.87}$$

Now on the right-hand side one has

$$\begin{aligned}\left\langle E_1(\mathbf{\Gamma})\right\rangle_{\rm ne} &= \int {\rm d}\mathbf{\Gamma}\, \frac{e^{-\beta_0 E_0(\mathbf{\Gamma})-\beta_1 E_1(\mathbf{\Gamma})+\tilde{S}_{\rm r,dyn}(\mathbf{\Gamma})/k_{\rm B}}}{\tilde{Z}(\beta_0,\beta_1)} E_1(\mathbf{\Gamma})\\ &= \int {\rm d}\mathbf{\Gamma}\, \frac{e^{-\beta_0 E_0(\mathbf{\Gamma})}\left[1-\beta_1 E_1(\mathbf{\Gamma})+\tilde{S}_{\rm r,dyn}(\mathbf{\Gamma})/k_{\rm B}+\mathcal{O}(\beta_1^2)\right]}{Z_{\rm MB}(\beta_0)+\mathcal{O}(\beta_1^2)} E_1(\mathbf{\Gamma})\\ &= -\beta_1\left\langle E_1(\mathbf{\Gamma})^2\right\rangle_{\rm MB}+\mathcal{O}(\beta_1^2)\\ &= \frac{1}{T_1}S^{-1}+\mathcal{O}(\beta_1^2). \end{aligned} \tag{9.88}$$

This is an identity that is equivalent to Eq. (9.31). In obtaining the penultimate equality, Eq. (9.83) has been used, as well as the fact that $\langle E_1(\mathbf{\Gamma})\rangle_{\rm MB}=0$, and also the expansion for the partition function, $\tilde{Z}(\beta_0,\beta_1) = Z_{\rm MB}(\beta_0)+\mathcal{O}(\beta_1^2)$, which follows because it has to be insensitive to the sign of the temperature gradient.

Similarly expanding the left-hand side, and using the fact that $\dot{E}_1^0$ has odd phase space parity, one has

$$\begin{aligned}\left\langle \dot{E}_1^0(\mathbf{\Gamma})\right\rangle_{\rm ne} &= \left\langle \frac{1-\beta_1 E_1(\mathbf{\Gamma})+S_{\rm r,dyn}(\mathbf{\Gamma},t)/k_{\rm B}+\mathcal{O}(\beta_1^2)}{1+\mathcal{O}(\beta_1^2)}\dot{E}_1^0(\mathbf{\Gamma})\right\rangle_{\rm MB}\\ &= k_{\rm B}^{-1}\left\langle S^{\rm odd}_{\rm r,dyn}(\mathbf{\Gamma},t)\dot{E}_1^0(\mathbf{\Gamma})\right\rangle_{\rm MB}+\mathcal{O}(\beta_1^2)\\ &= k_{\rm B}^{-1}\left\langle \tilde{S}^{\rm odd}_{\rm r,dyn}(\mathbf{\Gamma})\dot{E}_1^0(\mathbf{\Gamma})\right\rangle_{\rm MB}+\mathcal{O}(\beta_1^2). \end{aligned} \tag{9.89}$$

Equating the leading-order of the expansion on both sides, gives the thermal conductivity as

$$\begin{aligned}\lambda &= \frac{T_1}{Vk_{\rm B}T_0^2}\left\langle \tilde S^{\rm odd}_{\rm r,dyn}(\mathbf{\Gamma})\dot E_1^0(\mathbf{\Gamma})\right\rangle_{\rm MB} \qquad (9.90)\\ &= \frac{1}{2Vk_{\rm B}T_0^2}\int_0^t {\rm d}t' \left\langle \left[\dot E_1^0(\overline{\mathbf{\Gamma}}(t'|\mathbf{\Gamma},t)) - \dot E_1^0(\overline{\mathbf{\Gamma}}(t'|\mathbf{\Gamma}^\dagger,t))\right]\dot E_1^0(\mathbf{\Gamma})\right\rangle_{\rm MB}\\ &= \frac{1}{2Vk_{\rm B}T_0^2}\int_{-\infty}^0 {\rm d}t' \left\langle \left[\dot E_1^0(\overline{\mathbf{\Gamma}}(t'|\mathbf{\Gamma},0)) - \dot E_1^0(\overline{\mathbf{\Gamma}}(t'|\mathbf{\Gamma}^\dagger,0))\right]\dot E_1^0(\mathbf{\Gamma})\right\rangle_{\rm MB}.\end{aligned}$$

This shows that only the odd projection of the dynamic part of the reservoir entropy contributes to the thermal conductivity (c.f. §8.5). Comparing this with Eq. (9.77), with $\tau = -\infty$ and the isolated system average replaced by a Maxwell-Boltzmann average, one must have

$$\begin{aligned}&\int_{-\infty}^0 {\rm d}t' \left\langle \left[\dot E_1^0(\overline{\mathbf{\Gamma}}(t'|\mathbf{\Gamma},0)) - \dot E_1^0(\overline{\mathbf{\Gamma}}(t'|\mathbf{\Gamma}^\dagger,0))\right]\dot E_1^0(\mathbf{\Gamma})\right\rangle_{\rm MB}\\ &\quad= \int_{-\infty}^0 {\rm d}t' \left\langle \left[\dot E_1^0(\mathbf{\Gamma}^0(t'|\mathbf{\Gamma},0)) - \dot E_1^0(\mathbf{\Gamma}^0(t'|\mathbf{\Gamma}^\dagger,0))\right]\dot E_1^0(\mathbf{\Gamma})\right\rangle_{\rm MB}. \qquad (9.91)\end{aligned}$$

On the left-hand side the most likely non-equilibrium trajectory appears, in contrast to the right-hand side where the adiabatic or Hamiltonian trajectory appears. This is in essence the mathematical content of Fig. 8.2 on p. 278, and is Eq. (8.181) applied to the present problem of steady heat flow.

9.5.3 Relation with Odd Projection

The preceding result, which is based upon the entirely plausible assertion that the thermal conductivity of an isolated system is identical to that measured for steady heat flow, is obviously related to the odd projection of the dynamic part of the reservoir entropy and its expression as an integral over the adiabatic trajectory that was discussed in general in §8.5. It is worthwhile then to revisit the latter argument explicitly for the present problem.

The analysis will be based upon the operation that reverses all the z-coordinates and z-momenta. That is, if $\mathbf{\Gamma} = \{q_x^N, q_y^N, q_z^N, p_x^N, p_y^N, p_z^N\}$, then

$$\mathbf{\Gamma}^\ddagger = \{q_x^N, q_y^N, (-q_z)^N, p_x^N, p_y^N, (-p_z)^N\}. \qquad (9.92)$$

Assuming that in the absence of the applied temperature gradient the system is symmetric about the mid-plane, one can readily confirm the symmetries

$$E_0(\mathbf{\Gamma}^\ddagger) = E_0(\mathbf{\Gamma}),\;\; E_1(\mathbf{\Gamma}^\ddagger) = -E_1(\mathbf{\Gamma}),\;\; \dot E_1^0(\mathbf{\Gamma}^\ddagger) = -\dot E_1^0(\mathbf{\Gamma}). \qquad (9.93)$$

One also has

$$\mathbf{\Gamma}^0(t'|\mathbf{\Gamma}^\ddagger,t) = \mathbf{\Gamma}^0(t'|\mathbf{\Gamma},t)^\ddagger, \text{ and } \overline{\mathbf{\Gamma}}_{\rm MB}(t'|\mathbf{\Gamma}^\ddagger,t) = \overline{\mathbf{\Gamma}}_{\rm MB}(t'|\mathbf{\Gamma},t)^\ddagger. \qquad (9.94)$$

These results for the adiabatic and for the most likely canonical equilibrium trajectory follow from the symmetry of the system about the mid-plane.

In such a symmetric system the non-equilibrium probability must be invariant with respect to reversing all the z-coordinates, including the applied temperature gradient,

$$\wp(\mathbf{\Gamma}|\beta_0, \beta_1) = \wp(\mathbf{\Gamma}^{\ddagger}|\beta_0, -\beta_1). \tag{9.95}$$

It follows that the partition function has the expansion $Z(\beta_0, \beta_1) = Z_{\rm MB}(\beta_0) + \mathcal{O}(\beta_1^2)$, and that the reservoir entropy, the exponent, has an expansion of the form

$$S_{\rm r}(\mathbf{\Gamma}|\beta_0, \beta_1) = S_0(\mathbf{\Gamma}) + \beta_1 S_1(\mathbf{\Gamma}) + \mathcal{O}(\beta_1^2), \tag{9.96}$$

with the z-parities of the expansion coefficients being even and odd, respectively

$$S_0(\mathbf{\Gamma}^{\ddagger}) = S_0(\mathbf{\Gamma}) \text{ and } S_1(\mathbf{\Gamma}^{\ddagger}) = -S_1(\mathbf{\Gamma}). \tag{9.97}$$

The reservoir entropy is $S_{\rm r}(\mathbf{\Gamma}|\beta_0, \beta_1) = S_{\rm r,st}(\mathbf{\Gamma}|\beta_0, \beta_1) + S_{\rm r,dyn}(\mathbf{\Gamma}|\beta_0, \beta_1)$, with the static part being $S_{\rm r,st}(\mathbf{\Gamma}|\beta_0, \beta_1) = -E_0(\mathbf{\Gamma})/T_0 - E_1(\mathbf{\Gamma})/T_1$. Obviously, $S_0(\mathbf{\Gamma}) = -E_0(\mathbf{\Gamma})/T_0$, and $-E_1(\mathbf{\Gamma})/T_1$ belongs to $\beta_1 S_1(\mathbf{\Gamma})$. As indicated above, $E_0(\mathbf{\Gamma})$ has even z-parity and $E_1(\mathbf{\Gamma})$ has odd z-parity, so the parity of the expansion is satisfied by these. It remains to analyse in detail the contribution of the dynamic part of the reservoir entropy to $\beta_1 S_1(\mathbf{\Gamma})$. One has

$$\begin{aligned} S_{\rm r,dyn}(\mathbf{\Gamma}|\beta_0, \beta_1)/k_{\rm B} &= \beta_1 \int_0^t {\rm d}t'\; \dot{E}_1^0(\overline{\mathbf{\Gamma}}(t'|\mathbf{\Gamma}, t; \beta_0, \beta_1)) \\ &= \beta_1 \int_0^t {\rm d}t'\; \dot{E}_1^0(\overline{\mathbf{\Gamma}}_{\rm MB}(t'|\mathbf{\Gamma}, t; \beta_0)) + \mathcal{O}(\beta_1^2) \\ &\approx \beta_1 \int_0^t {\rm d}t'\; \dot{E}_1^0(\overline{\mathbf{\Gamma}}^0(t'|\mathbf{\Gamma}, t)). \end{aligned} \tag{9.98}$$

The second equality follows by setting $\beta_1 = 0$ to extract the leading-order term, and this converts the most likely trajectory in the non-equilibrium system to the canonical equilibrium most likely trajectory. The final approximation assumes that the latter is much the same as the adiabatic trajectory. In view of the symmetries given above one can show explicitly that this has odd z-parity,

$$\begin{aligned} S_{\rm r,dyn}(\mathbf{\Gamma}^{\ddagger}|\beta_0, \beta_1)/k_{\rm B} &= \beta_1 \int_0^t {\rm d}t'\; \dot{E}_1^0(\overline{\mathbf{\Gamma}}^0(t'|\mathbf{\Gamma}^{\ddagger}, t)) + \mathcal{O}(\beta_1^2) \\ &= \beta_1 \int_0^t {\rm d}t'\; \dot{E}_1^0(\overline{\mathbf{\Gamma}}^0(t'|\mathbf{\Gamma}, t)^{\ddagger}) \\ &= -\beta_1 \int_0^t {\rm d}t'\; \dot{E}_1^0(\overline{\mathbf{\Gamma}}^0(t'|\mathbf{\Gamma}, t)) \\ &= -S_{\rm r,dyn}(\mathbf{\Gamma}|\beta_0, \beta_1)/k_{\rm B}. \end{aligned} \tag{9.99}$$

Of course it also follows that this has odd z-parity if the canonical equilibrium most likely trajectory were to be used. This is a pleasing confirmation of the consistency of this argument based upon the symmetry of the probability distribution. One can therefore conclude two things about the dynamic part of the reservoir entropy: Eq. (9.98) shows that it can be expressed as an integral over the adiabatic trajectory, and Eq. (9.99) shows that it has odd parity with respect to this particular z-symmetry operation. Both conclusions are valid only to linear order in the temperature gradient.

These two results are similar to, but not as strong as, the general results argued for in §8.5. This present analysis is more rigorous. The present results are clearly similar to the Green-Kubo result established at the end of the preceding subsection. Of course 'odd parity' here refers to the ‡ operation (reversal of z-coordinates and z-momenta) whereas it refers to the † conjugation operation (reversal of all momenta) in the preceding subsection and elsewhere. However, Eq. (9.91) from the end of the preceding subsection follows directly from the present adiabatic result, Eq. (9.98), and from the fact that the adiabatic rate of change of the first energy moment has odd conjugation parity, $\dot{E}_1^0(\mathbf{\Gamma}^\dagger) = -\dot{E}_1^0(\mathbf{\Gamma})$.

9.6 Shear Flow

In order to illustrate the way in which these results for heat flow are typical of thermodynamic flows, the key results for shear flow will be given here. As shown in Eq. (5.34), the velocity is the thermodynamic conjugate of the momentum,

$$\frac{\partial \sigma(\mathbf{r},t)}{\partial \mathbf{p}(\mathbf{r},t)} = \frac{-1}{T}\mathbf{v}(\mathbf{r},t). \tag{9.100}$$

Here $\sigma(\mathbf{r},t)$ is the entropy density, $\mathbf{p}(\mathbf{r},t) = mn(\mathbf{r},t)\mathbf{v}(\mathbf{r},t)$, is the momentum density, m is the molecular mass (a single component system is assumed), $n(\mathbf{r},t)$ is the number density, $\mathbf{v}(\mathbf{r},t)$ is the velocity, and T is the temperature, assumed uniform and constant. Also $\mathbf{r} = \{x, y, z\}$ is the position and t is the time.

From this identification of the conjugate variables, one could immediately transcribe the results for shear flow from those for heat flow by making the identifications $E_1 \Rightarrow P_{x1}$, $T_1^{-1} \Rightarrow -v_{x1}/T$, and $T_0 \Rightarrow T$. This is in essence what is now done, with slightly more detail.

Consider two reservoirs with boundaries located at $z = \pm L/2$, moving with velocities $v_{x\pm}$. The zeroth and first velocities exerted by the reservoirs on the sub-system are

$$v_{x0} \equiv \frac{1}{2}\left[v_{x+} + v_{x-}\right], \text{ and } v_{x1} \equiv \frac{1}{L}\left[v_{x+} - v_{x-}\right]. \tag{9.101}$$

The zeroth velocity is the average or mid velocity, and the first velocity is the velocity gradient. It is straightforward to show, (see §4.2), that the sub-system velocities are thermodynamically conjugate to the zeroth and first sub-system momentum moments,

$$v_{sx0} \equiv -T\frac{\partial S_s(P_{x0}, P_{x1})}{\partial P_{x0}}, \text{ and } v_{sx1} \equiv -T\frac{\partial S_s(P_{x0}, P_{x1})}{\partial P_{x1}}. \tag{9.102}$$

The momentum moments here label a constrained sub-system macrostate; the other state variables are not shown explicitly. In the present case the moments are measured in the z-direction,

$$P_{x0} \equiv A \int_{-L/2}^{L/2} \mathrm{d}z\, p_x(z), \text{ and } P_{x1} \equiv A \int_{-L/2}^{L/2} \mathrm{d}z\, z p_x(z), \tag{9.103}$$

where A is the cross-sectional area of the sub-system.

The conductive part of the momentum flux is the viscous pressure tensor, $\underline{\underline{J}}_{\mathrm{P}}^{0}(\mathbf{r},t) = \underline{\underline{\Pi}}(\mathbf{r},t)$, as was derived in §5.1.4. This is the momentum transport that is not due to convection and that is not due to pressure work. The component $\Pi_{xz} = \Pi_{zx}$ is the x-momentum per unit area per unit time crossing a plane perpendicular to the z-axis by conduction. From momentum conservation, the rate of change of the momentum density is just the negative divergence of the flux, which in the present one-dimensional case is its gradient, $\dot{p}_x^0(z) = -\mathrm{d}\Pi_{xz}(z)/\mathrm{d}z$. Hence considering the sub-system as isolated (so that $\Pi_{xz}(\pm L/2) = 0$, and $\dot{P}_{x1} = \dot{P}_{x1}^0$), the adiabatic rate of change of the first momentum moment is,

$$\begin{aligned}
\dot{P}_{x1}^0 &= A \int_{-L/2}^{L/2} \mathrm{d}z\, z \dot{p}_x^0(z) \\
&= A \int_{-L/2}^{L/2} \mathrm{d}z\, z \frac{-\mathrm{d}\Pi_{xz}(z)}{\mathrm{d}z} \\
&= -Az\Pi_{xz}(z)\Big|_{-L/2}^{L/2} + A \int_{-L/2}^{L/2} \mathrm{d}z\, \Pi_{xz}(z) \\
&= V\overline{\Pi}_{xz},
\end{aligned} \tag{9.104}$$

where $V = AL$ is the volume of the sub-system. In the final equality, $\overline{\Pi}_{xz} \equiv L^{-1}\int_{-L/2}^{L/2} \mathrm{d}z\, \Pi_{xz}(z)$ is the flux averaged over the sub-system. In the steady state, the flux must be uniform, which means that the spatially averaged flux is the same as the flux itself.[10] This expression is formally exact and holds in non-optimum constrained states as well as in the optimum steady state.

The linear constitutive relation from hydrodynamics relates the most likely value of the traceless part of the viscous pressure tensor to the traceless symmetric part of the velocity gradient tensor, Eq. (5.72),

$$\overline{\underline{\underline{\Pi}}}^{*}(\mathbf{r},t) = -2\eta \left[\nabla \mathbf{v}(\mathbf{r},t)\right]^{*,\mathrm{sym}}, \tag{9.105}$$

where η is the shear viscosity. The fluid is here at rest except for the x-component, which only varies in the z-direction. This means that $\nabla \cdot \mathbf{v}(\mathbf{r},t) = 0$. Hence in the present problem the relevant component is

$$\overline{\Pi}_{xz} = -\eta \frac{\partial v_x(z)}{\partial z} = -\eta v_{x1}. \tag{9.106}$$

[10] The objection that there is a contradiction between this argument that $\Pi_{xz}(z)$ is uniform and using an apparently non-zero $\mathrm{d}\Pi_{xz}(z)/\mathrm{d}z$ in the integrand can be resolved by noting that the latter is formally a δ-function at the boundaries. See footnote 2 on p. 102.

The task is now to derive this relation from the second entropy theory. As a bonus, the Green-Kubo expression for the shear viscosity will also emerge.

9.6.1 Second Entropy for Shear Flow

For an isolated sub-system, the first entropy may be written as a quadratic form in the first momentum moment,

$$S_{\rm s}(P_{x1}) = \frac{1}{2} S P_{x1}^2, \quad S = -k_{\rm B} / \left\langle P_{x1}^2 \right\rangle . \tag{9.107}$$

The first velocity is essentially the thermodynamic force,

$$v_{{\rm s}x1}(P_{x1}) = -T \frac{\partial S_{\rm s}(P_{x1})}{\partial P_{x1}} = -T S P_{x1}. \tag{9.108}$$

The pure parity second entropy for the isolated sub-system is (see §4.3),

$$S_{\rm s}^{(2)}(P'_{x1}, P_{x1}|\tau) = \frac{1}{2} A(\tau)[P'^2_{x1} + P^2_{x1}] + B(\tau) P'_{x1} P_{x1}, \tag{9.109}$$

where P_{x1} is at time t, and P'_{x1} is at time $t+\tau$. This is maximised with respect to P'_{x1} by

$$\overline{P'_{x1}} \equiv \overline{P_{x1}}(P_{x1}, \tau) = -A(\tau)^{-1} B(\tau) P_{x1}. \tag{9.110}$$

Using the short time expansions this is

$$\overline{P}'_{x1} = P_{x1} + \frac{|\tau|\Lambda}{2} S P_{x1} + \mathcal{O}(\tau^2). \tag{9.111}$$

With this and the reduction condition the second entropy may be rewritten

$$\begin{aligned} S^{(2)}(P'_{x1}, P_{x1}|\tau) &= \frac{1}{2} S P_{x1}^2 - \frac{1}{2\Lambda|\tau|} \left[P'_{x1} - P_{x1} - \frac{|\tau|\Lambda}{2} S P_{x1} \right]^2 + \mathcal{O}(\tau^2) \\ &= \frac{1}{2} S P_{x1}^2 - \frac{|\tau|}{2\Lambda} \left[\overset{\circ}{P}_{x1} - \frac{\hat{\tau}\Lambda}{2} S P_{x1} \right]^2 , \end{aligned} \tag{9.112}$$

which is valid for small time intervals. Here $\hat{\tau} \equiv \text{sign}(\tau)$.

The adiabatic rate of change of momentum flux is the volume times the viscous pressure tensor $\overset{\circ}{P}{}_{x1}^{0} \equiv [\overline{P}'_{x1} - P_{x1}]/\tau = V\Pi_{xz}$. Hence the most likely value of this component of the viscous pressure tensor is

$$\overline{\Pi}_{xz} = \frac{1}{V} \overline{\overset{\circ}{P}}{}_{x1}^{0} = \frac{\hat{\tau}\Lambda}{2V} S P_{x1} = \frac{-\hat{\tau}\Lambda}{2VT} v_{{\rm s}x1}. \tag{9.113}$$

In view of the linear constitutive relation above, which applies forward in time, one can identify the shear viscosity as

$$\eta = \frac{\Lambda}{2VT}. \tag{9.114}$$

The time correlation function is (see Eq. 2.22),

$$\begin{aligned}
Q(\tau) &= k_{\rm B}^{-1}\langle P_{x1}(t+\tau)P_{x1}(t)\rangle_0 \\
&= -k_{\rm B}^{-1}A(\tau)^{-1}B(\tau)\langle P_{x1}P_{x1}\rangle_0 \\
&\sim -\left[1+\frac{|\tau|\Lambda}{2}S\right]S^{-1}+\mathcal{O}(\tau^2) \\
&= -S^{-1}-|\tau|VT\eta. \qquad (9.115)
\end{aligned}$$

This does not prove that the correlations in the first momentum moment are Markovian, since it only holds on small time scales. Nevertheless, by analogy with the result for the first energy moment, one might guess that they are indeed Markovian and that the time correlation function decays exponentially on longer time scales. If one imagines that this truncated expansion formally holds at all times such that this defines a time dependent shear viscosity conductivity $\eta(\tau)$, then one can rewrite this as

$$\begin{aligned}
\eta(\tau) &= \frac{-1}{|\tau|VT}\left[Q(\tau)-Q(0)\right] \\
&= \frac{-1}{|\tau|Vk_{\rm B}T}\langle\left[P_{x1}(t+\tau)-P_{x1}(t)\right]P_{x1}(t)\rangle_0 \\
&= \frac{-1}{|\tau|Vk_{\rm B}T}\int_0^\tau {\rm d}t'\,\left\langle \dot{P}^0_{x1}(t'+t)P_{x1}(t)\right\rangle_0 \\
&= \frac{-1}{|\tau|Vk_{\rm B}T}\int_0^\tau {\rm d}t'\,\left\langle \dot{P}^0_{x1}(\mathbf{\Gamma}^0(t'|\mathbf{\Gamma},0))P_{x1}(\mathbf{\Gamma})\right\rangle_0. \qquad (9.116)
\end{aligned}$$

The right-hand side becomes independent of τ for τ in the short time regime. Hence taking τ over to the other side and differentiating with respect to τ gives an alternative expression,

$$\begin{aligned}
\eta(\tau) &= \frac{-\hat{\tau}}{Vk_{\rm B}T}\left\langle \dot{P}^0_{x1}(\mathbf{\Gamma}^0(\tau|\mathbf{\Gamma},0))P_{x1}(\mathbf{\Gamma})\right\rangle_0 \\
&= \frac{1}{2Vk_{\rm B}T}\int_{-\tau}^{\tau} {\rm d}t'\,\left\langle \dot{P}^0_{x1}(\mathbf{\Gamma})\dot{P}^0_{x1}(\mathbf{\Gamma}^0(t'|\mathbf{\Gamma},0))\right\rangle_0. \qquad (9.117)
\end{aligned}$$

(See §9.5 above for the steps leading to the final equality.) A molecular expression for the adiabatic flux is given next. These are two Green-Kubo expressions for the shear viscosity. They are equivalent in the small time regime, but the second as a function of τ might possibly converge faster to its asymptotic value.

These results are for an isolated system. The treatment for shear flow caused by reservoirs moving with velocities $v_{x\pm}$ is entirely analogous to that for heat flow, §9.1.3.

9.6.2 Phase Space Probability Density

The general expression given in Ch. 8 for the non-equilibrium phase space probability distribution will now be applied to shear flow. With p_{xi} the x-component of the momentum of the ith molecule, and z_i its z-coordinate, the zeroth and first momentum moments are

$$P_{x0}(\mathbf{\Gamma}) = \sum_{i=1}^{N} p_{xi}, \text{ and } P_{x1}(\mathbf{\Gamma}) = \sum_{i=1}^{N} z_i p_{xi}. \tag{9.118}$$

The zeroth moment is a conserved variable, and hence its adiabatic rate of change is zero. The adiabatic rate of change of the first momentum moment is

$$\begin{aligned}
\dot{P}^0_{x1}(\mathbf{\Gamma}) &= \dot{\mathbf{\Gamma}}^0 \cdot \nabla P_{x1}(\mathbf{\Gamma}) \\
&= \sum_{i=1}^{N} \left[\dot{\mathbf{p}}^0_i \cdot \frac{\partial P_{x1}(\mathbf{\Gamma})}{\partial \mathbf{p}_i} + \dot{\mathbf{q}}^0_i \cdot \frac{\partial P_{x1}(\mathbf{\Gamma})}{\partial \mathbf{q}_i} \right] \\
&= \sum_{i=1}^{N} \left[z_i \dot{p}^0_{xi} + \dot{q}^0_{zi} p_{xi} \right] \\
&= \sum_{i=1}^{N} \left[\frac{p_{zi} p_{xi}}{m} - z_i \sum_{j=1}^{N(j \neq i)} u'(q_{ij}) \frac{x_i - x_j}{q_{ij}} \right] \\
&= \sum_{i=1}^{N} \frac{p_{zi} p_{xi}}{m} - \sum_{i<j}^{N} u'(q_{ij}) \frac{[z_i - z_j][x_i - x_j]}{q_{ij}}.
\end{aligned} \tag{9.119}$$

The penultimate equality assumes that the potential energy consists solely of central pair terms. The first term is the contribution to the rate of momentum moment change due to molecular diffusion, and the second term is the contribution from intermolecular forces.

The phase space probability distribution is essentially the exponential of the reservoir entropy. As usual the reservoir entropy for the non-equilibrium system is formally $S_{\rm r}(\mathbf{\Gamma}, t) = S_{\rm r,st}(\mathbf{\Gamma}) + S_{\rm r,dyn}(\mathbf{\Gamma}, t)$. Choose the reservoir velocities to be equal and opposite, $v_{x+} = -v_{x-}$, so that $v_{x0} = 0$ and $v_{x1} = 2v_{x+}/L$. In this case the static part of the reservoir entropy is

$$\begin{aligned}
S_{\rm st}(\mathbf{\Gamma}) &= \frac{-\mathcal{H}(\mathbf{\Gamma})}{T} + \frac{v_{x1}}{T} P_{x1}(\mathbf{\Gamma}) \qquad (9.120) \\
&= \frac{-1}{T} \sum_{i\alpha} \frac{\left[p_{i\alpha} - m_i v_{x1} z_i \delta_{x\alpha}\right]^2}{2m_i} + \frac{v_{x1}^2}{2T} \sum_i m_i z_i^2 - \frac{1}{T} \sum_{i<j}^{N} u(q_{ij}).
\end{aligned}$$

The first term on the right-hand side of the first equality arises from the exchange of energy with the reservoirs, and the second term from the exchange of momentum. The first term on the right-hand side of the second equality is the peculiar kinetic energy, which is the kinetic energy in the frame of reference moving with the local flow. The second term is second-order in the shear rate and is usually negligible.

The dynamic part of the reservoir entropy in the present case is

$$\begin{aligned} S_{\rm dyn}(\mathbf{\Gamma},t) &\equiv -\int_0^t {\rm d}t'\, \dot{S}^0_{\rm r,st}(\overline{\mathbf{\Gamma}}(t'|\mathbf{\Gamma},t),t') \\ &= \frac{-v_{x1}}{T}\int_0^t {\rm d}t'\, \dot{P}^0_{x1}(\overline{\mathbf{\Gamma}}(t'|\mathbf{\Gamma},t)). \end{aligned} \tag{9.121}$$

The explicit time dependence here reflects the entropy produced in the reservoirs up to the present time due to the momentum transferred between them via the sub-system.

Because this is a steady state system, the most likely value of the momentum flux is independent of time, $\overline{\dot{P}^0_{x1}}$. Also, the most likely trajectory is invariant with respect to translations in time, $\overline{\mathbf{\Gamma}}(t'|\mathbf{\Gamma},t) = \overline{\mathbf{\Gamma}}(t'+\tau|\mathbf{\Gamma},t+\tau)$. Finally, in the long time limit, on the most likely trajectory the heat flux must tend to its most likely value whatever its current value, $\dot{P}^0_{x1}(\overline{\mathbf{\Gamma}}(t'|\mathbf{\Gamma},t)) \rightarrow \overline{\dot{P}^0_{x1}}$, $|t'-t| \rightarrow \infty$. These three facts mean that the partial time derivative of the dynamic part of the reservoir entropy is

$$\begin{aligned} &\frac{\partial S_{\rm dyn}(\mathbf{\Gamma},t)}{\partial t} \\ &= \frac{-v_{x1}}{T\Delta_t}\left\{\int_0^{t+\Delta_t} {\rm d}t'\, \dot{P}^0_{x1}(\overline{\mathbf{\Gamma}}(t'|\mathbf{\Gamma},t+\Delta_t)) - \int_0^t {\rm d}t'\, \dot{P}^0_{x1}(\overline{\mathbf{\Gamma}}(t'|\mathbf{\Gamma},t))\right\} \\ &= \frac{-v_{x1}}{T\Delta_t}\left\{\int_{-\Delta_t}^{t} {\rm d}t''\, \dot{P}^0_{x1}(\overline{\mathbf{\Gamma}}(t''|\mathbf{\Gamma},t)) - \int_0^t {\rm d}t'\, \dot{P}^0_{x1}(\overline{\mathbf{\Gamma}}(t'|\mathbf{\Gamma},t))\right\} \\ &= \frac{-v_{x1}\overline{\dot{P}^0_{x1}}}{T}. \end{aligned} \tag{9.122}$$

This is in agreement with the general steady state result, Eq. (8.161).

The phase space probability for the canonical non-equilibrium problem

$$\begin{aligned} \wp(\mathbf{\Gamma}|v_{x1},T,t) &= \frac{1}{Z(v_{x1},T,t)} e^{S_{\rm r}(\mathbf{\Gamma},t)/k_{\rm B}} \\ &= \frac{1}{Z(v_{x1},T,t)} e^{-\beta\mathcal{H}(\mathbf{\Gamma})+\beta v_{x1}P_{x1}(\mathbf{\Gamma})} e^{S_{\rm r,dyn}(\mathbf{\Gamma},t)/k_{\rm B}}, \end{aligned} \tag{9.123}$$

where $\beta \equiv 1/k_{\rm B}T$. The partition function normalises this to unity, $Z(t) = \int {\rm d}\mathbf{\Gamma}\, e^{S_{\rm r}(\mathbf{\Gamma},t)/k_{\rm B}}$. Since this is a steady state system one must have $\partial\wp(\mathbf{\Gamma},t)/\partial t = 0$, which implies

$$\frac{1}{Z(t)}\frac{{\rm d}Z(t)}{{\rm d}t} = \frac{\partial S_{\rm r,dyn}(\mathbf{\Gamma},t)}{k_{\rm B}\partial t} = -\beta v_{x1}\overline{\dot{P}^0_{x1}}. \tag{9.124}$$

Hence $Z(t) = \tilde{Z}\exp -\beta v_{x1}\overline{\dot{P}^0_{x1}}t$. The exponent could be added to the dynamic part of the reservoir entropy to make it independent of time and the integrand short-ranged,

$$\tilde{S}_{\rm dyn}(\mathbf{\Gamma}) \equiv \frac{-v_{x1}}{T}\int_{-\tau}^0 {\rm d}t'\, \left\{\dot{P}^0_{x1}(\overline{\mathbf{\Gamma}}(t'|\mathbf{\Gamma},0)) - \overline{\dot{P}^0_{x1}}\right\}. \tag{9.125}$$

The logarithm of the partition function times Boltzmann's constant is the total entropy of the total system (or, more precisely, the sub-system dependent part thereof), $S_{\mathrm{tot}}(v_{x1}, T, t) = k_{\mathrm{B}} \ln Z(v_{x1}, T, t)$. This is related to the maximum value of the total entropy,

$$\begin{aligned} S_{\mathrm{tot}}(v_{x1}, T, t) &> S_{\mathrm{tot}}(\overline{\mathcal{H}}, \overline{P}_{x1}|v_{x1}, T, t) \\ &\equiv S_{\mathrm{s}}(\overline{\mathcal{H}}, \overline{P}_{x1}) - \frac{\overline{\mathcal{H}}}{T} + \frac{v_{x1}\overline{P}^{(\mathrm{r})}_{x1}(t)}{T}. \end{aligned} \tag{9.126}$$

Here $P^{(\mathrm{r})}_{x1}(t) = -P_{\mathrm{r}x1}(t)$ is the total reservoir contribution to the first momentum moment of the sub-system to the present time. The thermodynamic potential or free energy for steady shear flow, which is the analogue of the Helmholtz free energy, is

$$\begin{aligned} F(v_{x1}, T, t) &= -TS_{\mathrm{tot}}(\overline{\mathcal{H}}, \overline{P}_{x1}|v_{x1}, T, t) \\ &\approx -TS_{\mathrm{tot}}(v_{x1}, T, t) \\ &= -k_{\mathrm{B}}T \ln Z(v_{x1}, T, t). \end{aligned} \tag{9.127}$$

The time derivative of the thermodynamic potential is proportional to the logarithmic derivative of the partition function, which is proportional to that of the total entropy. From Eq. (9.122) above one has

$$\frac{\partial S_{\mathrm{tot}}(v_{x1}, T, t)}{\partial t} = \frac{-v_{x1}\overline{\dot{P}^0_{x1}}}{T}, \text{ and } \frac{\partial F(v_{x1}, T, t)}{\partial t} = v_{x1}\overline{\dot{P}^0_{x1}}. \tag{9.128}$$

This implies $\mathrm{d}\overline{P}^{(\mathrm{r})}_{x1}(t)/\mathrm{d}t = -\overline{\dot{P}^0_{x1}}$, which is to say that most likely the rate of relaxation of the first momentum moment of the sub-system due to the reservoirs is equal and opposite to the rate of its change due to internal influences.

9.6.3 Most Likely Trajectory

The stochastic dissipative equations of motion for shear flow are identical to those given for steady heat flow, §9.3, with the exception of the formula for the most likely reservoir force. That is one has

$$\mathbf{\Gamma}_2 = \mathbf{\Gamma}_1 + t_{21}\dot{\mathbf{\Gamma}}^0(\mathbf{\Gamma}) + \mathbf{R}(\mathbf{\Gamma}, t_{21}) + \mathcal{O}(t_{21}^2). \tag{9.129}$$

The random force, which only has momentum components, is the sum of a stochastic, $\tilde{\mathbf{R}}$, and a dissipative, $\overline{\mathbf{R}}$, force. The stochastic forces have variance

$$\left\langle \tilde{R}_{i\alpha}\tilde{R}_{j\gamma} \right\rangle = \sigma\delta_{ij}\delta_{\alpha\gamma}. \tag{9.130}$$

The variance represents the stochastic influence of the momentum reservoirs, $\sigma \sim |t_{21}|$. The dissipative force is given by Eq. (8.94),

$$\begin{aligned}
\overline{R}_{i\alpha}(\mathbf{\Gamma}, t_{21}) &= \frac{\sigma}{2k_{\mathrm{B}}} \frac{\partial S_{\mathrm{r,st}}(\mathbf{\Gamma})}{\partial p_{i\alpha}} + \frac{\sigma}{2k_{\mathrm{B}}} (\hat{t}-1)\, \overline{S'_{\mathrm{p}i\alpha}} \\
&= \frac{-\sigma}{2k_{\mathrm{B}}} \left[\frac{1}{T} \frac{\partial \mathcal{H}(\mathbf{\Gamma})}{\partial p_{i\alpha}} - \frac{v_{x1}}{T} \frac{\partial P_{x1}(\mathbf{\Gamma})}{\partial p_{i\alpha}} \right] + \frac{\sigma}{2k_{\mathrm{B}}} (\hat{t}-1)\, \overline{\frac{\partial S_{\mathrm{r,st}}(\mathbf{\Gamma})}{\partial p_{i\alpha}}} \\
&= \frac{-\sigma}{2k_{\mathrm{B}}} \left[\frac{p_{i\alpha}}{T m_i} - \frac{v_{x1}}{T} z_i \delta_{\alpha x} \right] - \frac{\sigma}{2k_{\mathrm{B}}} (\hat{t}-1) \left[\frac{\overline{p}_{i\alpha}}{T m_i} - \frac{v_{x1}}{T} \overline{z}_i \delta_{\alpha x} \right] \\
&= \frac{-\sigma}{2m_i} \beta \left[p_{i\alpha} - m_i v_{x1} z_i \delta_{\alpha x} \right]. \qquad (9.131)
\end{aligned}$$

One can recognise the bracketed term as the peculiar momentum, which is on average zero. This represents a drag or friction force. It is of the same form as the Langevin dissipation, but for the local fluctuating velocity of the atom.

Constraints

As discussed in §8.3.6, one can impose constraints on the reservoir perturbation so as to minimise the influence on the physical properties of interest. In the case of shear flow, $\Pi_{xz} = \dot{P}^0_{x1}/V$, one one can make the dissipative force orthogonal to the flux gradient,

$$\overline{\mathbf{R}}(\mathbf{\Gamma}, t_{21}) \cdot \nabla_{\mathrm{p}} \dot{P}^0_{x1}(\mathbf{\Gamma}) = 0. \qquad (9.132)$$

Denoting the unconstrained or free force given above by $\overline{\mathbf{R}}^{\mathrm{f}}(\mathbf{\Gamma}, t_{21})$ and introducing a Lagrange multiplier, $\mu(\mathbf{\Gamma})$, the constrained most likely force becomes

$$\overline{\mathbf{R}}(\mathbf{\Gamma}, t_{21}) = \overline{\mathbf{R}}^{\mathrm{f}}(\mathbf{\Gamma}, t_{21}) + \mu(\mathbf{\Gamma}) \nabla_{\mathrm{p}} \dot{P}^0_{x1}(\mathbf{\Gamma}), \qquad (9.133)$$

with the Lagrange multiplier being

$$\mu(\mathbf{\Gamma}) = \frac{-\overline{\mathbf{R}}^{\mathrm{f}}(\mathbf{\Gamma}, t_{21}) \cdot \nabla_{\mathrm{p}} \dot{P}^0_{x1}(\mathbf{\Gamma})}{\nabla_{\mathrm{p}} \dot{P}^0_{x1}(\mathbf{\Gamma}) \cdot \nabla_{\mathrm{p}} \dot{P}^0_{x1}(\mathbf{\Gamma})}. \qquad (9.134)$$

Using the expression for the adiabatic rate of change of first momentum moment given above, one has explicitly

$$\frac{\partial \dot{P}^0_{x1}(\mathbf{\Gamma})}{\partial p_{i\alpha}} = \frac{p_{xi}}{m_i} \delta_{z\alpha} + \frac{p_{zi}}{m_i} \delta_{x\alpha}. \qquad (9.135)$$

9.6.4 Equipartition Theorem

It is an exercise to show that the local equipartition theorem holds for the peculiar momentum,

$$\left\langle [p_{i\alpha} - m_i v_{x1} z_i \delta_{\alpha x}]^2 \right\rangle_{\mathrm{ne}/i} = m_i k_{\mathrm{B}} T + \mathcal{O}(v_{x1}^2). \qquad (9.136)$$

Recall that the notation $\langle \ldots \rangle_{/i}$ means that the coordinate z_i is fixed and not averaged over. This result can also be seen directly from the expression given above for the static part of the reservoir entropy that involves the peculiar kinetic energy.

Chapter 10

Generalised Langevin Equation

Brownian motion is the archetypal non-equilibrium system in thermodynamics and statistical mechanics. It encompasses not only the original application of the movement of small particles in a fluid, either free or driven by external forces, but also the fluctuations of thermodynamic variables. What these systems have in common is that they can be divided into slow and fast coordinates. The slow or macroscopic coordinates are treated explicitly, while the fast or microscopic coordinates are projected out of the problem and their influence is included via probabilistic methods.

In Ch. 3 the three simplest types of Brownian motion were treated: a free Brownian particle, a Brownian particle in a fixed potential trap, and a Brownian particle in a steadily moving trap. These results all invoked the diffusion constant (equivalently, the drag coefficient), which determines that the mean square displacement of a free particle grows linearly with time, and that the most likely velocity of a driven particle is linearly proportional to the constant applied force. The results were formally exact in the case of an external potential in uniform motion. For more general external potentials with arbitrary time and space dependence, two related exact methods were given, namely the non-equilibrium probability distribution for the particle, and the stochastic, dissipative equations of motion that characterise its trajectory. These were derived more formally from the principles of non-equilibrium statistical mechanics in Ch. 8.

In the more general case of an accelerating trap or a trap in arbitrary motion, the expressions that were obtained for the most likely position and velocity of the particle in uniform motion no longer hold exactly. One has now to account for not only the inertia of the particle, which causes a lag compared to the uniform motion case, but also the more challenging phenomenon that reflects the memory of the solvent for the past motion of the particle, which modifies the diffusion or drag constant that is to be applied.

Such memory effects signify non-Markov behaviour. The Einstein result obtained in Ch. 3 for the diffusion of the free Brownian particle, namely that the square of the displacement grows linearly with time, is predicated on Markovian behaviour, which indeed was explicitly invoked in §3.6.3 in the derivation of the most likely position and velocity. A Markov system has no memory beyond the immediately preceding state, and it becomes the limiting case if successive states are separated by a large time interval. However, the decomposition of a trajectory into large time steps is only valid when the motion varies slowly in time and in space. For example, in obtaining the Markovian exponential decay in Eq. (3.77) from the coarse velocity in Eq. (3.76), it can be seen that the second derivative of the displacement is neglected. For uniform motion of a parabolic trap one can make the time step arbitrarily large and still satisfy this condition.

The point is worth repeating: the restriction of the earlier analysis is that the diffusion constant necessitates time scales that are long compared to the molecular time scales that characterise the fast motion of the hidden variables. Accordingly, the analysis breaks down when the competing requirements—a small time interval so that the variation of the potential is negligible and a large time interval so that memory effects are negligible—can not be satisfied simultaneously. In such cases one has to go beyond the Markov approximation implicit in the diffusion constant.

Of course the full problem has already been formally solved in Ch. 8, where the probability distribution and the stochastic, dissipative equations of motion were given for phase space for an arbitrary mechanical or thermodynamic non-equilibrium system. In this formulation the Brownian particle is treated alongside the solvent as just one of the components of phase space. However, whilst this analysis has the virtue of being formally exact, it does not seek to exploit the separation in time scales that characterise Brownian motion, namely retaining explicitly only the slow coordinates (the particle) and projecting out the fast coordinates (the solvent). The twin disadvantages of the formally exact approach are that one is restricted to time steps that are small on molecular scales, which means very, very small on macroscopic scales, and one needs to track the coordinates in phase space of hundreds or thousands of molecules during their interaction and evolution, even though only the coordinates of the Brownian particle or thermodynamic variables are of direct interest.

A perturbation approach to the problem is the primary focus of this chapter. The equations of motion are formulated for the Brownian particle in the presence of an arbitrary external potential. This requires the time correlation function of the 'bare' Brownian particle, which is the free particle in the absence of any external potential. This is used as a response function to the external potential of interest. It will be seen that the influence of prior states progressively weakens as they are further removed in time.

The perturbation theory is used in §10.3.4 to generalise the Langevin equation to include memory effects. That is, the stochastic dissipative equations of motion are given that include the influence of prior states on the current deterministic and stochastic forces. The correlation between stochastic forces and its relation with the dissipative forces is a version of the fluctuation dissipation theorem for memory effects. It will be shown that the memory function is just the velocity autocorrelation function of the bare Brownian particle. In §10.4 linear response theory is presented for an external force acting on the Brownian particle in an adiabatic system that is initially in thermal equilibrium, and it is shown that the susceptibility that emerges is the same as that obtained in the non-equilibrium perturbation theory. Numerical tests of the perturbation theory for driven Brownian motion are given in §10.5.

The perturbation theory is generalised to mixed parity thermodynamic variables in §10.6. In this case two additional effects not present for pure parity Brownian motion occur, namely that the memory function has an additional term that represents indirect correlations between the velocity and the position of the trajectory, and also that there is now instantaneous coupling between position and velocity. An explicit formula for the memory function is given in terms of the time correlation functions of the unperturbed system. In §10.7, a summary of the projector operator formalism is given, and it is shown that the generalised Langevin equation that emerges from that approach is formally identical to the one given by the perturbation theory. In §10.8, Brownian motion in a harmonic oscillator model of the solvent (the Caldeira-Leggett model) is analysed and an analytic expression for the memory function is obtained.

10.1 Free Brownian Particle

Let x denote the position of a Brownian particle, let $\dot{x}$ be its velocity. In the simplest case only one-dimensional motion is considered, and x is a scalar of dimension $N = 1$. More generally the following analysis can be applied to three-dimensional motion, and x can be a vector of dimension $N = 3$. Even more generally, x can be considered to represent the positions of many Brownian particles or the values of a set of multiple thermodynamic coordinates. In these more general cases the multiplications below have to be considered as matrix multiplications, and the elements of the various time correlation and fluctuation matrices are themselves $N \times N$ matrices.

There is a distinction between the pure parity case, where all the components have the same parity, and the mixed parity case, where each component of x is either pure odd or pure even, but both cases occur. For clarity most of this section will be restricted to the pure parity case, and will be discussed in terms of the position of a single Brownian particle. Some of the present analysis can be used in the mixed parity case, as is briefly summarised in §10.1.2. The general mixed parity analysis will be given in §10.6.

A free particle is equally likely to be found anywhere within the volume, and so the first entropy is a constant with respect to position that can be set equal to zero, $S_0^{(1)}(x) = 0$. The kinetic energy of the Brownian particle is $m\dot{x}^2/2$, where m is its mass. Hence the first entropy of the free particle is

$$S_0^{(1)}(X) = \frac{-m}{2T}\dot{x}^2, \tag{10.1}$$

where $X \equiv \{x, \dot{x}\}$, and T is the temperature of the solvent, which acts as a thermal reservoir.

10.1.1 Time Correlation Functions

The position self correlation function (also called the position autocorrelation function) can be defined as[1]

$$q_x(\tau) \equiv k_B^{-1} \langle x(t+\tau)x(t) \rangle_0 . \tag{10.2}$$

This diverges at $\tau = 0$, $q_x(0) \equiv k_B^{-1} \langle x(t)x(t) \rangle_0 = \infty$. More precisely, it scales with the square of the size of the system, which diverges in the thermodynamic limit. In the general case that x is a vector, then this is an $N \times N$ dyadic matrix. In the present pure parity case, this is an even, symmetric matrix, $q_x(\tau) = q_x(-\tau) = q_x(\tau)^{\mathrm{T}}$. The fact that it is even follows because both elements of the dyadic matrix have the same time parity, and so both change sign in the same way when the velocity is reversed. This fact and time homogeneity imply that the matrix is symmetric, $\langle x(t+\tau)x(t) \rangle_0 = \langle x(t-\tau)x(t) \rangle_0 = \langle x(t)x(t+\tau) \rangle_0$, the second equality following upon replacing t by $t+\tau$.

Because of this divergence of $\langle x^2 \rangle_0$, it is more convenient to work with displacements, or differences in position. Accordingly one can define

$$\begin{aligned} q_0^{\mathrm{s}}(\tau) &\equiv \frac{1}{2k_B} \langle [x(t+\tau) - x(t)][x(t+\tau) - x(t)] \rangle_0 \\ &= [q_x(0) - q_x(\tau)]. \end{aligned} \tag{10.3}$$

This is an even function of the time interval τ, $q_0^{\mathrm{s}}(\tau) = q_0^{\mathrm{s}}(-\tau)$, and a symmetric matrix, $q_0^{\mathrm{s}}(\tau) = q_0^{\mathrm{s}}(\tau)^{\mathrm{T}}$. The superscript s here and below signifies the symmetrised correlation function, which distinguishes the present analysis from the more general analysis in §10.6.

The position autocorrelation function at large times grows linearly in time, as found by Einstein.[2] The result defines the diffusion tensor D,

$$\lim_{\tau \to \infty} q_0^{\mathrm{s}}(\tau) = |\tau| k_B^{-1} D. \tag{10.4}$$

In general, because $q_0^{\mathrm{s}}(\tau) = q_0^{\mathrm{s}}(\tau)^{\mathrm{T}}$, the diffusion matrix must be a symmetric matrix. For a single Brownian particle in three-dimensional space the diffusion tensor is generally a constant times the identity matrix.

Taking the time derivative, one obtains the symmetrised position velocity correlation function,

$$\begin{aligned} \dot{q}_0^{\mathrm{s}}(\tau) &= \frac{-1}{2} \left[\dot{q}_x(\tau) + \dot{q}_x(\tau)^{\mathrm{T}} \right] \\ &= \frac{-1}{2k_B} \left[\langle \dot{x}(t+\tau)x(t) \rangle_0 + \langle x(t)\dot{x}(t+\tau) \rangle_0 \right]. \end{aligned} \tag{10.5}$$

[1] The reader should be aware that time correlation functions are conventionally defined without Boltzmann's constant. Also, time correlation functions generally refer to the correlation of the departures from the average value, $\langle [x(t+\tau) - \langle x \rangle_0] \, [x(t) - \langle x \rangle_0] \rangle_0$. In the present analysis the average is zero, $\langle x \rangle_0 = 0$.

[2] Einstein, A. (1905), Ann. Phys. **17**, 549.

This is a symmetric matrix and an odd function of τ (since $q_0^{\rm s}(\tau)$ is an even function), which implies that it must vanish at $\tau = 0$. This may be seen directly by differentiating the first equality in the definition of $q_0^{\rm s}(\tau)$,

$$\begin{aligned}\dot{q}_0^{\rm s}(\tau) &= \frac{1}{2k_{\rm B}} \left[\left\langle \dot{x}(t+\tau) \left\{ x(t+\tau) - x(t) \right\} \right\rangle_0 \right. \\ &\quad \left. + \left\langle \left\{ x(t+\tau) - x(t) \right\} \dot{x}(t+\tau) \right\rangle_0 \right]. \end{aligned} \tag{10.6}$$

The two terms in braces clearly vanish at $\tau = 0$. In the pure parity case there is no instantaneous coupling of x and $\dot{x}$, and so the individual terms in Eq. (10.5) vanish, $\langle \dot{x}(t)x(t)\rangle_0 = 0$. In the opposite limit one has $\lim_{\tau\to\infty} \dot{q}_0^{\rm s}(\tau) = \hat{\tau} k_{\rm B}^{-1} D$, where $\hat{\tau} \equiv \text{sign}(\tau)$.

Taking a further derivative with respect to τ gives the symmetrised position acceleration correlation function,

$$\begin{aligned}\ddot{q}_0^{\rm s}(\tau) &= \frac{-1}{2} \left[\ddot{q}_x(\tau) + \ddot{q}_x(\tau)^{\rm T} \right] \\ &= \frac{-1}{2k_{\rm B}} \left[\langle \ddot{x}(t+\tau)x(t)\rangle_0 + \langle x(t)\ddot{x}(t+\tau)\rangle_0 \right]. \end{aligned} \tag{10.7}$$

This is in fact equal to the velocity self correlation (or autocorrelation) function, as may be seen from the fact that $\dot{q}_0^{\rm s}(\tau)$ is a symmetric matrix and an odd function of τ,

$$\begin{aligned}\ddot{q}_0^{\rm s}(\tau) &= \frac{-1}{2k_{\rm B}} \frac{\mathrm{d}}{\mathrm{d}\tau} \left[\langle \dot{x}(t+\tau)x(t)\rangle_0 + \langle x(t)\dot{x}(t+\tau)\rangle_0 \right] \\ &= \frac{1}{2k_{\rm B}} \frac{\mathrm{d}}{\mathrm{d}\tau} \left[\langle \dot{x}(t-\tau)x(t)\rangle_0 + \langle x(t)\dot{x}(t-\tau)\rangle_0 \right] \\ &= \frac{1}{2k_{\rm B}} \frac{\mathrm{d}}{\mathrm{d}\tau} \left[\langle \dot{x}(t)x(t+\tau)\rangle_0 + \langle x(t+\tau)\dot{x}(t)\rangle_0 \right] \\ &= \frac{1}{2k_{\rm B}} \left[\langle \dot{x}(t)\dot{x}(t+\tau)\rangle_0 + \langle \dot{x}(t+\tau)\dot{x}(t)\rangle_0 \right] \\ &= \frac{1}{k_{\rm B}} \langle \dot{x}(t+\tau)\dot{x}(t)\rangle_0 . \end{aligned} \tag{10.8}$$

This is evidently a symmetric matrix, and it is an even function of τ, since it consists of the product of two factors with the same parity. This means that the two terms on the right-hand side are equal, $\langle \dot{x}(t+\tau)\dot{x}(t)\rangle_0 = \langle \dot{x}(t-\tau)\dot{x}(t)\rangle_0 = \langle \dot{x}(t)\dot{x}(t+\tau)\rangle_0$, the second equality following upon using time homogeneity to replace t by $t+\tau$; this accounts for the final equality. This means that in the present pure parity case, the symmetrised velocity autocorrelation function is the same as the ordinary velocity autocorrelation function, $\ddot{q}_0^{\rm s}(\tau) = \ddot{q}_0(\tau)$. The fact that $\ddot{q}_0^{\rm s}(0) > 0$ implies that $q_0^{\rm s}(0)$ is a local minimum. In fact it is a global minimum because $q_0^{\rm s}(0) = 0$ and $q_0^{\rm s}(\tau) \geq 0$.

At large times the initial and final velocities must become uncorrelated. Hence the velocity autocorrelation function must go to zero,

$$\ddot{q}_0^{\rm s}(\tau) \to \frac{1}{k_{\rm B}} \langle \dot{x}(t+\tau)\rangle_0 \langle \dot{x}(t)\rangle_0 = 0. \tag{10.9}$$

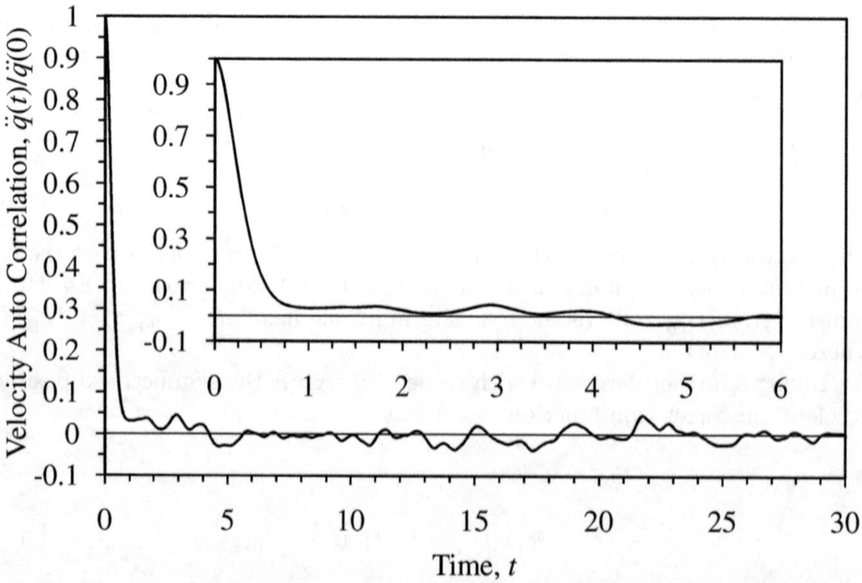

Figure 10.1: Relative velocity autocorrelation function for a free Brownian particle in a canonical equilibrium system. The horizontal line at 0 is a guide to the eye. The inset is a magnification of the short time data. Data from Attard and Gray-Weale, (2008).

The velocity is typical of thermodynamic fluctuations about zero, and this long time decorrelation can be expected quite generally.

Conversely, from general considerations of the nature of thermodynamic fluctuations, one expects that the velocity autocorrelation function is maximal at $\tau = 0$. (The discussion in this paragraph refers to the scalar case, $N = 1$.) The argument for this is as follows. It is certainly the case that since variables of opposite time parity are instantaneously uncorrelated, $\dddot{q}^{\mathrm{s}}(0) \equiv k_{\mathrm{B}}^{-1} \langle \ddot{x}(0)\dot{x}(0)\rangle_0 = 0$, which implies that $\ddot{q}_0^{\mathrm{s}}(\tau)$ must be an extremum at $\tau = 0$. The velocity autocorrelation function is positive at $\tau = 0$, and it must decay to zero at large times. It must be a local maximum, because non-zero values of the velocity are unlikely, and in the future the system must most likely be returning to zero (just as it most likely came from zero in the past). The decay from $\ddot{q}^{\mathrm{s}}(0)$ to the long time uncorrelated state must be either monotonic, or else damped oscillatory; in both cases one concludes that the maximum in the velocity autocorrelation function at $\tau = 0$ is a global maximum.

Figure 10.1 shows the velocity autocorrelation function for a free Brownian particle in a soft-sphere fluid.[3] The function is normalised by its zero time value,

[3] Attard, P. and Gray-Weale, A. (2008), J. Chem. Phys. **128**, 114509. The soft-sphere interaction potential was $u_{\alpha\gamma}(r) = [\sigma_{\alpha\gamma}/r]^{12}$. The solvent had a mass $m_1 = 1$, a length scale $\sigma_{11} = 0.5$, and a potential cut-off of $R_{11} = 1.7$. The solute had a mass $m_0 = 10$, it interacted with the solvent via $\sigma_{10} = 2.25$, and a potential cut-off of $R_{10} = 5.2$. The temperature was

which in this case is $\ddot{q}^{\rm s}(0) = T/m = 0.3$. The velocity autocorrelation function clearly has a maximum at $t = 0$, and decays fairly rapidly until $t \approx 1$, after which it has a rather slowly decaying tail. The negative values on the interval $0.45 \lesssim t \lesssim 5.5$ appear to be statistically significant, with $\ddot{q}^{\rm s}(5)/\ddot{q}^{\rm s}(0) = -0.03$, with the standard error on the mean being 0.01. This appears to be due to the general phenomenon of backscattering, whereby at high densities collisions with neighbouring molecules tend to reverse the velocity, which leads to negative regions in velocity autocorrelation function, such as that apparent in the figure.[4]

10.1.2 Mixed Parity Digression

Many of the results given in the preceding section also hold for the mixed parity case with minor modifications. The general mixed parity analysis will be given in §10.6, but here a brief detour is undertaken to summarise the above results. The difference between these results and the results to be given in §10.6 is that the present results are restricted to symmetrised time correlation matrices.

The parity matrix ε is diagonal with elements equal to ± 1, signifying the parity of the corresponding component of x. Under velocity reversal of the underlying phase space, $x \Rightarrow \varepsilon x$.

The position autocorrelation function, $q_x(\tau) \equiv k_{\rm B}^{-1} \langle x(t+\tau)x(t)\rangle_0$, defined in Eq. (10.2), but here considered as of mixed parity, has the general symmetry,

$$q_x(\tau) = q_x(-\tau)^{\rm T}, \tag{10.10}$$

as follows from time homogeneity, $t \Rightarrow t-\tau$, $\langle x(t+\tau)x(t)\rangle_0 = \langle x(t)x(t-\tau)\rangle_0$. This implies that the matrix is symmetric at $\tau = 0$, $q_x(0)^{\rm T} = q_x(0)$. Invoking the parity matrix and velocity reversal one must also have

$$q_x(-\tau) = \varepsilon q_x(\tau)\varepsilon. \tag{10.11}$$

Hence one also has $q_x(\tau)^{\rm T} = \varepsilon q_x(\tau)\varepsilon$.

The autocorrelation function for displacements in this mixed parity case is

$$\begin{aligned} q_0^{\rm s}(\tau) &\equiv \frac{1}{2k_{\rm B}} \langle [x(t+\tau)-x(t)][x(t+\tau)-x(t)]\rangle_0 \\ &= \frac{1}{2}[2q_x(0) - q_x(\tau) - q_x(\tau)^{\rm T}]. \end{aligned} \tag{10.12}$$

This is evidently a symmetric matrix, $q_0^{\rm s}(\tau) = q_0^{\rm s}(\tau)^{\rm T}$, and, since $q_x(-\tau) = q_x(\tau)^{\rm T}$, it is an even function of the time interval τ. Also, $q_0^{\rm s}(0) = 0$.

$k_{\rm B}T = 3$, which was maintained by a Nosé-Hoover chain thermostat. A bulk system with periodic boundary conditions was simulated, with a cube length of 16.48, and 3095 solvent atoms, giving a solvent density of 0.69. The unit of time corresponds to about 1 picosecond for a realistic liquid.

[4] Rahman, A. (1964), Phys. Rev. **136**, A405. Levesque, D. and Verlet, L. (1970), Phys. Rev. A **2**, 2514. Hansen, J.-P. and McDonald, I. R. (1986), *Theory of Simple Liquids*, (Academic Press, London).

Because this symmetrised function is an even function of time, in the mixed parity case one still has the Einstein result,

$$\lim_{\tau\to\infty} q_0^{\rm s}(\tau) = |\tau| k_{\rm B}^{-1} D. \tag{10.13}$$

Even in this case the diffusion matrix is a symmetric matrix.

Taking the time derivative, one obtains the symmetrised position velocity correlation function,

$$\begin{aligned}\dot q_0^{\rm s}(\tau) &= \frac{-1}{2}\left[\dot q_x(\tau) + \dot q_x(\tau)^{\rm T}\right] \\ &= \frac{-1}{2k_{\rm B}}\left[\langle \dot x(t+\tau) x(t)\rangle_0 + \langle x(t)\dot x(t+\tau)\rangle_0\right].\end{aligned} \tag{10.14}$$

This is a symmetric matrix and an odd function of τ (since $q_0^{\rm s}(\tau)$ is an even function), which implies that $\dot q_0^{\rm s}(0) = 0$.

This result, $\dot q_0^{\rm s}(0) = 0$, is a consequence of the symmetrised formulation and may be seen directly from the result

$$\langle \dot x(t+\tau) x(t)\rangle_0 = -\langle x(t+\tau)\dot x(t)\rangle_0 . \tag{10.15}$$

This 'dot transfer' is an example of a more general result that can be seen as follows. Because of time homogeneity, the time derivative of any equilibrium average must vanish. Hence it follows that

$$\frac{\rm d}{{\rm d}t}\Big\langle a(t'+t) b(t)\Big\rangle_0 = \Big\langle \dot a(t'+t) b(t)\Big\rangle_0 + \Big\langle a(t'+t)\dot b(t)\Big\rangle_0 = 0. \tag{10.16}$$

Hence one can conclude that the two terms in Eq. (10.14) cancel each other in the limit $\tau = 0$,

$$\langle \dot x(t) x(t)\rangle_0 = -\langle x(t)\dot x(t)\rangle_0 . \tag{10.17}$$

Generally in the mixed parity case neither side of this equation is zero, because the time derivative of components on one parity instantaneously couple to components of the opposite parity. In the pure parity case there is no instantaneous coupling of x and $\dot x$, and so the individual terms vanish, $\langle \dot x(t) x(t)\rangle_0 = 0$.

In the opposite limit in the general mixed parity case one has $\lim_{\tau\to\infty} \dot q_0^{\rm s}(\tau) = \hat\tau k_{\rm B}^{-1} D$, where $\hat\tau \equiv {\rm sign}(\tau)$. Finally, since $\langle x(t-\tau) x(t)\rangle_0 = \varepsilon \langle x(t+\tau) x(t)\rangle_0 \varepsilon$, then $\langle \dot x(t-\tau) x(t)\rangle_0 = -\varepsilon \langle \dot x(t+\tau) x(t)\rangle_0 \varepsilon$, and it follows that

$$\dot q_0^{\rm s}(-\tau) = -\varepsilon \dot q_0^{\rm s}(\tau)\varepsilon. \tag{10.18}$$

Taking a further derivative with respect to τ gives the symmetrised position acceleration correlation function,

$$\begin{aligned}\ddot q_0^{\rm s}(\tau) &= \frac{-1}{2}\left[\ddot q_x(\tau) + \ddot q_x(\tau)^{\rm T}\right] \\ &= \frac{-1}{2k_{\rm B}}\left[\langle \ddot x(t+\tau) x(t)\rangle_0 + \langle x(t)\ddot x(t+\tau)\rangle_0\right].\end{aligned} \tag{10.19}$$

This is in fact equal to the symmetrised velocity self correlation (or autocorrelation) function, since, as mentioned above, transferring a 'dot' from one term to the other gives a minus sign,

$$\ddot{q}_0^{\mathrm{s}}(\tau) = \frac{1}{2k_{\mathrm{B}}} \left[\langle \dot{x}(t+\tau)\dot{x}(t)\rangle_0 + \langle \dot{x}(t)\dot{x}(t+\tau)\rangle_0 \right]. \tag{10.20}$$

This is clearly an even function of τ, as it must be. The matrices in the brackets are not individually symmetric, but their sum evidently is. As in the pure parity case, at large times the initial and final velocities must become uncorrelated and this goes to zero.

10.1.3 Diffusion Constant

In practice, a useful way of obtaining the diffusion constant is as follows. One starts with an expression for the displacement as an integral of the velocity,

$$x(t) - x(0) = \int_0^t \mathrm{d}t' \, \dot{x}(t'). \tag{10.21}$$

Forming the dyadic of this with itself and taking the free particle average gives the symmetrised position self correlation function,

$$\begin{aligned} q_0^{\mathrm{s}}(t) &= \frac{1}{2k_{\mathrm{B}}} \langle [x(t)-x(0)][x(t)-x(0)] \rangle_0 \\ &= \frac{1}{2k_{\mathrm{B}}} \int_0^t \mathrm{d}t' \int_0^t \mathrm{d}t'' \, \langle \dot{x}(t'')\dot{x}(t') \rangle_0 \\ &= \frac{1}{2} \int_0^t \mathrm{d}t' \int_0^t \mathrm{d}t'' \, \ddot{q}_0^{\mathrm{s}}(t''-t') \\ &= \int_0^t \mathrm{d}t' \int_0^{t'} \mathrm{d}t''' \, \ddot{q}_0^{\mathrm{s}}(t'''), \quad t''' = t' - t'' \\ &= \int_0^t \mathrm{d}t''' \int_0^t \mathrm{d}t' \, \ddot{q}_0^{\mathrm{s}}(t''') \\ &= \int_0^t \mathrm{d}t''' \, \ddot{q}_0^{\mathrm{s}}(t''')[t - t''']. \end{aligned} \tag{10.22}$$

Since the integral is over a square, the integrand in the second equality can be written in symmetrised form, which leads to the third equality. In view of the Einstein limit, Eq. (10.4), one can define a diffusion function,

$$D(t) \equiv \frac{k_{\mathrm{B}}}{|t|} q_0^{\mathrm{s}}(t) = k_{\mathrm{B}} \int_0^t \mathrm{d}t' \, \ddot{q}_0^{\mathrm{s}}(t') \left[\hat{t} - \frac{t'}{|t|} \right], \tag{10.23}$$

where $\hat{t} \equiv \mathrm{sign}(t)$. This goes to the diffusion constant for large times, $D(t) \to D$, $t \to \infty$. Since the velocity self correlation function goes to zero for large times, the second term in the integrand makes a decreasing contribution to the diffusion

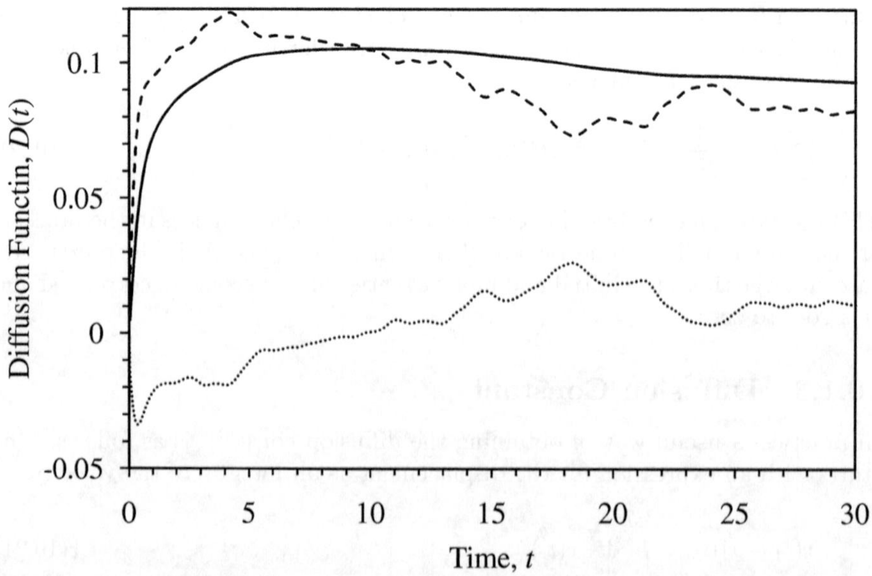

Figure 10.2: Diffusion function of a free Brownian particle in a soft sphere solvent. The dashed and dotted curves are the running integral of the first and second terms, respectively, on the right-hand side of Eq. (10.23), and the solid curve is their sum, which equals the left-hand side of that equation. Data from Attard and Gray-Weale, (2008).

constant as t increases. This is an example of a Green-Kubo formula, whereby a transport coefficient is related to a time correlation function.

The diffusion function for a soft sphere Brownian particle and fluid is plotted in Fig. 10.2. This corresponds to the velocity autocorrelation function shown in Fig. 10.1. This has a broad plateau with a maximum of $D(t) = 0.105$ at $t \approx 10$. It is clear that the left-hand side of Eq. (10.23) is a smoother function of time than are the individual terms on the right-hand side, which argues for the necessity of retaining both terms.

The peak of the diffusion function is generally taken to be 'the' diffusion constant. There is some evidence that the slow decay beyond the peak is dependent upon the periodic boundary conditions and the finite size of the simulation cell.

10.1.4 Trajectory Entropy and Correlation

A trajectory may be discretised with time step τ, and nodes $t_j = j\tau$. The velocities on an n-step trajectory may be specified as $\dot{\mathbf{x}}^{(n)} = \{\dot{x}_1, \dot{x}_2, \ldots, \dot{x}_n\}$. By convention, all trajectory vectors and matrices are ordered with time increasing for positive τ. For a small enough time step, successive positions are given by $x_j = x_{j-1} + \tau\dot{x}_{j-1}$, which will be assumed throughout. Hence in order to fully specify a trajectory one only has to give one position, say the terminal one x_n, plus the n velocities, $\mathbf{X}^{(n)} \equiv \{x_n, \dot{\mathbf{x}}^{(n)}\}$.

The free particle trajectory entropy is a quadratic form,[5]

$$\begin{aligned} S_0^{(n)}(\mathbf{X}^{(n)};\tau) &= \frac{1}{2} S_0^{(n)}(\tau) : \mathbf{X}^{(n)}\mathbf{X}^{(n)} \qquad (10.24)\\ &= \frac{1}{2} S_{0;xx}^{(n)}(\tau) : x_n^2 + S_{0;x\dot{x}}^{(n)}(\tau) : \dot{\mathbf{x}}^{(n)} x_n + \frac{1}{2} S_{0;\dot{x}\dot{x}}^{(n)}(\tau) : \dot{\mathbf{x}}^{(n)}\dot{\mathbf{x}}^{(n)}. \end{aligned}$$

The total fluctuation matrix $S_0^{(n)}$, which has dimension $(n+1)N \times (n+1)N$, is symmetric and consists of four sub-matrices: $S_{0;xx}^{(n)} = [S_{0;xx}^{(n)}]^{\mathrm{T}}$, which is $N \times N$, $S_{0;\dot{x}\dot{x}}^{(n)} = [S_{0;\dot{x}\dot{x}}^{(n)}]^{\mathrm{T}}$, which is $nN \times nN$, $S_{0;x\dot{x}}^{(n)}$, which is $N \times nN$, and $S_{0;\dot{x}x}^{(n)}(\tau) = S_{0;x\dot{x}}^{(n)}(\tau)^{\mathrm{T}}$.

The correlation matrix is

$$Q_0^{(n)}(\tau) \equiv k_{\mathrm{B}}^{-1} \left\langle \mathbf{X}^{(n)}\mathbf{X}^{(n)} \right\rangle_0, \qquad (10.25)$$

and it likewise consists of four sub-matrices:

$$Q_{0;xx}^{(n)} \equiv k_{\mathrm{B}}^{-1} \langle x(t)x(t)\rangle_0, \qquad (10.26)$$

$$Q_{0;\dot{x}\dot{x}}^{(n)}(\tau) \equiv k_{\mathrm{B}}^{-1} \left\langle \dot{\mathbf{x}}^{(n)}\dot{\mathbf{x}}^{(n)} \right\rangle_0 = \ddot{Q}_0^{(n)}(\tau), \qquad (10.27)$$

$$Q_{0;x\dot{x}}^{(n)}(\tau) \equiv k_{\mathrm{B}}^{-1} \left\langle x_n \dot{\mathbf{x}}^{(n)} \right\rangle_0, \qquad (10.28)$$

and $Q_{0;\dot{x}x}^{(n)}(\tau) = Q_{0;x\dot{x}}^{(n)}(\tau)^{\mathrm{T}}$. For the case of the free Brownian particle, $Q_{0;xx}^{(n)}$ diverges, and this is invoked below.

The elements of the velocity autocorrelation matrix are

$$\ddot{Q}_{0;ij}^{(n)}(\tau) = k_{\mathrm{B}}^{-1} \langle \dot{x}(t+i\tau)\dot{x}(t+j\tau)\rangle_0 = \ddot{q}_0((i-j)\tau). \qquad (10.29)$$

This is an $N \times N$ matrix. In the pure parity case, it is a symmetric matrix, $\ddot{q}_0(t) = \ddot{q}_0(t)^{\mathrm{T}}$, and an even function of time, $\ddot{q}_0(t) = \ddot{q}_0(-t)$. (In the mixed parity case $\ddot{Q}_0^{(n)}$ is itself symmetric: $\ddot{Q}_{0;i\alpha,j\gamma}^{(n)} = \ddot{Q}_{0;j\gamma,i\alpha}^{(n)}$, where Roman letters range over the nodes and Greek letters range over the components, but $\ddot{q}_0(t)$ is not.) The fact that in the pure parity case $\ddot{q}_0(t)$ is an even function of time means that $\ddot{Q}_0^{(n)}(\tau) = \ddot{Q}_0^{(n)}(-\tau)$, which in turn means that $\ddot{Q}_0^{(n)}(\tau)$ is a persymmetric matrix, $\ddot{Q}_{0;i,j}^{(n)}(\tau) = \ddot{Q}_{0;n-i+1,n-j+1}^{(n)}(\tau)$. One can conclude that in the present pure parity case the correlation matrix is a persymmetric Toeplitz matrix. This simplifies considerably its numerical inversion.

The fluctuation matrix is the negative inverse of the correlation matrix $Q_0^{(n)}(\tau)\, S_0^{(n)}(\tau) = -\mathrm{I}^{(n+1)}$. It is straightforward to show that the block matrix equation

$$\begin{pmatrix} A & B \\ B^{\mathrm{T}} & C \end{pmatrix} \begin{pmatrix} D & E^{\mathrm{T}} \\ E & F \end{pmatrix} = \mathrm{I}, \qquad (10.30)$$

[5]The superscript s ought to be appended to the trajectory matrices that appear here in order to signify that they are based upon the symmetrised time correlation function. It has been omitted in order to simplify the notation, with the hope that no confusion will arise.

with A, C, D and F symmetric matrices, has solution

$$\begin{aligned} D &= [I - A^{-1}BC^{-1}B^{\mathrm{T}}]^{-1}A^{-1} \\ F &= [I - C^{-1}B^{\mathrm{T}}A^{-1}B]^{-1}C^{-1} \\ E &= -[I - C^{-1}B^{\mathrm{T}}A^{-1}B]^{-1}C^{-1}B^{\mathrm{T}}A^{-1}. \end{aligned} \tag{10.31}$$

In the present case, one has $A = Q^{(n)}_{xx}$, $B = Q^{(n)}_{x\dot{x}}$ $C = Q^{(n)}_{\dot{x}\dot{x}}$, $D = -S^{(n)}_{xx}$, $E = -S^{(n)}_{\dot{x}x}$, and $F = -S^{(n)}_{\dot{x}\dot{x}}$. The minus sign arises because it is the negative inverse in the present case. Hence in the present case the fluctuation sub-matrices are given by

$$\begin{aligned} S^{(n)}_{xx} &= -\left[1 - [Q^{(n)}_{xx}]^{-1}Q^{(n)}_{x\dot{x}}[Q^{(n)}_{\dot{x}\dot{x}}]^{-1}[Q^{(n)}_{x\dot{x}}]^{\mathrm{T}}\right]^{-1}[Q^{(n)}_{xx}]^{-1} \\ &= -[Q^{(n)}_{xx}]^{-1},\ Q^{(n)}_{xx} \to \infty, \end{aligned} \tag{10.32}$$

$$\begin{aligned} S^{(n)}_{\dot{x}\dot{x}} &= -\left[\mathrm{I}^{(n)} - [Q^{(n)}_{\dot{x}\dot{x}}]^{-1}Q^{(n)}_{\dot{x}x}[Q^{(n)}_{xx}]^{-1}Q^{(n)}_{x\dot{x}}\right]^{-1}[Q^{(n)}_{\dot{x}\dot{x}}]^{-1} \\ &= -[Q^{(n)}_{\dot{x}\dot{x}}]^{-1},\ Q^{(n)}_{xx} \to \infty, \end{aligned} \tag{10.33}$$

and

$$\begin{aligned} S^{(n)}_{\dot{x}x} &= \left[\mathrm{I}^{(n)} - [Q^{(n)}_{\dot{x}\dot{x}}]^{-1}Q^{(n)}_{\dot{x}x}[Q^{(n)}_{xx}]^{-1}Q^{(n)}_{x\dot{x}}\right]^{-1}[Q^{(n)}_{\dot{x}\dot{x}}]^{-1}Q^{(n)}_{\dot{x}x}[Q^{(n)}_{xx}]^{-1} \\ &= [Q^{(n)}_{\dot{x}\dot{x}}]^{-1}Q^{(n)}_{\dot{x}x}[Q^{(n)}_{xx}]^{-1},\ Q^{(n)}_{xx} \to \infty. \end{aligned} \tag{10.34}$$

The limiting form holds for the free Brownian particle; it may or may not be an acceptable approximation for thermodynamic fluctuations, and in any case using the full form does not appear to impose significant difficulties (see §10.3.4 below). Because $Q^{(n)}_{\dot{x}\dot{x}} = \ddot{Q}^{(n)}_0$ is a Toeplitz matrix, its inverse is relatively easy to calculate.

Most Likely Velocity

It is not necessary to actually give the most likely trajectory for the present free particle, but is is useful to take a small digression and to establish one particular result that will prove useful in §10.3.4.

Differentiating the trajectory entropy with respect to the velocities yields the bare particle most likely trajectory velocities,

$$\begin{aligned} \overline{\dot{\mathbf{x}}}^{(n)} &= -[S^{(n)}_{\dot{x}\dot{x}}]^{-1}S^{(n)}_{\dot{x}x}x_n \\ &= Q^{(n)}_{\dot{x}\dot{x}}\left[\mathrm{I}^{(n)} - [Q^{(n)}_{\dot{x}\dot{x}}]^{-1}Q^{(n)}_{\dot{x}x}[Q^{(n)}_{xx}]^{-1}Q^{(n)}_{x\dot{x}}\right] \\ &\quad \times \left[\mathrm{I}^{(n)} - [Q^{(n)}_{\dot{x}\dot{x}}]^{-1}Q^{(n)}_{\dot{x}x}[Q^{(n)}_{xx}]^{-1}Q^{(n)}_{x\dot{x}}\right]^{-1}[Q^{(n)}_{\dot{x}\dot{x}}]^{-1}Q^{(n)}_{\dot{x}x}[Q^{(n)}_{xx}]^{-1}x_n \\ &= Q^{(n)}_{\dot{x}x}[Q^{(n)}_{xx}]^{-1}x_n. \end{aligned} \tag{10.35}$$

This result is exact and does not invoke the limiting form. Multiplying both sides of this by x_n and taking the equilibrium average yields for the left-hand side

$$\begin{aligned}\left\langle \overline{\dot{\mathbf{x}}}^{(n)} x_n \right\rangle_0 &= \left\langle \dot{\mathbf{x}}^{(n)} x_n \right\rangle_0 \\ &= k_{\rm B} \left\{ \dot{q}_0(t_n - t_1)^{\rm T}, \dot{q}_0(t_n - t_2)^{\rm T}, \ldots, \dot{q}_0(0)^{\rm T} \right\}, \end{aligned} \tag{10.36}$$

and for the right-hand side $Q_{\dot{x}x}^{(n)} [Q_{xx}^{(n)}]^{-1} \langle x_n x_n \rangle_0$. Hence one has

$$Q_{\dot{x}x}^{(n)} [Q_{xx}^{(n)}]^{-1} = \left\langle \dot{\mathbf{x}}^{(n)} x_n \right\rangle_0 \langle x_n x_n \rangle_0^{-1}. \tag{10.37}$$

In particular, the terminal element of the vector is

$$Q_{\dot{x}x;n}^{(n)} [Q_{xx}^{(n)}]^{-1} = \langle \dot{x}(t) x(t) \rangle_0 \langle x(t) x(t) \rangle_0^{-1} = -\dot{q}_0(0)^{\rm T} q_0(0)^{-1}. \tag{10.38}$$

These are *not* the symmetrised correlation functions, but this is more important for later sections than here. In fact, for the present pure parity case of free Brownian motion, $\langle \dot{x}(t) x(t) \rangle_0 = 0$, which means that this term vanishes. It needs to be retained for the mixed parity analysis of §10.3.4.

The result is important because it tells how the current position affects the current velocity. One reason for choosing the terminal position as the one specified, is that the generalised Langevin equation that emerges from this choice is the same as the one that is given by the projector operator approach in §10.7. It would have been possible to specify the trajectory by all the velocities and a preceding position, x_k say, rather than the terminal position x_n. The analysis could be performed for this case, and it turns out that the memory function to be given below also depends upon this choice. Nothing appears to be gained by this. Presumably the choice of specified position affects the two terms in such a way that the predicted terminal velocity is unaffected.

Nature of Trajectory Entropy

An explicit representation of the free particle fluctuation matrix will not be required beyond what has just been given, but it is worth making one point about its general properties. For a single node it has already been seen in Eq. (10.1) that $S_0^{(1)}(\dot{x}, x) = -m\dot{x}^2/2T$. This result follows because only the kinetic energy of the Brownian particle contributes to the total energy, and the solvent enters solely as a heat reservoir. This result necessarily assumes that all possible trajectories leading to the current velocity have been integrated over. In contrast, for a trajectory one cannot simply relate the trajectory entropy to the kinetic energy or its change. Even at the two node level, for example,

$$S_0^{(2)}(\dot{x}_1, \dot{x}_2; \tau) \neq \frac{1}{2} \left[S_0^{(1)}(\dot{x}_1) + S_0^{(1)}(\dot{x}_2) \right], \tag{10.39}$$

or similar. The reason is that although the change in kinetic energy of the free particle is exactly equal and opposite to the change in energy of the solvent, the latter is not sufficient to give the change in entropy; the specific trajectory corresponds to a particular spatial distribution of the energy, number, and momentum density fields of the solvent in the vicinity of the particle, and these affect its entropy. These perturbations on a uniform distribution occur over length scales of similar size to the Brownian particle itself and cannot be neglected. (In contrast, in the general non-equilibrium reservoir formalism, the size of the inhomogeneous interfacial region between the sub-system and the reservoir is small compared to the size of the sub-system.) The entropy for $n = 1$ corresponds to all prior trajectories being integrated over, in which case all such spatial distributions are averaged out leaving only the total energy of the solvent as the relevant variable for the entropy. One cannot carry out such an averaging process for a specified trajectory, and in this case the particular microscopic arrangement of the solvent and its consequent entropy cannot be expressed simply as changes in the kinetic energy.

Bare Particle Equations of Motion

The trajectory of the free Brownian particle in the solvent can be written in the form of equations of motion,

$$\begin{aligned} x(t+\tau) &= x(t) + \tau\dot{x}(t) \\ \dot{x}(t+\tau) &= \dot{x}(t) + \tau F_{\mathrm{s0}}(t)/m, \end{aligned} \tag{10.40}$$

where $F_{\mathrm{s0}}(t)$ is the force exerted by the solvent on the particle. This force at all prior times $t' \leq t$ is what is required to give the specific trajectory. This force is part of the contribution that changes the entropy of the solvent, and is incorporated into the fluctuation matrix above. The existence of the solvent forces, their rôle in determining the trajectory, and their influence on the trajectory entropy, provide the conceptual basis for the perturbation theory that is shortly developed.

The time correlation function $\ddot{q}_0(t)$, given above, can be expressed simply as an average over the trajectory. For example,

$$\begin{aligned} \ddot{q}_0^{\mathrm{s}}((n-j)\tau) &= k_{\mathrm{B}}^{-1}\left\langle \dot{x}(n\tau)\dot{x}(j\tau)\right\rangle_0 \\ &= \frac{k_{\mathrm{B}}^{-1}\int \mathrm{d}\dot{\mathbf{x}}^{(n)}\, e^{S^{(n)}_{\dot{x}\dot{x}}(\dot{\mathbf{x}}^{(n)};\tau)/k_{\mathrm{B}}}\,\dot{x}_n\dot{x}_j}{\int \mathrm{d}\dot{\mathbf{x}}^{(n)}\, e^{S^{(n)}_{\dot{x}\dot{x}}(\dot{\mathbf{x}}^{(n)};\tau)/k_{\mathrm{B}}}} \\ &= \frac{k_{\mathrm{B}}^{-1}\int \mathrm{d}\dot{x}_n\mathrm{d}\dot{x}_j\, e^{S^{(2)}_0(\dot{x}_n,\dot{x}_j;(n-j)\tau)/k_{\mathrm{B}}}\,\dot{x}_n\dot{x}_j}{\int \mathrm{d}\dot{x}_n\mathrm{d}\dot{x}_j\, e^{S^{(2)}_0(\dot{x}_n,\dot{x}_j;(n-j)\tau)/k_{\mathrm{B}}}}. \end{aligned} \tag{10.41}$$

This shows how the higher-order fluctuation matrix reduces to a lower-order matrix when the irrelevant nodes are integrated out.

10.2 Langevin and Smoluchowski Equations

Now a time dependent external potential, $U(x,t)$, is applied to the otherwise free Brownian particle. In the first instance the Smoluchowski and Langevin approximations for the problem will be given.

Langevin originally developed a stochastic, dissipative equation for the motion of a Brownian particle.[6] The dissipative force is taken to be a drag force $F_{\text{drag}} = -\gamma\dot{x}$. By setting the driving force to zero and solving the stochastic, dissipative equation of motion to obtain the mean square displacement of the free particle, it can be shown that the drag coefficient is related to the diffusion constant by $\gamma \equiv 1/\beta D$. For the present case of a driven Brownian particle, a mechanical force, $F(x,t) = -\partial U(x,t)/\partial x$ acts in addition to the dissipative force. Setting the stochastic force to zero gives Newton's equation for the most likely trajectory,

$$m\ddot{\overline{x}}(t) = F(\overline{x}(t),t) - \gamma\dot{\overline{x}}(t). \tag{10.42}$$

This is the deterministic form of Langevin's equation.

Smoluchowski independently obtained a somewhat simpler equation that is applicable in the large drag limit.[7] In this over damped case, Smoluchowski effectively set the acceleration to zero in the Langevin equation. In consequence, the drag force is equal and opposite to the applied force,

$$\dot{\overline{x}}(t) = \gamma^{-1}F(\overline{x}(t),t) = \beta D F(\overline{x}(t),t). \tag{10.43}$$

This is the Smoluchowski equation, and it applies in the large drag (equivalently, small diffusion) regime.

10.3 Perturbation Theory

10.3.1 Most Likely Velocity

Again impose upon the Brownian particle a time dependent external potential $U(x,t)$. The system belongs to the class of mechanical non-equilibrium systems, which was treated in Ch. 8, where a formally exact expression was given for the trajectory entropy (see §§8.2.1 and 8.6). As will now be seen, the trajectory entropy of the driven Brownian particle is just that of the free particle, as discussed above, §10.1.4, plus that due to the work done by the external potential over the trajectory.

The equations of motion for the driven particle are of the form,

$$\begin{aligned} x(t+\tau) &= x(t) + \tau\dot{x}(t) \\ \dot{x}(t+\tau) &= \dot{x}(t) + \frac{\tau}{m}\left[F_{\text{s0}}(t) + F_{\text{sU}}(t) + F(x(t),t)\right]. \end{aligned} \tag{10.44}$$

Here $F_{\text{s0}}(t')$ is the force due to the solvent that would give the trajectory if the particle were free, $F_{\text{sU}}(t')$ is the additional solvent force that has arisen because of the external potential, and $F(x',t') = -\partial U(x',t')/\partial x'$ is the external force. In order for F_{s0} to be the same here as in Eq. (10.40) for the same trajectory,

[6] Langevin, P. (1908), C. R. Acad. Sci. Paris, **146**, 530.

[7] von Smoluchowski, M. (1906), Ann. Phys. **21**, 756.

the additional solvent force has to be equal and opposite to the applied force, $F_{\rm sU} = -F$. In other words, the total solvent force, $F_{\rm s} = F_{\rm s0} + F_{\rm sU}$, plus the external force, F, gives the trajectory, and the trajectory is the same as that of a free particle when the total force, $F_{\rm s} + F$ equals that of the solvent acting on the bare particle, $F_{\rm s0}$.

The rate of change of energy of the particle due to this additional solvent force is $\dot{E}_{\rm p} = -\dot{x}F_{\rm sU}$. By energy conservation, this is equal and opposite to the rate of energy change of the solvent, $\dot{E}_{\rm s} = \dot{x}F_{\rm sU}$. Hence the additional rate of change of solvent entropy is $\dot{S}_{\rm sU} = -\dot{x}F_{\rm sU}/T = \dot{x}F/T$. Using this formulation, the entropy of the trajectory of the driven particle is equal to that of the free particle for the same trajectory plus the change in entropy of the solvent that arises from these additional solvent forces,

$$\begin{aligned} S^{(n)}(\mathbf{X}^{(n)};\tau,U) &= S_0^{(n)}(\mathbf{X}^{(n)};\tau) - \frac{1}{T}\int_0^t \mathrm{d}t'\, \dot{x}(t')F_{\rm sU}(t') \\ &= S_0^{(n)}(\mathbf{X}^{(n)};\tau) + \frac{1}{T}\int_0^t \mathrm{d}t'\, \dot{x}(t')F(x(t'),t'). \end{aligned} \tag{10.45}$$

The contribution of the free particle solvent forces, $F_{\rm s0}(t')$, has been incorporated into the free particle trajectory entropy, $S_0^{(n)}(\mathbf{X}^{(n)};\tau)$, as was discussed in relation to Eq. (10.40). Recall that the trajectory is specified by the terminal position and the velocities, $\mathbf{X}^{(n)} \equiv \{x_n, \dot{\mathbf{x}}^{(n)}\}$.

The discrete version of this is

$$S^{(n)}(\mathbf{X}^{(n)};\tau,U) = \frac{1}{2}S_0^{(n)}(\tau) : \mathbf{X}^{(n)}\mathbf{X}^{(n)} + \frac{\tau}{T}\dot{\mathbf{x}}^{(n)} \cdot \mathbf{F}^{(n)}. \tag{10.46}$$

The force vector is defined on the same nodes as the velocity vector, $\mathbf{F}^{(n)} \equiv \{F_1, F_2, \ldots, F_n\}$, with $F_j \equiv F(x_j, t_j)$. Obviously there are more refined ways in which the integral could be discretised, but these make no difference to linear order in the time step.

Differentiating this with respect to the velocities gives

$$\begin{aligned} &\frac{\partial S^{(n)}(\mathbf{X}^{(n)};\tau,U)}{\partial \dot{\mathbf{x}}^{(n)}} \\ &= S_{\dot{x}\dot{x}}^{(n)}(\tau)\dot{\mathbf{x}}^{(n)} + S_{\dot{x}x}^{(n)}(\tau)x_n + \frac{\tau}{T}\mathbf{F}^{(n)} - \frac{\tau^2}{T}\mathcal{U}^{(n)}[\nabla^{(n)}\mathbf{F}^{(n)}]\dot{\mathbf{x}}^{(n)} \\ &= S_{\dot{x}\dot{x}}^{(n)}(\tau)\dot{\mathbf{x}}^{(n)} + S_{\dot{x}x}^{(n)}(\tau)x_n + \frac{\tau}{T}\mathbf{F}^{(n)} + \mathcal{O}(U^2). \end{aligned} \tag{10.47}$$

Here $\mathcal{U}^{(n)}$ is an upper triangular matrix consisting of ones above the diagonal. Neglecting the final term of the first equality is justified because $\overline{\dot{\mathbf{x}}}^{(n)}$ is linear in the force. Accordingly, the most likely trajectory predicated upon x_n is

$$\begin{aligned} \overline{\dot{\mathbf{x}}}^{(n)}(x_n) &= -S_{\dot{x}\dot{x}}^{(n)}(\tau)^{-1}S_{\dot{x}x}^{(n)}(\tau)x_n - \frac{\tau}{T}S_{\dot{x}\dot{x}}^{(n)}(\tau)^{-1}\mathbf{F}^{(n)} \\ &= Q_{\dot{x}x}^{(n)}[Q_{xx}^{(n)}]^{-1}x_n + \frac{\tau}{T}\ddot{Q}_0^{(n)}(\tau)\mathbf{F}^{(n)},\ Q_{xx}^{(n)} \to \infty. \end{aligned} \tag{10.48}$$

The second equality holds for the free Brownian particle, in which case $S^{(n)}_{\dot{x}\dot{x}}(\tau)^{-1} = -\ddot{Q}^{(n)}_0(\tau)$, as discussed in connection with Eq. (10.33). The first term in the second equality is zero for the free Brownian particle, but it is retained here in order to facilitate the later discussion of mixed parity, thermodynamic fluctuations.

Differentiating the trajectory entropy with respect to the terminal velocity

$$\frac{\partial S^{(n)}(\mathbf{X}^{(n)};\tau,U)}{\partial \dot{x}_n} = \sum_{k=1}^{n} S^{(n)}_{0;nk}(\tau)\dot{x}_k + \frac{\tau}{T}F_n. \tag{10.49}$$

This neglects $S^{(n)}_{\dot{x}x;n}(\tau)x_n$ because, as discussed in connection with Eq. (10.38), its contribution to the most likely terminal velocity vanishes in the pure parity case of Brownian motion, $\langle \dot{x}(t)x(t)\rangle_0 = 0$. Terms $\dot{x}_j(\partial F_j/\partial x_j)(\partial x_j/\partial \dot{x}_n) = -\tau\dot{x}_j(\partial F_j/\partial x_j)$, $j < n$, have also been neglected for the reasons given above. Setting the derivative to zero yields the most likely terminal velocity conditional upon the preceding velocities and x_n,

$$\overline{\dot{x}}_n(\dot{\mathbf{x}}^{(n-1)}, x_n) = -S^{(n)}_{0;nn}(\tau)^{-1}\left[\sum_{k=1}^{n-1} S^{(n)}_{0;nk}(\tau)\dot{x}_k + \frac{\tau}{T}F_n\right] + \mathcal{O}(U^2). \tag{10.50}$$

The difference between the optimised trajectory, Eq. (10.48) and the optimised terminal point, Eq. (10.50) is not great in practical terms, as both can be used to calculate the most likely trajectory by stepping forward one node at a time. That is Eq. (10.48) can be written as,

$$\overline{\dot{\mathbf{x}}}^{(n/j)}(x_j) = Q^{(n)}_{\dot{x}x}[Q^{(n)}_{xx}]^{-1}x_j + \frac{\tau}{T}\ddot{Q}^{(n)}_0(\tau)\mathbf{F}^{(n/j)}, \tag{10.51}$$

and evaluated for the terminal node,

$$\overline{\dot{x}}_j(x_j) = \frac{\tau}{T}\left\{\ddot{Q}^{(n)}_0(\tau)\mathbf{F}^{(n/j)}\right\}_n = \frac{\tau}{T}\sum_{k=j-n+1}^{j} \ddot{q}^{\mathrm{s}}_0(t_j - t_k)F_k. \tag{10.52}$$

Here the velocities at the n nodes up to and including the node j are $\dot{\mathbf{x}}^{(n/j)} \equiv \{\dot{x}_{j-n+1}, \dot{x}_{j-n+2}, \ldots, \dot{x}_j\}$, and similarly for $\mathbf{F}^{(n/j)}$. The final element of the vector that is the first term on the right-hand side of Eq. (10.48) vanishes in the present pure parity case, $Q^{(n)}_{\dot{x}x;n}[Q^{(n)}_{xx}]^{-1} = 0$, as was discussed following Eq. (10.38). From the previous velocities $\dot{\mathbf{x}}^{(n/j-1)}$ and forces $\mathbf{F}^{(n/j-1)}$, the next position may be obtained, $\overline{x}_j = \overline{x}_{j-1} + \tau\overline{\dot{x}}_{j-1}$. Using this for the force F_j, one may find $\overline{\dot{x}}_j$, and then move to node $j+1$.

Equation (10.52) has one advantage over Eq. (10.50) in that the velocity autocorrelation function can be used directly. In Eq. (10.50), the velocity autocorrelation function has to be inverted to obtain the fluctuation matrix $S^{(n)}_0(\tau)$. This is inconvenient rather than prohibitive because, as has been mentioned, the velocity autocorrelation function is a positive definite Toeplitz matrix, at least in the present symmetrised case, and there exist efficient $\mathcal{O}(n^2)$ algorithms for their inversion.

Evaluating Eq. (10.52) for the terminal velocity, and using the fact that $\ddot{Q}^{(n)}_{0;ij}(\tau) = \ddot{q}^{\rm s}_0(t_i - t_j)$, the continuum version is

$$\overline{\dot{x}}(t) = \frac{1}{T}\int_0^t \mathrm{d}t'\, \ddot{q}^{\rm s}_0(t-t')F(x(t'),t'). \tag{10.53}$$

This result goes beyond the Markov approximation. It shows how previous values of the applied force influence the current motion. It also shows how this influence decreases over time, since $\ddot{q}^{\rm s}_0(t) \rightarrow 0$, $t \rightarrow \infty$. Hence the integrand goes to zero for $t - t'$ sufficiently large, which means that the lower limit of the integral can be replaced by $t - t^*$, for some relaxation time t^* large enough.

A major finding of the perturbation theory is that the kernel or memory function, which gives the current influence of the prior force, is just the velocity autocorrelation function. This explicit result is exact within the perturbation approach.

A second major outcome of the perturbation theory is that the difficult part of the calculation, which is to obtain the bare particle velocity autocorrelation function, need be done only once, and the result can be used to obtain the particle motion in the presence of any external force by a relatively trivial one-dimensional quadrature.

In practice, one generally works with the discretised version. One needs $n\tau$ sufficiently large that $\ddot{q}^{\rm s}_0(n\tau) \approx 0$, and τ sufficiently small that the discretisation approximation is valid. Various criteria can be used for the latter, such as that it is an accurate approximation to take $\overline{\dot{x}}_i \approx [\overline{x}_i - \overline{x}_{i-1}]/\tau$, or $\overline{U}_{i+1} \approx \overline{U}_i + \tau\overline{\dot{x}}_i\overline{F}_i + \tau\partial U(\overline{x}_i, t_i)/\partial t_i$.

Approximating the quadrature by just two terms gives

$$\begin{aligned}\overline{\dot{x}}_{j+1} &= \frac{\tau}{2T}\left[F_{j+1}\,\ddot{q}^{\rm s}_0(0) + F_j\,\ddot{q}^{\rm s}_0(\tau)\right] \\ &\approx \frac{1}{T}F_j\,\dot{q}^{\rm s}_0(\tau) \\ &\approx \beta D F_j. \end{aligned}\tag{10.54}$$

This is the Smoluchowski result. If τ is large enough to justify the two term expansion, then it must also be large enough for the velocity position correlation function to have reached its asymptotic limit. Differentiating the Einstein asymptote, Eq. (10.4), this is $\lim_{\tau\rightarrow\infty}\dot{q}^{\rm s}_0(\tau) = k_{\rm B}^{-1}D$. Using such a large value of τ is only valid when the force is slowly varying, which is the condition for the Smoluchowski approximation to be valid.

This limiting result can be directly confirmed when the force is constant, $F(x', t') = F$. In this case the continuum expression, Eq. (10.53), yields directly

$$\overline{\dot{x}}(t) = \frac{F}{T}\left[\dot{q}^{\rm s}_0(t) - \dot{q}^{\rm s}_0(0)\right] \stackrel{t\rightarrow\infty}{\rightarrow} \beta D F. \tag{10.55}$$

10.3.2 Alternative Derivation

There is another way to obtain the perturbation result that illustrates a rather general linearisation procedure that is common in statistical mechanics. In the first instance one ought to regard the external force as a function of time but independent of the particle position, $F(t)$. The reason for this will be discussed further below, but here it may be briefly pointed out that taking the external force to be independent of the particle position allows it to be taken outside of the average. Linearising the exponential, the average of the trajectory is may then be seen to be

$$\begin{aligned}
\left\langle \dot{\mathbf{x}}^{(n)} \right\rangle_U &= \frac{\int \mathrm{d}\dot{\mathbf{x}}^{(n)}\ e^{S^{(n)}(\dot{\mathbf{x}}^{(n)};\tau,U)/k_\mathrm{B}}\dot{\mathbf{x}}^{(n)}}{\int \mathrm{d}\dot{\mathbf{x}}^{(n)}\ e^{S^{(n)}(\dot{\mathbf{x}}^{(n)};\tau,U)/k_\mathrm{B}}} \\
&= \frac{\int \mathrm{d}\dot{\mathbf{x}}^{(n)}\ e^{S_0^{(n)}(\dot{\mathbf{x}}^{(n)};\tau)/k_\mathrm{B}}\left[1+\beta\tau\dot{\mathbf{x}}^{(n)}\cdot\mathbf{F}^{(n)}+\mathcal{O}(U^2)\right]\dot{\mathbf{x}}^{(n)}}{\int \mathrm{d}\dot{\mathbf{x}}^{(n)}\ e^{S_0^{(n)}(\dot{\mathbf{x}}^{(n)};\tau)/k_\mathrm{B}}\left[1+\beta\tau\dot{\mathbf{x}}^{(n)}\cdot\mathbf{F}^{(n)}+\mathcal{O}(U^2)\right]} \\
&= \frac{\left\langle \dot{\mathbf{x}}^{(n)} \right\rangle_0 + \beta\tau\left\langle \dot{\mathbf{x}}^{(n)}\dot{\mathbf{x}}^{(n)}\mathbf{F}^{(n)} \right\rangle_0}{1+\beta\tau\left\langle \dot{\mathbf{x}}^{(n)}\cdot\mathbf{F}^{(n)} \right\rangle_0} \\
&= \beta\tau\left\langle \dot{\mathbf{x}}^{(n)}\dot{\mathbf{x}}^{(n)}\mathbf{F}^{(n)} \right\rangle_0 + \mathcal{O}(U^2) \\
&= \beta\tau\left\langle \dot{\mathbf{x}}^{(n)}\dot{\mathbf{x}}^{(n)} \right\rangle_0 \mathbf{F}^{(n)} + \mathcal{O}(U^2) \\
&= \frac{\tau}{T}\ddot{Q}_0^{(n)}(\tau)\mathbf{F}^{(n)}.
\end{aligned} \tag{10.56}$$

The second equality is the linearisation with respect to the external potential, in which procedure the quadratic terms are neglected. The third equality follows from the definition of the free particle trajectory probability and averages. The fourth equality follows because $\langle \dot{x}(t) \rangle_0 = 0$. The penultimate equality follows because the external force here is only a function of time, not of the particle position on the trajectory, and so it may be taken outside of the average, as was discussed above. Obviously one can take the average trajectory to be equal to the most likely trajectory.

This result holds for an arbitrary time dependent external force, $F(t)$. This means that having obtained the result, one can choose the force to be a function of the most likely position on the trajectory, $F(t') \equiv F(\overline{x}(t'), t')$, $t' \leq t$. Here the most likely trajectory may be a function of the terminal position, here x_n, above x_j, and the most likely velocities. One can also use this freedom of choice for the force to evaluate this for the average terminal velocity, $\langle \dot{x}_n \rangle$, predicated upon the actual trajectory and actual force leading up to the current time, $F(x(t'), t')$, $t' \leq t$.

10.3.3 Most Likely Position

The velocity result can be integrated to obtain the most likely position,

$$\begin{aligned}
\overline{x}(t) &= \overline{x}(0) + \int_0^t \mathrm{d}t'\, \overline{\dot{x}}(t') \\
&= \overline{x}(0) + \frac{1}{T}\int_0^t \mathrm{d}t' \int_0^{t'} \mathrm{d}t''\, \ddot{q}_0^{\mathrm{s}}(t'-t'')F(x(t''),t'') \\
&= \overline{x}(0) + \frac{1}{T}\int_0^t \mathrm{d}t''\, F(x(t''),t'') \int_{t''}^t \mathrm{d}t'\, \ddot{q}_0^{\mathrm{s}}(t'-t'') \\
&= \overline{x}(0) + \frac{1}{T}\int_0^t \mathrm{d}t''\, F(x(t''),t'') \int_0^{t-t''} \mathrm{d}t'''\, \ddot{q}_0^{\mathrm{s}}(t'''),\ \ t'''=t'-t'' \\
&= \overline{x}(0) + \frac{1}{T}\int_0^t \mathrm{d}t''\, \left[\dot{q}_0^{\mathrm{s}}(t-t'') - \dot{q}_0^{\mathrm{s}}(0)\right] F(x(t''),t''). \qquad (10.57)
\end{aligned}$$

In both the pure parity case and the mixed parity case the symmetrised position velocity time correlation function vanishes at $t=0$, $\dot{q}_0^{\mathrm{s}}(0)=0$. The full correlation function does not vanish in the mixed parity case in this limit (see §10.6 below).

In the discrete case, this can simply be written $\overline{x}(t) = \overline{x}(t-\tau) + \tau\overline{\dot{x}}(t-\tau) + \mathcal{O}(\tau^2)$. These results allow the trajectory to be obtained by marching forward one node at a time. The length of the time steps, and the number of nodes retained for the backward time integration for the velocity must be chosen from the natures of the applied force and the time correlation function, respectively.

10.3.4 Stochastic Dissipative Equations of Motion

The preceding analysis gave the most likely velocity of a Brownian particle when it is driven by a time and space varying external potential, and when memory effects of the solvent are important. It is desirable to develop the stochastic equations for the trajectory since in some cases these yield information that is not available from the most likely trajectory. For example, one often wants the average of a 'phase function', $f(X,t)$, or a trajectory function, $f(\mathbf{X}^{(n)})$. In general, the most likely value of a function is not equal to the function evaluated at the most likely value of the trajectory, and similarly for average values. An algorithm for generating stochastic trajectories in the case that memory effects are non-negligible allows averages of such phase functions to be taken.

The stochastic dissipative equations will now be developed in a form that allows one to march along a trajectory remembering the previous n nodes. As in the preceding section, the results are restricted to the pure parity case and the symmetrised velocity autocorrelation function will be used. The general mixed parity analysis will be given in §10.6.

As above, the vector of the n most recent velocities up to and including the one at t_j is $\dot{\mathbf{x}}^{(n/j)} \equiv \{\dot{x}_{j-n+1}, \dot{x}_{j-n+2}, \ldots, \dot{x}_j\}$, and similarly for the vector of forces, $\mathbf{F}^{(n/j)}$. The terminal position plus the velocities were used to specify a trajectory, $\mathbf{X}^{(n/j)} \equiv \{x_j, \dot{\mathbf{x}}^{(n/j)}\}$. These and all vectors and matrices are ordered with time increasing when $\tau > 0$. As slightly modified from Eq. (10.46), the trajectory entropy for the current and preceding nodes is

$$S^{(n)}(\mathbf{X}^{(n/j)};\tau,U) = \frac{1}{2}S_0^{(n)}(\tau) : \mathbf{X}^{(n/j)}\mathbf{X}^{(n/j)} + \frac{\tau}{T}\dot{\mathbf{x}}^{(n/j)} \cdot \mathbf{F}^{(n/j)}. \quad (10.58)$$

The most likely trajectory maximises this, and its terminal node was given by Eq. (10.52),

$$\overline{\dot{x}}_j = \frac{\tau}{T}\left\{\ddot{Q}_0^{(n)}(\tau)\mathbf{F}^{(n/j)}\right\}_n = \frac{\tau}{T}\sum_{k=j-n+1}^{j} \ddot{q}_0^{\mathrm{s}}(t_j - t_k)F_k. \quad (10.59)$$

The force $F_i \equiv F(x_i, t_i)$ is here taken to be the force on the actual prior trajectory, whether or not this is equal to the most likely prior trajectory. Combining this with $\overline{x}_j = x_{j-1} + \tau\overline{\dot{x}}_{j-1}$ allows the most likely trajectory to be stepped out.

The physical meaning of this result can be clarified by looking at the change in the most likely terminal velocity,

$$\begin{aligned}
\overline{\dot{x}}_j - \overline{\dot{x}}_{j-1} &= \frac{\tau}{T}\sum_{i=j-n+1}^{j} \ddot{q}_0^{\mathrm{s}}(t_j - t_i)F_i - \frac{\tau}{T}\sum_{i=j-n}^{j-1} \ddot{q}_0^{\mathrm{s}}(t_{j-1} - t_i)F_i \\
&= \frac{\tau}{T}\ddot{q}_0^{\mathrm{s}}(0)F_j - \frac{\tau}{T}\ddot{q}_0^{\mathrm{s}}(n\tau)F_{j-n} \\
&\quad + \frac{\tau}{T}\sum_{i=j-n+1}^{j-1} \left[\ddot{q}_0^{\mathrm{s}}(t_j - t_i) - \ddot{q}_0^{\mathrm{s}}(t_{j-1} - t_i)\right] F_i \\
&= \frac{\tau}{m}F_j - \frac{\tau}{T}\ddot{q}_0^{\mathrm{s}}(n\tau)F_{j-n} + \frac{\tau^2}{T}\sum_{i=j-n+1}^{j-1} \dddot{q}_0^{\mathrm{s}}(t_j - t_i)F_i. \quad (10.60)
\end{aligned}$$

The first term, which arises from the fact that $\ddot{q}_0^{\mathrm{s}}(0) \equiv k_{\mathrm{B}}^{-1}\langle\dot{x}(0)\dot{x}(0)\rangle_0 = (T/m)\mathrm{I}$, is the usual Newtonian acceleration, and so this represents the adiabatic contribution to the change in velocity. The second term is negligible for $n\tau$ large enough, $\ddot{q}_0^{\mathrm{s}}(t) \to 0$, $t \to \infty$. The remaining terms give the dependence of the change in velocity on the history of the external force. These are non-Hamiltonian, deterministic terms, and their sum may be regarded as a generalised dissipative force for the case of memory.

One can simply add the random force to the equation for the most likely terminal velocity to obtain the stochastic dissipative equations of motion in perturbation approximation,

$$\begin{aligned}
x_j &= x_{j-1} + \tau\dot{x}_{j-1} \\
\dot{x}_j &= R_j + \frac{\tau}{T}\sum_{i=j-n+1}^{j} \ddot{q}_0^{\mathrm{s}}(t_j - t_i)F_i. \quad (10.61)
\end{aligned}$$

This is the generalised Langevin equation for position. The continuum version of the second of these is obvious,

$$\dot{x}(t) = R(t) + \frac{1}{T}\int_0^t \mathrm{d}t'\, \ddot{q}_0^{\mathrm{s}}(t-t')F(x(t'),t'). \tag{10.62}$$

Since the velocity autocorrelation function is short-ranged, the distant prior history contributes with ever decreasing weight, and so the integral becomes independent of the lower limit. The evolution of functions of x and t, including the probability density, can be obtained by expanding over a time step, retaining terms to second-order in the random force, and averaging. As mentioned in connection with the Fokker-Planck equation, in footnote 18 on p. 89, the difference between the Ito and Stratonovich stochastic calculus disappears when the variance is *not* a function of x, as is the case here and throughout. (See also §11.3.3 below.)

The random forces form a vector $\mathbf{R}^{(n/j)}$, and they have covariance matrix[8]

$$\begin{aligned}\left\langle \mathbf{R}^{(n/j)}\mathbf{R}^{(n/j)}\right\rangle_U &= \left\langle \left[\dot{\mathbf{x}}^{(n/j)} - \overline{\dot{\mathbf{x}}}^{(n/j)}\right]\left[\dot{\mathbf{x}}^{(n/j)} - \overline{\dot{\mathbf{x}}}^{(n/j)}\right]\right\rangle_U \\ &= -k_{\mathrm{B}}S_0^{(n)}(\tau)^{-1} \\ &= k_{\mathrm{B}}\ddot{Q}_0^{(n)}(\tau).\end{aligned} \tag{10.63}$$

The ikth element of this matrix is $\langle R(t_{j-i+1})R(t_{j-k+1})\rangle_U = k_{\mathrm{B}}\ddot{q}_0^{\mathrm{s}}(|i-k|\tau)$. It is significant that the random force for the current time step is correlated with the preceding random forces. The fact that the velocity autocorrelation function of the bare particle gives both the covariance of the random forces and the dissipative force may be regarded as the fluctuation dissipation theorem for a system with memory.

The probability distribution of the last n random forces is just the multivariate Gaussian

$$\wp(\mathbf{R}^{(n/j)}) = \frac{1}{\left|2\pi k_{\mathrm{B}}\ddot{Q}_0^{(n)}(\tau)\right|^{1/2}} e^{S_0^{(n)}(\tau):\mathbf{R}^{(n/j)}\mathbf{R}^{(n/j)}/2k_{\mathrm{B}}}. \tag{10.64}$$

Maximising the exponent with respect to R_j yields the most likely value conditional upon the preceding $n-1$ random forces,

$$\overline{R}_j(\mathbf{R}^{(n-1/j-1)}) = -S_{0;nn}^{(n)}(\tau)^{-1}\sum_{k=1}^{n-1} S_{0;nk}^{(n)}(\tau)R_{j-n+k}. \tag{10.65}$$

Hence the conditional probability distribution of the final random force is

$$\wp(R_j|\mathbf{R}^{(n-1/j-1)}) = \frac{e^{S_{0;nn}^{(n)}(\tau)[R_j-\overline{R}_j(\mathbf{R}^{(n-1/j-1)})]^2/2k_{\mathrm{B}}}}{[-2\pi k_{\mathrm{B}}S_{0;nn}^{(n)}(\tau)^{-1}]^{1/2}}. \tag{10.66}$$

---

[8]Note the distinction between the unconditional most likely velocity, $\overline{\dot{x}}_j$, which appears here in the $\overline{\dot{\mathbf{x}}}^{(n/j)}$, and the conditional most likely velocity, $\overline{\dot{x}}_j(\dot{\mathbf{x}}^{(n-1/j-1)}, x_{j-n+1})$, which is the left-hand side of Eq. (10.61) with $R_j = 0$. The unconditional most likely position and velocity satisfy the deterministic equations that are Eq. (10.59) with $F_i = F(\overline{x}_i, t_i)$, plus $\overline{x}_j = \overline{x}_{j-1} + \tau\overline{\dot{x}}_{j-1}$.

This equation has the virtue that it allows the random forces to be calculated one at a time from a Gaussian distribution for a single variable. Each random force has a non-zero mean that is a weighted average of the preceding $n-1$ random forces. It does require the inverse of the velocity autocorrelation function matrix, which, as mentioned above, can be efficiently obtained by exploiting its Toeplitz structure (in the present symmetrised case).

The present stochastic differential equation for driven Brownian motion can be compared with the Langevin equation. The Langevin equation is appropriate when one can take a time step that is large on the time scales of the solvent such that memory effects are negligible, in which regime the diffusion constant is truly constant. In this regime the dissipative force is simply the drag force, and the random forces are uncorrelated with each other and have a variance that is proportional to the diffusion constant—the fluctuation dissipation theorem. The Langevin equation is appropriate when the force is slowly varying in time and space such that a large time step is justified.

In contrast, rapidly varying forces necessitate the present approach that takes into account the memory of the solvent. In this case historical values of the force contribute directly to the current change in the velocity of the particle, and the current random force is correlated with the previous random forces. The non-Hamiltonian deterministic forces may be called the dissipative forces, and the stochastic forces may be called the fluctuation. It is the velocity autocorrelation function of the bare particle that determines the magnitude of the dissipation and its dependence on previous history. It also determines the covariance of the random forces. The relation between the two may be called the fluctuation dissipation theorem with memory.

10.3.5 Generalised Langevin Equation for Velocity

Differentiating the continuous version of the most likely velocity, Eq. (10.53), one obtains

$$\ddot{\overline{x}}(t) = m^{-1}F(x(t),t) + \frac{1}{T}\int_{-\infty}^{t} \mathrm{d}t'\ \dddot{q}_0^{\,\mathrm{s}}(t-t')F(x(t'),t'). \tag{10.67}$$

This uses the fact that $\ddot{q}_0^{\mathrm{s}}(0) = Tm^{-1}$, where m is the mass, a diagonal matrix in the multi-particle case, and the fact that $\ddot{q}_0^{\mathrm{s}}(t) = 0$ for t sufficiently large. The lower limit has been extended to $-\infty$, which is permissible because the integrand is short-ranged.

One can add a random force to this to obtain a version of the generalised Langevin equation for momentum. Noting that the most likely velocity is linearly proportional to the force (in an integral sense, Eq. (10.53)), it is conventional, however, to replace the force that appears in the integrand by the velocity, and to replace the present proportionality factor, the memory function $\dddot{q}_0^{\,\mathrm{s}}(t-t')$, by a new proportionality factor, the as yet undetermined memory function $K(t-t')$. Adding a random force, the generalised Langevin equation for the velocity then becomes

$$\begin{aligned} m\ddot{x}(t) &= F(x(t),t) + R_{\rm p}(t) - \int_{-\infty}^{t} {\rm d}t'\, K(t-t')\dot{x}(t') \\ &= F(x(t),t) + R_{\rm p}(t) - \int_{-\infty}^{\infty} {\rm d}t'\, \tilde{K}(t-t')\dot{x}(t'), \end{aligned} \tag{10.68}$$

where the truncated memory function is $\tilde{K}(t) = \Theta(t)K(t)$, with the Heaviside step function being $\Theta(t) = 1$, $t > 0$, and $\Theta(t) = 0$, $t < 0$. In the limit that the memory function is infinitely short-ranged, $K(t) = 2\gamma\delta(t)$, this becomes $m\ddot{x}(t) = F(x(t),t) + R_{\rm p}(t) - \gamma\dot{x}(t)$. One may recognise the term $-\gamma\dot{x}(t)$ as the friction or drag force, and this as the stochastic equation originally given by Langevin.

The Fourier transform of the generalised Langevin equation for the velocity is[9]

$$-i\omega m\dot{x}(\omega) = F(\omega) + R_{\rm p}(\omega) - \tilde{K}(\omega)\dot{x}(\omega), \tag{10.69}$$

with solution

$$\dot{x}(\omega) = \frac{1}{\tilde{K}(\omega) - i\omega m}\left[F(\omega) + R_{\rm p}(\omega)\right]. \tag{10.70}$$

Setting the random force to zero and comparing this with the Fourier transform of the most likely velocity, Eq. (10.53), $\bar{\dot{x}}(\omega) = T^{-1}\tilde{\ddot{q}}_0^{\rm s}(\omega)F(\omega)$, one can see that the memory function and the velocity autocorrelation function are related by

$$\frac{1}{T}\tilde{\ddot{q}}_0^{\rm s}(\omega) = \frac{1}{\tilde{K}(\omega) - i\omega m}. \tag{10.71}$$

This can be rearranged to give the velocity autocorrelation function in terms of the memory function,[10]

$$\begin{aligned} \ddot{q}_0^{\rm s}(\omega) &= 2{\rm Re}\left\{\tilde{\ddot{q}}_0^{\rm s}(\omega)\right\} \\ &= 2T{\rm Re}\left\{\frac{1}{\tilde{K}(\omega) - i\omega m}\right\} \\ &= \frac{TK(\omega)}{\left|\tilde{K}(\omega) - i\omega m\right|^2}, \end{aligned} \tag{10.72}$$

---

[9]The Fourier transform is defined as $f(\omega) = \int_{-\infty}^{\infty} {\rm d}t\ e^{i\omega t} f(t)$, with inverse $f(t) = (2\pi)^{-1}\int_{-\infty}^{\infty} {\rm d}\omega\ e^{-i\omega t} f(\omega)$. The Fourier transform of $\dot{f}(t)$ is $-i\omega f(\omega)$. A convolution in real space gives a product in Fourier space. Many results can be proven using the fact that $\int_{-\infty}^{\infty} {\rm d}t\ e^{i\omega t} = 2\pi\delta(\omega)$. As is common in this book, distinct but related functions are distinguished by their arguments; here ω and t distinguish the Fourier pair.

[10]Let $f_\pm(t) = f(t)$, $t > 0$, and $f_\pm(t) = \pm f(-t)$, $t < 0$. Then $\tilde{f}(t) \equiv \Theta(t)f(t) = [f_+(t) + f_-(t)]/2$. Hence $\tilde{f}(\omega) = [f_+(\omega) + f_-(\omega)]/2$, with $f_+(\omega)$ pure real, and $f_-(\omega)$ pure imaginary. Hence if $f(t)$ is even, then $f(\omega)$ is pure real, and so $f(\omega) = 2{\rm Re}\{\tilde{f}(\omega)\}$.

or *vice versa*,

$$\begin{aligned} K(\omega) &= 2\mathrm{Re}\left\{\tilde{K}(\omega)\right\} \\ &= 2\mathrm{Re}\left\{i\omega m + T/\tilde{\ddot{q}}_0^{\mathrm{s}}(\omega)\right\} \\ &= \frac{T\ddot{q}_0^{\mathrm{s}}(\omega)}{\left|\tilde{\ddot{q}}_0^{\mathrm{s}}(\omega)\right|^2}. \end{aligned} \tag{10.73}$$

10.3.6 Fluctuation Dissipation Theorem

The fluctuation dissipation theorem may be derived from Eq. (10.70) by setting the external force to zero, forming the product of the respective sides evaluated at ω and its complex conjugate at ω', and taking the equilibrium average,

$$\langle \dot{x}(\omega)\dot{x}^*(\omega')\rangle_0 = \frac{\left\langle R_{\mathrm{p}}(\omega)R_{\mathrm{p}}^*(\omega')\right\rangle_0}{\left[\tilde{K}(\omega) - i\omega m\right]\left[\tilde{K}^*(\omega') + i\omega' m\right]}. \tag{10.74}$$

Since the equilibrium system is homogeneous in time, each side of this contains a factor of $\delta(\omega - \omega')$. That is

$$\begin{aligned} \langle \dot{x}(\omega)\dot{x}^*(\omega')\rangle_0 &= \int_{-\infty}^{\infty} \mathrm{d}t \int_{-\infty}^{\infty} \mathrm{d}t'\; e^{i\omega t} e^{-i\omega' t'} \langle \dot{x}(t)\dot{x}(t')\rangle_0 \\ &= \int_{-\infty}^{\infty} \mathrm{d}t \int_{-\infty}^{\infty} \mathrm{d}t'\; e^{i\omega[t-t']} e^{-i[\omega'-\omega]t'} \langle \dot{x}(t-t')\dot{x}(0)\rangle_0 \\ &= \langle \dot{x}(\omega)\dot{x}(t=0)\rangle_0\, 2\pi\delta(\omega' - \omega) \\ &= 2\pi k_{\mathrm{B}} \ddot{q}_0^{\mathrm{s}}(\omega)\delta(\omega' - \omega). \end{aligned} \tag{10.75}$$

Similarly, $\left\langle R_{\mathrm{p}}(\omega)R_{\mathrm{p}}^*(\omega')\right\rangle_0 = 2\pi \left\langle R_{\mathrm{p}}(\omega)R(t=0)\right\rangle_0 \delta(\omega - \omega')$. Integrating out this δ-function on both sides of Eq. (10.74) gives

$$\ddot{q}_0^{\mathrm{s}}(\omega) = \frac{1}{k_{\mathrm{B}}} \frac{\langle R_{\mathrm{p}}(\omega)R_{\mathrm{p}}(t=0)\rangle_0}{\left|\tilde{K}(\omega) - i\omega m\right|^2}. \tag{10.76}$$

Comparing this with Eq. (10.72) shows that

$$\langle R_{\mathrm{p}}(\omega)R_{\mathrm{p}}(t=0)\rangle_0 = k_{\mathrm{B}} T K(\omega). \tag{10.77}$$

Taking the inverse Fourier transform yields the fluctuation dissipation theorem for the random force for the velocity,

$$\langle R_{\mathrm{p}}(t)R_{\mathrm{p}}(0)\rangle_0 = k_{\mathrm{B}} T K(t). \tag{10.78}$$

10.3.7 Weiner-Khintchine Theorem

These results are often derived by invoking the Weiner-Khintchine theorem.[11] This theorem broadly relates the autocorrelation function to the inverse Fourier transform of the noise or power spectrum.[12] In fact, Eq. (10.75) is related to the Weiner-Khintchine theorem.

The historical significance of the Weiner-Khintchine theorem is linked to the fact that the Fourier transform of a stochastic process, in this case $\dot{x}(t)$, does not exist except as a generalised function. One can truncate the process by setting $\dot{x}_\tau(t) = 0$, $|t| > \tau/2$, for which the transform exists. The theorem relates the power spectrum, which is the average of the product of the Fourier transforms, and which depends upon the truncation, to the Fourier transform of the time correlation function, which is independent of the truncation,[13]

$$\begin{aligned} S(\omega) &= \frac{1}{\tau} \left\langle \dot{x}_\tau(\omega) \dot{x}_\tau^*(\omega) \right\rangle \\ &= \int \mathrm{d}t\, e^{i\omega t} \left\langle \dot{x}_\tau(t) \dot{x}_\tau(0) \right\rangle \\ &= \int \mathrm{d}t\, e^{i\omega t} \left\langle \dot{x}(t) \dot{x}(0) \right\rangle . \end{aligned} \tag{10.79}$$

The non-trivial part of the theorem is contained in the second equality, which follows from the general result that

$$\begin{aligned} \int_{-\tau/2}^{\tau/2} \mathrm{d}t \int_{-\tau/2}^{\tau/2} \mathrm{d}t'\, f(t-t') &= \int_{-\tau}^{0} \mathrm{d}t'' \int_{-\tau/2}^{t''+\tau/2} \mathrm{d}t\, f(t'') \\ &\quad + \int_{0}^{\tau} \mathrm{d}t'' \int_{t''-\tau/2}^{\tau/2} \mathrm{d}t\, f(t'') \\ &= \tau \int_{-\tau}^{\tau} \mathrm{d}t''\, f(t''). \end{aligned} \tag{10.80}$$

Also the second equality interchanges the operations of Fourier transformation and statistical averaging. The third equality follows because the time correlation function is short-ranged, and so the correlation of the truncated series is the same as that of the full series. The power spectrum is an experimentally measurable object, and the truncation time is an experimental specification.

The present derivation of the fluctuation dissipation theorem invoked directly the properties of the Fourier transform, and the Weiner-Khintchine theorem was not required. This appears to be the general rule for other mathematical results in the field of stochastic processes. Generalised functions are not nowadays as challenging as they were in the early part of the twentieth century,

[11] Pottier, N. (2010), *Non-equilibrium Statistical Physics: Linear Irreversible Processes*, (Oxford University Press, Oxford). See §§10.4 and 10A.3, and also §§12.6 and 14.3.

[12] Wiener, N. (1930), Acta Math. **55**, 117. Khintchine, A. (1934) Math. Ann. **109**, 604.

[13] Le Bellac, M., Mortessagne, F., and Batrouni, G. G., (2004), *Equilibrium and Non-equilibrium Statistical Thermodynamics*, (Cambridge University Press, Cambridge). See Eqs. (9.189) and (9.190).

and this is perhaps the reason that it can be difficult to appreciate the import of the Weiner-Khintchine theorem.

10.3.8 Exponentially Decaying Memory Function

A common approximation is to assume that the memory kernel is exponential (Pottier, 2010),

$$K(t) = \gamma\omega_c e^{-\omega_c|t|}. \tag{10.81}$$

The relevant Fourier transforms of this are

$$K(\omega) = \frac{2\gamma\omega_c^2}{\omega^2+\omega_c^2}, \text{ and } \tilde{K}(\omega) = \frac{\gamma\omega_c}{\omega_c - i\omega}. \tag{10.82}$$

From the expression for the response function, the complex Fourier transform of the one-sided velocity autocorrelation function is therefore

$$\tilde{\ddot{q}}_0^s(\omega) = \frac{T}{m}\left[\frac{\gamma\omega_c}{\omega_c - i\omega} - i\omega\right]^{-1}. \tag{10.83}$$

In the short memory or low friction case, $\omega_c > 4\gamma$, there are simple poles at $-i\omega_\pm$, with

$$\omega_\pm \equiv \frac{\omega_c}{2}\left[1 \pm [1-4\gamma/\omega_c]^{1/2}\right]. \tag{10.84}$$

These give an exponentially decaying velocity autocorrelation function,

$$\ddot{q}_0^s(t) = \frac{T}{m}[1-4\gamma/\omega_c]^{-1/2}\left[\frac{\omega_+}{\omega_c}e^{-\omega_-|t|} - \frac{\omega_-}{\omega_c}e^{-\omega_+|t|}\right]. \tag{10.85}$$

Because $\omega_+ > \omega_-$, this is strictly positive and monotonic decaying. In the $\omega_c < 4\gamma$ case, which is the long memory or large friction case, one gets damped oscillatory behaviour. The velocity autocorrelation function for a Brownian particle in a soft sphere fluid, shown in Fig. 10.1 on p. 334, is clearly not pure exponential; to a crude approximation it might be fitted by a damped sinusoid.

In the large cut-off limit, $\omega_c \to \infty$, the memory function becomes a δ-function, $K(t) \to 2\gamma\delta(t)$. Since $\omega_+ \to \omega_c$ and $\omega_- \to \gamma$, the velocity autocorrelation function becomes a single exponential, $\ddot{q}_0^s(t) \to (T/m)e^{-\gamma|t|}$. In this limit the generalised Langevin equation reduces to

$$m\ddot{x}(t) = F^{\text{ext}}(x(t), t) + R_p(t) - \gamma\dot{x}(t), \tag{10.86}$$

which is just the ordinary Langevin equation without memory. The quantity γ is the friction or drag coefficient.

10.4 Adiabatic Linear Response Theory

Now a linear response approach is used as an alternative to the perturbation theory. In the present case the adiabatic evolution of an isolated total system is used, whereas in the perturbation theory the particle was treated explicitly and the reservoir was treated statistically. In the present case the initial distribution is taken to be the canonical equilibrium distribution, whereas in the perturbation theory the exact non-equilibrium probability distribution was implicit. In spite of these differences, the adiabatic linear response theory, like the perturbation theory, shows that the velocity autocorrelation function is the susceptibility that links the velocity to the applied force. A number of such linear response relationships between susceptibilities and time correlation functions were obtained by Green,[14] and by Kubo.[15]

For simplicity, a single Brownian particle is considered and only one dimension of its motion is treated explicitly. Let $\mathbf{X}$ be a point in the total system phase space, and write the total Hamiltonian $\mathcal{H}(\mathbf{X}, t) = \mathcal{H}_0(\mathbf{X}) + U(x, t)$, where x is the position of the Brownian particle. The total system phase space $\mathbf{X}$ includes x and $\dot{x}$ as well as the solvent positions and momenta. The total system is isolated (apart from the applied external potential) and evolves adiabatically according to Hamilton's equations.

Let the system be Boltzmann distributed at the initial time $t = t_0$,

$$\wp(\mathbf{X}, t_0) = \frac{1}{Z(t_0)} e^{-\beta \mathcal{H}(\mathbf{X}, t_0)}, \tag{10.87}$$

where $\beta \equiv 1/k_\mathrm{B}T$. The system evolves adiabatically according to the equations of motion for the total Hamiltonian; the trajectory may be written $\mathbf{X}_U(t|\mathbf{X}_1, t_0)$. Since Hamilton's equations are incompressible, the corresponding probability at time t is

$$\wp(\mathbf{X}, t) = \wp(\mathbf{X}_U(t_0|\mathbf{X}, t), t_0) = \frac{1}{Z(t_0)} e^{-\beta \mathcal{H}(\mathbf{X}_U(t_0|\mathbf{X}, t), t_0)}. \tag{10.88}$$

The total energy at time t equals the initial energy plus the work done on the trajectory,

$$\begin{aligned}
\mathcal{H}(\mathbf{X}, t) &= \mathcal{H}(\mathbf{X}_U(t_0|\mathbf{X}, t), t_0) + \int_{t_0}^{t} \mathrm{d}t' \, \frac{\partial \mathcal{H}(\mathbf{X}', t')}{\partial t'}, \quad \mathbf{X}' \equiv \mathbf{X}_U(t'|\mathbf{X}, t), t') \\
&= \mathcal{H}_0(\mathbf{X}_U(t_0|\mathbf{X}, t)) + U(x_U(t_0|\mathbf{X}, t), t_0) + \int_{t_0}^{t} \mathrm{d}t' \, \frac{\partial U(x', t')}{\partial t'} \\
&= \mathcal{H}_0(\mathbf{X}_U(t_0|\mathbf{X}, t)) + U(x, t) - \int_{t_0}^{t} \mathrm{d}t' \left\{ \frac{\mathrm{d}U(x', t')}{\mathrm{d}t'} - \frac{\partial U(x', t')}{\partial t'} \right\} \\
&= \mathcal{H}_0(\mathbf{X}_U(t_0|\mathbf{X}, t)) + U(x, t) + \int_{t_0}^{t} \mathrm{d}t' \, \dot{x}' F^{\mathrm{ext}}(x', t'),
\end{aligned} \tag{10.89}$$

[14] Green, M. S. (1954), J. Chem. Phys. **22**, 398.

[15] Kubo, R. (1966), Rep. Progr. Phys. **29**, 255. Kubo, R., Toda, M., and Hashitsume, N. (1978), *Statistical Physics II. Non-equilibrium Statistical Mechanics*, (Springer-Verlag, Berlin).

where the external force is $F^{\rm ext}(x,t) = -\partial U(x,t)/\partial x$, and $x' \equiv x_U(t'|\mathbf{X},t)$. Hence the non-equilibrium probability can be written as

$$\begin{aligned}\wp(\mathbf{X},t) &= \frac{1}{Z(t_0)} \exp -\beta \left[\mathcal{H}_0(\mathbf{X}) - \int_{t_0}^{t} \mathrm{d}t'\, \dot{x}' F^{\rm ext}(x',t') \right] \\ &= \frac{1}{Z(t_0)} e^{-\beta \mathcal{H}_0(\mathbf{X})} \left[1 + \beta \int_{t_0}^{t} \mathrm{d}t'\, \dot{x}' F^{\rm ext}(x',t') + \mathcal{O}(U^2) \right].\end{aligned} \tag{10.90}$$

The second equality is the linearisation with respect to the external potential. Because quadratic terms are neglected, and because the external force is linear in the potential, one can take $x' = x_0(t'|\mathbf{X},t) + \mathcal{O}(U)$. The partition function is unchanged to linear order,

$$\begin{aligned}Z(t) &= \int \mathrm{d}\mathbf{X}\, e^{-\beta \mathcal{H}_0(\mathbf{X},t)} \left[1 + \beta \int_{t_0}^{t} \mathrm{d}t'\, \dot{x}' F^{\rm ext}(x',t') + \mathcal{O}(U^2) \right] \\ &= Z(t_0) \left[1 + \beta \int_{t_0}^{t} \mathrm{d}t'\, \left\langle F^{\rm ext}(x(t'),t') \dot{x}(t') \right\rangle_0 \right] \\ &= Z(t_0) + \mathcal{O}(U^2).\end{aligned} \tag{10.91}$$

The last equality follows because the average exactly vanishes, there being no instantaneous correlation between variables of opposite parity in the equilibrium system. (Note that $\mathrm{d}\mathbf{X} = \mathrm{d}\mathbf{X}'$ and $\mathcal{H}_0(\mathbf{X}) = \mathcal{H}_0(\mathbf{X}')$.)

The average velocity induced by the external potential is[16]

$$\begin{aligned}\langle \dot{x}(t) \rangle_U &= \int \mathrm{d}\mathbf{X}\, \wp(\mathbf{X},t) \dot{x}(t) \\ &= \int \mathrm{d}\mathbf{X}\, \frac{1}{Z(t_0)} e^{-\beta \mathcal{H}_0(\mathbf{X})} \beta \int_{t_0}^{t} \mathrm{d}t'\, F^{\rm ext}(x(t'),t') \dot{x}(t') \dot{x}(t) \\ &= \beta \int_{t_0}^{t} \mathrm{d}t'\, \left\langle F^{\rm ext}(x(t'),t') \dot{x}(t') \dot{x}(t) \right\rangle_0 \\ &= \beta \int_{t_0}^{t} \mathrm{d}t'\, F^{\rm ext}(x(t'),t') \left\langle \dot{x}(t') \dot{x}(t) \right\rangle_0 .\end{aligned} \tag{10.92}$$

In the final equality the external force, which depends upon the actual position of the Brownian particle, has been taken outside the average. The argument for doing this is that the correlation between $x(t')$ and $\dot{x}(t)$ is linear in the external force, and so this contributes a second-order term (c.f. the similar discussion in the non-equilibrium perturbation theory in §10.3, specifically the justification for Eq. (10.47)). Alternatively, the force could be replaced by its most likely

[16] Equilibrium, denoted by the subscript 0, here means an average in the adiabatic or isolated system in the absence of the external force. In the preceding sections it denoted an average in the canonical equilibrium system that could exchange energy with a thermal reservoir, also in the absence of the external force. The two averages are expected to be equal. Also, the symmetrised velocity autocorrelation function is $\ddot{q}_0^{\rm s}(\tau) = \left[\langle \dot{x}(t+\tau)\dot{x}(t) \rangle_0 + \langle \dot{x}(t)\dot{x}(t+\tau) \rangle_0 \right] / 2k_{\rm B}$. In the present pure parity case this is equal to the ordinary velocity autocorrelation function, $\ddot{q}_0^{\rm s}(\tau) = \ddot{q}_0(\tau) = \langle \dot{x}(t+\tau)\dot{x}(t) \rangle_0 / k_{\rm B}$.

value, $\overline{F}^{\rm ext}(t') \equiv F^{\rm ext}(\overline{x}(t'), t')$, as was done in §10.3.2. The result shows that the equilibrium velocity autocorrelation function times the inverse temperature gives the susceptibility for the velocity induced by a time dependent external force.

In terms of the Heaviside step function, which is $\Theta(t) = 1,\ t > 0,\ \Theta(t) = 0,\ t < 0$, one can define $\tilde{\ddot{q}}_0(t) \equiv \Theta(t)\ddot{q}_0(t)$, with the velocity autocorrelation function being $\ddot{q}_0(t) = k_{\rm B}^{-1}\left\langle \dot{x}(t)\dot{x}(0)\right\rangle_0$. In the present section this is a scalar. With this the average induced velocity may be written

$$\left\langle \dot{x}(t)\right\rangle_U = \frac{1}{T}\int_{-\infty}^{\infty} {\rm d}t'\ F^{\rm ext}(x(t'), t')\tilde{\ddot{q}}_0(t-t'). \tag{10.93}$$

The lower limit on the integral has been extended to $-\infty$ because $\ddot{q}_0(t)$ is short-ranged. This has Fourier transform

$$\left\langle \dot{x}(\omega)\right\rangle_U = \frac{1}{T}\tilde{\ddot{q}}_0(\omega)F^{\rm ext}(\omega). \tag{10.94}$$

One often sees in the literature a susceptibility or linear response to the external force, $\left\langle \dot{x}(\omega)\right\rangle_U = \mathcal{A}(\omega)F^{\rm ext}(\omega)$. The present result shows that the response function is here $\mathcal{A}(t) = \Theta(t)\ddot{q}_0(t)/T$.

This result may be inverse transformed (or else use Eq. (10.92) directly), and the random force added to obtain the generalised Langevin equation for the position,

$$\dot{x}(t) = R_x(t) + \frac{1}{T}\int_{-\infty}^{t} {\rm d}t'\ \ddot{q}_0(t-t')F^{\rm ext}(t'). \tag{10.95}$$

Set the external force to zero, $F^{\rm ext}(t') = 0$, multiply the respective sides of this by their value at $t = t'$, and take the average to obtain the fluctuation dissipation theorem for this particular random force,

$$\left\langle R_x(t)R_x(t')\right\rangle_0 = \left\langle \dot{x}(t)\dot{x}(t')\right\rangle_0 = k_{\rm B}\ddot{q}_0(t-t'). \tag{10.96}$$

Of course R_x has units of velocity and as such it is a scaled force. These last two results and Eq. (10.92) agree with the non-equilibrium perturbation theory results in §10.3.4, namely Eq. (10.59), Eq. (10.61), and Eq. (10.63).

10.5 Numerical Results for a Brownian Particle in a Moving Trap

The perturbation theory will now be tested for a Brownian particle in a parabolic potential in harmonic motion. That is, the external potential is

$$U(x,t) = \frac{\kappa}{2}[x - b(t)]^2, \tag{10.97}$$

with the trap being located at

$$b(t) = B\cos\omega t. \tag{10.98}$$

The applied force is $F(x,t) = -\kappa[x - b(t)]$.

In the linear regime the most likely trajectory can be expected to be of the form,

$$\overline{x}(t) = A\cos(\omega t + \phi), \quad \dot{\overline{x}}(t) = -A\omega \sin(\omega t + \phi). \tag{10.99}$$

The approximate theories will be tested against computer simulation results for the relative amplitude A/B, and the phase lag ϕ.

10.5.1 Langevin Theory

The deterministic part of the Langevin equation adds a dissipative drag force, $F_{\text{drag}} = -\gamma\dot{x}$, to the mechanical force in Newton's equation,

$$m\ddot{\overline{x}}(t) = F(\overline{x}(t), t) - \gamma\dot{\overline{x}}(t). \tag{10.100}$$

The drag coefficient is $\gamma \equiv 1/\beta D$.

In the present case of a parabolic trap, with a harmonic time dependence and response, $\overline{x}(t) = A\cos(\omega t + \phi)$, the Langevin equation becomes

$$\begin{aligned} -Am\omega^2\cos(\omega t + \phi) &= -\kappa\left[A\cos(\omega t + \phi) - B\cos(\omega t)\right] \\ &\quad + \frac{\omega A}{\beta D}\sin(\omega t + \phi). \end{aligned} \tag{10.101}$$

Equating the coefficients of $\sin\omega t$ gives

$$Am\omega^2 \sin\phi = \kappa A \sin\phi + \frac{\omega A}{\beta D}\cos\phi, \tag{10.102}$$

or

$$\tan\phi = \frac{\omega/m\beta D}{\omega^2 - \kappa/m}. \tag{10.103}$$

This predicts a change in phase at the mechanical resonance of $\omega = \surd(\kappa/m)$. Equating the coefficients of $\cos\omega t$ gives

$$-Am\omega^2\cos\phi = -\kappa\left[A\cos\phi - B\right] + \frac{\omega A}{\beta D}\sin\phi, \tag{10.104}$$

or

$$A = \frac{B\kappa}{[\kappa - m\omega^2]\cos\phi - (\omega/\beta D)\sin\phi}. \tag{10.105}$$

This has the appearance of a damped harmonic oscillator.

10.5.2 Smoluchowski Theory

In the over damped case, the acceleration is zero and the drag force is equal and opposite to the applied force. This is the Smoluchowski equation,

$$\dot{\overline{x}} = \gamma^{-1} F(x) = \beta D F(x), \tag{10.106}$$

which becomes in this case

$$-A\omega \sin(\omega t + \phi) = -\beta D\kappa \left[A \cos(\omega t + \phi) - B \cos(\omega t) \right]. \tag{10.107}$$

Expanding these, the coefficient of $\sin \omega t$ is

$$A\omega \cos \phi = -A\beta D\kappa \sin \phi, \text{ or } \tan \phi = \frac{-\omega}{\beta D\kappa}. \tag{10.108}$$

The coefficient of $\cos \omega t$ is

$$A\omega \sin \phi = \beta D\kappa \left[A \cos \phi - B \right], \text{ or } A = \frac{\beta D\kappa B}{\beta D\kappa \cos \phi - \omega \sin \phi}. \tag{10.109}$$

Since the Smoluchowski theory is only valid for slowly varying forces, it is arguable that one ought to take the low frequency limit of this, which gives an amplitude that decreases quadratically with frequency,

$$A = B \left[1 - \frac{\omega^2}{2[\beta D\kappa]^2} \right] + \mathcal{O}(\omega^4). \tag{10.110}$$

10.5.3 Computer Simulations

Two types of simulations were performed for the driven Brownian particle: conventional equilibrium molecular dynamics, and non-equilibrium stochastic molecular dynamics, which will be described in Ch. 11.

The conventional molecular dynamics simulations are detailed by Attard and Gray-Weale (2008). These used an equilibrium Nosé-Hoover chain thermostat with thermostat relaxation of 20. The thermostat appeared adequate as far as controlling the temperature was concerned, since there was no evidence of local heating of the solvent. In these simulations 3095 solvent molecules were used, in a cube of edge length 16.48, with periodic boundary conditions. The trap constant was $\kappa = 16.8$, and the drive amplitude was either $B = 2.5$ or $B = 5$. The position and velocity of the particle were averaged at set points in the drive cycle and at the conclusion of the simulation a least squares fit of these yielded the relative amplitude and phase lag. All other parameters for the solvent and solute were as described for the simulation of the velocity autocorrelation function (see footnote 3 on p. 334).

The stochastic molecular dynamics simulations detailed by Attard (2009a)[17] used a simulation cell that was a cube of edge length 11.31, with 1000 solvent

[17] Attard, P. (2009a), J. Chem. Phys. **130**, 194113.

atoms and density of 0.69, and periodic boundary conditions. The drive amplitude was $B = 2.5$, and the trap constant was $\kappa = 16.81$. By varying the amplitude it was confirmed that the results were in the linear regime. Typically, 300 drive cycles were simulated using a time step of $\Delta_t = 10^{-3}$. All other parameters for the solvent and solute were as described for the simulation of the velocity autocorrelation function, (see footnote 3 on p. 334).

To obtain the amplitude and phase lag, in the stochastic molecular dynamics simulations the period of the harmonic potential was broken into 20 nodes, and the average position and velocity of the Brownian particle at each node were obtained, as well as the standard error on the mean. A least squares fit to these average values at the nodes yielded the relative amplitude and phase lag. This procedure is not exactly the same as obtaining directly the average relative amplitude and the average phase lag.

10.5.4 Perturbation Algorithm

The perturbation theory described in §10.3 was tested against the simulations. In the first instance only the most likely trajectory was generated, with the random force set to zero. The implementation was straightforward. The velocity autocorrelation function shown in Fig. 10.1 on p. 334, was used as input and interpolated onto the nodes used for the discrete trajectory. Values of $n \in [20, 100]$ were tested, and $\tau \in [0.05, 0.5]$. The most likely next velocity and position were generated from the current values on the n preceding nodes, as indicated in Eq. (10.52). Smoluchowski theory was used for the initial values, but it was found that the results were insensitive to this choice. The trajectory was followed until $t_j = \max(50, 10\pi/\omega)$. At the conclusion of the calculation, the amplitude and phase were obtained by a least squares fit to the trajectory on the n most recent nodes.

The perturbation theory is computationally trivial compared with the simulations (once the velocity autocorrelation function of the bare particle has been obtained). The computer program comprises on the order of 250 lines, and it takes about one second to obtain results for 50 frequencies on a personal computer.

10.5.5 Relative Amplitude and Phase Lag

Figure 10.3 shows the relative amplitude of the driven Brownian particle, as given by the various simulations and approximations. In general, the response amplitude decreases as the drive frequency increases, as one might expect for a damped harmonic oscillator. Indeed, it can be seen that the Langevin theory, which predicts just such an over damped oscillator, smoothly interpolates between the low frequency limit and the mechanical resonance at $\omega = \sqrt{(\kappa/m)} = 1.3$. The Langevin theory tends to underestimate the response amplitude in the range $\omega \in (0.2, 0.8)$, and to overestimate it in the range $\omega \in (0.9, 2)$.

The natural frequency that occurs in Smoluchowski theory is $\beta D \kappa = 0.59$. For frequencies somewhat less than this one might expect Smoluchowski theory

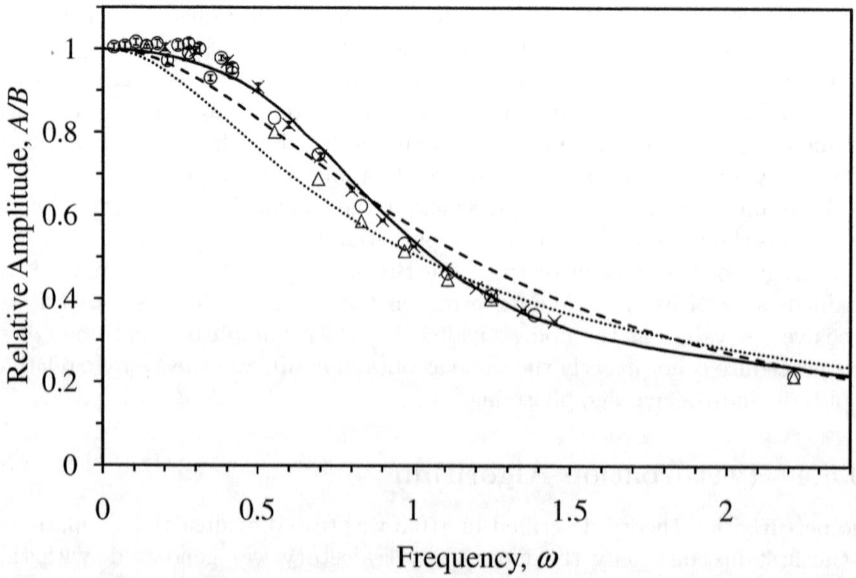

Figure 10.3: Relative amplitude as a function of drive frequency of a trapped Brownian particle. The circles ($B = 2.5$) and triangles ($B = 5$) are Nosé-Hoover equilibrium molecular dynamics simulations (Attard and Gray-Weale, 2008), and the crosses are non-equilibrium stochastic molecular dynamics simulations with $B = 2.5$ (Attard, 2009a). In these the statistical error is smaller than the size of the symbols. The solid line is the perturbation theory for the most likely trajectory, using $n = 50$, $\tau = 0.07$, and $t_{\max} = 50$, and the velocity autocorrelation function shown in Fig. 10.1 on p. 334. The dashed line is the Langevin result, Eq. (10.105), and the dotted line is the Smoluchowski result, Eq. (10.109), both using $D = 0.105$.

to be accurate. In fact, Smoluchowski theory only appears accurate for the response amplitude for $\omega \lesssim 0.1$, and it is probably somewhat of a coincidence that it comes back to reasonable agreement with the simulations for higher frequencies.

The two types of simulations—the Nosé-Hoover equilibrium molecular dynamics and the non-equilibrium stochastic molecular dynamics—are in quite good agreement with each other for the relative amplitude. This suggests that possibly the system is only weakly non-equilibrium, or that the thermostatting properties of the explicit solvent are efficient enough that the actual thermostat used does not significantly perturb the motion of the solute.

In so far as the simulated relative amplitude for $B = 5$ and $B = 2.5$ are in agreement, the results can be said to be in the linear regime, which is necessary for the validity of the perturbation theory. There is perhaps a hint of non-linearity in the range $\omega \in (0.5, 1)$, with the higher drive amplitude having a

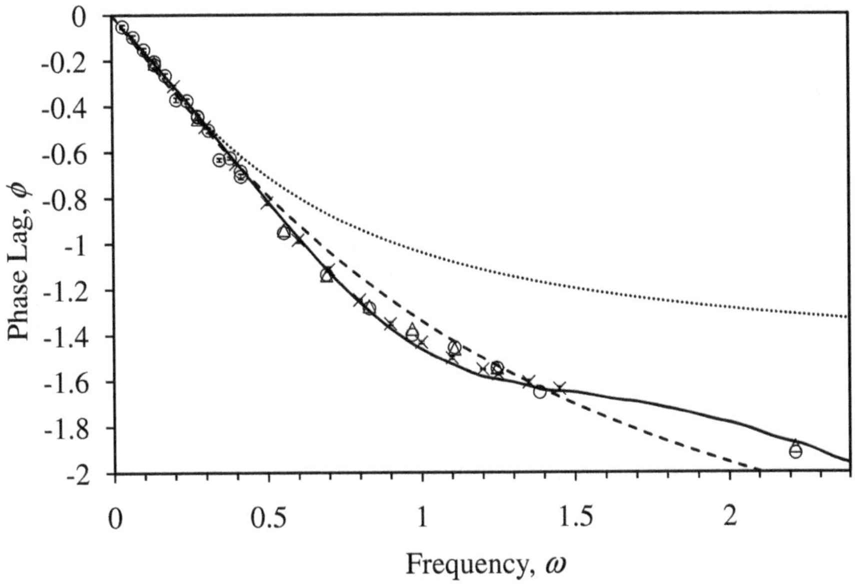

Figure 10.4: Phase lag as a function of drive frequency of a trapped Brownian particle. Symbols and curves are the same as in the preceding figure.

statistically significant lower response.

The perturbation theory for the most likely trajectory performs quite well, capturing the sigmoidal shape evident in the simulation data. At the highest frequency, $\omega = 2.5$, the time step used, $\tau = 0.07$, gives $\omega\tau/2\pi = 0.03$, or approximately 35 nodes per period, which is more than adequate. The number of nodes times the time step, $n\tau = 3.5$, is sufficient for the velocity autocorrelation function to have decayed significantly (see Fig. 10.1 on p. 334). In general varying the number of nodes or the time step made no significant change to the results. However the case $n = 50$, $\tau = 0.05$, which had the shortest memory of $n\tau = 2.5$, performed noticeably worse (i.e. an understated sigmoidal shape, rather like the Smoluchowski result) than the rest.

The phase lag is shown in Fig. 10.4. It can be seen that this is negative, which corresponds to the motion of the particle lagging that of the trap. The change from lag to lead, $\phi = \pm\pi/2$, occurs somewhere in the range $\omega \in (1.25, 1.39)$, according to the Nosé-Hoover molecular dynamics simulations (Attard and Gray-Weale, 2008). This is quite close to the mechanical resonance at $\omega = \sqrt{(\kappa/m)} = 1.3$, and explains to some extent the relatively good performance of the Langevin theory, since this smoothly interpolates between the low frequency results, where use of the diffusion constant and neglect of memory effects are exact, and the high frequency regime, where the particle behaves ballistically.

The Smoluchowski theory performs slightly better for the phase lag than for

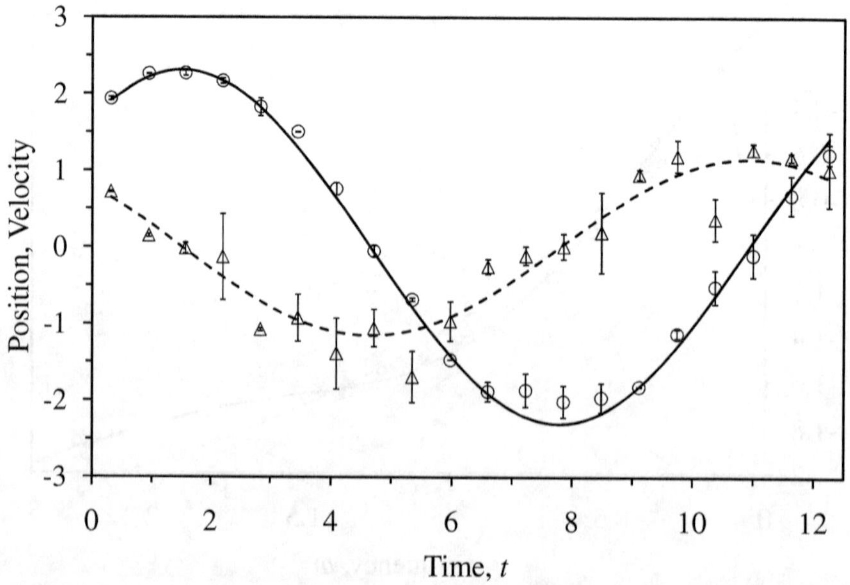

Figure 10.5: The average position (circles, solid curve) and velocity (triangles, dashed curve) of the driven Brownian particle over a period for $\omega = 0.5$ and $B = 2.5$. The symbols result from the stochastic dissipative equations, and the curves are best fits, $x(t) = A\cos(\omega t + \phi)$ and $\dot{x}(t) = -A\omega \sin(\omega t + \phi)$. The total length of the error bars is twice the standard error on the mean. The number of nodes was $n = 100$ and the time step was $\tau = 0.1$. All other parameters are as in the preceding figures.

the amplitude, and does not become inaccurate until $\omega \gtrsim 0.3$.

The perturbation for the most likely trajectory appears to be accurate over the whole regime. The small discrepancies between the perturbation theory and the simulation data may likely be attributed to systematic errors in the velocity autocorrelation function (such as, for example, those due to the periodic boundary conditions) or to the use of biased estimators (using the phase lag of the most likely trajectory rather than the most likely phase lag). There is little evidence for non-linear effects in the simulations.

10.5.6 Stochastic Trajectory

The algorithm for the stochastic dissipative equations resulting from the perturbation theory is essentially the same as that described in §10.5.4 for the most likely trajectory. The main difference is the addition of the random force at each step on the trajectory. Two sets of averages were stored: the amplitude and the phase lag, and the position and the velocity. In both cases the first three cycles were used for equilibration and then averages were taken until $t = 50$

Table 10.1: Amplitude and phase lag at $\omega \approx 0.5$ and at $\omega \approx 0.8$

ω	A/B	ϕ	Approach
0.50	$0.908 \pm .008$	$-0.820 \pm .009$	non-equil., stochastic MD[‡]
0.50	$1.01 \pm .01$	$-0.77 \pm .01$	non-equil., MC[♯]
0.55	$0.83 \pm .01$	$-0.95 \pm .01$	equil. MD[†]
0.50	0.921	-0.762	perturbation, most likely trajectory
0.50	$0.929 \pm .005$	$-0.753 \pm .006$	perturbation, stochastic trajectory
0.80	$0.661 \pm .004$	$-1.247 \pm .004$	non-equil. stochastic MD[‡]
0.80	$0.68 \pm .02$	$-1.19 \pm .02$	non-equil., MC[♯]
0.83	$0.62 \pm .01$	$-1.28 \pm .02$	equil. MD[†]
0.80	0.717	-1.29	perturbation, most likely trajectory
0.80	$0.717 \pm .004$	$-1.295 \pm .006$	perturbation, stochastic trajectory

[‡] Attard (2009a).
[♯] Attard (2009c), Phys. Rev. E **80**, 041126.
[†] Attard and Gray-Weale (2008).

(or until five cycles, whichever was greater). No attempt was made to optimise these. For the amplitude and phase, every $n/2$ nodes, a least squares fit was made to the trajectory on the preceding n nodes and stored for averaging. For the position and velocity, a period was divided into 20 points, and the position and velocity were stored each time the trajectory crossed one of these points in the cycle. The square of the quantities was also recorded and at the conclusion of the trajectory this was used to calculate the standard error on the mean. The time taken to run the stochastic algorithm differed immaterially from that taken for the deterministic algorithm.

The direct calculations of the amplitude and phase lag using the stochastic dissipative equations are not shown because they gave results indistinguishable from those obtained with the most likely trajectory and shown in Fig. 10.3 and in Fig. 10.4. Instead, Fig. 10.5 shows the position and velocity over a cycle for the case $\omega = 0.5$. Obviously the statistical error could have been reduced by collecting averages over more cycles. The relative amplitude obtained from a least squares fit to these data was $A/B = 0.923$ and the phase lag was $\phi = -0.752$. These can be compared to the averages obtained directly, $A/B = 0.906 \pm 0.013$ and $\phi = -0.763 \pm 0.033$. The values obtained from the deterministic calculation, the most likely trajectory, were $A/B = 0.921$ and $\phi = -0.762$ (for the same parameters, $n = 100$ and $\tau = 0.1$). In this particular case of driven Brownian motion, it appears that the amplitude and phase of the most likely trajectory were indistinguishable from the average amplitude and phase, and also from the amplitude and phase of the average trajectory.

Table 10.1 compares the results of the perturbation theory with three different simulation methods. The non-equilibrium stochastic molecular dynamics method and the non-equilibrium Monte Carlo method will be detailed in Ch. 11.

Of the three, the stochastic molecular dynamics results have the lowest statistical error and appear the most reliable. The perturbation theory results for a stochastic trajectory in Table 10.1 were obtained from a trajectory of length 1000.

There is some evidence for a weak non-linear dependence of the amplitude response at higher drive amplitudes. For example, at $\omega = 0.8$, in the case of non-equilibrium stochastic molecular dynamics simulations, drive amplitudes of 5, 2.5, 1, and 0.5 give $A/B = 0.624 \pm 0.002$, 0.657 ± 0.002, 0.664 ± 0.005, and 0.675 ± 0.006, respectively. The non-equilibrium Monte Carlo simulations for drive amplitudes of 1, 0.5, and 0.2 give 0.68 ± 0.02, 0.78 ± 0.05, and 0.76 ± 0.05. The phase lag showed less variability with drive amplitude.

10.6 Generalised Langevin Equation in the Case of Mixed Parity

10.6.1 Equilibrium System

The preceding perturbation theory used the symmetrised velocity autocorrelation function of the bare system as the response function. This is not a restriction in the pure parity case. However, in the general mixed parity case the velocity autocorrelation function contains antisymmetric components, and it is therefore worthwhile developing the perturbation theory for this more general case.

In §10.1.1 the time correlation functions of the bare Brownian particle were given, including the position autocorrelation function, Eq. (10.2),

$$q_x(\tau) \equiv k_{\mathrm{B}}^{-1} \left\langle x(t+\tau)x(t)\right\rangle_0 , \tag{10.111}$$

and the symmetrised displacement autocorrelation function,

$$\begin{aligned} q_0^{\mathrm{s}}(\tau) &\equiv \frac{1}{2k_{\mathrm{B}}} \left\langle [x(t+\tau)-x(t)][x(t+\tau)-x(t)]\right\rangle_0 \\ &= \frac{1}{2}[2q_x(0) - q_x(\tau) - q_x(\tau)^{\mathrm{T}}]. \end{aligned} \tag{10.112}$$

The latter matrix is by design symmetric, even in the mixed parity case, as was discussed in §10.1.2. It was from the symmetry of this matrix that the symmetrised form of the velocity autocorrelation function arose as the response function. The reason for starting the analysis from the symmetric displacement autocorrelation function, rather than from the position autocorrelation function, is that the latter scales like the square of the volume at $\tau = 0$, $q_x(0) \equiv k_{\mathrm{B}}^{-1} \left\langle x(t)x(t)\right\rangle_0 = V^2$. Hence the square root of this divided by the volume does not go to zero in the thermodynamic limit.

In a mathematical sense this divergence is not catastrophic, and even for the case of the bare Brownian particle one could formally proceed with the analysis based on the position autocorrelation function. In any event there can be no objections to the analysis for the case of thermodynamic fluctuations since such fluctuations occur about some fixed value and are bounded by a value that usually scales with the system size. Hence the root mean square fluctuation divided by the average value goes to zero in the thermodynamic limit.

There are qualitative differences in the evolution of the trajectory depending whether or not the position autocorrelation function diverges at $\tau = 0$, $\langle x(t)x(t)\rangle_0 \to \infty$. The fact that for the free Brownian particle, $Q_{0;xx} \to \infty$, but for a thermodynamic fluctuation, $Q_{0;xx} < \infty$, has consequences for the form of the response function (c.f. Eq. (10.33)), as will be detailed here.

A second difference between the pure parity Brownian motion case and the present mixed parity thermodynamic fluctuation case concerns the position velocity time correlation function. In the pure parity case, this vanishes at $t = 0$, $\langle \dot{x}(t)x(t)\rangle_0 = 0$, whereas this is non-zero in the mixed parity case. This also has consequences for the trajectory evolution.

Let x be a vector of N fluctuating thermodynamic variables of a sub-system. More precisely, let it measure the departure from the average so that its equilibrium average vanishes, $\langle x\rangle_0 = 0$. The sub-system, which can be isolated or in contact with a reservoir, has no macroscopic time dependence, and neither does the reservoir. Let the individual variables each be of pure parity, but let the vector as a whole have mixed parity, with diagonal parity matrix ε: $x(\mathbf{\Gamma}^\dagger) = \varepsilon x(\mathbf{\Gamma})$, where $\mathbf{\Gamma}$ is a point in the phase space of the sub-system and $\dagger$ denotes the conjugate point with all velocities of the total system reversed.

The velocity autocorrelation function of the mixed parity system is the $N\times N$ equilibrium time correlation dyadic matrix,[18]

$$\ddot{q}_0(\tau) \equiv k_{\mathrm{B}}^{-1} \langle \dot{x}(t+\tau)\dot{x}(t)\rangle_0 . \tag{10.113}$$

Since this matrix may have an antisymmetric component, it now has no superscript in order to distinguish it from the symmetrised version used in the preceding sections. Compare, for example, this expression for the velocity autocorrelation function with the mixed parity, symmetrised expression given above in Eq. (10.20). From the parity rules one has

$$\ddot{q}_0(-\tau) = \varepsilon\ddot{q}_0(\tau)\varepsilon = \ddot{q}_0(\tau)^{\mathrm{T}}, \tag{10.114}$$

and also

$$\begin{aligned}
\ddot{q}_0(\tau) &= k_{\mathrm{B}}^{-1}\frac{\mathrm{d}}{\mathrm{d}\tau}\langle x(t+\tau)\dot{x}(t)\rangle_0 \\
&= -k_{\mathrm{B}}^{-1}\varepsilon\frac{\mathrm{d}}{\mathrm{d}\tau}\langle x(t-\tau)\dot{x}(t)\rangle_0\,\varepsilon \\
&= -k_{\mathrm{B}}^{-1}\varepsilon\frac{\mathrm{d}}{\mathrm{d}\tau}\langle x(t)\dot{x}(t+\tau)\rangle_0\,\varepsilon \\
&= -k_{\mathrm{B}}^{-1}\varepsilon\,\langle x(t)\ddot{x}(t+\tau)\rangle_0\,\varepsilon \\
&= -k_{\mathrm{B}}^{-1}\,\langle \ddot{x}(t+\tau)x(t)\rangle_0 .
\end{aligned} \tag{10.115}$$

[18]Generally one is dealing with fluctuations in physical quantities, which are real valued variables. Complex variables can arise if one takes Fourier transforms, for example, and in such a case the time correlation functions are usually defined to include a complex conjugate.

The minus sign appears at the second equality because $-\varepsilon$ is the parity matrix of $\dot{x}$. The third equality follows from time homogeneity of an equilibrium system. In fact the final result could have been written down directly from the time homogeneity of the equilibrium system, Eq. (10.16). The reason for spelling out the derivation is to illustrate certain common manipulations of time correlation functions. It should be noted that all of the parity rules hold not just for an equilibrium adiabatic system, but also for an equilibrium sub-system in contact with an equilibrium reservoir, since forward and backward motion is equally likely for such equilibrium systems.

The position velocity correlation function is

$$\begin{aligned} \dot{q}_0(\tau) &= k_{\mathrm{B}}^{-1} \langle x(t+\tau)\dot{x}(t)\rangle_0 \\ &= -k_{\mathrm{B}}^{-1} \langle \dot{x}(t+\tau)x(t)\rangle_0 . \end{aligned} \tag{10.116}$$

Differentiating the first equality with respect to τ gives Eq. (10.113), and differentiating the second equality gives Eq. (10.115). The integration constant is chosen to be zero so that $\dot{q}_0(\tau) \to 0$, $\tau \to \infty$. (This is another difference between the present thermodynamic fluctuations and free Brownian motion.) The parity rules for this are

$$\dot{q}_0(-\tau) = -\varepsilon\dot{q}_0(\tau)\varepsilon = -\dot{q}_0(\tau)^{\mathrm{T}}. \tag{10.117}$$

In the analysis of Brownian motion above, §§10.1.1 and 10.1.2, the fact that $\dot{q}_0(\tau)$ was a symmetric matrix meant that the final equality here implied that $\dot{q}_0(0) = 0$. In the present case where the time correlation matrices have an antisymmetric part, this last equation shows that the position velocity correlation function is a strictly antisymmetric matrix at $\tau = 0$, $\dot{q}_0(0) = -\dot{q}_0(0)^{\mathrm{T}} \neq 0$. This result will be of use below.

The position autocorrelation function is

$$q_0(\tau) = -k_{\mathrm{B}}^{-1} \langle x(t+\tau)x(t)\rangle_0 . \tag{10.118}$$

Differentiating this yields the second equality in Eq. (10.116). Again unlike Brownian motion, the integration constant is chosen to be zero so that $q_0(\tau) \to 0$, $\tau \to \infty$. The parity rules for this are

$$q_0(-\tau) = \varepsilon q_0(\tau)\varepsilon = q_0(\tau)^{\mathrm{T}}. \tag{10.119}$$

Note the negative sign in the definition of $q_0(\tau)$. It is not possible to define both the velocity and the position autocorrelation functions with positive signs. (At least it is not possible to do so if the two are related by differentiation and go to zero at large times.) Since if the position autocorrelation function is a global maximum at $\tau = 0$, then the velocity autocorrelation function, its second derivative, must be negative and therefore a global minimum. Conversely, if one takes the position autocorrelation function to be a global minimum at $\tau = 0$, then the velocity autocorrelation function must be positive and therefore a global maximum. The convention chosen here is that the velocity autocorrelation function is defined with a positive sign.

Trajectory

As above, define $\dot{\mathbf{x}}^{(n/j)} \equiv \{\dot{x}_{j-n+1}, \dot{x}_{j-n+2}, \ldots \dot{x}_j\}$ to be the velocities on a trajectory of n nodes ending at node j, with $\dot{x}_k \equiv \dot{x}(t_k)$, and $t_k \equiv k\tau$. Let x_j be the terminal position and denote the trajectory by $\mathbf{X}^{(n/j)} \equiv \{x_j, \dot{\mathbf{x}}^{(n/j)}\}$, with the understanding that $x_{k-1} = x_k - \tau\dot{x}_k$, $k \in [2, n]$. (In §10.6.2 that follows, an example will be given that uses instead the initial position to fix the trajectory.)

As in §10.1.4, the free particle trajectory entropy is a quadratic form,

$$\begin{aligned} S_0^{(n)}(\mathbf{X}^{(n/j)};\tau) &= \frac{1}{2}S_0^{(n)}(\tau) : \mathbf{X}^{(n/j)}\mathbf{X}^{(n/j)} \\ &= \frac{1}{2}S_{0;xx}^{(n)}(\tau) : x_j^2 + S_{0;x\dot{x}}^{(n)}(\tau) : \dot{\mathbf{x}}^{(n/j)}x_j \\ &\quad + \frac{1}{2}S_{0;\dot{x}\dot{x}}^{(n)}(\tau) : \dot{\mathbf{x}}^{(n/j)}\dot{\mathbf{x}}^{(n/j)}. \end{aligned} \tag{10.120}$$

The total fluctuation matrix $S_0^{(n)}$, which has dimension $(n+1)N \times (n+1)N$, is symmetric and consists of four sub-matrices: $S_{0;xx}^{(n)} = [S_{0;xx}^{(n)}]^{\mathrm{T}}$, which is $N \times N$, $S_{0;\dot{x}\dot{x}}^{(n)} = [S_{0;\dot{x}\dot{x}}^{(n)}]^{\mathrm{T}}$, which is $nN \times nN$, $S_{0;x\dot{x}}^{(n)}$, which is $N \times nN$, and $S_{0;\dot{x}x}^{(n)} = [S_{0;x\dot{x}}^{(n)}]^{\mathrm{T}}$. The elements of the last 'vector' are $N \times N$ matrices, and this last symmetry means $\left\{S_{0;\dot{x}x;k}^{(n)}\right\}_{\alpha\beta} = \left\{S_{0;x\dot{x};k}^{(n)}\right\}_{\beta\alpha}$, where k refers to $\dot{x}_k$ and x refers to x_n.

The correlation matrix is

$$Q_0^{(n)}(\tau) \equiv k_{\mathrm{B}}^{-1}\left\langle \mathbf{X}^{(n)}\mathbf{X}^{(n)} \right\rangle_0, \tag{10.121}$$

and it likewise consists of four sub-matrices:

$$Q_{0;xx}^{(n)} \equiv k_{\mathrm{B}}^{-1}\left\langle x(t)x(t)\right\rangle_0 = -q_0(0), \tag{10.122}$$

$$Q_{0;\dot{x}\dot{x}}^{(n)}(\tau) \equiv k_{\mathrm{B}}^{-1}\left\langle \dot{\mathbf{x}}^{(n)}\dot{\mathbf{x}}^{(n)} \right\rangle_0 = \ddot{Q}_0^{(n)}(\tau), \tag{10.123}$$

$$Q_{0;x\dot{x}}^{(n)}(\tau) \equiv k_{\mathrm{B}}^{-1}\left\langle x_n\dot{\mathbf{x}}^{(n)} \right\rangle_0, \tag{10.124}$$

and $Q_{0;\dot{x}x}^{(n)} = [Q_{0;x\dot{x}}^{(n)}]^{\mathrm{T}}$.

The velocity autocorrelation matrix has elements

$$\ddot{Q}_{0;ik}^{(n)}(\tau) = k_{\mathrm{B}}^{-1}\left\langle \dot{x}(t_i)\dot{x}(t_k)\right\rangle_0 = \ddot{q}_0(t_i - t_k). \tag{10.125}$$

In the present mixed parity case $\ddot{Q}_0^{(n)}$ itself is symmetric: $\ddot{Q}_{0;i\alpha,j\gamma}^{(n)} = \ddot{Q}_{0;j\gamma,i\alpha}^{(n)}$, where Roman letters range over the nodes and Greek letters range over the components. However, $\ddot{q}_0(t)$ is not symmetric: $\ddot{q}_{0;\alpha\gamma}(t) \neq \ddot{q}_{0;\gamma\alpha}(t)$. In fact, $\ddot{q}_0(t) \neq \ddot{q}_0(-t)$, but $\ddot{q}_0(t) = \ddot{q}_0(-t)^{\mathrm{T}}$.

The position velocity time correlation vector (actually $N \times nN$ matrix) has elements

$$Q_{0;x\dot{x};k}^{(n)}(\tau) \equiv k_{\mathrm{B}}^{-1}\left\langle x(t_n)\dot{x}(t_k)\right\rangle_0 = \dot{q}_0(t_n - t_k). \tag{10.126}$$

Again these elements are $N \times N$ matrices, with $\left\{Q^{(n)}_{0;\dot{x}x;k}\right\}_{\alpha\beta} = \left\{Q^{(n)}_{0;x\dot{x};k}\right\}_{\beta\alpha} = \dot{q}_{0;\alpha\beta}(t_n - t_k) = -\dot{q}_{0;\beta\alpha}(t_k - t_n)$.

The fluctuation matrix is the negative inverse of the correlation matrix $Q_0^{(n)}(\tau)\, S_0^{(n)}(\tau) = -\mathrm{I}^{(n+1)}$. As shown in §10.1.4, this gives the fluctuation submatrices as

$$S^{(n)}_{0;xx} = -\left[1 - [Q^{(n)}_{0;xx}]^{-1} Q^{(n)}_{0;x\dot{x}} [Q^{(n)}_{0;\dot{x}\dot{x}}]^{-1} [Q^{(n)}_{0;x\dot{x}}]^{\mathrm{T}}\right]^{-1} [Q^{(n)}_{0;xx}]^{-1}, \tag{10.127}$$

$$S^{(n)}_{0;\dot{x}\dot{x}} = -\left[\mathrm{I}^{(n)} - [Q^{(n)}_{0;\dot{x}\dot{x}}]^{-1} Q^{(n)}_{0;\dot{x}x} [Q^{(n)}_{0;xx}]^{-1} Q^{(n)}_{0;x\dot{x}}\right]^{-1} [Q^{(n)}_{0;\dot{x}\dot{x}}]^{-1}, \tag{10.128}$$

and

$$S^{(n)}_{0;\dot{x}x} = \left[\mathrm{I}^{(n)} - [Q^{(n)}_{0;\dot{x}\dot{x}}]^{-1} Q^{(n)}_{0;\dot{x}x} [Q^{(n)}_{0;xx}]^{-1} Q^{(n)}_{0;x\dot{x}}\right]^{-1} [Q^{(n)}_{0;\dot{x}\dot{x}}]^{-1} Q^{(n)}_{0;\dot{x}x} [Q^{(n)}_{0;xx}]^{-1}. \tag{10.129}$$

The last two give

$$-[S^{(n)}_{0;\dot{x}\dot{x}}]^{-1} S^{(n)}_{0;\dot{x}x} = Q^{(n)}_{0;\dot{x}x} [Q^{(n)}_{0;xx}]^{-1}. \tag{10.130}$$

For the terminal node this is $Q^{(n)}_{0;\dot{x}x;n} [Q^{(n)}_{0;xx}]^{-1} = -\dot{q}_0(0)^{\mathrm{T}} q_0(0)^{-1}$, which will be used below.

The inverse of the velocity fluctuation matrix will play the rôle of the memory function for the perturbation theory below, and it is of some interest to attempt to interpret it physically. The memory matrix is

$$M^{(n)} \equiv [S^{(n)}_{0;\dot{x}\dot{x}}]^{-1} = -Q^{(n)}_{0;\dot{x}\dot{x}} + Q^{(n)}_{0;\dot{x}x} [Q^{(n)}_{0;xx}]^{-1} Q^{(n)}_{0;x\dot{x}}. \tag{10.131}$$

The first term represents the contribution to the entropy from the direct correlation between the velocities at a pair of nodes. This contribution occurs irrespective of the location of the terminal position. The second term represents the indirect contribution that arises from the correlation between velocities as it is affected by the location of the terminal position. In so far as the thermodynamic fluctuation is localised, this term is non-zero. For the case of a free Brownian particle, $Q^{(n)}_{0;xx} \to \infty$, leaving $M^{(n)} = -Q^{(n)}_{0;\dot{x}\dot{x}}$.

In component form one has

$$M^{(n)}_{ik} = -\ddot{q}_0(t_i - t_k) - \dot{q}_0(t_n - t_i)^{\mathrm{T}} q_0(0)^{-1} \dot{q}_0(t_n - t_k). \tag{10.132}$$

Each of these is an $N \times N$ matrix. The memory matrix itself is in total symmetric, $M^{(n)}_{i\alpha,k\gamma} = M^{(n)}_{k\gamma,i\alpha}$.

The case $i = n$ will be required below, and so one can define the memory 'vector' with components

$$M_k \equiv M^{(n)}_{nk} = -\ddot{q}_0(t_n - t_k) - \dot{q}_0(0)^{\mathrm{T}} q_0(0)^{-1} \dot{q}_0(t_n - t_k). \tag{10.133}$$

In the continuum limit this is $M(t-t') \equiv -\ddot{q}_0(t-t') - \dot{q}_0(0)^{\mathrm{T}} q_0(0)^{-1} \dot{q}_0(t-t')$. The symmetry of the memory matrix manifests itself as $M_{\alpha\gamma}(t-t') = M_{\gamma\alpha}(t'-t)$. The memory function is short-ranged, $M(t) \rightarrow 0$, $|t| \rightarrow \infty$, due to the short-range nature of the velocity autocorrelation function and of the position velocity correlation function for thermodynamic fluctuations. This means that either $n\tau$ in the discrete case or the integration interval in the continuum case can be relatively short.

It should be noted that the actual memory function depends upon the choice of the terminal node x_j as the location of the position that fixes the trajectory. As a result of this choice, the zero time position velocity correlation function, $\dot{q}(0)$, appears here and in the results below. If one were to choose another position, x_k say, then the form of the memory function would change, as would the first term in the equation for the most likely next velocity, but, presumably, the predicted most likely next velocity $\overline{\dot{x}}_j$ would remain unchanged. An example of this is now given, and it is also further discussed on p. 377 below.

10.6.2 Regression of Fluctuation

Above the terminal position x_n was used in conjunction with the velocities $\dot{\mathbf{x}}^{(n)}$ to specify a trajectory. This is somewhat arbitrary, with one motivation for this specific choice being to match the perturbation results with the projector operator results in §10.7 below. An alternative choice is the initial position just prior to the start of the trajectory, x_0, which accounts for the prior state of the system before the external force is applied.

For the general case one could in addition specify the prior velocity, $\dot{x}_0$, but this complicates the analysis with little gain. For the case of spontaneous regressions of fluctuations in an equilibrium system, one begins in the dynamically disordered state so that $\dot{x}_0 = 0$ anyway.

In view of this, in this section the trajectory is $\mathbf{X}^{(n)} \equiv \{x_0, \dot{\mathbf{x}}^{(n)}\}$. The trajectory entropy is formally the same as Eq. (10.120), except that $S_{0;x\dot{x}}^{(n)}$ now couples the velocity to the position x_0,

$$\begin{aligned} S_0^{(n)}(\mathbf{X}^{(n)};\tau) &= \frac{1}{2} S_0^{(n)}(\tau) : \mathbf{X}^{(n)}\mathbf{X}^{(n)} \\ &= \frac{1}{2} S_{0;xx}^{(n)}(\tau) : x_0^2 + S_{0;x\dot{x}}^{(n)}(\tau) : \dot{\mathbf{x}}^{(n)} x_0 \\ &\quad + \frac{1}{2} S_{0;\dot{x}\dot{x}}^{(n)}(\tau) : \dot{\mathbf{x}}^{(n)}\dot{\mathbf{x}}^{(n)}. \end{aligned} \tag{10.134}$$

The derivative of the trajectory entropy with respect to the velocities is

$$\frac{\partial S_0^{(n)}(\mathbf{X}^{(n)};\tau)}{\partial \dot{\mathbf{x}}^{(n)}} = S_{0;\dot{x}x}^{(n)}(\tau) x_0 + S_{0;\dot{x}\dot{x}}^{(n)}(\tau) \dot{\mathbf{x}}^{(n)}. \tag{10.135}$$

Hence the most likely trajectory given the starting fluctuation (and $\dot{x}_0 = 0$) is

$$\begin{aligned} \overline{\dot{\mathbf{x}}}^{(n)} &= -[S_{0;\dot{x}\dot{x}}^{(n)}(\tau)]^{-1} S_{0;x\dot{x}}^{(n)}(\tau)^{\mathrm{T}} x_0 \\ &= Q_{0;\dot{x}x}^{(n)} [Q_{0;xx}^{(n)}]^{-1} x_0. \end{aligned} \tag{10.136}$$

This has to be the same as what can be obtained directly. Right multiplying this by x_0 and taking the average gives

$$\text{LHS} = \left\langle \overline{\dot{\mathbf{x}}}_n x_0 \right\rangle_0 = \langle \dot{\mathbf{x}}_n x_0 \rangle_0 = k_{\text{B}} \dot{q}_0(t_0 - t_n)^{\text{T}}, \tag{10.137}$$

and

$$\text{RHS} = -\dot{q}_0(t_0 - t_n)^{\text{T}} q_0(0)^{-1} \langle x_0 x_0 \rangle_0 = k_{\text{B}} \dot{q}_0(t_0 - t_n)^{\text{T}}. \tag{10.138}$$

The equality of these is consoling.

The final element of $\overline{\dot{\mathbf{x}}}_n$, is the most likely terminal velocity, and this is explicitly

$$\overline{\dot{x}}_n = -\dot{q}_0(t_0 - t_n)^{\text{T}} q_0(0)^{-1} x_0 = \dot{q}_0(t_0 - t_n)^{\text{T}} S x_0, \tag{10.139}$$

since $Q^{(n)}_{0;xx} = -q_0(0) = -S^{-1}$. Here the fluctuation form for the equilibrium entropy has been used,

$$S^{(1)}(x) = \frac{1}{2} S : xx, \;\; S = k_{\text{B}} \langle xx \rangle_0^{-1} = -q_0(0)^{-1}. \tag{10.140}$$

Hence one sees that this is just the Onsager regression of the fluctuation, with $F = \partial S/\partial x = Sx$ being the thermodynamic force.

Initial Velocity

To add the initial velocity, one can take $X_0 \equiv \{x_0, \dot{x}_0\}$,

$$Q^{(n)}_{0;XX} \equiv k_{\text{B}}^{-1} \begin{pmatrix} \langle x_0 x_0 \rangle_0 & \langle x_0 \dot{x}_0 \rangle_0 \\ \langle \dot{x}_0 x_0 \rangle_0 & \langle \dot{x}_0 \dot{x}_0 \rangle_0 \end{pmatrix} = \begin{pmatrix} -q_0(0) & \dot{q}_0(0) \\ \dot{q}_0(0)^{\text{T}} & \ddot{q}_0(0) \end{pmatrix}, \tag{10.141}$$

and

$$Q^{(n)}_{0;X\dot{x};n} \equiv k_{\text{B}}^{-1} \begin{pmatrix} \langle x_0 \dot{x}_n \rangle_0 \\ \langle \dot{x}_0 \dot{x}_n \rangle_0 \end{pmatrix} = \begin{pmatrix} \dot{q}_0(t_0 - t_n) \\ \ddot{q}_0(t_0 - t_n) \end{pmatrix}. \tag{10.142}$$

Hence one has in this case

$$\begin{aligned} \overline{\dot{x}}_n &= -\left\{ [S^{(n)}_{0;\dot{x}\dot{x}}(\tau)]^{-1} S^{(n)}_{0;\dot{x}X}(\tau)^{\text{T}} \right\}_{n\cdot} X_0 \\ &= Q^{(n)}_{0;\dot{x}X;n} [Q^{(n)}_{0;XX}]^{-1} X_0 \\ &= \begin{pmatrix} \dot{q}_0(t_0 - t_n)^{\text{T}} \\ \ddot{q}_0(t_0 - t_n)^{\text{T}} \end{pmatrix} \cdot \begin{pmatrix} -q_0(0) & \dot{q}_0(0) \\ \dot{q}_0(0)^{\text{T}} & \ddot{q}_0(0) \end{pmatrix}^{-1} \begin{pmatrix} x_0 \\ \dot{x}_0 \end{pmatrix}. \end{aligned} \tag{10.143}$$

Right multiplying by X_0 and taking the average yields an identity. Clearly this vanishes as $|t_n - t_0| \to \infty$, which shows that the most likely terminal velocity becomes independent of the initial conditions.

10.6.3 Time Dependent Perturbation

Return to the original representation of the trajectory, $\mathbf{X}^{(n/j)} \equiv \{x_j, \dot{\mathbf{x}}^{(n/j)}\}$. In the case of driven Brownian motion, it was shown in Eq. (10.45) that the trajectory entropy was

$$\begin{aligned} & S^{(n)}(\mathbf{X}^{(n/j)}; \tau, U) \\ &= S_0^{(n)}(\mathbf{X}^{(n/j)}; \tau) + \frac{1}{T}\int_0^t \mathrm{d}t'\, \dot{x}(t') F(x(t'), t') \\ &= \frac{1}{2} S_0^{(n)}(\tau) : \mathbf{X}^{(n/j)}\mathbf{X}^{(n/j)} + \frac{\tau}{T}\dot{\mathbf{x}}^{(n/j)} \cdot \mathbf{F}^{(n/j)}. \end{aligned} \tag{10.144}$$

For simplicity, the integral is written for $t_j = t$ and $t_{j-n+1} = 0$. This is the same as Eq. (10.45) except that here the trajectory ends at j. In this case the second term is the entropy produced over the trajectory, since it may be written

$$\frac{1}{T}\int_0^t \mathrm{d}t'\, \dot{x}(t') F(x(t'), t') = \int_0^t \mathrm{d}t'\, \dot{x}(t') \nabla S_{\mathrm{st}}(X(t'), t'), \tag{10.145}$$

where the static part of the entropy in this mechanical case is just $S_{\mathrm{st}}(X, t) = -\mathcal{H}(X, t)/T$, and $F = -\nabla\mathcal{H}$. In obtaining this result it was necessary that the additional solvent forces on the particle due to the presence of the perturbing potential were equal and opposite to the applied perturbing force. This had to be the case in order for the trajectory in the perturbed system to be identical to that in the bare system, so that the bare trajectory entropy could be invoked. It is worth noting that the memory effects arose from the trajectory entropy of the bare system, and that only the simplest instantaneous entropy arose in direct response to the external force via the induced internal force.

Linear versus Non-Linear Force

In the present case of mixed parity thermodynamic fluctuations, the sub-system first entropy is given by the usual fluctuation formula. It is the linear regime that is of most interest, and in this case the entropy is

$$S(x, t) = \frac{1}{2} S : xx. \tag{10.146}$$

Consequently the force is

$$F_{\mathrm{TD}}(x, t) = Sx. \tag{10.147}$$

In this case the time dependence is indirect and occurs only via the current position $x(t)$. The fluctuation matrix is of course the negative inverse of the correlation matrix, which is just the position autocorrelation function at $\tau = 0$,

$$S = -k_{\mathrm{B}} \langle x(0)x(0)\rangle_0^{-1} = q_0(0)^{-1}. \tag{10.148}$$

Note the positive sign on the right-hand side of the final equality (c.f. the discussion on p. 368), which is consistent with the fact that the fluctuation matrix has to be negative definite.

The static part of the entropy is an equilibrium property and as such it does not couple variables of opposite time parity. Hence the static fluctuation matrix commutes with the parity matrix and is consequently symmetric,

$$\varepsilon S \varepsilon = S, \text{ and } S = S^{\mathrm{T}}. \tag{10.149}$$

In the non-linear regime, one can still invoke a fluctuation form for the entropy in terms of the departure from the most likely state,

$$S(x,t) = \frac{1}{2} S(t) : [x - \overline{x}(t)][x - \overline{x}(t)]. \tag{10.150}$$

Consequently the force is

$$F_{\mathrm{TD}}(x,t) = S(t)[x - \overline{x}(t)]. \tag{10.151}$$

The time dependence of $S(t)$, as discussed in §3.6.3 on p. 78, arises because it is a material property that may depend upon the most likely current state, $S(\overline{x}(t))$,

$$S(t) \equiv -k_{\mathrm{B}} \left\langle [x - \overline{x}(t)][x - \overline{x}(t)] \right\rangle_{\mathrm{st}}^{-1}. \tag{10.152}$$

In the non-linear case $\overline{x}(t)$ can differ significantly from $\langle x \rangle_0 = 0$, which in turn means that $S(t)$ can differ significantly from $S \equiv k_{\mathrm{B}} \langle xx \rangle_0^{-1}$. The fluctuations about $\overline{x}(t)$ are always relatively small, which justifies the quadratic form even in the non-linear regime.

In what follows results will be presented in terms of this generic non-linear thermodynamic force, $F(x,t)$, and also the linear case where $F(x,t) = Sx(t)$. The time correlation matrices defined above, $\ddot{q}_0(\tau)$, $\dot{q}_0(\tau)$, and $q_0(\tau)$, depend only upon the interval τ and not upon the time t, and so are most consistent with the linear case (c.f. the formula $S = q_0(0)^{-1}$). They could of course be redefined in the event that the system is non-linear.

Internal versus External Force

In the case of driven Brownian motion, the external mechanical force, $F(x,t)$, appeared explicitly. This was justified by saying that it caused an equal and opposite internal solvent force, and that it was this internal force that changed the entropy of the solvent. The nett result was that $F(x,t)/T \Rightarrow \nabla S(x,t)$. This was the instantaneous form for the entropy, and, in the perturbation approximation, it was simply added to the trajectory entropy of the bare system which incorporated all the memory effects.

In the present thermodynamic problem, it is most common to specify the preceding trajectory, $x(t')$, $t' < t$, without being explicit about the external force that caused this trajectory. The reasons for this will be discussed in the following paragraph. The generalised Langevin equation below will give the current velocity that arises from the past positions of the sub-system, without needing to specify how those past positions came to be. This means that the thermodynamic force $F_{\rm TD}(x,t) = \nabla S(x,t)$ that appears is to be interpreted as the additional entropy that has appeared as a result of this implied external force. Although $\nabla S(x,t)$ is the internal entropy gradient and depends upon the actual trajectory $x(t)$, it nevertheless represents additional entropy in the system due to the external force. Like the case of Brownian motion, it is the instantaneous entropy, and all memory effects are contained in the bare system trajectory entropy.

The reason for avoiding being explicit about the external force that causes the thermodynamic trajectory $x(t')$, $t' < t$, arises from practical considerations of the time dependent problem. In the rest of this book, the state of the sub-system is determined by an external reservoir whose field variables are taken as the independently specified variables. One might attempt to formulate the present problem in similar terms, by specifying a time dependent reservoir field variable or force, $\lambda_{\rm r}(t)$ and calculating the sub-system response $\lambda_{\rm s}(t) = Sx(t)$. If one were to do this, then one needs the internal sub-system velocity, $\dot{x}^0(t)$ and the reservoir induced sub-system velocity, $\dot{x}^{(\rm r)}(t)$, such that the actual velocity is the sum of these, $\dot{x}(t) = \dot{x}^0(t) + \dot{x}^{(\rm r)}(t)$. One would need some prescription for the reservoir induced velocity. If one were to naively attempt to maximise the total trajectory entropy with respect to the reservoir induced velocity without limitation, then one would obtain the result that the internal thermodynamic force is equal to the reservoir force at each instant, $\overline{\lambda}_{\rm s}(t) = \lambda_{\rm r}(t)$. In most real world problems this result is only correct in the steady state or slowly varying force limit. More generally, the internal force lags the reservoir force because there is a limit to the propagation of the reservoir changes through the sub-system that is more appropriately described by hydrodynamics than by the single moment description that was used, for example, to describe steady heat flow. The complications imposed by these practical considerations preclude a generic treatment of the time dependent reservoir problem.

Maximisation of the Trajectory Entropy

In view of the discussion above, the general perturbation expression for the trajectory entropy is

$$\begin{aligned} S^{(n)}(\mathbf{X}^{(n/j)};\tau,\mathbf{F}) &= S_0^{(n)}(\mathbf{X}^{(n/j)};\tau) + \int_0^t \mathrm{d}t'\, \dot{x}(t')\cdot\nabla S(x(t'),t') \\ &= \frac{1}{2} S_0^{(n)}(\tau) : \mathbf{X}^{(n/j)}\mathbf{X}^{(n/j)} + \tau\dot{\mathbf{x}}^{(n/j)}\cdot\mathbf{F}_{\rm TD}^{(n/j)}, \quad (10.153) \end{aligned}$$

where the thermodynamic force is $F_{\rm TD}(x,t) \equiv \nabla S(x,t) = \partial S(x,t)/\partial x$. This thermodynamic force differs from the usual force in a mechanical system by a factor of inverse temperature. This is the time dependent perturbation that is *not* contained in the bare equilibrium system.

Differentiating the general expression for the trajectory entropy with respect to the velocities gives

$$\frac{\partial S^{(n)}(\mathbf{X}^{(n/j)};\tau,\mathbf{F})}{\partial\dot{\mathbf{x}}^{(n/j)}} = S_{0;\dot{x}\dot{x}}^{(n)}\dot{\mathbf{x}}^{(n/j)} + S_{0;\dot{x}x}^{(n)}x_j + \tau\mathbf{F}_{\rm TD}^{(n/j)} + \mathcal{O}(S^2). \quad (10.154)$$

This neglects a term $\tau^2[[\mathcal{U}^{(n)}]^{\mathrm{T}}\nabla^{(n)}\mathbf{F}_{\mathrm{TD}}^{(n)}]\dot{\mathbf{x}}^{(n)}$, since $\overline{\dot{\mathbf{x}}}^{(n)}$ is linear in the force. (The matrix $\mathcal{U}^{(n)}$ is upper triangular with ones above the diagonal.) Accordingly, the most likely trajectory is

$$\begin{aligned}\overline{\dot{\mathbf{x}}}^{(n/j)} &= -[S^{(n)}_{0;\dot{x}\dot{x}}]^{-1}S^{(n)}_{0;\dot{x}x}x_j - \tau[S^{(n)}_{0;\dot{x}\dot{x}}]^{-1}\mathbf{F}_{\mathrm{TD}}^{(n/j)} \\ &= Q^{(n)}_{0;\dot{x}x}[Q^{(n)}_{0;xx}]^{-1}x_j - \tau M^{(n)}\mathbf{F}_{\mathrm{TD}}^{(n/j)}.\end{aligned} \tag{10.155}$$

This uses Eq. (10.130) and the memory matrix given in Eq. (10.131), $M^{(n)} \equiv [S^{(n)}_{0;\dot{x}\dot{x}}]^{-1}$. This is essentially the same as the result for Brownian motion, Eq. (10.48), except that in that case the position dependent term that appears here vanished, as did the indirect correlation term in the memory function, so that the latter was exactly the velocity autocorrelation function.

In fluctuation approximation the vector of forces is

$$\mathbf{F}_{\mathrm{TD}}^{(n/j)} = S\mathbf{x}^{(n/j)}. \tag{10.156}$$

This assumes that the static fluctuation matrix is independent of time. In this case the static fluctuation matrix is a scalar as far as the trajectory nodes are concerned, but an $N \times N$ matrix as far as the thermodynamic components are concerned.

Evaluating the most likely trajectory for the terminal velocity one obtains

$$\begin{aligned}\overline{\dot{x}}_j &= -\dot{q}_0(0)^{\mathrm{T}}q_0(0)^{-1}x_j - \tau\sum_{k=1}^{n} M_k F_{\mathrm{TD};j-n+k} \\ &= -\dot{q}_0(0)^{\mathrm{T}}q_0(0)^{-1}x_j - \tau\sum_{k=j-n+1}^{j} M_{j-k+1}F_{\mathrm{TD};k} \\ &= -\dot{q}_0(0)^{\mathrm{T}}q_0(0)^{-1}x_j - \tau\sum_{k=j-n+1}^{j} M_{j-k+1}Sx_k, \text{ (fluc.).}\end{aligned} \tag{10.157}$$

One could rewrite the first term using the fact that $\dot{q}_0(0)^{\mathrm{T}} = -\dot{q}_0(0)$. The continuum version of this is

$$\begin{aligned}\overline{\dot{x}}(t) &= -\dot{q}_0(0)^{\mathrm{T}}q_0(0)^{-1}x(t) - \int_0^t \mathrm{d}t'\, M(t-t')F_{\mathrm{TD}}(x(t'),t') \\ &= -\dot{q}_0(0)^{\mathrm{T}}q_0(0)^{-1}x(t) - \int_0^t \mathrm{d}t'\, M(t-t')Sx(t'), \text{ (fluc.).}\end{aligned} \tag{10.158}$$

The equilibrium entropy fluctuation matrix is $S = q_0(0)^{-1} = -k_{\mathrm{B}}\left\langle x(t)x(t)\right\rangle_0^{-1}$, the position velocity correlation function is $\dot{q}_0(0) = k_{\mathrm{B}}^{-1}\left\langle x(t)\dot{x}(t)\right\rangle_0$, and the memory function given by Eq. (10.133) is

$$\begin{aligned}M(t-t') &\equiv -\ddot{q}_0(t-t') - \dot{q}_0(0)^{\mathrm{T}}q_0(0)^{-1}\dot{q}_0(t-t') \\ &= -k_{\mathrm{B}}^{-1}\left\langle \dot{x}(t)\dot{x}(t')\right\rangle_0 \\ &\quad + k_{\mathrm{B}}^{-1}\left\langle \dot{x}(t)x(t)\right\rangle_0 \left\langle x(t)x(t)\right\rangle_0^{-1}\left\langle x(t)\dot{x}(t')\right\rangle_0.\end{aligned} \tag{10.159}$$

The memory function is short-ranged, $M(t) \to 0$, $|t| \to \infty$ (since for a thermodynamic fluctuation, $\ddot{q}_0(t) \to 0$ and $\dot{q}_0(t) \to 0$, when $|t| \to \infty$), which means that the integral can be truncated after a fixed interval greater than the relaxation time.

Non-Terminal Fixed Node

As mentioned above, it is possible to specify the trajectory by the velocities and the position at any node, say $x(t')$. In this case the most likely terminal velocity is

$$\begin{aligned}\overline{\dot{x}}(t) &= -\dot{q}_0(t'-t)^{\mathrm{T}} q_0(0)^{-1} x(t') - \int_0^t \mathrm{d}t''\, M(t-t'';t') F_{\mathrm{TD}}(x(t''),t'') \\ &= -\dot{q}_0(t'-t)^{\mathrm{T}} q_0(0)^{-1} x(t') - \int_0^t \mathrm{d}t''\, M(t-t'';t') S x(t''), \quad \text{(fluc.)}.\end{aligned} \tag{10.160}$$

The memory function in this case is

$$\begin{aligned}M(t-t'';t') &\equiv -\ddot{q}_0(t-t'') - \dot{q}_0(t'-t)^{\mathrm{T}} q_0(0)^{-1} \dot{q}_0(t'-t'') \\ &= -k_{\mathrm{B}}^{-1} \left\langle \dot{x}(t)\dot{x}(t'') \right\rangle_0 \\ &\quad + k_{\mathrm{B}}^{-1} \left\langle \dot{x}(t) x(t') \right\rangle_0 \left\langle x(t') x(t') \right\rangle_0^{-1} \left\langle x(t') \dot{x}(t'') \right\rangle_0 .\end{aligned} \tag{10.161}$$

10.6.4 Generalised Langevin Equation

The stochastic dissipative equations of motion that follow are essentially the same as for Brownian motion, Eq. (10.61),

$$\begin{aligned}x_j &= x_{j-1} + \tau \dot{x}_{j-1} \\ \dot{x}_j &= R_j - \dot{q}_0(0)^{\mathrm{T}} q_0(0)^{-1} x_j - \tau \sum_{k=j-n+1}^{j} M_{j-k+1} F_{\mathrm{TD};k} \\ &= R_j - \dot{q}_0(0)^{\mathrm{T}} q_0(0)^{-1} x_j - \tau \sum_{k=j-n+1}^{j} M_{j-k+1} S x_k, \quad \text{(fluc.)}.\end{aligned} \tag{10.162}$$

The first term may be called the random force and the final sum may be called the dissipative force (although in the mechanical case the $k = j$ term is just the adiabatic force). The middle term is only non-zero in the mixed parity case.

The random forces have covariance matrix

$$\begin{aligned}\left\langle \mathbf{R}^{(n/j)} \mathbf{R}^{(n/j)} \right\rangle_{\mathrm{st}} &= \left\langle \left[\dot{\mathbf{x}}^{(n/j)} - \overline{\dot{\mathbf{x}}}^{(n/j)} \right] \left[\dot{\mathbf{x}}^{(n/j)} - \overline{\dot{\mathbf{x}}}^{(n/j)} \right] \right\rangle_{\mathrm{st}} \\ &= -k_{\mathrm{B}} S^{(n)}_{0;\dot{x}\dot{x}}(\tau)^{-1} \\ &= -k_{\mathrm{B}} M^{(n)}.\end{aligned} \tag{10.163}$$

The covariance matrix is symmetric, $M^{(n)}_{i\alpha,j\gamma} = M^{(n)}_{j\gamma,i\alpha}$. In the linear case this is just an average in the bare system,

$$\left\langle \mathbf{R}^{(n/j)}\mathbf{R}^{(n/j)}\right\rangle_0 = \left\langle \dot{\mathbf{x}}^{(n/j)}\dot{\mathbf{x}}^{(n/j)}\right\rangle_0 = -k_{\mathrm{B}}M^{(n)}. \tag{10.164}$$

The fact that the memory function of the equilibrium system gives both the covariance of the random forces and the dissipative force may be regarded as the fluctuation dissipation theorem for a system with memory. The probability distribution of the last n random forces is just the multivariate Gaussian.

The continuum version of the generalised Langevin equation is

$$\begin{aligned}\dot{x}(t) &= R(t) - \dot{q}_0(0)^{\mathrm{T}} q_0(0)^{-1} x(t) - \int_0^t \mathrm{d}t'\ M(t-t')F_{\mathrm{TD}}(x(t'),t') \\ &= R(t) - \dot{q}_0(0)^{\mathrm{T}} q_0(0)^{-1} x(t) - \int_0^t \mathrm{d}t'\ M(t-t')Sx(t'),\ \ \text{(fluc.)}.\end{aligned} \tag{10.165}$$

The covariance of the random forces is given by the elements of the covariance matrix

$$\langle R(t)R(0)\rangle_0 = -k_{\mathrm{B}}M(t), \tag{10.166}$$

where the memory function is given by Eq. (10.159). The memory function is actually an $N \times N$ matrix that satisfies $M(-t) = M(t)^{\mathrm{T}}$, which is to say that $M_{\alpha\gamma}(-t) = M_{\gamma\alpha}(t)$. In the pure parity case this is a symmetric matrix.

The generalised Langevin equation can be rewritten in terms of the time correlation functions and the random force covariance,

$$\begin{aligned}\dot{x}(t) &= R(t) + \langle \dot{x}(t)x(t)\rangle_0 \langle x(t)x(t)\rangle_0^{-1} x(t) \\ &\quad - \int_0^t \mathrm{d}t'\ \langle R(t)R(t')\rangle_0 \langle x(t)x(t)\rangle_0^{-1} x(t'),\ \ \text{(fluc.)}.\end{aligned} \tag{10.167}$$

This form will facilitate comparison with the generalised Langevin equation that emerges from the projector operator formalism.

10.7 Projector Operator Formalism

This section presents an introduction to the projector operator approach and compares the generalised Langevin equation that emerges from that formalism with the above perturbation result based upon fluctuation theory. The two are shown to be formally identical when the memory function in the perturbation theory is written as the covariance of the random force. In so far as the perturbation theory (and the linear response theory) gives an explicit expression for the memory function in terms of time correlation functions, then it may be said to go beyond the projector operator formalism.

Projector operator methods were developed by Zwanzig and by Mori.[19] One aim was to provide a more fundamental justification for the Langevin equation that proceeds from systematic approximations to molecular level statistical mechanics. A second aim was to obtain tractable approximations for the time correlation functions, with the hope that the memory function has a simpler structure and is easier to approximate than the time correlation function itself. Certain formal relationships involving the memory function, such as the fluctuation dissipation theorem with memory, emerge from the projector operator formalism.

The following account is based upon that given by Zwanzig.[20] The focus is upon getting to the generalised Langevin equation as quickly as possible, whilst at the same time identifying the similarities and differences between this approach and the approach taken in the rest of this book.

Let $\mathbf{\Gamma}$ be a point in the phase space of the total system, which is isolated and therefore evolves by Hamilton's equations,

$$\dot{\mathbf{\Gamma}}^0_{\mathrm{q}} = \frac{\partial \mathcal{H}(\mathbf{\Gamma})}{\partial \mathbf{\Gamma}_{\mathrm{p}}}, \text{ and } \dot{\mathbf{\Gamma}}^0_{\mathrm{p}} = \frac{-\partial \mathcal{H}(\mathbf{\Gamma})}{\partial \mathbf{\Gamma}_{\mathrm{q}}}. \tag{10.168}$$

For simplicity an equilibrium total system is considered and the total Hamiltonian is not explicitly dependent upon time.

The use of an isolated system and adiabatic evolution contrasts with the treatment in most of this book, where the total system is divided into a sub-system and a reservoir, and the evolution of the sub-system contains a deterministic Hamiltonian sub-system part, a deterministic dissipative part, and a stochastic part, the latter two accounting for the reservoir contribution in a probabilistic fashion. The advantage of the reservoir formalism is that it focuses on the region of direct interest and treats it in molecular detail, while accounting for the influence of the 'far' part of the system in a gross or probabilistic fashion. The far region is irrelevant but for the fact that it sets the thermodynamic state of the system of interest and it determines the conditions at the boundary of the region of direct interest

Let $x(\mathbf{\Gamma})$ be a dynamical variable, or a set of macroscopic thermodynamic variables. For simplicity no explicit time dependence is considered. Also for simplicity its current value will often be written, $x(t) \equiv x(\mathbf{\Gamma}^0(t|\mathbf{\Gamma}_0))$, where the initial point of the trajectory, $\mathbf{\Gamma}_0$ at $t = 0$, is understood.

Although $\mathbf{\Gamma}$ represents a point in the phase space of the total system, x can readily represent thermodynamic variables defined on the sub-system of direct interest. In this case only some of the coordinates of $\mathbf{\Gamma}$ directly affect its instantaneous value. However all of the coordinates of $\mathbf{\Gamma}$ are required to predict the evolution of x.

[19] Zwanzig, R. (1961), in *Lectures in Theoretical Physics*, (W. E. Britton, B. W. Downs, and J. Downs, eds), Vol. III, p. 135, (Wiley Interscience, New York). Mori, H. (1965), Progr. Theor. Phys. **33**, 423 and **34**, 399.

[20] Zwanzig, R. (2001), *Non-equilibrium Statistical Mechanics*, (Oxford University Press, Oxford).

The rate of change of x is

$$\begin{aligned}\dot{x}(t) &\equiv \frac{\mathrm{d}x(t)}{\mathrm{d}t} \\ &= \left\{ \dot{\mathbf{\Gamma}}^0_{\mathrm{q}} \cdot \frac{\partial}{\partial \mathbf{\Gamma}_{\mathrm{q}}} + \dot{\mathbf{\Gamma}}^0_{\mathrm{p}} \cdot \frac{\partial}{\partial \mathbf{\Gamma}_{\mathrm{p}}} \right\} x(\mathbf{\Gamma}), \quad \mathbf{\Gamma} \equiv \mathbf{\Gamma}^0(t|\mathbf{\Gamma}_0) \\ &\equiv \mathcal{L} x(\mathbf{\Gamma}). \end{aligned} \tag{10.169}$$

The final equality defines the Liouville operator. For the projector operator formalism it is essential that the time evolution is given by a linear operator. This in turn explains why the derivation is given for the total system, because the Liouville operator only applies to adiabatic equations of motion.

The linearity that enters the projector operator formalism here is the counterpart of the fluctuation formulation of the perturbation theory treated above. The quadratic form used there leads directly to a thermodynamic force that is linear in the thermodynamic variable.

The formal operator solution to the evolution of x is

$$x(t) = e^{t\mathcal{L}} x(\mathbf{\Gamma}_0). \tag{10.170}$$

This again shows the linear nature of the formulation.

The heart of the projector operator formalism is the existence of an inner product on phase space. As inner product, the equilibrium time correlation function is chosen,

$$(A(t), B(0)) \equiv \langle A(t)B(0)\rangle_0 \equiv \int \mathrm{d}\mathbf{\Gamma}_0 \wp(\mathbf{\Gamma}_0) A(\mathbf{\Gamma}(t|\mathbf{\Gamma}_0)) B(\mathbf{\Gamma}_0). \tag{10.171}$$

One can think of phase space as a vector space, and $\mathbf{\Gamma}$ as the index of the vector.

The weight for the inner product, which is the probability distribution $\wp(\mathbf{\Gamma})$, is somewhat problematic. Strictly speaking, since one is dealing with an isolated system the energy is fixed and one ought to have $\wp(\mathbf{\Gamma}) = \delta(\mathcal{H}(\mathbf{\Gamma}) - E_0)$. This means that the average is not, for example, a canonical equilibrium average. However, in a mathematical sense there is nothing to prevent choosing the initial probability distribution as one likes without being overly concerned with its physical realisation. One can, for example, choose the Maxwell-Boltzmann distribution, $\wp(\mathbf{\Gamma}) = Z(T)^{-1} \exp -\beta\mathcal{H}(\mathbf{\Gamma})$, in which case the inner product is the same as the canonical equilibrium average. Because Hamilton's equations are incompressible, $\mathrm{d}\mathbf{\Gamma}(t) = \mathrm{d}\mathbf{\Gamma}_0$, the probability density is conserved, $\wp(\mathbf{\Gamma}, t) = \wp(\mathbf{\Gamma}^0(-t|\mathbf{\Gamma}))$. Because energy is conserved on an adiabatic trajectory, this means that for the Maxwell-Boltzmann distribution, $\wp(\mathbf{\Gamma}, t) = \wp(\mathbf{\Gamma})$. This is necessary for time homogeneity, which is invoked in some of the following analysis.

One can use the inner product to project any phase function $B(\mathbf{\Gamma})$ onto the sub-space spanned by the thermodynamic variables of interest at $t = 0$,

$$\mathcal{P}B(t) = \langle B(t)x(0)\rangle_0 \langle x(0)x(0)\rangle_0^{-1} x(0). \tag{10.172}$$

The normalisation factor preserves the initial value, $\mathcal{P}x(0) = x(0)$. The complement of the projector is $1 - \mathcal{P}$. Obviously $\mathcal{P}^2 = \mathcal{P}$ and $[1 - \mathcal{P}]^2 = 1 - \mathcal{P}$.

The random force is defined as that part of the evolution that occurs in the sub-space orthogonal to the initial thermodynamic variable,

$$R(t) \equiv e^{t(1-\mathcal{P})\mathcal{L}}(1-\mathcal{P})\mathcal{L}x(0). \tag{10.173}$$

Since $(1-\mathcal{P})\mathcal{L}$ is the time derivative operator in the sub-space orthogonal to $x(0)$, this represents the orthogonal evolution of the thermodynamic variable. This means that the random force remains orthogonal during its evolution,

$$\langle R(t)x(0)\rangle_0 = 0. \tag{10.174}$$

The force is random in the sense that it arises from outside the sub-space of interest, and therefore it is not determined solely by the sub-space of interest.

Following Zwanzig (2001),[21] write the Liouville operator as $\mathcal{L} = \mathcal{P}\mathcal{L} + (1-\mathcal{P})\mathcal{L}$, and note the operator identity,

$$e^{t\mathcal{L}} = e^{t(1-\mathcal{P})\mathcal{L}} + \int_0^t \mathrm{d}t'\; e^{(t-t')\mathcal{L}}\,\mathcal{P}\mathcal{L}\,e^{t'(1-\mathcal{P})\mathcal{L}}, \tag{10.175}$$

which may be verified using repeated differentiation to make a Taylor expansion about $t=0$ of both sides. Now multiply both sides of this by $(1-\mathcal{P})\mathcal{L}x(0) = \dot{x}(0) - \mathcal{P}\dot{x}(0)$. The left-hand side becomes

$$\begin{aligned} \text{LHS} &= e^{t\mathcal{L}}(1-\mathcal{P})\mathcal{L}x(0) \\ &= e^{t\mathcal{L}}\left[\dot{x}(0) - \langle \dot{x}(0)x(0)\rangle_0 \langle x(0)x(0)\rangle_0^{-1} x(0)\right] \\ &= \dot{x}(t) - \langle \dot{x}(0)x(0)\rangle_0 \langle x(0)x(0)\rangle_0^{-1} x(t). \end{aligned} \tag{10.176}$$

This is the evolved velocity, less the projected component of the initial velocity carried in the evolved subspace. With the random force the right-hand side is

$$\begin{aligned} \text{RHS} &= R(t) + \int_0^t \mathrm{d}t'\; e^{(t-t')\mathcal{L}}\,\mathcal{P}\mathcal{L}\,R(t') \\ &= R(t) + \int_0^t \mathrm{d}t'\; e^{(t-t')\mathcal{L}}\,\langle \mathcal{L}\,R(t')x(0)\rangle_0 \langle x(0)x(0)\rangle_0^{-1} x(0) \\ &= R(t) + \int_0^t \mathrm{d}t'\; \langle \mathcal{L}\,R(t')x(0)\rangle_0 \langle x(0)x(0)\rangle_0^{-1} x(t-t'), \end{aligned} \tag{10.177}$$

the final equality following because the time evolution operator commutes with the time correlation functions, which are independent of $\mathbf{\Gamma}$.

It is traditional to define a frequency matrix,

$$\mathrm{i}\Omega \equiv \langle \dot{x}(0)x(0)\rangle_0 \langle x(0)x(0)\rangle_0^{-1}, \tag{10.178}$$

[21] who followed Hynes, J. and Deutch, J. (1975), in *Physical Chemistry, an Advanced Treatise*, edited by H. Eyring, D. Henderson, and W. Jost, Vol. XI, (Academic Press, New York).

which is the projection of the initial velocity that occurs on the left-hand side, and a memory or kernel matrix,

$$\begin{aligned} K(t) &\equiv -\left\langle [\mathcal{L}\, R(t)]\, x(0)\right\rangle_0 \left\langle x(0)x(0)\right\rangle_0^{-1} \\ &= \left\langle R(t)\mathcal{L}x(0)\right\rangle_0 \left\langle x(0)x(0)\right\rangle_0^{-1} \\ &= \left\langle R(t)R(0)\right\rangle_0 \left\langle x(0)x(0)\right\rangle_0^{-1}, \end{aligned} \tag{10.179}$$

which occurs in the integrand on the right-hand side. The second equality follows from the time homogeneity of the equilibrium system, Eq. (10.16). The final equality follows because $R(t)$ lies in the orthogonal sub-space, and therefore only the orthogonal part of the second term contributes, $\mathcal{L}x(0) \Rightarrow (1-\mathcal{P})\mathcal{L}x(0) = R(0)$. Equating the above left and right-hand sides, inserting these definitions, and rearranging gives the generalised Langevin equation,

$$\begin{aligned} \dot{x}(t) &= R(t) + \mathrm{i}\Omega x(t) - \int_0^t \mathrm{d}t'\; K(t')x(t-t') \\ &= R(t) + \left\langle \dot{x}(0)x(0)\right\rangle_0 \left\langle x(0)x(0)\right\rangle_0^{-1} x(t) \\ &\quad - \int_0^t \mathrm{d}t''\; \left\langle R(t)R(t'')\right\rangle_0 \left\langle x(0)x(0)\right\rangle_0^{-1} x(t''), \end{aligned} \tag{10.180}$$

where in the second equality the substitution $t'' = t - t'$ has been made. It may be seen that this is identical to the perturbation result, Eq. (10.167).

It should be emphasised that the projector operator formalism leading to the generalised Langevin equation is formally exact and no approximations have been made. The physical interpretation of the formalism is another matter. The definition of the random force as the orthogonal evolution of the original vector is plausible, although whether or not it is precisely the same as the reservoir force in the perturbation theory remains to be clarified. Similarly questionable is the physical basis for the position $x(t'')$ that appears in the integrand (i.e. how to relate it in an actual physical problem to a time dependent external mechanical or thermodynamic force). In the light of the perturbation theory one could interpret the term $\left\langle x(0)x(0)\right\rangle_0^{-1} x(t'')$ as the internal thermodynamic force. It is not immediately obvious how to extend the projector operator method into the non-linear regime, whereas in the perturbation theory the non-linear extensions appear well defined in principle (c.f. the discussion on p. 373).

There has been a deal of effort to develop practical approximations for the memory function that arises in the projector operator formalism (Hansen and McDonald, 1986; Zwanzig, 2001; Das, 2004).[22] The explicit expression for the memory function given by the perturbation theory in terms of the time correlation functions of the unperturbed system, Eq. (10.159), which is the same as that given by the linear response theory, §10.4, has been shown to perform well in the case of driven Brownian motion, §§ 10.5 and 10.5.6 above, and for the regression of fluctuations, §10.6.2. To some extent the two theories have been used to address different problems: the projector operator method

[22] Das, S. P. (2004), Rev. Modern Phys. **76**, 785.

has been concerned with obtaining the equilibrium time correlation functions from simple approximations to the memory function, whereas the perturbation theory has been concerned with obtaining the behaviour of the non-equilibrium system using the given equilibrium time correlation functions as the memory function.

10.8 Harmonic Oscillator Model for the Memory Function

A particular molecular model for Brownian motion that yields the generalised Langevin equation and an analytic expression for the memory function, takes the solvent or heat reservoir to be a collection of harmonic oscillators of arbitrary frequency and coupling. The model has the virtue of being simple enough that analytic expressions can be obtained for the time evolution of the Brownian particle and the various time correlation functions. Its limitations are discussed as they arise below and in the concluding paragraphs of this section.

The harmonic oscillator model of the heat reservoir for Brownian motion has come to be called the Caldeira-Leggett model after its use by these authors to describe quantum tunnelling with dissipation.[23] However, a deal of work with the model, both in quantum dissipation and in classical Brownian motion, predates this.[24] Useful summaries of the model may be found in Zwanzig (2001) and Pottier (2010).

In the model, the Brownian particle has position x, momentum p, and mass m_0, and experiences a time dependent external potential $U(x,t)$. The solvent or heat reservoir consists of particles with position x_n, momentum p_n, and mass m_n, with $n = 1, 2, \ldots$ For simplicity the one-dimensional case is considered, and the coordinates are collectively denoted $\mathbf{X}$. The Hamiltonian is

$$\mathcal{H}(\mathbf{X},t) = \frac{p^2}{2m_0} + U(x,t) + \sum_n \left\{ \frac{p_n^2}{2m_n} + \frac{m_n\omega_n^2}{2}\left[x_n - \frac{c_n}{m_n\omega_n^2}x \right]^2 \right\}. \quad (10.181)$$

An essential feature of the model is that the coupling between the particle and the reservoir is bilinear in the positions, with terms of the form $-c_n x x_n$. Hence c_n measures the strength of the coupling with the nth oscillator, which has frequency ω_n. Note that even in the absence of the external potential, the term $x^2 \sum_n c_n^2/2m_n\omega_n^2$, acts as a parabolic trap pinning the Brownian particle about the origin. It is emphasised that this is the Hamiltonian of the total system and it completely determines the evolution.

The interaction between the particle and the reservoir depends only upon the distances $x_n(t) - x(t)$ (as can be seen by rearranging the Hamiltonian), even as the Brownian particle moves through the solvent. This means that the oscillators of the reservoir do not represent physical solvent molecules. Rather

[23] Caldeira, A. O. and Leggett, A. J. (1983), Ann. Phys. **149**, 374.

[24] Senitzky, I. R. (1960) Phys. Rev. **119**, 670. Ford, G. W., Kac, M., and Mazur, P. (1965) J. Math. Phys. **6**, 504. Ullersma, P. (1966), Physica **32**, 27, 56, 74, and 90.

they represent modes of interaction with the surroundings or effective statistical entities. Because these modes are effective quantities rather than physical oscillators, the masses m_n (and the momenta p_n), should not be taken overly literally. It is often convenient to set all the oscillator masses equal to the same value.

10.8.1 Generalised Langevin Equation

The equations of motion for the Brownian particle are

$$\frac{\mathrm{d}x}{\mathrm{d}t} = \frac{p}{m_0}, \text{ and } \frac{\mathrm{d}p}{\mathrm{d}t} = F(x,t) + \sum_n c_n \left[x_n - \frac{c_n}{m_n\omega_n^2} x \right], \tag{10.182}$$

where the external force is $F(x,t) = -\partial U(x,t)/\partial x$. Those for the solvent particles are

$$\frac{\mathrm{d}x_n}{\mathrm{d}t} = \frac{p_n}{m_n}, \text{ and } \frac{\mathrm{d}p_n}{\mathrm{d}t} = -m_n\omega_n^2 x_n + c_n x. \tag{10.183}$$

The equations of motion of the solvent can be solved in terms of the trajectory of the Brownian particle up to the present time and the initial configuration at time t_0,

$$\begin{aligned} x_n(t) &= x_n(t_0)\cos(\omega_n(t-t_0)) + \frac{p_n(t_0)}{m_n\omega_n}\sin(\omega_n(t-t_0)) \\ &\quad + c_n \int_{t_0}^{t} \mathrm{d}t' \, \frac{\sin(\omega_n(t-t'))}{m_n\omega_n} x(t'). \end{aligned} \tag{10.184}$$

Integrating by parts, this may be rearranged as

$$\begin{aligned} & x_n(t) - \frac{c_n}{m_n\omega_n^2} x(t) \\ &= \left[x_n(t_0) - \frac{c_n}{m_n\omega_n^2} x(t_0) \right] \cos(\omega_n(t-t_0)) + \frac{p_n(t_0)}{m_n\omega_n}\sin(\omega_n(t-t_0)) \\ &\quad - c_n \int_{t_0}^{t} \mathrm{d}t' \, \frac{\cos(\omega_n(t-t'))}{m_0 m_n\omega_n^2} p(t'). \end{aligned} \tag{10.185}$$

The right-hand side of this depends only upon the initial configuration of the reservoir and the past trajectory of the Brownian particle. Using this, the equation of motion for the acceleration of the particle can be written in the form of the generalised Langevin equation,

$$\begin{aligned} \frac{\mathrm{d}p(t)}{\mathrm{d}t} &= F(x(t),t) + \sum_n c_n \left[x_n(t) - \frac{c_n}{m_n\omega_n^2} x(t) \right] \\ &= F(x(t),t) + R(t) - \int_{t_0}^{t} \mathrm{d}t' \, K(t-t') \frac{p(t')}{m_0}. \end{aligned} \tag{10.186}$$

The memory function that appears here has the explicit microscopic representation,

$$K(t) \equiv \sum_n \frac{c_n^2}{m_n \omega_n^2} \cos \omega_n t. \tag{10.187}$$

It is conventional to display the factor of m_0 explicitly in the integrand even though it could be incorporated into the definition of the memory function. The random force is defined as

$$\begin{aligned} R(t) &\equiv \sum_n c_n \left[x_n(t_0) - \frac{c_n}{m_n \omega_n^2} x(t_0) \right] \cos(\omega_n(t - t_0)) \\ &\quad + \sum_n \frac{c_n p_n(t_0)}{m_n \omega_n} \sin(\omega_n(t - t_0)). \end{aligned} \tag{10.188}$$

The force is random in the sense that it depends upon the initial configuration of the reservoir, but not upon the current coordinates of the Brownian particle.

The random force is the sum of a large number of independent terms, and hence by the central limit theorem it is Gaussian distributed.[25] The distribution is therefore characterised by its first and second moments. Although one is dealing with a single isolated system that evolves adiabatically, one may nevertheless specify how the initial configuration is determined. It particular, it can be drawn randomly according to a specified law. If the initial distribution is Maxwell-Boltzmann,

$$\wp(\mathbf{X}, t_0) = \frac{1}{Z(t_0)} \exp[-\mathcal{H}(\mathbf{X}, t_0)/k_B T], \tag{10.189}$$

then it is a Gaussian in the variables $x_n(t_0) - c_n x(t_0)/m_n \omega_n^2$ and $p_n(t_0)$. Hence their first moments vanish,

$$\left\langle x_n(t_0) - \frac{c_n}{m_n \omega_n^2} x(t_0) \right\rangle_0 = 0, \text{ and } \langle p_n(t_0) \rangle_0 = 0, \quad n = 1, 2, \ldots \tag{10.190}$$

These hold irrespective of the external potential, $U(x, t_0)$. It should be noted that without loss of generality one can shift the origin so that the initial external potential is such that $\langle x(t_0) \rangle_0 = 0$, but this is not required here. Because the random force at time t is a linear combination of these initial first moments, then $\langle R(t) \rangle_0 = 0$. The second moments are

$$\left\langle \left[x_n(t_0) - \frac{c_n}{m_n \omega_n^2} x(t_0) \right]^2 \right\rangle_0 = \frac{k_B T}{m_n \omega_n^2}, \quad n = 1, 2, \ldots, \tag{10.191}$$

and

$$\left\langle p_n(t_0)^2 \right\rangle_0 = m_n k_B T, \quad n = 1, 2, \ldots \tag{10.192}$$

---

[25] In fact, for the Maxwell-Boltzmann distribution that is about to be invoked, the individual terms are Gaussian distributed. In this case the Gaussian distribution for the random force holds irrespective of the number of modes.

Since there is no correlation between different modes one has

$$\begin{aligned}
\langle R(t)R(t')\rangle_0 &= \sum_n \frac{k_\mathrm{B}Tc_n^2}{m_n\omega_n^2}\cos(\omega_n(t-t_0))\cos(\omega_n(t'-t_0)) \\
&\quad + \sum_n \frac{k_\mathrm{B}Tc_n^2}{m_n\omega_n^2}\sin(\omega_n(t-t_0))\sin(\omega_n(t'-t_0)) \\
&= \sum_n \frac{k_\mathrm{B}Tc_n^2}{m_n\omega_n^2}\cos(\omega_n(t-t')) \\
&= k_\mathrm{B}TK(t-t'). \qquad (10.193)
\end{aligned}$$

That the covariance of the random force is related to the memory function is the generalised fluctuation dissipation theorem.

The use of the Maxwell-Boltzmann distribution as the initial distribution is another limitation of the model. It is generally chosen for three reasons: it is simple, it is familiar, and it leads directly to the generalised fluctuation dissipation theorem. In the case that the external potential is not explicitly dependent on time for $t < t_0$, then it is indeed the correct probability distribution. More generally, however, the non-equilibrium probability distribution ought to be invoked.

The memory function is a functional of the distribution of mode frequencies and the strength of the couplings. This represents a further weakness of the model in that it is unclear how to go from a real physical system to a specific distribution of modes and couplings, and hence to the memory function of the physical system.

Because the spectrum of oscillators is divorced from the physical nature of the reservoir, the focus tends to shift to the memory function itself. One often chooses a distribution of modes to give a desired memory function. For example, in the continuum limit the sum over modes can be replaced by an integral over ω, with mode density $g(\omega)$, so that the memory function becomes

$$K(t) \equiv \int_0^\infty \mathrm{d}\omega\; g(\omega)\frac{c(\omega)^2}{m(\omega)\omega^2}\cos\omega t. \qquad (10.194)$$

Hence the memory function is essentially a Fourier transform of the mode coupling distribution. The general results given in §10.3.5 relate the memory function to the velocity autocorrelation function.

The choice of a Lorentzian,

$$\frac{g(\omega)c(\omega)^2}{m(\omega)\omega^2} = \frac{2\gamma}{\pi\left[\omega^2+\omega_\mathrm{c}^2\right]}, \qquad (10.195)$$

gives an exponentially decaying memory function,

$$K(t) = \gamma\omega_\mathrm{c}e^{-\omega_\mathrm{c}|t|}. \qquad (10.196)$$

This shows that the cut-off for the modes, ω_c, gives the inverse of the relaxation time, which tells how far back the system remembers.

In the limit of infinitely short relaxation time, $\omega_c \to \infty$, the mode coupling distribution becomes constant, and the memory function becomes a δ-function,

$$\frac{g(\omega)c(\omega)^2}{m(\omega)\omega^2} \to \frac{2\gamma}{\pi\omega_c^2}, \quad K(t) \to 2\gamma\delta(t), \quad \omega_c \to \infty. \tag{10.197}$$

In this limit the system is Markovian (i.e. it has no memory), and the generalised Langevin equation reduces to the usual Langevin equation without memory, with γ being the usual friction coefficient.

10.8.2 Modified Random Force

A slightly different formulation of the random force illustrates an interesting connection with the perturbation theory given above. That part of the random force that depends upon the initial position $x(t_0)$ of the Brownian particle can be written $-x(t_0)\sum_n(c_n^2/m_n\omega_n^2)\cos(\omega_n(t-t_0)) = -K(t-t_0)x(t_0)$. Hence the generalised Langevin equation can be written in the form,

$$\frac{\mathrm{d}p(t)}{\mathrm{d}t} = F(x(t),t) - K(t-t_0)x(t_0) + \tilde{R}(t) - \int_{t_0}^{t} \mathrm{d}t'\, K(t-t')\frac{p(t')}{m_0}, \tag{10.198}$$

with the memory function being as given above and the random force now being defined as

$$\tilde{R}(t) \equiv \sum_n c_n x_n(t_0)\cos(\omega_n(t-t_0)) + \sum_n \frac{c_n p_n(t_0)}{m_n\omega_n}\sin(\omega_n(t-t_0)). \tag{10.199}$$

In view of this, one can define a modified Hamiltonian,

$$\tilde{\mathcal{H}}(\mathbf{X},t) = \frac{p^2}{2m_0} + U(x,t) + \sum_n \left\{ \frac{p_n^2}{2m_n} + \frac{m_n\omega_n^2}{2}x_n^2 \right\}, \tag{10.200}$$

which has no coupling between the Brownian particle and the reservoir. The initial probability distribution can be taken to be the Boltzmann factor of this uncoupled Hamiltonian, $\tilde{\wp}(\mathbf{X},t_0) = Z(t_0)^{-1}\exp -\tilde{\mathcal{H}}(\mathbf{X},t)\,/k_BT$. Obviously the reason for doing this is to make the initial distribution a Gaussian in the variables that now define the random force. The averages are taken over this modified initial distribution. With this, all of the above analysis remains unchanged and the generalised fluctuation dissipation theorem holds for the modified random force,

$$\left\langle \tilde{R}(t)\tilde{R}(t') \right\rangle_{\tilde{0}} = k_BTK(t-t'). \tag{10.201}$$

The memory function remains unchanged in this modified analysis. The initial probability distribution is arguably physical if the coupling between the particle and the reservoir could somehow be turned on only for $t \geq t_0$.

This particular digression shows that by altering the definition of the random force, the generalised Langevin equation can be altered to explicitly show a term dependent on the initial position of the particle. This is reminiscent of the discussion of the choice of the position in the definition of the trajectory in the perturbation theory (c.f. §10.6.2 and also p. 377). In the perturbation theory, the memory function depended upon the choice of the position that was used to fix the trajectory, whereas in the present harmonic oscillator model the memory function is unchanged when the initial position is removed from the definition of the random force.

10.8.3 Discussion

As mentioned at the beginning of this section, the chief merit of the harmonic oscillator model for the interaction of the Brownian particle and the reservoir is that it has analytic solutions that explicitly yield the generalised Langevin equation and an analytic expression for the memory function. The distribution of the frequencies and couplings between the particle and the reservoir is arbitrary, which is both a strength—analytic solutions can be obtained in the generic case—and a weakness—results may correspond to distributions that inadvertently violate fundamental statistical requirements. Likewise the fact that the averages can be taken over an arbitrary initial probability distribution is a mixed blessing: on the one hand one can use an equilibrium distribution such as the Maxwell-Boltzmann distribution, which is both familiar and simple, but on the other hand one ought to use arguably the proper non-equilibrium probability distribution, which is not given directly by the model. Finally, in the harmonic oscillator model, the force between the reservoir and the particle is always linear in the distance between the particle and the centre of each oscillator of the reservoir, no matter where the particle moves. Under such circumstances the oscillators must be regarded as effective statistical entities rather than physical solvent particles, and this creates challenges in the quantitative application of the model to actual physical systems.

Chapter 11

Non-Equilibrium Computer Simulation Algorithms

Molecular level computer simulations have revolutionised statistical mechanics over the past several decades. One can identify three generic contributions: they have provided insight into the molecular mechanisms that cause observed phenomena, they have been used to establish benchmark results to test quantitatively approximation schemes and to verify hypotheses of a more fundamental thermodynamic or statistical mechanical nature, and they have enabled predictions to be made of the behaviour of systems in a more timely and cost effective manner than can be achieved in a laboratory or in regimes that might be dangerous or impossible to explore in the real world.

The essential requirement of computer simulations is that they generate exact results for the molecular model chosen. Issues such as the statistical accuracy of the results, or the influence of the finite size of the system can be addressed by systematically increasing the size of the simulation. The reality of the molecular model can also, in principle, be dealt with by either comparison with experiment or by quantum calculations of molecular interactions. Beyond these there remains the simulation algorithm itself, and if this is not formally exact, then most of the benefits of computer simulation summarised in the preceding paragraph are lost.

This is a major problem in the simulation of time dependent systems, where most of the results that have been generated to date have been obtained with algorithms that have been developed for equilibrium systems. For example, many molecular dynamics simulations use thermostats that are based upon the Maxwell-Boltzmann probability distribution. In other cases, artificial equations of motion have been invoked to yield certain non-equilibrium flows, whilst sacrificing fundamental thermodynamic requirements. What is known is that such algorithms are not exact; what is unknown is the degree to which the equilibrium assumption or the artificial equations of motion affect the results. It is unfortunate that many non-equilibrium simulations are accepted uncritically, as though they were as reliable as simulations of equilibrium systems.

There are two main simulation techniques for equilibrium systems: Monte Carlo and molecular dynamics. The Monte Carlo algorithm is based upon a transition probability that has as the stationary solution the equilibrium probability distribution, either the Maxwell-Boltzmann distribution in the case of a temperature reservoir, or else the related Gibbs distributions in the case of pressure or chemical potential reservoirs. The obvious analogue for time dependent systems is to develop an algorithm that preserves the non-equilibrium probability distribution as derived in Ch. 8. Because these probability distributions are of relatively recent discovery, there have been few attempts to use Monte Carlo techniques for non-equilibrium problems (apart from some studies that naively invoke the Maxwell-Boltzmann distribution using the instantaneous value of the potential).

In contrast there have been many simulations of time dependent systems using molecular dynamics, probably because even equilibrium molecular dynamics invokes time explicitly in the equations of motion. Consequently these have been applied unchanged, with little thought and no foundation, to non-equilibrium mechanical systems in which the potential varies explicitly with time. However even in equilibrium molecular dynamics there is the issue of the thermostat to consider, and the problem is potentially exacerbated in the non-equilibrium situation where there may be no real reason to suppose that the temperature of the system is constant, or that it is uniform in the region of work, or that the statistical effects of an artificial thermostat are negligible. Often, for example, constraints are introduced into the equations of motion to hold the kinetic energy constant. But since this fails even in the equilibrium case (the Maxwell-Boltzmann distribution is not stationary), an isokinetic thermostat also produces spurious results in the non-equilibrium case. The problem for non-equilibrium molecular dynamics is in essence no different to that for non-equilibrium Monte Carlo: whichever thermostatted equations of motion are invoked one must guarantee that they correspond to the correct non-equilibrium probability distribution.

A third simulation technique that has been applied to non-equilibrium systems is Brownian dynamics. This is really an approximation, at the same level as the perturbation theory of Ch. 10, which is suited to complex heterogeneous non-equilibrium systems. Because of the great range of time scales due to the presence of both molecular and macroscopic components, these may be intractable for an exact simulation. Such systems may be practically important in laboratory, technological, or industrial applications, and a more pragmatic approach can be justified compared with, for example, the benchmark results that are sought from simulations of simple systems.

This chapter is divided into three sections, one for each of the three simulation techniques, with most emphasis placed on molecular dynamics, §11.1, and Monte Carlo, §11.2. For each of these, both equilibrium and non-equilibrium algorithms are given, as well as numerical results for a mechanical non-equilibrium system (driven Brownian motion), and for a thermodynamic non-equilibrium system (steady heat flow). These sections are concerned more with applications and results than with foundations or fundamental principles. Because these non-equilibrium simulation algorithms are relatively recent, the focus here is mostly on the recipes for implementing the algorithms, the details of actual computations, and the tests and comparisons of the simulation results. The final section of this chapter, §11.3, reviews the Brownian dynamics simulation technique. Some speculation on systematically improving the Brownian dynamics algorithm via a perturbation expansion is offered in §11.3.2.

11.1 Stochastic Molecular Dynamics

11.1.1 Equilibrium Systems

At the simplest level, one might attempt to simulate an equilibrium system by discretising Hamilton's equations of motion and stepping along a trajectory of the system,

$$\begin{aligned} q_{i\alpha}(t+\Delta_t) &= q_{i\alpha}(t) + \Delta_t p_{i\alpha}(t)/m, \\ p_{i\alpha}(t+\Delta_t) &= p_{i\alpha}(t) + \Delta_t F_{i\alpha}(t). \end{aligned} \tag{11.1}$$

These are the adiabatic equations of motion. Here one considers a total isolated system of N identical atoms, labelled $i = 1, 2, \ldots, N$, in three-dimensional space, $\alpha = x, y, \text{or } z$, with kinetic energy $\mathcal{K}(\mathbf{\Gamma}) = \sum_{i\alpha} p_{i\alpha}^2/2m$, momentum $p_{i\alpha} = m\dot{q}_{i\alpha}$, and force that is the derivative of the potential, $F_{i\alpha}(\mathbf{\Gamma}) = -\partial U(\mathbf{\Gamma})/\partial q_{i\alpha}$. Generalisations of this include mixtures, where the species have different masses and interaction forces, and molecular systems, where the orientation and angular momentum have to be included in the phase space coordinates.

In the simplest case the potential is the sum of central pair-wise interactions,

$$U(\mathbf{\Gamma}) = \sum_{i<j} u(q_{ij}), \quad q_{ij} = |\mathbf{q}_i - \mathbf{q}_j|, \tag{11.2}$$

in which case the force is

$$F_{i\alpha}(\mathbf{\Gamma}) = -\sum_{j=1}^{N}{}^{(j\neq i)}\, u'(q_{ij}) \frac{q_{i\alpha} - q_{j\alpha}}{q_{ij}}. \tag{11.3}$$

Inhomogeneous systems can be modelled by invoking in addition a singlet potential, $\sum_i u^{(1)}(\mathbf{q}_i)$, and in such cases the pair potential might not be central. Less commonly, triplet, $\sum_{i<j<k} u^{(3)}(\mathbf{q}_i, \mathbf{q}_j, \mathbf{q}_k)$, and high order many body potentials can be invoked. For macromolecular systems, one might distinguish between the intra- and the inter-molecular potentials.

For an isolated system, Hamilton's equations of motion conserve the total energy, $\dot{\mathcal{H}}(\mathbf{\Gamma}) = 0$. However the discretised equations of motion given above are only exact to $\mathcal{O}(\Delta_t)$, and one finds in practice that, due to the accumulation of errors, the energy increases more or less monotonically at a rate dependent upon the size of the time step. The reason that the energy increases is that entropy increases with energy, and so random numerical errors tend to find high energy states rather than low energy states.

One partial solution to the problem is to use higher-order equations of motion, which reduce the error in each step (or else allow a larger time step for a given error). The simplest such formula is the central difference formula, which exploits the time reversibility of Hamilton's equations. By taking the difference of the time step expressions for $q_{i\alpha}(t \pm \Delta_t)$ (and changing $t \Rightarrow t - \Delta_t$), and $p_{i\alpha}(t \pm \Delta_t)$, one obtains the Verlet leap frog algorithm,[1]

$$\begin{aligned} q_{i\alpha}(t) &= q_{i\alpha}(t - 2\Delta_t) + 2\Delta_t p_{i\alpha}(t - \Delta_t)/m, \\ p_{i\alpha}(t + \Delta_t) &= p_{i\alpha}(t - \Delta_t) + 2\Delta_t F_{i\alpha}(t). \end{aligned} \tag{11.4}$$

Since the even terms vanish, these are exact to $\mathcal{O}(\Delta_t^2)$. Higher-order Gear predictor corrector algorithms are also used.[2] The main reason for using such higher-order algorithms is that they enable a larger time step to be used, which allows a larger region of phase space to be covered in a given amount of computer time. This in turn leads to more reliable statistical averages for two reasons: first, more representative regions of phase space are likely to be visited, and second, since adjacent points on a trajectory are highly correlated, and correlated points add no new information to averages, the number of distinct contributions to an average increases with the length of the time step.

Whilst such higher-order algorithms have benefits in terms of statistical efficiency, the inevitable energy (equivalently temperature) increase is only delayed, not prevented. For this reason almost all equilibrium molecular dynamics simulations invoke some form of thermostat that keeps the temperature constant. The rationale or physical interpretation of adding a thermostat to Hamilton's equations of motion is that the simulations represent a canonical equilibrium system. Under such an interpretation not all thermostats are acceptable. In fact since the discrete equations of motion represent a transition from one microstate of the sub-system to another, then they must be a manifestation of the transition probability for the canonical equilibrium system. This determines the nature and the specific form of the thermostat that can be used.

At a minimum the thermostat must ensure that the equilibrium probability distribution, the Maxwell-Boltzmann distribution, is stationary. Some thermostats commonly in use, such as velocity rescaling and isokinetic (constant kinetic energy) thermostats, fail this criterion.

Another essential requirement of the thermostat is that it should contain a stochastic element. This requirement stems from the fact that the non-Hamiltonian terms in the equations of motion represent the interactions between the sub-system and the thermal reservoir, with the reservoir coordinates projected out. As discussed in §1.3.4, such projection operations are the origin of randomness in classical mechanics, since the future behaviour is not uniquely determined by the current phase space point of the sub-system alone. Implicit

---

[1] Verlet, L. (1967), Phys. Rev. **159**, 98.

[2] Allen, M. P. and Tildesley, D. J. (1987), *Computer Simulations of Liquids*, (Oxford University Press, Oxford).

in the idea that the thermostat represents the hidden interactions between the sub-system and the reservoir is that it must also contain dissipative elements, and that these must be related by the equilibrium fluctuation dissipation theorem, as discussed next. (This is another reason that the isokinetic thermostat is unrealistic: as a purely deterministic thermostat it includes dissipation without fluctuation.)

Although they don't explicitly address the fluctuation dissipation theorem, there are three equilibrium stochastic thermostats worth mentioning because they preserve the Maxwell-Boltzmann distribution. One such thermostat is the Nosé-Hoover thermostat.[3] This is a deterministic thermostat in an extended system that preserves the canonical equilibrium probability distribution. This thermostat is stochastic in the sense that the evolution of the sub-system is not determined by the phase space of the sub-system alone. This is particularly true for the Nosé-Hoover chain thermostat, where the randomness increases with the links in the chain. A second stochastic algorithm proceeds by replacing a randomly selected particle's velocity by one chosen from a Maxwell distribution.[4] Such replacement is made after a number of Hamiltonian (adiabatic) time steps. This ensures that the probability distribution of the simulation is the Maxwell-Boltzmann distribution. A third equilibrium molecular dynamics algorithm invokes a transition probability for each atom at each time step in addition to the adiabatic motion.[5] This method has also been applied to grand canonical (constant chemical potential, varying particle number) simulations.

Stochastic Thermostat

In fact, the stochastic dissipative equations of motion for the canonical equilibrium transition probability were given in Ch. 7, and in §7.4.5 it was shown that the Maxwell-Boltzmann distribution was stationary under their action. In the present notation, Eq. (7.86) reads

$$\begin{aligned} q_{i\alpha}(t+\Delta_t) &= q_{i\alpha}(t) + \Delta_t p_{i\alpha}(t)/m, \\ p_{i\alpha}(t+\Delta_t) &= p_{i\alpha}(t) + \Delta_t F_{i\alpha}(t) - \frac{\sigma}{2mk_{\rm B}T} p_{i\alpha}(t) + \tilde{R}_{i\alpha}(t). \end{aligned} \tag{11.5}$$

This holds for positive or negative values of the time step. The momentum equation has the form of a simple Langevin equation, with the dissipative force, $\overline{R}_{i\alpha}(t) = -\sigma p_{i\alpha}(t)/2mk_{\rm B}T$, being of the form of a drag or friction force, and the random force, $\tilde{R}_{i\alpha}(t)$, having mean zero and variance σ, $\left\langle \tilde{R}_{i\alpha}(t)\tilde{R}_{j\gamma}(t) \right\rangle = \sigma\delta_{ij}\delta_{\alpha\gamma}$. The two are related by the equilibrium fluctuation dissipation theorem. The variance σ can be chosen arbitrarily, with the proviso that it be of the same

---

[3]Nosé, S. (1984), Mol. Phys. **52**, 255. Hoover, W. G. (1985), Phys. Rev. A **31**, 1695. Tuckerman, M. E. and Martyna, G. J. (2000), J. Phys. Chem. B **104**, 159.

[4]Andersen, H. C. (1980), J. Chem. Phys. **72**, 2384. Haile, J. M. and Gupta, S. (1983), J. Chem. Phys. **79**, 3067. Andrea, T. A., Swope, W. C., and Anderson, H. C. (1983), J. Chem. Phys. **79**, 4576. Heyes, D. M. (1983), J. Chem. Phys. **82**, 285.

[5]Attard, P. (2002b), J. Chem. Phys. **116**, 9616. Boinepalli, S. and Attard, P. (2003), J. Chem. Phys. **119**, 12769.

order as the time step, since the equations of motion are predicated upon an expansion in Δ_t and σ. It is important to note that the dissipative force does not depend upon the sign of the time step. There is no memory for the random force.

With these equations of motion, the change in energy in a time step is

$$\begin{aligned}\Delta\mathcal{H}(\mathbf{\Gamma}) &= \sum_{i\alpha}\left[\frac{-\sigma}{2mk_{\mathrm{B}}T}p_{i\alpha}(t)+\tilde{R}_{i\alpha}(t)\right]\frac{p_{i\alpha}(t)}{m}+\frac{1}{2m}\sum_{i\alpha}\tilde{R}_{i\alpha}(t)^2\\ &= \frac{-\sigma}{2m^2k_{\mathrm{B}}T}\sum_{i\alpha}p_{i\alpha}(t)^2+\frac{3N\sigma}{2m}\\ &= \frac{3N\sigma}{2m}\left[1-\frac{\mathcal{T}(\mathbf{\Gamma})}{T}\right]. \end{aligned}\tag{11.6}$$

Terms $\mathcal{O}(\Delta_t^2, \Delta_t\sigma, \sigma^2)$ have been neglected here. The second equality follows after averaging over the random force. In view of the equipartition theorem, the kinetic temperature has been defined as $\mathcal{T}(\mathbf{\Gamma}) \equiv 2\mathcal{K}(\mathbf{\Gamma})/3Nk_{\mathrm{B}} = \sum_{i\alpha} p_{i\alpha}(t)^2/3Nmk_{\mathrm{B}}$. From this one sees how the stochastic thermostat works: on average the energy of the sub-system decreases when the kinetic temperature is higher than the reservoir temperature, and *vice versa.*

One way to develop second-order equations of motion is to symmetrise the adiabatic and dissipative forces with respect to t and $t + \Delta_t$. That is

$$\begin{aligned}q_{i\alpha}(t+\Delta_t) &= q_{i\alpha}(t)+\frac{\Delta_t}{2m}\left[p_{i\alpha}(t)+p_{i\alpha}(t+\Delta_t)\right]\\ &= q_{i\alpha}(t)+\frac{\Delta_t}{m}p_{i\alpha}(t)+\frac{\Delta_t^2}{2m}F_{i\alpha}(t)-\frac{\Delta_t\sigma}{4mk_{\mathrm{B}}T}p_{i\alpha}(t)+\frac{\Delta_t}{2m}\tilde{R}_{i\alpha}(t),\end{aligned}\tag{11.7}$$

and

$$\begin{aligned}p_{i\alpha}(t+\Delta_t) &= p_{i\alpha}(t)+\frac{\Delta_t}{2}\left[F_{i\alpha}(t)+F_{i\alpha}(t+\Delta_t)\right]\\ &\quad -\frac{\sigma}{4mk_{\mathrm{B}}T}\left[p_{i\alpha}(t)+p_{i\alpha}(t+\Delta_t)\right]+\tilde{R}_{i\alpha}(t)\\ &= \frac{1-s}{1+s}p_{i\alpha}(t)+\frac{\Delta_t}{2[1+s]}\left[F_{i\alpha}(t)+F_{i\alpha}(t+\Delta_t)\right]+\frac{1}{[1+s]}\tilde{R}_{i\alpha}(t),\end{aligned}\tag{11.8}$$

where $s \equiv \sigma/4mk_{\mathrm{B}}T$. Note that it is $\tilde{R}_{i\alpha}(t)$, the random force of zero mean and variance σ, that appears here. The intention with this reformulation is to allow a larger time step than otherwise to be taken.

It would appear that for the physical realisation of canonical equilibrium systems there is only one acceptable form for the equations of motion, and that is the form given here. First, the reservoir formalism necessitates some form of stochastic equations, and these must appear as non-Hamiltonian terms in the equations of motion for the sub-system. Conversely, any equations of motion that contain non-Hamiltonian terms must arise from interactions with the reservoir, and such interactions are necessarily dissipative *and* stochastic. Second, the transition probability (and consequently the random force distribution) must be Gaussian (i.e. the second entropy must be a quadratic form), because it is the sum of a large number of independent, random reservoir perturbations. Third, maximisation of the second entropy leads uniquely to the dissipative force having the drag or friction form, with the friction coefficient being equal to the variance of the random force.

Equilibration

The point in phase space chosen for the start of the trajectory requires some discussion. If the initial point is chosen 'at random', by which is meant without regard to the proper equilibrium probability distribution, then one ought to observe an equilibration phase during which the system evolves into a likely equilibrium state, and during which variables should not be accumulated for averaging. The duration of the equilibration phase is often a matter of trial and error and experience. The signature of equilibration is often a more or less monotonic change in thermodynamic variables such as the kinetic temperature, energy, or pressure, the latter often being the most sensitive.

Alternatively, one might choose the initial point according to the appropriate equilibrium probability distribution. If this is done well, then no further equilibration is required for the molecular dynamics trajectory. The Metropolis algorithm, which is discussed in the Monte Carlo section below, §11.2.1, generates points chosen from the equilibrium probability distribution. However even in this algorithm a sequence of points needs to be generated before the system equilibrates. The Monte Carlo method offers the opportunity to optimise the rate of equilibration, particularly because one is not constrained to use a small time step to the same extent that one is in the molecular dynamics method. From the point of view of ease of programming, a not insignificant consideration, one is probably best advised to use the same method for equilibration as for collecting averages.

11.1.2 Mechanical Non-Equilibrium System

The stochastic molecular dynamics algorithm is readily extended to a mechanical non-equilibrium system. In such cases the Hamiltonian is explicitly time dependent, $\mathcal{H}(\mathbf{\Gamma}, t) = \mathcal{K}(\mathbf{\Gamma}_{\rm p}) + U(\mathbf{\Gamma}_{\rm q}, t)$. It is often the case that the time dependent part of the potential is an external potential that is added to the pre-existing time independent intermolecular potential, and it may act on all the atoms of the system, or on a particular sub-set, as in driven Brownian motion that will be treated next. The equations of motion are formally the same as given above in the equilibrium case, Eq. (11.5), except that now the force is $F_{i\alpha}(t) = -\partial U(\mathbf{\Gamma}_{\rm q}, t)/\partial q_{i\alpha}$. The equations give the evolution of a sub-system that can exchange energy with a thermal reservoir of temperature T.

One of the major aims of simulations is to obtain the average value of a phase function, $\langle f(\mathbf{\Gamma}, t)\rangle$. This is relatively straightforward in the case of a harmonic (i.e. periodic in time) system, such as the driven Brownian particle treated next, or in the steady state case, such as heat flow, treated below. However in the general non-equilibrium case it can be a challenge.

One way of proceeding is to generate M trajectories, which at time t yield M points in phase space, $\mathbf{\Gamma}_1(t), \mathbf{\Gamma}_2(t), \ldots, \mathbf{\Gamma}_M(t)$, where $\mathbf{\Gamma}_I(t) \equiv \mathbf{\Gamma}(t|\mathbf{\Gamma}_{I,0}, t_0)$, $\mathbf{\Gamma}_{I,0}$ being the initial point of the Ith trajectory. Provided that the system is equilibrated by time t (i.e. either the current time is sufficiently long after the start of the trajectories for the stochastic dissipative equations of motion to have converged on the correct non-equilibrium probability at the current time, or else the initial point of each trajectory has been chosen from the correct non-equilibrium probability distribution at the initial time), then the average is

$$\langle f(\mathbf{\Gamma}, t) \rangle = \frac{1}{M} \sum_{I=1}^{M} f(\mathbf{\Gamma}_I(t), t). \tag{11.9}$$

11.1.3 Driven Brownian Motion

The stochastic molecular dynamics algorithm and results will now be given for a Brownian particle in a moving potential trap, which is the same model as treated analytically in Ch. 3, and used to test the perturbation theory in Ch. 10.

System Details

The sub-system consisted of one solute atom, the Brownian particle, and 1000 solvent atoms. A soft sphere pair potential was used for the interactions between the solvents, and between the solute and solvent,

$$u_{\alpha\gamma}(r) = \begin{cases} [d_{\alpha\gamma}/r]^{12}, & r \le R^{\rm cut}_{\alpha\gamma}, \\ 0, & r > R^{\rm cut}_{\alpha\gamma}. \end{cases} \tag{11.10}$$

The solvent had a mass $m_1 = 1$, a length scale $d_{11} = 0.5$, and a potential cut-off of $R^{\rm cut}_{11} = 1.2$. The solute had a mass $m_0 = 10$, the solute-solvent length scale was $d_{10} = 2.25$, and solute-solvent cut-off was $R^{\rm cut}_{10} = 5.2$. The temperature was 3 in these units.

As in the earlier chapters, a one-dimensional parabolic potential along the x-axis was applied to the solute, with its minimum in oscillatory motion along that axis,

$$U(x, t) = \kappa[x - b(t)]^2/2, \tag{11.11}$$

with $\kappa = 16.81$, and the location of the trap being $b(t) = B \cos \omega t$. This potential was applied to the solute alone, and x is the solute position. Unless stated otherwise, the amplitude of oscillation was $B = 2.5$.

Periodic boundary conditions were used to simulate a bulk system. A cube of edge length 11.31 was used, and the solvent density was 0.69. A spatial neighbour table with cubic cells of side length ≈ 0.6 was used, which is about three times more efficient than the more common spatial neighbour table that uses $R^{\rm cut}$ as the size of the neighbour cells.

Results were obtained for a range of frequencies, $\omega \in [0.1, 1.5]$. One simulation was carried out for each radial frequency, with typically 300 oscillation cycles being simulated each time. The time step was $\Delta_t = 10^{-3}$, so that the simulation ran for $\approx 2 \times 10^6/\omega$ times steps. The variance was set at $\sigma = 10^{-3}$ for both the solvent and the solute. The second-order equations of motion, Eqs. (11.7) and (11.8), were used. A cycle (i.e. a period $2\pi/\omega$) was divided into 20 nodes, and the position and velocity of the solute was accumulated for averaging at each node. At the conclusion of the simulation, these data were fitted in a least squares sense using $\overline{x}(t) = A\cos[\omega t + \phi]$ and $\overline{\dot{x}}(t) = -A\omega \sin[\omega t + \phi]$, where ϕ is the phase lag, and A is the oscillation amplitude. The simulation was broken into 10 blocks, and 10 independent fits were made. From the fluctuation in these, the standard error on the mean was estimated.

Notice that this method of collecting averages for the present harmonic system is somewhat simpler than in the most general non-equilibrium case discussed in the preceding subsection, §11.1.2. In the present case one simply generates a single trajectory, and after an initial equilibration period, one accumulates averages each time the system passes through the particular point in the drive cycle that one is interested in. The number of drive cycles in the total trajectory determines the number of points that contribute to each average and hence the statistical accuracy.

Results

The relative response of the Brownian particle to the driving potential is shown in Fig. 11.1. The results of the present stochastic molecular dynamics algorithm (Attard, 2009a) are compared with conventional Nosé-Hoover simulations that employ a chain thermostat, which ensures the stationarity of the equilibrium Maxwell-Boltzmann distribution for the instantaneous value of the Hamiltonian, (Attard and Gray-Weale, 2008). There is good agreement between the two simulation methods, which is perhaps a little surprising considering that one is an equilibrium method and the other is a non-equilibrium method. Possibly the reason is that the ratio of solvent to solute is 1000:1, and the solvent, which is treated explicitly in both methods, acts as a thermal reservoir (in addition to the thermal reservoir that is represented implicitly by the respective thermostats). It would be interesting to systematically decrease the solvent solute ratio with a view to quantifying the extent to which the two simulation methods are insensitive to the system size. After all, one of the motivations and a rationale for the reservoir formalism is that it focuses on the sub-system of interest, which reduces the computational effort required. Presumably, since the non-equilibrium algorithm specifically accounts for the non-equilibrium effects of the reservoir, it does not require as much solvent. (Both methods require some solvent in order to produce the generalised diffusion constant for the particle.)

One might expect that the response ratio, A/B, would be less than one, but both simulation methods show that this is slightly greater than one at low frequencies. The matter is not resolved unambiguously by the data, because low frequencies are the most challenging to simulate, and although the effect is

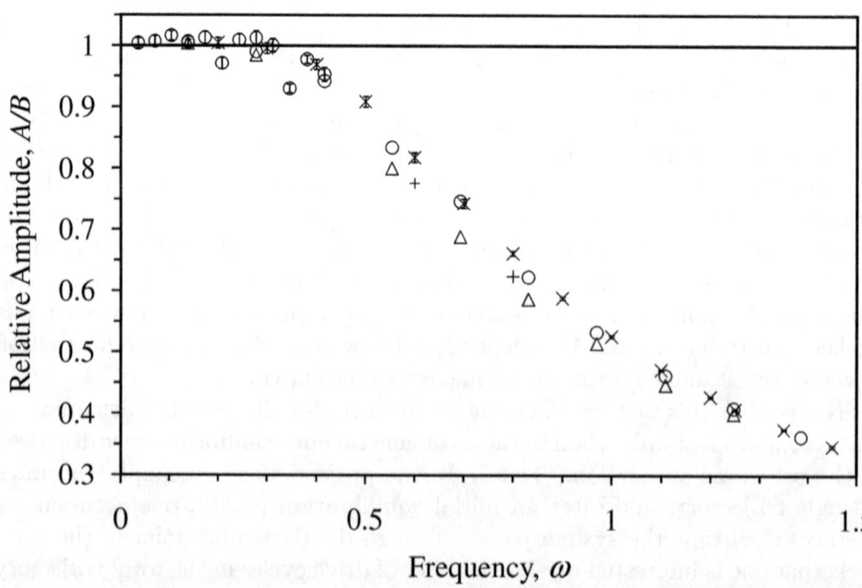

Figure 11.1: Relative amplitude of the driven Brownian particle in a soft sphere solvent. The crosses ($B = 2.5$) and plus symbols ($B = 5$) are non-equilibrium stochastic molecular dynamics simulations, [Attard, P. (2009a), J. Chem. Phys. **130**, 194113]. The circles ($B = 2.5$) and triangles ($B = 5$) are molecular dynamics simulations with equilibrium Nosé-Hoover chain thermostat, [Attard, P. and Gray-Weale, A. (2008), J. Chem. Phys. **128**, 114509]. The line is a guide to the eye. The error bars are in total twice the standard error on the mean, and are in most cases smaller than the symbols.

larger than the estimated statistical error, one cannot rule out systematic effects such as those due to the finite size of the system. It can be seen that after the initial slight increase in the ratio from unity at $\omega = 0$, the response amplitude decreases with further increase in frequency.

The diffusion constant that corresponds to this solvent solute combination has memory. This can be seen from the departure from the Langevin equation with $D = 0.105$, shown by the data in Figs 10.3 and 10.4 on pp. 362 and 363. The macroscopic description embodied in the Langevin equation with constant transport coefficient assumes Markovian behavior, which has limited validity in this case. The equations of motion in the present phase space molecular dynamics simulations are Markovian, but they yield non-Markovian behavior when projected onto the macroscopic solute coordinate space.

Increasing the amplitude by a factor of two, from $B = 2.5$ (circles and crosses) to $B = 5$ (triangles and plusses), tended to slightly decrease the relative response amplitude. The relative fluctuations shown in Fig. 3.3 on p. 73, for $\omega = 0.7$ and $B = 2.5$, show a hint of systematic departure from unity over the period of oscillation, which also suggests non-linear corrections. These are

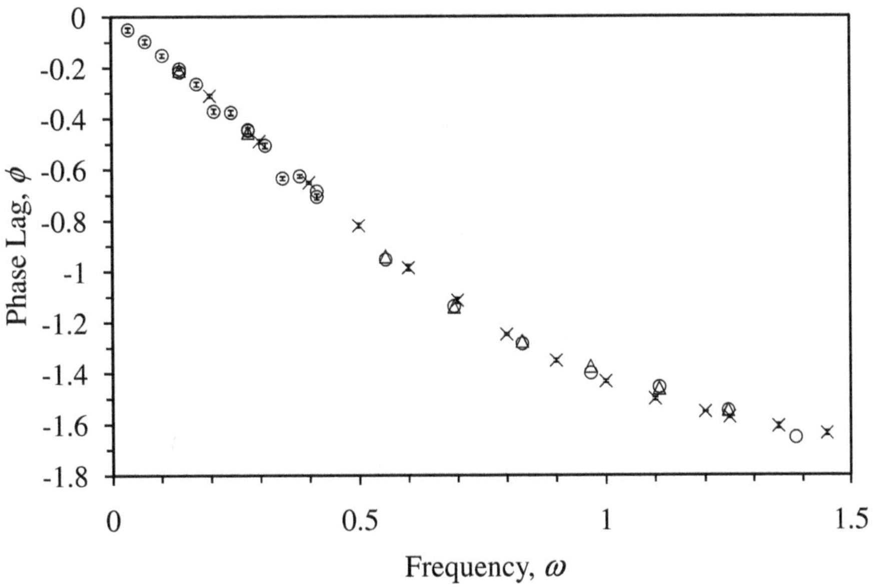

Figure 11.2: Phase lag of the driven solute. As in the preceding figure, the crosses ($B = 2.5$) and plus symbols ($B = 5$) are non-equilibrium stochastic molecular dynamics simulations (Attard, 2009a), and the circles ($B = 2.5$) and triangles ($B = 5$) are molecular dynamics simulations with equilibrium Nosé-Hoover chain thermostat (Attard and Gray-Weale, 2008).

rather small non-linear effects, but they nevertheless ought to be reliable as no linearisation has been assumed in the non-equilibrium stochastic molecular dynamics algorithm.

Figure 11.2 shows the phase lag of the solute as a function of the drive frequency. The solute oscillates almost in phase with the trap at low frequencies, and increasingly lags behind the trap as the frequency is increased, going through $\phi = \pi/2$ at approximately $\omega = 1.3$.

In the simulations summarised in the two figures, the stochastic thermostat performed well. The second-order equations of motion, Eqs. (11.7) and (11.8), were used, together with a time step of $\Delta_t = 10^{-3}$ and a variance $\sigma = 10^{-3}$. The average solvent kinetic temperature was less than 1.5% greater than the reservoir temperature over the whole frequency range with these parameters. At $\omega = 0.7$, increasing both by a factor of five did not change any results. However, increasing the time step by a factor of five and leaving the variance unchanged gave a temperature 6% above the nominal temperature, and increasing the variance by a factor of five and leaving the time step unchanged decreased the response amplitude by 11%. Obviously a lower drive frequency or a smaller drive amplitude would allow a smaller variance to be used to attain the desired temperature with a reduced perturbation of the dynamics.

11.1.4 Steady Heat Flow

Equations of Motion

The stochastic molecular dynamics algorithm for steady heat flow is based on the usual equations of motion,

$$\begin{aligned} q_{i\alpha}(t+\Delta_t) &= q_{i\alpha}(t) + \Delta_t \dot{q}^0_{i\alpha}(t) \\ &\equiv q^0_{i\alpha}(t+\Delta_t), \\ p_{i\alpha}(t+\Delta_t) &= p_{i\alpha}(t) + \Delta_t \dot{p}^0_{i\alpha}(t) + \overline{R}_{i\alpha}(t) + \tilde{R}_{i\alpha}(t) \\ &\equiv p^0_{i\alpha}(t+\Delta_t) + \overline{R}_{i\alpha}(t) + \tilde{R}_{i\alpha}(t). \end{aligned} \tag{11.12}$$

The superscript 0 denotes the adiabatic velocities, $\dot{q}^0_{i\alpha} = \partial\mathcal{H}/\partial p_{i\alpha} = p_{i\alpha}/m$, and $\dot{p}^0_{i\alpha} = -\partial\mathcal{H}/\partial q_{i\alpha} = F_{i\alpha}$, and the adiabatic evolution.

As shown in Chs 4 and 9, the static part of the reservoir entropy depends upon the zeroth and first energy moments of the sub-system and the zeroth and first temperatures of the reservoir,

$$S_{\mathrm{r,st}}(\mathbf{\Gamma}) = \frac{-E_0(\mathbf{\Gamma})}{T_0} - \frac{E_1(\mathbf{\Gamma})}{T_1}. \tag{11.13}$$

If the boundaries of the sub-system are located at $z = \pm L/2$, and if the temperatures of the respective reservoirs are $T_\pm$, then the zeroth and first temperatures of the reservoir are given by

$$\frac{1}{T_0} \equiv \frac{1}{2}\left[\frac{1}{T_+} + \frac{1}{T_-}\right], \text{ and } \frac{1}{T_1} \equiv \frac{1}{L}\left[\frac{1}{T_+} - \frac{1}{T_-}\right]. \tag{11.14}$$

For future reference is defined $\beta_n \equiv 1/k_{\mathrm{B}}T_n$, $n = 0, 1$, and also $\beta(z) \equiv \beta_0 + \beta_1 z$, which may be thought of as the local inverse temperature. The zeroth energy moment of the sub-system is just the total energy or Hamiltonian of the sub-system. In the present case a one body potential representing confining walls, $w(z)$ will be included, as well as a central pair potential, and Eq. (9.33) reads,

$$E_0(\mathbf{\Gamma}) = \mathcal{H}(\mathbf{\Gamma}) = \frac{1}{2m}\sum_{i=1}^{N} p_i^2 + \sum_{i=1}^{N} w(z_i) + \frac{1}{2}\sum_{i,j}^{N}{}' u(q_{ij}), \tag{11.15}$$

where the prime on the summation indicates that the $i = j$ term is excluded, and $z_i \equiv q_{iz}$. The first energy moment in the z-direction given by Eq. (9.34) in the present case reads,

$$E_1(\mathbf{\Gamma}) = \frac{1}{2m}\sum_{i=1}^{N} z_i p_i^2 + \sum_{i=1}^{N} z_i w(z_i) + \frac{1}{2!}\sum_{i,j}^{N}{}' \frac{z_i + z_j}{2} u(q_{ij}). \tag{11.16}$$

The adiabatic rate of change of the first energy moment is given by Eq. (9.35), and, in the present case, by Eq. (9.36). It is

$$\begin{aligned}\dot{E}_1^0(\mathbf{\Gamma}) &= \sum_{l=1}^N z_l \frac{\dot{\mathbf{p}}_l^0 \cdot \mathbf{p}_l}{m_l} + \sum_{l=1}^N \left[z_l\, \dot{z}_l^0 \frac{\partial w(z_l)}{\partial z_l} + \dot{z}_l^0 w(z_l) \right] \\ &\quad + \frac{1}{2!} \sum_{l,j}^{N}{}' \left[\{z_l + z_j\}\, \dot{\mathbf{q}}_l^0 \cdot \frac{\partial u(q_{lj})}{\partial \mathbf{q}_l} + \dot{z}_l^0 u(q_{lj}) \right] \\ &= \sum_{i=1}^N \sum_{\alpha=x,y,z} \kappa_{i\alpha} p_{i\alpha}, \end{aligned} \tag{11.17}$$

with

$$\kappa_{i\alpha} \equiv \frac{\epsilon_i}{m_i} \delta_{\alpha,z} - \sum_{j=1}^{N} {}^{j\neq i}\, u'(q_{ij}) \frac{q_{iz} - q_{jz}}{2 m_i q_{ij}} [q_{i\alpha} - q_{j\alpha}], \tag{11.18}$$

where $\delta_{\alpha,z}$ is a Kronecker delta, and ϵ_i is the energy of the ith atom,

$$\epsilon_i = \frac{p_i^2}{2m_i} + w(z_i) + \frac{1}{2} \sum_{j=1}^{N} {}^{j\neq i}\, u(q_{ij}). \tag{11.19}$$

In some cases below, a bulk system will be modelled by setting the wall potential to zero, $w(z) = 0$. Notice that κ_i does not depend upon the momentum except in the kinetic energy contribution to the energy per molecule, which means that the momentum derivative of the adiabatic rate of change of the first energy moment is

$$\frac{\partial \dot{E}_1^0(\mathbf{\Gamma})}{\partial p_{i\alpha}} = \kappa_{i\alpha} + \frac{p_{i\alpha} p_{iz}}{m_i^2}. \tag{11.20}$$

This result will be invoked in the constraint below.

Maximising the second entropy without constraint, the most likely force for the present problem of steady heat flow was given as Eq. (9.55),

$$\begin{aligned}\overline{R}_{i\alpha}^{\mathrm{f}}(\mathbf{\Gamma}, t_{21}) &= \frac{\sigma}{2k_{\mathrm{B}}} \frac{\partial S_{\mathrm{r,st}}(\mathbf{\Gamma})}{\partial p_{i\alpha}} + \frac{\sigma}{2k_{\mathrm{B}}} (\hat{t} - 1)\, \overline{S'_{\mathrm{p}i\alpha}} \\ &= \frac{-\sigma}{2m_i} \beta(z_i) p_{i\alpha} - \frac{\sigma}{2m_i} (\hat{t} - 1)\, \overline{\beta(z_i) p_{i\alpha}}, \end{aligned} \tag{11.21}$$

where $\hat{t} \equiv \mathrm{sign}(t_{21})$.

As was discussed in the general non-equilibrium case in §8.3.6, and specifically for heat flow in §9.3, p. 309, the reservoir formalism is idealised in the sense that the interactions occur throughout the sub-system, and this can disorder the internal sub-system flow in a way that does not occur with realistic boundary driven flows. One way to minimise this artifact is to make the reservoir perturbation orthogonal to the energy flux, so that $\overline{\mathbf{R}}(\mathbf{\Gamma}, t_{21}) \cdot \nabla_{\mathrm{p}} \dot{E}_1^0(\mathbf{\Gamma}) = 0$.

Maximising the second entropy subject to this constraint gives Eq. (9.57),[6]

$$\overline{\mathbf{R}}(\mathbf{\Gamma},t_{21}) = \overline{\mathbf{R}}^{\mathrm{f}}(\mathbf{\Gamma},t_{21}) + \mu(\mathbf{\Gamma})\nabla_{\mathrm{p}}\dot{E}_1^0(\mathbf{\Gamma}), \tag{11.22}$$

with $\mu(\mathbf{\Gamma}) = -\overline{\mathbf{R}}^{\mathrm{f}}(\mathbf{\Gamma},t_{21})\cdot\nabla_{\mathrm{p}}\dot{E}_1^0(\mathbf{\Gamma})/\nabla_{\mathrm{p}}\dot{E}_1^0(\mathbf{\Gamma})\cdot\nabla_{\mathrm{p}}\dot{E}_1^0(\mathbf{\Gamma})$ being the Lagrange multiplier. In component form this is

$$\overline{\mathbf{R}}_{i\alpha} = \frac{-\sigma}{2m_i}\beta(z_i)p_{i\alpha} - \frac{\sigma}{2m_i}\left(\hat{t}-1\right)\overline{\beta(z_i)p_{i\alpha}} + \mu\left[\kappa_{i\alpha} + \frac{p_{i\alpha}p_{iz}}{m_i^2}\right], \tag{11.23}$$

with

$$\mu = \frac{\sigma\sum_{i\alpha}\left[\beta(z_i)p_{i\alpha} + (\hat{t}-1)\overline{\beta(z_i)p_{i\alpha}}\right]\left[\kappa_{i\alpha} + p_{i\alpha}p_{iz}/m_i^2\right]/2m_i}{\sum_{i\alpha}\left[\kappa_{i\alpha} + p_{i\alpha}p_{iz}/m_i^2\right]^2}. \tag{11.24}$$

The stochastic molecular dynamics algorithm was based upon the above equations of motion. Each time step consisted of the adiabatic step, followed by the reservoir (dissipative plus stochastic) step. The dissipative force was evaluated at the position after the adiabatic step, $\overline{\mathbf{R}}(\mathbf{\Gamma}^0(t+\Delta_t))$, rather than at the initial point of the transition, $\overline{\mathbf{R}}(\mathbf{\Gamma}^0(t))$.

System Details

Non-equilibrium stochastic molecular dynamics were used to simulate heat flow in a Lennard-Jones fluid (Attard, 2009a). The pair potential was

$$u(r) = \begin{cases} 4\epsilon_{\mathrm{LJ}}\left[\left(\frac{\sigma_{\mathrm{LJ}}}{r}\right)^{12} - \left(\frac{\sigma_{\mathrm{LJ}}}{r}\right)^{6}\right] - u(R_{\mathrm{cut}}^-), & r \le R^{\mathrm{cut}} \\ 0, & r > R^{\mathrm{cut}}. \end{cases} \tag{11.25}$$

Results are reported in dimensionless form by invoking the well depth ϵ_{LJ}, the diameter σ_{LJ}, and the time constant $\tau_{\mathrm{LJ}} = \sqrt{(m_{\mathrm{LJ}}\sigma_{\mathrm{LJ}}^2/\epsilon_{\mathrm{LJ}})}$, where m_{LJ} is the mass, and by setting Boltzmann's constant to unity. The pair potential was cut-off at $R_{\mathrm{cut}} = 2.5$. No tail correction was used. The constant subtracted from the potential within the cut-off makes it zero approaching the cut-off, which is necessary for consistency between the Monte Carlo and the molecular dynamics results.

A spatial neighbour table with cubic cells of side length ≈ 0.6 was used.[7] The neighbour cells were defined as all those cells at least partially within the cut-off (plus the step length) of a given cell. Prior to the commencement of

[6] In Attard, (2009a) and in Attard, P. (2009b), Annu. Rep. Prog. Chem., Sect. C **105**, 63, an additional constraint was invoked, $\overline{\mathbf{R}}\cdot\nabla_{\mathrm{p}}S_{\mathrm{dyn}} = -\Delta_t k_{\mathrm{B}}\dot{Z}(t)/Z(t)$, which stems from Eqs (8.23) (with $\gamma = 0$), (8.26), (8.94), and (8.139). It turns out that for steady heat flow the associated Lagrange multiplier is negligible, $\eta \approx 0$, and results of simulations with this particular constraint imposed are indistinguishable from those without it. In contrast, neglecting the constraint $\overline{\mathbf{R}}(\mathbf{\Gamma},t_{21})\cdot\nabla_{\mathrm{p}}\dot{E}_1^0(\mathbf{\Gamma}) = 0$, treated explicitly here, leads to a measurable underestimate of the thermal conductivity, as will be shown in the results below.

[7] Attard, P. (2004), J. Chem. Phys. **121**, 7076.

the simulation, a list of such neighbour cells was calculated and stored. The advantage of such a neighbour list is that the neighbourhood volume composed of such small neighbour cells closely approximates the cut-off sphere. This is better than the conventional neighbour lists that use cubes of side length equal to the potential cut-off (Allen and Tildesley, 1987). Specifically, the present small neighbour cells reduce the enveloping neighbourhood volume from 27 large cubes (each of size $R_{\rm cut}$, neighbourhood volume $27R_{\rm cut}^3$), to approximately 667 small cubes giving a neighbourhood volume of the order of $(4\pi/3)(R_{\rm cut}+0.6)^3$. This is an improvement of almost a factor of three. That is, with these small cells each force or potential calculation requires the computation of one third the number of pair interactions that are required by conventional cells of length $R_{\rm cut}$.

Generally, 800 atoms were used in the stochastic molecular dynamics simulations, with the lateral width being $L_x = 5.87$, and, in the case of the density being $\rho = 0.8$, the nominal distance between the walls being $\tilde{L}_z = 29.87$, which gave a fluid phase width of $L_z = 28.55$. Only a slit pore was simulated with the stochastic molecular dynamics algorithm, since it was found that the bulk system with periodic boundary conditions gave unreliable results. The origin of the problem in the latter case is suspected to lie with the unrealistic behaviour of the first energy moment when an atom leaves the system at $z = +L_z/2$ and re-enters at $z = -L_z/2$; whether formulating the problem with itinerant coordinates (i.e. applying the periodic boundary conditions to $\mathbf{q}_{ij}$ rather than to $\mathbf{q}_i$) was not addressed. For this case, the temperatures of the reservoirs were $T_- = 2.25$ and $T_+ = 1.80$, which correspond to $T_0 = 2$ and $\beta_1 = 3.7 \times 10^{-3}$.

The slit pore had uniform Lennard-Jones walls perpendicular to the z-axis,

$$\begin{aligned} w(z) &= 8\pi\rho\epsilon_{\rm LJ}\left\{\left[\frac{\sigma_{\rm LJ}^{12}}{90((\tilde{L}_z/2)-z)^9} - \frac{\sigma_{\rm LJ}^{6}}{12((\tilde{L}_z/2)-z)^3}\right]\right. \\ &\quad \left. + \left[\frac{\sigma_{\rm LJ}^{12}}{90((\tilde{L}_z/2)+z)^9} - \frac{\sigma_{\rm LJ}^{6}}{12((\tilde{L}_z/2)+z)^3}\right]\right\}. \end{aligned} \tag{11.26}$$

The separation and number of atoms were varied by trial and error until the uniform density in the centre of the cell was equal to the nominal bulk density. The density of the fluid went rapidly but smoothly to zero approaching each wall. The effective width of the slit pore, which was required to calculate the volume and hence the thermal conductivity, was taken as the region where the density was non-zero (c.f. as mentioned above, for $\rho = 0.8$, the nominal distance between the walls was $\tilde{L}_z = 29.87$, and the fluid phase width was $L_z = 28.55$). Periodic boundary conditions and the minimum image convention were used in the lateral directions.

Most of the results presented below were obtained with a time step of $\Delta_t = 5 \times 10^{-3}$. No change was observed upon reducing this to 2×10^{-3}. A time step larger than 7.5×10^{-3} for these systems with these equations of motion causes the kinetic energy to increase significantly on an adiabatic trajectory.

A dimensionless variance was defined, $\sigma^* \equiv 2\sigma/m_{\rm LJ}k_{\rm B}T_0$, where the denominator is the average of the square of a component of the momentum. Values of $\sigma^* \sim 1$ mean that the stochastic change in a momentum component is of the same order as the momentum component itself. Results were obtained with $\sigma^* \in [5 \times 10^{-4}, 10^{-1}]$. The influence of the variance on the results is discussed in detail below.

A time step consists of an adiabatic part and a reservoir perturbation, as described above. In most of the results reported here, averages were collected every 100 time steps. In some cases this was reduced to every 10 time steps, which appeared more efficient (i.e. it gave smaller statistical error for little increase in computer time). In general terms, there is a small overhead in gathering averages, and since consecutive points are highly correlated, there is a cost to collecting averages too frequently. No effort was made to optimise the calculation of the averages or the frequency with which they were collected.

In the stochastic molecular dynamics method for steady state systems such as the present heat flow, averages mean simple time averages over the single trajectory that is simulated. This is simpler than collecting averages in the general non-equilibrium case discussed in §11.1.2, or in the harmonic case discussed above in the context of driven Brownian motion, §11.1.3. In the general steady state case one has

$$\langle f \rangle = \frac{1}{M} \sum_{I=1}^{M} f(\mathbf{\Gamma}(t_I)), \tag{11.27}$$

where the t_I are regularly spaced time nodes along the trajectory, and $f(\mathbf{\Gamma})$ is a function of the sub-system phase space. The standard error is estimated by breaking the simulation into Q blocks, and taking the averages separately in each block. With $\overline{f}_J$ the average of block J, and $\langle f \rangle = Q^{-1} \sum_{J=1}^{Q} \overline{f}_J$ the over all average, the standard error on the mean is

$$E_f = \frac{1}{Q^{1/2}} \sigma_f^{1/2} = \frac{1}{Q^{1/2}} \sqrt{\frac{1}{Q} \sum_{J=1}^{Q} \left[\overline{f}_J - \langle f \rangle\right]^2}. \tag{11.28}$$

The variance of f measured by the blocks is σ_f, and so $\sigma_f^{1/2}$ is an estimate of the variation of $\overline{f}_J$ each time a new block is simulated. Hence $\sigma_f^{1/2}/Q^{1/2}$ is an estimate of the overall accuracy of the simulation once the Q blocks have been combined. Often the Q in the denominator is replaced by $\sqrt{Q(Q-1)}$, but for typical values, $Q \approx 10$–20, this makes little difference. Once again there ought to be an equilibration period before one begins collecting averages.

It can be shown using Fourier's law that when the first energy moment relaxes in an adiabatic system, then the temperature profile has to have zero gradient at the walls, because there is no heat flux across the boundaries.[8] Hence in the slit pore of the present stochastic molecular dynamics simulations, the adiabatic step mitigates against having a linear temperature profile throughout the system. This also affects the dynamics in the vicinity of the walls. For this

[8] Attard, P. (2005a), J. Chem. Phys. **122**, 154101.

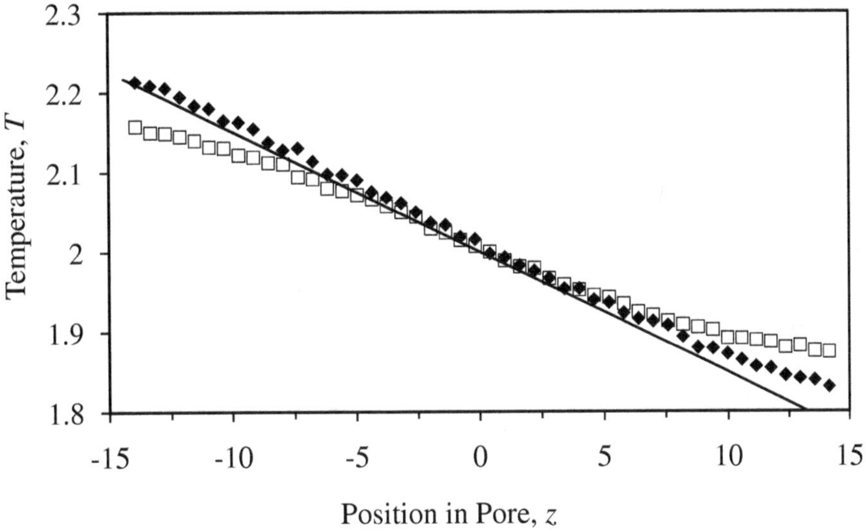

Figure 11.3: Induced temperature profile from stochastic molecular dynamics simulations for a slit pore with $\rho = 0.8$, $T_+ = 2.25$, $T_- = 1.80$, and $L_z = 29.87$, which give $T_0 = 2$, and $\beta_1 = 3.7 \times 10^{-3}$. The time step was $\Delta_t = 5 \times 10^{-3}$. The filled diamonds use $\sigma^* = 5 \times 10^{-3}$, and the open squares use $\sigma^* = 5 \times 10^{-4}$. The solid line is the applied temperature from the reservoirs, $T(z) = T_0 + (T_+ - T_-)z/L_z$. Data from Attard (2009a).

reason, when calculating the rate of change of energy moment, and also when calculating the μ, only the contribution from atoms in the central half of the slit pore was used.

Results

Figure 11.3 shows the induced temperature profile in a slit pore obtained with stochastic molecular dynamics for two cases of the variance. The equipartition theorem was used to obtain the local temperature $T(z)$ from the local average kinetic energy. It can be seen that the data lie on sigmoidal curves, which have approximately zero gradient at the boundaries. As discussed above, this shape is due to the absence of any heat flux across the walls during the adiabatic motion, and so this is most prominent at low values of the variance. The reservoir perturbation acts in part like a thermostat that induces the temperature profile to be equal to the applied profile. One can interpret the stochastic molecular dynamics algorithm as a competition between the adiabatic evolution and the reservoir perturbation, with the variance controlling their relative strengths. This explains why the sigmoidal shape is less pronounced at high variance.

In the central half of the system the simulated temperature profiles have approximately the same slope as the applied temperature profile from the reser-

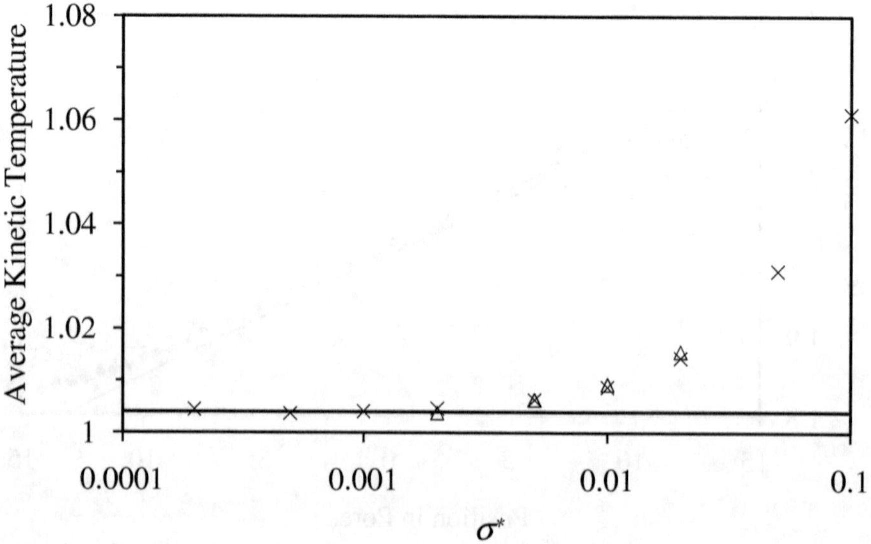

Figure 11.4: The kinetic temperature averaged across the whole system, $\langle p^2 \rangle / mkBT_0$, as a function of variance for the same case as the preceding figure. The crosses are for a time step $\Delta_t = 5 \times 10^{-3}$, and the triangles are for a time step $\Delta_t = 2 \times 10^{-3}$. The line is $\overline{T}_0/T_0 = [1 + (\beta_1^2/\beta_0^2)L_z^2/12] = 1.004$. Data from Attard (2009a).

voirs. Closer inspection of Fig. 11.3 shows that in this region the induced temperature gradient is slightly less than the applied temperature gradient, particularly for the lower variance. Again this is a result of the competition between the adiabatic step and the reservoir perturbation. During the adiabatic evolution heat flows from the high energy region of the sub-system to the low energy region as the first energy moment internally relaxes. The rate of internal heat flow is proportional to the magnitude of the first energy moment, which in turn is proportional to the induced temperature gradient. The reservoir perturbation attempts to counter this internal adiabatic heat flow and to restore the moment to what it was before the adiabatic evolution. Obviously the amount of heat that can be supplied by the reservoir is proportional to the magnitude of the variance, which represents the strength of the coupling between the reservoir and the sub-system. For small variances, the average moment decreases to a value such that the adiabatic heat flow is reduced to an amount that can be compensated by the reservoir. This effect is manifest in the reduced induced temperature gradient in the figure.

Figure 11.4 shows the kinetic temperature averaged over the whole system relative to T_0, which one would expect to be unity. The data show that at high values of the variance the kinetic temperature is significantly larger than the expected value. In deriving the transition probability that underlies the stochastic molecular dynamics algorithm, various expansions were carried out

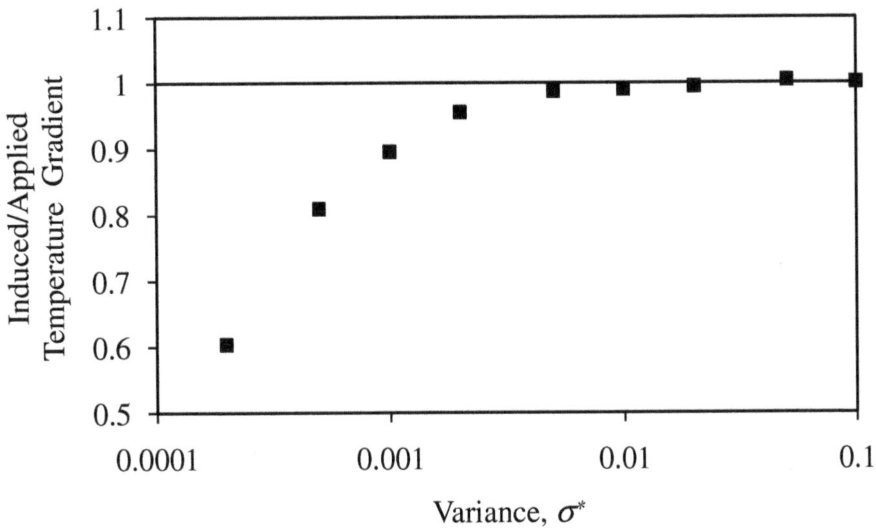

Figure 11.5: Ratio of the induced temperature gradient in the central half of the slit pore to the applied temperature gradient as a function of variance for the same case as the preceding figures ($\Delta_t = 5 \times 10^{-3}$). The line is a guide to the eye. Data from Attard (2009a).

to linear order in the variance. Presumably then, the heating up of the system for large values of σ is a consequence of neglecting the higher-order terms.

Two different time steps were tested in the simulations in Fig. 11.4, and it can be seen that they are in quantitative agreement. This indicates that the results can be relied upon for these values.

Small values of the variance lead to a kinetic temperature ratio close to unity, but slightly above unity. This is a second-order effect that arises from averaging the applied temperature profile over the whole system. The applied temperature has expansion

$$k_{\mathrm{B}}T(z) = \frac{1}{\beta_0 + z\beta_1} = \beta_0^{-1}[1 - z\beta_1/\beta_0 + z^2\beta_1^2/\beta_0^2 \ldots]. \tag{11.29}$$

Integrating this over z yields $\overline{T}_0 = T_0[1 + (\beta_1^2/\beta_0^2)L_z^2/12] = 1.004T_0$. This agrees with the low variance data in Fig. 11.4.

Figure 11.5 shows how the gradient of the temperature induced in the sub-system depends upon the variance. As already discussed in connection with Fig. 11.3, at small variance the coupling between the sub-system and the reservoir is too weak to supply heat fast enough to compensate for the adiabatic relaxation of the energy moment. Hence the system adopts a reduced energy moment, which is equivalent to a reduced temperature gradient. At this reduced value the internal flux can be matched to the flux from the reservoir and the energy moment can be held steady.

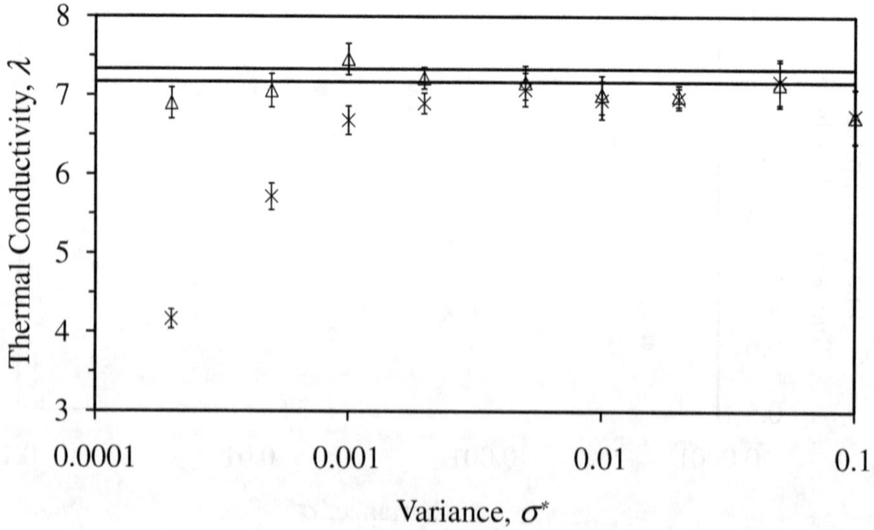

Figure 11.6: Thermal conductivity as a function of variance for the same case as the preceding figures ($T_0 = 2$, $\rho = 0.8$). The crosses use the applied temperature gradient, and the open triangles use the induced temperature gradient in the central half of the system. The total length of each error bar is twice the standard error on the mean. The two lines represent the range of values obtained with Green-Kubo and non-equilibrium molecular dynamics methods in the literature for this state point [Daivis, P. J. and Evans, D. J. (1993), Phys. Rev. E **48**, 1058. Evans, D. J. (1986) Phys. Rev. A **34**, 1449]. Data from Attard (2009a).

The effect just identified is very important in calculating the thermal conductivity, as is shown in Fig. 11.6. The thermal conductivity is essentially the ratio of the adiabatic rate of change of the first energy moment to the temperature gradient,

$$\lambda = \frac{\left\langle \dot{E}_1^0 \right\rangle}{\beta_1 V k_{\mathrm{B}} T_0^2}. \tag{11.30}$$

This is a simple time average over the stochastic molecular dynamics trajectory. Because the temperature gradient depends on the variance for the reasons just discussed, it makes a difference whether the β_1 that appears here is the induced or the applied temperature gradient, particularly at lower values of the variance. It can be seen in Fig. 11.6 that using the induced temperature gradient gives the same value of the thermal conductivity for over three orders of magnitude of range in σ. Note that the induced temperature gradient, the adiabatic rate of change of the first energy, and the volume were all taken in the central half of the slit pore.

The stochastic molecular dynamics results for the thermal conductivity for this case ($T_0 = 2$, $\rho = 0.8$) are in good agreement with the literature values shown in Fig. 11.6. Of the two methods that are commonly used in the literature to obtain the thermal conductivity, the equilibrium Green-Kubo method is the most reliable. This obtains the time correlation function for the heat flux from molecular dynamics simulations of an equilibrium system with thermostat, such as the Nosé-Hoover chain thermostat. From the long time behaviour of the time correlation function the thermal conductivity is obtained. The limitation of the Green-Kubo method is that as an equilibrium method the simulations need to be quite long in order to get reliable statistics for the fluctuations. Although quantitative tests have not been performed, one would expect that the stochastic molecular dynamics algorithm would be perhaps an order of magnitude more computationally efficient than the Green-Kubo method, because the thermal conductivity results directly from a non-equilibrium average rather than from an equilibrium fluctuation.

The second literature method is called the non-equilibrium molecular dynamics algorithm, and this relies on certain artificial non-Hamiltonian equations of motion that have been developed to mimic heat flow (Evans, 1986; Daivis and Evans, 1993). The problem with these deterministic equations of motion is that there is no fundamental thermodynamic basis for the non-Hamiltonian terms, and consequently the fluctuation dissipation theorem is violated. Although there are fundamental problems with the non-equilibrium molecular dynamics method, these do not appear to effect the thermal conductivity itself, judging by the agreement with the non-equilibrium stochastic molecular dynamics results and with the Green-Kubo results. One should be cautious about drawing broader conclusions about the artificial equations that underpin the non-equilibrium molecular dynamics algorithm.

The main point that can be taken from Fig. 11.6 is that the present non-equilibrium stochastic molecular dynamics algorithm produces results for the transport coefficient that are in agreement with literature values. Further comparison of the stochastic molecular dynamics algorithm for the thermal conductivity with literature values and with non-equilibrium Monte Carlo results is given below.

11.2 Non-Equilibrium Monte Carlo

11.2.1 Equilibrium Systems

As mentioned in the introduction to the chapter, there are two main simulation methods: molecular dynamics and Monte Carlo. Arguably, of the two, molecular dynamics is the more direct, brute force approach. This is because one follows, and one is restricted to following, the physical trajectory in time of the subsystem through its phase space. By comparison, the Monte Carlo method is more refined because it allows any convenient path, even non-physical paths, to be followed, and these can be optimised to increase the computational efficiency or to exploit the physical structure of the system in ways that are beyond the molecular dynamics method.

In mathematical terms, the Monte Carlo method is a quadrature method for evaluating integrals, based upon a uniform random sampling of the integration volume and a simple average of the integrand. In this raw form it is suited for multidimensional integrals in which the integrand is a relatively smooth function that almost completely fills the region of integration. This raw form is not suited for statistical mechanics because almost all of phase space has zero weight due to the overlap of atoms that arises when points are chosen at random. The actual hypersurface that contributes significantly to canonical equilibrium averages is a negligible fraction of the total phase space, and accordingly one has to focus the points chosen for the quadrature on this region.

To achieve this goal, the Monte Carlo simulation method for a canonical equilibrium system consists of generating a sequence of points in phase space,[9] $\mathbf{\Gamma}_1, \mathbf{\Gamma}_2, \ldots$, selected from the Maxwell-Boltzmann distribution,

$$\wp_{\mathrm{MB}}(\mathbf{\Gamma}) = \frac{1}{Z(N,V,T)} e^{-\beta \mathcal{H}(\mathbf{\Gamma})}, \tag{11.31}$$

where $\beta \equiv 1/k_{\mathrm{B}}T$, k_{B} being Boltzmann's constant and T being the temperature. The sequence of points in phase space is *not* a sequence in time. Because the points are already chosen with the proper weight, the canonical equilibrium average of a phase function is just a simple average over the sequence,

$$\langle f(\mathbf{\Gamma}) \rangle = \frac{1}{M} \sum_{I=1}^{M} f(\mathbf{\Gamma}_I). \tag{11.32}$$

This assumes that appropriate equilibration has taken place before commencement of the averaging process.

In general, the probability distribution, which is the exponential of an extensive variable, is a much more rapidly varying function of phase space than any function that one typically seeks to average. This is the reason why the probability distribution alone is used to bias the selection of phase space points. Umbrella sampling is a useful variant of this, which is discussed below.

Metropolis Algorithm

The key to selecting points in a sequence with a chosen weight lies in the choice of transition probability, $\wp(\mathbf{\Gamma}_{I+1}|\mathbf{\Gamma}_I)$. For Monte Carlo simulations, the most common choice for the transition probability is the Metropolis algorithm. This algorithm divides the transition into two stages: first the generation of a trial phase space point, $\wp_{\mathrm{trial}}(\tilde{\mathbf{\Gamma}}_{I+1}|\mathbf{\Gamma}_I)$, and second the decision whether to accept

[9] In an equilibrium system the momenta play a trivial rôle and are generally integrated out of the problem, so that only points in configuration space are generated, $\mathbf{\Gamma}_{\mathrm{q}1}, \mathbf{\Gamma}_{\mathrm{q}2}, \ldots$, with weight $\exp -\beta U(\mathbf{\Gamma}_{\mathrm{q}})$. In the present description of the equilibrium algorithm, the full phase space is used in order to make a more direct connection with the non-equilibrium problem, which requires the momenta.

this trial point as the next point in the sequence, $\mathbf{\Gamma}_{I+1} = \tilde{\mathbf{\Gamma}}_{I+1}$, or else to reject it and to retain the original point as the next point, $\mathbf{\Gamma}_{I+1} = \mathbf{\Gamma}_I$.

The trial transition most commonly consists of selecting an atom i, at random or in turn, and moving it by a random amount,

$$\tilde{\mathbf{\Gamma}}_{I+1;j} = \begin{cases} \mathbf{\Gamma}_{I;i} + \mathbf{D}\boldsymbol{\xi}_{I+1}, & j = i, \\ \mathbf{\Gamma}_{I;j}, & \text{otherwise}, \end{cases} \tag{11.33}$$

where upper case Roman indeces label the phase space point, and the lower case Roman indeces label the atom. It is sometimes useful to move several atoms simultaneously, but this is not discussed here. Here $\mathbf{D}$ is a diagonal matrix (six-dimensional in phase space; three-dimensional in the more usual configuration space) that is fixed at the outset and that determines the maximum size of the random trial move, which is typically a fraction of an atomic diameter (so that the energy change is on average a fraction of $k_{\mathrm{B}}T$). The vector $\boldsymbol{\xi}_{I+1}$ consists of six (or three) independent random numbers drawn from a uniform distribution on $[-1, 1]$. As will become clearer shortly, what is important for the algorithm is that the trial transition probability is symmetric,

$$\wp_{\mathrm{trial}}(\mathbf{\Gamma}'|\mathbf{\Gamma}) = \wp_{\mathrm{trial}}(\mathbf{\Gamma}|\mathbf{\Gamma}'). \tag{11.34}$$

The Metropolis recipe certainly satisfies this criterion because positive and negative values of $\boldsymbol{\xi}_{I+1}$ are equally likely, and because the atom to be moved is chosen independent of the point in phase space. Algorithms that do not have a symmetric trial probability are possible, but the analysis is somewhat simplified when this condition is obeyed.

For greatest generality the transitions can be made based on an arbitrary probability distribution $\wp(\mathbf{\Gamma})$. This is most commonly taken to be the Maxwell-Boltzmann distribution, $\wp_{\mathrm{MB}}(\mathbf{\Gamma})$, but there can sometimes be merit in using a different distribution, which is called umbrella sampling. The decision whether to select or reject the trial point depends on whether the probability increases or decreases in the trial move. Specifically,

$$\mathbf{\Gamma}_{I+1} = \begin{cases} \tilde{\mathbf{\Gamma}}_{I+1}, & \zeta_{I+1} \leq \wp(\tilde{\mathbf{\Gamma}}_{I+1})/\wp(\mathbf{\Gamma}_I), \\ \mathbf{\Gamma}_I, & \text{otherwise}, \end{cases} \tag{11.35}$$

where ζ_{I+1} is a random number uniformly distributed on $[0, 1]$. The effect of this is that if the probability increases after the trial move, then the trial point is certainly accepted as the next point in the sequence. If the trial probability decreases by only a small amount, then the trial point is likely accepted, whereas if the trial probability decreases by a large amount, then it is likely that the trial point will be rejected and the next point in the sequence will be set equal to the preceding point. The nett effect of this transition rule is that the phase space points in the sequence tend to be points of high probability, but not exclusively so. Note that because the ratio of probabilities appears on the right-hand side, unnormalised probabilities, $\omega(\mathbf{\Gamma})$, can be used.

The Metropolis algorithm corresponds to the transition probability,

$$\wp(\mathbf{\Gamma}'|\mathbf{\Gamma}) = \begin{cases} \wp_{\mathrm{trial}}(\mathbf{\Gamma}'|\mathbf{\Gamma}), & \wp(\mathbf{\Gamma}') \geq \wp(\mathbf{\Gamma}), \\ \wp_{\mathrm{trial}}(\mathbf{\Gamma}'|\mathbf{\Gamma})\wp(\mathbf{\Gamma}')/\wp(\mathbf{\Gamma}), & \text{otherwise}. \end{cases} \tag{11.36}$$

Because the trial transition probability is symmetric, $\wp_{\text{trial}}(\mathbf{\Gamma}'|\mathbf{\Gamma}) = \wp_{\text{trial}}(\mathbf{\Gamma}|\mathbf{\Gamma}')$, the transition can readily be shown to satisfy statistical symmetry,

$$\wp(\mathbf{\Gamma}'|\mathbf{\Gamma})\wp(\mathbf{\Gamma}) = \wp(\mathbf{\Gamma}|\mathbf{\Gamma}')\wp(\mathbf{\Gamma}'). \tag{11.37}$$

As mentioned above, the sequence of points in phase space is not a sequence in time, and so one should *not* interpret this in terms of molecular motion. This is the rule for joint probabilities that comes from the laws of probability, from which one can conclude that the transition rule visits points in phase space according to the probability distribution $\wp(\mathbf{\Gamma})$.[10]

A canonical equilibrium average for this algorithm is given by,

$$\langle f(\mathbf{\Gamma})\rangle = \frac{\sum_{J=1}^{M} f(\mathbf{\Gamma}_J)\omega_{\text{MB}}(\mathbf{\Gamma}_J)/\omega(\mathbf{\Gamma}_J)}{\sum_{I=1}^{M} \omega_{\text{MB}}(\mathbf{\Gamma}_J)/\omega(\mathbf{\Gamma}_J)}. \tag{11.38}$$

Typically an average is taken every nN Monte Carlo trial steps, $J = nNI$, where N is the number of atoms and $n = 10$–100. This is because adjacent points on the Monte Carlo sequence are highly correlated and contribute no new information to the average. Here the unnormalised probabilities appear, since the normalisation factors cancel top and bottom. In the case that the transition probability invokes the Maxwell-Boltzmann canonical form, this reduces to the simple average given above, Eq. (11.32). More generally, choosing a slightly different weight, $\omega(\mathbf{\Gamma}) \neq \omega_{\text{MB}}(\mathbf{\Gamma})$, is called umbrella sampling, and it is convenient when, for example, one wants to simulate multiple thermodynamic states simultaneously.

Almost always there is an equilibration period following the initiation of a Monte Carlo path. During this time the distribution of the points generated approaches the designated probability distribution $\wp(\mathbf{\Gamma})$. Averages should not be collected during this equilibration period (c.f. the discussion on p. 395).

11.2.2 Non-Equilibrium Systems

As just mentioned, Monte Carlo simulations of equilibrium systems depend upon generating a sequence of points with a weight corresponding to a known probability distribution. The well-known Metropolis algorithm achieves this goal by using a transition probability that has a known probability (usually equal to, or at least close to, the equilibrium probability) as its stationary solution.

A transition probability with the correct non-equilibrium probability distribution as its stationary state has already been invoked above: the stochastic dissipative equations of motion used in the molecular dynamics algorithm. These represent the physical transition probability of the system and accordingly they yield a sequence of points distributed according to the proper non-equilibrium probability distribution. Although in this sense the non-equilibrium stochastic

[10] Another way of seeing this is to note that integrating over $\mathbf{\Gamma}'$ yields $\wp(\mathbf{\Gamma})$ on the left-hand side. This says that $\wp(\mathbf{\Gamma})$ is an eigenfunction of $\wp(\mathbf{\Gamma}'|\mathbf{\Gamma})$ with eigenvalue unity. This means that a long sequence of applications of the transition probability will tend to $\wp(\mathbf{\Gamma})$, whatever the starting probability.

molecular dynamics algorithm is similar to the equilibrium Monte Carlo method, it does differ qualitatively in the sense that it follows the physical trajectory of the non-equilibrium system, and one does not have the freedom to utilise any alternative paths, which may have computational advantages, or which may exploit the physical behaviour of the system (c.f. the remarks at the beginning of the preceding subsection, §11.2.1).

The development of a non-equilibrium Monte Carlo algorithm that retains the same advantages as its equilibrium counterpart, is, in a formal sense, rather trivial. There are however two challenges in formulating the algorithm in a practical form suitable for actual computation. One challenge is actually to evaluate the non-equilibrium probability, which has been formally derived in general in Ch. 8. The second challenge is to give an efficient transition probability, and for this a form of umbrella sampling will be required.

Here the most general non-equilibrium case is treated. The algorithm simplifies somewhat for the case of a harmonic system and for the case of a steady state system. These simpler but more specialised algorithms are discussed below.

Metropolis Algorithm

For the present, let $\wp_{\mathrm{NE}}(\mathbf{\Gamma}, t)$ be the non-equilibrium probability distribution for the system of interest, and let $\wp(\mathbf{\Gamma}, t)$ be the arbitrary probability distribution that will be used in the transition probability (umbrella sampling). Let $\mathbf{\Gamma}_1(t), \mathbf{\Gamma}_2(t), \ldots$, be a sequence of points in phase space all at time t. These are generated by a transition probability of the Metropolis form. That is, a trial move is made on atom i,

$$\tilde{\mathbf{\Gamma}}_{I+1;j}(t) = \begin{cases} \mathbf{\Gamma}_{I;i}(t) + \mathbf{D}\boldsymbol{\xi}_{I+1}(t), & j = i, \\ \mathbf{\Gamma}_{I;j}(t), & \text{otherwise}, \end{cases} \tag{11.39}$$

where again $\mathbf{D}$ is a diagonal matrix, that determines the maximum move of atom i and the random vector $\boldsymbol{\xi}_{I+1}(t)$ has elements drawn from a uniform distribution on $[-1, 1]$. Evidently the trial transition probability is symmetric. $\wp_{\mathrm{trial}}(\mathbf{\Gamma}'(t)|\mathbf{\Gamma}(t)) = \wp_{\mathrm{trial}}(\mathbf{\Gamma}(t)|\mathbf{\Gamma}'(t))$. The actual transition is

$$\mathbf{\Gamma}_{I+1}(t) = \begin{cases} \tilde{\mathbf{\Gamma}}_{I+1}(t), & \zeta_{I+1}(t) \leq \wp(\tilde{\mathbf{\Gamma}}_{I+1}(t), t)/\wp(\mathbf{\Gamma}_I(t), t), \\ \mathbf{\Gamma}_I(t), & \text{otherwise}, \end{cases} \tag{11.40}$$

where $\zeta_{I+1}(t)$ is a random number uniformly distributed on $[0, 1]$. Again unnormalised probabilities, $\omega(\mathbf{\Gamma}, t)$, can be used.

This non-equilibrium analogue of the Metropolis algorithm corresponds to the transition probability

$$\wp(\mathbf{\Gamma}'(t)|\mathbf{\Gamma}(t); t) = \begin{cases} \wp_{\mathrm{trial}}(\mathbf{\Gamma}'(t)|\mathbf{\Gamma}(t)), & \wp(\mathbf{\Gamma}'(t), t) \geq \wp(\mathbf{\Gamma}(t), t), \\ \wp_{\mathrm{trial}}(\mathbf{\Gamma}'(t)|\mathbf{\Gamma}(t)) \dfrac{\wp(\mathbf{\Gamma}'(t), t)}{\wp(\mathbf{\Gamma}(t), t)}, & \text{otherwise}. \end{cases} \tag{11.41}$$

Again the symmetry of the trial transition probability means that the transition satisfies statistical symmetry,

$$\wp(\mathbf{\Gamma}'(t)|\mathbf{\Gamma}(t);t)\wp(\mathbf{\Gamma}(t),t) = \wp(\mathbf{\Gamma}(t)|\mathbf{\Gamma}'(t);t)\wp(\mathbf{\Gamma}'(t),t). \tag{11.42}$$

Again the sequence of points in phase space at time t is not a sequence in time, and so one should *not* interpret this in terms of molecular motion. As in the equilibrium case, this is the rule for joint probabilities that comes from the laws of probability, from which one can conclude that the transition rule visits points in phase space according to the probability distribution $\wp(\mathbf{\Gamma},t)$.

The non-equilibrium average for points generated with this algorithm is

$$\langle f(\mathbf{\Gamma},t)\rangle = \frac{\sum_{J=1}^{M} f(\mathbf{\Gamma}_J(t),t)\,\omega_{\rm ne}(\mathbf{\Gamma}_J(t),t)/\omega(\mathbf{\Gamma}_J(t),t)}{\sum_{I=1}^{M} \omega_{\rm ne}(\mathbf{\Gamma}_J(t),t)/\omega(\mathbf{\Gamma}_J(t),t)}, \tag{11.43}$$

assuming equilibration has taken place. The number of points retained for this average is typically about one or two orders of magnitude less than the number of trial moves in the Monte Carlo sequence, $J = nNI$, where N is the number of atoms and $n = 10$–100.

It bears repeating that this is the average at a single time t. For the most general mechanical non-equilibrium case with external potential of arbitrary time variation, the whole procedure has to be carried out at each time that one wants an average. For the case of steady state systems, a single time suffices.

Non-Equilibrium Probability

Still in this general time dependent case, two points remain to be addressed: calculation of the non-equilibrium weight, and the choice of umbrella weight.

In Ch. 8 the non-equilibrium probability density was shown to be proportional to the exponential of the reservoir entropy, $\wp_{\rm NE}(\mathbf{\Gamma},t) \propto \exp S_{\rm r}(\mathbf{\Gamma},t)/k_{\rm B}$ It was shown that the latter consisted of a static and a dynamic part,

$$S_{\rm r}(\mathbf{\Gamma},t) = S_{\rm r,st}(\mathbf{\Gamma},t) + S_{\rm r,dyn}(\mathbf{\Gamma},t). \tag{11.44}$$

This neglects the time dependent constant $\overline{S}_{\rm r}(t)$, which can be incorporated into the normalising partition function. Hence the exponential of the right-hand side divided by Boltzmann's constant can be taken as the unnormalised weight, $\omega_{\rm NE}(\mathbf{\Gamma},t) = \exp[S_{\rm r,st}(\mathbf{\Gamma},t) + S_{\rm r,dyn}(\mathbf{\Gamma},t)]/k_{\rm B}$.

The static part of the reservoir entropy can be derived from equilibrium considerations and is relatively trivial to write down in a given case. It is essential to note that it depends only upon the instantaneous configuration of the sub-system, and not upon the previous history.

In contrast, the dynamic part of the reservoir entropy depends upon the most likely trajectory of the sub-system leading to the current point,

$$S_{\rm r,dyn}(\mathbf{\Gamma},t) = -\int_{t_0}^{t} {\rm d}t\ \ \dot{S}^0_{\rm r,st}(\overline{\mathbf{\Gamma}}(t'|\mathbf{\Gamma},t),t'). \tag{11.45}$$

It is this term that poses a challenge to non-equilibrium Monte Carlo simulations because as a one-dimensional integral it requires, say, L nodes to evaluate it. Each time step on the backwards trajectory requires the force on all atoms, usually an N^2 operation for N atoms, but this an be reduced to $\mathcal{O}(N)$ with the use of neighbour tables. If one has M' points on the Monte Carlo sequence (typically $M' = (10\text{–}100)M$), then in the general case this is an $\mathcal{O}(NLM')$ operation for *each* time t. This is computationally demanding (it is L times the number of operations required to evaluate $S_{\mathrm{r,st}}$) and it is clear that one ought to minimise the number of evaluations of $S_{\mathrm{r,dyn}}$.

One obvious way of proceeding is to use umbrella sampling, with the umbrella weight simply the static part of the reservoir entropy,

$$\omega(\mathbf{\Gamma}, t) = \exp S_{\mathrm{r,st}}(\mathbf{\Gamma}, t)/k_{\mathrm{B}}. \tag{11.46}$$

This is about a factor of L less demanding to compute than the full $\omega_{\mathrm{ne}}(\mathbf{\Gamma}, t)$. For this to be useful the number of points used for the average, M, ought to be very much less than the total number of points in the Monte Carlo sequence, M', with typically $M \ll M' \lesssim LM$. The reason is that adjacent points on the Monte Carlo sequence are highly correlated, and they contribute no new information to the average. With this umbrella weight, generating a new point in the sequence is much less expensive than evaluating the weight of the point for the average. Typically, an average is accumulated once every $100N$ trial moves (i.e. an attempt has been made to move each atom 100 times). The algorithm is now $\mathcal{O}(NLM, NM')$ for each time t.

This umbrella weight has even parity, $S_{\mathrm{r,st}}(\mathbf{\Gamma}^\dagger, t) = S_{\mathrm{r,st}}(\mathbf{\Gamma}, t)$, whereas the non-equilibrium probability distribution must have an odd parity component. In view of the fact that the adiabatic rate of change of the static part of the reservoir entropy has an odd parity component, and since one expects $\overline{\dot{S}^0_{\mathrm{r,st}}(t)} < 0$, an alternative possibility for the umbrella weight is

$$\omega(\mathbf{\Gamma}, t) = \exp\left[S_{\mathrm{r,st}}(\mathbf{\Gamma}, t) - c\dot{S}^0_{\mathrm{r,st}}(\mathbf{\Gamma}, t)\right]/k_{\mathrm{B}}, \tag{11.47}$$

where $c > 0$ is a time constant chosen by experience or by trial and error. Usually one tries to achieve equality between the average value of $\dot{S}^0_{\mathrm{r,st}}$ taken with this instantaneous weight alone with that taken with the full non-equilibrium weight; linearising, this criterion is $c \approx -k_{\mathrm{B}} \left\langle \dot{S}^0_{\mathrm{r,st}} \right\rangle_{\mathrm{ne}} \Big/ \left\langle [\dot{S}^0_{\mathrm{r,st}}]^2 \right\rangle_{\mathrm{st}}$. (It should be stressed that even with this choice of c, the weight $\omega(\mathbf{\Gamma}, t)$ is not the exact non-equilibrium weight.) Of course the final average is independent of the umbrella weight, since this is replaced by the actual non-equilibrium weight, $\omega_{\mathrm{ne}}(\mathbf{\Gamma}, t)$, when the averages are accumulated; it is only the computational efficiency required to obtain the average with a given variance that is affected by the actual form of the umbrella weight. In general terms, the closer the umbrella weight to the actual non-equilibrium weight, the more efficient is the simulation. The particular form of umbrella weight discussed here is, in principle, better than the static weight because it biases the Monte Carlo path so that it visits regions of phase space that have the same parity as those that dominate the non-equilibrium system.

Odd Parity Approximation

In §8.5, the odd parity projection of the dynamic part of the reservoir entropy was analysed. It was argued that in many circumstances only this odd projection was required (e.g. obtaining transport coefficients), that the odd projection was computationally easier to evaluate than the full quantity (because adiabatic trajectories could be used, and because the integrand was short-ranged), and that the even parity projection was in any case negligible (because the intermolecular interactions of the sub-system dominated the entropy, and these were contained in $S_{\mathrm{r,st}}$, which has even parity, $S_{\mathrm{r,st}}(\mathbf{\Gamma}, t) \gg S^{\mathrm{even}}_{\mathrm{r,dyn}}(\mathbf{\Gamma}, t)$). These arguments led to Eq. (8.169),

$$S_{\mathrm{r}}(\mathbf{\Gamma}, t) \approx S_{\mathrm{r,st}}(\mathbf{\Gamma}, t) + S^{\mathrm{odd}}_{\mathrm{r,dyn}}(\mathbf{\Gamma}, t). \tag{11.48}$$

As noted in the discussion leading up to Eq. (8.181), the odd projection could be approximated by an integration over adiabatic trajectories,[11]

$$\begin{aligned} S^{\mathrm{odd}}_{\mathrm{r,dyn}}(\mathbf{\Gamma}, t) &\equiv \frac{1}{2}\left[S_{\mathrm{r,dyn}}(\mathbf{\Gamma}, t) - S_{\mathrm{r,dyn}}(\mathbf{\Gamma}^{\dagger}, t)\right] \\ &\approx \frac{-1}{2}\int_{t-\tau}^{t} \mathrm{d}t' \left[\dot{S}^{0}_{\mathrm{r,st}}(\mathbf{\Gamma}^{0}(t'|\mathbf{\Gamma}, t), t') - \dot{S}^{0}_{\mathrm{r,st}}(\mathbf{\Gamma}^{0}(t'|\mathbf{\Gamma}^{\dagger}, t), t')\right] \\ &\equiv S^{\mathrm{odd;0}}_{\mathrm{r,dyn}}(\mathbf{\Gamma}, t). \end{aligned} \tag{11.49}$$

The computational advantage of this expression for the dynamic part of the reservoir entropy is twofold. First the integrand is short-ranged, because the two terms tend to the same asymptotic value, $\overline{\dot{S}^{0}_{\mathrm{r,st}}(t')}$, $t - t' \to \infty$, and so the integral can be truncated after a fixed and relatively short time interval τ. Second, the original formulation required the most likely backward trajectory, $\overline{\mathbf{\Gamma}}(t'|\mathbf{\Gamma}, t)$, $t' < t$, which in turn required the most likely reservoir force for a negative time step, $\overline{R}(\mathbf{\Gamma}', t', \Delta_t < 0) = (\sigma/2k_{\mathrm{B}})\left[\nabla_{\mathrm{p}} S_{\mathrm{r,st}}(\mathbf{\Gamma}', t') - 2\overline{S'_{\mathrm{r,st}}(t')}\right]$. The final term is the most likely value of the gradient of the static part of the reservoir entropy at time t', and an efficient algorithm for the computation of this quantity is not immediately obvious.[12]

There is one potential problem with the use of adiabatic trajectories, namely that without a thermostat the numerical errors could lead to a temperature increase. This is ameliorated somewhat by the fact that relatively short trajectories are needed. Nevertheless, one should monitor this potential problem and keep in mind the options of short time steps or higher-order equations of motion. Another possibility is adding an equilibrium thermostat, although the

---

[11] Note that although the integrand is a total differential, there is no point in integrating this analytically. The upper limit vanishes, $S_{\mathrm{r,st}}(\mathbf{\Gamma}, t) - S_{\mathrm{r,st}}(\mathbf{\Gamma}^{\dagger}, t) = 0$, and value at the lower limit, $S_{\mathrm{r,st}}(\mathbf{\Gamma}^{0}(t-\tau|\mathbf{\Gamma}, t), t-\tau) - S_{\mathrm{r,st}}(\mathbf{\Gamma}^{0}(t-\tau|\mathbf{\Gamma}^{\dagger}, t), t-\tau)$ is not known explicitly other than by numerical integration.

[12] In the discussion of the regression of fluctuations on the most likely reservoir perturbed trajectory on p. 420, it is pointed out that one can evaluate the dynamic part of the reservoir entropy on the future most likely trajectory, which is relatively easy to evaluate. This also obviates the problem with using unthermostatted equations of motion.

derivation has not been done for this case and so the fundamental implications of doing this have yet to be established.

This argument for using only the odd projection of the dynamic part of the reservoir entropy rests on the general observation that the reservoir entropy is dominated by the static entropy, and that any part of the dynamic entropy that breaks a symmetry in the static entropy *needs* to be retained, and need *only* be retained. The most obvious symmetry is time reversal, but, as will now be seen, additional symmetries can also be exploited. The odd projection of the dynamic part of the reservoir entropy that results from this procedure can then be shown to be calculable on adiabatic trajectories rather than most likely backward trajectories.

11.2.3 Driven Brownian Motion

The model of driven Brownian motion consisting of a moving harmonic trap that has been invoked in Chs 3 and 10 and in § 11.1.3 above will now be used as a generic model for Monte Carlo simulations of a mechanical non-equilibrium system. The Hamiltonian of the sub-system consists of the kinetic energy, $\mathcal{K}(\mathbf{\Gamma}_{\rm p})$, the intermolecular potential energy, $U(\mathbf{\Gamma}_{\rm q})$ (here and throughout $\mathbf{\Gamma}$ contains both the solvent atoms and Brownian particle), and an external time dependent potential of parabolic form acting only on the Brownian particle,

$$U^{\rm ext}(x,t) = \frac{\kappa}{2}[x - b(t)]^2. \tag{11.50}$$

That is, $\mathcal{H}(\mathbf{\Gamma},t) = \mathcal{K}(\mathbf{\Gamma}_{\rm p}) + U(\mathbf{\Gamma}_{\rm q}) + U^{\rm ext}(x,t)$.

As in the general case of a mechanical non-equilibrium system treated in Ch. 8, the reservoir entropy is $S_{\rm r}(\mathbf{\Gamma},t) = S_{\rm r,st}(\mathbf{\Gamma},t) + S_{\rm r,dyn}(\mathbf{\Gamma},t)$, with the static entropy being

$$S_{\rm r,st}(\mathbf{\Gamma},t) \equiv \frac{-\mathcal{H}(\mathbf{\Gamma},t)}{T} \tag{11.51}$$

and the dynamic entropy being

$$S_{\rm r,dyn}(\mathbf{\Gamma},t) \equiv \frac{1}{T}\int_{t_0}^{t} {\rm d}t'\, \dot{\mathcal{H}}^0(\overline{\mathbf{\Gamma}}(t'|\mathbf{\Gamma},t),t'). \tag{11.52}$$

The adiabatic rate of change of the Hamiltonian is of course

$$\dot{\mathcal{H}}^0(\mathbf{\Gamma},t) = \frac{\partial U^{\rm ext}(x,t)}{\partial t} = -\kappa\dot{b}(t)[x - b(t)]. \tag{11.53}$$

Odd Projection of the Dynamic Entropy

For this trapped Brownian particle there are two symmetries that leave the value of the Hamiltonian unchanged. The first symmetry operation is velocity reversal, $\mathbf{\Gamma}^\dagger = \{\mathbf{\Gamma}_{\rm q}, -\mathbf{\Gamma}_{\rm p}\}$. This can apply to all components of the velocity or to just the x-component, but it must apply to all molecules of the sub-system,

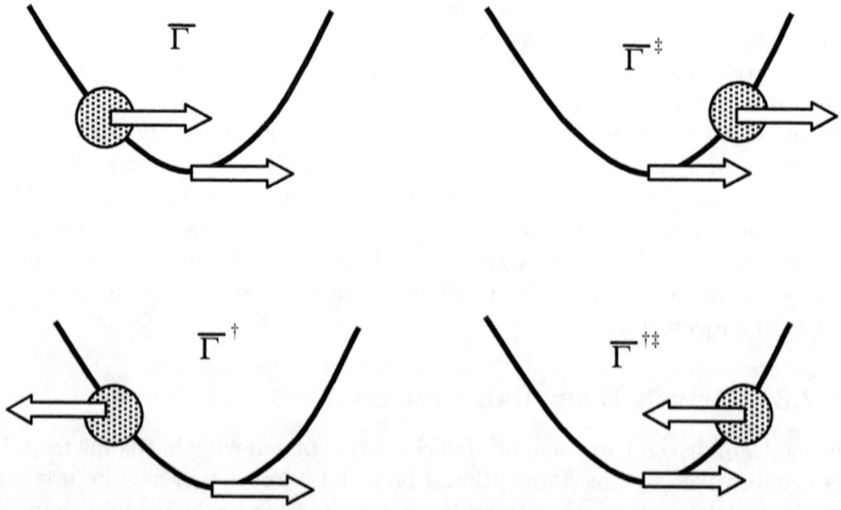

Figure 11.7: Four configurations of a Brownian particle in a potential trap moving to the right. The most likely configuration is denoted $\overline{\mathbf{\Gamma}}$. Particle and solvent velocity reversal, $\mathbf{\Gamma}_{\mathrm{p}} \Rightarrow -\mathbf{\Gamma}_{\mathrm{p}}$, is denoted by a dagger. Reflection of the particle and solvent positions in the trap minimum, $\mathbf{\Gamma}_{\mathrm{q}} \Rightarrow 2b(t)\hat{\mathbf{x}} - \mathbf{\Gamma}_{\mathrm{q}}$, is denoted by a double dagger. The static part of the reservoir entropy is the same for all four configurations.

not just the Brownian particle. The second symmetry operation is reflection in the trap minimum at time t, denoted by $\mathbf{\Gamma}^{\ddagger} = \{2b(t)\hat{\mathbf{x}} - \mathbf{\Gamma}_{\mathrm{q}}, \mathbf{\Gamma}_{\mathrm{p}}\}$. Again this can apply to either all components of the position or to just the x-component, and again it applies to all molecules in the sub-system, not just the Brownian particle. (An achiral solvent and particle and homogeneous sub-system is assumed.) By design, the Hamiltonian is invariant under these operations,

$$\mathcal{H}(\mathbf{\Gamma}, t) = \mathcal{H}(\mathbf{\Gamma}^{\dagger}, t) = \mathcal{H}(\mathbf{\Gamma}^{\ddagger}, t) = \mathcal{H}(\mathbf{\Gamma}^{\dagger\ddagger}, t). \tag{11.54}$$

The static part of the reservoir entropy, $S_{\mathrm{r,st}}(\mathbf{\Gamma}, t) = -\mathcal{H}(\mathbf{\Gamma}, t)/T$, obviously displays the same symmetries.

Four configurations related by these symmetry operations are sketched in Fig. 11.7. For the moving trap they are not equally likely. Since they all have the same energy and therefore the same static entropy, it is the dynamic part of the reservoir entropy that distinguishes them. As remarked in the previous sub-section on p. 417, since the reservoir entropy is dominated by the static entropy, one can reduce the dynamic part of the reservoir entropy to the symmetry breaking part. One could immediately write down the doubly odd projection without further discussion, but the following remarks may give physical insight into the mathematical result.

For the moving trap at a given time t, one can divide phase space into likely and unlikely regions. If $\mathbf{\Gamma}$ is a likely point in phase space, then $\mathbf{\Gamma}^{\dagger}$, $\mathbf{\Gamma}^{\ddagger}$, and $\mathbf{\Gamma}^{\dagger\ddagger}$ are all unlikely (c.f. Fig. 11.7). That is,

$$\wp(\mathbf{\Gamma},t) \gg \wp(\mathbf{\Gamma}^{\dagger},t) \approx \wp(\mathbf{\Gamma}^{\ddagger},t) \gg \wp(\mathbf{\Gamma}^{\dagger\ddagger},t). \tag{11.55}$$

Because of the invariance of the static part of the reservoir entropy under these operations, this ordering of the probabilities must be carried by the dynamic part of the reservoir entropy,

$$S_{\text{r,dyn}}(\mathbf{\Gamma},t) \gg S_{\text{r,dyn}}(\mathbf{\Gamma}^{\dagger},t) \approx S_{\text{r,dyn}}(\mathbf{\Gamma}^{\ddagger},t) \gg S_{\text{r,dyn}}(\mathbf{\Gamma}^{\dagger\ddagger},t). \tag{11.56}$$

In order to identify that portion of the dynamic part of the reservoir entropy that carries these distinctions, one needs to generalise the notion of odd parity to two variables by defining

$$\begin{aligned} & S^{ab}_{\text{r,dyn}}(\mathbf{\Gamma},t) \equiv \\ & \quad \frac{1}{4}\left[S_{\text{r,dyn}}(\mathbf{\Gamma}) + aS_{\text{r,dyn}}(\mathbf{\Gamma}^{\dagger}) + bS_{\text{r,dyn}}(\mathbf{\Gamma}^{\ddagger}) + abS_{\text{r,dyn}}(\mathbf{\Gamma}^{\dagger\ddagger})\right], \quad a,\, b = \pm 1. \end{aligned} \tag{11.57}$$

The projection $S^{++}_{\text{r,dyn}}$ is even with respect to both operations. The projection $S^{-+}_{\text{r,dyn}}$ is odd with respect to velocity reversal and even with respect to spatial reflection. The projection $S^{+-}_{\text{r,dyn}}$ is even with respect to velocity reversal and odd with respect to spatial reflection. The projection $S^{--}_{\text{r,dyn}}$ exclusively changes sign upon velocity reversal alone or upon spatial reflection alone, but remains unchanged when both are reversed.

The dynamic part of the reservoir entropy is the sum of its four projections,

$$S_{\text{r,dyn}}(\mathbf{\Gamma},t) = S^{++}_{\text{r,dyn}}(\mathbf{\Gamma},t) + S^{-+}_{\text{r,dyn}}(\mathbf{\Gamma},t) + S^{+-}_{\text{r,dyn}}(\mathbf{\Gamma},t) + S^{--}_{\text{r,dyn}}(\mathbf{\Gamma},t). \tag{11.58}$$

In view of the ordering of the dynamic part of the reservoir entropy deduced above from the broken symmetry of the probability, the two singly odd projections dominate this. Hence with negligible statistical error one can write,

$$S_{\text{r,dyn}}(\mathbf{\Gamma},t) \approx S^{\text{odd}}_{\text{r,dyn}}(\mathbf{\Gamma},t), \tag{11.59}$$

with the odd projection defined as

$$\begin{aligned} S^{\text{odd}}_{\text{r,dyn}}(\mathbf{\Gamma},t) &\equiv S^{-+}_{\text{r,dyn}}(\mathbf{\Gamma},t) + S^{+-}_{\text{r,dyn}}(\mathbf{\Gamma},t) \\ &= \frac{1}{2}\left[S_{\text{r,dyn}}(\mathbf{\Gamma},t) - S_{\text{r,dyn}}(\mathbf{\Gamma}^{\dagger\ddagger},t)\right] \\ &= \frac{1}{2T}\int_{t_0}^{t} \mathrm{d}t' \Big[\dot{\mathcal{H}}^0(\overline{\mathbf{\Gamma}}(t'|\mathbf{\Gamma},t),t') - \dot{\mathcal{H}}^0(\overline{\mathbf{\Gamma}}(t'|\mathbf{\Gamma}^{\dagger\ddagger},t),t')\Big]. \end{aligned} \tag{11.60}$$

It is worth noting that the explicit expression for the dynamic part of the reservoir entropy derived for the case of a Brownian particle in a trap in uniform motion, Eq. (3.43), is the sum of two terms: the first term is even with respect to reflection and odd with respect to velocity reversal, and the second term is even with respect to velocity reversal and odd with respect to reflection. In other words, it is exactly the odd projection in the sense that has just been defined.

Asymptotic Behaviour

At large times, the adiabatic rate of change of the energy asymptotes to its most likely value,

$$\dot{\mathcal{H}}^0(\overline{\mathbf{\Gamma}}(t'|\mathbf{\Gamma}, t), t') \sim \overline{\dot{\mathcal{H}}^0}(t'), \ |t - t'| \to \infty. \tag{11.61}$$

This is independent of the starting position of the trajectory, so that the same asymptote holds for $\mathbf{\Gamma}^{\dagger\ddagger}$. This means that the integrand goes to zero at large times and so the integral for $S^{\mathrm{odd}}_{\mathrm{r,dyn}}$ can be taken over a relatively short fixed time interval, with lower limit $t_0 \Rightarrow t - \tau$. This has obvious computational advantages, as will be shown explicitly below.

Regression of Fluctuation

By adding and subtracting the asymptote, the integrand of the odd dynamic entropy can be written as the difference between the regression of two fluctuations,

$$\begin{aligned} &\dot{\mathcal{H}}^0(\overline{\mathbf{\Gamma}}(t'|\mathbf{\Gamma}, t), t') - \dot{\mathcal{H}}^0(\overline{\mathbf{\Gamma}}(t'|\mathbf{\Gamma}^{\dagger\ddagger}, t), t') \\ &\quad = \left[\dot{\mathcal{H}}^0(\overline{\mathbf{\Gamma}}(t'|\mathbf{\Gamma}, t), t') - \overline{\dot{\mathcal{H}}^0}(t')\right] - \left[\dot{\mathcal{H}}^0(\overline{\mathbf{\Gamma}}(t'|\mathbf{\Gamma}^{\dagger\ddagger}, t), t') - \overline{\dot{\mathcal{H}}^0}(t')\right]. \end{aligned} \tag{11.62}$$

Just as in the equilibrium case, fluctuations in a non-equilibrium system are most likely even functions of time,

$$\begin{aligned} &\dot{\mathcal{H}}^0(\overline{\mathbf{\Gamma}}(t'|\mathbf{\Gamma}, t), t') - \overline{\dot{\mathcal{H}}^0}(t') = \\ &\quad \dot{\mathcal{H}}^0(\overline{\mathbf{\Gamma}}(2t - t'|\mathbf{\Gamma}, t), 2t - t') - \overline{\dot{\mathcal{H}}^0}(2t - t'), \end{aligned} \tag{11.63}$$

and similarly for the trajectory starting at $\mathbf{\Gamma}^{\dagger\ddagger}$. This equality is meant in a statistical sense. This result is true for a future Hamiltonian extended by reflection, $\tilde{\mathcal{H}}(\mathbf{\Gamma}, t'; t) = \mathcal{H}(\mathbf{\Gamma}, 2t - t')$, $t' > t$, as is discussed next.

With this result the integral for the dynamic part of the reservoir entropy, originally over the backward trajectory on the past interval $[t - \tau, t]$, can be rewritten over the forward trajectory on the future interval, $[t, t + \tau]$,

$$\begin{aligned} &S^{\mathrm{odd}}_{\mathrm{r,dyn}}(\mathbf{\Gamma}, t) \\ &\quad = \frac{1}{2T} \int_t^{t+\tau} \mathrm{d}t' \left[\dot{\mathcal{H}}^0(\overline{\mathbf{\Gamma}}(t'|\mathbf{\Gamma}, t), t') - \dot{\mathcal{H}}^0(\overline{\mathbf{\Gamma}}(t'|\mathbf{\Gamma}^{\dagger\ddagger}, t), t')\right]. \end{aligned} \tag{11.64}$$

The asymptote can be added to and subtracted from this if desired, but it makes no real difference. This expression for the dynamic part of the reservoir entropy appears computationally feasible, since the explicit formula for the most likely future trajectory is straightforward (c.f. footnote 12 on p. 416). To date, it has not been numerically implemented.

Future Adiabatic Behaviour

The original motivation for rewriting the dynamic part of the reservoir entropy as an integral over the forward most likely trajectory was to be able to exploit adiabatic trajectories. In general, the future regression of a fluctuation is determined by the adiabatic evolution of the isolated system. In contrast, the past behaviour leading up to a fluctuation in structure, the backward regression, differs markedly in an isolated and in a reservoir-driven sub-system. For an isolated system, the structure has arisen by a spontaneous fluctuation. For a sub-system of a reservoir, the structure has resulted from an external force or an imposed thermodynamic gradient. That the future regression, more precisely the adiabatic part of the future regression, is the same in an isolated system as in a sub-system of a reservoir, is part of the content of Onsager's regression hypothesis.

In view of these remarks, one can say that in the future, the rate of energy production is approximately the same on the adiabatic trajectory and on the most likely trajectory in the presence of the reservoir,

$$\dot{\mathcal{H}}^0(\mathbf{\Gamma}^0(t'|\mathbf{\Gamma},t),t') \approx \dot{\mathcal{H}}^0(\overline{\mathbf{\Gamma}}(t'|\mathbf{\Gamma},t),t'), \quad t'>t. \tag{11.65}$$

(Again this result assumes a future Hamiltonian, $\tilde{\mathcal{H}}(\mathbf{\Gamma},t';t) = \mathcal{H}(\mathbf{\Gamma},2t-t')$, $t'>t$, as is discussed shortly.) This result also holds for $\mathbf{\Gamma}^{\dagger\ddagger}$. Hence one can write $S^{\rm odd}_{\rm r,dyn}(\mathbf{\Gamma},t) \approx S^{\rm odd;0}_{\rm r,dyn}(\mathbf{\Gamma},t)$, with

$$\begin{aligned} & S^{\rm odd;0}_{\rm r,dyn}(\mathbf{\Gamma},t) \\ & \quad = \frac{1}{2T}\int_t^{t+\tau} \mathrm{d}t' \left[\dot{\mathcal{H}}^0(\mathbf{\Gamma}^0(t'|\mathbf{\Gamma},t),t') - \dot{\mathcal{H}}^0(\mathbf{\Gamma}^0(t'|\mathbf{\Gamma}^{\dagger\ddagger},t),t')\right]. \end{aligned} \tag{11.66}$$

Now the reversibility of the adiabatic equations of motion will be established with a view toward transforming this integral to one over the past adiabatic trajectory. For a Hamiltonian that is independent of time, $\mathcal{H}(\mathbf{\Gamma})$, then Hamilton's equations obey microscopic reversibility: if $\mathbf{\Gamma}_2 = \mathbf{\Gamma}^0(t'|\mathbf{\Gamma}_1,t)$ then $\mathbf{\Gamma}_1^\dagger = \mathbf{\Gamma}^0(t'|\mathbf{\Gamma}_2^\dagger,t)$. Because a Hamiltonian trajectory is uniquely specified by any point and time on it, and because of time homogeneity, this second result may be written $\mathbf{\Gamma}_2^\dagger = \mathbf{\Gamma}^0(t|\mathbf{\Gamma}_1^\dagger,t') = \mathbf{\Gamma}^0(2t-t'|\mathbf{\Gamma}_1^\dagger,t)$. In summary,

$$\mathbf{\Gamma}_2 = \mathbf{\Gamma}^0(t'|\mathbf{\Gamma}_1,t) \Leftrightarrow \mathbf{\Gamma}_2^\dagger = \mathbf{\Gamma}^0(2t-t'|\mathbf{\Gamma}_1^\dagger,t). \tag{11.67}$$

Since the operation $\ddagger$ reflects all the positions, then it has the effect of reversing all the forces. Consequently one also has,

$$\mathbf{\Gamma}_2 = \mathbf{\Gamma}^0(t'|\mathbf{\Gamma}_1,t) \Leftrightarrow \mathbf{\Gamma}_2^\ddagger = \mathbf{\Gamma}^0(2t-t'|\mathbf{\Gamma}_1^\ddagger,t). \tag{11.68}$$

This last result implicitly assumes a spatially homogeneous sub-system (i.e. no external force). These two results individually give the future behaviour of the trajectory based upon the past behaviour.

Neither of these two hold as such for a Hamiltonian that is explicitly time dependent, such as the present moving trap, because this breaks time and space homogeneity. However one can combine them to give future behaviour that is an exact reflection of past behaviour, provided that one extends the external potential into the future by reflecting its past. Hence, define the extended Hamiltonian as

$$\tilde{\mathcal{H}}(\mathbf{\Gamma}, t'; t) = \begin{cases} \mathcal{H}(\mathbf{\Gamma}, t'), & t' \leq t, \\ \mathcal{H}(\mathbf{\Gamma}, 2t - t'), & \text{otherwise.} \end{cases} \tag{11.69}$$

In the above the tilde has not been shown explicitly; rather this expression has been implicitly invoked whenever $t' > t$. Similarly the trajectories $\overline{\mathbf{\Gamma}}(t'|\mathbf{\Gamma}, t)$, and $\mathbf{\Gamma}^0(t'|\mathbf{\Gamma}, t)$ implicitly use $\tilde{\mathcal{H}}$ whenever $t' > t$. This expression for the future Hamiltonian is equivalent to reversing at time t all the velocities in the universe external to the sub-system, since ultimately any time dependent potential arises from molecular motion. For the present moving trap, this definition implies

$$b(t') = b(2t - t'), \text{ and } \dot{b}(t') = -\dot{b}(2t - t'). \tag{11.70}$$

With this extended Hamiltonian, at time t the velocity of the trap is reversed. Since the † operation reverses all the velocities of the sub-system, and the ‡ operation reverses all the forces in the sub-system including that due to the trap, one can see that

$$\mathbf{\Gamma}_2 = \mathbf{\Gamma}^0(t'|\mathbf{\Gamma}_1, t) \Leftrightarrow \mathbf{\Gamma}_2^{\dagger\ddagger} = \mathbf{\Gamma}^0(2t - t'|\mathbf{\Gamma}_1^{\dagger\ddagger}, t). \tag{11.71}$$

Here $\mathbf{\Gamma}_1^{\ddagger}$, means reflection of the positions in $b(t)$, and $\mathbf{\Gamma}_2^{\ddagger}$ means reflection of the positions in $b(t') = b(2t - t')$.

This establishes the reversibility of the adiabatic equations of motion in the presence of a time varying potential. Accordingly, the dynamic part of the reservoir entropy can be transformed from an integral over the future adiabatic trajectory to an integral over the past adiabatic trajectory,

$$\begin{aligned} & S_{\mathrm{r,dyn}}^{\mathrm{odd};0}(\mathbf{\Gamma}, t) \\ &= \frac{1}{2T} \int_t^{t+\tau} \mathrm{d}t' \left[\dot{\mathcal{H}}^0(\mathbf{\Gamma}^0(t'|\mathbf{\Gamma}, t), t') - \dot{\mathcal{H}}^0(\mathbf{\Gamma}^0(t'|\mathbf{\Gamma}^{\dagger\ddagger}, t), t') \right] \\ &= \frac{1}{2T} \int_{t-\tau}^{t} \mathrm{d}t'' \left[\dot{\mathcal{H}}^0(\mathbf{\Gamma}^0(t''|\mathbf{\Gamma}^{\dagger\ddagger}, t)^{\dagger\ddagger}, t'') - \dot{\mathcal{H}}^0(\mathbf{\Gamma}^0(t''|\mathbf{\Gamma}, t)^{\dagger\ddagger}, t'') \right] \\ &= \frac{1}{2T} \int_{t-\tau}^{t} \mathrm{d}t' \left[\dot{\mathcal{H}}^0(\mathbf{\Gamma}^0(t'|\mathbf{\Gamma}, t), t') - \dot{\mathcal{H}}^0(\mathbf{\Gamma}^0(t'|\mathbf{\Gamma}^{\dagger\ddagger}, t), t') \right]. \end{aligned} \tag{11.72}$$

The final equality invokes the result that $\dot{\mathcal{H}}^0(\mathbf{\Gamma}^{\dagger\ddagger}, t) = -\dot{\mathcal{H}}^0(\mathbf{\Gamma}, t)$. Recalling the arguments leading successively to $S_{\mathrm{r,dyn}}(\mathbf{\Gamma}, t) \approx S_{\mathrm{r,dyn}}^{\mathrm{odd}}(\mathbf{\Gamma}, t) \approx S_{\mathrm{r,dyn}}^{\mathrm{odd};0}(\mathbf{\Gamma}, t)$, this expression represents a computationally convenient form for the dynamic part of the reservoir entropy.

System Details

The system used for the non-equilibrium Monte Carlo simulations of driven Brownian motion is almost identical to that described for the stochastic molecular dynamics simulations in §11.1.3. The soft sphere potential was the same, as was the temperature, $T = 3$ and the density, $\rho = 0.69$. A bulk system was simulated, in the form of a cube with periodic boundary conditions and an edge length 8.98 or 11.31 (for N = 500 or 1000 solvent atoms, respectively). The small cell spatial neighbour was used with cells of edge length ≈ 0.3 (see p. 402).

Each simulation was performed for fixed drive frequency ω. The Metropolis algorithm and umbrella sampling were used. The umbrella potential was based upon the static part of the reservoir entropy (see below). Each simulation consisted of 500–1650 Monte Carlo cycles.

A Monte Carlo cycle was as follows. First the position of a solvent atom was changed as a trial with the Brownian particle fixed at the origin. Each trial move was accepted or rejected, based upon the usual Metropolis criteria, using the solvent-solvent and solvent-solute potentials *only* (no external potential). This was repeated successively for each solvent in turn as part of one cycle. The step length was adjusted to give an acceptance rate of $\lesssim 50\%$. This was repeated 10–25 times (i.e. 10–25 moves per solvent atom per cycle). Then the solvent and solute velocities were chosen randomly from the Maxwell-Boltzmann distribution, Gaussians of variance $\sqrt{k_{\mathrm{B}}T/m_1}$ and $\sqrt{k_{\mathrm{B}}T/m_0}$, respectively. Then, with the solute the y- and z-coordinates left unchanged at the origin, the solute x-coordinate was chosen from a Gaussian of variance $\sqrt{k_{\mathrm{B}}T/\kappa}$. After this, all the solvent x-coordinates were shifted by this value of the solute x-coordinate, so that their position relative to the solute remained unchanged. At this stage of the cycle, the procedure is equivalent to choosing the phase space position of the solute and solvent from an equilibrium Boltzmann distribution with the external potential being fixed at the origin (i.e. the position of the trap minimum being located at the origin). This phase space coordinate (solute plus solvent), $\mathbf{\Gamma}_\alpha$, is stored, and after the following loop of trajectories and averages, it is used as the starting point for the next Monte Carlo cycle, at which time the solute and solvent were shifted by the same amount, the negative of the current solute x-coordinate. Hence at the start of the next cycle, the solute was back at the origin, and the solvent molcecules were in the same positions relative to each other and to the solute as at the end of the current cycle.

Now, as the next stage in the current cycle, a loop was performed over a set of 10 time nodes, $t_1, t_2, \ldots, t_{10}$, typically evenly spaced on the interval $[0, 2\pi/\omega]$. Each time node defined the location of the external potential, $b_i = B\cos\omega t_i$. For each time node, the solute and the solvent x-coordinates were shifted by the amount $b_i\hat{\mathbf{x}}$, so that their position relative to the trap minimum was maintained. This phase space coordinate was denoted $\mathbf{\Gamma}_{\alpha,i}$. For each such coordinate, two adiabatic trajectories were calculated backward in time, $\mathbf{\Gamma}^0(t'|\mathbf{\Gamma}_{\alpha,i}, t_i)$ and $\mathbf{\Gamma}^0(t'|\mathbf{\Gamma}^{\dagger\ddagger}_{\alpha,i}, t_i)$, for $t' \in [t_i, t_i - \tau]$. The maximum time interval was $\tau = 7$. The second-order Verlet leap frog algorithm, Eq. (11.4), with a time step of 2×10^{-2} was used for the trajectory calculation. These trajectory calculations were in the presence of the time varying external potential, $U^{\mathrm{ext}}(x', t')$. A running integral of the partial time derivative of the external potential was performed over each trajectory, so that effectively integrals over successively longer intervals, $[t_i, t_i - \tau_k]$, $k = 1, 2, \ldots, 20$, with $\tau_{20} = 7$, were obtained. Taking the difference between the integrals for the two trajectories for a given τ_k gave $S^{\text{odd};0}_{\text{r,dyn}}(\mathbf{\Gamma}_{\alpha,i}; t_i, \tau_k)$. This was stored for use as the non-equilibrium weight for the subsequent averages.

This loop over the time nodes was the final stage in each Monte Carlo cycle. Before the start of the next cycle phase functions were evaluated, weighted, and accumulated for later averaging. For any given phase function 200 distinct averages were taken, $\langle f(\mathbf{\Gamma}, t)\rangle_{i,k}$: one for each point in the drive cycle t_i, and one for each interval used for the integration of the dynamic part of the reservoir entropy, τ_k. Values for the averages was taken once per Monte Carlo cycle α, and the simulation consisted of 500–1650 such cycles. In view of the umbrella sampling (the umbrella weight is the equilibrium Boltzmann factor, equivalently the static part of the reservoir entropy), an average value is

$$\langle f(\mathbf{\Gamma}, t)\rangle_{t_i,\tau_k} = \frac{\sum_\alpha f(\mathbf{\Gamma}_{\alpha,i}, t_i) e^{S^{\text{odd};0}_{\text{r,dyn}}(\mathbf{\Gamma}_{\alpha,i};t_i,\tau_k)/k_{\text{B}}}}{\sum_\alpha e^{S^{\text{odd};0}_{\text{r,dyn}}(\mathbf{\Gamma}_{\alpha,i};t_i,\tau_k)/k_{\text{B}}}}. \tag{11.73}$$

Running totals for these were added to, and stored as a function of, k and i at the end of each cycle.

The two main quantities averaged in each simulation were the solute position $x(t)$ and the solute velocity, $\dot{x}(t)$, and from them the amplitude and phase lag of the response from the functions $\langle x(t_i)\rangle_{\tau_k} = A\cos(\omega t_i + \phi)$, $\langle \dot{x}(t_i)\rangle_{\tau_k} = -A\omega\sin(\omega t_i + \phi)$. For each t_i and τ_k, these can be inverted to obtain the amplitude and phase lag, $A_i = \sqrt{x_i^2 + \dot{x}_i^2/\omega^2}$, and $\phi_i = \tan^{-1}([\dot{x}_i + \omega x_i]/[\dot{x}_i - \omega x_i])$. Values of ϕ in the first quadrant were shifted to the third, so that $\phi \in [-\pi, 0]$. For each τ_k, these were simply averaged over the 10 t_i to obtain the average amplitude and phase, A_{τ_k}, and ϕ_{τ_k}, and an estimate of the standard error in the mean of this average. Statistical error creates problems with the phase lag for small values of the lag due to the branch cut. Hence an alternative approach was also used, in which the two trigonometric response functions were fitted in a least squares sense to all the data at the 10 t_i to obtain A_{τ_k} and ϕ_{τ_k}. Except for the smallest values of τ_k (see discussion), the two methods gave very similar results.

A minor improvement on the umbrella sampling algorithm was explored. The static Maxwell-Boltzmann distribution is insensitive to velocity reversal and to spatial reflection, since $\mathcal{H}(\mathbf{\Gamma}^{\dagger\ddagger}, t) = \mathcal{H}(\mathbf{\Gamma}, t)$. By definition of the odd projection of the dynamic part of the reservoir entropy one has, $S^{\text{odd};0}_{\text{r,dyn}}(\mathbf{\Gamma}; t, \tau) = -S^{\text{odd};0}_{\text{r,dyn}}(\mathbf{\Gamma}^{\dagger\ddagger}; t, \tau)$. These two results mean that one can calculate the phase space average for both $\mathbf{\Gamma}_{\alpha,i}$ and $\mathbf{\Gamma}^{\dagger\ddagger}_{\alpha,i}$ with no additional work (two configurations per t_i per cycle). This ought to reduce the statistical error by about 40%.

Results

Non-equilibrium Monte Carlo simulation results are shown in Fig. 11.8. These are for the drive frequency $\omega = 0.8$ and the drive amplitude $B = 0.2$, and use the longest time interval for the integration, $\tau = 7$, but are fairly typical

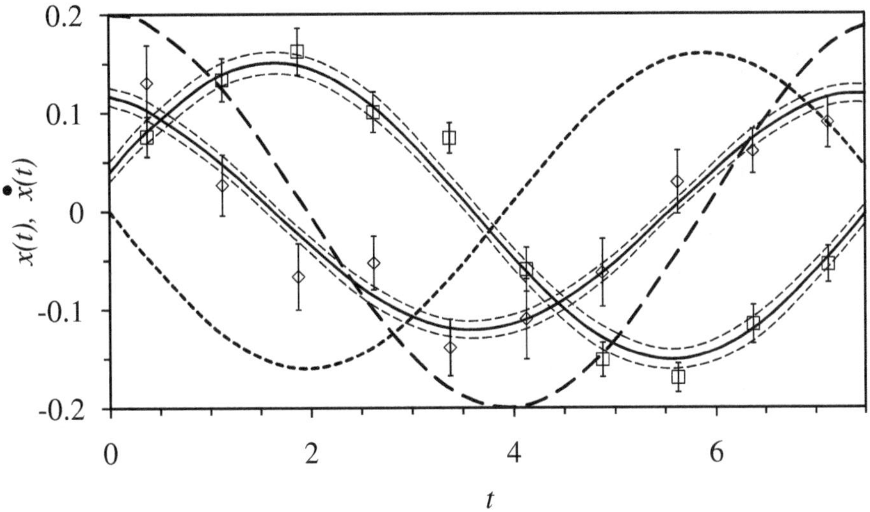

Figure 11.8: Non-equilibrium Monte Carlo simulation results for the average position (open squares) and velocity (open diamonds) of the solute at 10 times t_i during the drive cycle ($\omega = 0.8$, $B = 0.2$, $\tau = 7$). The solid curves are the trigonometric functions, $x(t) = A\cos(\omega t + \phi)$, and $\dot{x}(t) = -A\omega\sin(\omega t + \phi)$, where A and ϕ are obtained from a simple average of the 10 $A(x(t_i), \dot{x}(t_i))$ and $\phi(x(t_i), \dot{x}(t_i))$. The enveloping dashed curves are a guide to the estimated error. The bold dashed and dotted curves are the position $b(t)$ and velocity $\dot{b}(t)$ of the trap, respectively. Data from Attard, P. (2009c) Phys. Rev. E **80**, 041126.

of all the simulation results that were obtained. The figure shows the average position and velocity of the solute at 10 different times in the drive cycle. The error bars are the standard error on the mean, which was obtained by breaking the simulation into in this case 25 blocks, and obtaining the variance of the averages. As can be seen, trigonometric curves are well defined by the data, and a least squares fit to them readily yields the phase lag and the response amplitude.

Using the static Boltzmann distribution for the umbrella sampling yielded $\langle A/B \rangle = 0.736 \pm 0.007$ and $\langle \phi \rangle = -1.23 \pm 0.01$. Some alternative umbrella weights were explored, but these were found to be less efficient (i.e. required more trajectories to obtain a specified statistical error) than the present static entropy weight (Attard, 2009c). It was however demonstrated that the averages were independent of the umbrella weighting.

It should be understood that *all* of the non-zero phase lag and and non-unity relative amplitude comes from the dynamic part of the reservoir entropy in the Monte Carlo algorithm. The static part of the entropy alone, which was used as the umbrella weight, and which would be equivalent to using the usual equilibrium Monte Carlo algorithm with the instantaneous Maxwell-Boltzmann factor, would give $\langle \phi \rangle = 0$ and $\langle A \rangle = B$. This is just another way of saying that

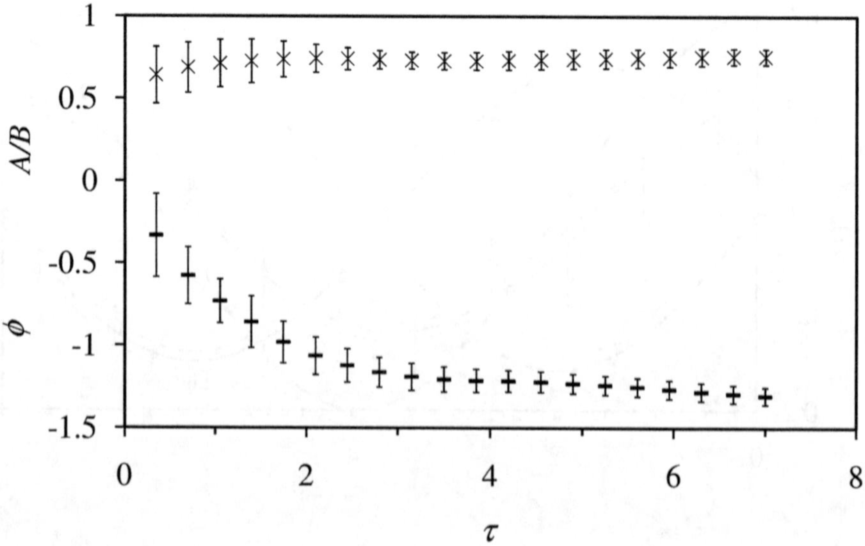

Figure 11.9: Non-equilibrium Monte Carlo simulation results for the average relative amplitude, $\langle A/B \rangle_\tau$, (crosses) and phase lag, $\langle \phi \rangle_\tau$, (bars) of the solute as a function of the time interval τ used for the averages, ($\omega = 0.8$, $B = 0.2$). Data from Attard, (2009c).

the Maxwell-Boltzmann distribution is not the correct probability distribution for a non-equilibrium system. The non-equilibrium part of the reservoir entropy in the present non-equilibrium probability distribution is an essential addition to the Boltzmann factor to produce a non-trivial response to the time varying force.

The non-equilibrium averages depend upon the dynamic part of the reservoir entropy, and this in turn depends upon τ, the time interval over which it is calculated in the Monte Carlo simulations. (This was denoted τ_k above in the description of the algorithm.) This is explored in Fig. 11.9, where the non-equilibrium Monte Carlo averages for the phase lag and for the amplitude ratio are plotted as a function of the time interval (i.e. $\langle \phi \rangle_\tau$, and $\langle A/B \rangle_\tau$). It can be seen that the averages asymptote to constant values as the time interval is increased. This confirms the asymptotic analysis above, which found that the integrand of the odd projection of the dynamic part of the reservoir entropy was a short-range function, and that therefore the integral was insensitive to the value of the lower limit once the integrand had reached its asymptotic regime. In this case, the figure shows that the averages have reached their asymptotic values for $\tau_k \gtrsim 3$.

As $\tau \to 0$, the dynamic part of the reservoir entropy vanishes. As discussed above, this limit is equivalent to using the instantaneous Maxwell-Boltzmann distribution as the probability distribution, in which case one would obtain $\langle x(t) \rangle_{\text{st}} = B \cos \omega t$ and $\langle \dot{x}(t) \rangle_{\text{st}} = 0$. The obvious inconsistency of these two

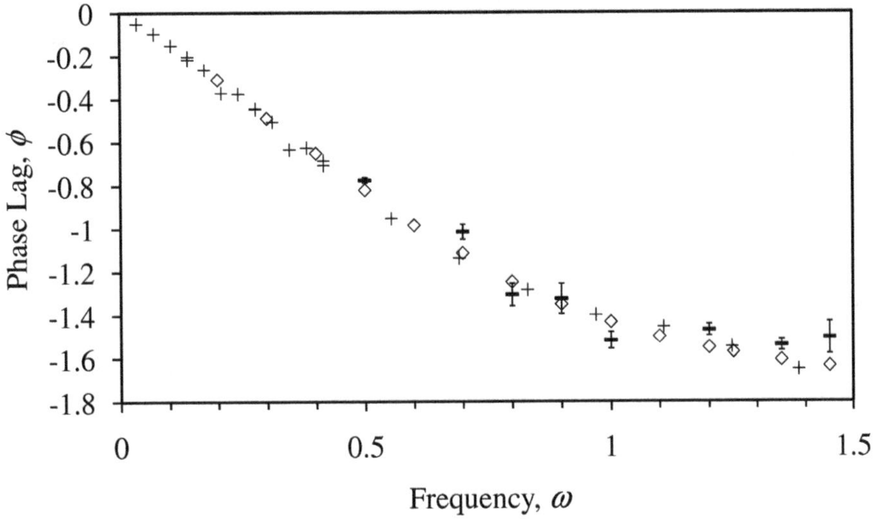

Figure 11.10: Phase lag of the solute as a function of the drive frequency. The plus symbols are MD simulations with the equilibrium Nosé-Hoover chain thermostat (Attard and Gray-Weale, 2008), the diamonds are non-equilibrium stochastic molecular dynamics simulations (Attard, 2009a), and the bars are the present non-equilibrium Monte Carlo simulations ($\tau = 7$). The error bars are the standard error on the mean, and are in most cases smaller than the symbols. Data from Attard, (2009c).

results (the second is not the time derivative of the first) confirms that the Maxwell-Boltzmann distribution is inappropriate for a non-equilibrium system. This inconsistency explains in part why the data in Fig. 11.9 at small τ are less reliable. Part of this is due to the fact that the statistical noise makes estimates of ϕ for small ϕ problematic due to the branch cut. But for the most part the unreliability at small τ arises from this inconsistency in the fitted functions (i.e. fitting to position alone would give $A = B$ and fitting to velocity alone would give $A = 0$). The figure graphically illustrates the need to use a large enough value of τ for the data to have attained its asymptotic value. Recording averages as a function of τ in order to retrospectively confirm that the data have indeed reached their asymptotes creates no extra computation time because the $\omega_{\mathrm{r,dyn}}^{\mathrm{odd};0}$ can be saved as a running integral of τ_k.

Figure 11.10 shows the phase lag , and Fig. 11.11 shows the relative response amplitude, both as a function of the drive frequency. It can be seen that the solute oscillates almost in phase with the trap at low frequencies, and increasingly lags the trap as the frequency is increased, and also that the response amplitude decreases with increasing frequencies. In both figures the results of the present non-equilibrium Monte Carlo algorithm are compared with Nosé-Hoover simulations that employ an equilibrium chain thermostat (Attard and Gray-Weale,

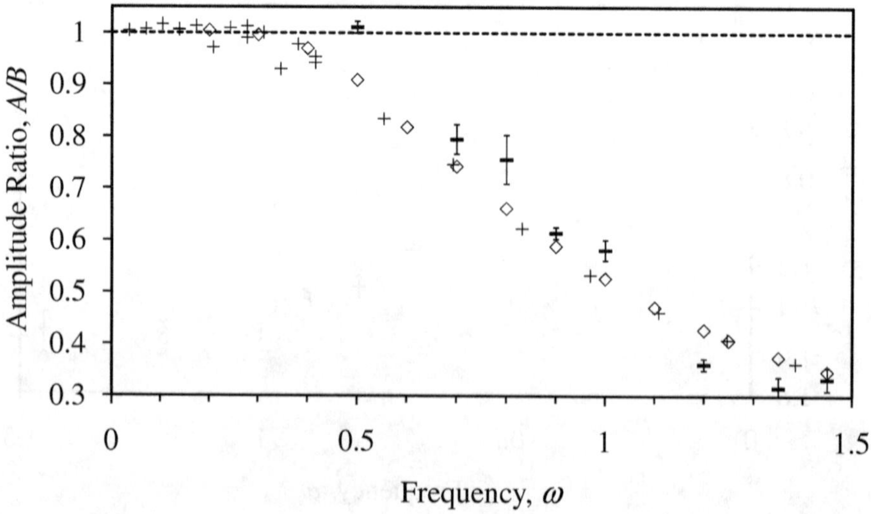

Figure 11.11: Relative amplitude of the driven solute. Symbols and curves as in the preceding figure. The line is a guide to the eye.

2008), and with the non-equilibrium stochastic molecular dynamics algorithm (Attard, 2009a), which was discussed and presented above in §11.1.3. The data represent a significant test of the non-equilibrium Monte Carlo algorithm as it is a qualitatively different approach to the molecular dynamics approaches. The Monte Carlo algorithm relies upon the correct non-equilibrium probability density (i.e. the correct weighting of phase space), whereas the stochastic molecular dynamics algorithm relies upon the correct non-equilibrium transition probability. The fact that the phase lag obtained in this approach is the same as that given by the molecular dynamics simulations is strong evidence for the validity of the non-equilibrium probability distribution that is invoked. Furthermore, it also confirms the arguments given for the dominance of the odd projection of the dynamic part of the reservoir entropy, and for adiabatic trajectories, $S_{\mathrm{r,dyn}}(\mathbf{\Gamma},t) \approx S_{\mathrm{r,dyn}}^{\mathrm{odd}}(\mathbf{\Gamma},t) \approx S_{\mathrm{r,dyn}}^{\mathrm{odd;0}}(\mathbf{\Gamma},t)$. One can conclude from the data in Fig. 11.10 and in Fig. 11.11 that the non-equilibrium Monte Carlo algorithm and the underlying non-equilibrium probability distribution are valid over the whole temporal regime from steady state to transient.

As mentioned in the discussion of Fig. 11.1, there is a weak non-linear dependence of the amplitude response. For example, at $\omega = 0.8$, in the case of non-equilibrium stochastic molecular dynamics simulations, drive amplitudes of 5, 2.5, 1, and 0.5 give $\langle A/B \rangle = 0.624 \pm 0.002$, 0.657 ± 0.002, 0.664 ± 0.005, and 0.675 ± 0.006, respectively. The non-equilibrium Monte Carlo simulations for drive amplitudes of 1, 0.5, and 0.2 give 0.68 ± 0.02, 0.78 ± 0.05, and 0.76 ± 0.05. The phase lag showed less variability with drive amplitude.

As far as computational efficiency is concerned, for the present case of a driven Brownian particle, the non-equilibrium stochastic molecular dynamics method was better than the present implementation of the non-equilibrium Monte Carlo method. For example, at $\omega = 0.8$, the stochastic molecular dynamics simulations used a total of 4×10^6 time steps to give $\langle A/B \rangle = 0.675 \pm 0.006$ and $\langle \phi \rangle = -1.27 \pm 0.01$. The non-equilibrium Monte Carlo algorithm at $\omega = 0.8$ and for $\tau = 3$ used 625 trajectories, each of length 3000 time steps, a total of 1.9×10^6 time steps, giving $\langle A/B \rangle = 0.73 \pm 0.05$ and $\langle \phi \rangle = -1.19 \pm 0.08$. (The data in Fig. 11.9 is for $\tau = 7$.) Hence the error is reduced by about a factor of 10 for twice as many time steps, which means that stochastic molecular dynamics is about 50 times more efficient than non-equilibrium Monte Carlo in this case.

This relative inefficiency in the non-equilibrium Monte Carlo method may be a little misleading, since this is the first implementation of the algorithm and there are opportunities to optimise it. It is clear that there is a great deal of redundancy in using 10 time nodes per cycle to get the amplitude and phase lag, and that this number could be reduced. No doubt there is a more optimal balance between the accuracy at each node and the number of nodes used for the fitting procedure. Also, different, more efficient forms of umbrella sampling could be developed with a view to reducing the variance. In more challenging non-equilibrium cases, such as polymeric or glassy systems, the non-equilibrium Monte Carlo method offers much greater potential for optimisation and for broad coverage of phase space than the stochastic molecular dynamics method.

11.2.4 Steady Heat Flow

Algorithm and System Details

The non-equilibrium statistical mechanics for steady heat flow was described above in the context of stochastic molecular dynamics, §11.1.4, and the molecular expressions for the various phase functions are given there. In terms of the zeroth, $E_0(\mathbf{\Gamma})$, and first, $E_1(\mathbf{\Gamma})$, energy moments of the sub-system, and the zeroth β_0 and first β_1 inverse temperatures of the reservoirs, the static part of the reservoir entropy is

$$k_{\mathrm{B}}^{-1} S_{\mathrm{r,st}}(\mathbf{\Gamma}) = -\beta_0 E_0(\mathbf{\Gamma}) - \beta_1 E_1(\mathbf{\Gamma}), \tag{11.74}$$

and the dynamic part of the reservoir entropy is

$$k_{\mathrm{B}}^{-1} S_{\mathrm{r,dyn}}(\mathbf{\Gamma}) = -\beta_1 \int_{-\tau}^{0} \mathrm{d}t\, \dot{E}_1^0(\overline{\mathbf{\Gamma}}(t|\mathbf{\Gamma}, 0)). \tag{11.75}$$

As was discussed in detail in §8.5, this is dominated by the odd projection, $S_{\mathrm{r,dyn}}(\mathbf{\Gamma}) \approx S_{\mathrm{r,dyn}}^{\mathrm{odd}}(\mathbf{\Gamma}) \equiv [S_{\mathrm{r,dyn}}(\mathbf{\Gamma}) - S_{\mathrm{r,dyn}}(\mathbf{\Gamma}^\dagger)]/2$ where the dagger means reversing the velocity, $\mathbf{\Gamma}^\dagger = \{\mathbf{\Gamma}_{\mathrm{q}}, -\mathbf{\Gamma}_{\mathrm{p}}\}$. By adding and subtracting the asymptote, one could see that this could be written as the difference in the regression of two fluctuations, and that consequently the integrand was short-ranged and an even function of time. The latter meant that the most likely trajectory going forward in time could be invoked, $[-\tau, 0] \Rightarrow [0, \tau]$, and that, in turn, this could be replaced by adiabatic trajectories. Finally, invoking the reversibility of Hamilton's equations one ends up with,

$$\begin{aligned} k_\mathrm{B}^{-1} S_{\mathrm{r,dyn}}(\mathbf{\Gamma}) &\approx k_\mathrm{B}^{-1} S_{\mathrm{r,dyn}}^{\mathrm{odd},0}(\mathbf{\Gamma}) \\ &\equiv \frac{-\beta_1}{2} \int_{-\tau}^{0} \mathrm{d}t \left[\dot{E}_1^0(\overline{\mathbf{\Gamma}}(t|\mathbf{\Gamma},0)) - \dot{E}_1^0(\overline{\mathbf{\Gamma}}(t|\mathbf{\Gamma}^\dagger,0)) \right] \\ &= \frac{-\beta_1}{2} \int_{-\tau}^{\tau} \mathrm{d}t \, \dot{E}_1^0(\overline{\mathbf{\Gamma}}(t|\mathbf{\Gamma},0)). \end{aligned} \tag{11.76}$$

In the following non-equilibrium Monte Carlo results this is used to compute the non-equilibrium probability, in conjunction with umbrella sampling as described below. The structure of the non-equilibrium Monte Carlo algorithm for steady heat flow is more or less the same as described above for driven Brownian motion, §11.2.3.

The Lennard-Jones potential used here was as described for the stochastic molecular dynamics simulations, §11.1.4. Both a uniform bulk fluid (periodic boundary conditions, minimum image convention) and a confined fluid were simulated. The latter was in the form of a slit pore, terminated in the z-direction by uniform Lennard-Jones walls, and with periodic boundary conditions in the lateral directions (see p. 403 for further system details).

The simulations were performed in $6N$-dimensional phase space, with $N =$ 120–500 atoms. The Metropolis algorithm was used with umbrella weight density

$$\omega(\mathbf{\Gamma}) = e^{-\beta_0 E_0(\mathbf{\Gamma})} e^{-\beta_1 E_1(\mathbf{\Gamma})} e^{c\beta_1 \dot{E}_1^0(\mathbf{\Gamma})}. \tag{11.77}$$

As discussed in conjunction with Eq. (11.47), the static part of the reservoir entropy has even parity, so that if it alone were used as the umbrella weight, the flux visited in the Monte Carlo sequence would be as likely positive as negative. Although these would be given the correct weight in the final averaging, about half the simulated points would be wasted. Including the final term biases the points visited in the Monte Carlo sequence to those with a flux in the right direction. This term obviously approximates $S_{\mathrm{r,dyn}}(\mathbf{\Gamma})/k_\mathrm{B}$, but is about a factor of 400 faster to evaluate. In the simulations reported below c was fixed at 0.08. It would be possible to optimise this choice, or to determine c on the fly.[13]

A trial move of an atom consisted of a small displacement in its position and momentum simultaneously. The step lengths were chosen to give an acceptance rate of $\lesssim 50\%$. A Monte Carlo cycle consisted of one trial move of each atom in turn, or N trial moves in total.

The full non-equilibrium weight was calculated once every 50 Monte Carlo cycles, and quantities for later averaging were accumulated at this time. This reduces the correlation between consecutive, costly trajectories, by inserting

[13] It is a moot point whether it is better to bias the umbrella sampling in this way with the parameter c, or whether to use the static part of the reservoir entropy alone for the umbrella weight, and to include both $\mathbf{\Gamma}$ and $\mathbf{\Gamma}^\dagger$ in the average, since $S_{\mathrm{r,dyn}}^{odd;0}(\mathbf{\Gamma}^\dagger) = -S_{\mathrm{r,dyn}}^{odd;0}(\mathbf{\Gamma})$, or to do both. The relative computational efficiency of these options was not explored.

many cheap, umbrella steps. Labelling the current configuration used for an average $\mathbf{\Gamma}_i$, the Hamiltonian trajectory $\mathbf{\Gamma}^0(t|\mathbf{\Gamma}_i, 0)$ was generated forward and backward in time using a second-order rule and a time step of $\Delta_t = 10^{-2}$, which gave satisfactory energy conservation. Prior to the generation of each trajectory the velocities of the particles were scaled and shifted at constant kinetic energy to give zero total z-momentum. In the inhomogeneous system, a constraint force was added to keep the total z-momentum zero on a trajectory.

The running integral for $S^{\mathrm{odd},0}_{\mathrm{r,dyn}}(\mathbf{\Gamma}_i; t)$ was calculated along the trajectory using both the trapezoidal rule and Simpson's rule, with indistinguishable results. The average flux was calculated as a function of the time interval,

$$\left\langle \dot{E}_1^0 \right\rangle_\tau = \frac{\sum_i \dot{E}_1^0(\mathbf{\Gamma}_i) e^{-c\beta_1 \dot{E}_1^0(\mathbf{\Gamma}_i)} e^{S^{\mathrm{odd},0}_{\mathrm{r,dyn}}(\mathbf{\Gamma}_i;\tau)/k_{\mathrm{B}}}}{\sum_i e^{-c\beta_1 \dot{E}_1^0(\mathbf{\Gamma}_i)} e^{S^{\mathrm{odd},0}_{\mathrm{r,dyn}}(\mathbf{\Gamma}_i;\tau)/k_{\mathrm{B}}}}. \tag{11.78}$$

The umbrella weight used in the Metropolis scheme is obviously cancelled here. The thermal conductivity is reported below as $\lambda(\tau) = \left\langle \dot{E}_1^0 \right\rangle_\tau / \beta_1 V k_{\mathrm{B}} T_0^2$. Of the order of 50,000 trajectories were generated for each case studied.

Compared to using the exact non-equilibrium weight, the umbrella method is orders of magnitude faster in generating configurations. Furthermore, it allows results as a function of τ to be collected, with no additional computational burden. This provides a check that the integration range for the dynamic part of the reservoir entropy is long enough for the integrand to have reached its asymptotic limit. It also allows results to be obtained for different thermal gradients simultaneously, since the costly weight $S^{\mathrm{odd},0}_{\mathrm{r,dyn}}$ is linearly proportional to β_1. Thus non-linear effects in the thermal conductivity could be obtained in a single simulation.

Results

Non-equilibrium Monte Carlo results for the thermal conductivity are given in Fig. 11.12, where the time interval used to calculate $S^{\mathrm{odd},0}_{\mathrm{r,dyn}}(\mathbf{\Gamma}; \tau)$ is tested. The thermal conductivity was taken from the ratio of the average adiabatic rate of change of the first energy moment, the heat flux, to the imposed thermal gradient. As can be seen, $\lambda(\tau)$ is independent of the integration limit for $\tau \gtrsim 1$. This asymptotic or plateau value is 'the' thermal conductivity. The value of τ required to reach the respective plateaus here is comparable to equilibrium Green-Kubo calculations,[14] but the present non-equilibrium Monte Carlo simulations used about one third the number of trajectories for comparable statistical error.

The simulation results in Fig. 11.12 were obtained in a bulk system with periodic boundary conditions. In particular, the adiabatic heat flux, $\dot{E}_1^0$, that was integrated to get the non-equilibrium weight and which was averaged to get the thermal conductivity, depends upon the separation between pairs of atoms, not their absolute position, Eq. (11.18). For the periodic boundary

[14] Attard, P. (2005b), J. Chem. Phys. **122**, 244105.

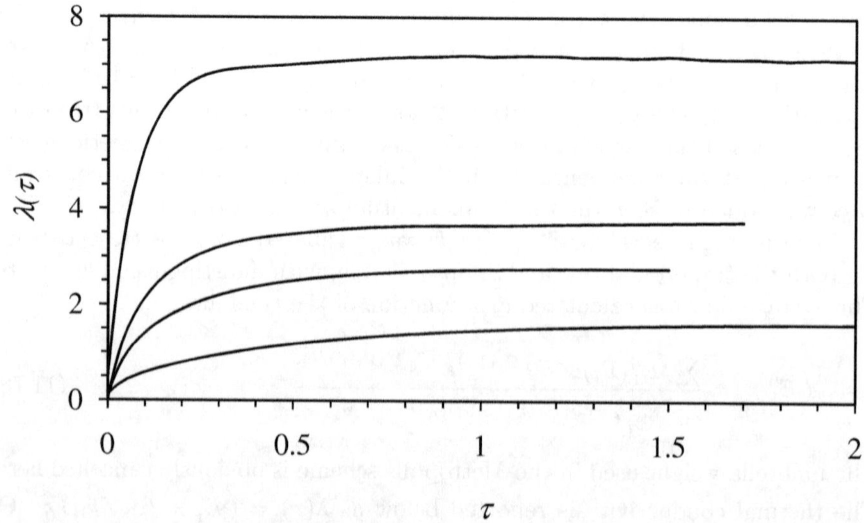

Figure 11.12: Non-equilibrium Monte Carlo results for the thermal conductivity as a function of the time interval used for the dynamic part of the reservoir entropy, $\lambda(\tau) = \left\langle \dot{E}_1^0 \right\rangle_\tau / V k_\mathrm{B} T_0^2 \beta_1$. The temperature is $T_0 = 2$, and from top to bottom the densities are 0.8, 0.6, 0.5, and 0.3. A bulk system with periodic boundary conditions was used. Data from Attard, P. (2006), J. Chem. Phys. **124**, 024109.

conditions these separations were evaluated using the nearest image convention. Figure 11.13 shows non-equilibrium Monte Carlo results for the slit pore. In this case there are no periodic boundaries in the z-direction, and the actual z-coordinate and separation are used. It can be seen from the data in the figure that as the width of the fluid increases, the extent of the plateau region for the conductivity increases and the eventual rate of decay slows. These data suggest that the extent of the plateau region scales with the size of the system, and that periodic systems are effectively infinite, at least on the time scales used in these simulations.

The stochastic molecular dynamics simulations presented in §11.1.4 above were for a slit pore. As discussed there, the thermal conductivity was obtained from the average adiabatic flux and the induced temperature gradient in the central half of the pore. This was necessary because the competition between the adiabatic and the reservoir induced evolution led to artifacts in the flux and in the temperature profile near the walls. In the non-equilibrium Monte Carlo simulations, this was not an issue, with the induced temperature profile being quite linear throughout. The data in Fig. 11.13 were obtained from the average adiabatic flux of the whole system and the applied temperature gradient. Of course the data in the figure do indicate that there is some dependence on the width of the slit, and for this reason the widest slit should be taken as an

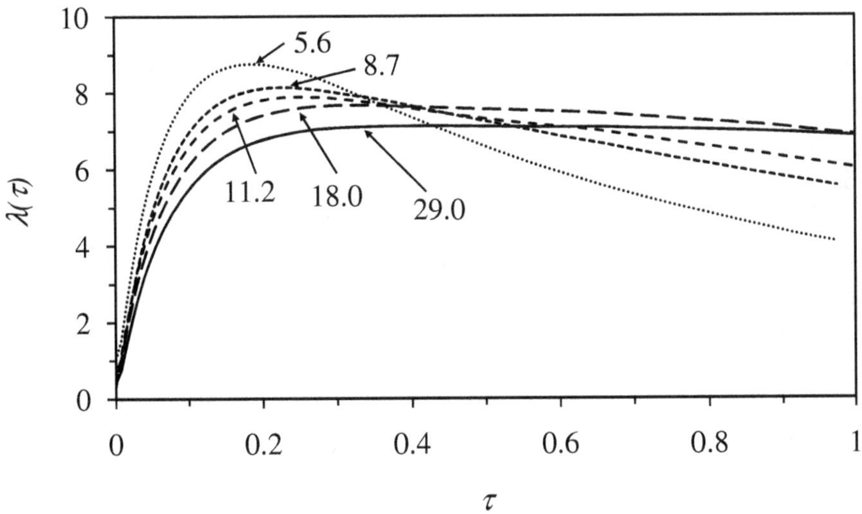

Figure 11.13: Non-equilibrium Monte Carlo results for the thermal conductivity $\lambda(\tau)$ obtained in a slit pore as a function of τ, ($\rho = 0.8$, $T_0 = 2$). The width of the pore, indicated by the numbers, increases from top to bottom at the peak. Data from Attard, (2006).

approximation to the thermal conductivity of the bulk system.

Figure 11.14 compares the thermal conductivity for a bulk system and for a slit pore, both case using non-equilibrium Monte Carlo. The value at the peak of the plateau was taken in all cases, and, in the case of the slit pore, the largest width that was studied. It can be seen there is good agreement between the results obtained in the two geometries over the range of densities. This indicates that the density inhomogeneity induced by the walls has little effect on the thermal conductivity.

Non-equilibrium stochastic molecular dynamics results are also given in Fig. 11.14. These were obtained for a slit pore with the algorithm described in §11.1.4. It can be seen that there is quite good agreement between the two non-equilibrium simulation algorithms. In terms of computational efficiency, it appears that the stochastic molecular dynamics algorithm is more efficient than the non-equilibrium Monte Carlo algorithm. For the case $\rho = 0.8$, ($\Delta_t = 0.005$, $\sigma^* = 0.002$), the standard error on the mean was 2%, and this took 4×10^6 time steps, with averages being collected every 10 steps. This is about a factor of 5 more efficient than the non-equilibrium Monte Carlo algorithm. Alternatively, the non-equilibrium Monte Carlo simulations required about 3×10^7 time steps (typically 2×10^5 independent trajectories, each of about 75 time steps forward and backward to get to the plateau). This obviously depends upon the size of the applied thermal gradient (the statistical error decreases with increasing gradient), but appears comparable to that required by deterministic molecular

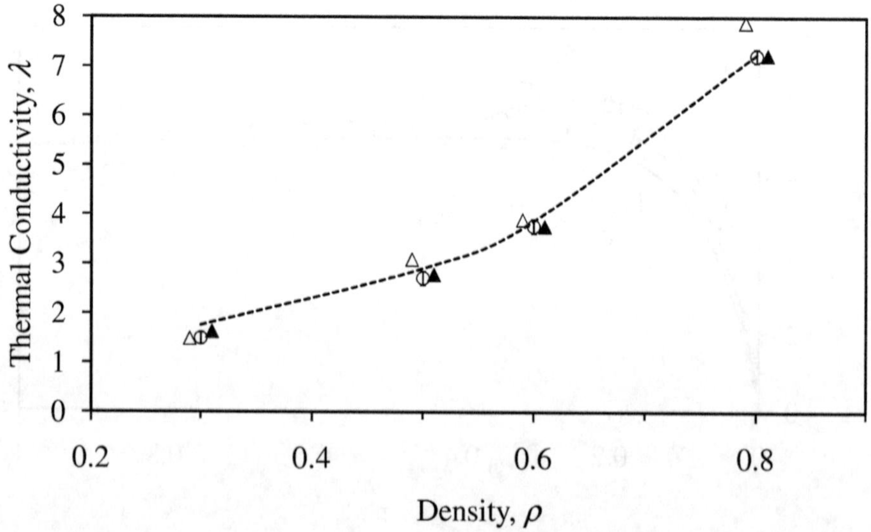

Figure 11.14: Thermal conductivity as a function of density at $T_0 = 2$. The filled triangles are non-equilibrium Monte-Carlo results for a bulk system (shifted horizontally by $+0.01$), and the open triangles are non-equilibrium Monte-Carlo results for a slit pore (shifted horizontally by -0.01) (Attard, 2006). The open circles are the stochastic molecular dynamics results, obtained with $\Delta_t = 5 \times 10^{-3}$ and $\sigma^* = 0.002$, (Attard, 2009a). The dotted line is deterministic molecular dynamics results, (Evans, 1986; Daivis and Evans, 1993).

dynamics simulations. No doubt the non-equilibrium Monte Carlo algorithm could be made more efficient by optimising the number of Monte Carlo cycles between trajectory evaluations or the value of the umbrella parameter.

Both non-equilibrium simulation algorithms, the stochastic molecular dynamics method discussed above and the present Monte Carlo method, are inherently non-linear methods. Unlike, for example, Green-Kubo methods, or deterministic molecular dynamics methods with artificial non-equilibrium equations of motion, no linearisation has been invoked. A non-linear effect that cannot be explored with such linear methods is the dependence of the thermal conductivity on the applied thermal gradient. However, this variation can be obtained with the non-equilibrium Monte Carlo algorithm, and results are shown in Fig. 11.15. The increase in λ with increasing β_1 is due primarily to the coupling of the induced density gradient to the heat flux. The non-linear effects appear greater for $\rho = 0.6$ than for $\rho = 0.8$, which is perhaps surprising. One should be cautious in invoking these results. The largest gradient shown corresponds to 10^{11} K/m in argon, which may be difficult to achieve in the laboratory. In this case the temperature discontinuity across the periodic z-boundaries is 1 $k_{\text{B}}/\epsilon_{\text{LJ}}$. It is not clear how this discontinuity and other finite size effects affect the non-linear conductivity.

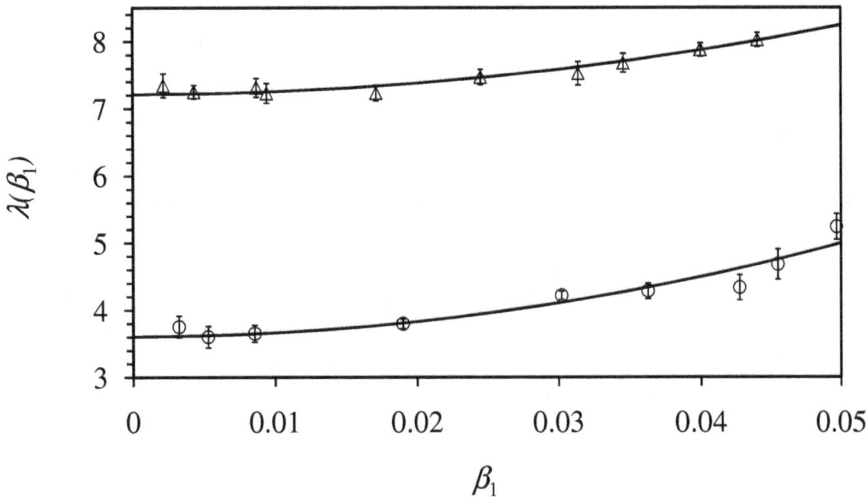

Figure 11.15: The non-linear thermal conductivity, $\lambda(\beta_1) = \left\langle \dot{E}_1^0 \right\rangle_{\beta_1} / V k_{\mathrm{B}} T_0^2 \beta_1$, as a function of the imposed thermal gradient for $T_0 = 2$. The symbols are non-equilibrium Monte Carlo simulation data, and the curves are fitted quadratics. The triangles are $\rho = 0.8$ and $\lambda(\beta_1) = 7.21 + 413\beta_1^2$, and the circles are $\rho = 0.6$ and $\lambda(\beta_1) = 3.61 + 554\beta_1^2$. Data from Attard (2006).

11.3 Brownian Dynamics

Brownian dynamics is an approximate form of non-equilibrium computer simulation that is most like the stochastic molecular dynamics method described earlier in this chapter, but which differs qualitatively from it in a number of important respects. In fact it is an implementation of Langevin's equation, or, more rarely, the generalised Langevin equation as derived by perturbation theory and by the projector operator formalism in Ch. 10. Like these approximations, it is designed for a system that can be divided into components that have fast and slow relaxation times. Typical examples include the simulation of polymers, biomolecules, or other macromolecules, colloids or other heavy solutes dispersed in a solvent, and monolayer, membrane, micelle, and microemulsion self-assembly. In such cases the solvent typically relaxes on much shorter time scales than the macro-component of the system, and it is this that limits the maximum time step that can be used in the stochastic molecular dynamics. In fact the separation in time scales can be so large that thousands of solvent steps are required before any measurable change in macrostructure occurs. Furthermore, the number of solvent atoms can be comparable to, or even many times larger than, the number of atoms or sites that comprise the macro-component, and so most of a molecular dynamics simulation can consist of solvent evaluations and moves. This is somewhat perverse since generally the primary motivation in simulating such systems is to characterise the structure and the dynamics of the macro-component, with the behaviour of the solvent being of secondary interest. Brownian dynamics seeks to overcome this problem by removing the explicit solvent and incorporating it implicitly via effective forces and interactions for the macro-components.

There are typically two types of problems that are addressed by Brownian dynamics. In one case one is interested in the spontaneous evolution of the structure of the macro-component. A typical example would be the folding or the unfolding of a protein following a change, such as solvent exchange, or such as the binding of an ion or antigen to a receptor site. This type of problem corresponds to the regression analysis of §10.6.2, and to the projector operator formalism of §10.7, where the perturbation theory for the spontaneous relaxation of a fluctuation was presented. A second type of problem concerns driven motion due to an external force. Examples include the electrophoretic mobility or sedimentation of colloids, ion or solute transport through membrane pores or microchannels, and the viscoelastic deformation of gels. This type of problem corresponds to the generalised Langevin equation, Eq. (10.61) or Eq. (10.165).

Brownian dynamics in typical implementation is an approximation, often modified with *ad hoc* corrections. This is in contrast to the non-equilibrium stochastic molecular dynamics and the non-equilibrium Monte Carlo algorithms presented above, which are formally exact.[15] The spirit of Brownian dynamics is focused on gaining an understanding of the behaviour of otherwise intractable heterogeneous systems, rather than on establishing benchmark results or on validating generic thermodynamic formulae or theorems. To this end an approximate approach is more defensible than it is for simulations of simple systems.

As mentioned above, the approximation in Brownian dynamics consists of subsuming the solvent into effective solute forces and interactions, so that only the solute or macro-component is simulated explicitly. In §11.3.1, which immediately follows, the simplest form of Brownian dynamics is discussed. This is the no memory limit of the generalised Langevin equation and perturbation theory of §10.3, and is typical of the field. Going beyond this to include memory effects is a more sophisticated form of Brownian dynamics simulation. Results have already been given for this type of algorithm in the form of the perturbation computations for driven Brownian motion presented in §10.5; these will not be repeated here. In §11.3.2, a way of systematically improving Brownian dynamics for more concentrated dispersions and for macromolecules is described, again in the context of perturbation theory. This section is somewhat speculative and no computational results are presented. In §11.3.3 an introduction to stochastic calculus is given, and its relevance to Brownian dynamics and to non-equilibrium statistical mechanics more generally is discussed.

[15]There are of course limitations in the non-equilibrium algorithms, primarily the model of the system studied, the size of the system, and the length of the simulation. These are practical restrictions that can be systematically checked and overcome, rather than fundamental approximations in the algorithm itself.

11.3.1 Elementary Brownian Dynamics

In §10.3.5, the generalised Langevin equation for velocity was derived from non-equilibrium perturbation theory for the pure parity case. With $x = \{x_1, x_2, \ldots\}$ a vector with components representing the positions of each macro-component (e.g. monomer or site of the macromolecule, colloid particle of the dispersion, etc.), in the continuum limit, the generalised Langevin equation (10.68) is

$$m\ddot{x}(t) = F^{\text{ext}}(x(t), t) + R_{\text{p}}(t) - \int_{-\infty}^{t} \mathrm{d}t' \, K(t-t')\dot{x}(t'). \tag{11.79}$$

Here m is the mass, $F^{\text{ext}}(x,t)$ is the external force, and $R_{\text{p}}(t)$ is the random force for the velocity, which satisfies the fluctuation dissipation theorem, Eq. (10.78),

$$\langle R_{\text{p}}(t) R_{\text{p}}(0) \rangle_0 = k_{\text{B}} T K(t). \tag{11.80}$$

The memory kernel is simply related to the velocity autocorrelation function in Fourier space by Eq. (10.73),

$$K(\omega) = \frac{T \ddot{q}_0(\omega)}{\left| \tilde{\ddot{q}}_0(\omega) \right|^2}. \tag{11.81}$$

Here the one-sided velocity autocorrelation function is $\tilde{\ddot{q}}_0(t) = \ddot{q}_0(t)$, $t \geq 0$, and $\tilde{\ddot{q}}_0(t) = 0$, $t < 0$. This is a pure parity case, and so the symmetrised velocity autocorrelation function is the same as the ordinary velocity autocorrelation function, $\ddot{q}_0(t) \equiv \langle \dot{x}(t)\dot{x}(0) \rangle_0 / k_{\text{B}}$.

The simplest approximation used for Brownian dynamics simulations is to take the memory function to be a δ-function, (Allen and Tildesley, 1987),

$$K(t) = 2\gamma\delta(t). \tag{11.82}$$

With this, the generalised Langevin equation reduces to the ordinary Langevin equation,

$$m\ddot{x}(t) = F(x(t), t) + R_{\text{p}}(t) - \gamma\dot{x}(t). \tag{11.83}$$

Evidently γ is the friction or drag coefficient, which is related to the diffusion constant by $\gamma \equiv k_{\text{B}} T D^{-1}$. Because one is generally dealing with a multi-component system, $x \equiv \{x_1, x_2, \ldots\}$, in this simplest case there is no coupling between different components and γ is a diagonal matrix with friction coefficient appropriate to each component on the diagonal. This δ-function memory function corresponds to an exponentially decaying velocity autocorrelation function, $\ddot{q}_0(t) = T m^{-1} e^{-\gamma m^{-1}|t|}$, §10.3.8.

Alternatively, and more commonly, one proceeds from the continuum version of the generalised Langevin equation for position, Eq. (10.62),

$$\begin{aligned}\dot{x}(t) &= R(t) + \frac{1}{T}\int_0^t \mathrm{d}t'\, \ddot{q}_0(t-t') F(x(t'),t') \\ &= R(t) + m^{-1}\int_0^t \mathrm{d}t'\, e^{-m^{-1}\gamma|t-t'|} F(x(t'),t') \\ &= R(t) + \frac{1}{k_{\mathrm{B}}T} D F(x(t),t). \end{aligned} \tag{11.84}$$

The second equality holds in the no memory limit, and the third equality holds in the large friction, $\gamma \to \infty$, or, equivalently, in the slowly varying force limit. The fluctuation dissipation theorem for the random force for the position follows from Eq. (10.63),

$$\begin{aligned}\langle R(t)R(t')\rangle &= k_{\mathrm{B}}\ddot{q}_0(t-t') \\ &= k_{\mathrm{B}}Tm^{-1}e^{-k_{\mathrm{B}}TD^{-1}m^{-1}|t_{kl}|} \\ &= 2D\,\delta(t-t'). \end{aligned} \tag{11.85}$$

The successive equalities correspond to the short memory and high friction limits, respectively.

In practice in Brownian dynamics simulations one discretises time, $t_k = k\Delta_t$, so that the Langevin equation for position becomes (short memory, high friction limit),[16]

$$x_{k+1} = x_k + \tilde{R}_k + \frac{\Delta_t}{k_{\mathrm{B}}T} D F_k. \tag{11.86}$$

Here the random force has been rescaled, $\tilde{R}_k = \Delta_t R_k$, so that the covariance is

$$\left\langle \tilde{R}_k \tilde{R}_l \right\rangle = 2|\Delta_t| D\,\delta_{kl}. \tag{11.87}$$

The Dirac δ-function of the continuum has been interpreted as a 'top-hat' function, of width Δ_t and height Δ_t^{-1}, $\delta(t_k - t_l) \Rightarrow \Delta_t^{-1}\delta_{kl}$, leading to the present Kronecker δ for the discrete time step. It is significant for the discussion of the stochastic calculus in §11.3.3 below that the covariance of the random force is proportional to the width of the time step. In the mathematics literature on stochastic calculus, this property is used as the definition of a Brownian motion, namely that the quadratic variation of the increments is proportional to the step length.

In component form the covariance is

$$\left\langle \tilde{R}_{I\alpha k} \tilde{R}_{J\beta l} \right\rangle = 2|\Delta_t| D_{I\alpha,J\beta}\,\delta_{kl}. \tag{11.88}$$

[16] This is equivalent to, for example, Eqs (9.33) and (9.34) of Allen and Tildesley, (1987). The extra term $\Delta_t \nabla D(x_i)$ that appears in Eq. (9.33) that doesn't appear here results from a configuration dependent diffusion tensor (see next), and from the stochastic calculus (see §11.3.3), neither of which, it is therein argued, are justified for physical systems.

Here upper case Roman indices refer to the Cartesian coordinate, lower case Greek indices refer to the solute species, and lower case Roman indices refer to the time node. In the most usual case, there is no coupling between the diffusion tensor for different species, $\mathbf{D}_{\alpha\beta} = \mathbf{D}_\alpha^{(3)} \mathrm{I}_{\alpha\beta}$, nor between different Cartesian components, $\mathbf{D}_{\alpha;JK}^{(3)} = D_{\alpha J} \mathrm{I}^{(3)}$. For a spherically symmetric solute, this is the same for all Cartesian components. Either one can obtain the diffusion constant from a simulation of the velocity autocorrelation function of a single solute in the solvent, or else one might assume that it is simply given by Stokes hydrodynamic drag, $\mathbf{D}_\alpha = k_\mathrm{B}T/6\pi\eta\rho_\alpha$, where η is the viscosity of the solvent and ρ_α is the Stokes radius of the monomer.

As mentioned above, the major application of Brownian dynamics is to complex systems, such as macromolecules or colloid dispersions. In such cases one expects that the interactions between the monomers or solutes strongly affect their diffusion. The simplest attempt to take such interactions into account is based on macroscopic hydrodynamics, and it invokes the Oseen tensor for the pair-wise interaction,[17]

$$\mathbf{D}_{\alpha\beta} = \begin{cases} \dfrac{k_\mathrm{B}T}{6\pi\eta\rho_\alpha} \mathrm{I}^{(3)}, & \alpha = \beta, \\ \dfrac{k_\mathrm{B}T}{8\pi\eta x_{\alpha\beta}} \left[\mathrm{I}^{(3)} + \dfrac{\mathbf{x}_{\alpha\beta}\mathbf{x}_{\alpha\beta}}{x_{\alpha\beta}^2} \right], & \text{otherwise}, \end{cases} \tag{11.89}$$

where $\mathbf{x}_{\alpha\beta} \equiv \mathbf{x}_\alpha - \mathbf{x}_\beta$ and $x_{\alpha\beta} \equiv |\mathbf{x}_{\alpha\beta}|$. Note that this expression represents *only* the hydrodynamic interaction; it does *not* include any interaction potentials between the monomers or solutes.

The status of this approach is a little unclear. This might be interpreted as the first term in a systematic expansion, with higher-order terms involving three body, four body, etc. interactions that can be obtained from macroscopic hydrodynamics. But whether this term and the higher-order terms are consistent with neglecting memory effects remains an issue. There is also an issue with the apparent dependence of the diffusion constant on the configuration. This is not trivial because such a dependence creates an ambiguity that turns on the difference between the Itô and Stratonovich stochastic calculus (see Ermak and McCammon, (1978), Allen and Tildesley, (1987), Mazo, (2008), which were mentioned above, and others[18]). This was mentioned in connection with the Fokker-Planck equation, in footnote 18 on p. 89, and will be analysed in detail in §11.3.3 below.

This brief review of Brownian dynamics simulations raises two related questions. The first concerns the use of more sophisticated diffusion constants that

[17]Ermak, D. L. and McCammon, J. A. (1978), J. Chem. Phys. **69**, 1352. Mazo, R. M. (2008), *Brownian Motion: Fluctuations, Dynamics, and Applications*, (Oxford University Press, Oxford).

[18]Gardiner, C. W. (1983), *Handbook of Stochastic Methods*, (Springer, New York). Risken, H. (1984), *The Fokker-Planck Equation*, (Springer-Verlag, Berlin). Tough, R. J. A., Pusey, P. N., Lekerkerker, H. N. W., and van den Broeck, C. (1986), Mol. Phys. **59**, 595. Keizer, J. (1987), *Statistical Thermodynamics of Non-equilibrium Processes*, (Springer-Verlag, New York).

include memory effects and possibly configuration dependence, with the aim of developing systematic expansions that lead to controlled approximations and tractable algorithms. The second question concerns the stochastic calculus, and the difference between Brownian stochastic processes, such as one finds in the mathematics and financial engineering literature, and the more realistic formulations of Brownian dynamics that are suitable for physical systems. The first question is discussed in the following section, which approaches the problem in the spirit of the perturbation theory that was used to derive the generalised Langevin equation in Ch. 10. The issue of the stochastic calculus is addressed in §11.3.3.

11.3.2 Perturbative Brownian Dynamics

In one sense an exact algorithm for Brownian dynamics for complex systems has already been given: it is the generalised Langevin equation for position with the velocity autocorrelation function providing the memory kernel, Eq. (10.62). This equation is exact in the context that the external force can be treated as a perturbation. The problem is, however, that obtaining the full velocity autocorrelation function of the solutes or monomers in the solvent is a challenging computational task; once it has been obtained the consequent simulations for different external forces are relatively trivial. This section seeks to simplify the computation of the solute velocity autocorrelation function and to place the many body diffusion tensor mentioned above on a more formal basis.

Consider a macromolecule in a solvent. Let the positions of the monomers comprising the macromolecule be denoted by $\mathbf{X} = \{\mathbf{x}_1, \mathbf{x}_2, \ldots\}$, with $\mathbf{x}_\alpha = \{r_{x\alpha}, r_{y\alpha}, r_{z\alpha}\}$. (This formulation can as well apply to a concentrated colloid dispersion, with the Greek indices labelling the colloids.) There are five types of potential energies relevant to the problem: there is the solvent-solvent potential, the solvent-monomer potential, the non-bonding monomer-monomer potential, the bonding monomer-monomer potential, and the external potential on the monomers. The bonding potential, $U_{\rm b}(\mathbf{X})$, is short-ranged and consists of pair terms between nearest neighbours, which describe the connectivity of the macromolecule and the length of the bonds, and three body and higher-order potentials, which describe bond bending and bond rotation effects. (The bonding potential is not present for a colloid dispersion.) The non-bonding monomer-monomer potential is non-specific compared to the bonding potential, and it contains short-range excluded volume terms, for example hard-sphere or Lennard-Jones interactions, and longer range terms such as Coulomb interactions, if present.

In the simplest case the macromolecule is a linear chain, with identical monomers labelled in order along the chain, with the bonding potential acting only between adjacent pairs, $U_{\rm b}(\mathbf{X}) = \sum_\alpha u_{\rm b}(x_{\alpha,\alpha+1})$, where $x_{\alpha\beta} \equiv |\mathbf{x}_{\alpha\beta}|$, and $\mathbf{x}_{\alpha\beta} \equiv \mathbf{x}_\alpha - \mathbf{x}_\beta$. Typically, $u_{\rm b}(x)$ might represent a stiff spring, with possibly a non-harmonic component. Short-range repulsions and excluded volume effects are probably best included in the non-bonding part of the monomer-monomer potential, as typically these are unsuited for a perturbation treatment.

Conditional Correlations and Response

There is a fundamental difference between the perturbation theory that was developed in Ch. 10 and the perturbation theory that will be developed here. In the perturbation theory of §10.3, the velocity autocorrelation function of the bare or equilibrium system was invoked, §10.1.1, as well as the trajectory entropy, §10.1.4. The trajectory in that case was specified by velocities at the time nodes, $\dot{\mathbf{X}}^{(n/j)} \equiv \{\dot{\mathbf{X}}(t_{j-n+1}), \dot{\mathbf{X}}(t_{j-n+2}), \ldots, \dot{\mathbf{X}}(t_j)\}$, and one position, usually the final position $\mathbf{X}(t_j)$, but sometimes instead the initial position $\mathbf{X}(t_{j-n+1})$. (In the present notation, the position of the solutes is denoted $\mathbf{X}$.) The velocity autocorrelation function that emerged as the memory function, $\ddot{q}_0^{\rm s}(t) \equiv k_{\rm B}^{-1}\left\langle \dot{\mathbf{X}}(t)\dot{\mathbf{X}}(0)\right\rangle_0$, was independent of the initial position. This corresponds to integrating over all initial positions, and to treating the positions explicitly in the quadratic form for the trajectory entropy, §10.1.4. In the case that $\mathbf{X}$ represents a single Brownian particle, or that it represents a set of Brownian particles at low concentration in the solvent, this is appropriate.

When $\mathbf{X}$ represents the monomers on a macromolecule, or the colloids in a high concentration dispersion, more accurate results can be anticipated when the initial positions of the particles are taken into account, so that one has the conditional velocity autocorrelation function,

$$\begin{aligned} \ddot{q}_0(t,t'|\mathbf{X}(0)) &\equiv k_{\rm B}^{-1}\left\langle \dot{\mathbf{X}}(t)\dot{\mathbf{X}}(t')\middle|\, \mathbf{X}(0)\right\rangle_0 \\ &\quad - k_{\rm B}^{-1}\left\langle \dot{\mathbf{X}}(t)\middle|\, \mathbf{X}(0)\right\rangle_0 \left\langle \dot{\mathbf{X}}(t')\middle|\, \mathbf{X}(0)\right\rangle_0 . \end{aligned} \tag{11.90}$$

(The superscript s is dropped because only the pure parity case is analysed here, and there is no distinction between the symmetrised and the full correlation function in this case. The meaning of the subscript 0 in the present context will be defined shortly; until then, it can be taken to signify the equilibrium system.) The reasons that this is important in the present problem is that unlike a dilute dispersion, there may be a high probability of $\mathbf{x}_\alpha(t)$ and $\mathbf{x}_\beta(t)$ being close together, and the consequent interactions significantly affect the time correlations of the monomer velocities. The dependence of a particular velocity autocorrelation function $\ddot{q}_{0;\alpha\beta}(t|\mathbf{X}(0))$ on the initial positions is not restricted to the initial separation of the pair, $\mathbf{x}_{\alpha\beta}(0)$, but may include the separations of other monomers initially in the vicinity (three body and higher-order effects).

This conditional velocity autocorrelation function $\ddot{q}_{0;\alpha\beta}(t|\mathbf{X}(0))$ contains the so-called hydrodynamic interactions. Neglecting memory effects, at the pair level it gives the Oseen form for the diffusion tensor, Eq. (11.89), with two significant improvements: it includes the contributions from the non-bonding monomer-monomer interactions (or else these can be added to the perturbation potential), and it is $\mathbf{x}_{\alpha\beta}(0)$ that appears here, which is arguably more correct than the $\mathbf{x}_{\alpha\beta}(t)$ that appears in the Oseen equation. The significance of this is that as far as the stochastic calculus is concerned, the transport coefficient (equivalently the velocity autocorrelation function) is independent of the stochastic variable, $\partial\ddot{q}_0(t;\mathbf{X}(0))/\partial\mathbf{X}(t) = \partial D(\mathbf{X}(0))/\partial\mathbf{X}(t) = 0$. This is consistent with the remarks made in footnote 18 on p. 89 on the relevance of the Itô and Stratonovich stochastic calculus to the Fokker-Planck equation, and will be discussed in detail in §11.3.3 below.

Bond Potential as a Perturbation

The non-trivial problem is to obtain the conditional solute velocity autocorrelation function in a computationally efficient fashion, with the eventual aim of using it in a form of Brownian dynamics to simulate the evolution of the structure of the macromolecule, $\mathbf{X}(t)$, under the influence of an external force $F^{\text{ext}}(\mathbf{X}, t)$, or of a non-equilibrium (unlikely) starting position $\mathbf{X}(0)$. As mentioned in the introduction the separation of time scales of solvent motion and macromolecule relaxation makes the full simulation of the solute velocity autocorrelation function challenging. Hence a solution is sought in terms of Brownian dynamics of the macromolecule that invokes time correlation functions of the *non-bonded* macromolecule.

The bonding force on monomer α is

$$\begin{aligned} F_{\mathrm{b};\alpha}(\mathbf{X}) &= \frac{-\partial U_{\mathrm{b}}(\mathbf{X})}{\partial \mathbf{x}_\alpha} \\ &= -u_{\mathrm{b}}'(x_{\alpha,\alpha+1})\frac{\mathbf{x}_{\alpha,\alpha+1}}{x_{\alpha,\alpha+1}} - u_{\mathrm{b}}'(x_{\alpha,\alpha-1})\frac{\mathbf{x}_{\alpha,\alpha-1}}{x_{\alpha,\alpha-1}}. \end{aligned} \tag{11.91}$$

The second equality holds in the simplest case of a linear chain with nearest neighbour bonds only.

Now consider the bonding potential as an external potential, and develop a perturbation theory using the non-bonded macromolecule as the reference system. As is well known in equilibrium statistical mechanics, there is no fundamental distinction between an internal potential and an external potential. Hence the total perturbation force may be taken to be

$$\mathbf{F}(\mathbf{X}, t) = \mathbf{F}_{\mathrm{b}}(\mathbf{X}) + \mathbf{F}^{\text{ext}}(\mathbf{X}, t). \tag{11.92}$$

The bare or unperturbed system consists of the solvent and macromolecule without the external and bonding potentials. It is an equilibrium system and is denoted by a subscript 0. The bare trajectory entropy, conditional upon $\mathbf{X}(0) \equiv \mathbf{X}_0$, is a quadratic form that is modified from Eq. (10.24),

$$S_0^{(n)}(\dot{\mathbf{X}}^{(n)}|\mathbf{X}_0) = S_{0;x\dot{x}}^{(n)}(\mathbf{X}_0) \cdot \dot{\mathbf{X}}^{(n)} + \frac{1}{2} S_{0;\dot{x}\dot{x}}^{(n)}(\mathbf{X}_0) : \dot{\mathbf{X}}^{(n)}\dot{\mathbf{X}}^{(n)}. \tag{11.93}$$

This is maximised by the most likely trajectory velocities,

$$\overline{\dot{\mathbf{X}}}^{(n)}(\mathbf{X}_0) = -[S_{0;\dot{x}\dot{x}}^{(n)}(\mathbf{X}_0)]^{-1} S_{0;x\dot{x}}^{(n)}(\mathbf{X}_0). \tag{11.94}$$

The velocity autocorrelation function is related to the fluctuation matrix in the usual fashion,

$$\begin{aligned}
\ddot{Q}_0^{(n)}(\mathbf{X}_0) &= k_\mathrm{B}^{-1}\left\langle \left[\dot{\mathbf{X}}^{(n)} - \overline{\dot{\mathbf{X}}}^{(n)}(\mathbf{X}_0)\right]\left[\dot{\mathbf{X}}^{(n)} - \overline{\dot{\mathbf{X}}}^{(n)}(\mathbf{X}_0)\right] \middle| \mathbf{X}_0 \right\rangle_0 \\
&= k_\mathrm{B}^{-1}\left\langle \dot{\mathbf{X}}^{(n)}\dot{\mathbf{X}}^{(n)} \middle| \mathbf{X}_0 \right\rangle_0 - k_\mathrm{B}^{-1}\left\langle \dot{\mathbf{X}}^{(n)} \middle| \mathbf{X}_0 \right\rangle_0 \left\langle \dot{\mathbf{X}}^{(n)} \middle| \mathbf{X}_0 \right\rangle_0 \\
&= -[S_{0;\dot{x}\dot{x}}^{(n)}(\mathbf{X}_0)]^{-1}. \qquad (11.95)
\end{aligned}$$

One also has that

$$S_{0;x\dot{x}}^{(n)}(\mathbf{X}_0) = -S_{0;\dot{x}\dot{x}}^{(n)}(\mathbf{X}_0)\overline{\dot{\mathbf{X}}}^{(n)}(\mathbf{X}_0) = [\ddot{Q}_0^{(n)}(\mathbf{X}_0)]^{-1}\left\langle \dot{\mathbf{X}}^{(n)} \middle| \mathbf{X}_0 \right\rangle_0. \qquad (11.96)$$

These averages are in the absence of the bond and external potentials, and with the initial positions of the monomers at $\mathbf{X}_0$. The elements of the velocity autocorrelation function matrix are

$$\begin{aligned}
\{\ddot{Q}_0^{(n)}(\mathbf{X}_0)\}_{ij} &= \ddot{q}_0(t_i, t_j|\mathbf{X}_0) \\
&= k_\mathrm{B}^{-1}\left\langle \dot{\mathbf{X}}(t_i)\dot{\mathbf{X}}(t_j) \middle| \mathbf{X}_0 \right\rangle_0 \\
&\quad - k_\mathrm{B}^{-1}\left\langle \dot{\mathbf{X}}(t_i) \middle| \mathbf{X}_0 \right\rangle_0 \left\langle \dot{\mathbf{X}}(t_j) \middle| \mathbf{X}_0 \right\rangle_0. \qquad (11.97)
\end{aligned}$$

In general this is not homogeneous in time. But for large times it does become homogeneous, $\ddot{q}_0(t_i, t_j|\mathbf{X}_0) \sim \ddot{q}_0(t_{ij})$, $t_i \to \infty$. In terms of monomers this is

$$\begin{aligned}
\ddot{q}_{0;\alpha\beta}(t, t'|\mathbf{X}_0) &= k_\mathrm{B}^{-1}\left\langle \left[\dot{\mathbf{X}}_\alpha(t) - \overline{\dot{\mathbf{X}}}_\alpha(t|\mathbf{X}_0)\right]\left[\dot{\mathbf{X}}_\beta(t') - \overline{\dot{\mathbf{X}}}_\beta(t'|\mathbf{X}_0)\right] \right\rangle_0 \\
&= k_\mathrm{B}^{-1}\left\langle \dot{\mathbf{X}}_\alpha(t)\dot{\mathbf{X}}_\beta(t') \middle| \mathbf{X}_0 \right\rangle_0 \\
&\quad - k_\mathrm{B}^{-1}\left\langle \dot{\mathbf{X}}_\alpha(t) \middle| \mathbf{X}_0 \right\rangle_0 \left\langle \dot{\mathbf{X}}_\beta(t') \middle| \mathbf{X}_0 \right\rangle_0. \qquad (11.98)
\end{aligned}$$

Adding now the total force, $\mathbf{F}^{(n)}$, the trajectory entropy of the perturbed system is formally the same as in Eq. (10.46), with the conditional dependence on the starting positions now exhibited explicitly,

$$S^{(n)}(\dot{\mathbf{X}}^{(n)}|\mathbf{X}_0, \mathbf{F}^{(n)}) = S_0^{(n)}(\dot{\mathbf{X}}^{(n)}|\mathbf{X}_0) + \frac{\tau}{T}\dot{\mathbf{X}}^{(n)} \cdot \mathbf{F}^{(n)}. \qquad (11.99)$$

The subscript zero here signifies either the non-bonded macromolecule in the case of $S_0^{(n)}$, or else the initial position in the case of $\mathbf{X}_0$. The force here is the sum of the bonding force and the external force, $\{\mathbf{F}^{(n)}\}_i \equiv F(\mathbf{X}(t_i), t_i) = F_\mathrm{b}(\mathbf{X}(t_i)) + F^\mathrm{ext}(\mathbf{X}(t_i), t_i)$. The trajectory has been uniformly discretised, $\tau = t_{i+1} - t_i$.

As discussed in connection with Eq. (10.47), to leading-order in the perturbing force, one may neglect the dependence of $\mathbf{F}^{(n)}$ on $\dot{\mathbf{X}}^{(n)}$. (Or else one may regard the perturbing force as a given that is later set equal to the actual force on the most likely trajectory, §10.3.2.) In this case the most likely trajectory velocities are given by

$$\begin{aligned}\overline{\dot{\mathbf{X}}}^{(n)}(\mathbf{X}_0, \mathbf{F}^{(n)}) &= \ddot{Q}_0^{(n)}(\mathbf{X}_0)\left[S_{0;x\dot{x}}^{(n)}(\mathbf{X}_0) + \frac{\tau}{T}\mathbf{F}^{(n)}\right] \\ &= \overline{\dot{\mathbf{X}}}^{(n)}(\mathbf{X}_0) + \frac{\tau}{T}\ddot{Q}_0^{(n)}(\mathbf{X}_0)\mathbf{F}^{(n)}. \end{aligned} \tag{11.100}$$

A subscript 0 could be added to the first term here, $\overline{\dot{\mathbf{X}}}_0^{(n)}(\mathbf{X}_0)$, to make it clear that this is the most likely trajectory velocity in the unbonded system. Here and above, $\mathbf{X}_0 \equiv \mathbf{X}(0)$ is the initial position; it is *not* the unbonded trajectory $\mathbf{X}_0^{(n)}$.

One can evaluate this for the terminal position, transform to the continuum, and add the random force,

$$\begin{aligned}\dot{\mathbf{X}}(t|\mathbf{X}_0, \mathbf{F}) &= \overline{\dot{\mathbf{X}}}(t|\mathbf{X}_0) + \mathbf{R}(t) \\ &\quad + \frac{1}{T}\int_0^t \mathrm{d}t'\ \ddot{q}_0(t, t'|\mathbf{X}_0)\mathbf{F}(\mathbf{X}(t'), t'). \end{aligned} \tag{11.101}$$

The random force has zero mean and covariance

$$\langle \mathbf{R}(t)\mathbf{R}(t')\rangle = k_{\mathrm{B}}\ddot{q}_0(t, t'|\mathbf{X}_0). \tag{11.102}$$

Obviously, because the memory function is short-ranged, one can set the lower limit of the integral to $t - t^*$ for some fixed t^* greater than the relaxation time, and set $\mathbf{X}_0 \Rightarrow \mathbf{X}(t - t^*)$. Also, $\overline{\dot{\mathbf{X}}}(t|\mathbf{X}_0) = \left\langle \dot{\mathbf{X}}(t) \middle| \mathbf{X}(0) \right\rangle_0$ (i.e. it is the unbonded conditional most likely position), and this must have been obtained at the same level of systematic approximation as $\ddot{q}_0(t, t'|\mathbf{X}_0)$. By systematic approximation is meant that for each component of the velocity autocorrelation only a small sub-set of monomers are fixed at their initial positions, for example $\ddot{q}_{0;\alpha\beta}(t, t'|\mathbf{x}_\alpha(0), \mathbf{x}_\beta(0), \ldots)$, where the ellipsis represents the closest neighbours of α and β. Obviously if one can compute the velocity autocorrelation function $\langle \dot{\mathbf{x}}_\alpha(t)\dot{\mathbf{x}}_\beta(t')|\, \mathbf{X}_0\rangle_0$ as a function of t, t', and a sub-set of $\mathbf{X}_0$, then one can even more easily compute $\langle \dot{\mathbf{x}}_\alpha(t)|\, \mathbf{X}_0\rangle_0$ as a function of t and a sub-set of $\mathbf{X}_0$.

There are three reasons for presenting these results here even though there are no numerical results to illustrate them or to confirm their utility. The first reason is to show how one should go beyond the simple Langevin equation to the generalised Langevin equation with memory to reduce the approximation invoked in Brownian dynamics simulations. The second reason is to show how a configuration dependent diffusion tensor enters the problem in a formal sense, as opposed to postulating its influence on the basis of macroscopic hydrodynamics. The third reason is that there may be computational advantages of this formulation of the problem for macromolecules. The hope is that the time correlation functions of the unbonded system are more tractable than those of the bonded system. In part this is because one expects the relaxation time of a non-bonded macromolecule to be shorter than that of a bonded macromolecule. And in part it is because, both in the case of a macromolecule and in the case of a concentrated colloid dispersion that has no bonding potentials, one can systematically obtain in turn the velocity correlation functions conditioned on no monomers, $\ddot{q}_{0;\alpha\beta}(t)$, two monomers, $\ddot{q}_{0;\alpha\beta}(t;\mathbf{x}_{\alpha\beta}(0))$, three monomers, $\ddot{q}_{0;\alpha\beta}(t;\mathbf{x}_{\alpha\beta}(0),\mathbf{x}_{\alpha\gamma}(0))$, etc., and truncate the series as dictated by physical reality, mathematical convenience, or computer limitations. The utility or otherwise of this approach remains to be determined, which omission doubtless will be rectified in time.

11.3.3 Stochastic Calculus

Now the question raised briefly at the end of §11.3.1 is addressed, namely the rôle of the stochastic calculus in realistic physical systems. In order to make the analysis as transparent as possible, the simplest form of the stochastic equations will be used that retains their essential features: they are discrete, and they consist of a continuous deterministic part (the most likely part), and a stochastic part.

First, begin with the generalised Langevin equation for velocity,

$$\begin{aligned} \dot{x}(t|x_0,F) &= \overline{\dot{x}}(t|x_0) + R(t) \\ &\quad + \frac{1}{T}\int_0^t \mathrm{d}t'\, \ddot{q}_0(t-t'|x_0) F(x(t'),t'). \end{aligned} \tag{11.103}$$

The random force for the velocity has zero mean and covariance,

$$\langle R(t)R(t')\rangle = k_{\mathrm{B}}\ddot{q}_0(t,t'|x_0). \tag{11.104}$$

Now integrate this to obtain the generalised Langevin equation for position,

$$\begin{aligned} x(t|x_0,F) &= \overline{x}(t|x_0) + \int_0^t \mathrm{d}t'\, R(t') \\ &\quad + \frac{1}{T}\int_0^t \mathrm{d}t' \int_0^{t'} \mathrm{d}t''\, \ddot{q}_0(t',t''|x_0) F(x(t''),t'') \\ &= \overline{x}(t|x_0) + \int_0^t \mathrm{d}t'\, R(t') \\ &\quad + \frac{1}{T}\int_0^t \mathrm{d}t'' \int_{t''}^{t} \mathrm{d}t'\, \ddot{q}_0(t',t''|x_0) F(x(t''),t'') \\ &= \overline{x}(t|x_0) + R_1(t) \\ &\quad + \frac{1}{T}\int_0^t \mathrm{d}t''\, \left[\dot{q}_0(t,t''|x_0) - \dot{q}_0(t'',t''|x_0)\right] F(x(t''),t''). \end{aligned} \tag{11.105}$$

Here $R_1(t) \equiv \int_0^t \mathrm{d}t'\, R(t')$. One expects $R_1(t)$ to be a Gaussian random variable, because it is the sum of Gaussian random variables.

The covariance of the random force for the position is

$$\begin{aligned}\langle R_1(t)R_1(t')\rangle &= \int_0^t \mathrm{d}s \int_0^{t'} \mathrm{d}s'\, \langle R(s)R(s')\rangle \\ &= k_\mathrm{B} \int_0^t \mathrm{d}s \int_0^{t'} \mathrm{d}s'\, \ddot{q}_0(s,s'|x_0) \\ &= \int_0^t \mathrm{d}s \int_0^{t'} \mathrm{d}s'\, \langle \dot{x}(s)\dot{x}(s')|\, x_0\rangle_0 \\ &= \langle [x(t)-x_0][x(t')-x_0]|x_0\rangle_0 . \end{aligned} \tag{11.106}$$

What will be required below is the variance of the increment in the random force. This can be obtained from,

$$\begin{aligned}\left\langle [R_1(t)-R_1(t')]^2\right\rangle &= \left\langle [x(t)-x_0]^2|x_0\right\rangle_0 + \langle [x(t)-x_0][x(t')-x_0]|x_0\rangle_0 \\ &\quad + \langle [x(t')-x_0][x(t)-x_0]|x_0\rangle_0 + \left\langle [x(t')-x_0]^2|x_0\right\rangle_0 \\ &= \left\langle [x(t)-x(t')]^2|x_0\right\rangle_0 \\ &= \left\langle [x(t)-x(t')]^2\right\rangle_0, \ t,t'\to\infty, \\ &= 2k_\mathrm{B} q_0^\mathrm{s}(t-t') \\ &\sim \begin{cases} k_\mathrm{B}\ddot{q}_0^\mathrm{s}(0)[t-t']^2, & |t-t'|\to 0, \\ 2D|t-t'|, & |t-t'|\gtrsim \tau_\mathrm{relax}. \end{cases} \end{aligned} \tag{11.107}$$

The third equality holds at long times, such that the initial state has negligible influence. In the molecular time interval limit, the upper equality on the final right-hand side, the variance of the difference is proportional to the square of the time interval because the symmetrised position autocorrelation function is an analytic, even function of time. (See, for example, Eq. (3.34). Also, $\ddot{q}_0^\mathrm{s}(0) = Tm^{-1}$, but this is not required here.) The short time limit, the lower equality on the final right-hand side, is the usual Einstein limit, in which the mean square displacement of the Brownian particle grows linearly with time.

It is emphasised that this is how the stochastic force in real physical systems behaves. In particular, the Einstein (or Brownian process) limit does *not* hold on infinitesimal time intervals. In the case of a Brownian particle in a soft sphere fluid shown in Fig. 10.2 on p. 338, it can be seen that $\tau_\mathrm{relax} \approx 5$. The difference between the time scales where a thermodynamic fluctuation shows quadratic and linear behaviour was discussed in reference to Fig. 2.1 on p. 40.

Simplest Stochastic Equation

In order to focus upon what is essential, one can set the perturbing force to zero, $F = 0$, and remove the dependence on the initial configuration by taking the limit $t, t' \to \infty$. In this case the discrete stochastic equation to be analysed is

$$x_i = \overline{x}(t_i) + R_{1,i}, \tag{11.108}$$

with the time discretised as $t_i = i\Delta_t$. By subtracting this from the version for $i+1$, it may be written in the form of a stochastic evolution equation,

$$x_{i+1} = x_i + \Delta_t \dot{\overline{x}}(t_i) + [R_{1,i+1} - R_{1,i}]. \tag{11.109}$$

This is in essence the same as the conventional form used for Brownian dynamics, Eq. (11.86), with the random force invoked there being $\tilde{R}_i \equiv [R_{1,i+1} - R_{1,i}]$.

The deterministic part $\overline{x}(t)$ and all its derivatives are assumed continuous, and it is not responsible for the unique features of the stochastic calculus that are discussed here. Hence one can interpret below $\overline{x}(t) = \overline{x}_i + [t - t_i]\dot{\overline{x}}_i + \ldots$

Focusing on the random force, the variance of the increment is given by the above expression with $t - t' = \Delta_t$,

$$\begin{aligned} \left\langle [R_{1,i+1} - R_{1,i}]^2 \right\rangle &= 2k_{\mathrm{B}} q_0^{\mathrm{s}}(\Delta_t) \\ &\sim \begin{cases} k_{\mathrm{B}} \ddot{q}_0{}^{\mathrm{s}}(0)\Delta_t^2 & \Delta_t \to 0, \quad \text{(memory)}, \\ 2D|\Delta_t|, & \Delta_t \gtrsim \tau_{\mathrm{relax}}, \quad \text{(no memory)}. \end{cases} \end{aligned} \tag{11.110}$$

The lower equality is the traditional no memory limit for the Langevin equation (see, for example, Eq. (11.86)). It embodies the constant drag force and for an equilibrium system yields the Einstein result for Brownian motion. Although the no memory result only holds for times greater than some relaxation time, in practice, particularly in mathematical and financial applications, the covariance of the random force is taken to be linear in the time step *and* the time step is regarded as infinitesimal on the scale of motion that is being considered. Virtually the whole field of stochastic calculus in mathematics is predicated on the so-called Brownian result, that the covariance of the random force is linear in the time step.

Internodal Behaviour

The key to understanding the origin of the stochastic calculus is to note that the discrete stochastic equations of motion are silent about the behaviour of the system between time steps, $t \in (t_i, t_{i+1})$. Consequently one has a certain freedom in defining the value between nodes, and this choice affects integrals of functions of the stochastic variable. For example, one could define

$$x(t) = \overline{x}(t) + R_{1,i}, \quad \text{(Itô)}. \tag{11.111}$$

Here and below, $t \in (t_i, t_{i+1})$, with $i \equiv \lfloor t/\Delta_t \rfloor$, where $\lfloor \ldots \rfloor$ means the largest integer that is smaller than or equal to the argument. This has the interpretation that the stochastic force is applied at the beginning of the time interval and is zero throughout. Consequently the position is discontinuous at the time nodes. This interpretation is equivalent to the Itô calculus.

A second possibility is

$$x(t) = \overline{x}(t) + [R_{1,i} + R_{1,i+1}]/2, \quad \text{(Stratonovich)}. \tag{11.112}$$

This is equivalent to the Stratonovich calculus. One objection to this interpretation is that the position at time t *anticipates* the force at time $t_{i+1} > t$, which appears to violate time causality. This objection need not be taken too seriously because in physical reality the random force arises from reservoir perturbations that are continuous in time, and so in principle it is possible to predict the future force from the present and past forces. Obviously other interpolation or extrapolation schemes are possible.

Quadrature

Now consider the integration of a function over the trajectory of the form,

$$\begin{aligned} G(t) &= \int_{x_0}^{x(t)} \mathrm{d}x(t')\, g(x(t')) \\ &= \sum_{i=0}^{n-1} [x_{i+1} - x_i]\, g(x(t')), \\ &= \sum_{i=0}^{n-1} [\Delta_t \dot{\overline{x}}(t_i) + R_{1,i+1} - R_{1,i}]\, g(x(t')). \end{aligned} \tag{11.113}$$

Here, $n \equiv \lfloor t/\Delta_t \rfloor$, $t' = t_i + \alpha\Delta_t$, and $\alpha \in [0, 1)$. In ordinary calculus, the result of the quadrature is independent of the position t' in the sub-interval at which the integrand is evaluated. More precisely, different choices of $\alpha \in [0, 1)$ change $g(x(t'))$ by an amount $\sim \Delta_t$, and the integral by an amount $\sim n\Delta_t^2 \sim \Delta_t$, which is negligible in the limit $\Delta_t \to 0$. In the stochastic calculus this is not the case.

The two common choices yield

$$g(x(t')) = \begin{cases} g(x_i) + [\alpha\Delta_t \dot{\overline{x}}(t_i) + R_{1,i}]g'(x_i) + \ldots, & \text{(Itô)}, \\ g(x_i) + [\alpha\Delta_t \dot{\overline{x}}(t_i) + (R_{1,i} + R_{1,i+1})/2]g'(x_i) + \ldots, & \text{(Strat.)} \end{cases} \tag{11.114}$$

The average of the difference between the respective quadratures is

$$\begin{aligned} &\left\langle G^{\text{Strat}}(t) - G^{\text{Ito}}(t) \right\rangle \\ &= \sum_{i=0}^{n-1} \left\langle [\Delta_t \dot{\overline{x}}(t_i) + R_{1,i+1} - R_{1,i}]\, [(R_{1,i+1} - R_{1,i})/2] g'(x_i) \right\rangle \\ &= \sum_{i=0}^{n-1} k_{\text{B}} q_0^{\text{s}}(\Delta_t) g'(x_i) \\ &= \sum_{i=0}^{n-1} g'(x_i) \begin{cases} k_{\text{B}} \ddot{q}_0^{\text{s}}(0)\Delta_t^2/2, & \text{(memory)}, \\ D|\Delta_t|, & \text{(no memory)}. \end{cases} \end{aligned} \tag{11.115}$$

For the no memory case, $\langle R_1(t)R_1(t')\rangle = 2D|t - t'|$, it can be seen that this is $G^{\text{Strat}}(t) - G^{\text{Ito}}(t) \sim n\Delta_t = t$, which cannot be neglected in the limit that the time step goes to zero. However for the case of memory, $\langle R_1(t)R_1(t')\rangle = 2k_B\ddot{q}_0^s(t-t') \sim k_B\ddot{q}_0^s(0)[t-t']^2$, $[t'-t]\to 0$, and hence $G^{\rm Strat}(t) - G^{\rm Ito}(t) \sim n\Delta_t^2 = t\Delta_t$, which is negligible when the time step goes to zero. Hence in the case of memory, the choice of stochastic calculus is irrelevant.

Conclusion

To summarise the stochastic calculus, one has to distinguish between mathematics and the physical sciences. As a mathematical discipline, it is perfectly legitimate to define a Brownian process as one in which the variance of the random force increment is linearly proportional to the time step for all values of the time step, and to pursue the consequences of different choices for the internode behaviour for the stochastic calculus.

In the physical sciences, one ought to recognise the real origins of the stochastic force, and to distinguish between large time steps, when the displacement of the Brownian particle is linear in the time step (the Einstein result), and small time steps, when it is quadratic. The following remarks address the relevance of the stochastic calculus to the physical sciences, and are not concerned with its value as a mathematical pursuit in its own right.

If one has rapidly varying external or perturbing forces, then one has no choice but to invoke small time steps, and under these circumstance the generalised Langevin equation with memory is required. In this case the stochastic calculus is not required.

A difficulty remains in the large time step, no memory case (slowly varying external force), where the stochastic calculus is required. There is no firm axiomatic basis for choosing between the Itô, Stratonovich, or other form for the internode value of the stochastic variable, but the value of a quadrature depends upon such a choice. It is axiomatic that physical values cannot have such an ambiguity. Perhaps the resolution of the problem lies in the observation that it is not only the external force that provides an upper bound on the time step, but also any function whose estimate is required. In particular, for an accurate quadrature, the variation of $g(x)$ in a time step must be relatively negligible. If one demands that $|Dg'(x(t))| \ll \left|\dot{\overline{x}}(t)g(x(t))\right|$, then the stochastic calculus is irrelevant. If the diffusion constant is not small enough to satisfy this, then one must instead use a small time step and the memory function.

In non-equilibrium statistical mechanics, this ambiguity in the stochastic calculus for the no memory case occurs in the relationship between the Fokker-Planck equation and the Langevin equation, with the differences depending upon the gradients of the diffusion constant, $\nabla\cdot D$.[19] (Once again it is stressed that this is not an issue for the generalised Langevin equation with memory.) In the present derivation of the Fokker-Planck equation, §3.7.3, the problem did not occur because the diffusion constant was independent of the stochastic variable. Similarly in Ch. 10, such gradients did not appear because the position variables were averaged out in obtaining the time correlation function

[19] See Eq. (11.86) *et seq.*, and also Ermak and McCammon, (1978), Gardiner, (1983), Risken, (1984), Tough et al. (1986), Keizer, (1987), Allen and Tildesley, (1987), and Mazo, (2008).

and the consequent diffusion constant. The advanced Brownian dynamics presented above, §11.3.2, invoked a time correlation function that was conditional upon the position of the Brownian particle at the beginning of the trajectory, $\langle [x(t) - x_0][x(t') - x_0] | x_0 \rangle$. The configuration dependence of the diffusion constant that results from this, $D(x_0)$, does not give rise to any dependence on the stochastic calculus, $\partial D(x_0)/\partial x(t) = 0$, since, it is arguable, that in the no memory limit, the dependence on the initial configuration ought to be forgotten.

References

Allen, M. P. and Tildesley, D. J. (1987), *Computer Simulations of Liquids*, (Oxford University Press, Oxford). [392, 403, 437– 439, 449][1]

Andersen, H. C. (1980), J. Chem. Phys. **72**, 2384. [393]

Andrea, T. A., Swope, W. C., and Anderson, H. C. (1983), J. Chem. Phys. **79**, 4576. [393]

Attard, P. (2002a), *Thermodynamics and Statistical Mechanics: Equilibrium by Entropy Maximisation*, (Academic Press, London). [35, 97, 141, 176, 193, 194]

Attard, P. (2002b), J. Chem. Phys. **116**, 9616. [291, 393]

Attard, P. (2004), J. Chem. Phys. **121**, 7076. [402]

Attard, P. (2005a), J. Chem. Phys. **122**, 154101. [146, 297, 404]

Attard, P. (2005b), J. Chem. Phys. **122**, 244105. [431]

Attard, P. (2006), J. Chem. Phys. **124**, 024109. [288, 291, 305, 432–435]

Attard, P. and Gray-Weale, A. (2008), J. Chem. Phys. **128**, 114509. [334, 338, 360–365, 398, 427]

Attard, P. (2009a), J. Chem. Phys. **130**, 194113. [73, 291, 360–365, 398– 408, 427, 434]

Attard, P. (2009b), Annu. Rep. Prog. Chem., Sect. C **105**, 63. [253, 402]

Attard, P. (2009c) Phys. Rev. E **80**, 041126. [365, 425–427]

Attard, P. (2011), AIP Advances **1**, 032146. [146]

Barker J. A. and Klein, M. L. (1973), Phys. Rev. B **7**, 4707. [176]

Le Bellac, M., Mortessagne, F., and Batrouni, G. G., (2004), *Equilibrium and Non-equilibrium Statistical Thermodynamics*, (Cambridge University Press, Cambridge). [354]

Bénard, H. (1900), Rev. Gen. Sci. Pures Appl. **11**, 1261. [145]

Bochkov, G. N. and Kuzovlev, Yu. E. (1981), Physica, **106A**, 443. [288, 289, 294]

Boinepalli, S. and Attard, P. (2003), J. Chem. Phys. **119**, 12769. [291, 393]

Boltzmann, L. (1872), Wien. Ber. **66**, 275. [2]

[1]Numbers in brackets give the pages on which the reference is made.

Boltzmann, L. (1877), Wien. Ber. **76**, 373. [2]

Brown, R. (1828), Phil. Mag. **4**, 171. [61]

Busse, F. H. (1967), J. Math. and Phys. **46**, 140. [154, 156]

Busse, F. H. and Whitehead, J. A. (1971), J. Fluid Mech. **47**, 305. [167, ??]

Caldeira, A. O. and Leggett, A. J. (1983), Ann. Phys. **149**, 374. [383]

Casimir, H. B. G. (1945), Rev. Mod. Phys. **17**, 343. [60]

Chatraei, Sh. and Macosko, C. W. (1981), J. Rheology **25**, 433. [140]

Clausius, R. (1850), Ann. Phys. **79**, 368, 500. English translation: Clausius, R. (1851), Phil. Mag. **2**, 1, 102. [1]

Clausius, R. (1865), Ann. Phys. **125**, 353. [1]

Cox, R. T. (1961), *The Algebra of Probable Inference*, (Johns Hopkins Press, Baltimore, MD). [9]

Cox, R. T. (1978), *Of Inference and Inquiry: An Essay in Inductive Logic*, in *The Maximum Entropy Formalism*, Levine, R. D. and Tribus, M. (eds), (MIT Press, Cambridge, MA). [9]

Daivis, P. J. and Evans, D. J. (1993), Phys. Rev. E **48**, 1058. [408, 434]

Das, S. P. (2004), Rev. Modern Phys. **76**, 785. [382]

de Groot, S. R. and Mazur, P. (1984), *Non-equilibrium Thermodynamics*, (Dover, New York). [60, 112, 119, 134]

Doob, J. L. (1942), Ann. Math. **43**, 351. [42]

Drazin P. G. and Reid, W. H. (1981), *Hydrodynamic Stability*, (Cambridge University Press, Cambridge). [143, 156, 166]

Dubois, M. and Bergé, P. (1978), J. Fluid Mech. **85**, 641. [156]

Einstein, A. (1905), Ann. Phys. **17**, 549. [61, 332]

Ermak, D. L. and McCammon, J. A. (1978), J. Chem. Phys. **69**, 1352. [439, 449]

Evans, D. J. (1986) Phys. Rev. A **34**, 1449. [408, 434]

Evans, D. J., Cohen, E. G. D., and Morriss, G. P. (1993), Phys. Rev. Lett. **71**, 2401. [288]

Evans, D. J. (2003), Molec. Phys. **101**, 1551. [288]

Faber, T. E. (1995), *Fluid Dynamics for Physicists*, (Cambridge University Press, Cambridge). [126]

Fokker, A. (1914), Ann. Phys. **43**, 810. [61]

Ford, G. W., Kac, M., and Mazur, P. (1965) J. Math. Phys. **6**, 504. [383]

Fox, R. F. and Uhlenbeck, G. E. (1970), Phys. Fluids, **13**, 1893 and 2881. [41, 122, 139]

Gardiner, C. W. (1983), *Handbook of Stochastic Methods*, (Springer, New York). [89, 439, 449]

Getling, A. V. (1998), *Rayleigh-Bénard Convection: Structure and Dynamics*,

(World Scientific, Singapore). [145]

Gibbs, J. W. (1902), *Elementary Principles in Statistical Mechanics Developed with Special Reference to the Rational Foundation of Thermodynamics*, (Yale Univ. Press, New Haven, CT). [19, 27]

Glauber, R. J. (1963), J. Math. Phys. **4**, 294. [291]

Green, M. S. (1954), J. Chem. Phys. **22**, 398. [46, 301, 356]

Haile, J. M. and Gupta, S. (1983), J. Chem. Phys. **79**, 3067. [393]

Hansen, J.-P. and McDonald, I. R. (1986), *Theory of Simple Liquids*, (Academic Press, London). [301, 335, 382]

Haw, M. D. (2002), J. Phys.: Condens. Matter, **14**, 7769. [61]

Heyes, D. M. (1983), J. Chem. Phys. **82**, 285. [393]

Hoover, W. G. (1985), Phys. Rev. A **31**, 1695. [393]

Hynes, J. and Deutch, J. (1975), in *Physical Chemistry, an Advanced Treatise*, edited by H. Eyring, D. Henderson, and W. Jost, Vol. XI, (Academic Press, New York). [381]

Jarzynski, C. (1997), Phys. Rev. Lett. **78**, 2690. [289, 294]

Jaynes, E. T. (1957), Phys. Rev. **106**, 620 and **108**, 171. [10]

Kawasaki, K. (1966), Phys. Rev. **145**, 145. [291]

Keizer, J. (1987), *Statistical Thermodynamics of Non-equilibrium Processes*, (Springer-Verlag, New York). [49, 122, 139, 439, 449]

Khintchine, A. (1934) Math. Ann. **109**, 604. [354]

Kita, T. (2006), J. Phys. Soc. Japan **75**, 114005 and 124005. [146]

Kondepudi, D. and Prigogine, I. (1998), *Modern Thermodynamics: From Heat Engines to Dissipative Structures*, (Wiley, Chichester). [5]

Kubo, R. (1966), Rep. Progr. Phys. **29**, 255. [46, 301, 356]

Kubo, R., Toda, M., and Hashitsume, N. (1978), *Statistical Physics II. Non-equilibrium Statistical Mechanics*, (Springer-Verlag, Berlin). [46, 301, 356]

Kuiken, G. D. C. (1984), *Thermodynamics for Irreversible Processes*, (John Wiley and Sons, Chichester, England). [135]

Landau, L. D. and Lifshitz, E. M. (1959), *Fluid Mechanics*, (Pergammon Press, Oxford). [121, 139]

Langer, J. S. (1969), Annals Phys. **54**, 258. [291]

Langevin, P. (1908), C. R. Acad. Sci. Paris, **146**, 530. [61, 343]

Levesque, D. and Verlet, L. (1970), Phys. Rev. A **2**, 2514. [335]

Malkus, W. V. R. and Veronis, G. (1958), J. Fluid Mech. **4**, 225. [145]

Mazo, R. M. (2008), *Brownian Motion: Fluctuations, Dynamics, and Applications*, (Oxford University Press, Oxford). [439, 449]

Metiu, H., Kitahara, K., and Ross, J. (1975), J. Chem. Phys. **63**, 5116. [291]

Mori, H. (1965), Progr. Theor. Phys. **33**, 423 and **34**, 399. [379]

Nosé, S. (1984), Mol. Phys. **52**, 255. [393]

Onsager, L. (1931), Phys. Rev. **37**, 405, and **38**, 2265. [45, 297, 303]

Ortiz de Zárate, J. M. and Sengers, J. V. (2001), Physica A **300**, 25. [166]

Ortiz de Zárate, J. M. and Sengers, J. V. (2006), *Hydrodynamic Fluctuations in Fluids and Fluid Mixtures*, (Elsevier, Amsterdam). [122, 139]

Perrin J. (1908), C. R. Acad. Sci. Paris, **146**, 967. [69]

Planck, M. (1916), Sitz. Ber. Preuss. Akad. Wiss. 324. [61]

Popper, K. R. (1959), Brit. J. Philos. Sci. **10**, 25. [11]

Pottier, N. (2010), *Non-equilibrium Statistical Physics: Linear Irreversible Processes*, (Oxford University Press, Oxford). [60, 112, 119, 354, 355, 383]

Rahman, A. (1964), Phys. Rev. **136**, A405. [335]

Risken, H. (1984), *The Fokker-Planck Equation*, (Springer-Verlag, Berlin). [88, 89, 439, 449]

Rosenkrantz, R. D. (1983), (Ed.), *E. T. Jaynes: Papers on Probability, Statistics, and Statistical Physics*, (D. Reidel, Dordrecht). [10]

Rugh, H. H. (1997), Phys. Rev. Lett. **78**, 772. [194]

Senitzky, I. R. (1960) Phys. Rev. **119**, 670. [383]

Shannon, C. E. and Weaver, W. (1949), *The Mathematical Theory of Communication*, (Univ. of Illinois Press, Urbana) [19]

von Smoluchowski, M. (1906), Ann. Phys. **21**, 756. [61, 343]

Tough, R. J. A., Pusey, P. N., Lekerkerker, H. N. W., and van den Broeck, C. (1986), Mol. Phys. **59**, 595. [439, 449]

Tuckerman, M. E. and Martyna, G. J. (2000), J. Phys. Chem. B **104**, 159. [393]

Ullersma, P. (1966), Physica **32**, 27, 56, 74, and 90. [383]

Verlet, L. (1967), Phys. Rev. **159**, 98. [392]

Wiener, N. (1930), Acta Math. **55**, 117. [354]

Yih, C.-S. (1977), *Fluid Mechanics: A Concise Introduction*, (West River Press, Ann Arbor). [143, 156, 166]

Zwanzig, R. (1961), in *Lectures in Theoretical Physics*, (W. E. Britton, B. W. Downs, and J. Downs, eds), Vol. III, p. 135, (Wiley Interscience, New York). [379]

Zwanzig, R. (2001), *Non-equilibrium Statistical Mechanics*, (Oxford University Press, Oxford). [46, 94, 301, 379–383]

Index

The manufacturer's authorised representative in the EU for product safety is Oxford University Press España S.A. of el Parque Empresarial San Fernando de Henares, Avenida de Castilla, 2 – 28830 Madrid (www.oup.es/en or product.safety@oup.com). OUP España S.A. also acts as importer into Spain of products made by the manufacturer.

www.ingramcontent.com/pod-product-compliance
Lightning Source LLC
LaVergne TN
LVHW062203200926
848484LV00006B/1310
* 9 7 8 0 1 9 9 6 6 2 7 6 0 *